World Sustainability Series

Series Editor

Walter Leal Filho, Wissenschaften Hamburg, Hochschule für Angewandte, Hamburg, Hamburg, Germany

Due to its scope and nature, sustainable development is a matter which is very interdisciplinary, and draws from knowledge and inputs from the social sciences and environmental sciences on the one hand, but also from physical sciences and arts on the other. As such, there is a perceived need to foster integrative approaches, whereby the combination of inputs from various fields may contribute to a better understanding of what sustainability is, and means to people. But despite the need for and the relevance of integrative approaches towards sustainable development, there is a paucity of literature which address matters related to sustainability in an integrated way.

Notes on the quality assurance and peer review of this publication

Prior to publication, the works published in this book are initially assessed and reviewed by an in-house editor. If suitable for publication, manuscripts are sent for further review, which includes a combined effort by the editorial board and appointed subject experts, who provide independent peer-review. The feedback obtained in this way was communicated to authors, and with manuscripts checked upon return before finally accepted. The peer-reviewed nature of the books in the "World Sustainability Series" means that contributions to them have, over many years, been officially accepted for tenure and promotion purposes.

Weisheng Zhou
Editor

Decarbonization Policy Engineering

Energy Transition and Regional Cooperation in East Asia

Springer

Editor
Weisheng Zhou
College of Policy Science
Ritsumeikan University
Ibaraki, Osaka, Japan

ISSN 2199-7373 ISSN 2199-7381 (electronic)
World Sustainability Series
ISBN 978-981-95-5952-7 ISBN 978-981-95-5953-4 (eBook)
https://doi.org/10.1007/978-981-95-5953-4

This Springer imprint is published by the registered company Springer Nature Singapore Pte Ltd.
The registered company address is: 152 Beach Road, #21-01/04 Gateway East, Singapore 189721, Singapore

Preface

One of the greatest challenges facing humanity in the 21st century is addressing climate change under conditions of profound uncertainty while striving to build a sustainable society. As the impacts of climate change intensify and the geopolitics of energy grow more complex, countries around the world are both cooperating and competing to set carbon neutrality targets and restructure their energy systems. In particular, East Asia stands at the center of global energy consumption and CO_2 emissions, while also serving as a hub of rapid economic growth and technological innovation. For this reason, realizing decarbonization in East Asia is not merely a regional issue—it also holds the key to forging a sustainable future on a global scale.

This book, *Decarbonization Policy Engineering: Energy Transition and Regional Cooperation in East Asia*, aims to provide a comprehensive and interdisciplinary analysis of the policy, technological, institutional, and behavioral dimensions of decarbonization in East Asia. Recognizing the uncertainty, complexity, and regional diversity inherent in climate change and the energy transition, it adopts a systems-based approach called "policy engineering" to design, integrate, and evaluate climate and energy policies and technologies.

The book is structured into three parts, systematically presenting both theoretical and empirical analyses of policy design, energy transition, and regional cooperation toward carbon neutrality in East Asia.

Strategies and Prospects for Carbon Neutrality in East Asia focuses on decarbonization strategies and policy prospects in East Asia, discussing the current state of decarbonization and the need for policy integration across the region. By examining the complexity of energy transitions faced by countries such as China, Japan, and South Korea, and by highlighting the potential for cross-border cooperation, it outlines a path toward achieving a sustainable green recovery.

System Evaluation of the Optimal Energy Mix concentrates on the system evaluation and design of optimal regional energy mixes. Incorporating climate uncertainty, it explores the harmonization of energy supply and demand using scientific and technological methods. Through case studies and scenario analyses from various countries, it offers practical and adaptable guidance for designing realistic energy transition strategies.

Innovations for a Decarbonized Society shines a light on technological innovation and the current challenges in policy implementation. It explores innovations supporting decarbonization from multiple angles, including advancements in carbon cycling technologies, analysis of consumer behavior, the role of climate finance, and the harmonization of environmental regulations in international trade. Bridging theory and practice, it presents concrete proposals that contribute to building a decarbonized society at the ground level.

A distinguishing feature of this book is its effort not only to analyze individual elements of policy and technology, but also to integrate them within a systematic framework of policy engineering to design decarbonization strategies for the entire region. Since energy policy is deeply intertwined with transformations in social structures and economic systems, regional cooperation and institutional design beyond national borders are essential. The diverse research findings and case studies included in this volume aim to serve as a practical and concrete compass for policymakers, researchers, businesses, and civil society in East Asia as they chart a path toward decarbonization.

In climate and energy policy, viewing uncertainty not as a risk but as an "opportunity" is key to fostering innovation and driving social transformation. While uncertainty has traditionally been considered a barrier to policy design, today it can act as a catalyst for breaking away from outdated systems. From this perspective, energy policy becomes more flexible and strategic—extending beyond technological choices to encompass the restructuring of socio-economic systems and governance. Especially in the case of long-term, high-uncertainty climate policies, the pursuit of both *optimality* and *robustness* is essential. Now more than ever, we need strategic thinking that harnesses uncertainty constructively to envision a sustainable and inclusive future.

In editing this volume, the collaboration between researchers and policy experts has enabled the sharing of practical insights into East Asia's energy transition and decarbonization policies—an achievement of great significance. These insights will serve as a vital foundation for advancing regional cooperation and fostering technological innovation.

Climate change is a multifaceted challenge that affects not only the environment, but also the economy, society, and security. Achieving a sustainable future requires evidence-based policy design, regional collaboration, and the promotion of innovation.

We hope that this book will contribute to providing an intellectual foundation for achieving these goals.

Part of this publication is based on the research outcomes of the project "Research on Green Recovery and the Realization of a Decarbonized Society in East Asia" (Principal Investigator: Professor Weisheng Zhou, College of Policy Science, Ritsumeikan University), which was supported by the Ritsumeikan Asia-Japan Research Organization's "Asia-Japan Research Promotion Program." We would like to express our sincere gratitude for this support.

On behalf of the authors

Kyoto, Japan
November 2025

Weisheng Zhou

Contents

About the Editor

Weisheng Zhou, Ph.D. in Engineering (Kyoto University), Emeritus Professor and Specially Appointed Professor at Ritsumeikan University; Qiushi Distinguished Professor at Zhejiang University; Director of the Global 3E Research Institute; and Foreign Member of the Engineering Academy of Japan (EAJ).

Dr. Zhou has held various positions, including Researcher at the New Energy and Industrial Technology Development Organization (NEDO); Senior Researcher and Research Advisor at the Research Institute of Innovative Technology for the Earth (RITE); Specially Appointed Professor at Osaka University; Visiting Professor at Kyoto University; and Founding Director of the Ritsumeikan Research Center for Sustainability Science.

His main fields of teaching and research include Earth Environmental Studies, Environmental Economics and Policy, Energy Systems Engineering, Sustainable Development Studies, and Policy Engineering. His key research topics encompass the East Asian Low-Carbon Community (EA-LCC), Global Resource Recycling System (GRS), East Asian Nuclear Power Safety System (EA-NPSS), East Asian Emissions Trading System (EA-ETS), Quantitative Happiness Studies, and Policy Engineering.

Dr. Zhou has published numerous monographs, including *The Earth Regeneration Plan: CO_2 Reduction Strategy* (Nikkan Kogyo Shimbun, 2000); *East Asian Low-Carbon Community* (Springer, 2021); *Sustainable Development Studies in the Era of SDGs* (Horitsu Bunka Sha, 2022); *Nuclear Power Safety and Governance* *in East Asia* (Routledge, 2023); *Creep-Fatigue Fracture: Analysis of Internal Damage* (Springer, 2024); and *Foundations of Quantitative Happiness: Economic Growth vs Life Satisfaction* (Springer Nature, 2025), among others, including co-authored works.

Strategies and Prospects for Carbon Neutrality in East Asia

The Framework of Decarbonization Policy Engineering and the Challenges in East Asia

Weisheng Zhou

1 Introduction

Global environmental issues caused by climate change represent one of the most pressing challenges of contemporary society. In particular, the reduction of greenhouse gas emissions and the realization of a decarbonized society are common goals for all countries; however, many uncertainties remain along the path toward these objectives. Reports by the Intergovernmental Panel on Climate Change (IPCC) not only clarify the mechanisms and severity of global warming scientifically but also indicate that future projections inherently contain uncertainties (IPCC 2023). These uncertainties significantly influence the formulation of climate policies and energy transition strategies, making it essential to properly understand their types.

In the era of decarbonization, it is necessary to outline flexible and effective pathways for energy transition that are grounded in scientific knowledge while taking into account social, economic, and policy constraints. East Asia, in particular, faces unique challenges, including rapid economic growth, high energy demand, and dependence on energy imports (Sovacool and Griffiths 2021; Zhou et al. 2021). Consequently, policy approaches for achieving carbon neutrality (CN) must incorporate region-specific perspectives.

Against this background, Chapter 1 first presents an overview of decarbonization policy engineering and surveys the specific challenges faced by East Asia. It clarifies the analytical direction and structure of this book, particularly by introducing methodological perspectives and analytical frameworks designed to support policy decision-making in complex and uncertain realities, with the goal of application in practical policy design. The chapter serves as a foundation for subsequent chapters, which will examine more concrete policy issues and technological strategies.

W. Zhou (✉)
College of Policy Science, Ritsumeikan University, Osaka, Japan
e-mail: zhou@sps.ritsumei.ac.jp

W. Zhou (ed.), *Decarbonization Policy Engineering*, World Sustainability Series,
https://doi.org/10.1007/978-981-95-5953-4_1

2 IPCC Reports and the Causes of Global Warming: Between Certainty and Uncertainty

The Intergovernmental Panel on Climate Change (IPCC) has published six comprehensive assessment reports since its First Assessment Report (AR1) in 1990. These reports have consistently identified human activities as the primary driver of global warming, particularly since the latter half of the twentieth century. As shown in Fig. 1, the IPCC has progressively increased its confidence that human influence is the main cause of global warming, culminating in the Sixth Assessment Report (IPCC 2021a, 2021b, 2021c) with the following key conclusions:

1. It is unequivocal that human influence has warmed the atmosphere, oceans, and land.
2. The scale of recent changes across the climate system is unprecedented over many centuries to many thousands of years.
3. Human-induced climate change is already affecting many weather and climate extremes in every region across the globe.

Nevertheless, scientists tend to avoid using absolute or categorical expressions such as "unequivocal" or "certain" unless there is conclusive evidence. In fact, the IPCC reports employ terminology based on quantified confidence levels or probabilities, such as "high confidence" or "very likely" (IPCC 2014b, 2021b).

The Chinese meteorologist Zhu Kezhen (1890–1974) conducted studies of China's historical climate using long-term historical documents, including agricultural records, famine and disaster reports, and phenological observations. Based on

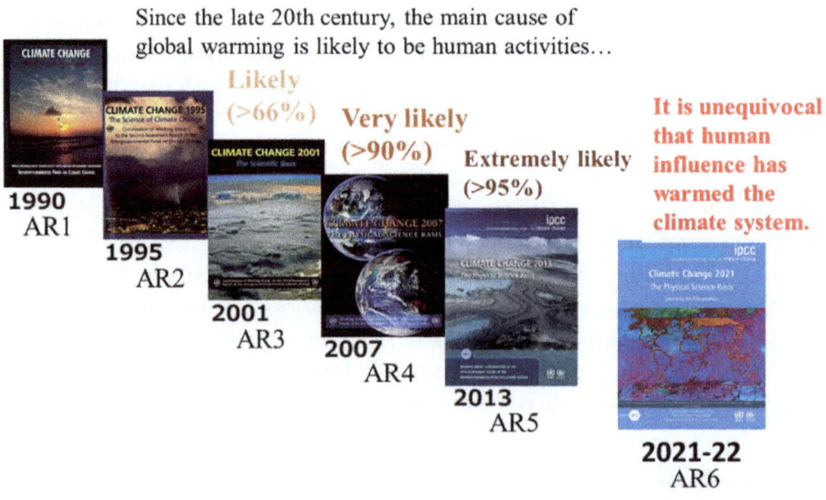

Fig. 1 The IPCC assessment report on the impacts of human activities on climate change

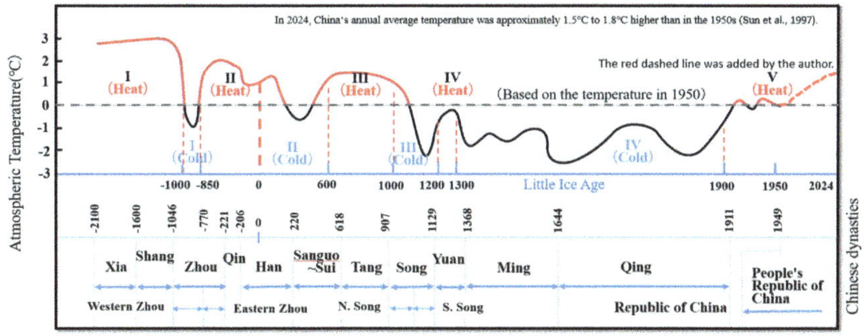

Fig. 2 Climatic trends and dynastic cycles in China over the Past 5,000 Years (Zhu 1973; Adapted by the author)

this research, he presented a curve depicting the temperature variations over approximately the past 5,000 years, now known as the "Zhu Kezhen Curve" (Zhu 1973) (Fig. 2).

The Zhu Kezhen Curve illustrates significant oscillations of warming and cooling over the past five millennia, with notable trends such as:

1. Relatively warm periods during the Han and Tang dynasties
2. Slightly cooler during the Song dynasty
3. Marked cooling during the Ming and Qing dynasties, corresponding to the Little Ice Age
4. Pronounced warming in the modern era

This curve also highlights how climate variability has historically influenced Chinese agricultural civilization and societal development. The dashed line in Fig. 2, added by the author based on Sun et al. (2021), shows that current temperatures are about 1.7 °C higher than in the 1950s, lower than during the Qin dynasty (221–206 BCE), and roughly at the same level as during the Tang dynasty (618–907 CE).

Throughout the long history of humanity prior to the First Industrial Revolution, the Earth experienced repeated periods of warming and cooling due to natural factors. These climate variations were clearly not caused by human activity.

The Paris Agreement, the international framework for climate policy, aims to limit the rise in global average temperature to well below 2 °C above pre-industrial levels, while pursuing efforts to limit it to 1.5 °C. In this context, "pre-industrial" is generally defined in IPCC and related literature as the average temperature from 1850 to 1900 (IPCC 2021b). However, this period corresponds to the final phase of the Little Ice Age (approximately the 14th–nineteenth centuries), during which the global climate remained relatively cold (see Fig. 2). Using this relatively cold period as a reference may make contemporary temperature increases appear relatively large. Therefore, the uncertainties inherent in selecting a base year, and the need for transparent explanation of such choices, are critically important for building trust between the scientific community and the public.

It is scientifically well established that greenhouse gases such as carbon dioxide and methane exist; however, the causal relationship between these gases and the observed climate changes, specifically global warming, is not yet fully understood. Therefore, climate change can be regarded as a highly uncertain issue.

Regarding climate change, significant uncertainties remain across multiple dimensions, including scientific knowledge, economic impacts, and societal responses. These uncertainties extend beyond academic interest, deeply influencing actual policy formulation, technology selection, investment decisions, and even citizen behavior. Consequently, it is crucial to systematically understand the nature of these uncertainties and incorporate their characteristics into policy design.

Below, uncertainties in climate change are categorized broadly into three aspects—scientific, economic, and other dimensions—and their characteristics and policy implications are discussed.

3 Types of Uncertainty in Climate Change

3.1 Four Levels of Uncertainty

Uncertainty refers to situations in which probabilities are unknown (Knight 1921). The Intergovernmental Panel on Climate Change (IPCC) often defines uncertainty as "future variables for which the probability distribution is unknown or can only be expressed subjectively," following Knight's definition (IPCC 2001).

As shown in Fig. 3, uncertainty can be classified into four levels (Courtney et al. 1997).

Level 1: Certainty—A Nearly Predictable Future

The future can be predicted with sufficient accuracy, and a single strategy is sufficient to address it.

Level 2: Risk—A Scenario-Based Future

Several scenarios can be envisioned, though it is uncertain which one will actually occur. It is possible to assign probabilities to each.

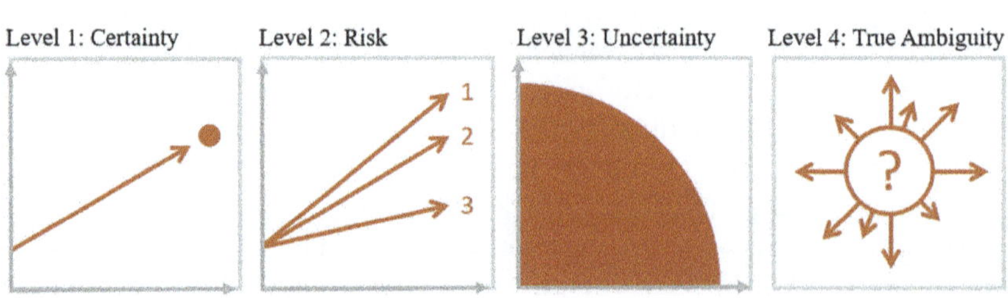

Fig. 3 Four levels of uncertainty (Adapted from Courtney et al. 1997)

Level 3: Uncertainty—A Directionally Clear Future

The future is bounded by several key variables, which provide a general direction, but the exact outcome within that range is unpredictable.

Level 4: True Ambiguity—A Completely Unforeseeable Future

It is not even possible to foresee what might happen in a given situation—or whether anything might happen at all.

In a broad sense, uncertainty refers to both [Level 2] Risk and [Level 3] True Uncertainty. An important recognition is that not all decisions face the same level of uncertainty. Decision theory categorizes choices into three types: (i) decisions under certainty, (ii) decisions under risk, and (iii) decisions under uncertainty. Climate change is a challenge that requires action under conditions of high uncertainty.

At the same time, climate change is a "complex system" issue that extends beyond the natural sciences to encompass social, economic, institutional, and cultural dimensions. Therefore, in discussions of climate policy and technological responses, "uncertainty" becomes an unavoidable and central element. Uncertainty does not merely imply a lack of information; it refers to situations in which future events or system behaviors are difficult to predict and the outcomes of choices are inherently limited in visibility. Here, uncertainties in climate change are categorized into four types—scientific, technological, economic, and social—and their characteristics and implications for policy and technology decisions are discussed in detail.

3.2 Scientific Uncertainty in Climate Change

Scientific uncertainty is an unavoidable challenge in formulating climate change policies. Here, "scientific uncertainty" refers to uncertainty arising from the limits of knowledge and prediction based on the natural sciences. Key factors include the incomplete understanding of the climate system, limitations of climate models, and the diversity and unpredictability of future greenhouse gas emission pathways.

3.2.1 Complexity of the Climate System and Natural Variability

The Earth's climate system is an extremely complex physical and biological system in which components such as the atmosphere, oceans, ice sheets, biosphere, and land surface interact. Nonlinear relationships exist among these components, and many phenomena cannot be explicitly understood in terms of cause and effect, significantly limiting prediction accuracy.

According to the IPCC Sixth Assessment Report (AR6) Working Group I (IPCC 2021b), natural variabilities such as the El Niño–Southern Oscillation (ENSO) and the North Atlantic Oscillation (NAO) occur cyclically over periods ranging from years to decades, exerting significant influence on temperature and precipitation patterns. However, accurately predicting the timing and magnitude of these events remains challenging with current scientific knowledge. Moreover, changes in cloud formation and properties, alterations in snow and ice albedo causing positive and negative feedbacks, and carbon cycle dynamics mediated by terrestrial ecosystems and oceans all contribute additional uncertainty to climate change progression and impacts.

Additionally, attention has been drawn to "tipping points", thresholds beyond which the climate system may undergo irreversible changes. For example, the melting of the Greenland ice sheet or methane release from Arctic permafrost could trigger accelerated warming if these critical thresholds are surpassed. While these phenomena are theoretically understood, predicting when and how they will manifest in reality remains highly uncertain.

3.2.2 Limitations and Structural Challenges of Climate Models

General Circulation Models (GCMs) play a central role in projecting future climate changes by simulating the Earth's climate system as a whole (IPCC 2021c). These models numerically represent the physical, chemical, and biological processes of the atmosphere, oceans, land, and biosphere, and are used to predict future climate conditions (Stouffer et al. 2007).

In recent years, numerical simulation techniques in physics have gained high scientific recognition. In 2021, Syukuro Manabe and colleagues were awarded the Nobel Prize in Physics for their work on climate modeling (Nobel Prize 2021). This recognition indicates that numerical simulations have transitioned from auxiliary tools to officially acknowledged methods in physics, becoming an essential component of scientific research alongside experimental and theoretical physics. Nevertheless, climate models have several intrinsic limitations.

First, there is the issue of parameterization. Micro-scale phenomena such as cloud formation, atmospheric turbulence, and precipitation processes cannot be physically resolved at every grid point with current computational capabilities, and are instead approximated statistically or empirically as parameters. Parameterization is a major source of differences among models, leading to variations in projected temperature increases and precipitation changes for the same greenhouse gas emission scenario (Flato et al. 2013).

Second, there is the problem of sensitivity to initial conditions. Climate models start calculations from a specific state of the atmosphere and oceans, but even small differences in initial conditions can lead to substantial long-term divergences in projections. This property, known as the butterfly effect, limits the predictability of the climate as a nonlinear system.

Third, spatial and temporal resolution constraints must also be considered. While broad trends, such as global mean temperature increases, can be predicted with reasonable confidence, deriving detailed climate impacts at the country or city level—such as the frequency of extreme weather, agricultural effects, or disaster planning applications—remains challenging. This limitation can undermine decision-making in policy practice, particularly regarding adaptation strategies and budget allocation.

Thus, although climate models represent a significant scientific achievement, the uncertainties arising from their structural limitations remain substantial, and caution is required when interpreting their predictive reliability.

3.2.3 Uncertainty in Greenhouse Gas Emission Pathways

To project future climate conditions, climate models require assumptions about greenhouse gas (GHG) emission scenarios (IPCC 2021c). However, these emission pathways themselves are highly uncertain. The root of this uncertainty lies in the future trajectory of human society. Factors such as global population trends, lifestyles, economic growth rates, energy demand, energy mix (fossil fuels vs. renewables and nuclear), the pace of technological innovation, the adoption of carbon pricing, and potential changes in international climate agreements (e.g., withdrawal from the Paris Agreement) all influence future emissions.

Accordingly, the IPCC uses multiple emission pathways, including Representative Concentration Pathways (RCPs) and Shared Socioeconomic Pathways (SSPs), each corresponding to different climate projections (IPCC 2021a). For example, RCP2.6 assumes stringent emission reductions, while RCP8.5 represents a "business-as-usual" high-emission scenario. However, it remains inherently unpredictable which path humanity will actually follow (IPCC 2021a).

Uncertainty in social behavior and institutional design is also crucial. While an increasing number of countries and companies declare carbon neutrality goals, these targets often lack legal enforceability and depend heavily on political and economic realities, making their actual achievement uncertain. For instance, under the Trump administration, the United States withdrew from the Paris Agreement, creating significant uncertainty about the realization of carbon neutrality goals. Therefore, climate scenarios are not merely a natural science problem but are deeply intertwined with social science and policy considerations.

3.2.4 Implications of Scientific Uncertainty for Policy

Scientific uncertainty exerts a complex influence on climate policy design. Given the broad range of predictions and their inherent limitations, policymakers are forced to make decisions under uncertainty.

3.3 Technological Uncertainty in Climate Change

There are two primary approaches to addressing climate change: mitigation and adaptation. Mitigation aims to slow the progression of climate change itself through the reduction or absorption of greenhouse gas emissions, whereas adaptation seeks to enable societies and ecosystems to respond flexibly to climate impacts that are already occurring or are inevitable.

Technological uncertainty refers to the difficulty in predicting the development, deployment, diffusion, and long-term effectiveness of mitigation and adaptation technologies. All such technological measures involve various uncertainties, posing significant challenges for policy decision-making and implementation.

3.3.1 Uncertainty in Mitigation Technologies

Uncertainty in mitigation measures can be primarily grouped into four aspects: technological maturity, cost, policy and institutional frameworks, and social acceptance. For example, advanced technologies such as Carbon Capture, Utilization, and Storage (CCUS) and green hydrogen hold great theoretical potential, yet most have not reached full commercial deployment. Only after demonstration results are obtained can these technologies become realistic options. Even when technically feasible, economic competitiveness is uncertain, as factors such as price fluctuations of key materials (e.g., lithium or nickel) significantly influence the cost of batteries, demonstrating the close interconnection between technology and markets.

The implementation of mitigation technologies is also closely linked to policy and institutional framework stability. Carbon pricing mechanisms or renewable energy incentives also carry the risk of being altered by political shifts or administrative transitions. Should these frameworks shift, private investment that depends on them may falter, hindering long-term technology deployment. Furthermore, from the perspective of social acceptance, even technically advanced measures may face delays due to local opposition, misunderstandings, or lack of trust. Typical examples include aesthetic and noise concerns with wind power, and persistent distrust of nuclear energy.

This uncertainty has a multi-layered structure. First, there is uncertainty associated with Technology Readiness Levels (TRLs) (Grübler et al. 2018). Renewable technologies such as solar PV and onshore wind that are already widely deployed differ significantly in reliability, scalability, and institutional acceptance compared with demonstration-stage technologies like Direct Air Capture (DAC) or negative emissions technologies (NETs). For instance, Carbon Capture and Storage (CCS) can theoretically achieve large-scale emission reductions, but the long-term safety of underground storage and local acceptance remain uncertain.

Second, complex interdependencies among technologies amplify uncertainty. Integrating variable renewable energy requires complementary technologies such as transmission and distribution infrastructure, batteries, smart grids, and demand-side management. However, the development stages and technical standards of these components are not always fully aligned, making it difficult to assess overall performance and cost–benefit ratios (IPCC 2014a).

3.3.2 Uncertainty in Adaptation Technologies

Adaptation measures also carry their own unique uncertainties. A particularly prominent source is the uncertainty associated with predicting climate change itself. While global-scale predictions achieve a certain level of accuracy, regional-scale changes—such as temperature and precipitation patterns at the level of cities or river basins—remain highly uncertain (IPCC 2021b). Consequently, it is difficult to determine the scale, location, and timing at which adaptation measures will be required. Moreover, adaptation technologies, such as disaster-resilient infrastructure or climate-tolerant crops, do not come with guarantees regarding their "temporal validity," i.e., how effective they will remain under environmental conditions several decades into the future (Smit and Wandel 2006).

Institutionally, adaptation often represents a preventive investment, which tends to be assigned lower fiscal priority. In particular, developing countries and local governments may lack the financial and human resources necessary to implement adaptation plans, resulting in cases where plans exist but are not operationalized. Adaptation also raises issues of inequality: wealthier individuals and urban areas can adopt adaptation technologies more flexibly, whereas poorer populations and peripheral regions may lack the capacity, leading to uneven impacts of climate change. This is not merely a technical problem but increasingly a matter of climate justice and distributional fairness. For example, infrastructure design for coastal flooding or urban flood control faces scientific uncertainty tied to future rainfall patterns, while the development of drought-resistant crops is influenced by diverse factors such as climate, pests, and ecosystem changes (Adger et al. 2005).

3.3.3 Complex Systems Nature and Interdependencies of Climate Technologies

A common feature of both mitigation and adaptation technologies is that climate action represents a typical complex systems problem. The introduction of a single technology can have ripple effects across energy systems, employment, land use, ecosystems, and urban infrastructure, making it difficult to predict overall impacts in advance (Sovacool et al. 2022).

Ensuring the effectiveness of technologies requires supporting data, monitoring, and evaluation frameworks, which are often underdeveloped and impede assessment of policy outcomes. Technological uncertainty cannot be fully eliminated. Therefore, building adaptive governance—which assumes uncertainty and emphasizes flexible institutional and technological adjustments in response to changing circumstances—is essential for 21st-century climate policy. This approach corresponds to the "adaptive policymaking" discussed in Sect. 10.2.3.2.

Uncertainty in climate technologies is not merely a matter of technical difficulty; it also encompasses the multi-layered challenges of institutional alignment, normative frameworks, and social acceptance inherent in technologies embedded within society.

Moving forward, there is an increasing need to systematize new policy knowledge and practical approaches that take uncertainty itself as a point of departure.

3.4 Economic Uncertainty in Climate Change

Economic uncertainty refers to the unpredictability associated with the costs and profitability of technological investments, resource prices, market mechanisms, and policy incentive structures (Pfenninger et al. 2014). Many climate technologies require high upfront investments while delivering long-term benefits that are uncertain, making investment decisions inherently risky. For example, the levelized cost of electricity (LCOE) for renewable energy has decreased significantly over the past decade, but it is unclear whether this trend will continue at the same pace. Moreover, the widespread adoption of batteries depends on the supply of critical metals such as lithium, nickel, and cobalt, which are subject to significant price fluctuations due to geopolitical risks and extraction conditions. High-cost technologies like CCUS and DAC face difficulties in attracting private investment if future carbon price expectations are unstable (Grübler et al. 2018).

Similarly, in adaptation measures, accurately predicting future disaster losses is challenging, and cost–benefit analyses for preventive investments (e.g., levees, elevated housing, water infrastructure) are highly sensitive to assumptions. On the international stage, trade policies, climate-related tariffs (e.g., CBAM, Carbon Border Adjustment Mechanism), and funding flows for developing countries (e.g., Green Climate Fund) further increase economic uncertainty. Market-based mechanisms such as carbon taxes or emissions trading systems are vulnerable to policy changes, and if long-term predictability is lacking, corporate and municipal investment decisions may be restrained.

3.5 Social Uncertainty in Climate Change

Social uncertainty refers to the unpredictability of individual, group, and societal values, perceptions, behaviors, and institutional responses regarding climate change and related technologies (O'Brien et al. 2007). This encompasses policy acceptance, public participation, risk perception, and ethical decision-making. For instance, public support for climate policies is highly sensitive to short-term increases in living costs (electricity bills, gasoline taxes) or restrictions on freedoms (e.g., private car regulations), and can quickly reverse with elections or regime changes. Technologies such as nuclear power, CCS, or large-scale renewable installations (e.g., offshore wind) may be technically promising but remain difficult to implement without local consensus. Building social trust in technological safety is a long-term process, yet accidents or incidents can instantly erode it.

Furthermore, the impacts of climate change and the distribution of adaptation measures are often uneven across social strata, giving rise to issues of climate justice. Vulnerable groups, including the elderly, people with disabilities, and low-income populations, may be left behind by adaptation measures, potentially perpetuating inequalities. In addition, the uneven distribution of information and misinformation through media and social networks can exacerbate social divisions and prompt misguided behavioral choices.

3.6 Addressing Complex, Dynamic, and Institutional Uncertainty

As discussed above, uncertainties related to climate change span multiple dimensions—scientific, technological, economic, and social—and are intricately interlinked, influencing each other in complex ways. Effectively addressing this multilayered uncertainty requires flexible and adaptive policy design as well as strategic responses grounded in a long-term perspective. Scientific uncertainty affects technology choices, technological indeterminacy increases economic investment risks, and economic instability can undermine social support—forming a multi-layered structure of interdependent uncertainties.

To respond to such uncertainty, policy design must prioritize adaptive capacity over the mere accuracy of decisions. This entails securing options that can respond to future changes, facilitating dialogue among stakeholders, ensuring transparency of information, and implementing multi-stage policy evaluation and revision processes (i.e., policy learning). Crucially, this approach reflects a shift in perspective: uncertainty is not an obstacle to be eliminated but a condition to coexist with, which serves as a critical starting point for policy engineering in the era of climate change.

The domain of climate change and energy transition is characterized by: (i) Scientific uncertainty; (ii) Diversity of policy objectives; (iii) Heterogeneity of stakeholders. The pace of technological progress, economic impacts, and international policy trends are difficult to predict. Additionally, assessments of benefits and risks vary across industry, civil society, and government actors. Under such circumstances, policy design must incorporate mechanisms to diversify risk and maintain flexibility capable of withstanding future uncertainty.

First, regarding complexity, energy supply structures, technology choices, economic systems, and environmental impacts are closely interrelated, meaning that a single policy decision can have cascading effects across domains. For instance, the deployment of renewable energy promotes greenhouse gas reduction but also requires coordination with grid stability and land use. Due to these strong interdependencies, analysis based on a single factor is insufficient for informed policy decisions.

Second, regarding uncertainty, the pace of scientific discovery, technological innovation, and international policy developments complicate predictions. While IPCC reports present long-term trends, regional impacts and economic consequences still show wide variability. Furthermore, international mechanisms such as carbon pricing or regulatory tightening depend on national political and economic contexts, making it difficult to establish a stable medium- to long-term outlook.

Third, regarding divergent interests, a diverse array of actors—including industry, civil society, government agencies, and international stakeholders—participate in policy formation. Energy-intensive industries prioritize competitiveness, while environmental NGOs and local communities emphasize environmental protection and health risk reduction. Conflicts may also arise between local government and central government objectives. Resolving such divergent interests requires institutional designs that go beyond mere consensus-building to foster collaborative governance (Zhang 2024).

Therefore, in policy environments characterized by complexity, uncertainty, and pluralistic interests, conventional linear and deterministic policy design is increasingly ineffective. From a policy engineering perspective, it is essential to design policies that distribute risk and possess flexibility to adapt to future uncertainties. Practically, this entails: evaluating policies under multiple scenarios, introducing feedback mechanisms based on empirical data, and establishing institutional frameworks that maintain continuous dialogue among stakeholders. Such approaches pave the way for sustainable and stable policy implementation, even in the long-term, uncertainty-laden task of decarbonization.

In short, high levels of uncertainty in climate change are not synonymous with ignorance. Rather, they necessitate the introduction of design principles and methods that enable the best possible decisions under conditions of limited knowledge. Effective policy formation requires a precise understanding of the nature of uncertainty and the implementation of appropriate risk management strategies—one concrete approach being Policy Engineering, as discussed in the following sections.

4 Proposal of Policy Engineering and Its Approach to Decarbonization

Contemporary social problems are inherently complex, intertwining politics, economics, and technology; short-term and long-term perspectives; as well as local and global dimensions. Accordingly, the policies designed to address these issues are themselves multi-dimensional and constitute complex systems. Integration of the humanities and sciences is required not only in terms of the problem domain but also methodologically, necessitating interdisciplinary fusion. While many advocate for such integration, concrete results have been limited. This is not merely an academic issue but an urgent societal challenge. In corporate contexts, intellectual property, ethics, environmental issues, and bioethics increasingly demand solutions that combine insights from both the humanities and sciences, and there is a recognized shortage of personnel capable of bridging these domains. New academic frameworks must therefore offer concrete pathways for integrating the humanities and sciences.

Decarbonization, as a mitigation strategy against climate change, represents a technical challenge centered on energy transition, but it is also a comprehensive issue involving transformation of social institutions and economic systems. Consequently, addressing it within the boundaries of a single discipline or policy domain is insufficient. A policy engineering approach—which seeks to understand complex interactions, mediate competing interests, and identify optimal solutions—is indispensable. The following sections outline the key elements of policy engineering methods and their application to decarbonization strategies.

4.1 Definition of Policy Engineering and Its Methodological Characteristics

Policy Engineering is an interdisciplinary approach that systematically addresses complex and highly uncertain social challenges by leveraging scientific knowledge and engineering methods—analytical, mathematical, and quantitative—to design, implement, and evaluate policies throughout the entire policy formation process. Its ultimate goal extends beyond partial optimization to achieve overall optimization of the social system or robust optimization that accounts for uncertainty. The academic character of policy engineering lies in its transdisciplinarity, going beyond conventional interdisciplinarity. It views society as a complex system and adopts a perspective that comprehensively analyzes the interactions among its constituent elements. In particular, in the domains of climate change and energy transition, it is essential to integrate knowledge across diverse fields such as natural sciences, economics, sociology, political science, and international relations. Policy engineering serves as the intellectual nexus for this integration (Zhou et al. 2014).

It is worth noting that, as shown in Table 1, the course titled *Introduction to Policy Engineering* has been offered for over ten years as a foundational liberal arts subject for undergraduate students at Ritsumeikan University.

The core of Policy Engineering lies in the logic of optimization, aiming to maximize social utility (the benefits enjoyed by society) while simultaneously minimizing costs and risks (such as resource constraints, environmental burdens, and political friction). In this process, an engineering approach is indispensable, with methods such as complex network analysis, mathematical optimization (including robust and stochastic optimization), as well as simulation techniques like system dynamics and agent-based modeling being actively utilized. As illustrated in Fig. 4, policy engineering is positioned as an optimization system based on design science, pursuing the dual objectives of increasing "work energy" (social utility) and reducing "loss energy" (costs and risks).

Furthermore, policy engineering is not merely a collection of technical tools; it is a design science grounded in discovery science (SCJ 2005) that seeks to integrate scientific knowledge and institutional design to address the complex, multidimensional challenges of contemporary society. In doing so, it treats the policy process

Table 1 Syllabus of the foundational undergraduate course "Introduction to Policy Engineering"

Session 1: Introduction: Why "Policy Engineering"? The utility of policy science and policy engineering, and the social utility of knowledge construction in policy engineering	**Session 8: Extraction of Policy Issues (Problem Discovery and Goal Setting): Uncertainty Handling** (1) Ensuring reliability of communication (2) Decision-making under uncertainty (3) Uncertainty in global warming issues
Session 2: Design Science and Discovery Science Exploration of the demand for "what ought to be" and "what is," scientific approach to human practical activities, and the science of human cognitive activities	**Session 9: Policy Formulation: Systems Thinking** (1) Systems Engineering (2) Black Box, White Box, and Grey Box Approaches (3) Analysis, synthesis, and integration
Session 3: Engineering Approaches and Optimization of Policy Systems Engineering, Systems Engineering, and Social Engineering	**Session 10: Policy Formulation: Optimization Methods** Scenario planning and linear programming, robust optimization, from partial optimization to overall optimization
Session 4: Policy Science and the Policy Life Cycle Recognition of problems (extraction of issues), policy formulation (drafting), implementation, and management (evaluation)	**Session 11: Policy Analysis and Evaluation Methods** Fundamentals of policy evaluation, grasping causal relationships, policy cycle
Session 5: Fundamentals of Policy Engineering, Basics of Mathematics Basic concepts of mathematics frequently used in the policy field, data science of social data, and ensuring data accuracy	**Session 12: Policy Analysis and Evaluation Methods** Simulation, strategic management systems
Session 6: Extraction of Policy Issues (Problem Discovery and Goal Setting), Basics of Decision-Making Creativity required for policy proposals, implementation of policy brainstorming, etc	**Session 13: Practice of Policy Engineering: Experimental Policy Studies** Experiments in natural sciences and social sciences, computer-based experiments
Session 7: Extraction of Policy Issues (Problem Discovery and Goal Setting), Mathematical Modeling (1) Linear programming models (2) Optimal models and suboptimal models (3) Linear models and nonlinear models	**Session 14: Complex Systems Theory and Policy Engineering** Insights from Nobel Prize–winning physicists, complex systems theory, cooperation and integration

Fig. 4 Schematic diagram of an optimization system aiming to maximize work energy (social utility) and minimize loss energy (costs or risks) (Author's creation)

as a lifecycle spanning from problem analysis to policy recommendations, thereby transforming conventional policymaking based on rules of thumb or intuition into a rational process founded on scientific and engineering methodologies.

Methodologically, policy engineering is characterized first by systems thinking. It does not treat domains such as energy, economy, or environment in isolation but emphasizes their dynamic interdependencies. For example, introducing renewable energy is linked not only to greenhouse gas reduction but also to industrial restructuring, community acceptance, and international technological competitiveness. This requires a holistic perspective that accounts for multi-dimensional impacts rather than focusing solely on a single policy goal.

Second, integrated assessment is essential. By combining climate predictions and energy supply–demand simulations, economic cost–benefit analyses, and considerations of social acceptance and ethics, policy engineering allows for the evaluation of overall policy validity. This ensures that institutional design achieves both scientific rationality and social legitimacy.

Third, empirical analysis is emphasized. Theoretical policy effects often produce unforeseen side effects when applied in real-world contexts. Therefore, it is critical to verify policy impacts and limitations through actual data and case studies, incorporating the results into the redesign and refinement of institutions. Policy engineering distinguishes itself from traditional policy research by embedding this design–empirical validation–redesign feedback loop.

Where traditional policy studies have focused on institutional design and political processes, policy engineering is notable for integrating natural sciences, engineering, and social sciences. In areas such as energy policy, where technological innovation, market mechanisms, institutional design, and citizen behavior interact, a single academic framework is insufficient. An integrated approach combining engineering methods with social science insights is essential.

First, policy engineering emphasizes the integration of policy, technology, institutions, and behavioral science. For instance, promoting the adoption of renewable energy requires not only technological advancement in solar and wind power but also reforms in the transmission grid, electricity market price signals, and citizen acceptance and behavioral changes. Systematically linking these multi-dimensional elements enables highly effective policymaking.

Second, this framework is applicable to energy transitions. In East Asia, energy transition involves multiple challenges, including reducing fossil fuel dependence, introducing renewable energy, and advancing electrification, hydrogen utilization, and digitalization. Applying a policy engineering perspective allows for simultaneous analysis of (i) the pace of technological innovation and cost fluctuations, (ii) country-specific regulatory and market differences, and (iii) behavioral changes in civil society and industry, enabling the design of strategic transition pathways.

Third, policy engineering provides methodologies to address uncertainty and complexity. The future of climate change and energy markets is inherently uncertain, with fluctuating technology costs and variable international policy coordination. By employing methods such as scenario planning, system dynamics, complex network analysis, and risk governance, multiple options can be evaluated, enabling flexible and adaptive policy design. In regions like East Asia, where energy security, economic growth, and environmental protection coexist as conflicting demands, comprehensive optimization through policy engineering is indispensable.

Overall, policy engineering is positioned as an academic field that integrates scientific analysis with institutional implementation to enable effective and sustainable policymaking in a society characterized by complexity, uncertainty, and diverse interests. Two methods are particularly noteworthy: robust optimization (pursuing solutions that remain viable under any future scenario rather than aiming for a single optimal outcome) (Ben-Tal et al. 2009), and adaptive policymaking (adjusting policies incrementally under uncertainty while learning during the implementation process) (Walker et al. 2013). See Sect. 10.2.3 for further details.

4.2 Application Examples of Policy Engineering

Policy engineering is not an academic pursuit for its own sake; it is inherently practical, aiming to apply scholarly knowledge to the real world, implement policies, and improve societal outcomes.

4.2.1 Necessity of an Integrated Approach and Illustrative Cases

Policy engineering emphasizes the integration of policy, technology, institutions, and behavior. For example, promoting renewable energy deployment involves not only technological development but also market design and public acceptance.

Japan: The Feed-in Tariff (FIT) system accelerated solar power adoption, but challenges arose regarding grid capacity and output variability. In response, the government promoted next-generation transmission networks (smart grids) and regional microgrids, combining engineering approaches (demand–supply forecasting, battery control) with institutional reforms.

South Korea: Under initiatives such as RE100 and the Korean New Deal, offshore wind and hydrogen infrastructure are being developed as national strategies. This combines industrial and energy policy to simultaneously foster domestic industries and reduce greenhouse gas emissions.

China: As the world’s largest renewable energy market, China has advanced large-scale deployment of solar, wind, and storage technologies, and enabled interregional power exchange through ultra-high-voltage (UHV) transmission networks. This represents a typical policy-engineering case that integrates technological innovation, institutional reform, and investment guidance.

4.2.2 Applications to Energy Transition

Energy transitions in East Asia involve simultaneous decarbonization, electrification, hydrogen adoption, and digitalization. Policy engineering provides a framework for integrated design in this context.

Hydrogen energy introduction:

1. Japan: Formulated the Basic Hydrogen Strategy in 2017, advancing hydrogen production, transport, and utilization infrastructure.
2. South Korea: Released the Hydrogen Economy Roadmap (2019), prioritizing fuel cell vehicles and power generation.
3. China: Issued the Medium- and Long-Term Plan for Hydrogen Industry Development (2022), promoting cost reduction and commercialization of green hydrogen.

These cases integrate technology assessment, institutional design, investment guidance, and public acceptance.

Coal phase-out strategies:

4. South Korea plans to reduce aging coal-fired power plants significantly by 2030 and has introduced a carbon emissions trading system (ETS).
5. Japan pursues a gradual "fade-out" while considering energy security, implementing co-firing and ammonia introduction technologies.
6. China, despite a high coal share, aims for coal peaking and clean technology deployment in the 2040s, representing long-term optimization combining market mechanisms, technology, and diplomacy.

4.2.3 Addressing Uncertainty and Complexity

Policy engineering includes methodologies to handle uncertain futures, such as:

Scenario planning: Japan's Strategic Energy Plan considers multiple electricity supply scenarios to evaluate pathways for achieving carbon neutrality by 2050.

Robust optimization: South Korea evaluates renewable energy investments and grid expansion using models that account for price fluctuations and technology cost uncertainties, comparing multiple investment scenarios.

Adaptive policymaking: China conducts decentralized energy pilot projects, using empirical data to incrementally adjust national strategies. For instance, renewable energy demonstration zones in Guangdong and Inner Mongolia function as "experimental grounds" for learning and adapting institutional and market arrangements.

4.2.4 Empirical Validation and Feedback

A distinguishing feature of policy engineering is its emphasis on the cycle of practice, validation, and redesign. Policies are not merely theoretical designs; they are tested, refined, and evolved through real-world implementation.

1. Japan: The Zero Energy Building (ZEB) pilot projects repeatedly implement design, operation, and evaluation cycles, feeding results back into building standards and subsidy schemes.
2. China: The Energy Internet initiative incorporates local-level pilot results into national policy, linking decentralized experimentation to strategic decision-making.
3. South Korea: The Smart City projects integrate transportation, energy, and ICT sectors, utilizing citizen behavior data to inform policy design.

These examples demonstrate that policy engineering generates a feedback loop between theory and practice, contributing to the development of sustainable and flexible social systems.

Policy engineering is a science of integrated and adaptive policy formation, simultaneously pursuing maximization of social utility and minimization of risks and costs. In East Asia's energy policy, for example, integrating technological innovation, institutional design, citizen behavior, and international cooperation enables the design of realistic and sustainable strategies. This book examines these practical cases to explore East Asia's strategy toward achieving carbon neutrality.

4.3 Approach to Decarbonization Using Policy Engineering Methods

Decarbonization is not limited to the mere introduction of greenhouse gas (GHG) reduction technologies; it requires multidimensional efforts, including the transformation of energy supply structures, minimization of environmental impacts, rationalization of economic costs, and establishment of institutional support. The deployment of renewable energy, for instance, necessitates advanced grid systems and market reforms, and carbon pricing must be considered in conjunction with industrial policy and social acceptance.

Such initiatives should be promoted through collaboration among multilayered actors: international organizations, national governments, local authorities, businesses, and civil society groups. Conflicting interests should not be treated as purely zero-sum but rather as opportunities for coordination and collaboration. Participatory workshops, social experiments, and stakeholder dialogues are essential tools for fostering consensus.

Furthermore, promoting decarbonization requires a framework that integrates technological innovation, institutional design, economic incentives, and socio-cultural contexts. For example, the adoption of electric vehicles depends not only on advancements in battery technology but also on charging infrastructure, subsidies and tax incentives, and changes in consumer awareness. Policy engineering models these factors as an integrated system, enabling both design and evaluation.

This integrated approach is implemented through a cyclical process:

1. Goal setting
2. Policy design
3. Impact prediction
4. Assessment of social acceptance
5. Implementation and feedback

Rather than one-off institutional measures, repeated empirical validation and adaptive adjustments optimize policy and enhance the effectiveness of decarbonization.

Policy engineering also utilizes diverse analytical methods to address future uncertainty and complexity:

1. Scenario analysis considers multiple potential futures to design flexible policy options.
2. Integrated assessment models combine climate, economic, and technological factors to evaluate long-term policy effects.
3. Robust optimization emphasizes solutions that remain resilient under various futures rather than pursuing a single "optimal" solution.
4. Adaptive policymaking provides a framework for incrementally adjusting policies and learning through the implementation process.

By combining integrated design with uncertainty-responsive analytical methods, decarbonization policies can advance more effectively, achieving both social acceptance and flexibility.

In summary, policy engineering provides an integrated and collaborative framework to tackle the complex and uncertain challenges of energy transition and decarbonization. Linking technological, institutional, economic, and social factors, and creating a feedback loop between analysis and practice, constitutes a pathway toward a sustainable low-carbon society.

5 Carbon Neutral Targets and the Specificity of Energy Transition in East Asia

The path to carbon neutrality is deeply rooted in the institutional, economic, and energy contexts of each country, and East Asia exhibits particularly notable diversity. The region is both the world's largest CO_2 emitter and an area of rapid growth in renewable energy investment. However, the substantive content of carbon neutrality pledges and the direction of energy policies vary significantly across countries. This section compares the decarbonization strategies of China, Japan, and South Korea, highlighting institutional challenges and trade-offs with energy security.

5.1 *Status and Differences Among China, Japan, and South Korea*

5.1.1 China: Top-Down Approach and Institution-Led Decarbonization

(1) National Targets and Current Energy Mix

China has set national targets to achieve carbon neutrality by 2060 and peak CO_2 emissions by 2030. To this end, the central government has implemented a top-down, centrally led policy framework, rapidly deploying renewable energy, particularly solar and wind. However, coal dependency remains high, and challenges such as regional disparities and delays in power system development persist.

In China's 2023 energy mix, coal-fired generation still accounts for roughly 60% of electricity, although non-fossil energy installed capacity (renewables, hydro, nuclear) reaches 58.2% on a capacity basis. By the end of 2024, cumulative installed solar capacity reached ~ 887 GW, and wind capacity ~ 521 GW, both among the largest in the world. In 2023 electricity generation, hydro accounted for 13.2%, wind 9.4%, solar 6.2%, and nuclear 4.6%.

In July 2025, electricity consumption reached a historical maximum of 1.02 trillion kWh, comparable to the annual consumption of all ASEAN countries or Japan (State Council 2025). This surge was driven primarily by strong demand in the production and consumption sectors. Record consumption reflects strong manufacturing and retail performance, along with extreme summer heat increasing peak air-conditioning demand. Despite this surge, large-scale blackouts were rare, reflecting technical advancements and stable supply infrastructure. However, the increase raises concerns regarding economic efficiency and environmental impacts.

China continues to build large-scale coal-fired power plants, with new capacity expanding in early 2025. Notably, the central government has promoted a policy-package approach, using coal as a backup while advancing green power market expansion and grid integration as part of a long-term strategic plan (Reuters 2025).

(2) Institutional Framework: Dual-Control Policy and Emissions Trading System (ETS)

Institutionally, China's dual-control policy is a hallmark. Initially targeting energy consumption (total and intensity), it has recently expanded to carbon emissions (total and intensity), assigning numerical targets to local governments to encourage regional reduction efforts. The decoupling of direct energy consumption control may partly explain the over 1 trillion kWh electricity consumption in July mentioned earlier.

China's national ETS, launched in 2021, currently covers the power sector, with plans to expand to heavy industry sectors such as steel, cement, and petrochemicals. A distinguishing feature is the rapid, centrally led development of institutional infrastructure and market mechanisms, highlighting China's unique approach to decarbonization.

5.1.2 Japan: Public–Private Partnership–Driven Transition and Institutional/Infrastructure Constraints

(1) National Goals and the Three-Pillar Strategy

Japan has set the achievement of carbon neutrality by 2050 as a national goal. Its energy policy is centered on three pillars: "expansion of renewable energy deployment," "reevaluation and reuse of nuclear power," and "practical and commercial application of next-generation fuels such as hydrogen and ammonia." Through these measures, the decarbonization of the energy mix is being promoted in stages, with plans to accelerate emissions reductions primarily in the power generation sector.

The 7th Strategic Energy Plan was formulated in alignment with Japan's national goal of achieving carbon neutrality by 2050, targeting a 73% reduction in greenhouse gas emissions by FY2040 compared with FY2013 levels. The plan also reaffirms the fundamental policy principles of "Safety," "Energy Security," "Economic Efficiency," and "Environment"—collectively known as "S + 3E".

Looking toward FY2040, Japan has set a long-term vision to raise the share of renewable energy to 40–50% and nuclear power to approximately 20%, resulting in a maximum of 70% non-fossil power generation. This indicates a transition away from fossil fuel dependence toward a power supply dominated by renewables and nuclear energy.

(2) Structural Challenges and Institutional Bottlenecks

However, Japan's energy transition faces several structural and institutional challenges. First, since the Great East Japan Earthquake, the direction of nuclear policy has been unclear, lacking long-term consistency, which leaves overall energy policy with uncertainty. Second, the development of transmission infrastructure necessary for expanding renewable energy has lagged, and weak interregional grid connections constrain renewable deployment, while geographic imbalances and mismatches remain problematic. Third, while liberalization of the electricity market has increased competition, it has also made long-term government-led investment guidance more difficult, reducing the relative policy inducement power.

(3) Progress in Private Sector–Led Decarbonization

Meanwhile, decarbonization efforts by the private sector have steadily advanced. Companies are enhancing disclosure in line with the Task Force on Climate-related Financial Disclosures (TCFD), expanding renewable energy adoption through RE100 and Power Purchase Agreements (PPAs), and implementing voluntary carbon offsets. These initiatives serve as effective means toward achieving carbon neutrality through public–private collaboration. Moving forward, further institutional development, strengthening of physical infrastructure, and ensuring long-term policy stability will be increasingly necessary.

Japan's institutional approach is characterized by "innovation-driven, public–private coordinated, and plan-focused" features. Unlike the strict regulatory approach seen in Europe, Japan aims for a transition compatible with economic growth, promoting gradual decarbonization by leveraging voluntary corporate initiatives and market mechanisms.

5.1.3 South Korea: European-Oriented Institutional Design and Corporate-Led Growth Strategy

(1) Energy Policy Restructuring and Nuclear Reevaluation

South Korea has set a national goal of achieving carbon neutrality by 2050 and formulated a long-term plan to have 70% of the power mix come from non-fossil energy (nuclear + renewable energy) by 2038. As of 2023, the share of carbon–neutral power sources was about 39%, but it is planned to rise to 70.7% by 2038.

While South Korea once pursued a "nuclear phase-out" path, the 10th Basic Energy Plan adopted in 2023 reevaluated the role of nuclear power, targeting a nuclear share of over 30% by 2030. Meanwhile, the renewable energy share was in the low 8% range in 2022, among the lowest in OECD countries, with site constraints, local conflicts, and delayed grid development hindering deployment.

(2) Carbon Pricing Policies and European-Oriented Institutional Design

In 2015, South Korea introduced the first nationwide emissions trading system in Asia (K-ETS), which currently covers about 700 entities. Future measures under consideration include a carbon budget system and carbon tax. The institutional design adopts a European-style approach heavily influenced by the EU-ETS, and domestic rules are being aligned with the Carbon Border Adjustment Mechanism (CBAM). South Korea's carbon neutrality strategy also draws on European frameworks for "hydrogen policy" and "renewable energy/ESG strategies," with many policies referencing European standards, indicators, and target-setting.

(3) Corporate-Led Green Investment and Decarbonization

Significant green investments by major companies such as Samsung, SK, LG, and Hyundai have been advancing, with notable expansion into renewable PPAs, hydrogen, and battery sectors. Especially in export-oriented manufacturing, participation in RE100, TCFD disclosure, and setting of GHG reduction targets have become more active, strengthening a corporate-led framework to respond to international decarbonization demands (Table 2).

5.2 *Leveraging the Latecomer Advantage*

The challenges facing East Asia can be analyzed across three interrelated dimensions: economic development, pollution abatement, and global environmental issues. Historically, advanced countries such as Japan have sequentially experienced these challenges. First, they reduced poverty through economic growth; next, they

Table 2 Differences in carbon neutral measures among China, Japan, and South Korea and future challenges

Perspective	China	Japan	South Korea
CN Target	2060	2050	2050
Main Actors	Central government, state-led	Government–civil cooperation	Government + enterprises
Institutional Characteristics	Top-down, policy package-type apprach, introduction of ETS, dual-control policy	Technology innovation-led, government–civil collaboration, voluntary initiatives	Legal institutionalization, ETS already in place, carbon pricing aligned with EU goals, president-led, centralized authority
Technological Strategy	Decarbonization of fossil fuels, simultaneous expansion of renewables and nuclear power	Renewable energy + nuclear power + hydrogen + ammonia	Nuclear power resurgence + hydrogen-centered strategy
Role of Private Sector	Guided by policy	Strong autonomy, market-responsive	Investment-led by large corporations

addressed severe industrial and urban pollution; and finally, they tackled global environmental issues, including climate change. This staged experience enabled the gradual establishment of institutional, technological, and regulatory frameworks.

In contrast, developing countries, including China and Southeast Asian nations, currently face all three challenges simultaneously. They must sustain economic growth while continuing poverty reduction, cope with severe pollution arising from rapid industrialization and urbanization, and meet international expectations for climate action and decarbonization. The coexistence of these triple challenges necessitates an integrated balancing of economic development, domestic environmental protection, and global climate responsibility, thereby increasing the complexity of policy design.

Figure 5 illustrates the carbon Kuznets curve for Japan, China, and South Korea, showing the temporal changes in per capita GDP and per capita CO_2 emissions. The figure indicates that up to a per capita GDP of approximately USD 10,000 (PPP), the three countries' CO_2 emissions followed almost identical upward trends. However, beyond the USD 10,000 threshold, China's emissions are lower than those of Japan and South Korea at the same per capita GDP level. This difference reflects not only the absolute level of emissions but also variations in industrial structure, energy composition, and technological level, suggesting the manifestation of a so-called "latecomer advantage."

Furthermore, Japan and South Korea reached their carbon peaks at a per capita GDP (PPP) of approximately USD 40,000. In China's case, as shown in Fig. 6, the government's target year for the carbon peak is 2030, by which time per capita GDP

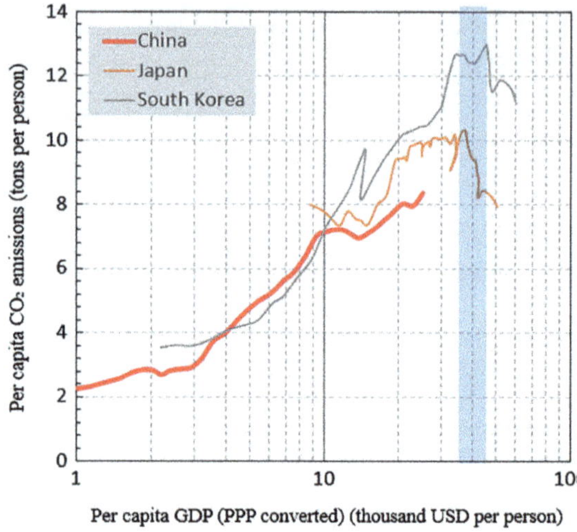

Fig. 5 Carbon Kuznets curve for Japan, China, and South Korea (IMF 2025; Ritchie et al. 2023)

(PPP) is projected to be around USD 40,000, indicating a peak at a level similar to that experienced by Japan and South Korea.

This analysis suggests that the stage of industrial development and the speed of technological innovation are critical factors in the relationship between economic growth and emissions. It also indicates that late-developing economies could potentially limit emissions at comparable stages of economic development through more efficient energy use and the adoption of low-carbon technologies. Therefore, the carbon Kuznets curve can serve not only as an empirical observation but also as a useful reference for formulating decarbonization strategies aligned with each country's industrial and technological policies.

Source: GDP and population data up to 2024 are actual values; subsequent figures are author's estimates.

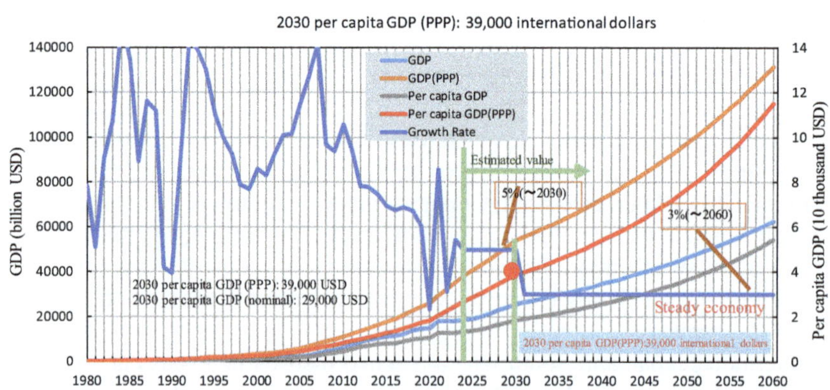

Fig. 6 Actual and projected values of China's economic scale (Created by the author)

Therefore, assuming the structural differences in experience between advanced and developing countries, achieving sustainable development across East Asia requires not only domestic policies in each country but also a multi-layered strategy that includes regional coordination. Policy design that integrates economic, environmental, and social dimensions will be key to achieving decarbonization and sustainability in the region.

Japan, China, and South Korea have adopted different decarbonization strategies according to their respective political, economic, and institutional contexts, but they share common challenges: "policy consistency" and "infrastructure development." The success of future carbon–neutral strategies will depend on how these challenges are overcome and implemented sustainably.

5.3 Energy Security and the Trade-Off with Economic Growth

Japan and South Korea rely heavily on imported energy, particularly crude oil and natural gas from the Middle East. According to the International Energy Agency (IEA), as of 2023, imported energy accounted for approximately 94% of Japan's total energy supply and 92% for South Korea (IEA 2023). This high import dependency exposes both countries to geopolitical risks, such as those in the Strait of Hormuz, and makes their economies sensitive to international energy price fluctuations, including surges in oil and LNG prices.

Under such circumstances, advancing a decarbonization transition entails short-term economic burdens. Renewable energy surcharges under the FIT (Feed-in Tariff) and FIP (Feed-in Premium) schemes raise electricity costs for households and companies. Massive investments required for the reconstruction and reinforcement of transmission networks also contribute to cost pass-through. In Japan, while renewable energy deployment in rural areas is progressing, transmission bottlenecks to urban centers limit the pace of decarbonization.

On the other hand, long-term prospects for economic growth through a green economy are promising. Investment in clean energy, development and export of decarbonization technologies, and the creation of new employment opportunities are expected to act as growth engines. Consequently, countries face a policy choice: whether to prioritize short-term stability and price containment or to pursue long-term transformation and growth.

In South Korea, the tension between implementing decarbonization policies and industrial pressure is particularly evident. The government initially planned to reflect renewable energy deployment costs in electricity rates, but strong demands from manufacturing industries have led to partial reconsideration of this approach. Maintaining the competitiveness of energy-intensive industries (steel, chemicals, semiconductors) is a key national policy objective, so the pass-through of green costs must be handled cautiously. Such trade-offs between accelerating decarbonization and economic rationality are common across many East Asian countries.

China faces similar challenges. As the world's largest energy consumer, China must balance energy security with economic growth. While promoting renewables and nuclear power, reliance on coal remains high, and a rapid shift to renewable energy carries risks of supply instability and higher electricity costs. In July 2025, for example, electricity consumption reached 10.2 trillion kWh, and regional transmission constraints caused localized supply shortages (State Council 2025). The Chinese government is attempting to balance these issues through centrally guided policies, including region-specific reduction targets and green power deployment plans, but reconciling short-term economic burdens with long-term decarbonization remains a challenge.

5.4 The Substantive Meaning and Institutional Challenges of Carbon Neutrality Declarations

In the United States, under President Trump's second term, the withdrawal from the Paris Agreement continued, effectively halting the implementation of greenhouse gas reduction targets. This temporarily reduced the U.S.'s international leadership in carbon neutrality efforts and impacted global climate cooperation.

Meanwhile, many countries have set carbon neutrality targets due to international pressure and domestic political needs, but there are significant differences in their substantive meaning and institutional progress. In some cases, the declaration is merely a political slogan, while in others, it is backed by a clear roadmap and institutional reforms.

In centralized systems like China, industrial and decarbonization policies can be integrated under national strategy, allowing for rapid implementation. However, challenges such as divergences with local governments and limited policy transparency remain.

Japan operates under a highly institutionalized decentralized system, where local governments play a significant role in energy policy and urban planning. Therefore, reforms toward carbon neutrality require consensus-building and institutional coordination between national and local governments.

South Korea, in contrast, has a relatively centralized system, with policy formulation and implementation led by the presidency and relevant ministries. While decision-making can be relatively fast, diverse implementation at the local level and societal consensus remain challenges. Key institutional priorities include reforming the transmission network, improving the effectiveness of emissions trading schemes, and designing incentives for renewable energy adoption.

Achieving carbon neutrality also requires long-term transformation of social systems, including not only energy supply structures but also urban planning, transportation, buildings, and lifestyles. This necessitates multi-layered governance, integrating national and regional strategies, cross-sectoral coordination, and collaboration with the private sector.

Carbon neutrality declarations are symbolic policies demonstrating responsible climate leadership internationally but require complex institutional coordination during implementation.

1. MRV (Monitoring, Reporting, and Verification) systems must be reliable. Challenges exist in China's ETS, Japan's GX League, and South Korea's K-ETS regarding transparency and compliance frameworks.
2. Transition planning and industrial restructuring must be reconciled. Just transition for regions (e.g., coal mining, fossil fuel power) and industries (steel, cement, petrochemicals) affected by decarbonization is critical.
3. Finance and international cooperation are also foundational. In ASEAN countries, mechanisms akin to an "Asian Green Deal" funded by Japan, China, and international development banks are needed. Mobilizing private capital, ESG finance, and green bonds is key.
4. Many countries still lack a mature integrated climate policy mix, with gaps among carbon pricing, subsidies, regulation, and R&D support creating bottlenecks. For example, emissions trading markets like the EU-ETS have not yet formed in East Asia (Zhou et al. 2021).

Decarbonization strategies in East Asia are shaped by differences in economic structures, political systems, and social acceptance. Even when countries formally declare the same goal of carbon neutrality, the meaning and policy instruments vary greatly. This highlights that achieving decarbonization requires policy design grounded in institutional, cultural, and political contexts, necessitating an integrated and evidence-based policy engineering approach.

6 Structure of This Book and Analytical Perspective

This book examines decarbonization challenges in East Asia using Policy Engineering methodology, analyzing strategies (decarbonization pathways), systems (optimal energy mix and institutional design), and innovation (technological and social transformation). Its aim is to provide practical, interdisciplinary scenarios for designing decarbonization policies in a complex reality characterized by uncertainty and multiple competing interests.

Traditionally, energy and environmental policy research has been pursued separately in economics, political science, engineering, and sociology. However, energy transition and decarbonization challenges cannot be addressed solely through technology or economic analysis, as diverse factors—institutions, regulation, citizen behavior, international relations—interact. An integrated approach is therefore essential. Policy engineering provides a methodology for optimizing policy design, institutional frameworks, and technology deployment in such complex and uncertain environments.

The book is structured as follows:

1. Part I provides an overview of decarbonization strategies and policies in East Asian countries.
2. Part II analyzes optimal energy mix and institutional design.
3. Part III presents perspectives on innovation, focusing on technological progress and social transformation.

Through these discussions, the book uses concrete East Asian cases to apply policy engineering methodology and offer comprehensive insights for building a sustainable and cooperative decarbonized society.

References

Adger WN, Arnell NW, Tompkins EL (2005) Successful adaptation to climate change across scales. Glob Environ Chang 15(2):77–86. https://doi.org/10.1016/j.gloenvcha.2004.12.005

Ben-Tal A, El Ghaoui L, Nemirovski A (2009) Robust optimization. Princeton University Press

Courtney HG, Kirkland J, Viguerie SP (1997) Strategy under uncertainty. Harv Bus Rev 75(6):67–79

Flato G et al. (2013) Evaluation of climate models. In: Stocker TF et al (eds) Climate change 2013: the physical science basis. Cambridge University Press, pp 741–866

Grübler A et al. (2018) A low energy demand scenario for meeting the 1.5 °C target and sustainable development goals without negative emission technologies. Nature Energy 3(6):515–527. https://doi.org/10.1038/s41560-018-0172-6

Grübler A, Wilson C, Bento N, Boza-Kiss B, Krey V, McCollum D, Riahi K, Samadi S, de Stercke, S (2018) A low energy demand scenario for meeting the 1.5 °C target and sustainable development goals without negative emission technologies. Nature Energy 3(6):515–527. https://doi.org/10.1038/s41560-018-0172-6

International Energy Agency (IEA). (2023). World Energy Balances 2023. IEA, Paris. Retrieved from https://www.iea.org/data-and-statistics/data-product/world-energy-balances

International Monetary Fund (2025) GDP per capita, current prices [Data set]. IMF DataMapper. Retrieved December 22, 2025, from https://www.imf.org/external/datamapper/PPPPC@WEO/OEMDC/ADVEC/WEOWORLD

IPCC. (2001). Climate change 2001: impacts, adaptation, and vulnerability. Contribution of working group II to the third assessment report of the Intergovernmental panel on climate change. Cambridge University Press, Cambridge, UK

IPCC (2014a) Climate Change 2014: mitigation of climate change. Contribution of working group III to the fifth assessment report of the intergovernmental panel on climate change. Cambridge University Press

IPCC (2014b) Climate change 2014: synthesis report. Contribution of working groups I, II and III to the fifth assessment report of the intergovernmental panel on climate change [Core Writing Team, R.K. Pachauri and L.A. Meyer (eds.)]. IPCC, Geneva, Switzerland, p 151

IPCC (2021a) Climate Change 2021: the physical science basis, Summary for policymakers. Cambridge University Press

IPCC (2021b) Climate Change 2021: The physical science basis. Contribution of working group I to the sixth assessment report of the intergovernmental panel on climate change. Summary for Policymakers, p 5
IPCC (2021c. Climate Change 2021: the physical science basis. Contribution of working group I to the sixth assessment report of the intergovernmental panel on climate change. Cambridge University Press. https://www.ipcc.ch/report/ar6/wg1/
IPCC (2023) Climate change 2023: synthesis report. Intergovernmental panel on climate change. https://www.ipcc.ch/report/ar6/syr/
Knight FH (1921) Risk, uncertainty and profit. Houghton Mifflin, Boston
Nobel Prize (2021) The nobel prize in Physics 2021. https://www.nobelprize.org/prizes/physics/2021/summary/
O'Brien K, Eriksen S, Schjolden A, Nygaard LP (2007) Why different interpretations of vulnerability matter in climate change discourses. Climate Policy 7(1):73–88. https://doi.org/10.1080/14693062.2007.9685639
Pfenninger S, Hirth L, Schlecht I (2014) The role of storage and flexibility in integrating renewable electricity. Energy Policy 69:467–478. https://doi.org/10.1016/j.enpol.2014.02.013
Reuters (2025, August 25) China's commissioning of coal power at 9-year high in first half, report says. Reuters. https://www.reuters.com/sustainability/boards-policy-regulation/chinas-commissioning-coal-power-9-year-high-first-half-report-says-2025-08-25/
Ritchie H, Rosado P, Roser M (2023) CO_2 and greenhouse gas emissions. Our World in Data. Retrieved June 24, 2025, from "Annual CO_2 emissions" page: https://archive.ourworldindata.org/20250624-125417/grapher/annual-co2-emissions-per-country.html
Science Council of Japan (SCJ) (2005) A new system of knowledge. Science council of Japan. https://www.scj.go.jp/ja/info/kohyo/pdf/kohyo-18-t995-60-2.pdf
Smit B, Wandel J (2006) Adaptation, adaptive capacity and vulnerability. Glob Environ Chang 16(3):282–292. https://doi.org/10.1016/j.gloenvcha.2006.03.008
Sovacool BK, Griffiths S (2021) Energy policy and the transition to low-carbon futures in Asia. Energy Policy 149:112011. https://doi.org/10.1016/j.enpol.2020.112011
Sovacool BK, Ryan SE, Stern PC, Janda K, Rochlin G, Spreng D, Pasqualetti MJ, Wilhite H, Lutzenhiser L (2022) The clean energy revolution and complex socio-technical systems. Energy Res Soc Sci 88:102543. https://doi.org/10.1016/j.erss.2022.102543
State Council (2025, August 21) China's power use sees strong growth in July. Retrieved from https://english.www.gov.cn/archive/statistics/202508/21/content_WS68a6dbb2c6d0868f4e8f4fe5.html
Stouffer RJ et al (2007) Coupled Climate Model Intercomparison Project (CMIP3) overview and results. J Clim 20(19):500–513
Sun Y, Zhang X, Ding Y, Chen D, Qin D, Zhai P (2021) Understanding human influence on climate change in China. Natl Sci Rev 9(3), nwab113. https://doi.org/10.1093/nsr/nwab113
Walker WE, Haasnoot M, Kwakkel JH (2013) Adapt or perish: a review of planning approaches for adaptation under deep uncertainty. Sustainability 5(3):955–979. https://doi.org/10.3390/su5030955
Zhang C (2024) Study on sustainable carbon reduction strategies based on collaboration and integration in China [Doctoral dissertation, Ritsumeikan University].
Zhou W, Qian X, Nakagami K (2021) East Asian low-carbon community : realizing a sustainable decarbonized society from technology and social systems. Springer, Cham, pp 61–81.
Zhou W, Ren H, Su X, Qian X, Yamazaki M, Ibano K, Sun F, Haga F (2014).Realizing the vision of an "East Asian Low-Carbon Community" and the creation of "Policy Engineering": a summary and prospects of the R-GIRO area-specific project "Fundamental technology development and strategic innovation for building a low-carbon society." Policy Science, 21(3).
Zhu K (1973) Overview of climatic changes in Chinese history. Science Press, Beijing

Various Pathways to Realize Carbon Neutrality

Kenji Yamaji

1 Development of Addressing Global Warming

1.1 Kyoto Protocol

After the collapse of the Cold War structure in the early 1990s, the world began to seriously address global climate change as a shared, important issue for humanity. In 1992, the United Nations Conference on Environment and Development (commonly known as the Earth Summit) was held in Rio de Janeiro, where the basic treaty for climate change, the Framework Convention on Climate Change (FCCC), was signed. The FCCC came into force in 1994. From the following year, the annual Conference of the Parties (COP) to the FCCC was held, and in 1997, the COP3 held in Kyoto adopted the Kyoto Protocol, which set greenhouse gas reduction targets for industrialized countries, including Russia, and measures to achieve these targets efficiently (such as emissions trading mechanisms known as Kyoto mechanisms). Negotiations on the treatment of developing countries, specific implementation details of the Kyoto mechanisms, and sanctions if numerical targets were not met continued, and the Kyoto Protocol finally entered into force in 2005. However, during this time, the United States, the largest emitter of greenhouse gases at the time, withdrew from the Kyoto Protocol. The reduction targets set by the Kyoto Protocol were aimed for the five years centered at 2010, and they were generally achieved, including by Japan, due to various factors such as the economic downturn caused by the 2008 Lehman Brothers collapse. However, most industrialized countries outside of Europe, including Japan, did not participate in the next framework for measures from 2013 to 2020 (the second commitment period of the Kyoto Protocol).

K. Yamaji (✉)
Research Institute of Innovative Technology for the Earth (RITE), Kyoto, Japan
e-mail: vyl05345@nifty.com

W. Zhou (ed.), *Decarbonization Policy Engineering*, World Sustainability Series,
https://doi.org/10.1007/978-981-95-5953-4_2

1.2 Paris Agreement

A major turning point in this situation came with the adoption of the Paris Agreement at COP21 in 2015. As shown in Fig. 1, the Paris Agreement defines the international framework for the post-2020 period and is based on two main pillars. One is the long-term goal for the entire world, which is to reduce the temperature rise since the industrial revolution to well below 2 °C, with an effort to limit it to 1.5 °C. The other pillar is the Nationally Determined Contributions (NDCs), where each country voluntarily sets and reports its own targets. Currently, most countries have submitted their 2030 targets, which are to be updated every five years, with the target for 2035 expected to be submitted in 2025. Under the Kyoto Protocol, emission reduction constraints were only assigned to developed countries, but under the Paris Agreement, each country submits its own reduction targets, allowing nearly all countries in the world to participate. Furthermore, as shown in the center of Fig. 1, a mechanism called the Global Stocktake has been established to align and adjust the bottom-up NDCs with the global long-term goal. The first Stocktake took place at COP28 in 2023. The Paris Agreement came into force smoothly in 2016, but at the same time, Donald Trump won the U.S. presidential election. Trump decided to withdraw the U.S. from the Paris Agreement, and the U.S. officially withdrew in 2020. However, in the same year's presidential election, Joe Biden won, and the U.S. rejoined the Paris Agreement when he assumed office the following year. Then, in November 2024, Donald Trump, who was re-elected, again declared the intention to withdraw from the Paris Agreement. Under the Paris Agreement, if a withdrawal is notified this time, it will be implemented one year later.

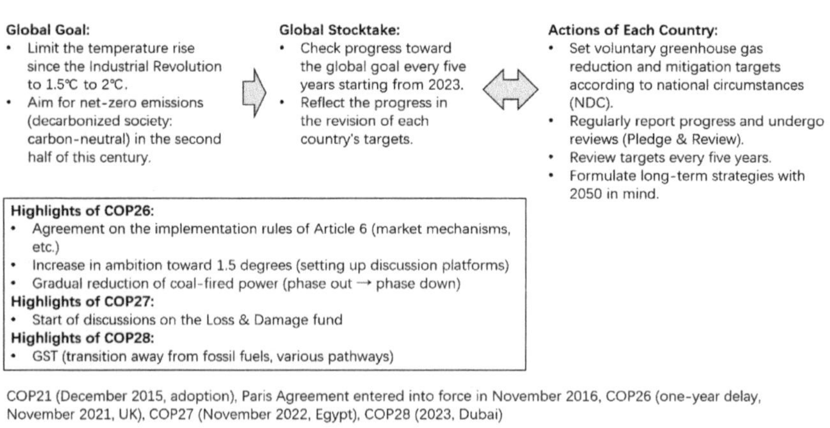

Fig. 1 Basic structure of the Paris agreement

1.3 *Carbon Neutrality by 2050*

Regarding the 1.5 °C effort target, which was included in the goal set under the Paris Agreement, the IPCC (Intergovernmental Panel on Climate Change) released the Special Report on 1.5 °C in 2018. This special report stated that achieving the 1.5 °C target requires a decarbonized society with net-zero greenhouse gas emissions by 2050, i.e., achieving carbon neutrality. This report served as a trigger for the European Union to take the lead in aiming for carbon neutrality by 2050. The ambition for achieving carbon neutrality by 2050 quickly spread worldwide, including the United States under the Biden administration. Japan's original long-term goal under the Paris Agreement was to achieve carbon neutrality later in this century, but in October 2020, just before the U.S. presidential election, Prime Minister Yoshihide Suga announced Japan's intention to achieve carbon neutrality by 2050, anticipating that Biden would win the election and the U.S. would also declare 2050 carbon neutrality. The aspirational target under the Paris Agreement is now gradually being transformed into this concrete target.

In April 2021, U.S. President Joe Biden, who had just taken office, held a climate summit, and at this summit, Japan revised its NDC target (a 26% reduction from 2013 levels by 2030) to a 46% reduction, in line with the 2050 net-zero goal. As shown in Fig. 2, this is completely consistent with achieving net-zero by 2050.

About 85% of Japan's greenhouse gases come from CO_2 originating from energy, so climate change measures and energy policies are inherently intertwined. The

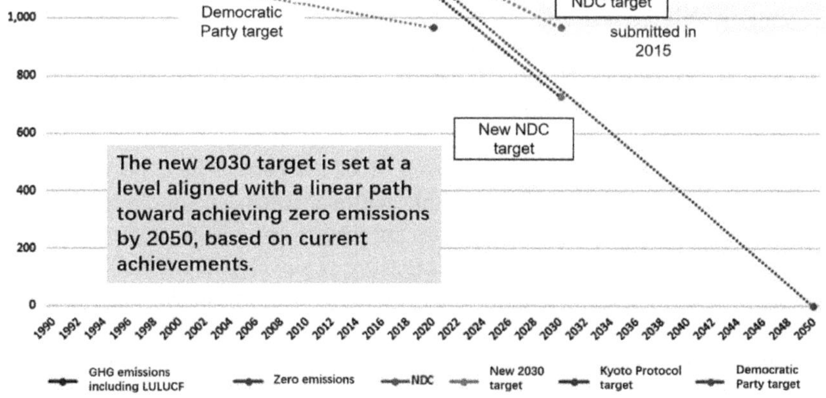

Fig. 2 Japan's greenhouse gas emissions and reduction targets

revision of the Strategic Energy Plan, which was under development at the time, was adjusted to align with the new NDC targets. In October 2021, the Sixth Strategic Energy Plan was adopted by the Cabinet, in alignment with the 46% reduction target, and this NDC was officially submitted to the Paris Agreement Secretariat. Also, the COP26, which was delayed by one year due to the COVID-19 pandemic, was held in Glasgow, UK. As noted in the "Highlights of COP26" in the lower-left section of Fig. 1, the ambition to achieve the 1.5 °C target, initially an aspirational goal, was clearly elevated. Furthermore, at recent COP meetings, developing countries have increased their demands for enhanced financial support from developed nations to cover the costs of climate adaptation, including the recovery of the damage & loss of climate change.

1.4 Restoring the Balance of the 3E Goals Due to Changes in International Circumstances

In October 2021, Japan's Prime Minister changed from Yoshihide Suga to Fumio Kishida. The new Prime Minister, Kishida, immediately set up a review committee for the Clean Energy Strategy and began discussions on various climate change countermeasures, including nuclear energy, of which discussions had been avoided until then after the Fukushima nuclear disaster in 2011, as well as hydrogen. Additionally, in January the following year, discussions began on a roadmap for the commercialization of CO_2 capture and storage (CCS) technology. These actions were integrated into the Green Transformation (GX) policy, which will be discussed later.

A major shift occurred in international circumstances. In February 2022, Russia's invasion of Ukraine began, causing fuel prices to rise, particularly for natural gas. Especially in Europe, the cessation of pipeline natural gas imports from Russia raised concerns about the stability of energy supplies. Japan also experienced surging fuel and electricity costs. Meanwhile, China began to face population decline and economic stagnation. The economic conflict between the U.S. and China deepened. In this international context, the presence of countries in the Global South, particularly India, grew more prominent. This is where Trump's resurgence in the U.S. presidential election in November 2024 comes in. The international situation has become increasingly unstable, and the world is on the verge of fragmentation. Addressing climate change, a common global challenge since the end of the Cold War, requires global cooperation, but with the current international situation, simply increasing the ambition of climate change goals risks moving further away from practical solutions. Although CO_2 emissions have been decreasing in developed countries, as shown in Fig. 3, global CO_2 emissions have continued to rise from 1990 to the present. The occasional dips are due to economic recessions triggered by events such as the Lehman Brothers collapse and the COVID-19 pandemic, but the general trend

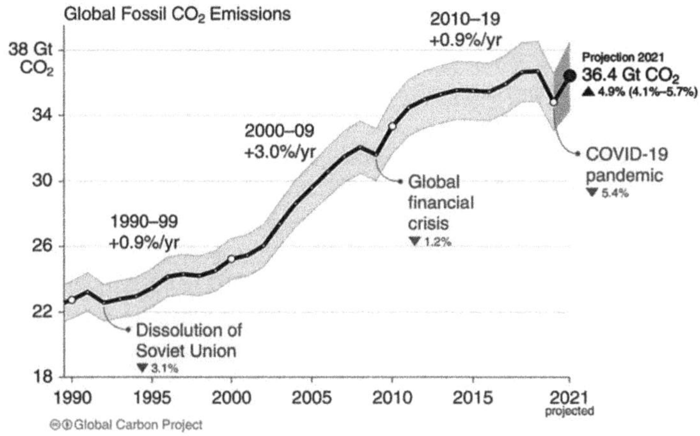

Fig. 3 Trends in global CO_2 emissions

remains upward. While the goal of addressing global warming remains decarbonization, it is critically important to aim for achievable targets based on past performance and current circumstances.

Japan's energy policy is based on the three E's: Energy security, Economic efficiency, and Environmental compatibility, with the goal of achieving them simultaneously. Under policies aimed at achieving the 2050 carbon neutrality target, the E for addressing climate change has recently received special emphasis. However, in response to global developments such as Russia's invasion of Ukraine, the importance of Energy security and Economic efficiency has also grown in recognition. The GX policy can be seen as a way to restore the three E's balance.

The following will discuss scenario analysis, the role of innovation, and the significance of societal system transformation toward achieving a decarbonized society.

2 RITE's Scenario Analysis for Achieving Decarbonization

This chapter introduces the scenario analysis by the Research Institute of Innovative Technology for the Earth (RITE) concerning the realization of carbon neutrality by 2050. Due to space constraints, the details regarding the models used, scenario settings, and quantitative results cannot be explained here. For more detailed data settings and analysis results, please refer to the materials published by RITE's systems analysis group on their official website (https://www.rite.or.jp/system/).

2.1 DNE21 + Model and Scenario Settings

The model used for scenario analysis is called "DNE21 + (Dynamic New Earth 21 plus)." The "New Earth 21" part of the model's name refers to the English nickname for Japan's "Earth Regeneration Plan" proposed at the 1990 G7 Summit held in Houston. RITE was established in 1990 as a research institute to advance the Earth Regeneration Plan, and this model is named accordingly. The prototype of the DNE21+ model originally named "New Earth 21" was developed in the laboratory of Professor Yoichi Koya (currently an advisor to RITE) at the University of Tokyo's Faculty of Engineering after the Houston Summit, in order to numerically depict the Earth Regeneration Plan. The author was involved in the development of this model as a visiting professor from the Central Research Institute of Electric Power Industry. The main developer at the time was Yasumasa Fujii, who was a graduate student (currently a professor at the University of Tokyo), and Keigo Akimoto, who was a graduate student when Fujii was a lecturer at Yokohama National University, is now the leader of RITE's Systems Analysis Group.

DNE21+ is an optimization model based on linear programming, which draws a system that optimizes (minimizes in the case of cost) the set objective function (usually system total cost) under various constraints. For the analysis of the 2050 carbon neutrality scenario, the most important constraint is to set CO_2 emissions to net zero by 2050. However, as DNE21+ treats the entire world with regional subdivisions, the way the net-zero constraint is applied differs by region. Typically, Japan and Western countries are each given a net-zero constraint by 2050 for their respective regions, while other regions are collectively given a net-zero constraint by 2050. In addition to greenhouse gas emission constraints, various other constraints representing the system structure, such as energy supply and demand balance, are also set. The original New Earth 21 model had about 10,000 variables, but it has now become a massive model with approximately 10 million variables and about 10 million constraints. The optimization time frame is set until 2100, and the entire world is divided into 77 regions, with Japan represented as one of those regions. The model incorporates the performance of various climate change mitigation technologies and energy technologies, and it also assumes future fuel prices. Specific technologies are included for major energy demand sectors such as steel and chemicals. Furthermore, international shipping and international air transport are also represented, and the lifespans of various facilities are considered.

One weakness of the DNE21+ model is that when dividing the world into regions and extending the analysis until 2100, the representation of power supply and demand balance inevitably becomes coarser. In the DNE21+ model, the annual load duration curve is approximated by four blocks (instantaneous peak, peak, intermediate, and baseload), and balance is maintained within these blocks. However, when renewable power generation, particularly wind and solar power with variable outputs, is introduced in large quantities, it becomes necessary to represent the power supply and demand balance at least on an hourly basis, and transmission capacity constraints must also be considered. To handle such variable renewable energy sources, the results of model analyses by the University of Tokyo and the Institute of Energy Economics, Japan are borrowed, and thus, the effect of increasing grid integration costs to balance supply and demand on the grid in proportion to the share of variable renewable energy sources in the electricity supply are incorporated into the DNE21+ model.

Moreover, it is common practice to set multiple scenarios in scenario analysis. In RITE's net-zero scenario analysis, various scenarios are set as well. Even when achieving net-zero emissions optimally, costs increase compared to the situation without a net-zero constraint (often called the baseline scenario). Within the net-zero scenarios, various scenarios are set beyond the standard case (often called the reference case), such as changing the upper limit constraints on the shares of nuclear energy (including cases where nuclear energy is prohibited) or altering hydrogen production costs. The analysis seeks the optimal solution under these various scenario settings. Here, the detailed scenario settings are omitted, and only the basic results will be presented.

2.2 *Basic Framework for Achieving Japan's Net-Zero Emissions*

Due to space constraints, the analysis results cannot be presented with specific numerical values. Instead, Fig. 4 presents the basic image of Japan's achievement of net-zero emissions based on the results of the scenario analysis.

As shown here, the first necessary step to achieve net-zero emissions is energy conservation. This includes energy demand reduction through the realization of a

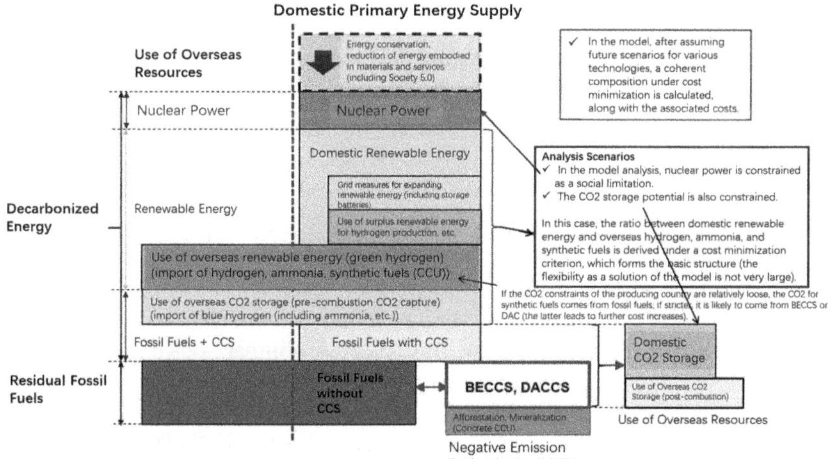

Fig. 4 Image of Japan's net-zero emissions

sharing and circular economy using digital technologies, such as the super-smart society (Society 5.0) mentioned later. On the energy supply side, decarbonized energy sources like nuclear power and domestic renewable energy will be maximized. As renewable energy generation is introduced in large quantities, it will be necessary to adjust the output of existing power sources and store electricity using batteries or hydrogen production in order to balance supply and demand. These are the basic measures to achieve decarbonization, but additional technologies will be employed to optimally achieve net-zero emissions.

According to scenario analysis, it will also be necessary to consider the option of producing green hydrogen (hydrogen produced by electrolysis of water using renewable energy sources without CO_2 emissions) overseas and importing it to Japan. Hydrogen could be converted to ammonia and imported as it would be easier to transport. Another possibility is to produce synthetic fuels using hydrogen and captured CO_2 and import them. Furthermore, the use of CCS (Carbon Capture and Storage) overseas is also a potential solution. Since overseas regions have greater economic CO_2 storage capacity, CCS may be implemented abroad to produce hydrogen (referred to as blue hydrogen) without CO_2 emissions from fossil fuels, which could then be imported. In fact, companies such as Kawasaki Heavy Industries and Electric Power Development Co., Ltd. are working on projects to produce hydrogen from lignite in southern Victoria, Australia, and import it as liquid hydrogen to Japan. Of course, there is also the option of using fossil fuels with CCS domestically. Since Japan's CO_2 storage capacity is limited, transporting CO_2 by ship to use CCS storage capacity abroad may also be considered.

The unavoidable CO_2 emissions will come from the use of fossil fuels without CCS. These unavoidable emissions are referred to as residual emissions. To address residual emissions, negative emissions technologies (NETs) such as BECCS, which combines CCS with biomass utilization, or DACCS, which combines CCS with direct air capture (DAC) to capture CO_2 from the atmosphere, can be utilized. CO_2 can also be captured from the atmosphere through afforestation or mineralization into carbonates. In other words, NETs are necessary to offset any unavoidable CO_2 emissions to achieve net zero. Using these multiple options, the DNE21+ model analysis provides a scenario for achieving Japan's net-zero greenhouse gas emissions.

2.3 Insights from the 2050 Net-Zero Scenario Analysis Results

The results of the scenario analysis involve extensive comparisons between the baseline, the reference case, and multiple other cases, making it difficult to present all the quantitative details in this document. For detailed results, please refer to the RITE website. Below is a summary of the key insights derived from RITE's 2050 net-zero scenario analysis.

- Achieving carbon neutrality requires a full deployment of various technologies, including energy conservation, renewable energy, nuclear power, CCS, hydrogen/ammonia, and CO_2 removal from the atmosphere (NETs). In the optimal solution, the amount of nuclear power introduced reaches the upper limit set in the scenarios.
- Electrification and decarbonization of the power sector are necessary in all scenarios. However, the cost of electricity generation (marginal cost) nearly doubles. If 100% renewable power is adopted, the cost doubles even further. A 100% renewable power scenario may hinder the promotion of electrification.
- In non-electric sectors, hydrogen and zero-emission synthetic fuels will be necessary. However, even with these measures, achieving zero emissions will not be possible without utilizing NETs to reach net-zero emissions.
- Direct Air Capture (DAC) is utilized in all scenarios. Although some of the captured CO_2 is used, the scale is not large. Japan's CO_2 storage capacity alone is insufficient, so overseas storage capacity is also utilized.
- The promotion of a sharing economy in a digital society could enable new energy-saving methods through the use of information, significantly reducing the costs of the energy system.
- In the scenario fully utilizing international carbon credits (which optimize for global net-zero emissions), Japan's CO_2 reduction in 2050 is limited to 63% compared to 2013 levels, but the cost of mitigation is significantly reduced.

3 Japan's Innovation Policies Toward Decarbonization

Achieving an optimal net-zero emissions scenario, as introduced in the previous chapter, requires both technological and societal innovation. This chapter explains Japan's innovation policies aimed at realizing decarbonization.

3.1 Formulation and Development of the Innovative Environmental Innovation Strategy

The importance of innovation in Japan's climate change policy has been recognized since the adoption of the Paris Agreement at COP21. Currently, these efforts have been consolidated into the Green Transformation (GX) policy. However, the first significant step in this direction was the publication of the Innovative Environmental Innovation Strategy in January 2020. Figure 5 provides an overview of this strategy.

As shown here, the Innovative Environmental Innovation Strategy first presents an Innovation Action Plan. This action plan is designed with a global perspective, promoting innovation across five key fields and sixteen challenges. The five fields include Energy Conversion, Transportation, Industry, Consumer, and Agriculture. The role of innovation in the energy, transportation, and industrial sectors has long been well recognized. The agricultural sector, however, contributes significant

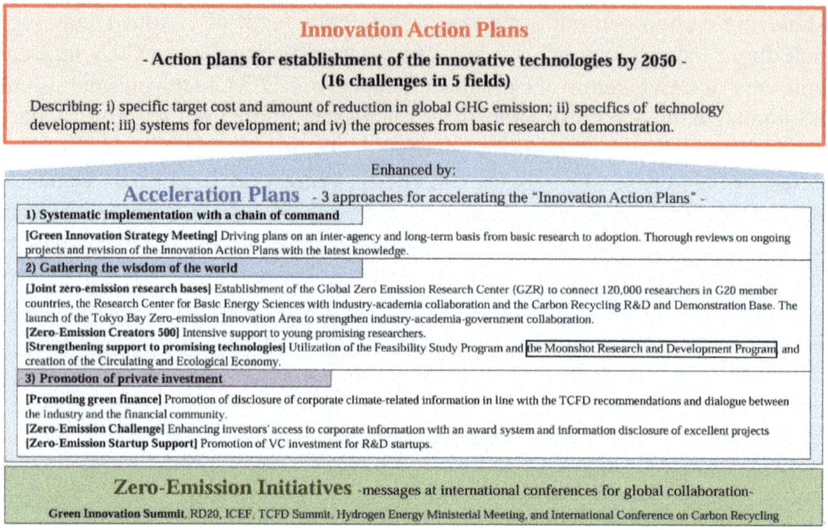

Fig. 5 Overview of the innovative environmental innovation strategy

amounts of non-CO_2 greenhouse gases, such as methane and N_2O. Additionally, agriculture serves as a CO_2 absorption source through vegetation and soil. The consumer sector, which includes households and cities, extends beyond conventional energy efficiency measures to incorporate societal system transformations and lifestyle changes to reduce greenhouse gas emissions. Furthermore, it is essential to focus on the common base technologies that support innovation across these sectors. These include digital technologies such as the Internet of Things (IoT), blockchain, and big data analytics, as well as power electronics and advanced materials. Innovation in these common base technologies plays a crucial role.

Figure 5 also shows the acceleration plans and zero-emission initiatives supporting the action plan. The Moonshot Research and Development Program listed within the acceleration plans began in 2020, and one of the programs (Moonshot Goal 4) has initiated research and development related to DAC.

The Innovative Environmental Innovation Strategy has evolved into the Green Growth Strategy through the creation of concrete implementation plans. The Green Growth Strategy views addressing climate change as an opportunity for growth and seeks to achieve benefits for the public in addition to decarbonization. While the details are omitted here, fourteen key sectors were selected based on industrial classifications. A green innovation fund with a scale of 2 trillion yen was introduced for related funding. This development took place during Prime Minister Shinzo Abe's time. Following this, Prime Minister Yoshihide Suga's declaration for carbon neutrality by 2050 was made, and the Clean Energy Strategy initiated by Prime Minister Fumio Kishida, as previously mentioned, continued this effort.

3.2 *Green Transformation (GX) Policy*

Immediately after the review of the Clean Energy Strategy began, Russia's invasion of Ukraine occurred. In the interim report compiled in May 2022, it was emphasized that energy security and the maximal use of high decarbonization-effect power sources, including nuclear power, should be prioritized. At the same time, the necessity of a Green Transformation (GX) for industry and decarbonization in regions and everyday life was highlighted. Within this framework, the government decided to issue 20 trillion yen in GX economic transition bonds, aiming to leverage these bonds as seed funding for private investments, ultimately guiding 150 trillion yen in total investments from both public and private sectors. This marked the beginning of the GX policy's development. In February 2023, the basic guidelines for achieving GX were approved by the Cabinet, and in May of the same year, the GX Promotion Act and the GX Decarbonization Electricity Act were passed in the Diet. Figure 6 presents an overview of these two acts related to GX.

As shown in Fig. 6, the GX Promotion Act includes 20 trillion yen in transition bonds and 150 trillion yen in public–private investments. Additionally, as a source of funds for the repayment of the 20 trillion yen in transition bonds, the act incorporates a growth-oriented carbon pricing system, which involves levies on fossil fuels and the auctioning of emission allowances in the emissions trading scheme. The issue of carbon pricing had been under discussion between the Ministry of the Environment and the Ministry of Economy, Trade, and Industry, but the GX policy helped

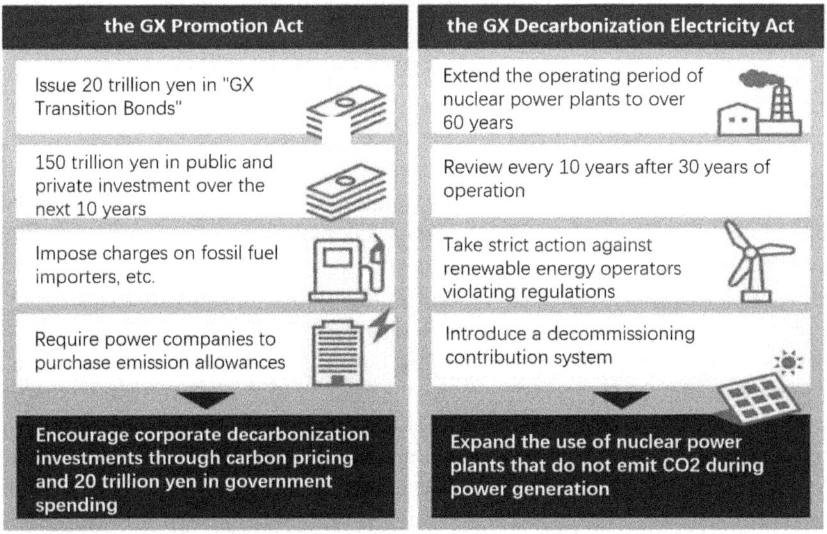

Fig. 6 Overview of the two acts related to GX policy

resolve the matter. Additionally, the GX Power Law introduced a system to effectively extend the operating life of nuclear power plants, clearly affirming a stance of active promotion of nuclear energy.

As previously mentioned, the GX policy is expected to restore the balance of the 3E goals, which had previously been tilted toward focusing solely on climate change mitigation. Furthermore, the GX policy also includes efforts to expand international decarbonization movements through international cooperation, particularly with Southeast Asian countries.

Expectations for the GX policy are high, but caution is also necessary. In general, policy instruments can be divided into carrots and sticks. In this case, carbon pricing is the stick, and GX economic transition bonds are the carrot. While it could be argued that the GX policy cleverly combines both carrots and sticks, we cannot be fully reassured until the specific details of the measures are finalized. The author has been involved in renewable energy policies, such as designing systems like Renewable Portfolio Standards (RPS) and Feed-in Tariffs (FIT) for renewable energy, and understands the importance of system details. As the saying goes, "God is in the details." In addressing issues that involve many stakeholders, the concrete details of policies are crucial.

4 The Importance of Social System Innovation

4.1 New Measures Addressing Global Warming in the Digital Age

Among innovation policies, social system innovation is a crucial area that has been recognized as important but has not yet been fully developed in terms of policy implementation. Figure 7 is an enlarged version of the lecture materials I used when I was a university professor, and in this figure, social system innovation is placed at the top of the "human activity" category. This will be explained further later, but the advancement of a digital society, utilizing information technology, is expected to create new possibilities for measures addressing global warming in this area.

Figure 7 illustrates the basic structure of measures addressing global warming, showing the pathway from "human activities" through various stages that ultimately lead to climate change impacts. Measures addressing global warming are structured according to these stages. Traditional measures begin with energy conservation and resource efficiency improvement, aimed at reducing energy and material demand generated by human activities. The next stage involves reducing CO_2 emissions from energy and material production, which is achieved through low-carbon technologies such as nuclear energy, renewable energy, and biomass. Beyond this, since CO_2 emissions do not cause the greenhouse effect unless they increase atmospheric CO_2 concentrations, the next step is Carbon Capture and Storage (CCS), which captures CO_2 before it is released into the atmosphere. Recently, there has been a

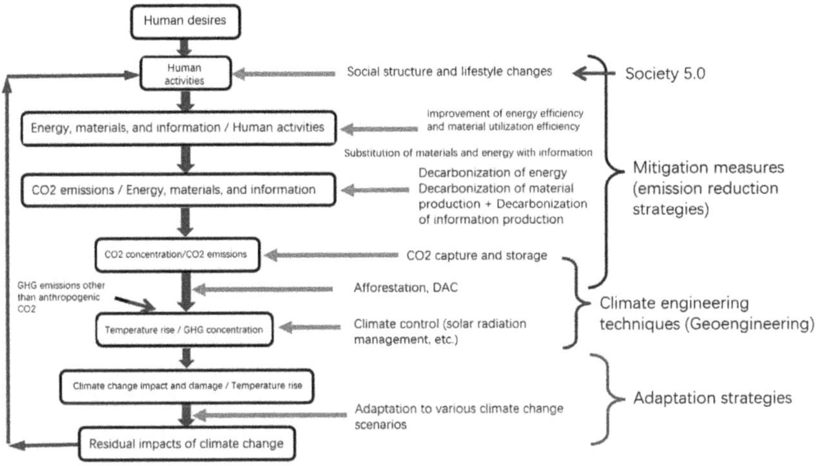

Fig. 7 Basic structure of measures addressing global warming

focus on Carbon Capture, Utilization, and Storage (CCUS), which includes utilizing captured CO_2. The next level includes negative emissions technologies (NETs) such as afforestation and Direct Air Capture (DAC), which remove CO_2 already present in the atmosphere. Although not currently implemented, another potential measure is climate engineering, which artificially offset the temperature rise caused by the greenhouse effects, including solar radiation management. Since such measures are specifically designed for climate change intervention, they are often referred to as geoengineering or climate engineering techniques. Lastly, "adaptation" is the process of taking measures to address climate change that has already occurred. Typical adaptation measures include preventing agricultural production losses due to global warming by developing industrialization of agriculture such as vegetable factories, or breeding improved crop varieties, as well as strengthening river management and disaster prevention to mitigate the effects of increasing natural disasters such as floods.

On the other hand, social system innovation involves transforming human activities, positioned at the top of Fig. 7, through changes in social structures and lifestyles, so that they contribute effectively to addressing global warming. The role of the super-smart society (Society 5.0), as pointed out in the Basic Plan for Science and Technology, is key to innovation in this domain. The focus here is on "information," which can be used to substitute for materials and energy. However, since the collection, processing, and analysis of information consumes large amounts of electricity, improving the efficiency and decarbonization of information production and processing becomes critically important.

4.2 Low Energy Demand (LED) Scenario

Recycling and resource recovery are being promoted as part of the formation of a circular economy. In addition to this, sharing resources such as cars, clothing, and rooms is also progressing. To realize this sharing and circular economy, the use of information is necessary, and the advancement of the digital society seems to make this a viable business model. In a sharing and circular economy, the demand for materials in human activities decreases, the energy required for material production decreases, and ultimately, CO_2 emissions are reduced.

This new possibility for a digital society was clearly presented in the Low Energy Demand (LED) scenario, as shown in Fig. 8, in the 2018 IPCC Special Report on 1.5 °C mentioned in Sect. 1.3. LED stands for Low Energy Demand. The researcher who compiled this section of the report was Anulf Grübler, a long-time acquaintance of the author, from the International Institute for Applied Systems Analysis (IIASA) in the suburbs of Vienna. Since the expression "Low" first in "LED" has a negative connotation, the name has recently been changed to HwL scenario (HwL stands for High Well-being with Low Energy Demand).

An excellent example of product-level sharing is the smartphone. This is referred to as digital convergence. As shown in Fig. 9, the power consumption of a smartphone is around 5W, but the smartphone's functions go beyond just making calls; it serves as a camera, an information terminal, a lighting device, and an audio device. If these functions were to be achieved by separate devices, 449W would be needed, which represents a difference of two orders of magnitude in energy consumption. Additionally, a smartphone weighs around 100 g, and its production requires 75 kWh of energy. However, if the smartphone's functions were implemented by separate devices, the total weight of the equipment would be 26 kg, and 1706 kWh would be required to produce them. In other words, using a smartphone significantly reduces both the energy consumed during use and the energy required for production.

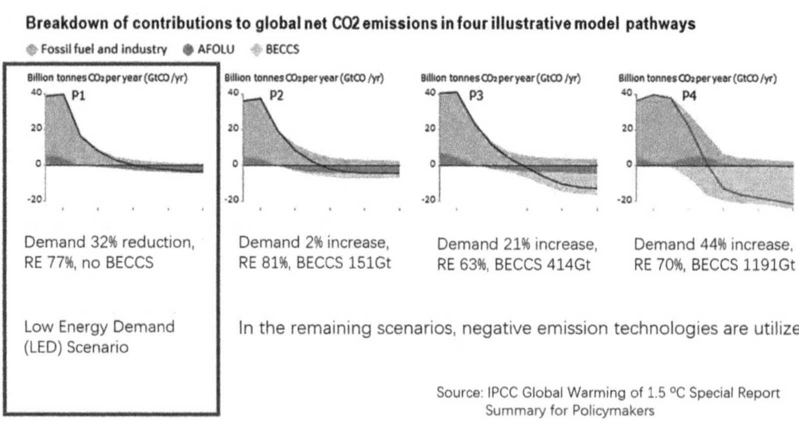

Fig. 8 Emergence of the LED scenario

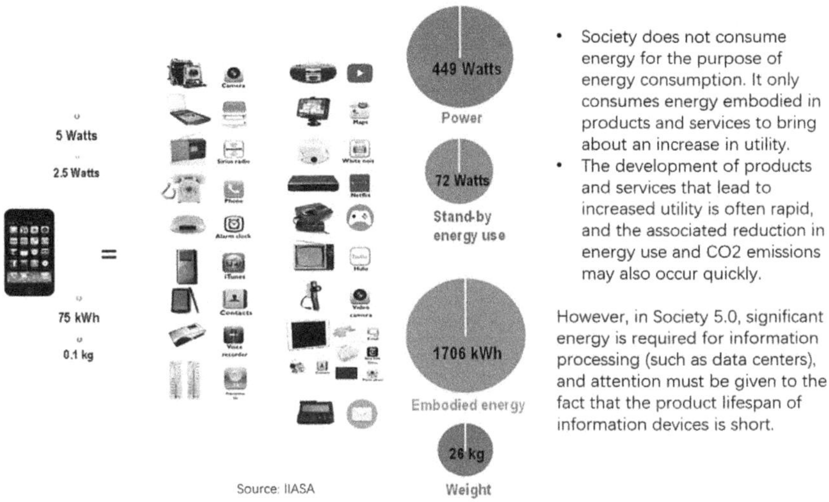

Fig. 9 Digital convergence through smartphones

However, it must be considered that the information processing in data centers and other systems associated with smartphone use requires significant energy. Additionally, since the product lifespan of smartphones is shorter than that of typical devices, the demand for replacements must also be taken into account. As previously mentioned, social system innovations that utilize information technology will need to address the growing energy demand related to information processing and other aspects.

4.3 The Role of Cities in Social Innovation

In social system innovation, the role of cities becomes crucial. Cities are where houses and buildings are concentrated and serve as the central stage for social activities. Energy demand in cities arises primarily from households and buildings, covering most of the consumer sector. As shown in Fig. 10, direct CO_2 emissions from Japan's consumer sector are relatively low, accounting for about 11% of total emissions. However, if we also consider indirect emissions from the energy production used by the consumer sector, the CO_2 emissions from the consumer sector rise to just over 33%, making it comparable in scale to emissions from the industries sector. Additionally, a large portion of the CO_2 emissions from the transport sector, which accounts for around 18% of total emissions, can also be attributed to cities.

There are many expectations for the role of cities in efforts addressing global warming. Insulation of buildings such as offices and homes, along with efficiency improvements and decarbonization of various devices such as air conditioners and

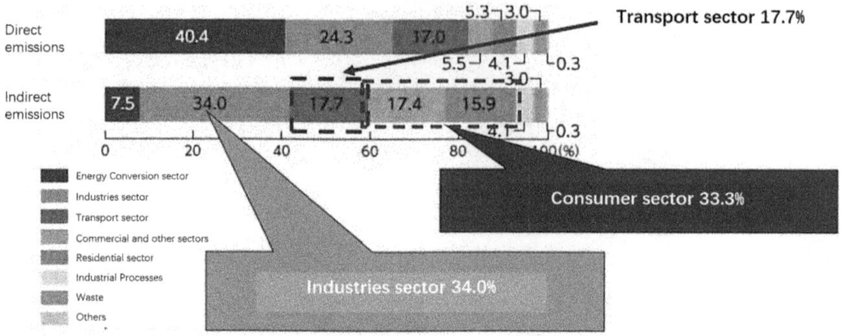

Note 1: Direct emissions refer to the CO2 emissions from energy sources associated with power generation and heat production, calculated as the emissions on the producer's side (before electricity and heat allocation).
2: Indirect emissions refer to the CO2 emissions from energy sources associated with power generation and heat production, allocated to each sector based on their consumption (after electricity and heat allocation).

Fig. 10 The importance of cities in addressing global warming

water heaters inside buildings, have already been implemented with positive results. District heating and cooling systems are also being used. These are areas in which Japan excels.

Going forward, innovations in city systems, such as transportation and communication infrastructure concentrated in cities, will become necessary. Cities are places where social innovations, like Society 5.0, are unfolding. Recently, the concept of City OS (Operating System) has emerged. Optimizing flexible power demand (including charging electric vehicles) to support the electrification of cars, automation of driving, and the large-scale integration of variable renewable energy, cities provide an ideal platform to realize these new challenges. Cities will play an extremely important role in achieving social system innovation.

5 Toward the Realization of a Decarbonized Society

To optimally realize a decarbonized society, various innovations in both technology and society must be mobilized. The central components are clean and efficiently usable secondary energy carriers, with electricity currently at the forefront and hydrogen expected to play a key role in fuel and thermal applications in the future. Figure 11 illustrates a framework for achieving carbon neutrality, centered around electricity and hydrogen, with various innovations incorporated.

Electricity and hydrogen can be produced from various resources, making it possible to achieve low carbon and decarbonization through technological innovation. If Carbon Capture, Utilization, and Storage (CCUS) technologies are considered, not only the use of nuclear energy and renewable energy, but also, fossil resources could also play an important role. However, with the expected large-scale introduction of power sources with fluctuating output, such as solar and wind power, the role of energy storage technologies, including electrolytic hydrogen storage

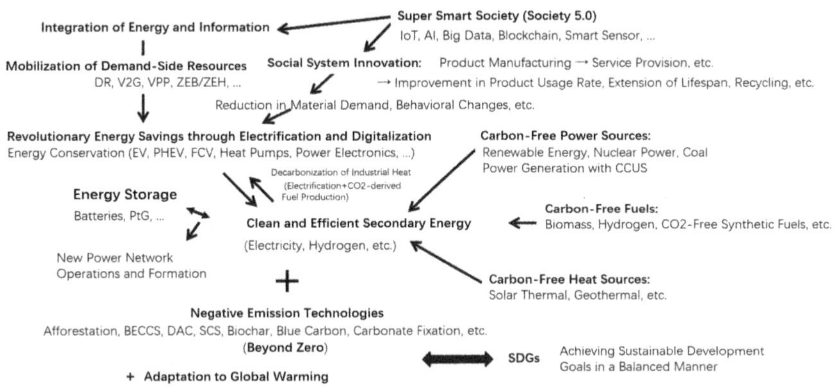

Fig. 11 Structure for realizing a decarbonized society

(PtG), will become crucial. Additionally, the power network connecting fluctuating renewable sources will require flexibility and resilience.

The use of electricity is expected to continue to increase with the advancement of digital society, the electrification of the transportation sector, and the growth of thermal supply via heat pumps. The utilization of decentralized energy sources, such as solar power generation and cogeneration, as well as storage devices on the demand side, like batteries in electric vehicles and storage tanks in heat pump water heaters, will also progress.

The system for the transmission and distribution sector is undergoing significant changes due to power system reforms, and it is expected that the traditional model of power supply matching demand will shift to a network formation and operation that integrates supply and demand. The utilization of digital technologies will support this network innovation.

Further electrification in the industrial sector is important, and the use of hydrogen is also expected. Hydrogen is expected to see increased demand not only for use in fuel cells but also for combustion power generation and thermal use, including its use as fuel ammonia, as well as its growing role in carbon recycling and as a raw material for various synthetic fuels and chemical synthesis. Of course, hydrogen used here must be produced without CO_2 emissions.

Furthermore, if the sharing and circular economy are realized through the super-smart society (Society 5.0), not only will there be a significant reduction in energy demand through the substitution of energy and materials by information, but the integration of energy and information systems will advance further. Decentralized resources such as equipment placed on the demand side will be used more efficiently. Information devices, like smartphones, which achieve various functions through application software, will continue to increase, and the substitution of energy and materials through information is also expected to progress at the device level. As a result, a revolutionary energy-saving potential through electrification and digitization can be envisioned. However, there is a need to respond to the increased power demand associated with information processing, such as blockchain, and collaborate with innovations in the information field, such as photonic devices and quantum computers.

When looking at human economic activities as a whole, such as steel production, cement manufacturing, chemical industries, and agriculture, it seems difficult to achieve zero greenhouse gas emissions even if all of the above measures are implemented. Therefore, it is also necessary to have technologies in place for capturing CO_2 from the atmosphere, such as afforestation, DAC (direct air CO_2 capture from the atmosphere), BECCS (biomass energy utilization with CCS), and mineral fixation of CO_2 using waste concrete.

On top of this, it is essential to balance the adaptation to potential global warming with the achievement of goals beyond the warming issue in the SDGs (United Nations Sustainable Development Goals). From a broader perspective, if societal innovations for achieving the SDGs lower the CO_2 emissions baseline, and technological innovations lead to the production of clean secondary energy such as electricity or hydrogen without CO_2 emissions, and systems for efficiently utilizing them are constructed, the prospects for solving the global warming issue will open up.

Author's note: This chapter is a revised and expanded English version of the author's article, "Various Pathways to Realize Carbon Neutrality", originally published in *Policy Science*, Vol. 32, No. 3 (2025).

The Realization of Carbon Neutrality by 2050 and Energy Security

Ken'ichi Nakagami

1 The Genealogy of Energy Security in Japan

1.1 The Transformation of Security Studies and Human Security

Since the beginning of the twenty-first century, the concept of security has been undergoing transformation. In today's situation, where the protection of national territorial integrity and independence does not necessarily guarantee the safety of citizens, there is a growing need to broaden the traditional understanding of security (Nakagami 2025).

From a historical perspective, discussions on security have long been dominated by military aspects. In recent years, however, non-military dimensions such as environment and energy have emerged as subjects of "ensuring safety," with further expansion into areas such us economy, natural resources and ideological culture.

Dr. Mahbub ul Haq (1934–1998), a Pakistani economist, first proposed the concept of "human security" in the 1994 UNDP Human Development Report (UNDP 1994). Human security was established as a reflective concept on traditional national security, placing the highest priority on protecting individuals rather than states.

At the UN Millennium Summit in September 2000, Secretary-General Kofi Annan emphasized the importance of achieving two overarching goals: "Freedom from Fear" and "Freedom from Want." These ideals were subsequently reflected in the "Millennium Development Goals (MDGs)" adopted later that year.

In response to these UN initiatives, and with support grom the Government of the Japan, the "Commission on Human Security" was established in 2001, co-chaired by Dr. Amartya Sen and Ms. Sadako Ogata. In 2003, the commission issued its

K. Nakagami (✉)
College of Policy Science, Ritsumeikan University, Osaka, Japan
e-mail: nakagami@sps.ritsumei.ac.jp

W. Zhou (ed.), *Decarbonization Policy Engineering*, World Sustainability Series,
https://doi.org/10.1007/978-981-95-5953-4_3

final report, "Human Security Now." (Commission on Human Security 2003). This report defined human security as "protecting the vital core of all human lives in ways that enhance human freedom and human fulfillment." It stressed the importance of strategies of protection and empowerment to secure people's survival, livelihood, and dignity.

This concept and message offer a transformative perspective on how security can be understood. In particular, human security provides an expanded framework that complements and enriches traditional discussions of energy security.

1.2 The Formation and Historical Evolution of the Concept of Energy Security

Over the past century, energy security has been a central concern in energy policy, particularly in relation to ensuring reliable fuel supplies for military operations during World War I and World War II. In times of war, energy security effectively becomes synonymous with national security, as maintaining a stable flow of oil is essential for powering battleships, tanks, and fighter planes (Energy Charter Secretariat 2015).

Since the 1950s, global energy demand has expanded rapidly, and the international oil supply system came to be dominated and managed by Western oil majors. In response to growing dissatisfaction with the inequitable distribution of revenues from oil export, oil-producing countries organized to assert greater control over their natural resources. This led to the establishment of the Organization of the Petroleum Exporting Countries (OPEC) in 1960 marking a significant turning point in the political economy of global energy.

The first oil shock in 1973, triggered by the oil embargo imposed by the Arab Petroleum Exporting Countries (OAPEC), undermined the energy security of oil-importing nations and highlighted the vulnerability of their energy systems. A few years later, between 1979 and to 1980, international oil prices approximately doubled, rising from about 12–15 USD per barrel to a peak of 30–35 USD per barrel, further intensifying global concerns about energy security and prompting major structural reforms in energy policy. This was a exceptionally high price level given global economic conditions at the time, and from 1981 to 1983, oil prices remained relatively elevated at around 25–30 USD per barrel. Growing concerns over the stability of oil supplies triggered of significant reactions in oil-countries. These included a slowdown in energy demand growth, a marked increase in inflation rates, and the effective end of the era of high economic growth in many Western economies. Importing countries began to implement various measures under the name of energy security policy, such as energy efficiency, diversification, stockpiling, and energy investments both domestically and internationally.

In 1988, the Intergovernmental Panel on Climate Change (IPCC) was established to provide scientific assessments on climate change, a development that has had important implications for energy security. The rising frequency of natural disasters and extreme weather events associated with climate change has increased the vulnerability of energy infrastructure, highlighting risks to the stability of energy systems. These challenges have encouraged a policy shift towards reducing dependence on fossil fuels and accelerating the adoption of renewable energy.

Throughout the 1990s, international measures to address global warming became increasingly institutionalized, and in 1997, the Kyoto Protocol was adopted as, the first international treaty to impose legally binding obligations on developed countries to reduce greenhouse gas emissions, was adopted.

In the early 2000s, concerns about energy insecurity re-emerged, and widely discussed again, and concerns about energy security expanded beyond oil supply.

Today, energy security is a critical issue affects that directly impacts the economy stability and safety of nations, not only in times of war but also in peacetime. Modern energy security is commonly defined as the continuous availability of diverse energy sources in sufficient quantities at reasonable prices (Energy Charter Secretariat 2015).

1.3 The Genealogy of Energy Security in Japan

Regarding the genealogy of energy security in Japan, Kazutomo IRIE and Keiji KANDA (Irie and Kanda 2002) classified the period up to 2000 into the following categories: (1) Emergence period (1961–1972), (2) Establishment period (1973–1982), (3) Transformation period 1 (1983–1988), and (4) Transformation period 2 (1989–).

In this paper, based on this classification, the periods of energy security from 1960 to the present are organized as follows:

1.3.1 Cradle Period (1961–1972)

The Ikeda Cabinet formulated the "Income Doubling Plan" in December 1960, setting a goal to double the real gross national product in ten years. As concrete measures for achieving rapid economic growth, the plan emphasized it the enhancement of social capital, the a structural shift toward to heavy and chemical industries centered on oil and steel, the expansion of exports, and the promotion of science and technology. These high growth policies drove a major transition in Japan's energy policy from coal to oil. While the development of the steel and petrochemical industries facilitated the domestic production of petroleum-based products, the import of raw materials such as coal from overseas increased. This led to a decline in the domestic coal industry and marked the beginning of Japan's "energy revolution". At the same time, the Organization of the Petroleum Exporting Countries (OPEC) was established in 1960. For Japan, which had been dependent on overseas oil, this represented a significant and potentially destabilizing factor for energy security.

From the stage of being concerned about oil prices, the necessity arose to expand the perspective to energy security centered on oil. The term "Energy Security" was first used in the report of the European Energy Policy Survey Mission (Ministry of International Trade and Industry) in July 1961. The report raised the issue, "If Japan relies almost entirely on overseas oil supply, what will happen if that supply were cut off?".

This concept was also adopted in the interim report of the Energy Roundtable (August 1961). The interim report stated, "Active cooperation in coal consumption will increase the burden on the electric power industry, but considering the supply responsibility imposed on the electric power industry, it should not be considered merely a cost issue in terms of achieving energy security as much as possible." It suggested a shift in energy policy by stating, "A certain percentage of the domestic oil market should be placed under national influence."

In June 1965, the Comprehensive Energy Survey Council was established as an advisory body to the Ministry of International Trade and Industry to establish a comprehensive energy policy that oversees the overall supply and demand of energy, and the policy of "securing low-cost and stable supply" was put forward (Hoshino 1964). The Comprehensive Energy Survey Council Establishment Law (Law No. 136 of June 28, 1965) was revised to "investigate and deliberate important matters concerning comprehensive and long-term measures to ensure the stable and rational supply of energy."

Thus, the discussion on energy security, which had been centered in Europe for nearly 100 years, began to emerge as a concept in Japan's energy policy.

1.3.2 Establishment Period (1973–1982)

Due to the risk of supply disruptions and price surges of imported oil, energy conservation progressed, particularly in reducing oil consumption and lowering oil dependency. Recognition that oil imports were unstable deepened, and the concept of energy security as an emergency response to sudden supply reductions or interruptions was established.

In August 1975, after experiencing the first oil crisis, the Comprehensive Energy Investigation Committee submitted a report titled "Energy Stabilization Policy of the 50s," which introduced a new policy philosophy prioritizing the stable supply of energy, leading to the establishment of the concept of energy security. The basic direction of energy supply and demand centered on oil, long-term energy measures including the energy supply structure, and other aspects were repeatedly examined, and on December 19, 1975, the "Basic Direction of Comprehensive Energy Policy" was decided (Ono 1976).

In May 1980, the Law for the Promotion of Development and Introduction of Alternative Energy to Oil was enacted. To set the supply targets for alternative energy to oil based on this law, the Supply and Demand Subcommittee of the Comprehensive Energy Investigation Committee compiled the "Basic Concept for Formulating Guidelines for the Introduction of Alternative Energy to Oil" in November of the same year. This "Basic Concept" pointed out that the significance of developing and introducing alternative energy to oil lies in ensuring energy security (Ayabe 2014).

1.3.3 Turning Point Period of the Energy Security Concept: 1983–1988

After the second oil crisis caused by the Iranian Revolution in 1978, global oil demand decreased, and in March 1983, the Organization of the Petroleum Exporting Countries (OPEC) had to lower the benchmark crude oil price for the first time since its establishment.

The Japanese economy, although significantly affected by the second oil crisis, began to recover from 1983. One of the factors was the decline in oil prices, which improved trade conditions and stabilized prices. The post-oil crisis Japanese economy became export-dependent, which also contributed to the bubble economy in the late 1980s (Kondoh [2011]).

By 1983, when the impact of the second oil crisis had lessened, a new point of discussion emerged in Japan's energy policy. The focus on securing supply in energy security now included the importance of oil prices. In other words, while ensuring a stable supply of oil was essential, the balance with costs needed to be considered, and this point was emphasized in the development and introduction of alternative energy sources. This was reflected in the long-term energy supply and demand outlook of November 1983.

In November 1986, the "21st Century Energy Vision" was compiled. As the basic philosophy of medium- to long-term energy policy, it emphasized energy security, cost, and the need to consider suitability (high quality, convenience, safety, cleanliness, etc.). It aimed for an optimal energy supply and demand structure balanced from these three perspectives.

In other words, the "21st Century Energy Vision" presented a long-term vision for Japan's energy policy, aiming for sustainable energy supply and environmental protection, reducing the use of fossil fuels, and promoting the introduction of renewable energy.

(a) Improvement of energy efficiency: Promotion of technological innovation and efficiency to reduce energy consumption. (b) Expansion of renewable energy: Increase the use of renewable energy such as solar, wind, and hydro power. (c) Environmental protection: Introduction of policies and technologies to minimize environmental impact.

The "21st Century Energy Vision" became the foundation of Japan's subsequent energy policy, serving as an important guideline for the direction of energy strategy and forming the framework of the energy security concept in Japan.

1.3.4 Period of Initiating Global Environmental Conservation and Asian Cooperation (1988–2002)

In November 1988, the Intergovernmental Panel on Climate Change (IPCC), a governmental body for collecting and organizing scientific research on global warming, was established. The purpose of the IPCC is to provide a scientific basis for the climate change policies of various governments. As of 2023, 195 countries and regions are participating, and it has been leading global warming policies since its establishment.

In discussions on energy security, environmental protection was considered an element of energy security in the "21st Century Energy Vision." However, with the establishment of the IPCC, it became a global situation where measures against global warming had to be incorporated.

In October 1990, the interim report of the Comprehensive Energy Survey Committee published "Challenges to the New Global Energy Trends," which stated that "the greatest challenge is to ensure the stable supply of energy, i.e., security," while also pointing out that "it is necessary to take maximum measures in energy policy to ensure sustainable economic development while balancing human activities and global environmental conservation," particularly regarding global warming issues.

The report "On Future Energy and Environmental Measures," compiled in November 1992, indicated that energy policy and environmental measures should be considered together as "energy and environmental measures," and that environmental conservation, economic growth, and energy supply stability should be tackled as a "trinity" in addressing global warming issues.

In June 1995, the International Energy Subcommittee of the Comprehensive Energy Survey Committee compiled the interim report "Asia Energy Vision," focusing on Asia, where energy conditions are changing significantly due to rapid economic growth and increasing energy demand. The report analyzed the current state of energy issues and examined the future intensification of energy security and environmental issues.

In April 2000, the Comprehensive Energy Survey Committee established the "Energy Security Working Group" (Fuji Research Institute 2003).

The June 2001 report raised new issues, such as whether price fluctuations arising from supply and demand conditions in the international oil market, rather than political decisions by oil-producing countries, should be considered a risk to energy security.

The report pointed out that "today's energy security" should be considered from the perspective of "the security or sustainability of humanity and the global environment."

1.3.5 Period of Enacting the Basic Energy Policy Law (2002–Present)

The "Basic Energy Policy Law" (Law No. 71 of 2002) was enacted by parliamentary legislation on June 14, 2002, with the aim of promoting long-term and comprehensive measures related to energy supply and demand. The purpose of the law is stated in Article 1: "This law aims to contribute to the conservation of the regional and global environment and to the sustainable development of the economic society of our country and the world by establishing basic policies related to measures concerning energy supply and demand, clarifying the responsibilities of the national and local governments, and promoting long-term, comprehensive, and planned measures concerning energy supply and demand, considering that energy is indispensable for the stability and improvement of national life and the maintenance and development of the national economy, and that its use has a significant impact on the regional and global environment."

The basic policy of energy policy is to achieve the 3Es (energy security, economic efficiency, and environmental compatibility) (Articles 2–4). Based on Article 12 of the law, the government is obliged to establish a basic plan concerning energy supply and demand, review it at least every three years, and make changes as necessary. The Minister of Economy, Trade and Industry is required to formulate a draft basic plan after hearing the opinions of the Comprehensive Resources and Energy Survey Committee, which consists of experts in energy and other fields, and seek a Cabinet decision (Kondoh 2012).

2 The Energy Basic Plan: Decarbonized Society, Asia Strategy, and Hydrogen Energy Technology

2.1 Energy Security

The Basic Act on Energy Policy stipulates in Article 12 the Basic Energy Plan. That is, the government must formulate a basic plan concerning the supply and demand of energy to promote measures related to the supply and demand of energy in a long-term, comprehensive, and planned manner.

In line with the basic policies of the Basic Act on Energy Policy, which are "ensuring stable supply," "environmental compatibility," and "utilization of market principles" while fully considering these, the government has decided to formulate and publish the "Basic Energy Plan" to promote measures related to the supply and demand of energy in a long-term, comprehensive, and planned manner. Since the enactment of the Basic Act on Energy Policy, seven Basic Energy Plans have been formulated. Below is an overview of the concept of energy security in the Basic Energy Plans.

2.1.1 The 1st Basic Energy Plan (October 2003)

The challenges of energy security are defined as "reducing risks related to energy supply and ensuring stable supply," "considering the approach to policies related to energy supply and demand, and preventing global warming," and "when utilizing market principles in the energy sector, ensuring that the autonomy and creativity of businesses are not impaired while the government appropriately intervenes." From the perspective of maintaining national life stability and smooth economic operation, oil stockpiling is considered an important pillar of Japan's energy security policy, supporting cooperation between consuming and producing countries for stable supply.

2.1.2 The 2nd Basic Energy Plan (March 2007)

Regarding energy security, it is recognized that "energy is once again recognized as a familiar issue for the public, and ensuring stable energy supply is the most important issue for the nation as the foundation of national life and economic activities," "a plan to achieve the goals of the Kyoto Protocol has been formulated, incorporating measures and policies to achieve the goals of the Kyoto Protocol, and strengthening and enhancing measures against global warming related to energy, and addressing environmental issues such as energy issues and climate change in an integrated manner," and "to ensure sustainable growth, Japan will actively and strategically utilize its technological capabilities, know-how, and experience in addressing energy issues, and play a leading role in strengthening the development foundation of the entire region and stabilizing the international energy market." Based on the current recognition of changes in the domestic and international environment surrounding energy, the "New National Energy Strategy" (2006) was formulated, incorporating particularly important policy programs toward 2030.

2.1.3 The 3rd Basic Energy Plan (June 2010)

Energy security is defined as building a system that ensures stable energy supply from resource production areas to final domestic consumers and minimizing risks that threaten it. To strengthen energy security, the following five policy goals were set to comprehensively ensure: (1) improvement of self-sufficiency rate, (2) energy conservation, (3) diversification of energy composition and supply sources, (4) maintenance of supply chains, and (5) enhancement of emergency response capabilities.

2.1.4 The 4th Basic Energy Plan (June 2014)

Due to the Great East Japan Earthquake on March 11, 2011, and the accident at the Tokyo Electric Power Company's Fukushima Daiichi Nuclear Power Plant, the environment surrounding energy has changed significantly both domestically and internationally, and energy policy has faced a situation requiring major adjustments. The goal of energy security is to ensure stable energy supply with safety as a prerequisite, achieve low-cost energy supply through improved economic efficiency, and make maximum efforts to achieve environmental compatibility, aiming to realize a "flexible energy supply and demand structure with multiple layers and diversity."

2.1.5 The 5th Basic Energy Plan (July 2018)

In addition to further strengthening efforts to achieve the energy mix for 2030, a challenge toward energy transition and decarbonization by 2050 was set as a new energy choice. Under the principle of "3E+S" (Energy Security, Economic Efficiency, Environment + Safety), the policy and corresponding actions will be steadily advanced to achieve the energy mix for 2030. The evaluation axis of "higher 3E+S" was set, prioritizing safety, improving self-sufficiency rate, challenging decarbonization, and strengthening the competitiveness of domestic industries.

2.1.6 The 6th Basic Energy Plan (October 2021)

The safety of all energy-related facilities is a prerequisite for energy policy. Especially for nuclear power, safety is prioritized above all else, and all efforts are made to alleviate public concerns. Given the challenges of limited negotiation power in resource procurement and the impact of changes in the situation in resource countries and sea lanes, the risk of facing supply insecurity is always present, and ensuring energy security remains a major issue for Japan.

2.1.7 The 7th Basic Energy Plan (February 2025)

The energy policy aims to simultaneously achieve stable energy supply, economic growth, and decarbonization. In May 2023, the GX Promotion Law and GX Decarbonization Power Law were enacted, and in May 2024, the Hydrogen Society Promotion Law and CCS Business Law were enacted. The increasing geopolitical risks, such as Russia's invasion of Ukraine and the intensification of the situation in the Middle East, have reaffirmed that it is difficult to consider Japan's energy security separately from the global energy situation. As issues such as climate change, which are difficult to solve by a single country, increase, an international perspective becomes more important.

2.2 *Decarbonized Society*

2.2.1 The 1st Basic Energy Plan (October 2003)

Achieving a 6% reduction in total greenhouse gas emissions compared to the base year during the first commitment period of the Kyoto Protocol (2008 to 2012) is an urgent issue. Since about 90% of greenhouse gases causing global warming are energy-related carbon dioxide, it is extremely important to consider the prevention of global warming when thinking about energy supply and demand policies.

2.2.2 The 2nd Basic Energy Plan (March 2007)

From the perspective of energy policy, we aim to lead discussions towards building an effective framework that enables all countries to work on emission reductions according to their capabilities, and encourages maximum reduction efforts by all major emitting countries, including the United States, China, and India, to achieve the ultimate goal of the Climate Change Framework Convention.

2.2.3 The 3rd Basic Energy Plan (June 2010)

To contribute to achieving medium- and long-term goals for reducing greenhouse gas emissions, it is necessary to promote a transformation in the energy supply and demand structure in national life, corporate activities, and local communities. Since CO_2 reduction in all sectors, including the residential sector, is urgent, it is essential to implement policies that delve into social systems and people's lives.

2.2.4 The 4th Basic Energy Plan (June 2014)

The IPCC Fifth Assessment Report indicates that there is no doubt about the warming of the climate system and that fundamental and continuous reductions in greenhouse gases are necessary to curb climate change. To fundamentally solve the global warming problem, it is urgent to significantly reduce greenhouse gas emissions worldwide, in addition to domestic emission reductions.

2.2.5 The 5th Basic Energy Plan (July 2018)

Considering the uncertainties associated with the 2050 scenario, lessons learned from the situation in leading countries, and Japan's unique energy environment, we adopt an "ambitious multi-track scenario aiming for energy transition and decarbonization" that pursues all options, including renewable energy, hydrogen, CCS, and nuclear power. It consists of realizing the long-term energy supply and demand outlook for 2030 and designing scenarios with an eye on 2050.

2.2.6 The 6th Basic Energy Plan (October 2021)

In October 2020, Japan declared its goal to achieve "carbon neutrality by 2050." In April 2021, a new policy was announced aiming to reduce greenhouse gas emissions by 46% from the 2013 level by 2030, with a further challenge towards a 50% reduction. To achieve carbon neutrality by 2050, it is necessary to advance an ambitious and fundamental shift in thermal power policy to make CO_2 emissions from thermal power generation virtually zero.

2.2.7 The 7th Basic Energy Plan (February 2025)

Based on the Innovative Environmental Innovation Strategy (January 2020), the "Green Growth Strategy" was formulated in June 2021 in collaboration with relevant ministries and agencies. It is believed that many companies in the industrial sector need to fundamentally change their business models and strategies with an eye on carbon neutrality by 2050. The costs of GX-related technologies, such as decarbonized power sources like renewable energy and nuclear power, energy-saving technologies like energy-saving semiconductors and photoelectric fusion technology, next-generation fuel technologies like hydrogen necessary for non-fossil conversion, and carbon capture and removal technologies like CCUS and DACCS, need to be significantly reduced.

2.3 Hydrogen Energy Technology

2.3.1 The 1st Basic Energy Plan (October 2003)

Although hydrogen is a zero-emission energy medium at the utilization stage, carbon dioxide and other emissions are produced when hydrogen is manufactured from fossil fuels. Therefore, improvements in hydrogen production technology through the reforming of fossil fuels will be promoted.

2.3.2 The 2nd Basic Energy Plan (March 2007)

Efforts will be made to establish safe, convenient, efficient, and low-cost technologies for the production, transportation, and storage of hydrogen. Additionally, infrastructure development for hydrogen supply, as well as the review of regulations related to hydrogen production, storage, transportation, and utilization, will be explored.

2.3.3 The 3rd Basic Energy Plan (June 2010)

Hydrogen derived from fossil fuels will be utilized to promote the effective use of fossil fuels, and by-product hydrogen from steel mills and other sources will be utilized. In the future, the establishment of hydrogen production technology that does not emit CO_2 by combining fossil fuel-derived hydrogen with CCS, and the development and utilization of non-fossil energy-derived hydrogen will be promoted. Support for the development of hydrogen supply infrastructure, such as hydrogen stations, will be promoted with the aim of starting dissemination from 2015.

2.3.4 The 4th Basic Energy Plan (June 2014)

To realize a society that fully utilizes hydrogen, a "hydrogen society," it is important to promote the development of a wide range of technologies and cost reduction under a strategy that takes an overview of the entire supply chain from hydrogen production to storage, transportation, and utilization. Technologies with high safety, convenience, economic efficiency, and environmental performance will be selected from among various technological possibilities.

2.3.5 The 5th Basic Energy Plan (July 2018)

To realize a "hydrogen society" ahead of the rest of the world, it is essential to reduce the procurement and supply costs of hydrogen, including its environmental value, to a level comparable to conventional energy. To achieve a "hydrogen society," there are still many challenges in terms of technology, cost, systems, and infrastructure. Therefore, based on the Hydrogen Basic Strategy formulated in December 2017, strategic efforts will be made to develop systems and infrastructure, promote the development of various technologies and cost reduction, and implement highly feasible technologies in society.

2.3.6 The 6th Basic Energy Plan (October 2021)

In anticipation of the carbon–neutral era, hydrogen is expected to contribute to the decarbonization of heat utilization, zero-emission power sources, decarbonization of the transportation and industrial sectors, production of synthetic fuels and synthetic methane, and efficient use of renewable energy. As Japan continues to lead internationally in this field, it is necessary to further strengthen efforts. In addition to resource diplomacy for securing and stable procurement of hydrogen and other resources, efforts will be made to strengthen relations with consumer countries with limited domestic resource potential and expected future hydrogen imports, and to promote the overseas deployment of Japan's excellent hydrogen utilization technologies. The Hydrogen Ministerial Meeting, a platform for dialogue among hydrogen ministers from various countries initiated by Japan, will continue to be utilized to lead international cooperation for the expansion of hydrogen utilization worldwide.

2.3.7 The 7th Basic Energy Plan (February 2025)

To "win with technology and win in business," efforts will continue to enhance competitiveness through advanced technology development under the Green Innovation Fund Project and other initiatives. To promote social implementation, strong support will be provided for the construction of large-scale supply chains for low-carbon hydrogen based on the Hydrogen Society Promotion Law. Policies integrating regulations and support will continue to be implemented to further promote large-scale supply and utilization of low-carbon hydrogen domestically and internationally, while reducing costs and expanding utilization.

2.4 Asia Strategy

2.4.1 The 1st Basic Energy Plan (October 2003)

It is expected that global energy demand will continue to increase, particularly in the rapidly growing Asian region. Oil is expected to remain the central source of global energy supply. Due to the decline in supply capacity from other oil-producing regions, dependence on the Middle East, which has a large production capacity, is likely to increase further.

2.4.2 The 2nd Basic Energy Plan (March 2007)

In the context of energy and environmental cooperation with Asian countries, the need to promote energy-saving cooperation is rapidly increasing as it significantly contributes to easing and stabilizing energy supply and demand and to global environmental conservation, including addressing global warming. In order to develop these collaborations, it is necessary to actively utilize regional cooperation through multilateral frameworks such as ASEAN + 3 (Japan, China, and South Korea), Asia–Pacific Economic Cooperation (APEC), and the East Asia Summit (EAS), in addition to bilateral cooperation.

2.4.3 The 3rd Basic Energy Plan (June 2010)

Energy demand in emerging countries in the Asia–Pacific region, including China and India, is expected to continue increasing. From the perspective of strengthening global energy security and addressing climate change issues, enhancing energy efficiency and promoting low-carbon development in the Asia–Pacific region will continue to be the pillars of Japan's international energy cooperation.

2.4.4 The 4th Basic Energy Plan (June 2014)

Japan's leading role in improving the stability of energy supply and demand in the rapidly growing Asian region will also improve Japan's energy security environment. The East Asia Summit (EAS) should be upgraded to a more effective framework for discussing multilateral energy security, with the Economic Research Institute for ASEAN and East Asia (ERIA) as the core institution.

2.4.5 The 5th Basic Energy Plan (July 2018)

Strengthening cooperation with emerging countries in Asia, where significant energy demand growth is expected, is an important issue for enhancing energy security in the Asian region and Japan, and for developing potential markets for the energy industry. The East Asia Summit (EAS) should be upgraded to a more effective framework for discussing multilateral energy security, with the Economic Research Institute for ASEAN and East Asia (ERIA) as the core institution.

2.4.6 The 6th Basic Energy Plan (October 2021)

With the intensification of US-China tensions, the importance of securing economic and energy security in the Asia–Pacific region has never been higher. To achieve global carbon neutrality, it is necessary to utilize diverse energy and technologies tailored to the circumstances of each country to simultaneously achieve sustainable economic growth and realistic energy transitions in emerging countries such as those in Asia, where energy demand is expected to grow.

2.4.7 The 7th Basic Energy Plan (February 2025)

The Asia Zero Emission Community (AZEC) aims to simultaneously achieve economic growth, energy security, and decarbonization, and to realize carbon neutrality through diverse pathways. In accordance with the "Action Plan for the Next Decade" agreed upon by leaders in October 2024, individual projects and rulemaking will be promoted as two wheels of a cart.

3 Energy Security for Achieving Carbon Neutrality by 2050

3.1 Framework of Energy Security

In discussions on energy security in Japan, ensuring a stable supply of energy has consistently been regarded as one of the most important national priorities, forming the basis of citizens' lives and economic activities.

Having experienced two oil crises, intensified global warming counter measures, the establishment of basic energy policies, and the nuclear power plant accident triggered by the Great East Japan Earthquake, contemporary energy policy requires the coexistence of decarbonization and energy security.

Before the oil crises, energy security primarily emphasized on the quantity of available energy supplied. However, during the two oil crises- crude oil prices surged- the appropriateness and stability of energy prices was also came to be understood as an integral component of energy security (Ishida 2010).

As concern for global warming grew, a broader notion of concern energy security emerged, one that includes environmental considerations and long-term sustainability. In addition, rising energy demand in rapidly developing Asian economies such as China and India has become a major factor influencing the global energy supply—demand balance. Geopolitical conditions similarly shape the international energy landscape.

The accident at the Fukushima Daiichi nuclear power station on March 11, 2011, has since been recognized as a serious and fundamental risk for Japan, a country that had actively promoted nuclear energy as a key policy pillar.

Risks to energy security include physical risks such as supply disruptions and economic risks such as price volatility surges. Physical risks may arise from accidents, labor disputes, natural disasters, terrorist attacks within the domestic or international energy supply chain, or political instability and regional conflicts in major energy-producing supplying countries (Mitchell 2000).

While Japan's energy security discourse has historically been framed around achieving national prosperity through a stable supply of energy, it typically lacks explicit reference to "human security" a concept central to broader international security discussions.

Rachel Schindler (2015) organizes the framework for energy security is organized as shown in Fig. 1, (2015), identifying the following interrelated dimentions:

Human Security: Protecting individual freedom and human rights, and ensuring social and economic stability.

National Security: Safeguarding national sovereignty and promoting sound governance for the protection of the nation's land and environment.

Climate Security: Addressing risks and disasters linked to extreme weather events.

Economic Security: Strengthening supply chain to support economic growth.

Energy Security: Ensuring a stable energy supply while reducing vulnerability to climate change.

Water Security: Conserving water resources and preserving the environment.

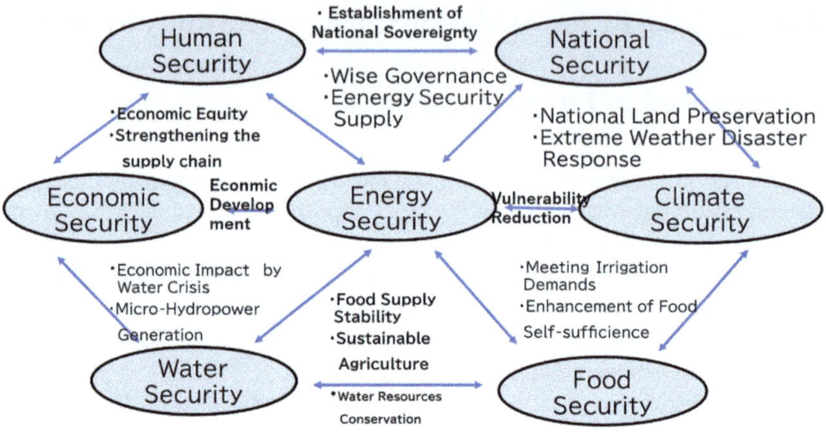

Fig. 1 The framework of energy security

Food Security: Achieving sustainable agriculture and improving food self-sufficiency rates.

Establishing energy security requires consideration of these interconnections and a deepened understanding of how of these elements interact and influence one another.

3.2 The Role of a Hydrogen Society in Achieving Carbon Neutrality by 2050

In the Fourth Basic Energy Plan, the government stated that "the time has come to comprehensively consider a 'hydrogen society' that utilizes hydrogen as an energy source."

In line with this direction, the "Hydrogen and Fuel Cell Strategy Roadmap" (June 2014) outlined the following phased approach:

Phase 1: Dramatic Expansion of Hydrogen Utilization (Present -) Rapidly expand the use of stationary fuel cells and fuel cell vehicccles (FCVs), which are already being commercialized, and strengthen Japan's leadership in the global hydrogen and fuel cell market.

Phase 2: Full-Scale Introduction of Hydrogen Power Generation and Establishment of Large-Scale Hydrogen Supply System (Realized in the late 2020s) Further increase hydrogen demand, diversify hydrogen sources to include unused or untapped energy, and construct a new secondary energy structure that adds "hydrogen" to the traditional pillars of electricity and heat.

Phase 3: Establishment of a Total CO_2-Free Hydrogen Supply System (around 2040) Combine hydrogen production with carbon capture and storge (CCS) or use hydrogen derived entirely from renewable energy to achieve a fully CO_2-free hydrogen supply chain.

The Japan Business Federation (Keidanren, 2021) has proposed that hydrogen and ammonia be actively utilized as carbon—free power sourced in the pursuit of carbon neutrality by 2050, targeting approximately 10% adoption through technological innovation.

Similarly, the Japan Chamber of Commerce and Industry (2022) has proposed the following measures to support Japan's progress toward carbon neutrality:

- Support for stable supply, including the development of hydrogen and ammonia supply chains.
- Improvements in supply efficiency, such as promoting the use of hydrogen across regions and industrial clusters and prioritizing storage and transportation infrastructure.
- Commercialization of next-generation ammonia technology capable of operating at low temperatures and pressures, and support for the deploying co-firing technology in coal-fired power plants to Southeast Asia and other regions.

At the same time, in response to these policy directions have offered critical perspectives. Japan's hydrogen strategy should urgently establish GHG emission standards for blue hydrogen, which is internationally recognized (Renewable Energy Institute 2022). The following points have been raised:

- Reconsider the role of hydrogen in Japan's decarbonization strategy and minimize the volume of hydrogen considered essential.
- Prioritized green hydrogen and establish a system to supply it through a combination of imports and domestic production.
- Reassess policies centered on gray hydrogen and blue hydrogen. Set GHG emission standards based on LCA and correct the treatment of hydrogen under the Energy Conservation Act and the Act on the Rational Use of Energy.
- Reevaluate the policy of constructing large-scale supply chain, particularly if they rely on carbon -incentive hydrogen pathways.

3.3 Japan-China-Korea Hydrogen Economic Zone and Energy Security

Within Japan's energy security strategy, the core policy objective defined in the Basic Energy Policy Law, which is "ensuring stable supply," remains central. At the same time, the realization of carbon neutrality by 2050, cooperation with Asia, particularly Japan, China, and South Korea, and the realization of a hydrogen-based society as a form of renewable energy have emerged as pressing practical issues.

Japan formulated, the "Green Growth Strategy for Carbon Neutrality by 2050" was formulated in 2021, in China released, the "Hydrogen Industry Development Plan" was formulated in 2022, and in South Korea, the "Hydrogen Economy Activation Roadmap" was formulated (Zhen 2022).

In Japan's hydrogen strategy under the "Green Growth Strategy," hydrogen is positioned as an important energy source for a decarbonized society, and the growth of hydrogen-related industries is promoted. Major initiatives includes:

- Establishment of a Large-Scale Hydrogen Supply System: To achieve an annual hydrogen supply of 3 million tons by 2030 and 20 million tons by 2050, hydrogen imports from overseas will be promoted, and the production of green hydrogen using renewable energy will be promoted domestically.
- Expansion of Hydrogen Power Generation: Introduce hydrogen and ammonia power generation will be introduced by 2030, and deploy large-scale hydrogen power plants as a substantial component of the energy mix from around 2040.
- Development of Hydrogen Fuel Cell Vehicles (FCVs) and Infrastructure Development: Target 800,000 hydrogen fuel cell vehicles and 1,000 hydrogen stations by 2030.
- Industrial Utilization of Hydrogen in the Industrial Sector: Promote hydrogen use in the steel, chemical, and cement industries to reduce CO_2 emissions and support the development of new hydrogen technologies.
- International Cooperation and Supply Chain Development: Strengthen partnerships with Australia, the United Arab Emirates (UAE), Saudi Arabia, and other countries to diversify hydrogen imports and enhance hydrogen transportation and storage technologies.

China's Medium- to Long-Term Plan for the Development for the Hydrogen Industry (2021–2035) China's hydrogen plan aims to accelerated the development of China's hydrogen energy industry and promote the use of clean energy to contribute to the realization of a decarbonized society by combining hydrogen energy with renewable energy, aiming for carbon neutrality (zero CO_2 emissions by 2060). Key policies include:

- Expansion of Green Hydrogen Production Using Renewable Energy: Significantly improve the supply capacity of hydrogen energy by 2030.
- Promote of Hydrogen Utilization: Expand the spread of hydrogen fuel cell vehicles, promote technological innovation and infrastructure development for the use of hydrogen in the industrial sector.
- Strengthen R&D of Hydrogen Production, Storage, and Transportation Technologies: Establish hydrogen stations nationwide.
- Strengthen Technical Cooperation with Countries Such as Japan and Europe: Actively engage in international standardization.

South Korea's Hydrogen Economy Activation Roadmap South Korea's roadmap positions hydrogen as a national strategy with the ambition of becoming "the world's first hydrogen economy". This roadmap aims for South Korea to become "the first country to realize a hydrogen economy." The South Korean government has also set the goal of "carbon neutrality by 2050," positioning hydrogen energy as an important pillar of a decarbonized society. The main policy issues of the "Hydrogen Economy Activation Roadmap" are as follows:

- Expansion of Clean Hydrogen Production and Imports by 2040: Improve the self-sufficiency rate of hydrogen by utilizing renewable energy in South Korea.
- Mass Development of Hydrogen Vehicles: Aim to have 200,000 hydrogen vehicles by 2025 and 2.9 million by 2040, and promote the introduction of hydrogen buses, trucks, and taxis.
- Establishment of 1,200 Hydrogen Stations by 2040: Develop hydrogen transportation and storage technologies.
- Securing 15 GW of Hydrogen Power Generation Capacity by 2040: Expand the use of hydrogen in the steel, chemical, and cement industries.
- Support the Development of Hydrogen Technologies by South Korean Companies: Foster the hydrogen industry as an export industry, promote international cooperation with Japan and Europe, and strengthen the global competitiveness of the hydrogen industry.

Although the annual targets and figures differ among Japan, China, and South Korea, the direction is consistent. To achieve this, it is necessary to solve the issues of hydrogen diffusion, such as the production process of blue hydrogen and the spread of water electrolysis hydrogen using renewable energy. Furthermore, it is necessary to solve the issues of high dependence on hydrogen imports and the limitations of domestic CO_2 storage capacity. To solve these issues, the construction of a "Japan-China-Korea Hydrogen Economic Zone" through cooperation among Japan, China, and South Korea will likely increase the possibility of achieving carbon neutrality by 2050 (Zhen 2022).

References

Ayabe H (2014, March). The background on renewable energy policies—Japanese development. National Diet Library Research and Legislative Reference Bureau, Research project: aspects surrounding renewable energy. https://dl.ndl.go.jp/view/download/digidepo_8563846_po_20130404.pdf?contentNo=1

Commission on Human Security (2003) Human security now: protecting and empowering people. United Nations

Energy Charter Secretariat (2015, March) International energy security: Common concept for energy producing, consuming and transit countries. https://www.energycharter.org/fileadmin/DocumentsMedia/Thematic/International_Energy_Security_2015_en.pdf

Fuji Research Institute. (2003, January) Study on the expansion of multidimensional evaluation indicators (II). Japan Nuclear Cycle Development Institute

Hoshino S (1964) Japan's energy issues. DIAMOND

Irie K, Kanda K (2002) Formation and transformation of energy security concept. Acad J EnerMix 81(5)

Ishida H (2010) Energy and security. J Int Secur 38(2)

Kondoh M (2011, March 14) The two oil crises and the trends in the Japanese economy. Cabinet Office Economic and Social Research Institute. https://www.esri.cao.go.jp/jp/esri/prj/sbubble/history/history_01/history_01.html

Kondoh K (2012, December 26) History and challenges of Japan's energy policy: based on discussions following the Fukushima Daiichi nuclear power plant accident. National Diet Library Research and Information Issue Brief, p 762

Mitchell JV (2000, November) Energy supply security: change in concepts. Royal Institute of International Affairs

Nakagami K (2025) The Genealogy of energy security and decarbonized society, Asia and hydrogen energy strategy. J Assoc Study of Polit Soc 10

Ono T (1976) Japan's new energy policy. J Fuel Soc Japan 55(8)

Renewable Energy Institute (2022, September) Re-examining Japan's hydrogen strategy: moving beyond the "hydrogen society" fantasy

Schindler R (2015, October 29) Eliminate hunger from the equation of poverty. United Nations University. https://jp.unu.edu/publications/articles/taking-hunger-out-of-the-poverty-equation.html

The Japan Business Federation (2021, March 5) Towards achieving carbon neutrality by 2050: commitment and actions of the business community

The Japan Chamber of Commerce and Industry (2022, May 12) Opinion on the clean energy strategy for achieving carbon neutrality by 2050

UNDP (1994) Human development report 1994. Oxford University Press

Zhen J (2022, August) The decarbonization and new energy strategies in Japan, China, and Korea: potential for cooperation in the hydrogen economy among Japan, China, and Korea. In Miyamoto Y (ed) East Asia risks and Japan-China relations (Asia Research Report [FY2021] 12)

An Analysis of Japan's Climate Policy Based on the Economic Philosophy of "Sustainable Development"

Kazuo Matsushita

1 Introduction

Japan experienced serious industrial pollution during its period of rapid economic growth in the 1960s and beyond, as exemplified by Minamata disease and the Yokkaichi air pollution incident.

The situation was so severe that Japan came to be known internationally as a country with advanced pollution damage. Subsequently, with the support of movements by pollution victims and advanced initiatives by local governments, the government strengthened pollution regulations, and with the progress of technological responses by industry, Japan came to be recognized as a leading country in pollution control. Then, after two oil shocks in the 1970s, Japan established its position as a leading country in energy conservation and environmental technology (Matsushita 2014).

However, Japan lags behind in current climate change countermeasures and measures to transition to a decarbonized economy, particularly in the expansion of renewable energy.

This is causing the Japanese economy to stagnate and damaging the international competitiveness of Japanese industries.

Looking back, Japan was once a world leader in the development and manufacture of solar panels, and hybrid vehicles were launched in Japan in conjunction with the adoption of the Kyoto Protocol at the Third Conference of the Parties to the United Nations Framework Convention on Climate Change (COP3, December 1997). Japanese manufacturers continue to dominate the global market. However, the market share of Japanese manufacturers in the solar panel sector, which once accounted for over half of the global market, has now significantly declined. Additionally, in the

K. Matsushita (✉)
Kyoto University, Kyoto, Japan
e-mail: matsushita@iges.or.jp

W. Zhou (ed.), *Decarbonization Policy Engineering*, World Sustainability Series,
https://doi.org/10.1007/978-981-95-5953-4_4

rapidly expanding global electric vehicle market, no Japanese manufacturers are among the top 10 companies.

On the other hand, looking at the change in the percentage of electricity supply from renewable energy sources over the past 23 years (2000 to 2023), the UK has increased by 45 points from 2.6% to 47.3%, and Germany has increased by 46 points from 6.6% to 53.0%, Japan has only increased from 9.2% to 22.1%, a mere 13-point increase. This is largely due to differences in policy rather than natural conditions. Japan has lacked a proper climate and energy policy.

Recently, Japan has been slow to implement the institutional reforms and investments necessary for decarbonization and the shift away from fossil fuels. The setting of ambitious climate change targets, institutional reforms to expand renewable energy, and the development of power grids have also been delayed. Furthermore, the full-scale introduction of carbon pricing, a core policy measure for decarbonization, has been significantly postponed.

In this paper, against the backdrop of the unfortunate current situation resulting from Japan's climate and energy policies to date, we will consider what constitutes a proper climate policy for Japan. As a prerequisite, we will review economic ideas related to "sustainable development," and then examine the "GX Strategy" and "GX Transition Bonds," which form the core of Japan's current decarbonization transition policy.

2 What Is a "proper climate policy"?

So, what is a "proper climate policy"? There are various opinions on this, and there is no consensus. Table 1 quotes from a lecture by Shuzo Nishioka, one of the initiators of the "Let's Create a Proper Climate Policy for Japan Now" campaign.

I basically agree with Nishio's definition, but I would also like to consider the elements shown in Table 2.

Examples of the latest and best scientific knowledge include numerous reports from the Intergovernmental Panel on Climate Change (IPCC) and others, as well as analyses from the International Energy Agency (IEA) and the International Renewable Energy Agency (IRENA) in the energy sector. In terms of international discussions, agreements reached at the United Nations Framework Convention on Climate Change (UNFCCC) Conference of the Parties (COP) and the Group of Seven (G7) summit are important.

For example, the IEA's report projects that approximately 90% of the global power supply mix by 2050 will be sourced from renewable energy (IEA 2021). Furthermore, in the vast body of scientific literature from the IEA and IPCC, it is a common understanding that wind power and solar power are the technologies most contributing to decarbonization, and their widespread adoption is crucial.

Table 1 What is a "proper climate policy"?

Proper climate policy
(1) Philosophy and goals:
· The climate is the foundation for the safe and secure survival and livelihood of humankind, and there is no substitute for it. · Climate change threatens this foundation, and stabilizing it must take priority over everything else. (The vanguard of sustainability policy) ➡Policies should aim to integrate nature and human society.
(2) Science-based
• Climate is a natural phenomenon, and its mechanisms are not yet fully understood. ➡ Policies should be based on the latest scientific knowledge that reflects the essence of "climate."
(3) Fairness and ethics-based
· Climate and weather are global public goods connected across the earth's surface. · A long-term challenge for human sustainability. · Cooperation on an international and intergenerational scale. ➡Stabilization is impossible without greenhouse gas (GHG) reductions through equitable burden-sharing among all countries and generations.
(4) Inclusivity
• All humanity depends on climate for survival and suffers from its variability. At the same time, we are the agents capable of altering climate through GHG emissions. • Therefore, we have the motivation, responsibility, and capacity to act as stakeholders in climate stabilization. ➡ Integrated efforts from both top-down and bottom-up approaches are effective.

From a lecture by Shuzo Nishioka (October 4, 2024)

Table 2 "Proper climate policy"

(1) Basic understanding that climate change is a human rights issue (related to intergenerational equity, north-south issues, gender, etc.)
(2) Shared awareness of the urgency of the climate crisis and the race against time
(3) Grounded in sound science (including both natural scientific foundations and social scientific insights)
(4) Aligned with the Paris agreement's 1.5 °C target pathway and consistent with international trends and guidelines
(5) Policy decision-making processes that ensure the participation of diverse stakeholders and transparency/openness
(6) Just transition

Author's supplement

3 Economic Thought Related to Sustainable Development

As a prerequisite for considering a sound climate policy, let us look back on the history of economic thought related to sustainable development.

The late 1960s was a period when the damage caused by industrial pollution, mainly in advanced industrialized countries, became apparent worldwide. In 1972, the first UN-sponsored environmental conference (the UN Conference on the Human Environment in Stockholm) was held at the initiative of the Nordic countries, particularly Sweden, which were suffering from acid rain damage caused by transboundary air pollution from the UK and Germany at the time.

In the same year, the Club of Rome published a book titled "The Limits to Growth," which shocked the world (Meadows et al. 1972). This book was written by Dennis Meadows of MIT, who was commissioned by the Club of Rome to use system dynamics methods and quantitative estimates to warn of future global trends. It pointed out that if the world continued to grow economically at the same rate as it was at the time, human society could face an uncontrollable crisis within the next 100 years in terms of population, food, resources, and pollution. This caused a great shock and sparked heated debate.

3.1 Recognition of a Closed Economy

Even before the publication of "The Limits to Growth," British-born American economist Kenneth E. Boulding was quick to point out that infinite economic growth is impossible within the finite closed system of the Earth.

In 1966, he wrote an essay titled "The Economics of Spaceship Earth," in which he criticized conventional economics for assuming that resources are infinite, calling it "cowboy economics (Boulding 1966)." "Cowboy economics" is an economy that aims to maximize consumption through repeated plundering and destruction of natural resources. Boulding stated, "The future 'closed economy' should be called the 'astronaut economy.' Earth has become a single spaceship, with no infinite reserves, no place to mine, and no place to pollute. Therefore, within this economy, humans must understand that they are part of a cyclical ecosystem.

" Against the backdrop of humanity's first view of Earth from space via satellite, Boulding's warning had a profound impact on many people, instilling a sense of reality. However, it did not lead to significant changes in actual economic activities. The concept of a "closed economy" has since been adopted by various fields, including environmental science, economics, and management.

Boulding is also known for saying, "Only a madman or an economist believes in exponential economic growth." Exponential economic growth refers to compound interest growth, which is fundamentally and essentially "unsustainable." Incidentally, if economic growth continues through compound interest, a 10% growth rate would double the economy in 7 years, a 7% growth rate in 10 years, and a 3.5% growth rate over 100 years would increase the economy by more than 30 times. Considering the physical and ecological limits of the Earth's environment, such a scenario is impossible.

3.2 Globalization and the Global Environment

More than half a century has passed since Boulding's remarks, but the governments and leaders of almost every country in the world still believe in the myth that "economic growth will solve all problems" and continue to focus on it as their primary concern.

Rather, with the globalization of capitalism today, the destruction of the global environment is accelerating even further due to the removal of barriers such as national borders. The rapid globalization of the economy since the late 1980s has led to the integration of economic activities on a global scale through the acceleration of trade, capital investment, and information transfer. The expansion of economic activities against the backdrop of population growth and globalization around the world has led to the worsening of diverse and complex environmental issues.

This means that at the local, national, regional, and global levels, economic activities have exceeded the carrying capacity of the ecosystems on which they depend, causing various types of damage to nature, people's lives, and health. Furthermore, there are an increasing number of cases in which the economic activities of certain countries and companies are affecting other countries and the global environment across national borders.

If Boulding's observation is correct, then the world today is filled with madmen and economists. Why has this happened?

It is important to note that "growth" and 'development' are fundamentally different concepts. As symbolized by the expansion of GDP, "growth" refers to quantitative expansion. GDP ignores the social costs associated with the consumption and depletion of resources that accompany economic activity, and in cases of environmental pollution, it adds the costs of pollution control measures to the calculation. On the other hand, "development" refers to qualitative change and does not necessarily mean quantitative expansion. The original policy goal should be the sustainable maintenance and development of people's welfare. Moreover, this should be achieved within a closed ecosystem, i.e., a sustainable ecosystem.

3.3 *Sustainable Development: Seeking Continuous Social Change*

Sustainable development was proposed with the intention of making economic development environmentally and socially sustainable. Regarding sustainable development, World Commission on Environment and Development "Our Common Future" (1987) established by the United Nations defined it as "development that meets the needs of the present generation without compromising the ability of future generations to meet their own needs." This definition clearly emphasizes intergenerational responsibility, stating that economic development must not threaten the development potential of future generations. Sustainable development inherently means development that maintains environmental, social, and economic sustainability, aiming to improve the quality of life for people and maintain the sustainability of ecosystems.

Behind this was the recognition and expectation that "economic growth and environmental conservation are not inherently contradictory, and that it is entirely possible to make economic development environmentally sustainable. Furthermore, it is also possible to achieve environmental and social justice within and between generations." However, in the world since then, the focus has been solely on the sustainability of economic growth, and there has been a strong tendency to emphasize technical and symptomatic approaches to environmental issues based on the premise of maintaining economic growth.

The Brundtland Report states that sustainable development is "a process of change in which the development of resources, investment directions, technological development trends, and institutional reforms are harmonized with current and future needs." This statement emphasizes that sustainable development is deeply connected to society's technology and institutions and that it is necessary to focus on the process of change.

Expanding on this definition, "sustainable development" can be described as an environmental philosophy that presents a desirable vision of the environment and society that should be realized, while at the same time intending to promote a policy process of continuous change toward that vision. In other words, it should be understood that the Brundtland Report expected countries and the international community to recognize the limits of the global environment through their collective political actions and policies, and to fundamentally change the existing patterns of economic development.

In this sense, achieving "sustainable development" means solving the various environmental problems that have arisen in the course of the advancement of advanced industrial societies, envisioning a "new environmental society" for post-advanced industrial societies, ensuring social equity, and continuously reforming and integrating systems, technologies, resource utilization, and investment methods to achieve this vision. This indicates that innovation in the social system itself and social management of social common capital are required.

3.4 Ecological Economics and Sustainability Indicators

When considering "sustainable development" from the perspective of people's lives, it can be rephrased as "development that can maintain and improve the well-being of people from the present to the future over the long term." The nature of economic development varies depending on assumptions such as policy objectives, indicators, and perceptions of environmental and resource constraints. Table 3 compares the basic characteristics of the current mainstream economic model, the green economic model, and the ecological economic model.

The assumptions of mainstream economic models were made at a time when the population and human economic activity were relatively small compared to the global environment. At that time, the constraints on economic development were artificial capital (roads, factories, machinery, etc.), and natural capital (air, water, forests, natural resources, etc.) was still abundant. There was little need to worry about the "externalities" of human activities on the environment. It was reasonable to focus on expanding gross domestic product (GDP) as the main means of improving human welfare and to aim to increase the quantity of goods and services traded in the market. However, the situation has changed dramatically.

However, in today's world, human activities have expanded and artificial capital such as infrastructure has become abundant. The increase in material production and consumption has pushed the global environment to its limits (planetary boundaries) (Stockholm Resilience Center 2024). The current expansion of material growth is already unproductive (uneconomical) in the sense that it has a negative impact on happiness, social capital, and natural capital, or will eventually become so. We must once again remember that the purpose of economic activity is to sustainably improve human happiness and quality of life, and that material consumption and GDP are merely means to that end.

Table 3's "green economy" model argues that desirable results can be achieved by making several policy adjustments to mainstream economic models. For example, introducing carbon taxes and emissions trading would put a price on carbon dioxide emissions (carbon pricing) and curb the depletion of natural capital by placing an adequate price on it. Additionally, it advocates prioritizing investments in natural capital. Through such government policy interventions aimed at internalizing externalities, it is claimed that economic growth can continue while addressing issues such as pollution and climate change.

On the other hand, the "ecological economy" model considers many of the policies advocated by the green economy model to be necessary but argues that they alone are not sufficient to achieve sustainable human happiness and calls for more fundamental changes (changes in objectives and paradigms).

Table 3 Basic characteristics of mainstream economic models, green economic models, and ecological economic models

	Mainstream economic Model	Green economic model	Ecological economic model
Main policy goal	More economic growth, measured by GDP. It assumes that growth ultimately solves all other problems.	More economic growth, but with less environmental impact. GDP growth decoupled from carbon and other material and energy consumption impacts.	Better, not merely more. It is necessary to shift the focus from mere economic growth recognizing that growth entails significant negative side effects to 'development' in the sense of enhancing sustainable human Well-being.
Main progression indicator	GDP	Still GDP, but with recognition of impacts on natural capital.	Sustainable economic welfare index (ISEW), genuine Progress indicator (GPI), and other improved indicators of true Well-being.
Scale/ environmental capacity	Assumes that markets can overcome any resource constraints through new technologies and that substitute resources will always be available.	Acknowledges the issue, but assumes it can be resolved through decoupling.	A primary concern as a determinant of ecological sustainability. Natural capital and ecosystem services cannot be infinitely substituted and are subject to real limits.
Economic productivity/ allocation	A primary concern, though generally limited to market goods and services (GDP) and the functioning of market mechanisms.	Recognizes the necessity of incorporating natural capital and embedding its value within market incentives.	A primary concern, encompassing the effects of both market and non-market goods and services. Emphasizes the necessity of incorporating the value of natural and social capital to enhance allocative efficiency.
Government role	Minimizes government intervention, replacing it with private sector involvement and market mechanisms.	Recognition of the necessity of government intervention to internalize natural capital.	The government plays a central role within new frameworks for the governance of common property resources.

Source World watch Institute 2014, Chap. 11, Fig. 11–1, edited by the authors

3.5 Ecological Economics

Herman Daly, a proponent of ecological economics, defined "sustainable development" as development that does not involve growth exceeding the carrying capacity of the environment (Daly 1997). Here, "development" refers to qualitative improvement, while "growth" refers to quantitative expansion. GDP expansion, defined as economic growth, is defined as the growth of quantitative components (resource throughput, He argued that economic growth, defined as an increase in GDP, can be decomposed into quantitative components (resource throughput, i.e., resources extracted and used in economic activities before being discharged as pollutants or waste) and qualitative components (improvements in resource efficiency). He contended that the primary cause of environmental degradation lies in the growth of total throughput, and that reducing throughput or improving resource efficiency is essential for environmental preservation.

The scale of human economic activity expands year by year, but it is ultimately dependent on nature, which consists of a closed ecosystem, and cannot expand indefinitely. In order for development to be sustainable, the level of economic activity must be kept within the limits of what the ecosystem can sustain.

Daly divided natural capital into renewable and non-renewable resources, viewed sustainable development from the perspectives of material cycles and ecosystems, and proposed the following three principles (Meadows et al. 1992).

The sustainable utilization rate of "renewable resources" such as soil, forests, water, and fish must not exceed the rate of regeneration. (For example, in the case of fish, if they are caught at a rate that can be replenished by the remaining fish reproducing, they are renewable.)

The sustainable utilization rate of "non-renewable resources" such as fossil fuels, high-quality minerals, and fossil water must not exceed the extent to which renewable resources can be utilized at a sustainable pace. (For example, in the case of oil, after depleting reserves, a portion of the profits from oil use should be automatically invested in solar collectors or plants to ensure that an equivalent amount of renewable energy can be obtained. This constitutes sustainable utilization.)

The sustainable emission rate of "pollutants" must not exceed the extent to which the environment can circulate, absorb, and neutralize such substances. (For example, when discharging sewage into rivers or lakes, the pollutants must be neutralized to the extent that the environment can handle them.)

In modern society, the economy has grown significantly larger than the ecosystem, and natural capital is scarce compared to artificial capital. Natural capital cannot be completely replaced by artificial capital, so natural capital has become a limiting factor in economic development. For example, even if high-speed fishing boats and advanced technology are developed, if the fish stock itself decreases, it will not be possible to increase the catch.

Traditional economics has focused on two main goals: (1) efficient resource allocation and (2) fair income distribution. However, Daly added a third policy goal: achieving a sustainable (optimal) economic scale based on the carrying capacity of natural ecosystems (environmental capacity). Under a free-market economy and a decentralized democratic system, it is even more challenging to achieve these policy goals simultaneously.

3.6 *Implications of Hirofumi Uzawa's Theory of Social Common Capital*

Hirofumi Uzawa was a pioneer in proposing the concept of social common capital as a theoretical framework for addressing issues such as climate change (Uzawa 2000).

According to Uzawa, social common capital can be broadly divided into three categories: natural environment (mountains, forests, rivers, oceans, atmosphere, etc.), social infrastructure (roads, public transportation, gas, water, etc.), and institutional capital (hospitals, schools, finance, judiciary, etc.). It refers to natural and social mechanisms that enable a country or specific region to maintain a prosperous economy, develop a rich culture, and sustain a socially attractive society in a stable and sustainable manner. It is managed and operated in accordance with social standards as a common asset of society as a whole.

Social common capital is what makes a society where everyone's human dignity is protected, people can be independent, and civil liberties are as secure as possible. In other words, it is a social stabilization mechanism that enables a country or society to harmonize with the natural environment, maintain a high cultural standard, engage in economic activities in a sustainable manner, and realize a stable society. Its specific composition is not determined based on a priori or logical criteria, but rather by giving full consideration to natural, historical, cultural, economic, social, and technological factors.

Social common capital generally consists of scarce resources that cannot be privately owned or managed, but in many cases, due to historical or social circumstances, it takes the form of private ownership or management. Uzawa states that regardless of the form of ownership, the management and operation of social common capital should not be governed by bureaucratic standards, nor should it be significantly influenced by market standards. He argues that each professional group involved in social common capital should manage and operate it based on their professional knowledge and ethical standards.

Based on the concept of "social common capital," Uzawa tackled the issue of pollution and published "The Social Costs of Automobiles" (Uzawa 1974). This book focused on the extent to which social common capital, such as the natural environment and social infrastructure, has been polluted or destroyed using automobiles, and attempted to calculate the social costs of automobiles. The key point was that it specifically calculated the social costs of automobiles from the perspective of ensuring citizens' basic rights, explored ways to internalize these costs, and suggested the ideal form of urban transportation. "The Social Costs of Automobiles" became a bestseller and had a significant impact on Japanese society.

Regarding global warming, he has proposed a "proportional carbon tax and international fund for atmospheric stabilization." He has advocated the institutionalization of a global carbon tax as a feasible climate stabilization policy. However, imposing a uniform carbon tax not only raises issues of international fairness, but also risks stifling the economic development of developing countries. Therefore, he proposed a "proportional carbon tax" based on each country's per capita income. Furthermore, he proposed the concept of an International Atmospheric Stabilization Fund to eliminate the economic gap between developed and developing countries.

Uzawa's proportional carbon tax and International Fund for Atmospheric Stabilization are not yet acceptable as actual policies.

The reason for this is that in today's international society, which is based on sovereign states, the right to levy taxes lies with individual countries. In the absence of a centralized world government, introducing international taxation to solve global issues is a major challenge for the international community. However, as with nuclear abolition, international taxation is theoretically possible if each sovereign state agrees, and persistent efforts must be made to realize such a system.

In reality, the European Union (EU), which is a supranational organization, has been discussing the introduction of a common EU carbon tax as a measure to combat global warming for many years, but no agreement has been reached. Instead, an emissions trading system has been introduced across all EU member states. At the same time, each country is implementing a combination of measures, including a global warming countermeasure tax (or carbon tax), measures to promote the introduction of renewable energy, and voluntary agreements between the government and industry.

The concept of "social common capital" proposed by Uzawa provides an important institutional and policy analysis framework for policy formulation and selection and has become a paradigm for opening up a new era. However, further consideration is needed regarding how to manage social common capital in order to realize a sustainable and stable society. In particular, the concrete management and institutional design of social common capital, including issues such as global warming, are important challenges that researchers, practitioners, and policymakers must address seriously and work to form a national consensus on.

3.7 New Indicators of Sustainability

Appropriate indicators are necessary for evaluating sustainable development. However, GDP, which is widely used as an indicator of economic growth, is a flow indicator that measures short-term economic fluctuations and does not necessarily correlate with the wealth of the general population. For this reason, many indicators have been developed to replace GDP. Among them, the "Inclusive Wealth Index (IWI)" (new national wealth) developed by Partha Dasgupta and others is attracting attention (Dasgupta et al. 2022).

Dasgupta defines sustainable development as development that achieves continuous improvement in social welfare (quality of life). He argues that if the productive base of a society is not shrinking in terms of per capita population, then that society's development is sustainable. The productive base is a combination of a society's capital stock, which creates quality of life, and the systems that utilize it.

Excessive emphasis on economic growth as indicated by GDP indicators leads to excessive use of resources and damages the development potential of future generations. Therefore, in order to measure long-term sustainable development, the IWI was developed with an emphasis on capital stock (quantity).

The IWI is an indicator that evaluates the well-being of people that cannot be assessed by GDP alone and aims to maintain it for future generations. Unlike flow concepts such as GDP and HDI (Human Development Index), which focus on income, it quantitatively evaluates sustainability based on capital assets that form the productive base.

The IWI attempts to quantify and measure the overall wealth of countries and cities using three indicators: "artificial capital," "natural capital," and "human capital." The IWI and welfare are related in terms of stock and flow, and the national wealth composed of these three types of capital forms the basis for welfare. Specifically, national wealth is used in the annual production activities of a given society. This national wealth is composed of three types of stock: artificial capital such as factories and machinery, natural capital such as forests and farmland, and human capital such as education and health. When used in productive activities, it generates the flow of welfare.

IWI has attracted the attention of the United Nations and various national governments, and its measurement and application to policy is progressing. The United Nations (UNEP) has published the Inclusive Wealth Report based on the IWI concept (UNEP 2023).

4 Development Environmental Measures Costs and Carbon Pricing

4.1 Principles Regarding Environmental Policy Costs

Throughout the history of environmental policy, several principles and concepts regarding environmental policy costs have been internationally accepted. The following are representative examples.

· Polluter Pays Principle (PPP).

The idea that the costs of pollution prevention and control should be borne by the polluters themselves. This principle was first proposed in the "Guiding Principles on the International Economic Aspects of Environmental Policy" adopted at a meeting of the Organization for Economic Cooperation and Development (OECD) in 1972, and was included as one of 27 principles in the "Rio Declaration on Environment and Development" adopted at the United Nations Conference on Environment and Development (Earth Summit) held in Rio de Janeiro, Brazil, in 1992. The Earth Summit also adopted the "Rio Declaration on Environment and Development," which established international principles on the environment and development. One of the 27 principles in the declaration is the PPP. ("Rio Declaration" (Principle 16): National authorities should strive to internalize environmental costs and promote the use of economic instruments, considering the principle that polluters should bear the costs of pollution, while giving due consideration to the public interest and without distorting international trade and investment.")

· Internalization of environmental external costs (external diseconomies).

When the activities of an economic entity affect a third party without going through the market, this is called an "externality." When the impact is negative, it is called an external diseconomy. "Pollution" is a typical example of external diseconomies. Approaches to internalizing external diseconomies include policies that use government taxes and subsidies to internalize externalities, and policies that use laws to set optimal environmental standards and impose sanctions or penalties on economic entities that fail to comply with those standards (direct regulation).

· Carbon pricing.

A policy measure introduced to change the behavior of emitters by placing a price on the carbon dioxide (CO2) emitted by companies and other entities. The main methods include "carbon taxes" and "emissions trading." It is also a method of internalizing environmental external costs (external diseconomies).

4.2 The Meaning and Current Status of Carbon Pricing

Carbon pricing refers to the act of assigning a price to carbon. Its purpose is to internalize the social external costs (such as various damages caused by climate change) resulting from CO2 emissions, which are a cause of climate change, by imposing some form of charge based on the amount of carbon contained in emitted CO2. By assigning a price to carbon, economic incentives for reducing emissions are created, thereby promoting measures to address climate change. This creates economic incentives for emissions reduction and promotes action against climate change. By assigning a price to carbon, CO2 emitters must choose between reducing emissions or paying for them. As a result, society as a whole can reduce CO2 emissions in a more flexible and economically efficient manner.

The marginal cost of carbon emissions is the present value, discounted at the social discount rate from the present to the future, of the marginal decrease in the value of the natural benefits of the atmosphere and the economic, social, and cultural aspects of human life resulting from a marginal increase of one unit of CO2 emissions (Uzawa 1995).

Specific carbon pricing methods include carbon taxes (environmental taxes) and emissions trading systems. Carbon pricing also provides incentives for investment in low-carbon technologies and market expansion (World Bank 2023).

According to the World Bank, as of April 2023, 73 carbon pricing measures have been introduced worldwide. Of these, eight countries and one region have implemented carbon taxes only, 21 countries and 25 regions have implemented emissions trading systems only, and 19 countries and 17 regions have implemented both carbon taxes and emissions trading systems. Carbon pricing measures cover approximately 23% of global greenhouse gas emissions. While new systems have been introduced and existing systems have been expanded to include new sectors (such as the aviation sector), some systems have been abolished or their expansion has been delayed due to the global energy crisis. While the number of systems introduced in emerging economies has increased, the number of systems introduced in developed economies remains higher.

Revenue from carbon pricing measures reached approximately $95 billion worldwide in 2022, an increase of more than 10% from 2021 levels and a record high. More than 76% of the increase was accounted for by increased revenue from the EU ETS. With pressure on government budgets intensifying, the importance of carbon pricing revenue is increasing.

However, according to the High-Level Committee on Carbon Pricing, in order to achieve the Paris Agreement targets, a carbon price of US$61 to US$122 per ton of CO2 will be necessary by 2030. On the other hand, it is estimated that carbon pricing levels consistent with the Paris Agreement targets cover less than 5% of global emissions.

5 Are the Green Transformation (GX) Strategy and GX Transition Bonds Reasonable?

Based on the above considerations, we will conclude by examining the GX Strategy and GX Transition Bonds, which are at the core of the Japanese government's current decarbonization policy.

5.1 What Are the GX Strategy and GX Transition Bonds: The GX Promotion Act and the GX Decarbonization Power Supply Act

The GX Promotion Act aims to promote GX investment in Japan with the goal of achieving carbon neutrality by 2050, strengthening industrial competitiveness, and realizing economic growth, against the backdrop of accelerating international investment competition toward green transformation (GX).

The main contents of the act are as follows.

5.1.1 Formulation and Implementation of GX Promotion Strategies

The government will formulate strategic plans for the systematic promotion of GX, and the Ministry of Economy, Trade and Industry will create targets, directions, and policy details, which will be decided by the Cabinet.

5.1.2 Issuance of GX Economic Transition Bonds

GX Economic Transition Bonds are a type of government bond limited to use for decarbonization projects. The government issues GX Economic Transition Bonds to investors, who then use the funds for decarbonization initiatives. There are various types of government bonds, including those for individuals, corporations and organizations, and those that can be purchased by both corporations and individuals, but GX Economic Transition Bonds are also planned to be issued for individual investors. The issuance of GX Economic Transition Bonds is projected to reach 20 trillion yen over a 10-year period from the 2023 fiscal year to the 2032 fiscal year, and the funds raised are scheduled to be repaid by 2050 using revenues from carbon pricing.

5.1.3 Introduction of Growth-Oriented Carbon Pricing

As mentioned above, carbon pricing is a policy measure intended to limit greenhouse gas emissions by placing a price on CO2 emissions from companies and collecting taxes and levies based on the amount of emissions.

Carbon pricing includes a "carbon tax" that taxes CO2 emissions from fossil fuels, and "emissions trading," which sets emission limits for each company and allows them to buy and sell excess or deficient amounts. Under growth-oriented carbon pricing, a fossil fuel levy will be introduced in 2028, following emissions trading in 2026. The fossil fuel levy involves collecting a levy from fossil fuel importers based on their CO2 emissions, while emissions trading involves allocating CO2 emission quotas to power generators on a partially paid basis and collecting a specific operator levy based on the allocated amount.

By combining carbon pricing regulations with support through upfront investments, the aim is to create an environment where companies actively engage in GX initiatives.

5.1.4 Establishment of the GX Promotion Organization

With the approval of the Ministry of Economy, Trade and Industry, a new organization related to GX (GX Promotion Organization) will be established and will perform the following duties as stipulated.

Support for GX investment by private companies.

Collection of fossil fuel levies and specific operator charges.

Operation of the emissions trading system, such as allocation of specific operator emission allowances and bidding.

5.1.5 Progress Evaluation and Necessary Revisions

Many of the specific measures under the GX Promotion Act are still under consideration as of 2023, and revisions are planned based on future circumstances. It is necessary to accurately grasp the implementation status of GX investments and evaluate points that require revision. In particular, measures must be taken as necessary in response to the impact on businesses, especially regarding fossil fuel levies and specific operator fees (Article 11 of the Supplementary Provisions stipulates that necessary legal measures must be taken within two years of the implementation of the GX Promotion Act).

Regarding GX Economic Transition Bonds, the government plans to issue GX Economic Transition Bonds (Decarbonization Growth-Type Economic Structure Transition Bonds) over a 10-year period starting in fiscal year 2023 as a leading investment toward the realization of the GX Promotion Strategy. The bonds are intended to support innovative technological development and capital investment that contribute to the decarbonization of energy and raw materials, as well as the improvement of profitability. This is based on the recognition that we have entered an era where the success or failure of decarbonization investments will determine the competitiveness of companies and nations. In fiscal year 2023, a total of 1.6 trillion yen in GX Economic Transition Bonds were issued.

5.2 GX Transition Strategy, Issues with GX Transition Bonds, and Future Prospects

The "growth-oriented carbon pricing" outlined in the GX Promotion Act aims to increase the relative added value of GX-related products and businesses by placing a price on CO2 emissions generated by businesses (carbon pricing).

Specifically, it consists of the following two measures.

5.2.1 Introduction of Fossil Fuel Levy

From fiscal 2028, a fossil fuel levy will be imposed on importers of fossil fuels, etc., based on the amount of CO2 emitted from the imported fossil fuels. The "levy" is effectively a tax.

5.2.2 Introduction of an Emissions Trading System

The emissions trading system is a mechanism designed to encourage companies to reduce their greenhouse gas emissions. It sets emission allowances for companies that emit a certain amount of CO2 or more, and allows companies to trade their allowances with each other. From April 2026, an emissions trading system will be introduced for companies that emit 100,000 tons of CO2 or more per year.

From fiscal 2033, power generators will be allocated CO2 emission allowances (quantities), some of which will be paid for, and a specific operator levy will be collected according to the quantity.

Looking at the contents of the GX Promotion Law, the introduction of a full-fledged carbon tax (equivalent to the introduction of a fossil fuel levy) and an emissions trading system has been postponed to the future. For the time being, 20 trillion yen in GX economic transition bonds will be issued over a 10-year period, and these bonds will be used as a source of funds to provide subsidies and support to companies for economic growth, strengthening international competitiveness, and reducing emissions. As a means of securing funds to repay these bonds by 2050, the government plans to utilize future revenues from fossil fuel levies and emissions trading system auctions (specific operator fees). This approach aims to ensure the sustainability of the GX transition while minimizing the burden on businesses.

The content of these GX economic transition bonds and carbon pricing presents many challenges when compared to the principles of environmental policy costs and precedents overseas. Furthermore, their consistency as policy measures for achieving the 2030 targets and 2050 decarbonization goals is unclear. Specifically, the following points should be noted.

First, the introduction of a fossil fuel levy will begin in 2028, and emissions trading will be left to the discretion of companies until 2026, which means that the greenhouse gas reduction targets for 2030 will not be met.

Second, the emissions trading system was introduced in fiscal year 2023 as a voluntary participation and voluntary target-setting system, but it will be left to the discretion of companies until at least 2026, with no legal binding force and reliance on corporate discretion.

Third, the fossil fuel surcharge and emissions trading fees for specific operators are not positioned as emission reduction measures. The revenue from the fossil fuel surcharge and specific operator fees is allocated to the repayment of GX Economic Transition Bonds, but these should be designed to achieve emission reduction effects, with revenue generation being a secondary consideration.

Since revenue generation is the primary objective, the unit price of these levies and fees is set only at the level necessary for the repayment of GX economic transition bonds. As a result, the original purpose of carbon pricing—emission reduction through price effects—is not realized, and its primary function—minimizing societal emission reduction costs through the equalization of marginal abatement costs (the additional cost required to reduce one unit) using market mechanisms, as well as providing strong incentives for transitioning to a decarbonized industrial structure—cannot be expected in the near term.

Fourth, the carbon price set by carbon pricing will be lower than international standards. The unit price of the fossil fuel levy and specific operator levy is capped at the amount reduced from the base year for the petroleum and coal tax and renewable energy levy. Based on the 20 trillion yen GX Economic Transition Bond, Japan's carbon price is estimated to be 2750 yen per ton of carbon dioxide on average from 2028 to 2050. According to the International Energy Agency (IEA), a carbon price of 130 dollars per ton is necessary in developed countries by 2030, and in the EU Emissions Trading System (EU-ETS), Emissions Trading System (EU-ETS), the trading price for emission allowances exceeded 100 euros (approximately 16,000 yen) on February 21, 2023. At 2750 yen per ton, the emission reduction effect is insufficient, and the unit price should be set at an internationally recognized level.

Fifth, the criteria for investment support should include requirements related to the contribution to emissions reduction. Since there are no legal requirements for support criteria, the government has a lot of discretion, and technologies that are unlikely to contribute to emissions reduction by 2030 may be included in the support targets. For example, hydrogen and ammonia derived from fossil fuels and CCUS have limited emissions reduction effects and may prolong the life of coal-fired power plants. This risk diverting funds that should be maximally utilized for existing renewable energy and energy-saving technologies.

Sixth, major power companies, which are large GHG emitters, are supported through subsidies using transition bonds as a funding source, with the intention of recovering these funds in the future through carbon taxes or similar measures. Ultimately, consumers will bear the burden based on their fossil fuel consumption, raising questions about consistency with the polluter-pays principle.

Seventh, the use of funds and the flow of funds from the "GX Economic Transition Bonds" are opaque, and the system does not allow for oversight or verification by the Japanese Diet. Appropriate criteria must be established to ensure that projects genuinely beneficial to the transition to a decarbonized growth-oriented economic structure are supported, along with transparency in the selection process, regular information disclosure, and evaluation and accountability by external experts.

As the world accelerates its transition to decarbonization, Japan's lagging efforts could damage the international competitiveness of its industries. It is desirable to promote decarbonization, encourage industrial renewal, and enhance the competitiveness of the Japanese economy.

The success or failure of decarbonization investments toward GX will determine Japan's path to zero emissions and the competitiveness of individual companies and the nation as a whole. Maximizing the use of Japan's technologically advantageous fields to accelerate GX, alongside stable energy supply, could serve as a catalyst to return the economy to a sustainable development trajectory. The key question is how to leverage Japan's technology, allocate government resources, and guide private investment in the right direction.

6 Conclusion

The current GX strategy lacks a sense of crisis and urgency regarding climate collapse and remains a traditional industrial policy that prioritizes support measures and consideration for specific industries, rather than the comprehensive climate policy that should be pursued. There is no clear sign of it being in line with the 1.5 °C goal, and there is no roadmap for transforming society itself and making a net-zero, sustainable society happen.

As a result, not only will the swift transition to a decarbonized society not be achieved, but it is also feared that this will negatively impact the international competitiveness of Japanese industries, delay the transition to a high-carbon production structure, and hinder decarbonization innovation.

What Japan now needs is a comprehensive strategy toward a decarbonized and sustainable society. In the face of frequent climate change-related disasters today, it is essential to reflect on the fundamental and structural causes, share a sense of urgency and crisis regarding decarbonization policies, rethink the role of energy and society, and accelerate the necessary reforms.

Author's note: This chapter is a revised and expanded English version of the author's article, "An Analysis of Japan's Climate Policy Based on the Economic Philosophy of "Sustainable Development"", originally published in *Policy Science*, Vol. 32, No. 3 (2025).

References

Boulding K E (1966) The economics of the coming spaceship Earth. http://arachnid.biosci.utexas.edu/courses/thoc/readings/boulding_spaceshipearth.pdf
Daly H (1997) Beyond growth: economics of sustainable development. Penguin Random House
Dasgupta P, Managi S, Kumar P (2022) The inclusive wealth index and sustainable development goals. Sustain Sci 17:899–903. https://doi.org/10.1007/s11625-021-00915-0
International Energy Agency (IEA) (2021) Net zero by 2050: A roadmap for the global energy sector. https://www.iea.org/reports/net-zero-by-2050
Matsushita K (2014) Consideration of Japan's sustainable development strategy. Environ Econ Policy Res 7(2):63–76
Meadows D H, Meadows D L, Randers J (1992) Beyond the limits: Confronting global collapse, envisioning a sustainable future (trans. by Kaya, Y.; Matsuhashi, R.; Murai, M.). Diamond Inc
Meadows D H, Meadows D L, Randers J, Behrens W W III (1972) The limits to growth. https://www.clubofrome.org/publication/the-limits-to-growth/
Stockholm Resilience Center (2024) Planetary boundaries. https://www.stockholmresilience.org/research/planetary-boundaries.html
United Nations Environment Programme (UNEP) (2023) Inclusive wealth report 2023: Measuring sustainability and equity. https://wedocs.unep.org/handle/20.500.11822/43131
Uzawa H (1974) Social costs of automobiles. Iwanami Shoten
Uzawa H (1995) The economics of global warming. Iwanami Shoten
Uzawa H (2000) Social Common Capital. Iwanami Shoten
World Bank (2023) State and trends of carbon pricing 2023

World Commission on Environment and Development (1987) Our common future. https://sustainabledevelopment.un.org/content/documents/5987our-common-future.pdf
World watch Institute (2014) State of the world 2014

Status and Prospects of Decarbonization in China: Energy Transition Progress, Policy Drivers, and Carbon Market

Anjun Jin, Yong Li, and Weisheng Zhou

1 Introduction—Chinese Imperative Green Transformation

Confronting the global climate crisis, China—as the world's largest emitter and renewable energy investor—has embarked on an unprecedented journey toward its "Dual Carbon" goals: peaking carbon emissions by 2030 and achieving carbon neutrality by 2060. This monumental shift demands a complete overhaul of its colossal energy system, driven by strategic state planning, aggressive policy frameworks, and innovative market mechanisms. "Chinese Green Energy Pathways" explores the nation's multi-faceted approach to decarbonization, analyzing the synergy between policy ambition, technological deployment, and market evolution that defines its energy transition.

1.1 Strategic Anchors: The 14th and 15th Five-Year Plans

Chinese transition is meticulously charted through its Five-Year Plans (FYPs). The 14th FYP (2021–2025) established binding targets: 20% non-fossil energy in primary consumption, 18% carbon intensity reduction, and massive wind/solar capacity

A. Jin
School of Resources and Safety Engineering, Chongqing University, Chongqing, China

Y. Li
School of Management Science and Engineering, Central University of Finance and Economics, Beijing, China

W. Zhou (✉)
College of Policy Science, Ritsumeikan University, Osaka, Japan
e-mail: zhou@sps.ritsumei.ac.jp

W. Zhou (ed.), *Decarbonization Policy Engineering*, World Sustainability Series,
https://doi.org/10.1007/978-981-95-5953-4_5

expansion. Crucially, it launched the "1+N" policy architecture, providing a top-level decarbonization blueprint. The forthcoming 15th FYP (2026–2030) will be decisive for achieving the 2030 emissions peak, likely mandating stricter efficiency standards, deeper coal cuts, and accelerated heavy-industry transitions.

1.2 Policy Drive and Carbon Markets

A robust policy toolkit enforces progress: stringent efficiency mandates, renewable portfolio standards, green finance incentives, and state-led industrial scaling. Central to this is Chinese national Emissions Trading Scheme (ETS), the world's largest carbon market by covered emissions. Initially targeting power generation, its expansion to steel, cement, and aluminum will amplify cost-effective abatement. Parallelly, Environmental, social, and governance (ESG) integration reshapes corporate governance, aligning investment with sustainability and reinforcing regulatory efforts.

1.3 Technology Deployment and Global Resonance

China leads globally in solar/wind deployment and manufacturing, enabled by plummeting costs and grid modernization. Energy storage and green hydrogen pilot projects address intermittency and hard-to-abate sectors. This domestic surge catalyzes a global Green Industrial Revolution, slashing clean-tech costs worldwide and influencing energy strategies across East Asia and emerging economies via initiatives like the green Belt and Road (National Development and Reform Commission et al. 2022).

Chinese pathway offers critical lessons for global climate action that balances growth with sustainability (Zhou et al. 2021b). This article examines its progress, challenges, and the implications of its choices for planetary sustainable development.

This article examines and undertakes the intricate interplay of these forces among the following that includes strategic planning, policy drive, market innovation, technological deployment, and industrial transformation. The above undertake defines Chinese monumental journey. and shapes its Dual Carbon future. It assesses the remarkable strides made, the substantial challenges that persist, and the profound implications of Chinese chosen pathways for both its own sustainable development and the global pursuit of climate stability.

Carbon emissions are directly related to the progression of climate change, and both the 1.5 °C and 2 °C pathways can be regarded as critical management thresholds under the Paris Agreement. Under the Paris Agreement, the international community aims to achieve a net-zero emission scenario through the 2 °C pathway.

China, as the world's largest emitter, accounts for approximately 30% of energy-related CO_2 emissions. Reducing coal consumption and decarbonizing the energy

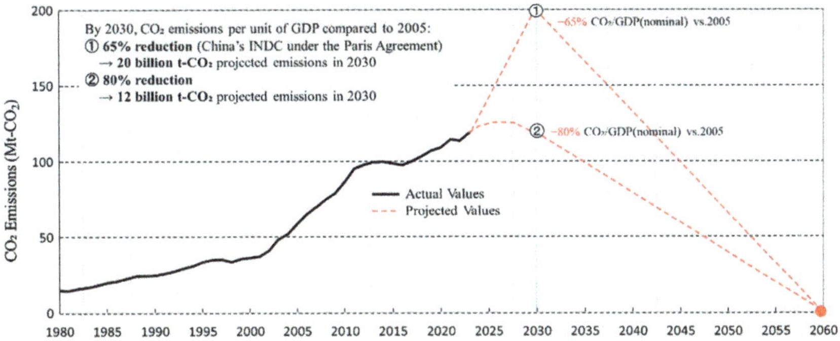

Fig. 1 China's actual CO_2 emissions (actual values) and "dual carbon" targets (compiled by the author). (*Source* China's CO_2 emissions up to 2023 are from Our World in Data (2025), while emissions from 2024 onward are projected by the author. Nominal GDP figures from 2024 onward are based on Fig. 1.6)

system are therefore key measures for achieving the targets set under the Paris Agreement.

However, as shown in Fig. 1, even if China achieves its NDC target under the Paris Agreement—which aims to reduce carbon intensity by more than 65% by 2030 compared to 2005—CO_2 emissions are projected to reach approximately 20 billion tons on a nominal GDP basis by 2030. Moreover, even if the target is raised to a more ambitious 80% reduction, emissions are still expected to amount to around 12 billion tons in 2030 (Zhou 2021; Zhou et al. 2021a).

The goal is to achieve net-zero emissions for these approximately 12 billion tons of CO_2 over the 30-year period from 2030 to 2060. To accomplish this, it is essential to significantly expand the deployment of new energy sources such as solar and wind power and to advance the green transformation of the energy system.

2 The Decarbonization Policy Drives Chinese Green Transition

Chinese ambitious "Dual Carbon" goals—peaking carbon dioxide emissions before 2030 and achieving carbon neutrality before 2060—represent a fundamental shift for the world's largest energy consumer and emitter. Achieving this transformation necessitates a powerful and multifaceted policy engine, meticulously designed to steer the colossal economy towards sustainability. Chinese decarbonization policy framework is characterized by its comprehensiveness, state-led coordination, and evolving sophistication, moving from initial subsidies towards a complex blend of regulatory mandates, economic incentives, industrial strategy, and market mechanisms, all orchestrated under the overarching "1+N" system.

2.1 The "1+N" Framework: The Strategic Backbone

The cornerstone of Chinese decarbonization policy architecture is the "1+N" policy framework, formally announced in late 2021. This structure provides the essential coherence and top-level direction for the entire endeavor.

The "1": Refers to the central, guiding document: "Working Guidance for Carbon Dioxide Peaking and Carbon Neutrality in Full and Faithful Implementation of the New Development Philosophy" issued by the Chinese Governmental Central Committee and State Council. This sets the overarching principles, main goals, and key tasks for the entire Dual Carbon journey, emphasizing systemic planning, a holistic approach, and embedding climate action within broader socio-economic development.

The "N": Represents a suite of specific action plans and implementation guidelines tailored to key sectors (e.g., energy, industry, transport, buildings) and critical enabling areas (e.g., technology, finance, carbon sinks). These documents translate the high-level guidance into concrete targets, timelines, and responsibilities for provinces, ministries, and industries. Examples include the Action Plan for Carbon Dioxide Peaking Before 2030 and sector-specific peaking plans for steel, cement, and non-ferrous metals. This framework ensures alignment and coordinated action across all levels of government and economic sectors, mitigating fragmentation and conflicting priorities.

2.1.1 Regulatory Mandates and Standards Set the Command-and-Control Instruments

The core policy encompasses critical built-in multi-pronged instruments to achieve dual-carbon goals. China deploys a diverse range of policy tools to drive decarbonization as follows.

Energy and Carbon Intensity Targets:

Binding targets set within Five-Year Plans (FYPs) compel provinces and industries to reduce energy consumed per unit of GDP (energy intensity) and carbon dioxide emitted per unit of GDP (carbon intensity). The 14th FYP (2021–2025) mandates an 18% reduction in carbon intensity and a 13.5% reduction in energy intensity by 2025.

Renewable Portfolio Standards (RPS)/Guaranteed Consumption Mechanisms:

Obligate grid companies and provinces to source an increasing minimum percentage of electricity from non-fossil sources (reaching around 33% by 2025 and rising further). Penalties for non-compliance create strong pressure for renewable integration.

Strict Energy Efficiency Standards:

Continuously tightened mandatory standards for appliances, industrial processes, vehicles, and buildings push technological improvement and reduce absolute energy demand.

Industrial Restructuring and Capacity Phase-outs:

Policies actively target the closure of outdated, inefficient, and highly polluting industrial capacity, particularly in coal power, steel, and cement. "Supply-side structural reform" aims to eliminate excess, dirty capacity while promoting advanced, cleaner production.

Coal Consumption Control:

Provincial coal consumption caps and policies promoting coal substitution (especially in industry and heating) aim to constrain and ultimately reduce absolute coal use, though balancing energy security remains a key challenge. As the State Council announced: "During the 14th Five-Year Plan period (2021–2025), strictly and reasonably control the growth of coal consumption, and gradually reduce it during the 15th Five-Year Plan period (2026–2030)."

2.1.2 Instrumentation of Economic Incentives and Financial Leverage

The instrumentation enforces the fund and fundamentals of the economic tasks.

Subsidies and Feed-in Tariffs (FiTs):

Historically the subsidy is crucial for kickstarting the renewable energy boom (especially wind and solar PV), providing guaranteed above-market prices for clean electricity. While being phased down for mature technologies (transitioning to auctions), targeted subsidies remain for emerging areas like offshore wind, distributed PV, and green hydrogen.

Tax Incentives:

This provides preferential value-added tax (VAT) rates, corporate income tax reductions, and import duty exemptions for key green technologies and equipment lower the cost of deployment and encourage investment.

Green Finance:

This provides rapidly expanding ecosystem channels capital towards sustainable projects:

Green Bonds:

China has become one of the world's largest green bond markets, with standards evolving to enhance credibility and prevent "greenwashing."

Green Loans:

The policy banks (e.g., China Development Bank) and commercial banks are encouraged and guided (e.g., via central bank relending facilities) to increase lending for green projects, often at preferential rates.

Green Equity Funds:

The government-backed fund and private venture capital increasingly targets clean tech innovation and deployment.

Central Bank Guidance:

The People's Bank of China (PBOC) incorporates climate risk into its financial stability framework and provides targeted support for green lending.

2.1.3 Industrial Policy and Strategic Scaling

The state plays a direct and proactive role in fostering world-leading green industries that is illustrated at below.

Massive Strategic Investment:

Direct state investment, low-cost financing from state-owned banks, and favorable land/regulatory treatment have propelled Chinese companies to dominate global manufacturing in solar PV, wind turbines, batteries, and electric vehicles (EVs).

Supply Chain Development:

Policies actively support the development of complete domestic supply chains for critical green technologies and minerals, reducing reliance on imports and enhancing competitiveness.

Economies of Scale and Cost Reduction: By creating vast domestic markets and supporting manufacturing scale-up, Chinese policy has been instrumental in driving down the global cost of key clean technologies, accelerating their adoption worldwide.

2.1.4 The Instrumentation of Market-Based Mechanisms

National Emissions Trading Scheme (ETS):

Launched in July 2021, the national ETS is a centerpiece of Chinese market-based approach (covered in detail in Part 3). The ETS initially covers the power sector (that contributes nearly 40% of national CO_2 emissions; it puts a price on carbon, incentivizing the most cost-effective emission reductions within the capped sector. Its planned expansion to other high-emitting industries (steel, cement, aluminum) is a major policy priority.

Green Electricity Certificates (GECs):

A tradable instrument representing the environmental attributes of 1 MWh of renewable electricity. While still evolving, GECs aim to create a voluntary market for corporates to demonstrate renewable energy use and meet sustainability goals, complementing the mandatory RPS.

2.2 *Implementation, Challenges, and Evolution*

The sheer scale and complexity of Chinese decarbonization pose immense implementation challenges as follows.

Balancing Act:

Reconciling rapid decarbonization with energy security (still heavily reliant on coal), economic growth, and regional development disparities requires constant policy calibration.

Local Implementation: Ensuring provincial and local governments fully embrace and effectively enforce central mandates, avoiding "campaign-style" compliance or local protectionism hindering national goals.

Grid Integration:

Accelerating the build-out of grid infrastructure (transmission, flexibility, storage) to accommodate the massive influx of variable renewables is critical.

Hard-to-Abate Sectors:

Developing and scaling viable decarbonization pathways for heavy industries (steel, cement, chemicals) and long-haul transport remains a significant hurdle, requiring breakthroughs in technologies like green hydrogen and CCUS.

ETS Effectiveness:

Enhancing the national ETS's liquidity, price signal strength, and regulatory robustness to truly drive cost-effective abatement across the economy.

Chinese decarbonization policy toolkit is not static; it is constantly evolving along with the targets' progress. The focus is shifting from initial subsidy-driven renewable expansion towards deeper systemic transformation: integrating markets (ETS, GECs), leveraging finance (green bonds, loans), tackling harder sectors (industry, transport), and emphasizing technological innovation for the next phase (CCUS, green hydrogen). The effectiveness of this intricate policy engine will ultimately determine the pace and success of Chinese historic green transition, with profound implications for global climate efforts.

3 Carbon Markets: Chinese Evolving Market-Based Carbon-Climate Driver

The establishment of a robust carbon market represents a pivotal shift in Chinese decarbonization strategy, moving beyond pure command-and-control measures to harness market forces for cost-effective emission reductions. The launch of the national Emissions Trading Scheme (ETS) in July 2021 marked a watershed moment, creating the world's largest carbon market by covered emissions volume. This section examines the structure, progress, challenges, and future trajectory of Chinese carbon market, alongside the synergistic rise of Environmental, Social, and Governance (ESG) principles, which are increasingly shaping corporate behavior and investment flows, reinforcing the carbon pricing signal.

3.1 Emissions Trading Scheme

The National ETS includes Structure and Initial Phases. More details are discussed later in this section.

3.1.1 Foundational Design & Scope

Sectoral Focus:

The initial phase targets the power sector, encompassing over 2,000 coal and gas-fired power plants responsible for approximately 40% of Chinese national CO_2 emissions (around 4.5 billion tons annually). This sector was chosen for its high emissions concentration, relative homogeneity, and robust data monitoring capabilities.

Cap-and-Trade Principle:

The scheme operates on a cap-and-trade model. A national emissions cap is set, declining over time to align with carbon intensity targets. Emission allowances are distributed to covered entities, who must surrender enough allowances annually to cover their verified emissions.

Allocation Mechanism:

Primarily employs a Benchmarking Approach combined with some Grandfathering. Free allowances are allocated based on the actual output (e.g., electricity generated, heat supplied) and predefined emissions intensity benchmarks per unit of product. Entities emitting below their benchmark can sell surplus allowances; those exceeding must buy allowances or face penalties. This incentivizes efficiency improvements within the sector. Grandfathering (based on historical emissions) plays a smaller, transitional role.

Trading Platform:

Centralized trading occurs on the Shanghai Environment and Energy Exchange (SEEE). Trading modalities include listed trading, block trading, and negotiated transfer.

Compliance & Enforcement:

Strict Monitoring, Reporting, and Verification (MRV) protocols are mandated. Non-compliance penalties include fines (up to 30,000 RMB per ton of CO_2e exceeding allowances, though this is considered low), public naming, and potential deduction of future allowances (International Energy Agency (IEA) 2024).

3.1.2 ETS Initial Phase in the Power Sector: Progress and Performance

Market Scale:

The sheer volume of covered emissions makes it globally significant, dwarfing the EU ETS in terms of tons covered.

Price Formation & Liquidity:

Carbon prices have generally ranged between 50–80 RMB per ton (approx. $7–11 USD). While providing a signal, liquidity has been lower than anticipated (International Energy Agency (IEA) 2024). Trading activity often spikes significantly around compliance deadlines ("compliance season"), indicating a market still driven primarily by regulatory obligation rather than continuous price discovery for hedging or speculation.

Achieving Compliance:

The first compliance cycles demonstrated high compliance rates (exceeding 99%), validating the basic operational framework. Covered entities successfully navigated allowance surrender.

Impact:

The primary impact so far has been heightened awareness and internal carbon cost accounting within the power sector. While driving some efficiency gains, the relatively low price and free allocation have limited transformative abatement investment triggered solely by the ETS thus far. Its effectiveness is intertwined with complementary policies like renewable mandates and efficiency standards.

3.1.3 Expansion and Evolution: The Next Critical Phase

The true potential of the ETS hinges on its expansion beyond the power sector. The sectoral expansion includes of "hard-to-abate" sectors that is a top priority under the 14th FYP and that will be crucial for the 15th FYP.

Key targets include the following sectors:

1) Steel: Major integrated steel mills and electric arc furnaces.
2) Cement: Large clinker production facilities.
3) Aluminum: Primary aluminum smelters (significant electricity consumers).

The ETS system enhancement include the following:

1) Tightening the Cap: Gradually reducing the overall cap to drive meaningful emission reductions and strengthen the price signal.
2) Increasing Auctioning: Gradually shifting from predominantly free allocation towards auctioning a percentage of allowances. This injects a direct cost for emissions, generates revenue for climate initiatives, and enhances market liquidity.
3) Market Stabilization Mechanisms: Implementing mechanisms like a price floor/ceiling or allowance reserves to manage excessive volatility and provide investor certainty.
4) Financialization: Exploring the introduction of derivatives (futures, options) to improve hedging capabilities and attract greater market participation (including financial institutions), deepening liquidity.
5) Linking: While politically complex, discussions on potential future linkages with other carbon markets (e.g., regional Asian schemes, EU ETS) exist for the long term, promoting global carbon price convergence.

3.1.4 The Synergistic Rise of ESG Integration

The carbon market does not operate in isolation. The parallel surge in ESG adoption in China creates a powerful reinforcing dynamic stated at below.

Regulatory Push for ESG Disclosure:

Regulators (e.g., China Securities Regulatory Commission—CSRC, stock exchanges) are mandating enhanced environmental and climate-related financial disclosures for listed companies, particularly in high-impact sectors. This includes reporting on carbon emissions, climate risks, and mitigation strategies.

ESG Investing Momentum:

Institutional investors (domestic and international) are increasingly integrating ESG factors, including carbon footprints and climate performance, into their investment decisions and stewardship activities. Funds explicitly labelled “ESG” or “green” are growing rapidly.

Corporate Response:

Companies face growing pressure from investors, regulators, customers, and employees to demonstrate sustainability leadership. Strong ESG ratings enhance reputation, reduce financing costs (via green bonds/loans), and attract investment.

Convergence with Carbon Markets:

This ESG wave directly complements the ETS:

Data Foundation:

ETS MRV systems provide high-quality, verified emissions data—a core component of environmental (E) disclosures within ESG.

Performance Metric:

A company's carbon intensity, compliance status, and strategy within the ETS become critical ESG performance indicators.

Financial Materiality:

Carbon costs (current under ETS or anticipated future costs with expansion/ tightening) and climate transition risks become material financial factors assessed under ESG frameworks.

Capital Allocation:

Green finance instruments (bonds, loans) increasingly link preferential terms to ESG performance, which includes carbon management aligned with the ETS trajectory. Companies performing well under the ETS and demonstrating strong climate governance attract cheaper capital.

Voluntary Action:

Companies seeking higher ESG ratings often go beyond compliance, investing in additional emission reductions or purchasing Green Electricity Certificates (GECs) or China Certified Emission Reductions (CCERs—see below) to offset residual emissions, creating demand in voluntary carbon markets.

3.1.5 The Role of CCERs and Voluntary Markets

Alongside the compliance ETS, China is revitalizing its China Certified Emission Reduction (CCER) program, a domestic voluntary carbon credit mechanism. Details refer to the following:

Purpose:

CCERs represent verified emission reductions from projects in sectors not covered by the ETS (e.g., forestry, renewable energy projects not under FiTs, methane capture). They can be used by ETS-regulated entities to offset a limited percentage (expected to be around 5%) of their compliance obligations. They also serve the voluntary market for corporate carbon neutrality claims.

Revival:

After a pause starting in 2017 due to concerns over credit quality, the program resumed accepting new project applications in 2023/2024 with revised, stricter methodologies focusing on integrity and additionality, and avoiding double counting.

Synergy:

The CCER market provides flexibility for compliance entities, channels finance to emission reduction projects outside the ETS cap, and supports corporate ESG goals through voluntary offsetting.

In summary, Chinese national ETS, coupled with the accelerating integration of ESG principles, represents a fundamental evolution in its climate policy arsenal. While the initial phase focused on establishing the operational infrastructure within the power sector, the coming years are critical for its expansion and deepening. Successfully integrating major industrial emitters, tightening the cap, enhancing market liquidity, and strengthening the price signal are paramount. The synergy with ESG is amplifying the impact, embedding carbon costs and climate considerations into the core of corporate finance and investment decisions. The effectiveness of this combined market and governance approach will be a major determinant of Chinese ability to achieve its Dual Carbon goals and solidify its position in the global green economy.

4 Energy in the Transition: Chinese Energy Technology Surge

Chinese ambitious dual-carbon goals are being propelled forwards by an unprecedented deployment of clean and green energy technologies, which transform its energy landscape at a scale and speed unmatched globally. This technological leapfrogging is not merely an adjunct to policy but is the physical manifestation of the transition process, enabling tangible progress towards peaking emissions and achieving carbon neutrality.

Exciting development of the alternative energy technologies in China propels the significant advancement toward its dual-carbon goals. Details refer to the references and illustration as follows.

4.1 Energy Storage

Energy storage (E.S.) technology refers to the process of converting energy into physical or chemical forms for storage, then releasing it when needed. Storage devices are commonly known as accumulators or batteries. Hydropower also constitutes energy storage, though hydropower will be discussed separately.

As long as generation capacity exists, output can be rapidly increased to compensate for fluctuations. This provides a highly consistent overall power supply with minimal energy loss, reducing reliance on hydropower. Where suitable hydropower (e.g., pumped storage) is unavailable, other forms of grid energy storage—such as compressed air storage and thermal energy storage—can be deployed. These systems store energy generated during peak solar/wind periods and release it on demand.

Stored energy enhances the economic value of photovoltaics (PV) and wind power by displacing higher-cost generation during peak demand. This revenue potential can offset storage costs and efficiency losses.

To achieve "carbon neutrality" at an early date, China is vigorously developing its new energy industry. Energy storage technology can reduce wind/solar curtailment rates, increase renewable energy utilization, and ensure the secure, stable operation of power systems.

4.1.1 Energy Storage: Enabling the Renewable Era or Called Sun Era

Recognizing the critical challenge of intermittency, China is investing heavily in energy storage solutions to ensure grid stability and maximize renewable utilization. Utility-scale battery storage deployment is accelerating rapidly, underpinned by Chinese leadership in battery manufacturing (dominating lithium-ion production). Pumped hydro storage (PHS), a mature large-scale solution, continues to expand significantly, with numerous large projects under development. Policy mandates and targets increasingly require new renewable projects to incorporate storage capacity. Research and pilot projects also explore alternative storage technologies like compressed air and flow batteries.

In 2024, China added 42.37 GW/1.01 GWh of new energy storage capacity nationwide, as shown in Fig. 2. The average energy storage duration reached 2.3 h, marking an increase of approximately 0.2 h compared to the end of 2023. Since the start of the 14th Five-Year Plan period, the energy storage duration has shown a consistent upward trend (International Energy Agency (IEA) 2024).

In 2024, eleven provincial-level regions in China each added over 1 GW of new energy storage capacity. Notably, Jiangsu, Xinjiang, Zhejiang, and Inner Mongolia registered particularly robust growth, with newly installed capacity exceeding 3 GW in each region—reaching 4.86 GW, 3.84 GW, 3.53 GW, and 3.05 GW respectively, as shown in Fig. 3.

In 2024, the utilization of new energy storage systems saw a significant increase compared to 2023, with an annual average equivalent utilization of 911 h—an increase of approximately 300 h over the previous year. The annual average equivalent charge–discharge cycles reached 221 times, up by about 59 times from 2023. Provinces and autonomous regions such as Zhejiang, Jiangsu, Chongqing, Xinjiang, Guangdong, Tibet, Hubei, and Ningxia all recorded annual equivalent utilization hours exceeding 1,000 h, as shown in Fig. 4.

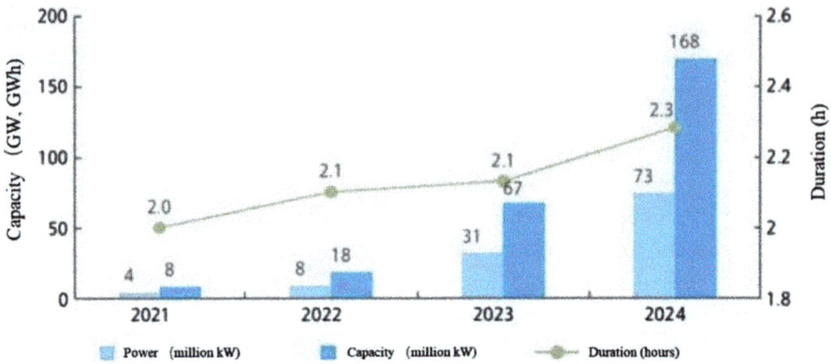

Fig. 2 Growth trends in new energy storage installations under China's 14th FYP (NDRC 2021)

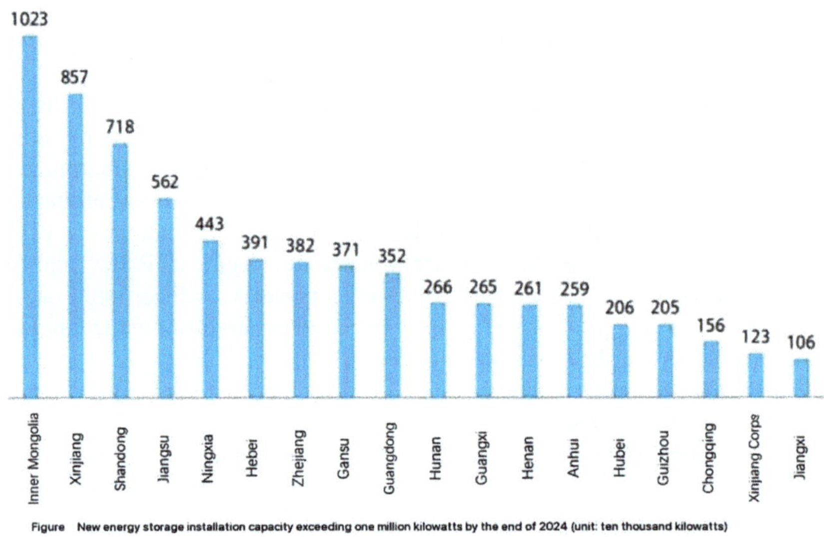

Fig. 3 Provincial deployment of advanced energy storage for regions exceeding 1 GW by end of 2024 (unit in 10 MW) (NDRC 2021)

4.2 Photovoltaic (PV) Power Generation Systems

The photovoltaic system generates electricity via the photovoltaic effect, converting solar energy into electrical power. It comprises PV modules (solar panels), inverters, mounting structures, energy storage devices (e.g., batteries), and monitoring/management systems.

PV systems have broad applications. Residential rooftop installations provide power, reduce grid dependence, and lower electricity costs. Commercial/industrial-scale PV plants deliver stable power while reducing carbon emissions.

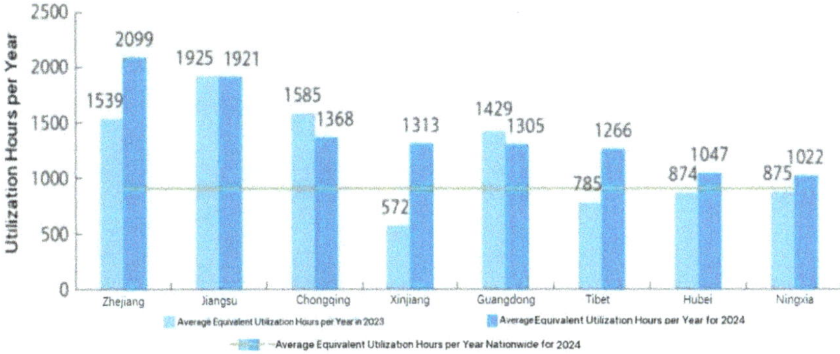

Fig. 4 Utilization status of new energy storage in key provinces (Unit: hours) (NDRC 2021)

Solar and Wind: Scaling Dominance

China stands as the undisputed global leader in both solar photovoltaic (PV) and wind power, driven by massive domestic manufacturing capacity, plummeting costs, and strong policy support. Deployment consistently exceeds Five-Year Plan targets. Gigantic utility-scale "clean energy bases," particularly in sun-drenched western provinces like Qinghai and Gansu and wind-rich regions like Inner Mongolia, are central to the strategy. Distributed solar, especially on commercial and industrial rooftops, is also surging. This dominance extends to manufacturing: Chinesc firms supply over 80% of global solar PV components and a major share of wind turbines, continuously driving down costs worldwide through economies of scale and relentless innovation.

A photovoltaic (PV) power generation system is a technological system that utilizes solar energy to generate electricity through the photovoltaic effect, also known as a PV system. It converts solar energy into electrical energy and primarily consists of PV modules (solar panels), inverters, mounting systems, energy storage devices (such as batteries), and monitoring and management systems.

The applications of PV systems are extensive. In households, rooftop PV systems can provide electricity for residents, reducing reliance on the traditional power grid and lowering electricity bills. In commercial and industrial sectors, large-scale PV power plants can supply stable electricity to businesses while reducing carbon emissions. Figure 5 shows photos of selected PV power station components. Figure 6 shows the trends in the cost of solar and wind energy technologies, as well as the in-stalled power capacity.

4.3 *Wind Power Generation Technology*

The wind power is a widely adopted clean energy source that has emerged as a vital alternative to traditional coal, oil, and gas. It offers significant commercial

Fig. 5 Photos of selected PV power station components (Jin et al. 2025)

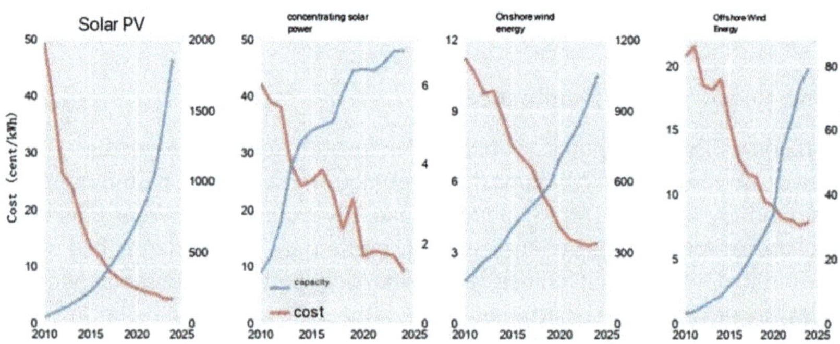

Fig. 6 The cost of solar and wind energy technologies worldwide is declining rapidly and the installed power capacity is growing rapidly (IEA 2025,b)

opportunities. Wind power converts kinetic energy into electricity using turbines. Wind rotates turbine blades, driving generators to produce power. As a clean, green energy source, it consumes no fuel, emits no pollutants, and is environmentally benign. Wind power plays a crucial role in achieving carbon neutrality and holds immense potential. Applications include onshore, offshore, and distributed wind power. Key advantages include the following.

1) Renewable Energy: Wind is an infinite, sustainable green resource that provides long-term clean energy.
2) Greenhouse Gas Reduction: Wind power produces near-zero emissions, curbs fossil fuel dependence, and advances low-carbon economies.
3) Cost-Effectiveness: Technological advances and economies of scale have progressively reduced costs, enhancing market competitiveness.
4) Wide Availability: Wind resources exist globally due to planetary atmospheric dynamics.

4.4 *Hydrogen Fuel Cells*

Interest in hydrogen fuel cells—a clean, efficient energy conversion technology—is growing. They generate electricity through electrochemical reactions between hydrogen and oxygen, emitting only water vapor. With high energy density, zero emissions, and rapid refueling, they are widely used in transportation, energy storage, and industry.

Fuel cells convert chemical energy from hydrogen and oxygen into electrical energy. Hydrogen is oxidized at the anode and oxygen reduced at the cathode, creating electron flow through an external circuit to produce electricity and water.

Green hydrogen (produced via electrolysis using renewable electricity) is identified as a crucial solution for decarbonizing "hard-to-abate" sectors like heavy industry (steel, chemicals) and long-haul transport. While still nascent compared to solar and wind, China is laying the groundwork for a future hydrogen economy. National and numerous provincial hydrogen strategies outline ambitious development pathways. Large-scale pilot projects for green hydrogen production, particularly co-located with renewable bases in the west, and for industrial use (e.g., hydrogen-based steel-making) are underway. Significant investment targets the entire value chain, from electrolyzer manufacturing to storage, transportation, and end-use applications.

4.5 *Smart Control Systems and the Integrated Energy*

Smart control systems enable real-time monitoring and management of Distributed Energy Resources (DERs), improving wind power efficiency and reliability. Innovations include AI-based controls and big-data forecasting systems.

4.5.1 AI in the Integrated Energy

To achieve optimal energy conversion efficiency in Distributed Energy Resources (DER) systems, artificial intelligence is a valuable tool. AI's capacity to forecast, optimize, and adapt makes it indispensable for unlocking the full potential of distributed energy resources, driving greater efficiency, resilience, and integration of renewables.

The value of smart integrated energy systems is demonstrated by empowering the user in the power decisions. The user can involve easily in the daily electricity decision-making processes that is illustrated in Fig. 7 as follows:

Smart power systems equipped with AI functions are becoming new and important push of technology that drives the electricity/power sector toward carbon neutrality. A Smart power system, commonly known as the smart grid, has huge applications toward the carbon neutrality. For example, AI is an enabling and emerging technology with significant share in energy market.

Fig. 7 AI applies to the decision-making processes in the smart integrated energy systems (Jin et al. 2022)

4.5.2 Smart Control Systems and Integrated Energies

This transformation, however, introduces complexity due to the variable nature of renewables, necessitating advanced frameworks for grid management and integration.

The EEE-model (or PECODER model) provides a critical framework for this transition. It is designed to be operated and moved forward by smart system in orchestrating diverse energy assets. Central to its function is the integration of a smart energy platform, which acts as a digital nervous system. This platform seamlessly combines centralized utility-scale renewables with decentralized, complementary energy supplies, including distributed energy resources (DERs) like rooftop solar, community batteries, electric vehicles, along with selected decarbonizing grid power.

By leveraging real-time data and predictive analytics, the platform optimizes the entire energy ecosystem—balancing generation with consumption, managing grid stability, and facilitating energy trading. This empowers users to make informed choices, allowing them to select their optimal balance of energy requirements, carbon emissions, and economic benefits.

Refer to Fig. 8, the main goal of EEE-model with both smart control systems and AI applications is to minimize the carbon emission coefficient of the integrated energies.

Chinese energy decarbonization deploys the innovative energy technology and the green energies to deliver the energy transition. The sheer scale of investment and installation in solar, wind, storage, and enabling infrastructure like UHV grids is unmatched, delivering measurable progress and shaping the global clean energy landscape. The strategic focus on emerging solutions like green hydrogen underscores its commitment to tackling the next frontier of decarbonization.

5 Green Industry Revolution in East Asia

Chinese monumental energy transition is far more than a domestic recalibration; it is the primary engine igniting a global Green Industrial Revolution (GIR) [24]. This revolution signifies a fundamental shift in industrial systems, supply chains,

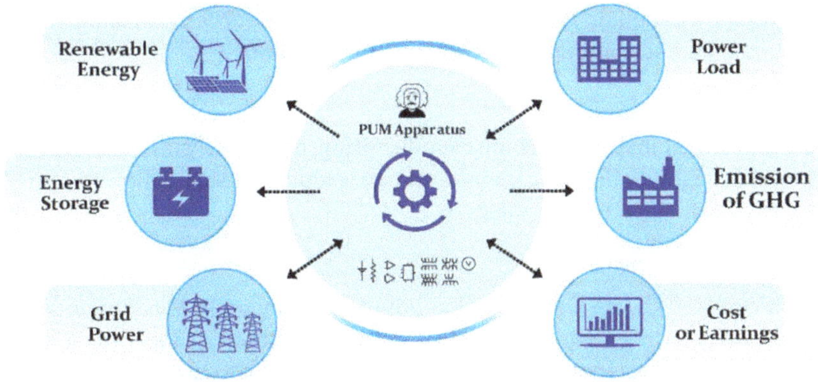

Fig. 8 Schematic diagram of multiple energy inputs and three output functions (known as the Energy-Emission of Carbon-Economy model, or EEE-Model) (Jin 2025)

and economic paradigms, driven by the imperative of decarbonization. By achieving unprecedented economies of scale in clean technology manufacturing and deployment, China has dramatically reshaped global markets, influenced regional strategies across East Asia, and accelerated the dissemination of green technologies and practices worldwide. This section examines Chinese catalytic role, the regional dynamics in East Asia, and the mechanisms through which the GIR is spreading globally.

5.1 China's Demonstration

During the course of the Green Industrial Revolution, renewable energy—including hydropower, wind, solar, and biomass—is projected to surpass coal-fired power by 2026, becoming the world's leading comprehensive source of electricity. For further information, please refer to Fig. 1 as illustrated previously.

China has achieved the Strategic State-Led Scaling. The Chinese rise was not accidental. It resulted from decades of concerted industrial policy: massive state investment (often via state-owned banks and enterprises), targeted subsidies (especially in the formative years), R&D support, and favorable regulatory frameworks. The "1+N" policy explicitly identifies strategic emerging industries like renewables, EVs, and batteries as national priorities.

5.1.1 Manufacturing Supremacy & Cost Collapse

Solar PV:

Chinese manufacturers supply over 80% of global solar PV polysilicon, wafers, cells, and modules. Massive scale, continuous process innovation, and vertical integration have driven down module costs by over 90% in the past decade, making solar the cheapest new-build electricity source in most of the world.

Wind Power:

China dominates global wind turbine manufacturing, particularly in onshore turbines, and is rapidly scaling offshore capabilities. Cost reductions have followed a similar, though less steep, trajectory as solar.

Batteries:

Chinese firms (CATL, BYD) lead global lithium-ion battery production, crucial for EVs and energy storage. China controls significant portions of the critical mineral refining and battery component supply chains (cathodes, anodes, separators).

Electric Vehicles (EVs):

China is the world's largest EV market and producer. Aggressive policy support (purchase subsidies, manufacturing incentives, ZEV mandates), massive investment in charging infrastructure, and fierce domestic competition have fostered globally competitive automakers and a vast supplier ecosystem.

5.1.2 Economies of Scale as Global Benefits

Chinese manufacturing scale has acted as a powerful deflationary force on clean tech prices globally. This "China Price Effect" has been arguably the single most significant factor in accelerating the global energy transition by making renewables and EVs economically viable far sooner than otherwise possible. It lowered the entry barrier for countries worldwide.

Beyond Hardware with System Integration & Digitalization

Chinese companies are increasingly leaders in integrated solutions—smart inverters, energy management systems, EV charging networks, and digital platforms for grid management and energy services. This ecosystem approach leverages hardware dominance with digital innovation.

5.2 East Asia Marching on Through Competition, Collaboration, and Regional Reconfiguration

Chinese GIR surge has profoundly reshaped the economic and energy landscape of East Asia, forcing adaptation and fostering complex interdependence. Brief analysis is shown at below.

Japan and South Korea:

Strategic Responses to the China Challenge: Facing intense competition from China in traditional manufacturing strongholds (e.g., batteries, displays, increasingly chips), Japan and South Korea have doubled down on:

High-Value Niche Leadership:

Focusing on advanced materials (e.g., battery chemistries like solid-state, hydrogen fuel cells, specialized semiconductors for EVs/energy), high-efficiency components, and sophisticated production equipment.

Hydrogen Economy:

Both nations have made hydrogen (particularly green/blue hydrogen) a cornerstone of their decarbonization and industrial strategy, investing heavily in R&D, supply chains, and international partnerships, seeking leadership where China is still building scale.

Supply Chain Resilience & "De-risking":

Actively diversifying supply chains (e.g., battery materials, rare earths) away from over-reliance on China, investing in domestic production and partnerships with other nations such as Australia, Canada, US allies. This involves significant government support and corporate restructuring.

EV Ambitions:

Korean (Hyundai/Kia) and Japanese (Toyota, Honda, Nissan) automakers are aggressively accelerating their EV rollouts and battery investments, seeking to counter Chinese EV dominance, particularly in global export markets.

Regional Supply Chain Integration (and Friction):

Despite geopolitical tensions, East Asia remains deeply integrated.

Component Dependence:

Chinese manufacturers rely heavily on advanced machinery, components, and specialty chemicals from Japan and South Korea. Conversely, Japanese/Korean battery and EV makers depend on Chinese raw materials and refined products.

Collaborative Ventures:

Joint ventures and technology licensing agreements persist, particularly in complex areas like battery production and semiconductor manufacturing for green tech.

ASEAN as Manufacturing Hub & Market:

Southeast Asian nations (Vietnam, Thailand, Indonesia, Malaysia) are becoming crucial manufacturing bases for both Chinese firms (diversifying production) and Japanese/Korean firms (seeking lower costs and market access). ASEAN is also a rapidly growing market for EVs, solar panels, and energy storage, attracting investment from all major East Asian players.

Policy Cross-Pollination:

The scale and ambition of Chinese policies have pushed Japan and South Korea to set more aggressive decarbonization targets and increase green industrial support. Competition drives ambition.

Global Diffusion:

How Chinese GIR Reshapes the World

The effects of Chinese GIR extend far beyond East Asia, accelerating the global energy transition through multiple channels.

Belt and Road Initiative (BRI)—The Evolving "Green BRI"

Infrastructure Exports:

Chinese firms are dominant players in building renewable energy projects (solar farms, wind parks, hydropower) globally, particularly in developing countries. This exports technology, expertise, and financing models.

Technology Transfer & Localization:

Increasingly, BRI projects involve local manufacturing partnerships or technology transfer components, helping build green industrial capacity in host countries (e.g., EV assembly plants in Southeast Asia, solar panel factories in the Middle East).

Financing:

Chinese policy banks (China Development Bank, Exim Bank) and commercial banks provide significant financing for overseas renewable projects, although scrutiny over debt sustainability and environmental/social standards continues. The push for "Green BRI" aims to address these concerns.

Global Cost Reduction & Market Acceleration:

As established, Chinese scale drives down global prices, making solar, wind, batteries, and EVs affordable for virtually all nations. This is the most pervasive impact.

Exporting the EV Revolution:

Chinese EV makers are rapidly expanding into international markets (Europe, Southeast Asia, Latin America, Middle East), offering competitively priced, technologically advanced models. This pressures legacy automakers globally to accelerate their own electrification plans and invest more heavily.

Supply Chain Diversification & New Geographies:

The drive for supply chain resilience (partly in response to over-reliance on China) is spurring massive investments in clean tech manufacturing globally.

US & Europe:

Policies like the US Inflation Reduction Act (IRA) and the EU Green Deal Industrial Plan offer massive subsidies to attract battery giga factories, EV plants, solar component manufacturing, and clean hydrogen projects. While aimed at reducing dependence on China, this globalizes manufacturing capacity, further accelerating the GIR, albeit with potential inefficiencies from duplication.

Resource Nations:

Countries rich in critical minerals (e.g., Australia, Chile, Indonesia, DRC) are moving beyond raw material exports to develop domestic refining and mid-stream processing capabilities, seeking a larger share of the green value chain, often with Chinese investment and technology.

Knowledge & Model Diffusion:

Chinese experience in rapidly deploying renewables at scale, building integrated supply chains, and implementing large-scale policy frameworks (including ETS pilots) provides valuable lessons (both positive and cautionary) for other developing and emerging economies navigating their own transitions. Chinese think tanks and companies actively engage in global knowledge exchange.

5.3 *Challenges and Contradictions Within the GIR*

The global diffusion of the GIR driven by China has significant challenges.

Geopolitical Friction & Protectionism:

US-China tensions, EU investigations into Chinese subsidies, and "de-risking" strategies create trade barriers, market fragmentation, and potential inefficiencies. This complicates global collaboration.

Overcapacity Concerns:

Massive investments, particularly in China, lead to concerns about global overcapacity in sectors like solar PV and batteries, potentially triggering trade disputes and bankruptcies.

Environmental and Social Footprint:

Scaling green tech requires vast resources. Ensuring responsible mining of critical minerals, managing manufacturing pollution, and upholding labor standards across global supply chains remain critical challenges. The "green" label must encompass the entire lifecycle.

Technology Access and Equity:

While costs fall, access to the latest technologies and financing remains uneven, potentially leaving the poorest nations behind in the GIR. Ensuring a just transition globally is paramount.

Carbon Leakage vs. Genuine Global Decarbonization:

While manufacturing shifts geographically, ensuring that this leads to a net global decrease in emissions, rather than just shifting them (carbon leakage), requires robust carbon accounting and international cooperation.

An Irreversible Force Reshaping the Global Economy

Chinese green energy surge has irrevocably launched a global GIR. Its unparalleled manufacturing scale has drastically reduced clean technology costs, making decarbonization economically feasible worldwide. Within East Asia, it has triggered a dynamic mix of fierce competition, strategic adaptation, and deep supply chain interdependence, accelerating the region's collective march towards a greener future. Globally, through exports, BRI projects, and the catalytic effect of its policies on others, Chinese GIR is reshaping industrial landscapes, energy systems, and geopolitical dynamics. While challenges of geopolitics, sustainability, and equity loom large, the direction is clear: the industrial foundation of the global economy is being fundamentally rewired towards sustainability, with China acting as the primary, albeit complex and contested, accelerator. The success of this revolution will depend not only on continued technological innovation and deployment but also on navigating the intricate web of international cooperation, competition, and shared responsibility for a stable climate.

6 Summary—China's Green Energy Transition and Global Implications

The Chinese energy transition has made significant progress in terms of its global implication [28]. This article provides a comprehensive analysis of Chinese multifaceted journey in energy sector toward decarbonization, examining the interplay of policy, market mechanisms, technological innovation, and global influence in its pursuit of the dual-carbon goals (carbon peak by 2030 and carbon neutrality by 2060).

Central to Chinese progress is its hierarchical policy architecture, currently anchored in the 14th and upcoming 15th Five-Year Plans, which integrate binding emission reduction targets, sector-specific action plans, and a mix of regulatory tools, economic incentives, and market mechanisms. The national carbon emissions trading scheme (ETS), expanded from regional pilots to cover the power sector and soon high-emission industries like steel and cement, has emerged as a key market-driven tool, complemented by ESG disclosure mandates that align corporate behavior with low-carbon goals. Coal is still the mainstay of Chinese grid. But the country has also been installing more clean energy than the rest of the world put together. Chinese carbon-dioxide emissions are so huge that, if the trend continues and they are shown to have actually peaked, that could alter the trajectory of global emissions.

Technological innovation has been a linchpin, with China leading global deployment in solar, wind, hydrogen, and energy storage—driving cost reductions and scaling renewable capacity to over 1.9 terawatts by the end of 2024. These advancements have catalyzed a domestic Green Industry Revolution (GIR), transforming traditional sectors like automotive and steel while fostering new green tech ecosystems.

Regionally, Chinese transition has positioned East Asia as a hub for green collaboration, with cross-border initiatives in hydrogen, carbon market linkage, and renewable integration. Globally, the GIR has spread through technology exports, green infrastructure investments (e.g., via the Belt and Road Initiative), and policy knowledge sharing, enabling developing nations to accelerate their own decarbonization.

While challenges—including regional disparities, critical mineral dependencies, and global policy fragmentation—persist, Chinese experience demonstrates that strategic policy alignment, technological scale, and market innovation can drive transformative energy transitions. Its pathways offer valuable lessons for global climate governance, emphasizing the need for inclusive, collaborative approaches to balance growth and sustainability (International Energy Agency [IEA] 2025a,b; Institute of Climate Change and Sustainable Development under Tsinghua University 2021).

References

BloombergNEF (BNEF) (2023) Chinese clean energy investment soars, driving global growth. https://about.bnef.com/blog/chinas-clean-energy-investment-soars-driving-global-growth/

BloombergNEF (BNEF) (2024a) 1H 2024 China market outlooks (energy storage, hydrogen, renewables). https://about.bnef.com/blog/

BloombergNEF (BNEF) (2024b) 1H 2024 global lithium-ion battery supply chain ranking. https://about.bnef.com/blog/

Climate Bonds Initiative, SynTao Green Finance (2023) China green bond market report 2023. https://www.climatebonds.net/resources/reports/china-green-bond-market-report-2023

Hillman J, McCreary E (2023) Chinese Belt and Road Initiative: Green or greenwashed? Center for Strategic and International Studies. https://www.csis.org/analysis/chinas-belt-and-road-initiative-green-or-greenwashed

ICAP (International Carbon Action Partnership) (2024a) Emissions trading worldwide: status report 2024. https://icapcarbonaction.com/en/ets-status-report-2024

ICAP (International Carbon Action Partnership) (2024b) Emissions trading worldwide: status report 2024—China chapter. https://icapcarbonaction.com/en/ets/china

IMF (2025) World economic outlook 2025

Institute of Climate Change and Sustainable Development, Tsinghua University et al (2021, May) China's long-term low-carbon development strategies and pathways

International Energy Agency (IEA) (2023a) Renewables 2023: Analysis and forecast to 2028. https://www.iea.org/reports/renewables-2023

International Energy Agency (IEA) (2023b). Update on Chinese energy policy: tracking progress in decarbonization and security. https://www.iea.org/reports/update-on-chinas-energy-policy
International Energy Agency (IEA) (2024). Global energy and climate model documentation—2024-4. IEA, Paris. https://www.iea.org
International Energy Agency (IEA) (2025a) Chinese energy transition: global implications. IEA, Paris
International Energy Agency (IEA) (2025b) International renewable energy source. IEA, Paris
Jin AJ (2023) Fundamental theory on multiple energy resources and related case studies. Sci Rep 13, Article 10965. https://doi.org/10.1038/s41598-023-37653-5
Jin AJ (2025) Einsteinian lens: unraveling energy to GHG conversion in optimized distributed energy systems. Carbon Neut Technol. https://doi.org/10.1016/j.cnt.2025.100003
Jin AJ, Li Y (2025) Carbon neutrality and green energy systems. Peking University Press, Beijing
Jin AJ, Li C, Su J, Tan J (2022) Fundamental studies of smart distributed energy resources along with energy blockchain. Energies 15:8067. https://doi.org/10.3390/en15218067
Lee J, Zhao F (2023) Chinese green industrial policy: the case of new energy vehicles. J Environ Econ Policy 12(4):421–438. https://doi.org/10.1080/21606544.2022.2090001
Ma J, Chen L (2023) Chinese green finance: policies, progress and challenges. Climate Policy 23(3):287–302. https://doi.org/10.1080/14693062.2022.2090891
Ministry of Ecology and Environment (MEE) (2023) Notice on the launch of the national carbon emission trading market (power generation industry) for the 2021 and 2022 compliance cycles. http://www.mee.gov.cn/xxgk2018/xxgk/xxgk03/202304/t20230412_1027079.html
National Development and Reform Commission (NDRC) (2021, October 24) Working guidance for carbon dioxide peaking and carbon neutrality in full and faithful implementation of the new development philosophy. https://en.ndrc.gov.cn/policies/202110/t20211027_1301020.html
National Development and Reform Commission, Ministry of Ecology and Environment, Ministry of Foreign Affairs, Ministry of Commerce (2022, March 28) Opinions on the joint implementation of green development in the Belt and Road Initiative
Our World in Data (2025) Annual CO_2 emissions. Our World in Data. https://archive.ourworldindata.org/20250624-125417/grapher/annual-co2-emissions-per-country.html
S&P Global Commodity Insights (2024) Chinese carbon market: expansion plans, price outlook and link to ESG. https://www.spglobal.com/commodityinsights/en/ci/research-analysis/china-carbon-market-expansion-plans-price-outlook.html
State Council (2021) Action plan for carbon dioxide peaking before 2030. https://www.gov.cn/zhengce/content/2021-10/26/content_5644984.htm
Su J, Li Z, Jin AJ (2021) Practical model for optimal carbon control with distributed energy resources. IEEE Access 9:161603–161612. https://doi.org/10.1109/ACCESS.2021.3132699
Tsinghua University Institute for Carbon Neutrality (2023) Annual report on Chinese carbon neutrality. https://lce.tsinghua.edu.cn/__local/4/66/C4/86E4A05F1FB60B0021BDAB56640_AD00CB96_3B6BB4.pdf
Tsinghua University Institute of Energy, Environment and Economy (2023) Annual report on Chinese low-carbon development. https://equalocean.com/analysis/2023092620252
Tsinghua University Institute for Climate Change and Sustainable Development. (2024). Chinese green technology exports and global climate governance. https://iccsd.tsinghua.edu.cn/en/research.html
Tsinghua University Institute for Climate Change and Sustainable Development (2025) Assessing Chinese dual-carbon progress: policy, markets, and technology. Tsinghua University Press, Beijing. https://www.icon.tsinghua.edu.cn/en/info/1074/1091.htm
Woodrow W. Clark II (2017) Sustainable cities and communities design handbook: green engineering, architecture, and technology. Butterworth-Heinemann. Paperback ISBN: 9780128139646; eBook ISBN: 9780128139653
Zhang D, Wang Y (2024) The evolution and effectiveness of Chinese industrial decarbonization policies: focusing on steel and cement. J Clean Prod 434:140132. https://doi.org/10.1016/j.jclepro.2023.140132

Zhang Z, Chen X, Strbac G (2023) Effectiveness and challenges of the Chinese national carbon emissions trading scheme: evidence from the power sector. Energy Econ 126:106972. https://doi.org/10.1016/j.eneco.2023.106972

Zhou W (2021, September 26) The East Asian low-carbon community concept and its realization: challenges for achieving China's "30–60" targets. Paper presented at the 2021 annual conference of the society for environmental economics and policy studies

Zhou W, Wang J, Ling Y, Qian L, Miyawaki N (2021a) Research on China's coal phase-out—Part 1: analysis of current coal consumption and influencing factors. Policy Sci 29(1)

Zhou W, Qian X, Nakagami K (2021b) East Asian low-carbon community: realizing a sustainable decarbonized society from technology and social systems. Springer, Cham

Attributing Historical Radiative Forcing and Temperature Change by Climate Forcer and Sector: Methodology and Global Assessment (1850–2020)

Xuanming Su and Weisheng Zhou

1 Introduction

Historical and attribution studies consistently show that long-lived greenhouse gases (LLGHGs), particularly carbon dioxide from fossil fuel combustion (CO_2-FF), dominate anthropogenic radiative forcing (ΔF) and temperature rise (ΔT) since the preindustrial era (Allen et al., 2018; Forster et al., 2025; Fu et al., 2021; Su et al., 2022; Su and Zhou 2025). Methane (CH_4) and nitrous oxide (N_2O) are the next most important contributors, with CH_4 exerting a strong near-term warming potential due to its high radiative efficiency and decadal lifetime. Short-lived climate forcers (SLCFs), including black carbon (BC), tropospheric ozone (O_3), and aerosols (sulfate, nitrate, and organic carbon (OC)), exert complex climate effects: some species are warming (BC, O_3), and others are cooling (sulfate, nitrate, and OC). Their short atmospheric lifetimes mean that mitigation can quickly alter the net forcing (Zheng and Unger 2021; Lin et al., 2022; Su et al., 2022; Su and Zhou 2025).

Global assessments (Forster et al., 2025; Su and Zhou 2025) indicate that by 2020, net anthropogenic radiative forcing reached 2.2–2.9 Wm^{-2}, with LLGHGs providing the bulk of the positive forcing. Aerosol cooling offset roughly one-third of GHG warming during the mid-twentieth century but has weakened since 2000 due to emission controls. On the one hand, most of the SLCFs are atmospheric pollutants that might cause environmental issues and affect human health (Shindell and Smith 2019; Nakajima et al., 2020; Zheng and Unger 2021; Li et al., 2023). On the other hand, despite the relatively short lifespan, SLCFs have a substantial impact

X. Su (✉)
Research Center for Environmental Modelling and Application/Research Institute for Global Change, Japan Agency for Marine-Earth Science and Technology (JAMSTEC), Yokohama, Japan
e-mail: suxuanming@jamstec.go.jp

W. Zhou
College of Policy Science, Ritsumeikan University, Osaka, Japan
e-mail: zhou@sps.ritsumei.ac.jp

W. Zhou (ed.), *Decarbonization Policy Engineering*, World Sustainability Series,
https://doi.org/10.1007/978-981-95-5953-4_6

on climate change, particularly due to their high global warming potential (GWP) (Shindell and Smith 2019; Xie et al., 2023). Reports suggest that suppressing SLCFs could contribute to a swift reduction in radiative forcing, thereby mitigating climate change (Bowerman et al., 2013; Shoemaker et al., 2013; Rogelj et al., 2015; Su et al., 2017, 2022; Nakajima et al., 2020).

In addition, atmospheric aerosols can mitigate global warming by obscuring some of the warming effects caused by greenhouse gas (GHG) emissions. Reducing these aerosols might result in an increase in the global mean ΔT. Modeling shows that 50% emission cuts in domestic, agriculture, and waste sectors could yield the largest $PM_{2.5}$ health gains globally, while also reducing ΔT by up to 0.085 K in the near term. Such integrated strategies are essential for aligning climate and public health goals, particularly in densely populated regions with high pollution burdens. It is therefore important to comprehensively assess the climate responses induced by the SLCFs. Mitigating SLCFs offers immediate co-benefits for air quality, health, and ecosystems (Zheng and Unger 2021).

Sectorally, the energy and industrial sectors dominate CO_2 and sulfate emissions, agriculture is the main CH_4 and N_2O source, and waste and transport contribute both GHGs and SLCFs. Natural forcings (solar, volcanic) add episodic variability but negligible net trends over the past century (Su et al., 2022; Gulev et al., 2021).

Furthermore, key uncertainties include the magnitude and spatial pattern of aerosol forcing, CH_4–O_3-aerosol chemical interactions, and the efficacy of specific mitigation measures across sectors. Differences in model treatment of atmospheric chemistry, land-climate feedbacks, and emission inventories contribute to the spread in ΔF and ΔT estimates. Improved observational constraints, harmonized inventories, and integrated modeling frameworks are needed to refine regional attribution and guide coordinated mitigation strategies.

This study applied a normalized marginal approach to the simple climate model for optimization version 3.3 (SCM4OPT v3.3) (Su et al., 2017, 2022, 2024; Su and Zhou 2025) to conduct a comprehensive assessment of the climate responses to the LLGHGs and SLCFs for the world. A comprehensive assessment that encompasses all long-lived and short-lived compounds and precursors could aid in evaluating detailed mitigation options to achieve the Paris Agreement. This agreement aims to limit global warming to well below 2 °C above pre-industrial levels, while also pursuing efforts to limit global warming to 1.5 °C above pre-industrial levels.

The paper was structured in the following manner. An updated version of the SCM4OPT v3.3 and the normalized marginal approach are presented in Sect. 2. These methods are utilized to quantify the individual climatic responses of different regional and sectoral emissions from the year 1850 to the year 2020. While this is going on, the emission datasets that were utilized in this investigation are described in Sect. 2. After that, the results of radiative forcing and temperature rises that were generated by both the LLGHGs and SLCFs are presented in Sect. 3 for the global context. The discussions and conclusions that are linked with the topic are presented in Sects. 4 and 5.

2 Methods

2.1 SCM4OPT V3.3

Using a comprehensive set of climate-related emissions and climate forcers, SCM4OPT v3.3 (Su et al., 2022, 2024) is able to evaluate the state of the global climate. These emissions and climate forcers include CO_2 from fossil fuels, CO_2 from land-use change, carbon monoxide (CO), nitrogen dioxide (N_2O), halogenated gases (16 Ozone-Depleting Substances (ODS) and 23 other fluorinated gases), aerosols and pollutants (carbon monoxide (CO), NO_x, VOCs, SO_x, BC, OC), land albedo, and natural forcings (such as volcanic and solar irradiance). Carbon cycles and temperature increases above pre-industrial levels were calibrated against the outputs of 26 coupled atmosphere–ocean general circulation models (AOGCMs) and Earth System Models (ESMs), which were part of the Coupled Model Intercomparison Project, Phase 6 (CMIP6) (Eyring et al., 2016). The most recent version of SCM4OPT, version 3.3, was updated from earlier versions (Su et al., 2017). Each set of parameters that simulated the behavior of the individual AOGCMs and ESMs was used to depict the uncertainties that were brought about by the carbon cycles and the climate system collectively. The radiative forcings of solar irradiance and solar irradiance were used to prescribe warmings that were generated by solar irradiance and volcanic activity for CMIP6 inputs (Zanchettin et al., 2016; Matthes et al., 2017). Within the context of temperature increases beyond pre-industrial levels, the most recent version of HadCRUT v5.0 was utilized as a restriction to calibrate the ΔT simulated in SCM4OPT v3.3. The Supplementary Materials in Su et al. (2022) provided a detailed description of the atmospheric physical and chemical processes.

2.2 Normalized Marginal Method

For the purpose of our research, we employed the normalized marginal technique, as described in earlier studies (Trudinger and Enting 2005; Li et al., 2016; Fu et al., 2021; Boucher et al., 2021; Su et al., 2022, 2024; Su and Zhou 2025). The ratio of the relative warming contribution of emissions that are specific to a region, sector, or climate forcer, which is denoted as *e*, to the total warming increase, which is defined as α_e^T, is proportional to the marginal effect of *e* divided by the magnitude of the entire marginal effect. To calculate α_e^T for every time series of *e*,, we carried out two simulations: (1) a simulation that included all emissions to ascertain the total warming increase, which was denoted as T_{all}; and (2) a parallel simulation in which emission *e* was reduced by a fraction $\epsilon = 0.001$, thereby subtracting $e \cdot \epsilon$ over the evaluation period (for example, 1850–2020) to derive the corresponding temperature increase, which was designated as $T_{e,\epsilon}$. Consequently, the relative contribution, denoted by α_e^T, can be determined in the following manner:

$$\alpha_e^T(t) = \frac{T_{all}(t) - T_{e,\epsilon}(t)}{\sum_{e'}\left(T_{all}(t) - T_{e',\epsilon}(t)\right)} \tag{1}$$

Therefore, the temperature increase T_e that takes place as a consequence of e can be computed in the following manner:

$$T_e(t) = T_{all}(t) \cdot \alpha_e^T(t) \tag{2}$$

To quantify the uncertainty in T_e, we conducted 200 pairs of runs for each source e, using randomized parameter sets for the climate system (see Table S4 in Su et al. (2022)). This allowed us to determine the extent of the uncertainties. We showed the radiative forcings and warming increases for each of the different climate forcers. The radiative forcings were also computed by employing the SCM4OPT v3.3 and the normalized marginal approach above.

2.3 Emission Datasets Used in This Study

In this research, the uncertainties in historical emissions, and uncertainties in climate models were all taken into consideration along with other uncertainties. The Community Emissions Data System (CEDS_v_2024_07_08) (Hoesly et al., 2018), the Emission Database for Global Atmospheric Research (EDGAR) v8.0_GHG 1970–2022 and EDGAR v8.1_AP 1970–2022 (Crippa et al., 2021, 2022), and the BB4CMIP (Marle et al., 2017) were some of the emission datasets that were utilized in order to accurately represent the uncertainties that were a consequence of the estimations of historical emissions. As a consequence of this, our findings are representative of the warming that is caused by the emission of GHGs, aerosols, and pollutants from a wide range of sectors, such as agriculture, commercial and residential, energy, industrial, transportation, and waste management. In addition, the data that is available in these datasets can be utilized to account for temperature fluctuations that are brought about by aerosols and pollutants that are produced by the burning of agricultural waste, residential biomass, or large-scale forest and grass fires. We used the same procedure as that in Su et al. (2022) when it came to compiling and processing the data. When a particular emission was not available, we utilized emission data from a separate dataset that was equivalent to the unavailable emission. The sectoral mapping for the main emission datasets was shown in Table 1. Emissions like these were utilized in the process of calculating atmospheric concentrations and radiative forcings, which ultimately led to the determination of ΔT.

Since around 1980, global anthropogenic emissions have shown strong but varied trajectories depending on the species, as shown in Fig. 1. CO_2 from fossil fuel combustion has continued to rise steadily, with only brief slowdowns during economic recessions (e.g., early 1980s, 2008–2009) and a possible plateau or reduced growth rate in the most recent decade. CH_4 emissions increased notably since the 1950s, slowed through the 1980s and 1990s, but resumed growth after 2007, reflecting

Table 1 Sectoral mapping for the emission datasets

Sector	CEDS_v_2024_07_08	EDGAR v8.0
Agriculture	3B_Manure-management 3D_Rice-Cultivation 3D_Soil-emissions 3E_Enteric-fermentation 3I_Agriculture-other	4A Enteric fermentation 4B Manure management 4C Rice cultivation 4D1 Direct soil emissions 4D2 Manure in pasture/range/paddock 4D3 Indirect N2O from agriculture 4D4 Other direct soil emissions
Agricultural Waste Burning		4F Agricultural waste burning
Domestic Housing and Commercial	1A4a_Commercial-institutional 1A4b_Residential 1A4c_Agriculture-forestry-fishing 1A5_Other-unspecified	1A4 Residential and other sectors
Energy	1A1a_Electricity-autoproducer 1A1a_Electricity-public 1A1a_Heat-production 1A1bc_Other-transformation 1B1_Fugitive-solid-fuels 1B2_Fugitive-petr 1B2b_Fugitive-NG-distr 1B2b_Fugitive-NG-prod 1B2d_Fugitive-other-energy 7A_Fossil-fuel-fires	1A1a Public electricity and heat production 1A1bc Other Energy Industries 1A2 Manufacturing Industries and Construction 1A5 Other Energy Industries 1B1 Fugitive emissions from solid fuels 1B2 Fugitive emissions from oil and gas 7A Fossil fuel fires

(continued)

Table 1 (continued)

Sector	CEDS_v_2024_07_08	EDGAR v8.0
Industry	1A2a_Ind-Comb-Iron-steel 1A2b_Ind-Comb-Non-ferrous-metals 1A2c_Ind-Comb-Chemicals 1A2d_Ind-Comb-Pulp-paper 1A2e_Ind-Comb-Food-tobacco 1A2f_Ind-Comb-Non-metalic-minerals 1A2g_Ind-Comb-Construction 1A2g_Ind-Comb-machinery 1A2g_Ind-Comb-mining-quarying 1A2g_Ind-Comb-other 1A2g_Ind-Comb-textile-leather 1A2g_Ind-Comb-transpequip 1A2g_Ind-Comb-wood-products 2A1_Cement-production 2A2_Lime-production 2Ax_Other-minerals 2B_Chemical-industry 2B2_Chemicals-Nitric-acid 2B3_Chemicals-Adipic-acid 2C1_Iron-steel-alloy-prod 2C3_Aluminum-production 2C4_Non-Ferrous-other-metals 2H_Pulp-and-paper-food-beverage-wood 7BC_Indirect-N_2O-non-agricultural-N	2A1 Cement production 2A2 Lime production 2A3 Limestone and dolomite use 2A4 Soda ash production and use 2A7 Production of other minerals 2B Production of chemicals 2C Production of metals 2G Non-energy use of lubricants/waxes (CO_2)
Industry Solvents	2D_ Chemical-products-manufacture-processing 2D_Degreasing-Cleaning 2D_Other-product-use 2D_Paint-application	3A Solvent and other product use: paint 3B Solvent and other product use: degrease 3C Solvent and other product use: chemicals 3D Solvent and other product use: other
Surface Transportation	1A3b_Road 1A3c_Rail 1A3di_Oil_Tanker_Loading 1A3dii_Domestic-navigation 1A3eii_Other-transp	1A3b Road transportation 1A3c Rail transportation 1A3d Inland navigation 1A3e Other transportation
Waste Treatment	5A_Solid-waste-disposal 5C_Waste-combustion 5D_Wastewater-handling 5E_Other-waste-handling 6A_Other-in-total 6B_Other-not-in-total	6A Solid waste disposal on land 6B Wastewater handling 6C Waste incineration 6D Other waste handling
Open Forest Burning		
Open Grassland Burning		

(continued)

Table 1 (continued)

Sector	CEDS_v_2024_07_08	EDGAR v8.0
Aviation	1A3ai_International-aviation 1A3aii_Domestic-aviation	1A3a Domestic aviation 1C1 International aviation
International Shipment	1A3di_International-shipping	1C2 International navigation

contributions from agriculture, waste, and fossil fuel extraction. N_2O emissions show a gradual upward trend, driven mainly by agricultural fertilizer use. For aerosol precursors, SO_2 emissions peaked globally around the late 1980s-early 1990s and have since declined sharply, especially in North America and Europe, with more recent reductions in East Asia due to air-quality policies. NO_x emissions continued to rise into the 2000s before leveling off or slightly decreasing, while NH_3 remained relatively constant or slightly increasing, reflecting agricultural sources. BC and OC emissions have decreased in some regions but remain significant globally, particularly from combustion in Asia and Africa. For industrial fluorinated gases, many CFCs and halons fell rapidly after the 1987 Montreal Protocol, while HCFCs rose temporarily as transitional substitutes before also beginning to decline. In contrast, HFCs (e.g., HFC-125, HFC-134a) and other synthetic gases (SF_6, NF_3, perfluorocarbons) have grown rapidly in recent decades due to refrigeration, air conditioning, and electronics manufacturing, despite their high global warming potential.

Overall, the recent decades show a divergence: LLGHGs, particularly CO_2, N_2O, CH_4, and industrial fluorinated gases, continue to increase; aerosol precursor emissions are declining; and the composition of SLCFs is shifting toward species with less cooling effect (e.g., sulfate reductions) and more warming effect (e.g., BC, CH_4). This shift implies a gradual unmasking of underlying GHG-driven warming.

3 Results

A thorough evaluation of the climatic responses of both LLGHGs and SLCFs was conducted for the world. We calculated the radiative forcings and temperature variations for particular LLGHGs and SLCFs for the world using the previously outlined methodology, and the subsequent two subsections present the results.

3.1 Radiative Forcing

Although the global industrial revolution began in the eighteenth century, the LLGHGs emissions before World War II were relatively small. Rapid industrialization and urbanization after the 1950s led to a significant increase in LLGHGs

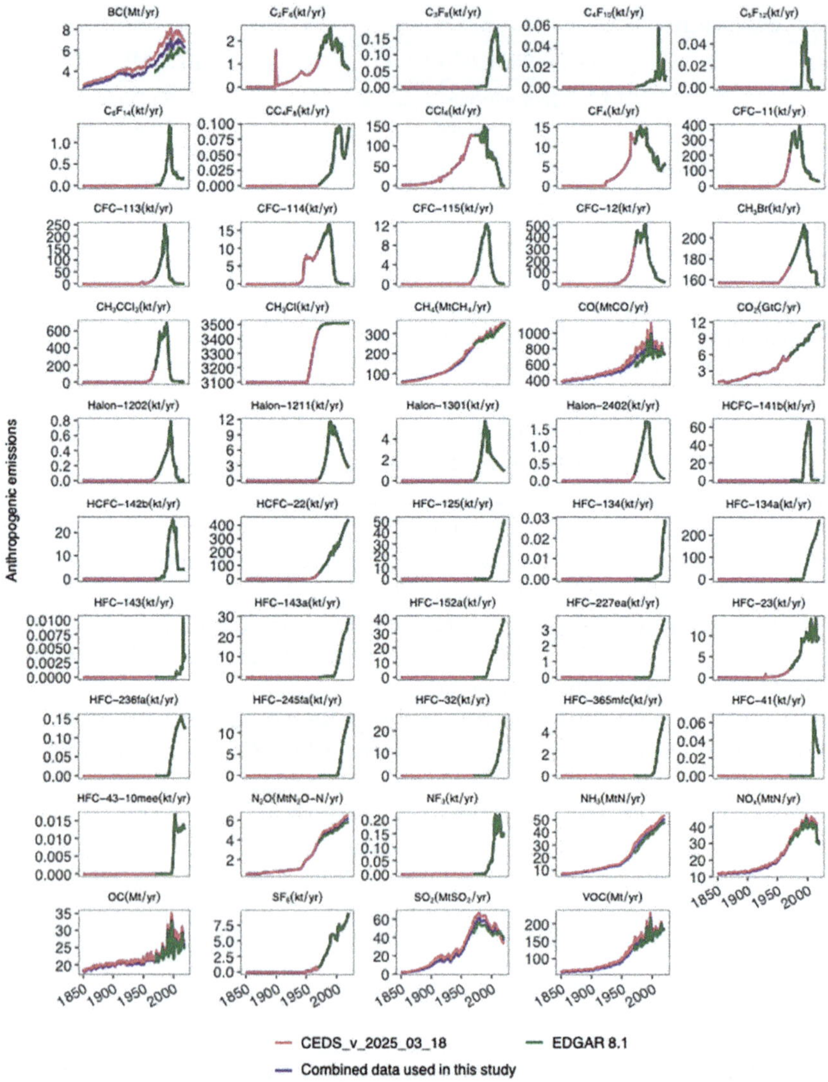

Fig. 1 World's emissions of greenhouse gases and aerosols. Data sources are from the Community Emissions Data System (CEDS_v_2024_07_08) (Hoesly et al., 2018) and the Emission Database for Global Atmospheric Research (EDGAR) v8.0_GHG 1970–2022 and EDGAR v8.1_AP 1970–2022 (Crippa et al., 2021, 2022). For the historical biomass burning, the BB4CMIP (Marle et al., 2017) (not shown) was used when the data was not available

emissions, and the associated radiative forcing has been a major concern for climate change, as shown in Fig. 2a. Among the LLGHGs emissions, CO_2 and CH_4 are the most significant contributors to the radiative forcing, accounting for 59% and 26% of the total LLGHGs radiative forcing, respectively, in recent years. LLGHGs emissions contribute most of the total positive radiative forcing in 2020, reaching up to 3.5 Wm^{-2}. However, the cooling effects of aerosols and pollutants have mitigated a significant portion of the warming. The sum of the radiative forcing of LLGHGs is comparable to that of total radiative forcing, since all of them are positive. However, the radiative forcing of SLCFs is relatively small, as the cooling effects of aerosols and pollutants cancel out a substantial portion of the radiative forcing.

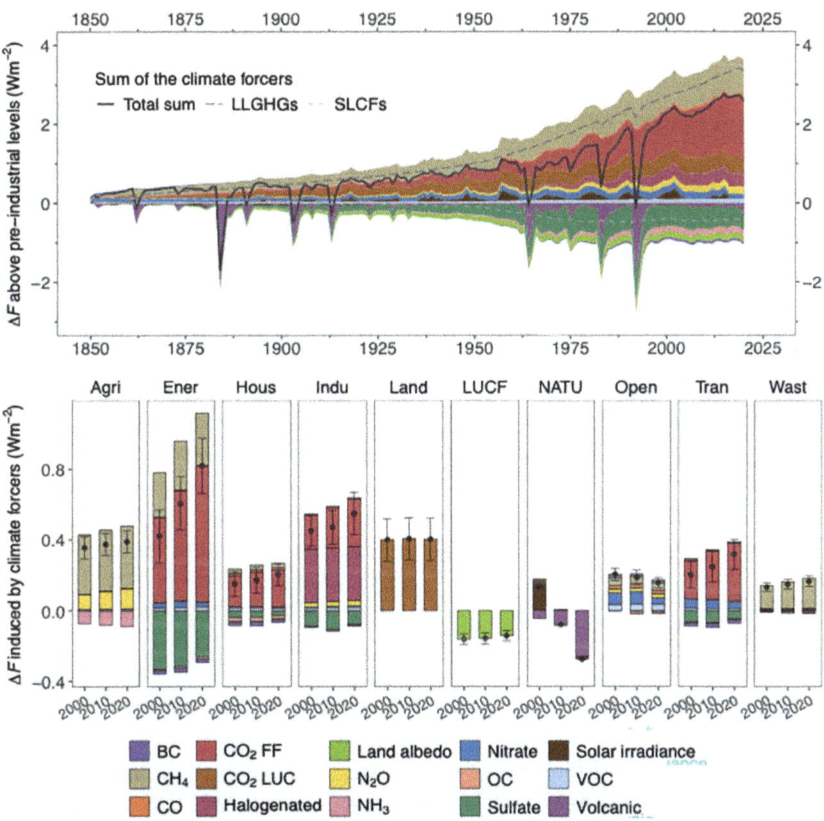

Fig. 2 World radiative forcing (ΔF) induced by climate forcers. **a** World radiative forcing above pre-industrial levels. The thick solid line shows the sum of the total; the thin dashed lines show the sums of LLGHGs and SLCFs climate forcers, respectively. **b** World radiative forcing induced by anthropogenic and natural climate forcers in 2000, 2010, and 2020. The points and error bars show the sums of the mean and one-sigma obtained from Monte Carlo simulations. Agri, Agriculture; Ener, Energy; Hous, Household; Indu, Industry; Tran, Transport; Wast, Waste; Open, Open burning; Land, Land use related emissions; LUCF, Land albedo; DUST, Dust; NATU, Natural forcing

The energy, residential (household), industrial, and transportation sectors are the main contributors to CO_2 forcing, whereas the agriculture, energy, and waste treatment sectors primarily drive CH_4 forcing (Fig. 2b). The primary contributors to negative radiative forcing include aerosols and pollutants resulting from fossil fuel combustion, along with land albedo effects and volcanic activities. A minor negative radiative forcing is produced from ammonia (NH_3) emissions related to agriculture, the residential sector, industrial processes, and the conventional transportation sector, which is largely reliant on fossil fuels. The radiative forcing of LLGHGs may reach up to 3.2–3.5 Wm^{-2}, while SLCFs exhibit a radiative forcing of –0.7 to –0.1 Wm^{-2}. Notably, the cooling effects attributed to aerosols and pollutants are significant, and there are considerable uncertainties associated with the forcing estimates for these components.

3.2 *Temperature Increases*

The trends in global warming driven by specific climate forcers typically align with their radiative forcing trends, as illustrated in Fig. 3a. However, the forcing signals display more significant annual variations compared to the warming signals, particularly in the cases of volcanic and solar irradiance forcings. This phenomenon can probably be attributed to ocean thermal inertia, resulting in a delay in temperature variations (Szopa et al., 2021). The LLGHGs exhibit a warming effect that is somewhat less than the overall warming, while SLCFs exerted a cooling effect which offset a part of the warming. By 2020, the primary contributors to positive warming included agriculture, energy, residential, industry, transportation, open burning and land use sectors, whereas the energy sector was the main contributor to negative warming (Fig. 3b). The overall increase in temperature was 1.2–1.6 °C, with the LLGHGs accounting for 1.6–1.8 °C and the SLCFs contributing –0.4 to 0 °C in the year 2020.

4 Discussion

Compared to existing attributions of climate responses (Shindell and Smith 2019; Nakajima et al., 2020; Zheng and Unger 2021; Li et al., 2023; Xie et al., 2023), we offered a more comprehensive attribution of the radiative forcing and warming to LLGHGs and SLCFs. The detailed attributions in this study might only cause relatively minor changes in the absolute values of the radiative forcing and warming. However, they could provide more insights into the relative contributions of different climate forcers to the warming under the global context, particularly when aggressive mitigation actions are implemented. The global attribution results are generally consistent with the earlier estimates (Szopa et al., 2021; Fu et al., 2021). Therefore,

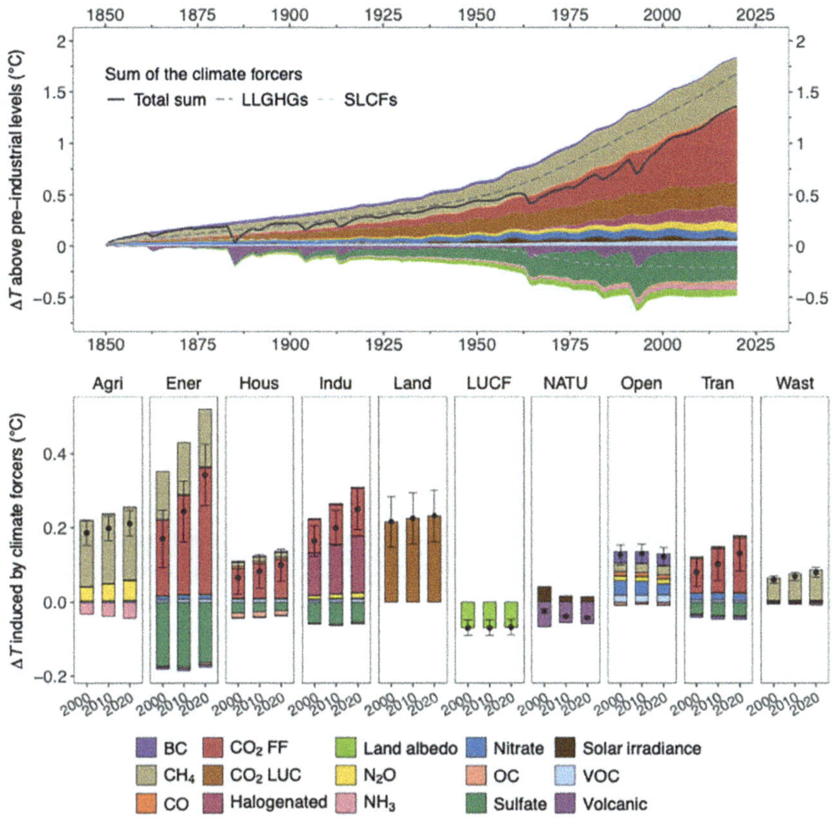

Fig. 3 World temperature increases (ΔT) induced by climate forcers. **a** World temperature increases above pre-industrial levels. The thick solid line shows the sum of the total; the thin dashed lines show the sums of LLGHGs and SLCFs climate forcers, respectively. **b** World temperature increases induced by anthropogenic and natural climate forcers in 2000, 2010, and 2020. The points and error bars show the sums of mean and one-sigma obtained from Monte Carlo simulations

the results here are quantitatively robust, allowing for detailed attribution assessment by regional and sectoral level of individual climate forcers.

This study presents specific caveats and limitations. First, the emission-based attribution is influenced by the administrative boundaries of a particular area, and the results of the attribution solely represent the global climate responses triggered by emissions within that administrative region. This study does not take into account the regional climate impacts, such as the cooling effect caused by aerosols and pollutants released from other areas. Second, the attribution results exhibit sensitivity to the emission data and climate models. This study employs Monte Carlo simulations for the uncertainty analysis, a widely recognized approach for estimating uncertainties. Nonetheless, the simulated uncertainties may still be underestimated due to the restricted number of realizations and the inadequate representation of intricate physical and chemical climate processes through simplified climate models (Su et al.,

2022). Third, this study does not differentiate between the climate responses associated with ozone, which is clearly defined in the SLCFs and includes both tropospheric and stratospheric ozone. The reason lies in the attribution approach utilized in this context.

While decreasing SLCFs may result in a rapid alleviation of global warming, the findings indicate a significant cooling impact from aerosols and pollutants. The emissions of aerosols and pollutants, especially sulfate, remain substantial in recent years. Reducing SLCFs or specific portions of them through environmental mitigation actions can enhance air quality, decrease exposure to outdoor air pollution, and lower the related health risks, along with benefiting land ecosystem health (Zheng and Unger 2021). Conversely, the cooling influence of aerosols and pollutants may counterbalance the warming impact of LLGHGs. A reduction in these cooling effects suggests that an additional warming effect will be triggered, potentially resulting in elevated temperatures, increased climate extremes, and greater risks of exposure for populations (Li et al., 2023). Consequently, the mitigation actions of SLCFs must be thoughtfully evaluated in relation to the current aerosols and pollutants. A thorough attribution of the climate responses of LLGHGs and SLCFs could assist global environmental experts and policymakers in grasping the trade-offs involved in the mitigation actions of individual LLGHGs and SLCFs. Nevertheless, implementing rigorous mitigation measures for both LLGHGs and SLCFs could be essential to address environmental pollution issues, lower health risks for individuals and ecosystems, and alleviate the effects of climate change.

5 Conclusions

This study presents a thorough attribution of the radiative forcing and warming associated with LLGHGs and SLCFs, utilizing a normalized marginal approach applied to the simple climate model, SCM4OPT v3.3. The findings offer a deeper insight into the climate responses elicited by specific climate forcers, especially regarding the implications of robust mitigation strategies. The findings indicate that global SLCFs have led to a radiative forcing of –0.7 to –0.1 W m^{-2} and a temperature rise of –0.4 to 0 °C. The global LLGHGs resulted in a radiative forcing of 3.2–3.5 W m^{-2} and an increase in warming of 1.6–1.8 °C. The results highlight the significance of accounting for the cooling effects of aerosols and pollutants in the assessment of SLCF impacts. The comprehensive attribution approach outlined here could be an essential resource for climate change researchers and policymakers to assess the trade-offs among various mitigation strategies and their potential effects on environmental issues, global warming and climate extremes. Future studies should concentrate on evaluating the challenges or economic implications that emerge from the implementation of strict mitigation strategies, to effectively identify prioritized measures for emission reduction.

Acknowledgements This work was supported by the Program for the advanced studies of climate change projection (SENTAN, Grant Number JPMXD0722681344) from the Ministry of Education, Culture, Sports, Science and Technology (MEXT), Japan. The computational resources were provided by the Japan Agency for Marine-Earth Science and Technology (JAMSTEC).

References

Allen MR et al (2018) Framing and context. In Masson-Delmotte V, Zhai P, Pörtner HO, Roberts D, Skea J, Shukla PR, Pirani A (eds), Global warming of 1.5 °C: An IPCC special report. Cambridge University Press, pp 49–92

Boucher O, Borella A, Gasser T, Hauglustaine D (2021) On the contribution of global aviation to the CO_2 radiative forcing of climate. Atmos Environ 267:118762. https://doi.org/10.1016/j.atmosenv.2021.118762

Bowerman NHA, Frame DJ, Huntingford C, Lowe JA, Smith SM, Allen MR (2013) The role of short-lived climate pollutants in meeting temperature goals. Nat Clim Chang 3(12):1021–1024. https://doi.org/10.1038/nclimate2034

Crippa M, Guizzardi D, Banja M, Solazzo E, Muntean M, Schaaf E, Pagani F et al (2022) CO_2 emissions of all world countries—JRC/IEA/PBL 2022 report. Publications Office of the European Union. https://doi.org/10.2760/730164

Crippa M, Guizzardi D, Solazzo E, Muntean M, Schaaf E, Monforti-Ferrario F, Banja M et al (2021) GHG emissions of all world: 2021 report. Publications Office of the European Union. https://doi.org/10.2760/173513

Eyring V, Bony S, Meehl GA, Senior CA, Stevens B, Stouffer RJ, Taylor KE (2016) Overview of the Coupled Model Intercomparison Project Phase 6 (CMIP6) experimental design and organization. Geosci Model Dev 9(5):1937–1958. https://doi.org/10.5194/gmd-9-1937-2016

Forster PM, Smith C, Walsh T, Lamb WF, Lamboll R, Cassou C, Hauser M et al (2025) Indicators of global climate change 2024: Annual update of key indicators of the state of the climate system and human influence. Earth Syst Sci Data 17(6):2641–2680. https://doi.org/10.5194/essd-17-2641-2025

Fu B, Li B, Gasser T, Tao S, Ciais P, Piao S, Balkanski Y et al (2021) The contributions of individual countries and regions to the global radiative forcing. Proc Natl Acad Sci 118(15):e2018211118. https://doi.org/10.1073/pnas.2018211118

Gulev SK et al (2021) Changing state of the climate system. In Masson-Delmotte V, Zhai P, Pirani A, Connors SL, Péan C (eds) Climate change 2021: the physical science basis. Cambridge University Press, pp 287–422. https://doi.org/10.1017/9781009157896.004

Hoesly RM, Smith SJ, Feng L, Klimont Z, Janssens-Maenhout G, Pitkanen T, Seibert JJ et al (2018) Historical (1750–2014) anthropogenic emissions of reactive gases and aerosols from the Community Emissions Data System (CEDS). Geosci Model Dev 11(1):369–408. https://doi.org/10.5194/gmd-11-369-2018

Li B, Gasser T, Ciais P, Piao S, Tao S, Balkanski Y, Hauglustaine D et al (2016) The contribution of China's emissions to global climate forcing. Nature 531(7594):357–361. https://doi.org/10.1038/nature17165

Li Y, Wang Z, Lei Y, Che H, Zhang X (2023) Impacts of reductions in non-methane short-lived climate forcers on future climate extremes and the resulting population exposure risks in eastern and southern Asia. Atmos Chem Phys 23(4):2499–2523. https://doi.org/10.5194/acp-23-2499-2023

Lin J, Khanna N, Liu X, Wang W, Gordon J, Dai F (2022) Opportunities to tackle short-lived climate pollutants and other greenhouse gases for China. Sci Total Environ 842:156842. https://doi.org/10.1016/j.scitotenv.2022.156842

Matthes K et al (2017) Solar forcing for CMIP6 (v3.2). Geosci Model Dev 10(6):2247–2302. https://doi.org/10.5194/gmd-10-2247-2017

Nakajima T, Ohara T, Masui T, Takemura T, Yoshimura K, Goto D, Hanaoka T et al (2020) A development of reduction scenarios of the short-lived climate pollutants (SLCPs) for mitigating global warming and environmental problems. Prog Earth Planet Sci 7(1):1–21. https://doi.org/10.1186/s40645-020-00351-1

Rogelj J, Meinshausen M, Schaeffer M, Knutti R, Keywan (2015) Impact of short-lived non-CO_2 mitigation on carbon budgets for stabilizing global warming. Environ Res Lett 10(7):075001. https://doi.org/10.1088/1748-9326/10/7/075001

Shindell D, Smith CJ (2019) Climate and air-quality benefits of a realistic phase-out of fossil fuels. Nature 573(7774):408–411. https://doi.org/10.1038/s41586-019-1554-z

Shoemaker JK, Schrag DP, Molina MJ, Ramanathan V (2013) What role for short-lived climate pollutants in mitigation policy? Science 342(6164):1323–1324. https://doi.org/10.1126/science.1240162

Su X, Shiogama H, Tanaka K, Tachiiri K, Hajima T, Watanabe M, Kawamiya M, Takahashi K, Yokohata T (2024) Reductions in atmospheric levels of non-CO_2 greenhouse gases explain about a quarter of the 1998–2012 warming slowdown. Commun Earth Environ 5(1):594. https://doi.org/10.1038/s43247-024-01723-x

Su X, Tachiiri K, Tanaka K, Watanabe M, Kawamiya M (2022) Identifying crucial emission sources under low forcing scenarios by a comprehensive attribution analysis. One Earth 5(12):1–13. https://doi.org/10.1016/j.oneear.2022.10.009

Su X, Takahashi K, Fujimori S, Hasegawa T, Tanaka K, Kato E, Shiogama H, Masui T, Emori S (2017) Emission pathways to achieve 2.0 °C and 1.5 °C climate targets. Earth Fut 5(6):592–604. https://doi.org/10.1002/2016EF000492

Su X, Zhou W (2025) Quantifying climate responses of short-lived climate forcers for China. Pol Sci 32(3):83–128. https://doi.org/10.34382/0002002520

Szopa S et al (2021). Short-lived climate forcers. In Masson-Delmotte V, Zhai P, Pirani A, Connors SL, Péan C (eds) Climate change 2021: the physical science basis. Cambridge University Press. pp. 817–922. https://doi.org/10.1017/9781009157896.008

Trudinger C, Enting I (2005) Comparison of formalisms for attributing responsibility for climate change: non-linearities in the Brazilian proposal approach. Clim Change 68(1):67–99. https://doi.org/10.1007/s10584-005-6012-2

van Marle MJE, Kloster S, Magi BI, Marlon JR, Daniau A-L, Field RD, Arneth A et al (2017) Historic global biomass burning emissions for CMIP6 (BB4CMIP). Geosci Model Dev 10(9):3329–3357. https://doi.org/10.5194/gmd-10-3329-2017

Xie B, Zhang H, Yu X, Wang Z (2023) Fast and slow responses of surface air temperature in China to short-lived climate forcers. Sci Total Environ 882:162888. https://doi.org/10.1016/j.scitotenv.2023.162888

Zanchettin D, Khodri M, Timmreck C, Toohey M, Schmidt A, Gerber EP, Hegerl G et al (2016) The Model Intercomparison Project on the climatic response to volcanic forcing (VolMIP): experimental design and forcing input data for CMIP6. Geosci Model Dev 9(8):2701–2719. https://doi.org/10.5194/gmd-9-2701-2016

Zheng Y, Unger N (2021) Reducing planetary health risks through short-lived climate forcer mitigation. GeoHealth 5(7):e2021GH000422. https://doi.org/10.1029/2021GH000422

Integrated Analysis of Carbon Reduction Strategies at the Regional Level: Evidence from Zhejiang, China

Xuanxuan Ming, Qiang Wang, Kun Luo, Liujie Zhang, and Jianren Fan

1 Regional Energy Transition and Carbon Reduction Strategies

1.1 China's Emission Goals

Climate change is a significant global challenge that has garnered widespread attention from the international community (Wu et al. 2018). Global warming since pre-industrial times has been primarily attributed to the increasing concentrations of CO_2 in the atmosphere, which mainly stems from the burning of fossil fuels (IPCC 2021). Recognized as a major contributor to annual CO_2 emissions, China has committed to peak its carbon emissions before 2030 and attain carbon neutrality before 2060 (Li et al. 2022). Carbon emissions are closely tied to energy consumption. China's carbon emissions increased by 750 million tonnes between 2019 and 2021, driving the rebound of global emissions above pre-pandemic levels. The increase was due

X. Ming · Q. Wang (✉) · K. Luo (✉) · L. Zhang · J. Fan
State Key Laboratory of Clean Energy Utilization, Department of Energy Engineering, Zhejiang University, Hangzhou, China
e-mail: zjuqw@zju.edu.cn

K. Luo
e-mail: zjulk@zju.edu.cn

X. Ming
e-mail: m17860628900@163.com

L. Zhang
e-mail: 22127007@zju.edu.cn

J. Fan
e-mail: fanjr@zju.edu.cn

Q. Wang · K. Luo · J. Fan
Zhejiang Key Laboratory of Clean Energy and Carbon Neutrality, Hangzhou, China

W. Zhou (ed.), *Decarbonization Policy Engineering*, World Sustainability Series,
https://doi.org/10.1007/978-981-95-5953-4_7

to a sharp rise in electricity demand that relied heavily on coal power, despite China also experiencing its largest-ever increase in renewable power output in 2021 (International Energy Agency 2022). Energy system transformation will play a decisive role in achieving carbon neutrality (Ma et al. 2023).

1.2 Limitations of National-Level Studies

Most studies focus on designing peak emission pathways at the national level. Zhang and Chen (2022) constructed the Integrated MARKAL-EFOM System model of China to elaborate and compare the technology options for decarbonizing the energy system and the synergistic effects on air pollutant reduction under different scenarios for achieving carbon neutrality. Fang et al. (2022) combined regression analysis with Monte Carlo simulation to explore the potential emission peak paths of industrial sectors in China. Additionally, scholars typically focus on in-depth analysis of specific industries or particular sectors of energy production in research on national energy transformation. According to statistics, specific industries are mainly in the construction industry (Yao et al. 2023), cement industry (Zhang et al. 2021), power sector with high penetration of renewable energy (Chen et al. 2020), coal production industry (Lee and Huh 2017) and so on.

Previous studies have demonstrated that energy use and carbon reduction strategies are typically developed at the national levels and in specific industries. Research focusing on the national level often fails to fully account for the variations among different regions in terms of economic development, environmental protection, and energy structure. This may result in the formulation of policies that are challenging to implement at the local level or have limited effectiveness, lacking in regional specificity and targeted approaches. Hence, regional-scale studies on energy transition and carbon emission reduction have gradually become a trend in research (Wei et al. 2022; Wu et al. 2021; Liu et al. 2020). Additionally, there is a lack of literature exploring economic performance in design scenarios at the regional level. When researching energy production, it is crucial to take both environmental and economic impacts into account. In the case of studying clean energy generation, some researchers focus on analyzing the wake flows between wind turbines to minimize their impact on the environment and reduce costs simultaneously (Wang et al. 2022a, b, 2023). Therefore, under the dual carbon goals, there is a lack of comprehensive analysis of energy transition design at the regional level in China.

1.3 Application of Energy Models in Regional Transition

Scholars frequently employ energy models to discuss the environmental impacts of energy transition, providing valuable insights for strategies to achieve goals. Based on different modeling methods, energy models can be roughly divided into top-down, bottom-up, and hybrid models. The CGE (Computable General Equilibrium) (Guo et al. 2021; Liang et al. 2022) model is the most prevalent tool employed in top-down analyses which is more suitable for assessing the impact of policies on economic sectors and agents from a macroscopic perspective. The bottom-up model begins with a technical perspective and simulates transportation, conversion, and end-use processes within the energy system. The bottom-up model mainly includes the LEAP (Low Emissions Analysis Platform) (Maduekwe et al. 2020) and MARKAL (Market Allocation Model) (Stoltmann et al. 2019). Comparing two types of models, the top-down approaches ignore specific technical details when measuring the impacts caused by energy policies, while the bottom-up approaches can assess the efficiency and costs of distinct technological options (Qi et al. 2023). Hybrid models (Shi et al. 2017) combine the above two types of models and are suitable for simulating macroeconomics, analyzing energy technology, and evaluating policies. Considering the characteristics of the aforementioned models, bottom-up models are more suitable for multiple sub-sectors modeling at the regional level, and they also can take economic issues into account. To this end, this study focuses on the urban agglomeration of Zhejiang as a case study, aiming to collect carbon reduction strategies with regional characteristics and provide a comprehensive analysis of various emission reduction measures implemented between 2020 and 2050.

2 Scenario Analysis for Achieving Decarbonization

This chapter introduces the methodology for analyzing different decarbonization scenarios in Zhejiang Province. The scenario analysis evaluates the long-term impacts of different policy measures, energy transitions, and technological advancements. Through this approach, policymakers and researchers can assess the effectiveness and costs of each strategy, helping to identify optimal solutions for achieving carbon neutrality while balancing economic and environmental objectives. The subsequent sections will detail the LEAP framework, calculation principles, and various decarbonization scenarios that have been set to guide this analysis.

2.1 LEAP Framework

The use of the LEAP enables an integrated economic, energy, and environmental analysis, allowing policymakers and researchers to evaluate the long-term impacts and costs of various policy options and technological choices (Heaps 2022). This includes the balancing of energy supply and demand, reduction of greenhouse gas emissions, mitigation of environmental impact, evaluation of economic costs, and identification of social benefits. The calculation process of total energy demand for the LEAP model is shown in Fig. 1.

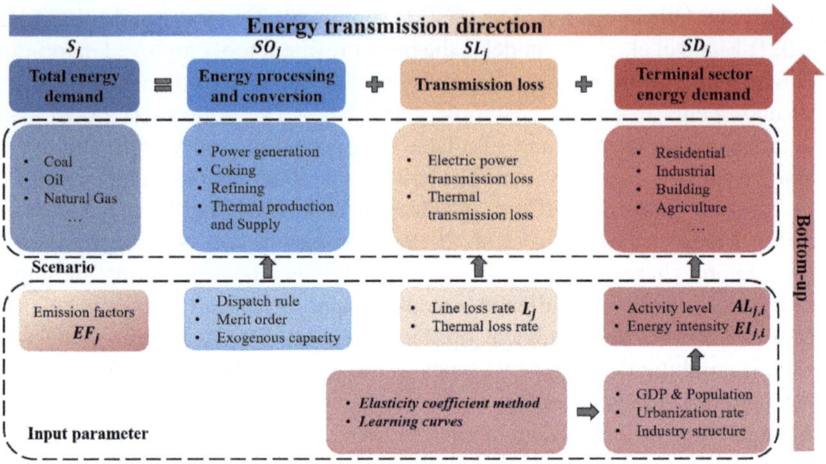

Fig. 1 Diagram of the calculation process for total energy demand (LEAP framework)

Total energy demand can be divided into three modules based on the direction of energy transmission. The terminal sector energy demand refers to the energy required by end users (such as residents, factories, or transportation) in production or daily life. This study calculates terminal sector energy demand based on the activity levels of end users and the corresponding energy intensities for various activities. Transmission loss refers to the energy loss that mainly occurs during the transmission and distribution of electric power, thermal power, and natural gas transportation. The energy processing and conversion module refers to the process of converting primary energy into secondary energy. It mainly includes power generation, refining, coking, and thermal production and supply in the sector of processing and conversion. The total energy demand is the sum of the energy consumption in the three modules mentioned above. At the same time, anthropogenic CO_2 emissions are set in the module of energy conversion and terminal sector energy demand.

2.2 *Calculation Principles*

2.2.1 Total Energy Demand

The terminal sector energy demand SD_i can be expressed as:

$$SD_j = \sum\nolimits_i AL_{j,i} \times EI_{j,i} \tag{1}$$

where $AL_{j,i}$ is the activity level of department i, as a measure of energy type j consumption for terminal sector. $E_{j,i}$ is the energy intensity corresponding to various levels of activity.

$$SL_j = \frac{SD_j}{E_j} \times L_j \tag{2}$$

$$E_j = 1 - L_j \tag{3}$$

where SL_j represents transmission loss of energy type j, where L_j is the loss rate. E_j is the transmission efficiency of energy.

$$SO_j = \frac{SD_j}{EI_j} - SD_j \tag{4}$$

$$S_j = PS_j + SS_j \tag{5}$$

where SO_j represents energy processing and conversion, and EI_j is the conversion rate of energy. S_j represents total energy demand. The energy available for local consumption includes PS_j (primary energy supply) which involves coal, oil, natural gas, etc. and SS_j (secondary energy supply) such as electricity and heat.

2.2.2 CO_2 Emission

To calculate *CE* (total carbon emissions) for the energy demand, the carbon emission factor for each energy source j is multiplied by its corresponding amount of usage. The calculation formula is as follows:

$$CE = \sum_i S_j \times EF_j \tag{6}$$

where S_j represents the total energy demand and EF_j is the emission factor. The present study references emission factors from the EFDB (Emission Factor Database) (Intergovernmental Panel on Climate Change 2023) of the IPCC (Intergovernmental Panel on Climate Change).

2.2.3 Elasticity Coefficient Method

The elasticity coefficient method is a predictive approach based on historical data, which can be used to forecast the output of a particular industry or region. It typically refers to the sensitivity of one economic indicator to another (Sanoh et al. 2022). We apply the method in the subsequent case study to forecast the GDP of sub-sectors within our research object under future scenarios. The elasticity coefficient γ can be calculated as:

$$\gamma = \frac{\alpha}{\beta} = \frac{\Delta I / I_0}{\Delta G / G_0} \tag{7}$$

where α represents the growth rate of industrial output value by sub-sectors, β indicates GDP growth rate, ΔI is the difference between industrial output values by sub-sectors in two adjacent years, and I_0 is the output value of in target year. Additionally, ΔG represents the change in GDP between two consecutive years. while G_0 indicates GDP for the target year. This study collects the GDP data of sub-sectors in Zhejiang Province from 2010 to 2019, combined with the elasticity coefficients method to predict industrial output value.

2.2.4 Learning Curves

According to the studies reported by McDonald and Schrattenholzer (2001) and Cai et al. (2022), the change in per unit electricity investment aligns with the learning curve model. This model explains that cost reduction is a result of accumulated experience, which can be used to illustrate the correlation between production costs and continuous cumulative outputs (Kumbaroğlu et al. 2008). Therefore, we use learning curves to predict the per unit electricity investment cost represented by *fci(k)* for different power generation methods. The learning curve is calculated as follows:

$$fc_i(k) = fc_i(1) \times N(k)^{-\varphi} \tag{8}$$

$$LR = 1 - PR = 1 - 2^{-\varphi} \tag{9}$$

where *N(k)* is the cumulative electricity installed capacity. $fc_i(1)$ represents the initial per unit capital investment cost. The elasticity of cumulative electricity installed capacity to unit capital investment cost is indicated by $-\varphi$. And *LR* indicates the learning rate. Finally, the declining rate of investment cost or technological progress rate can be determined by φ. The learning rate settings for different power generation methods in this article are shown in Table 1.

Table 1 Learning rate setting for different power generation methods

Power generation method	Learning rate (%)	References
Thermal power plant	1	Cai et al. (2022)
Wind power plant	8	Xu and Lin (2008)
Hydropower plant	1.4	Cai et al. (2022)
Solar PV	10	Cai et al. (2022)
Biomass	15	McDonald and Schrattenholzer (2001)
Nuclear power	5.8	Cai et al. (2022)
Coal with CCUS	7	Lohwasser and Madlener (2013)

2.2.5 Time Series Function

The process of electric power planning involves calculating values for installed capacity and cost using a time-series function for unspecified years.

$$\mathrm{V}_{iv} = \mathrm{V}_{fv} + \left[\mathrm{V}_{ey} - \mathrm{V}_{fy}\right] \cdot \left[\frac{\mathrm{Y}_{iy} - \mathrm{Y}_{fy}}{\mathrm{Y}_{ey} - \mathrm{Y}_{fy}}\right] \tag{10}$$

where V means values and Y means year. And, the *iy* represents the intermediate period, the value of which is to be interpolated. The *ey* is the end period and the *fy* is the first period used as the basis for the interpolation (Heaps 2022).

2.2.6 Total Power Costs

For a better understanding of the changes in power generation costs under different policies and energy regulations, this article introduces a method (Cai et al. 2022) for calculating the total power costs in the energy processing and conversion module. TC (total power costs) include CP (carbon costs), capital costs, fixed OM costs, variable costs, and E_i (fuel costs). The calculation formula is as follows:

$$TC = CP + \sum_{i}^{n} (fc_i \cdot Ca_i + foc_i \cdot Ca_i + fc_i \cdot P_i + E_i) \tag{11}$$

2.3 Scenario Settings and Description

2.3.1 Historical Scenario in Zhejiang

Here we take Zhejiang Province as an example to evaluate carbon reduction strategies at the regional level using an integrated economic, energy, and environmental analysis. Zhejiang Province is located in southeastern China, it is situated between 27°12′N~31°31′N and 118°1′E~123°47′E. There are 11 administrative units under its jurisdiction (Zhejiang Provincial Bureau of Statistics 2023). The resident population in 2019 was 63.75 million and the urbanization rate was 71.6%. The total and per capita carbon emissions in Ningbo are relatively high, at 127.63 million tonnes and 20.98 tonnes per person respectively. While Hangzhou is the provincial capital city of Zhejiang and ranks second in total emissions, its per capita carbon emissions remain reasonable due to the large population. On the other hand, Zhoushan has a smaller population but records the highest per capita carbon emissions among all cities in Zhejiang. The CO_2 emissions distribution, population distribution, and Per capita CO_2 emissions distribution can be found in Fig. 2.

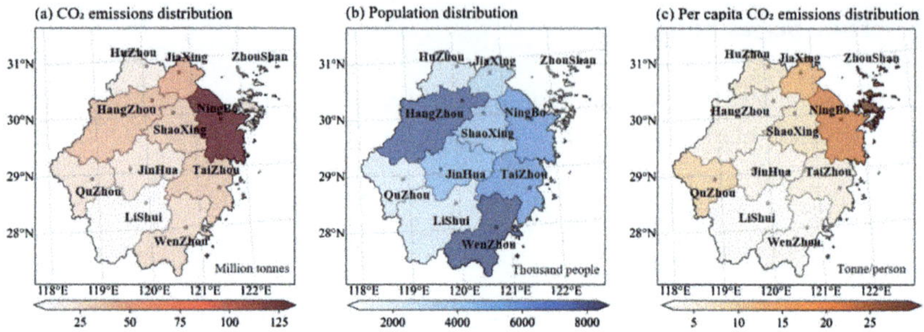

Fig. 2 Zhejiang province in 2019. **a** CO_2 emissions distribution. **b** Population distribution, and **c** Per capita CO_2 emissions distribution

We explore the industrial structure of cities with high per capita carbon emissions (Zhoushan, Ningbo, Jiaxing, Quzhou), and find that the proportion of their secondary industry in these cities is higher than that in Hangzhou.

Carbon Emission Accounts & Datasets (CEADs) (Guan et al. 2023) provide the latest energy and emission inventories for Zhejiang (Guan et al. 2021; Shan et al. 2016). We divide sub-sectors of the secondary industry in Zhejiang into I and II based on the energy consumption per unit of GDP which is calculated by the 2019 energy inventory of Zhejiang in CEADs and sub-sectors' GDP released by the Zhejiang Provincial Bureau of Statistics. According to industrial classification for national economic activities (GB/T 4754-2017) (Ministry of Civil Affairs of the People's Republic of China 2023), the corresponding relationships between sub-sectors and code are shown in Table 2.

Table 2 Sub-sectors of the secondary industry in Zhejiang Province

Classification	Sub-sectors	Code
I	Raw Chemical Materials and Chemical Products	C26
	Smelting and Pressing of Ferrous Metals	C31
	Production and Supply of Electricity and heat	D44
	Papermaking and Paper Products	C22
	Electric Equipment and Machinery	C38
	Textile Industry	C17
II	Petroleum Processing and Coking	C25
	Transportation Equipment	C37
	Smelting and Pressing of Nonferrous Metals	C30
	Medical and Pharmaceutical Products	C27
	Ordinary Machinery	C34
	Equipment for Special Purposes	C35
	Electronic and Telecommunications Equipment	C39
III	Other sub-sectors	

Additionally, comparing the carbon emissions of different industries in Zhejiang Province simulated by LEAP and offered by CEADs in 2019 will lend greater credibility to the model. The error is small, as indicated by R (Correlation Coefficient) of 0.99 and MAE (Mean Absolute Error) of 1.668. The calculation methods for the correlation coefficient and mean absolute error are as follows. Detailed data can be found in Table 3. Consequently, the model is suitable for subsequent analysis. Based on the correct results of the model, multiple comprehensive scenarios including the BAU scenario (Business As Usual), LCE scenario (Low Carbon Emission), CLT scenario (Clean Transformation), and CCUS scenario (Carbon Capture, Utilization, and Storage) are established.

Table 3 Carbon emissions simulated by LEAP and in CEADs

Sectors	Code	CEADs	LEAP
Farming, Forestry, Fishery and Water Conservancy (FFAFW)	A	7.12	7.53
Textile Industry	C17	2.51	2.90
Papermaking and Paper Products	C22	0.91	1.30
Petroleum Processing and Coking	C25	6.56	12.63
Raw Chemical Materials and Chemical Products	C26	2.81	6.90
Medical and Pharmaceutical Products	C27	0.16	0.36
Smelting and Pressing of Ferrous Metals	C31	13.74	16.40
Smelting and Pressing of Nonferrous Metals	C30	0.74	6.84
Ordinary Machinery	C34	1.02	1.22
Equipment for Special Purposes	C35	0.19	0.29
Transportation Equipment	C37	0.76	1.68
Electric Equipment and Machinery	C38	0.61	0.72
Electronic and Telecommunications Equipment	C39	0.21	0.22
Production and Supply of Electricity and Heat	D44	238.01	240.1
Construction	E	5.31	7.28
Transportation, Storage, Post, and Telecommunication Services (TSPTS)	G	27.37	29.39
Wholesale, Retail Trade, and Catering Services (WRTCS)	F	4.36	4.65
Urban	–	10.34	9.22
Rural	–	7.44	4.88

2.3.2 Business as Usual Scenario

The baseline scenario involves making minor adjustments to technological progress and energy structure. The case is aligned with Zhejiang Province's 14th Five-Year Plan to ensure that established economic development goals are met through the implementation of relevant policies. The average annual growth rate of regional GDP will be 5.5% by 2025 (The People's Government of Zhejiang Province 2021). Basic economic and population parameters as shown in Table 4. We make a population forecast for Zhejiang Province referring to (Jiang et al. 2014), and combine it with the population growth trend of the middle migration plan and a time-series function. It is predicted that there will be negative population growth starting from 2033. According to Zhejiang Province's 14th Five-Year Plan for National Economic and Social Development (The People's Government of Zhejiang Province 2022), urbanization will continue to improve and reach the target of 75% by 2035.

The primary focus of this article is on improving energy activities within Zhejiang Province's second industry by controlling traditional high-energy-consuming industries in terms of energy intensity and activity level, promoting the development of new manufacturing industries, and setting future values in sub-sectors based on the elasticity coefficient. For Industry I, the annual reduction rate of energy intensity is 2% and the activity level is represented by the annual output growth rate, which is derived from the sub-sectors' GDP. The setting method is as follows: (i) Collecting GDP growth rates in Zhejiang Province and sub-sectors' growth rates from 2010 to 2019; (ii) Calculating the elasticity coefficient of all sub-sectors during the five years and selecting the median within the range after excluding outliers; (iii) Setting the median as the growth rate in 2020–2025 and decreasing it continuously in subsequent years. The parameters for energy intensity and activity level in the second industry are shown in Table 5.

Table 4 Basic economic and population parameters

Key parameters	Unit	2020	2025	2030	2040	2050
Total population	Million	64.68	66.72	67.75	67.68	63.85
Urbanization rate	%	72.2	72.8	74.2	77.0	78.3
Urban GDP	Billion RMB	4665.08	Growth rate 5.5%			
Rural GDP	Billion RMB	1796.25				
Agroforestry and Fishery GDP	Billion RMB	5643.99				
Construction industry GDP	Billion RMB	2039.00				

Table 5 Energy intensity and energy activity parameters of sub-sectors in the BAU scenario

Energy intensity	Sub-sectors	Annual output growth rate			
		2020–2025 (%)	2026–2030 (%)	2031–2040 (%)	2041–2050 (%)
I Growth (−2%)	C26	5.30	4.10	3.50	3.10
	C31	3.32	2.58	2.20	1.98
	D44	2.65	2.06	1.75	1.57
II Growth (−0.5%)	C22	3.04	2.37	2.01	1.82
	C38	6.47	4.72	4.01	3.01
	C17	0.48	0.37	0.32	0.28
	C25	0.48	0.37	0.32	0.28
	C30	2.28	1.78	1.51	1.36
	C38	2.49	1.82	1.55	1.16
	C37	3.81	2.78	2.36	1.77
	C27	1.66	1.21	1.03	0.77
	C34	6.54	4.70	4.00	3.00
	C35	0.01	0.01	0.01	0.00
	C39	4.01	2.93	2.49	1.86

2.3.3 Low Carbon Emission Scenario

The low carbon scenario involves the implementation of advanced industrial technologies with reduced carbon emissions. In comparison to the BAU scenario, there is an annual reduction rate of 2.5% in energy intensity for industry I. Additionally, high-energy-consuming industries are limited to a growth rate of 3%, while low-energy-consuming industries are limited to a growth rate of 4.8%. The detailed parameters are shown in Table 6. The energy conversion process in the LCE scenario aligns with the BAU scenario and the time series function is used to calculate the power transmission and distribution loss. Besides, the loss rate in 2020 was 3.7%, and it is assumed to be 3.1% in 2050.

2.3.4 Clean Transformation Scenario

In the energy conversion section, this article focuses on studying the impact of changes in energy generation structure on the environment. The overall installed capacity of power plants will continue to increase from 2020 to 2050. According to the 14th Five-Year Plan for Energy Development in Zhejiang Province (The People's Government of Zhejiang Province 2022), the proportion of coal-fired power generation will decrease from 46.7 to 40%, which means coal-fired thermal power generation will gradually phase out in the future. Meanwhile, there will be a strong development of eco-friendly renewable energy, promoting centralized photovoltaic

Table 6 Energy intensity and energy activity parameters of sub-sectors in the LCE scenario

Energy intensity	Sub-sectors	Annual output growth rate			
		2020–2025	2026–2030	2031–2040	2041–2050
I	C26	2.89	2.23	1.91	1.69
	C31	1.81	1.41	1.20	1.08
	D44	1.45	1.13	0.96	0.86
Growth (−2.5%)	C22	1.66	1.29	1.10	0.99
	C38	1.36	0.99	0.85	0.63
	C17	0.26	0.20	0.17	0.15
II	C25	0.35	0.27	0.23	0.20
	C30	1.65	1.29	1.10	0.99
	C38	1.80	1.31	1.12	0.84
Growth (−0.5%)	C37	2.77	2.02	1.72	1.29
	C27	1.20	0.87	0.74	0.56
	C34	4.76	3.42	2.91	2.18
	C35	0.01	0.01	0.01	0.00
	C39	2.91	2.12	1.81	1.35

development and focusing on building million-kilowatt-level offshore wind power bases. It is estimated that by 2025, the photovoltaic installed capacity will reach up to 27.62 GW, and wind power installed capacity will exceed 6.41 GW in Zhejiang. Figure 3a demonstrates the structure and installed capacity parameters of power generation methods in the energy conversion section in the CLT scenario.

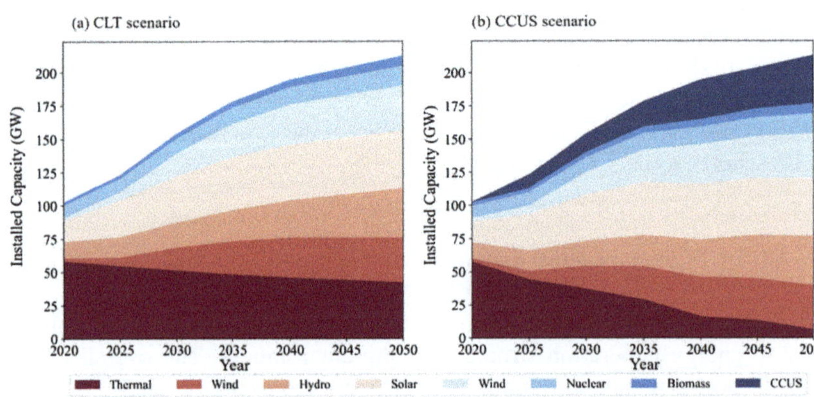

Fig. 3 Installed capacity of different power generation methods in **a** CLT scenario and **b** CCUS scenario

2.3.5 CCUS Scenario

Developing CCUS technologies to gradually innovate traditional coal-fired power generation is effective and will play a decisive role in achieving carbon reduction. Because CCUS technology is expected to capture 90% of CO_2 emissions from coal-fired power plants, resulting in a significant reduction in the emission factor (Liu et al. 2022a, b). In addition to outlining the 14th Five-Year Plan for Energy Development in Zhejiang Province, this article presents an alternative scenario for CCUS power generation. In this scenario, the planned installed capacity for CCUS power generation is in Fig. 3b. The setting of the power cost data is referenced from Zheng et al. (2020).

3 Energy Demand, Carbon Emissions, and Power Cost Analysis in Different Scenarios

This chapter focuses on the long-term trends in energy consumption, carbon emissions, and power costs, with a particular emphasis on the role of energy intensity reduction, changes in the power generation structure, and the implementation of Carbon Capture, Utilization, and Storage (CCUS) technologies. The results highlight how different scenarios impact energy demand, carbon emissions, and the economic viability of power generation technologies. Below, we present the findings from each of the specific analyses.

3.1 Total Energy Demand in Different Scenarios

Figure 4 shows the predicted total energy demand under three scenarios: LCE (1), BAU (2), and CLT (3). It is obvious that if no additional measures are taken to reduce energy intensity and development follows the conventional path of social and technological progress, the total energy demand in Zhejiang Province will continue to increase. This is due to the increasing demand for energy from industrial development, which mainly affects fossil fuel consumption. At the same time, higher end-use demands mean more losses during energy transmission. The CLT scenario mainly relies on changes in power generation structure and transportation technology innovation, with overall energy demand remaining unchanged.

However, the proportion of fossil fuel consumption decreases from 88.08 to 73.72% from 2020 to 2050, which will have positive effects on air pollution control and carbon reduction efforts. The LCE scenario mainly depends on further reducing industrial energy intensity. Although total energy demand increases year by year in LCE, it remains lower than that of other scenarios. Total energy demand analysis from three scenarios indicates that reducing energy intensity is more effective than

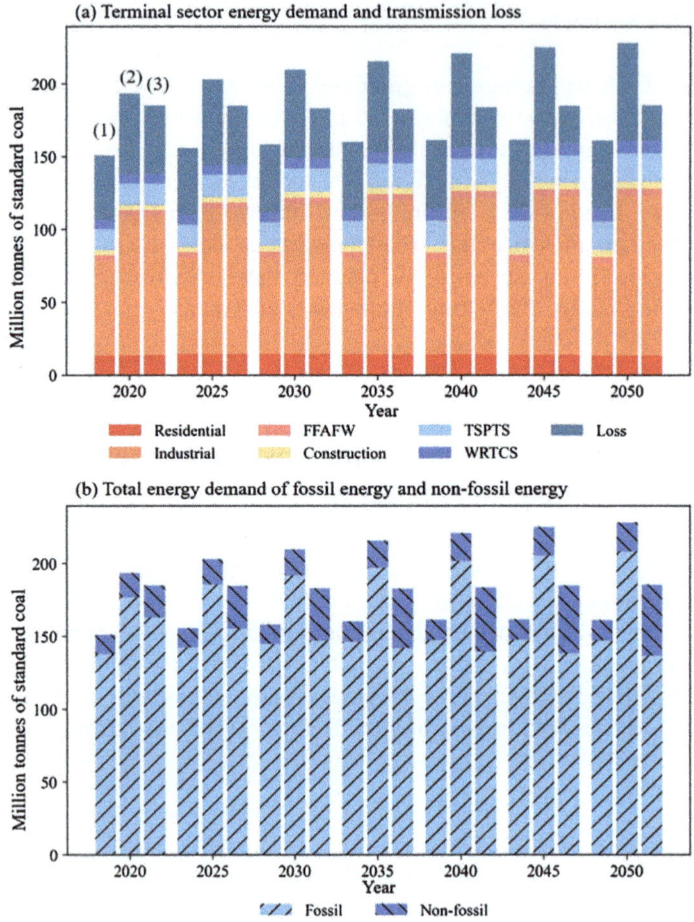

Fig. 4 Total energy demand in different scenarios: **a** Terminal sector energy demand and transmission loss and **b** Fossil energy and non-fossil energy

changing the power generation structure and innovating transportation technology from the perspective of overall energy consumption.

3.2 Carbon Emissions in Different Scenarios

Figure 5a shows the total carbon emissions in four scenarios. In the BAU (1) scenario, the total carbon emissions are on an upward trend, reaching 487.77 million tonnes in 2050, with electricity generation and industrial sectors being the main sources of carbon emissions. In the LCE (2) scenario, Zhejiang Province will achieve peak carbon emissions in 2044 at a total of 346.59 million tonnes, and carbon emissions will decrease to 344.50 million tonnes in 2050.

Although reducing energy intensity can effectively reduce CO_2 emissions, Zhejiang Province may find it challenging to achieve the goal of peaking carbon emissions by 2030 which was outlined in 2020. The total CO_2 emissions in CLT (3) and CCUS (4) scenarios will continue to decline and reach respectively at 293.59 million tonnes and 210.76 million tonnes in 2050 which are both lower than that in the LCE scenario. Figure 5b represents the changes in carbon emissions of per unit energy consumption. Reducing energy intensity may not invariably lower the emissions per unit energy consumption, whereas adjusting the power generation structure and promoting CCUS prove to be effective in Zhejiang Province. In 2050, the reduction will be 1.58 and 1.13 tonnes of CO_2/tonne of coal equivalent respectively in the scenarios of CLT and CCUS.

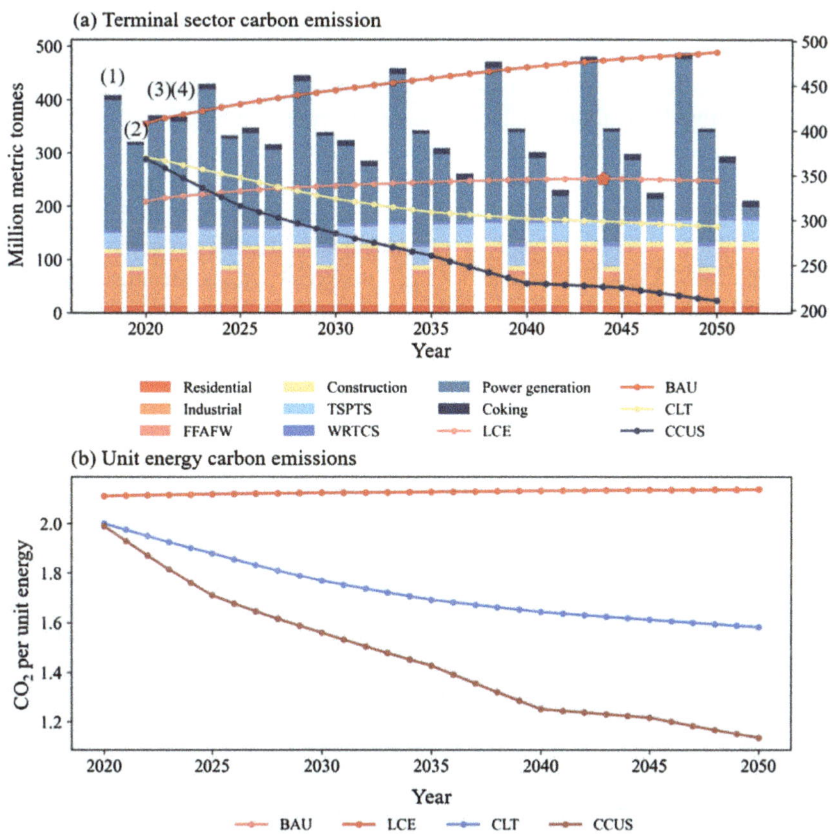

Fig. 5 Carbon emissions in different scenarios: **a** Terminal sector carbon emissions (The left coordinates is for the bar displayed every 5 years, while the right for the curve displayed every 1 year). **b** Unit energy carbon emissions

3.3 Power Cost Analysis

Analyzing the power cost is crucial in understanding the economic efficiency of various energy and power generation technologies (IEA and NEA 2020). The report points out that with the increase in carbon emission costs, CCUS power generation technology will also become competitive. The analysis serves as a foundation for optimizing power generation structures and ensuring economic benefits.

According to the annual transaction data in the national carbon market platform of the Shanghai Environment and Energy Exchange, the average price of carbon tax is 50 RMB/tonne. This article proposes a growth rate of 3.6% for the carbon tax, which is based on the average growth rate observed between 2020 and 2022. Based on Cai et al. (2022), another growth rate is set to 7.2%. Figure 6 presents the total cost in CLT and CCUS scenarios under different carbon tax growth rates.

With a growth rate of 3.6% in Fig. 6a, the total costs in CLT (1) and CCUS (2) scenarios in 2050 are 129.28 and 159.18 billion RMB respectively. In this hypothetical scenario, CCUS power generation technology does not have an economic advantage before 2050 and the highest proportion of CCUS cost is 34.67% (Nordhaus 2017) claimed that although the estimates carry great uncertainties, global social costs could exceed 100 USD/tonne of CO_2. When the carbon tax growth rate reaches 7.2%, the carbon tax price will be 431 RMB/tonne in 2050 which is lower than 100 USD/tonne. The power cost in scenario CCUS (2) will be lower than that in CLT (1) before 2050. This finding is consistent with Cai et al. (2022). The estimated costs are 167.77 billion RMB in CCUS and 171.45 billion RMB in CLT. Besides, the highest proportion of CCUS cost decreases to 33.74%. Despite its higher initial construction costs, CCUS proves to be a more economically advantageous option compared to traditional thermal power generation by 2050.

4 Key Findings and Recommendations for Regional Carbon Mitigation

This section summarizes the major findings from the scenario-based analysis of Zhejiang Province's energy transition and carbon reduction pathways. It also outlines practical policy recommendations aimed at promoting effective and economically viable emission reduction strategies at the regional level.

4.1 Summary of Scenario-Based Key Findings

The scenario analysis indicates that Zhejiang Province is unlikely to achieve its 2030 carbon peaking goal without substantial interventions to reduce energy consumption or improve the power generation structure. Under the BAU scenario, both total

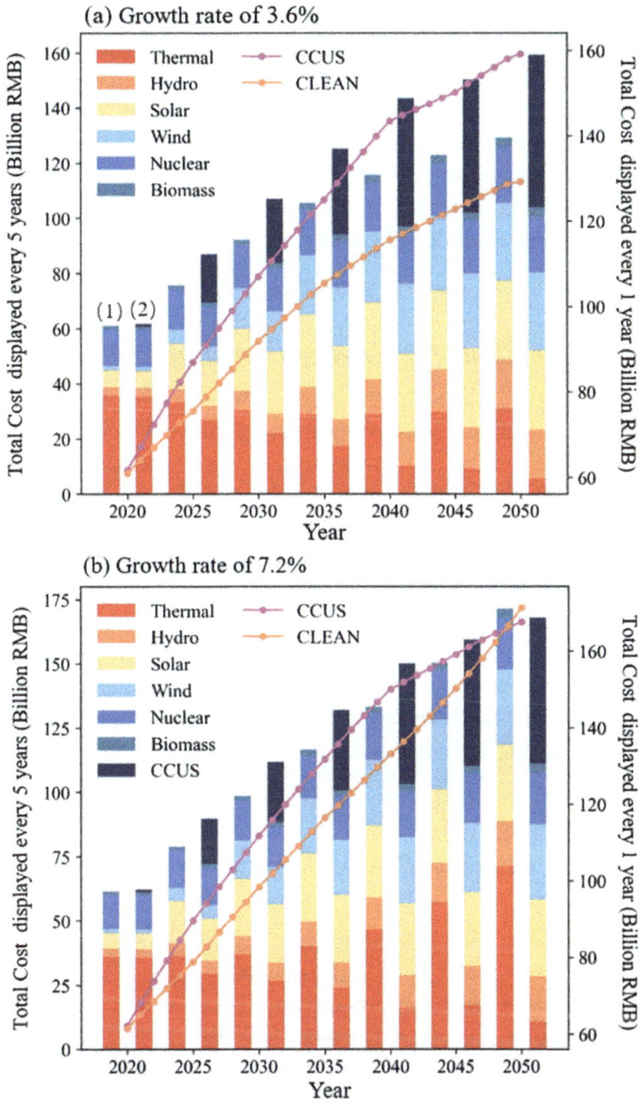

Fig. 6 Total cost in CLT and CCUS: **a** Growth rate of 3.6%. **b** Growth rate of 7.2% (The left coordinate is for the bar, while the right for the curve)

energy demand and carbon emissions increase significantly, reaching 228.06 million tonnes of coal equivalent and 487.76 million tonnes of CO_2 by 2050, respectively. Although transforming the power generation structure and improving transportation technology can reduce fossil fuel dependency from 88.08% in 2020 to 73.72% in 2050, these efforts alone are insufficient. The LCE scenario demonstrates that reducing energy intensity is more effective in managing total energy demand and can delay emissions growth. Specifically, carbon emissions peak in 2044 at 346.59 million tonnes. The CLT and CCUS scenarios further highlight the benefits of structural reform and advanced technology, with carbon intensity per unit of energy reduced to 1.58 and 1.13 tonnes of CO_2 per tonne of coal equivalent by 2050, respectively. However, the economic viability of CCUS largely depends on future carbon pricing. At a carbon tax growth rate of 7.2%, CCUS becomes more cost-effective than the clean transformation path, emphasizing the critical role of market-based mechanisms in enabling deep decarbonization.

4.2 Policy Implications for Energy Consumption Control

To effectively control energy consumption and achieve long-term carbon mitigation, Zhejiang must prioritize energy-saving strategies and structural adjustments. Advancing energy efficiency through technological innovation and reducing the energy intensity of industrial activities should be central to future policy design. Particular attention should be paid to high-energy-consuming sectors, which require strict management measures to phase out outdated production capacity and accelerate industrial upgrading. Additionally, reducing transmission and distribution losses in the energy supply chain is essential to improve overall system efficiency. Managing the growth of industrial activity and implementing differentiated control strategies across sub-sectors will also help contain rising energy demand while supporting sustainable economic development.

4.3 Pathways to Reshape Power Generation Structure

Achieving significant emission reductions requires a deep restructuring of the power generation system toward low-carbon and renewable sources. Zhejiang should accelerate the deployment of wind power, photovoltaic energy, and other clean technologies to increase the share of non-fossil energy in its power mix. This transition must be supported by proactive policy incentives and long-term planning to ensure large-scale, diversified, and stable development of renewable energy capacity. Furthermore, CCUS technology presents an important pathway for decarbonizing existing fossil fuel-based power generation. With adequate investment and a supportive carbon pricing regime, CCUS can offer a cost-effective solution for reducing emissions in hard-to-abate sectors. To ensure the success of this strategy, the government should establish a comprehensive market environment that includes carbon tax mechanisms, innovation support, and regulatory certainty for low-carbon technologies.

Author's note: This chapter is a revised and expanded version of the author's article, "An integrated economic, energy, and environmental analysis to optimize evaluation of carbon reduction strategies at the regional level: A case study in Zhejiang, China" published in *Journal of Environmental Management*, Vol. 351 (2024).

References

Cai L, Duan J, Lu X, Luo J, Yi B, Wang Y, Jin D, Lu Y, Qiu L, Chen S, Zhang H, Wang L (2022) Pathways for electric power industry to achieve carbon emissions peak and carbon neutrality based on LEAP model: a case study of state-owned power generation enterprise in China. Comput Ind Eng 170:108334. https://doi.org/10.1016/j.cie.2022.108334

Chen S, Liu P, Li Z et al (2020) Low carbon transition pathway of power sector with high penetration of renewable energy. https://doi.org/10.1016/j.rser.2020.109985

Fang K, Li C, Tang Y, He J, Song J (2022) China's pathways to peak carbon emissions: new insights from various industrial sectors. Appl Energy 306:118039. https://doi.org/10.1016/j.apenergy.2021.118039

Guan S, Shan Y, Liu Z (2023) Carbon emission accounts & datasets. https://www.ceads.net.cn/

Guan Y, Shan Y, Huang Q, Chen H, Wang D, Hubacek K (2021) Assessment to China's recent emission pattern shifts. Earth's Future 9:e2021EF002241. https://doi.org/10.1029/2021EF002241

Guo Z, Zhang X, Ding Y, Zhao X (2021) A forecasting analysis on China's energy use and carbon emissions based on a dynamic computable general equilibrium model. Emerg Mark Financ Trade 57(3):727–739. https://doi.org/10.1080/1540496X.2019.1597704

Heaps CG (2022) LEAP: the low emissions analysis platform (Version 2022.1.90) [Computer software]. Stockholm Environment Institute. https://leap.sei.org

International Energy Agency (2020) Projected costs of generating electricity 2020. IEA. https://www.iea.org/reports/projected-costs-of-generating-electricity-2020

International Energy Agency (2022) Global energy review: CO_2 emissions in 2021. IEA. https://www.iea.org/reports/global-energy-review-co2-emissions-in-2021-2

Intergovernmental Panel on Climate Change (2021) Summary for policymakers. In: Masson-Delmotte V, Zhai P, Pirani A, Connors SL, Péan C, Berger S, Caud N, Chen Y, Goldfarb L, Gomis MI, Huang M, Leitzell K, Lonnoy E, Matthews JBR, Maycock TK, Waterfield T, Yelekçi O, Yu R, Zhou B (eds) Climate change 2021: the physical science basis. Contribution of Working Group I to the Sixth Assessment Report of the Intergovernmental Panel on Climate Change. Cambridge University Press, pp 3–32. https://doi.org/10.1017/9781009157896.001

Intergovernmental Panel on Climate Change (2023) Emission factor database. https://www.ipcc-nggip.iges.or.jp/EFDB/main.php/

Jiang J, Wang W, Yin W, Yao Y (2014) Study on future population prediction and development trends in Zhejiang Province. http://zjjcmspublic.oss-cn-hangzhou-zwynet-d01-a.internet.cloud.zj.gov.cn/jcms_files/jcms1/web3077/site/attach/0/e2e0ccc37edd4265bdce90b7ec2bd841.pdf

Kumbaroğlu G, Karali N, Arıkan Y (2008) CO_2, GDP and RET: an aggregate economic equilibrium analysis for Turkey. Energy Policy 36(7):2694–2708. https://doi.org/10.1016/j.enpol.2008.03.026

Lee C-Y, Huh S-Y (2017) Forecasting new and renewable energy supply through a bottom-up approach: the case of South Korea. Renew Sustain Energy Rev 69:207–217. https://doi.org/10.1016/j.rser.2016.11.173

Li L, Zhang Y, Zhou T et al (2022) Mitigation of China's carbon neutrality to global warming. Nat Commun 13:5315. https://doi.org/10.1038/s41467-022-33047-9

Liang S, Lin X, Liu X, Pan H (2022) The pathway to China's carbon neutrality based on an endogenous technology CGE model. Int J Environ Res Public Health 19(10):6251. https://doi.org/10.3390/ijerph19106251

Liping X, Li L (2008) Study on the cost trends of wind power in China based on the learning curve. Electr Power Sci Eng 24*(3):1. https://kns.cnki.net/kcms2/article/abstract?v=QOR9nnE_NmIlPUDCgpPoUwU7Gj_AmV0_hU6B90TQBHJJ9R58MJzbUfkcDzmfYAXmA92JK-sITgpYeCm6KFbUyhyiVPMFqrqFqSTw0l8QJMEY0Hg1qLWHqj9J6fX41OQ6jgf9Qv9ObRO769TwtUnON1vZQ3uJjRC8&uniplatform=NZKPT

Liu G, Cai B, Li Q, Zhang X, Ouyang T (2022a) China's pathways of CO_2 capture, utilization and storage under carbon neutrality vision 2060. Carbon Manag 13(4):435–449. https://doi.org/10.1080/17583004.2022.2117648

Liu Y, Chen S, Jiang K, Kaghembega WS-H (2022b) The gaps and pathways to carbon neutrality for different type cities in China. Energy 244:122596. https://doi.org/10.1016/j.energy.2021.122596

Liu Z, Wang F, Tang Z, Tang J (2020) Predictions and driving factors of production-based CO_2 emissions in Beijing, China. Sustain Cities Soc 53:101909. https://doi.org/10.1016/j.scs.2019.101909

Lohwasser R, Madlener R (2013) Relating R&D and investment policies to CCS market diffusion through two-factor learning. Energy Policy 52:439–452. https://doi.org/10.1016/j.enpol.2012.09.061

Maduekwe M, Akpan U, Isihak S (2020) Road transport energy consumption and vehicular emissions in Lagos, Nigeria: an application of the LEAP model. Transp Res Interdiscip Perspect 6:100172. https://doi.org/10.1016/j.trip.2020.100172

Ma X, Peng T, Zhang Y, Wang L, Pan X (2023) Accelerating carbon neutrality could help China's energy system align with below 1.5°C. J Environ Manag 337:117753. https://doi.org/10.1016/j.jenvman.2023.117753

McDonald A, Schrattenholzer L (2001) Learning rates for energy technologies. Energy Policy 29(4):255–261. https://doi.org/10.1016/S0301-4215(00)00122-1

Ministry of Civil Affairs of the People's Republic of China (2023) National economic industry classification (GB/T4754-2017). https://www.mca.gov.cn/images3/www/file/201711/1509495881341.pdf

Nordhaus WD (2017) Revisiting the social cost of carbon. Proc Natl Acad Sci 114(7):1518–1523. https://doi.org/10.1073/pnas.1609244114

Sanoh O, Zhang Q, Wang D, Aurélien N (2022) Guinea's bauxite resources evaluation and forecasting using elasticity demand method. Open J Soc Sci 10(4):400–418. https://doi.org/10.4236/jss.2022.104029

Shan Y, Liu J, Liu Z, Xu X, Shao S, Wang P, Guan D (2016) New provincial CO_2 emission inventories in China based on apparent energy consumption data and updated emission factors. Appl Energy 184:742–750. https://doi.org/10.1016/j.apenergy.2016.03.073

Shi W, Ou Y, Smith SJ, Ledna CM, Nolte CG, Loughlin DH (2017) Projecting state-level air pollutant emissions using an integrated assessment model: GCAM-USA. Appl Energy 208:511–521. https://doi.org/10.1016/j.apenergy.2017.09.122

Stoltmann A, Jaskólski M, Bućko P (2019) Optimization of combined heat and power (CHP) market allocation: the case of Poland. In: IOP conference series: earth and environmental science. IOP Publishing, vol 214, p 012036. https://doi.org/10.1088/1755-1315/214/1/012036

Qi Y, Liu T, Jing L (2023) China's energy transition towards carbon neutrality with minimum cost. J Clean Prod 388:135904. https://doi.org/10.1016/j.jclepro.2023.135904

The People's Government of Zhejiang Province (2021) Zhejiang Province's 14th five-year plan for national economic and social development. https://www.zj.gov.cn/art/2025/3/4/art_1229019364_2546972.html

The People's Government of Zhejiang Province (2022) Zhejiang Province's 14th five-year plan for energy development. https://www.zj.gov.cn/art/2022/5/19/art_1229505857_2404396.html

Wang Q, Luo K, Wu C, Mu Y, Tan J, Fan J (2022a) Diurnal impact of atmospheric stability on inter-farm wake and power generation efficiency at neighboring onshore wind farms in complex terrain. Energy Convers Manage 267:115897. https://doi.org/10.1016/j.enconman.2022.115897

Wang Q, Luo K, Wu C, Zhu Z, Fan J (2022b) Mesoscale simulations of a real onshore wind power base in complex terrain: wind farm wake behavior and power production. Energy 241:122873. https://doi.org/10.1016/j.energy.2021.122873

Wang Q, Luo K, Wu C, Tan J, He R, Ye S, Fan J (2023) Inter-farm cluster interaction of the operational and planned offshore wind power base. J Clean Prod 396:136529. https://doi.org/10.1016/j.jclepro.2023.136529

Wei YM, Chen K, Kang JN, Chen W, Wang XY, Zhang X (2022) Policy and management of carbon peaking and carbon neutrality: a literature review. Engineering 14:52–63. https://doi.org/10.1016/j.eng.2021.12.018

Wu F, Geng Y, Tian X, Zhong S, Wu W, Yu S, Xiao S (2018) Responding climate change: a bibliometric review on urban environmental governance. J Clean Prod 204:344–354. https://doi.org/10.1016/j.jclepro.2018.09.067

Wu W, Zhang T, Xie X, Huang Z (2021) Regional low carbon development pathways for the Yangtze River Delta region in China. Energy Policy 151:112172. https://doi.org/10.1016/j.enpol.2021.112172

Xu L, Lin L (2008) Study on the cost trends of wind power in China based on the learning curve. https://kns.cnki.net/kcms2/article/abstract?v=QOR9nnE_NmIlPUDCgpPoUwU7Gj_AmV0_hU6B90TQBHJJ9R58MJzbUfkcDzmfYAXmA92JK-sITgpYeCm6KFbUyhyiVPMFqrqFqSTw0l8QJMEY0Hg1qLWHqj9J6fX41OQ6jgf9Qv9ObRO769TwtUnON1vZQ3uJjRC8&uniplatform=NZKPT

Yao H, Xu P, Wang Y, Chen R (2023) Exploring the low-carbon transition pathway of China's construction industry under carbon-neutral target: a socio-technical system transition theory perspective. J Environ Manage 327:116879. https://doi.org/10.1016/j.jenvman.2022.116879

Zhang CY, Yu B, Chen JM, Wei YM (2021) Green transition pathways for cement industry in China. Resour Conserv Recycl 166:105355. https://doi.org/10.1016/j.resconrec.2020.105355

Zhang S, Chen W (2022) China's energy transition pathway in a carbon neutral vision. Engineering 14:64–76. https://doi.org/10.1016/j.eng.2021.09.004

Zhejiang Provincial Bureau of Statistics (2023) Zhejiang Province statistical yearbook in 2020. http://zjjcmspublic.oss-cn-hangzhou-zwynet-d01-a.internet.cloud.zj.gov.cn/jcms_files/jcms1/web3077/site/flash/tjj/Reports1/2020-%E7%BB%9F%E8%AE%A1%E5%B9%B4%E9%89%B4/indexcn.html

Zheng B, Liu Z, Wang Z (2020) China's power industry almanac in 2020. https://data.cnki.net/yearBook/single?id=N2022080097

Decarbonization Policy Cooperation Toward a Carbon Neutral Society in China, Japan and South Korea: Toward the Establishment of a CJK Joint Carbon Market

Soocheol Lee

1 Introduction

Achieving carbon neutrality by mid-century is a global trend that no country can afford to ignore. According to the Climate Ambition Alliance, as of October 2022, 155 countries and regions have committed to carbon neutrality.

According to the International Energy Agency (IEA), China, Japan and South Korea (hereafter, CJK) will account for about 32, 2.7, and 1.8% of the 37.4 billion tons of global energy-derived carbon dioxide emissions in 2023, respectively, making these three countries alone responsible for about 36.8% of global emissions. The EU's share of global carbon dioxide emissions is about 8%, making CJK alone the world's largest carbon dioxide emitting region. It is clear that East Asian countries, especially CJK, have a tremendous responsibility to help the world move toward carbon neutrality by the middle of this century.

However, CJK are more dependent on fossil energy than other countries in the region, such as Europe, and achieving carbon neutrality by mid-century will be a very steep road. For example, the share of fossil energy in primary energy supply is about 87% in China, 85% in Japan, and 83% in Korea, according to the IEA (2023), all above the global average of 80%.

Unlike the EU, which has introduced common intra-regional energy conservation, renewable energy, and greenhouse gas targets and collaborative policies such as the EU-ETS, in the CJK, only individual country measures have been implemented, with little intra-regional policy cooperation or joint efforts. Collaborative intra-regional efforts toward carbon neutrality, as seen in the EU experience, can bring mutual

S. Lee (✉)
Faculty of Economics, Meijo University, Nagoya, Japan
e-mail: slee@meijo-u.ac.jp

W. Zhou (ed.), *Decarbonization Policy Engineering*, World Sustainability Series,
https://doi.org/10.1007/978-981-95-5953-4_8

benefits through spillover effects of low and decarbonized technologies and carbon cost savings within the region.

In this paper, we will use the IEEJ Outlook from the Institute of Energy Economics, Japan (IEEJ) to examine how much carbon dioxide can be reduced by 2050 based on current decarbonization-related policy scenarios and decarbonization-related technology development scenarios in CJK. We will then examine the institutional design and economic impact of Japan's efforts to achieve carbon neutrality as a case study. Finally, we will consider the necessity and benefits of establishing a joint carbon market and cooperating on decarbonization policies in Japan, China and South Korea in order to achieve carbon neutrality.

2 CJK's Decarbonization Outlook: Scenario Analysis of IEEJ OUTLOOK 2025

This section examines the path to decarbonization in Asia, particularly in CJK, using the IEEJ OUTLOOK 2025, published annually by the IEEJ, which uses two scenarios, a reference scenario and a technological progress scenario, to estimate energy, carbon dioxide emissions, and GDP in the world and Asian countries by 2050.

First, in the reference scenario, the future outlook is created according to past trends and current policies and technologies related to energy and the environment. Here, the effects of policies and technologies that are expected to be implemented in the future as an extension of the past are factored in. This does not mean that the current state of policies and technologies is fixed. However, it is assumed that radical energy conservation and low-carbon policies that deviate from the current trend will not be introduced.

The technological progress scenario assumes that in all countries of the world, energy conservation standards, low-carbon fuel standards, renewable energy introduction standards, feed-in tariffs, subsidies and subsidy programs, environmental taxes, emissions trading programs, and other measures that contribute to securing a stable energy supply, combating climate change, and strengthening air pollution control, etc., are implemented. It is assumed that energy and environmental policies will be strongly implemented and that they will be maximally successful. It is also assumed that advanced technologies on the demand and supply sides of energy will be introduced to the maximum extent possible, based on their applicability and acceptability in the real world. However, the energy demand-supply structure can be greatly affected not only by energy use technologies and energy and environmental policies, but also by social and economic factors such as population and economic growth, as well as energy prices, etc. Therefore, economic growth and population are assumed to be common to both the reference scenario and the technological progress scenario. The same assumptions are made for economic growth and population in both the reference scenario and the technological progress scenario.

Based on the above assumptions, the IEEJ OUTLOOK 2025 GDP growth forecast for the period 2022–2050 places the Asian region 0.9 percentage points above the global average annual GDP growth rate of 2.6%, positioning the region as a driver of global economic growth in the future (Table 1).

The outlook for carbon dioxide emissions from 2022 to 2050 according to IEEJ OUTLOOK 2025 is as follows: First, under the Reference Scenario, energy-derived carbon dioxide emissions are expected to remain at the current level in 2050 for both the world and Asia (Table 2). Within Asia, carbon dioxide emissions are expected to continue to decline, with Japan's emissions decreasing by 40.9% in 2050 compared to 2022 emissions, China's by 41.7%, and South Korea's by 22.0%. However, all of these countries are far from carbon neutral. On the other hand, under the Technological Progress Scenario, carbon dioxide emissions are expected to decrease significantly in most Asian countries. In particular, as shown in Table 2, CJK are expected to decrease their carbon dioxide emissions significantly by 86.2%, 92.7%, and 82.0%, respectively, in 2050 compared to 2022.

The above IEEJ OUTLOOK2025 shows that carbon dioxide emissions can be significantly reduced in Asia if progress is made in decarbonizing technologies and the diffusion of related energy-related facilities and equipment on both the demand and supply sides is expanded. However, assuming that the amount of carbon dioxide emissions absorbed by forests is at the level of 5% of current emissions, the study also indicates that carbon dioxide emissions under the technological progress scenario may still bc insufficient to achieve carbon neutrality by mid-century. In other words, achieving carbon neutrality by mid-century will require more decarbonizing technological innovations and related energy and equipment deployment than indicated in the IEEJ OUTLOOK2025 Technology Progress Scenarios, as well as the design and implementation of decarbonization policies that can stimulate these innovations and equipment deployment.

Table 1 GDP growth prospects in CJK by IEEJ OUTLOOK 2025 (In billions of dollars at 2015 prices)

	1990	2010	2022	2030	2040	2050	2050/2022 (annual average, %)
World	35,916	64,773	89,488	110,594	143,985	184,098	2.6
Asian	6,690	17,895	30,825	41,436	59,046	81,326	3.5
China	1,027	7,554	16,325	22,246	31,544	42,922	3.5
Japan	3,510	4,219	4,530	4,846	5,254	5,659	0.8
Korea	402	1,261	1,741	2,049	2,439	2,769	1.7

Note 2050/2022(%) is the average annual growth rate during the relevant period. *Source* Actual figures from World Bank (2023); forecasts from Institute of Energy Economics, Japan (IEEJ 2025)

Table 2 Outlook of energy-derived carbon dioxide emissions in CJK by IEEJ OUTLOOK 2025 (Unit: million ton CO_2)

	Scenarios	1990	2010	2022	2030	2040	2050	2050/2022 annual growth rate (%)
World	Reference	20,522	30,703	34,116	32,912	32,464	32,662	−0.2 (−4.3)
	Technological progress				29,605	20,487	12,924	−3.4 (−62.1)
Asian	Reference	4,700	13,032	17,077	16,717	16,388	16,210	−0.2 (−5.7)
	Technological progress				14,897	9,577	5,841	−3.8 (−65.8)
China	Reference	2,195	8,110	10,613	9,469	7,767	6,188	−1.9 (−41.7)
	Technological progress				8,518	3,953	1,465	−6.8 (−86.2)
Japan	Reference	1,056	1,137	974	768	664	576	−1.9 (−40.9)
	Technological progress				667	372	71	−8.9 (−92.7)
Korea	Reference	208	528	549	539	488	428	−0.9 (−22.0)
	Technological progress				468	253	99	−5.9 (−82.0)

Note 1: Figures in parentheses are the percentage increase from 2022 to 2050
2: The actual values for 1990, 2010, and 2022 are the same as those for the reference and technological progress scenarios
Source Actual figures are from IEA (2023); forecasts are from the Institute of Energy Economics, Japan (IEEJ OUTLOOK2025)

3 Institutional Design and Economic Impact Analysis Toward Carbon Neutrality: A Case Study of Japan

3.1 Setting a Decarbonization Policy Scenario

The above discussion shows that achieving carbon neutrality in the middle of this century is a steep road ahead. This section discusses the design of decarbonization policies (decarbonization policy scenarios) necessary for Japan to achieve carbon neutrality in 2050 and the analysis of the economic impact if these policies are implemented in reality, based on a paper (Kato et al. 2023) in which the author participated as an organizer. In this paper, the "decarbonization policy scenarios" are the Japanese government's 2030 GHG reduction target (46% reduction from 2013 levels) and the authors' own set of low- and decarbonization policies needed to achieve carbon neutrality by 2050 (Table 3). The simulations with policy scenario inputs are based on the E3ME macro-econometric model.

The decarbonization policy scenarios are then based on the following two scenarios. First, Decarbonization Policy Scenario 1 (S1) assumes that the power supply plan in 2030 follows the 6th Basic Energy Plan of the Japanese government

Table 3 Japan's decarbonization policy design to achieve carbon neutrality by 2050

Category		Scenario setting
Carbon tax		The tax rate applies to all sectors and is US$50/$CO_2$t in 2021 Increase to US$410/CO_2t in 2040 in stages Maintains US$410/$CO_2$t from 2040 to 2050
Power sector	Nuclear power	S1: Share of electricity generation: 20% in 2030 and 10% in 2050 are assumed S2: Phased out by 2040
	Coal-fired thermal power	S1: Assumes 19% share of power generation in 2030 S2: Phased out by 2040
	Renewable energy	S1: Japanese government's power mix plan for 2030 and 2050 S2: Endogenously determined by the model
Transport sector	Passenger car sales	Gasoline/diesel vehicle sales regulations from 2035
	EV Subsidy	Maintain purchase subsidy through 2025
	Biofuel mandates	Mandate the percentage of biofuels used in cargo vehicles and aircraft
Iron and steel sector		Zero emissions from blast furnaces by 2050
Building air conditioning		Phase out fossil fuel boilers by 2050

Note S1 and S2 represent policy scenarios 1 and 2, respectively. Unannotated sections apply to both S1 and S2
Source Compiled from Kato et al. (2023)

(METI 2021) and that the power supply plan in 2050 follows METI's assumptions. Under this scenario, the share of nuclear power in the power supply decreases from 6.2% in 2019 to 20% in 2030 and 10% in 2050. Meanwhile, renewable energy expands from 19.8% in 2019 to 38% in 2030 and 60% in 2050.

Decarbonization Policy Scenario 2 (S2) relies on the study's own setting, independent of the Japanese government's Power Supply Plan and Basic Energy Plan described above. Nuclear power plants are phased out by 2040, with no new construction after 2018, and phase-out is based on the number of years of operation (40 years). Similarly, coal-fired power plants will be phased out between 2021 and 2040, in order of oldest operating history. Other power sources are determined endogenously in the model. Thus, the simulation of the decarbonization policy scenario employed in this study consists of two cases: one in which the power generation sector follows the government's 2030 and 2050 power supply plans (S1) and one unique to this study (S2).

The common decarbonization policy design for the above two decarbonization policy scenarios to achieve carbon neutrality in 2050 is summarized in Table 3. First, a new carbon tax is introduced in addition to the current "global warming tax," with the tax rate increasing proportionally from US$50/$CO_2$t in 2025–2040 and remaining constant at US$410/$CO_2$t until 2050. Revenues generated from the carbon tax will be used to finance low-carbon and decarbonization investments, renewable energy promotion, and costs associated with phasing out thermal power generation. The use of this tax revenue is in accordance with the principle of tax revenue neutrality.

In the transportation sector, passenger cars that rely solely on internal combustion engines should not be sold starting in 2035 in order to achieve a 100% sales rate of new electric vehicles by 2035. The government will subsidize between US$8,000 and US$13,000 per electric vehicle, depending on battery capacity at the time of purchase, to be implemented by 2025. The use of biofuels is mandatory for cargo vehicles and aircraft until 2050.

In the industrial sector, the policy was developed only for the steel industry, which accounts for 12.7% of the economy-wide energy-related CO_2 emissions and about 40% of total emissions of industries. For the iron and steel industry, a regulatory scenario was set to achieve zero CO_2 emissions by 2050, assuming a method to reduce CO_2 emissions in the blast furnace sector by 100% by 2050. Finally, in the building sector (including households), it was decided to phase out fossil fuel boilers by 2050.

The IEEJ OUTLOOK reference scenario was used as the baseline scenario against which the above decarbonization policy scenarios were compared in the model simulations to 2050.

3.2 Carbon Dioxide Emission Pathways Under the Decarbonization Policy Scenario

The above two decarbonization policy scenarios for achieving carbon neutrality by 2050 (hereafter S1 and S2) and the baseline scenario (i.e., the IEEJ OUTLOOK reference scenario) were fed into the E3ME model and the carbon emission pathway to 2050 was Simulated.

First, the GHG emission pathway in the baseline scenario is estimated to be 34.7% lower in 2050 than in 2019, according to the IEEJ OUTLOOK reference scenario. The paper estimates that the GHG emission pathway in S1 will continue to decrease to 46% of 2013 emissions by 2030 and to about 65 million tons of CO_2 in 2050. The remaining 65 million tons of GHG emissions in 2050 will be absorbed by land use, land use change, forestry (LULUCF), etc. S2 then estimates that GHG emissions will continue to decline to 51% of 2013 levels by 2030 and to about 65 million tons of CO_2 by 2050, as in S1.

3.3 Economic Impacts of Carbon Neutrality

Based on the E3ME model simulations, implementation of policy scenarios S1 and S2 is expected to increase GDP by about 2% by 2030 and by another 3% by 2050 relative to the baseline scenario, as shown in Table 4. Both policy scenarios S1 and S2 are estimated to have a positive impact on GDP despite higher carbon costs, due to increased demand for low- and decarbonization-related investments and subsequent reductions in generation costs from renewable energy penetration. The increase in GDP in both scenarios relative to the baseline scenario is mainly due to the low and decarbonization-related investment demand needed to achieve carbon neutrality through the revitalization of various sectors, including power generation, industry, and transportation, as shown in Table 4, where investment demand leads to a marked increase of 6–7%.

And while the introduction of decarbonization policies will lead to higher energy costs by 2030, especially the cost of electricity generation, leading to various cost increases, the cost of generating renewable energy as an alternative to fossil fuels is projected to drop significantly by 2050, which will have a downward energy price effect.

In addition, increased demand for low-carbon and decarbonization investment will lead to the creation of new employment opportunities, resulting in an estimated 1.4–1.5% increase in employment by 2050 relative to the baseline scenario under both policy scenarios. Another major economic benefit of achieving carbon neutrality by 2050 is an improved trade balance. Exports are projected to decline slightly (−0.2 to −0.3%), mainly due to lower exports of certain capital goods. However, an improvement in the trade balance due to a significant decrease in demand for fossil energy imports is expected to contribute to GDP growth. For example, Japan's fossil energy imports, including crude oil, LNG, and coal, amounted to 22 trillion yen in 2022, equivalent to about 20% of total imports (Agency for Natural Resources and Energy 2024). Therefore, reducing this import value is important for improving the trade balance.

Table 4 Macroeconomic impacts of achieving carbon neutrality by 2050 (difference from baseline)

	2030		2050	
	S1	S2	S1	S2
GDP	2.0%	1.9%	3.1%	3.1%
Consumption	0.5%	0.5%	2.3%	2.5%
Investment	7.7%	7.5%	7.0%	6.0%
Export	−0.4%	−0.5%	−0.3%	−0.2%
Import	0.2%	0.1%	−2.1	−2.7%
Employment	1.1%	1.2%	1.4%	1.5%
Inflation	3.3%	3.3%	−0.3%	−1.4%

Source Compiled from Kato et al. (2023)

4 Decarbonization Policy Cooperation Among CJK: Establishment of a Joint Carbon Market

The discussion so far shows that achieving carbon neutrality in CJK is a long way off, and the IEEJ OUTLOOK scenario of technological progress, which assumes the development and diffusion of decarbonization technologies, shows that carbon emissions could be significantly reduced, but not to the point of carbon neutrality. Also, in the Japanese case, achieving carbon neutrality would require a considerably higher carbon cost (US\$50/CO_2t to US\$400/CO_2t in terms of carbon tax rate).

On the other hand, the EU has been building an internal carbon market (EU-ETS) since 2005, which has encouraged decarbonization innovation in the region and has also eased the burden of carbon costs: greenhouse gases in the EU have decreased by about 30% since the EU-ETS started in 2005 to 2023, while EU GDP grew 55.1% from US\$11.8 trillion to US\$18.3 trillion during the same period, so decoupling was achieved in the EU during the same period. However, while respective domestic carbon markets exist in CJK, there has been no movement yet to establish a cooperative carbon market in the region and to create an intra-regional emissions trading system based on such a market. The establishment of a cooperative carbon market in CJK will contribute to strengthening the competitiveness of domestic industries by taking advantage of economies of scale in reducing carbon dioxide emissions, encouraging innovation in low- and decarbonization technologies in each country, and helping to reduce the burden of carbon costs.

According to the International Energy Agency (IEA 2023), China emits 10.65 billion tons (31.7% of the global total), Japan 1.0 billion tons (3.0% of the global total), and South Korea 560 million tons (1.7% of the global total) of energy-related CO_2 emissions in 2021, accounting for 36.4% of the global total of 33.6 billion tons. In the case of the EU, the total emissions of the 27 EU member states amounted to 2.58 billion tons, or 7.7% of the global total. If a joint carbon market were to be established among CJK, it would certainly be the largest carbon market in the world. In South Korea, a national cap-and-trade emissions trading scheme has been in effect since 2015 (the scheme covers about 70% of greenhouse gases), while in China, a pilot emissions trading scheme at the regional level has been transitioned to a national scheme starting in 2021 (the scheme covers about 40% of greenhouse gases). Japan also plans to launch a national emissions trading scheme in 2026 as part of its green transformation (the scheme will cover about 60% of greenhouse gas emissions). It is recommended that the emissions trading schemes of the three countries also be designed and revised with a view to linking the schemes among the three countries in the future.

A joint CJK approach to emissions trading schemes offers a variety of advantages over other climate policies, including greater efficiency and flexibility in emissions reductions. In theory, under conditions of certainty of emission reductions, an emissions trading scheme could be as cost-effective as a Pigouvian carbon tax (Weitzman 1974). The more heterogeneous the marginal abatement costs of carbon dioxide of the partners participating in an emissions market, the more favorable market linkages will be for the partners (Green et al. 2014). Li and Duan (2021) found that a full CJK emissions trading scheme with participation through consolidation could provide the greatest welfare benefits for all participating countries.

In January 2022, the Japanese government announced the Asia Zero Emission Community (AZEC) initiative, a joint decarbonization-related platform comprising nine ASEAN countries (excluding Myanmar) and 11 Asian countries including Japan and Australia. Through initiatives such as the Japanese Ministerial and the Public–Private Investment Forum, AZEC aims to promote a virtuous cycle of decarbonized growth on both sides of the region, based on the principles of coexistence of decarbonization and energy security, economic growth, and diversification of pathways to carbon neutrality for the countries in the region. The initiative aims to promote a virtuous cycle of decarbonized growth in both Japan and ASEAN. The establishment of a joint carbon market and emissions trading system encompassing CJK, CJK + ASEAN, and Asia as a whole should proceed on the premise that each country has different growth trajectories, domestic challenges, and development issues, and that a long negotiation and consensus building process will be required.

The creation of a cross-border emissions trading scheme faces barriers of institutional resistance based on administrative challenges, barriers of shared governance reflecting the difficulty of resolving complex sovereignty issues, and barriers of environmental integrity based on lack of trust among regional partners. However, as shown in Table 5, the establishment of a joint carbon market will require harmonization of regional emission reduction efforts, promotion of decarbonization technology exchanges among participating countries, creation of new sources of climate change financing through auction revenues, linkage with international emissions trading schemes with other regions, support for carbon border adjustment mechanisms in other countries, and establishment of an emissions reporting framework. There are numerous benefits, including economic, environmental, and strategic benefits among participating countries, such as the establishment of a platform for emissions reporting. According to Zhou et al. (2018), the establishment of a carbon emissions trading market in CJK has advantages such as a wider, more distributable trading market and more efficient reductions, and the use of mandatory and compulsory participation schemes, an appropriate mix of free allocation and paid allocation schemes, unified cost mitigation measures and a compliance framework, and other diverse factors need to be considered when designing the system.

A phased approach is needed to establish a joint carbon market in CJK, starting with piloting cross-border trading and harmonizing rules. Then, emission rights and associated measurement and verification protocols need to be standardized, proceeding from cross-border industry sector-specific trading. Once the foundation is laid with the above two steps, the establishment of a joint carbon market in CJK may become feasible. The major hurdle, however, is the absence of a governing body necessary for a successful intra-regional emissions trading system; unlike the EU, CJK lacks such a centralized authority, and the road to establishment is long. The establishment of a joint carbon market and emissions trading system encompassing CJK, and in the future, Asia as a whole, must proceed on the premise that it

Table 5 The benefits of the East Asia joint carbon market

Benefits	Explanations
Economic benefits	
Efficiency	A bigger, more diverse system provides better access to low-cost emission reduction options
Liquidity	Reducing market volatility by bringing together more actors and more trading activity which leads to a higher level of liquidity in a system
Market power	Broadening market participation reduces volatility by diminishing the ability of large entities to wield disproportionate market-manipulating market-manipulating power
Competitiveness	Harmonized carbon pricing corrects divergences in carbon prices within the region, enhancing competitiveness and limiting exposure to other countries' carbon border adjustment mechanisms
Finance	A new source of climate finance through auctioning revenues
Environmental benefits	
Carbon leakage	Disincentivizing the movement of emitting activities from one economy to another within the joint market
Ambition	By reducing costs, linkage allows countries to adopt more ambitious policies
Co-benefits	Reducing transboundary air pollutions in the region
Strategic benefits	
Trade	The required degree of harmonization and the institutional framework required from linking carbon markets will further develop trade within the region
Regional confidence building	Through the required robust MRV system, a deep interaction among stakeholders across policies and business sectors will be developed
Global regional signal	Harmonization of regional emission reduction efforts and providing the region with more negotiating power at global climate submit
Platform for international collaboration	Ability to link with other international emission trading systems
Technology transfers	Through deeper trade and regional integration, low-carbon technologies will diffuse more rapidly, resulting in further cost reductions and accelerating the region's decarbonization efforts.

Source Developed from Ewing (2022), Roadmap to a Northeast Asian Carbon Market, ASPI and authors' own inputs will require a long negotiation and consensus building process, as each country has different growth trajectories, domestic challenges, and development issues.

5 Conclusion

The discussion in this paper reveals that for CJK to move toward a decarbonized society by mid-century, ambitious energy and climate change policies are needed, including the introduction and expansion of carbon pricing, such as carbon taxes and emissions trading schemes, and active support for decarbonized investments. While these policies can stimulate a wide variety of decarbonized investments and boost GDP growth by reducing fossil energy imports, they also face challenges that make them difficult to be chosen as actual policies due to the increased economic burden of carbon costs.

Therefore, this paper proposes the establishment of a joint carbon market in CJK and the introduction of a cross-border emissions trading system. The establishment of a CJK joint carbon market has a number of advantages, including the reduction of carbon costs in participating countries and the expansion of low- and decarbonization technology and investment exchanges and related business opportunities, linkage of emissions trading with the EU and other countries outside the region, and support for a carbon border adjustment mechanism outside the region.

However, in CJK, there is no governing body pertaining to the establishment and operation of a joint carbon market. In CJK, it is necessary to establish a mechanism for the establishment and operation of a joint carbon market among nations, even if it does not have the binding power to constrain the actions of each nation. In the climate change and climate adaptation sectors, the organization should also take on the role of measuring the impacts of climate change in the region, supporting greenhouse gas reduction targets and adaptation strategies, and supporting the development of a regional climate adaptation platform to provide information to advance preparation for climate risks. The role of the climate adaptation will also be to provide information to advance climate risk preparedness.

With the goal of achieving carbon neutrality by 2050 or 2060, CJK are expected to take the opportunity of building a joint carbon market to realize large-scale energy and environmental policy programs such as the Super Grid, the Green Deal, the Circular Economy Action Plan, and the establishment of a nuclear power safety regime, through close cooperation is expected.

Author's note: This chapter is a revised and expanded English version of the author's article, "Japan-China-Korea Decarbonization Policy Cooperation Toward a Carbon-Neutral Society in East Asia", originally published in *Policy Science*, Vol. 32, No. 3 (2025).

References

Agency for Natural Resources and Energy (2024) Energy White Paper 2024. Natural Energy Foundation website. https://www.renewable-ei.org/. Accessed 31 Oct 2024
Zhou H, Hu Y, Qian X, Nakagami K (2018) Designing an East Asian Carbon Emissions Trading Scheme (EA-ETS)—Considering the links among CJK from the EU case. Policy Sci 25(3):13–26
Green J, Sterner T, Wagner G (2014) A balance of bottom-up and top-down in linking climate policies. Nat Clim Chang 4(4):1064–1067
IEA (2023) World Energy Balances. IEA
IEA (2024) Greenhouse Gas Emissions from Energy. IEA
IEEJ (2023) IEEJ OUTLOOK 2024. IEEJ
IEEJ (2024) IEEJ OUTLOOK 2025. IEEJ
Kato S, Lee S, He Y, Yoshioka T, Morotomi T, Chewpreecha U (2023) Impact of carbon neutrality on the economy and industry assuming Japan's achievement of 2030 power mix plan: a 2050 perspective based on the E3ME macro-econometric model. Energies (MDPI) 16(18):6661, 1–18. https://doi.org/10.3390/en16186661. 17 Sept 2023
Lee S (2025) Japan-China-Korea decarbonization policy cooperation toward a carbon-neutral society in East Asia. Policy Sci 32(3)
Li M, Duan M (2021) Exploring linkage opportunities for China's emissions trading system under the Paris targets—EU-China and Japan-Korea-China cases. Energy Econ 102:105528
Ministry of Economy, Trade and Industry, Japan (2021) The Sixth Basic Energy Plan
Ministry of Economy, Trade and Industry Japan website Asian Zero Emissions Community (AZEC). https://www.meti.go.jp/policy/energy_environment/global_warming/azec.html
Weitzman ML (1974) Prices vs. Quantities. Rev Econ Stud 41(4):477–491
World Bank (2023) World Development Indicators. World Bank

Local Carbon Neutrality Scenarios and Social Acceptance: Evidence from Tochigi, Japan

Wakana Takahashi, Manabu Utagawa, Tetuji Miyake, Yoshihisa Imade, and Bernadett Kiss

1 Introduction—Research Motivation and Objectives

1.1 Carbon Neutrality as a Top Political Priority

A sustainable transition toward carbon neutrality (CN) has become an urgent political priority both domestically and internationally, with countries exhibiting high per capita emissions bearing a particular responsibility. In 2019, China ranked first in total CO_2 emissions, followed by the United States. However, this ranking changes when emissions are measured on a per capita basis. As shown in Fig. 1, U.S. per capita emissions have historically remained high—declining in recent years yet still reaching 14.30 tonnes in 2023. Japan's level, while lower than that of the United States, remains comparatively high among advanced economies at 7.95 tonnes in 2023, despite a gradual decline since the early 2010s. Notably, China, which was

W. Takahashi (✉)
School of International Studies, Utsunomiya University, Utsunomiya, Japan
e-mail: wakana.uu@gmail.com

M. Utagawa
National Institute of Advanced Industrial Science and Technology, Tokyo, Japan
e-mail: m.utagawa@aist.go.jp

T. Miyake · Y. Imade
NPO Utsunomiya Environmental Action Forum, Renewable Energy Subcommittee, Utsunomiya, Japan
e-mail: tm_ym@sky.ucatv.ne.jp

Y. Imade
e-mail: yo-imade@moon.ucatv.ne.jp

B. Kiss
The International Institute for Industrial Environmental Economics, Lund University, Lund, Sweden
e-mail: bernadett.kiss@iiiee.lu.se

W. Zhou (ed.), *Decarbonization Policy Engineering*, World Sustainability Series,
https://doi.org/10.1007/978-981-95-5953-4_9

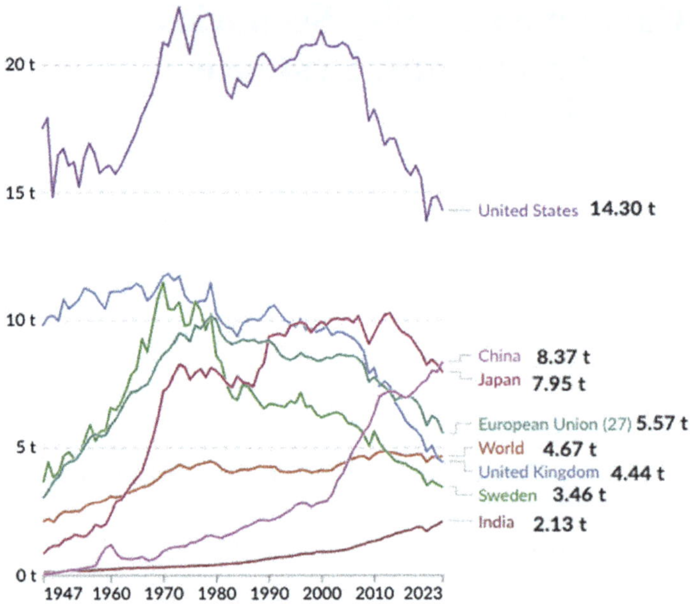

Fig. 1 Trends in CO_2 emissions per capita (1945–2023). *Source* Global Carbon Budget (2024); Population based on various sources (2024)—with major processing by Our World in Data

below the global average in 1995 (approximately 2.7 tonnes), has rapidly increased its emissions, reaching 8.37 tonnes in 2023.

A sustainable transition toward carbon neutrality (CN) has become an urgent political priority both domestically and internationally, with countries exhibiting high per capita emissions bearing a particular responsibility. In 2019, China ranked first in total CO_2 emissions, followed by the United States. However, this ranking changes when emissions are measured on a per capita basis. As shown in Fig. 1, U.S. per capita emissions have historically remained high—declining in recent years yet still reaching 14.30 tonnes in 2023. Japan's level, while lower than that of the United States, remains comparatively high among advanced economies at 7.95 tonnes in 2023, despite a gradual decline since the early 2010s. Notably, China, which was below the global average in 1995 (approximately 2.7 tonnes), has rapidly increased its emissions, reaching 8.37 tonnes in 2023 (Fig. 2).

1.2 Focus on Regional Cities

Following the adoption of the Paris Agreement in 2015, the Japanese government set a target of reducing greenhouse gas emissions by 46% from 2013 levels by 2030 (Ministry of the Environment 2021b). Consequently, all local governments are now expected to engage in CN initiatives.

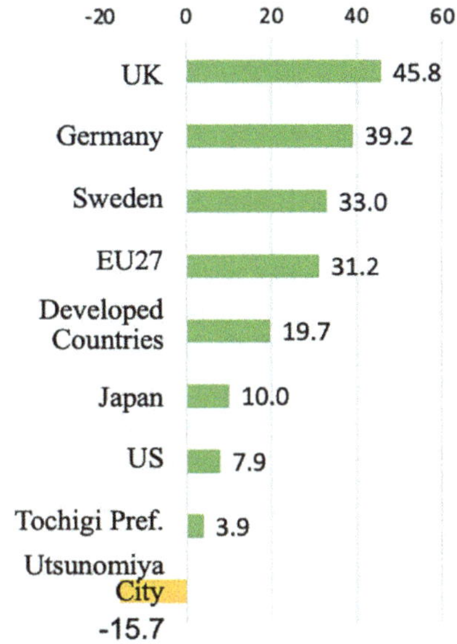

Fig. 2 CO_2 Emission Reduction Rates in Developed Countries and Japan, Tochigi Prefecture, and Utsunomiya City (1990–2021). *Source* Our World in Data based on the Global Carbon Project (2022) and various data from Tochigi Prefecture and Utsunomiya City

Large metropolitan areas, with their concentrations of population, industry, and wealth, are major sources of greenhouse gas emissions but also have access to the information, expertise, human resources, and financial capacity needed to implement transformative measures. For example, the Tokyo Metropolitan Government has mandated solar panel installation for new residential buildings, demonstrating the capacity of large cities to undertake independent initiatives.

Regional cities, however, face the different pressure to develop CN strategies. Approximately 68% of Japan's population lives in municipalities with fewer than 500,000 residents, which collectively constitute significant sources of emissions. Many of these cities are highly car-dependent, resulting in substantial transportation-sector emissions. Declining populations, aging infrastructure across industrial, commercial, and residential sectors, and generally lower energy efficiency are common. At the same time, untapped energy resources offer significant potential for reducing emission. Yet, these cities also face worsening heat conditions, increased flooding, and other climate-related hazards, heightening the urgency of adaptation measures.

By early 2023, over 870 municipalities across 45 prefectures—including Tochigi Prefecture and 13 of its municipalities—had declared CN targets (Ministry of the Environment 2023). Most have set interim targets of approximately 50% reduction by 2030 and have begun developing roadmaps. While some municipalities have undertaken pioneering efforts, many lack sufficient information, expertise, and financial resources. Their range of feasible mitigation options is often limited, and despite calls for greater action from residents and businesses, the feasibility of such measures is frequently low. Moreover, in some cases—such as the installation of large-scale solar farms—renewable energy projects have caused environmental damage, creating conflicts with local development goals.

To address these challenges, the Ministry of the Environment launched the "Decarbonization Leading Areas" program, offering grants to selected municipalities to serve as models and trigger an "implementation domino effect." As of March 2023, two selection rounds had been conducted, with Utsunomiya City and Haga Town, as well as Nasushiobara City in Tochigi Prefecture, among those selected (Ministry of the Environment 2022). Most of the 46 municipalities selected by November 2022 have set targets of achieving CN by 2050 and halving emissions by 2030.

Internationally, more ambitious targets exist. In the European Union, 100 cities have committed to becoming climate-neutral and smart by 2030, serving as hubs for experimentation and innovation so that the rest of Europe can follow by 2050 (EU 2022). These cities vary widely in climate, population size, industrial structure, and infrastructure, yet share the goal of achieving climate neutrality within less than a decade.

Against this backdrop, our research has examined the potential transferability of advanced European approaches to Tochigi Prefecture. Over the past two years, we have conducted an international, interdisciplinary, and socially collaborative research project with colleagues from Lund University and a local non-profit organization. As will be discussed later, our findings highlight several blind spots in Utsunomiya City's CN strategy: insufficient attention to energy efficiency improvements, limited utilization of renewable heat, and a lack of multifunctional, socially innovative measures such as nature-based solutions (NbS). The study also underscores the importance of participatory approaches to enhance social acceptance.

Building on these insights, this paper focuses specifically on energy efficiency improvements and presents multiple strategic scenarios for achieving CN in regional contexts. Furthermore, it evaluates their social acceptance through a social experiment conducted in Tochigi Prefecture via the Multicultural Public Sphere Forum.

2 Research Methods and Process

2.1 Previous Studies on Information Provision and Social Acceptance in Energy TransitionsInternational, Interdisciplinary, and Socially Co-Creative Research

Following the adoption of the Paris Agreement in 2015, the Japanese government set a target of reducing greenhouse gas emissions by 46% from 2013 levels by 2030 (Ministry of the Environment 2021b). Consequently, all local governments are now expected to develop and implement carbon neutrality (CN) initiatives.

Large metropolitan areas, with their concentration of population, industry, and wealth, are major sources of greenhouse gas emissions but also possess the information, expertise, human resources, and financial capacity to implement transformative measures. For example, the Tokyo Metropolitan Government has mandated the installation of solar panels on new residential buildings, illustrating the capacity of large cities to undertake independent initiatives.

Regional cities, however, face different pressures in developing CN strategies. Approximately 68% of Japan's population resides in municipalities with fewer than 500,000 residents, which collectively account for a substantial share of national emissions. Many of these cities are highly car-dependent, resulting in significant transportation-sector emissions. They also tend to experience declining populations, aging infrastructure across industrial, commercial, and residential sectors, and generally lower energy efficiency. At the same time, untapped energy resources offer substantial opportunities for emissions reduction. Yet these cities face worsening heat conditions, increased flooding, and other climate-related hazards, intensifying the need for adaptation measures.

In the 2020s, an increasing number of local governments have declared CN targets. As of 28 February 2023, 871 municipalities across 45 prefectures—including Tochigi Prefecture and 13 of its municipalities—had made such declarations (Ministry of the Environment 2023). Most have set interim targets of approximately 50% reduction by 2030 and have begun developing roadmaps. While some municipalities have initiated pioneering efforts, many lack sufficient information, expertise, and financial resources. Their feasible mitigation options are often limited, and despite calls for stronger action from residents and businesses, the practical viability of such measures remains low. Moreover, in some cases—such as the development of large-scale solar farms—renewable energy projects have caused environmental damage, generating conflicts with local development goals.

To address these challenges, the Ministry of the Environment launched the "Decarbonization Leading Areas" program, providing grants to selected municipalities to serve as models and catalyze an "implementation domino effect." As of March 2023, two selection rounds had been conducted, with Utsunomiya City, Haga Town, and Nasushiobara City in Tochigi Prefecture among those selected (Ministry of the Environment 2022). Most of the 46 municipalities selected by November 2022 have set targets of achieving CN by 2050 and halving emissions by 2030.

Internationally, more ambitious targets exist. In the European Union, 100 cities have committed to becoming climate-neutral and smart by 2030, serving as hubs for experimentation and innovation so that the rest of Europe can follow by 2050 (EU 2022). These cities vary widely in climate, population size, industrial structure, and infrastructure, yet share the goal of achieving climate neutrality in less than a decade.

Against this backdrop, our research examines the potential transferability of advanced European approaches to Tochigi Prefecture. Over the past two years, we have conducted an international, interdisciplinary, and socially collaborative project in partnership with colleagues from Lund University and a local non-profit organization. As discussed later, our findings highlight several blind spots in Utsunomiya City's CN strategy: insufficient attention to energy efficiency improvements, limited utilization of renewable heat, and a lack of multifunctional, socially innovative measures such as nature-based solutions (NbS). The study also emphasizes the importance of participatory approaches to enhance social acceptance.

Building on these insights, this paper focuses specifically on energy efficiency improvements and presents multiple strategic scenarios for achieving CN in regional contexts. It also evaluates their social acceptance through a social experiment conducted in Tochigi Prefecture via the Multicultural Public Sphere Forum.

2.2 International, Interdisciplinary, and Socially Co-Creative Research

Building on the theoretical insights from the previous section regarding the importance of information provision and social acceptance, our study sought to operationalize these concepts in the context of regional energy transitions. The aim was to bridge the gap between quantitative modeling of carbon–neutral scenarios and the social processes that determine their feasibility.

To this end, we developed a research framework that is:

- International, drawing on experiences and practices from multiple countries;
- Interdisciplinary, combining perspectives from engineering, social sciences, and policy studies; and
- Socially co-creative, engaging diverse local stakeholders throughout the research process.

This framework integrates two complementary components:

1. Quantitative scenario analysis—assessing the technical and economic potential of various carbon–neutral pathways; and
2. Participatory engagement—examining the social viability and acceptance of these pathways through dialogue and co-design with local actors.

The participatory dimension was embedded as a core element of the research design, ensuring that the scenarios reflected both technical constraints and community priorities. Methodologically, two key elements were central to implementing this approach: the *mobile lab*, which facilitated interdisciplinary field-based research, and *action research* within the multicultural public sphere, which enabled sustained stakeholder dialogue and collaborative problem-solving.

2.2.1 Mobile Lab

A mobile lab is a collaborative research approach in which an interdisciplinary team investigates local phenomena and related controversies. The process includes joint preparation, discussion, fieldwork, data collection, analysis, and synthesis of results. Each researcher's individual expertise, combined with the personal and institutional experiences of other stakeholders (e.g., citizens, government officials), deepens collective understanding of the subject matter.

Operating as a team-based activity, the mobile lab strengthens group cohesion, fosters continuous learning among researchers, and provides a platform for exchanging experiences, building shared values, and enhancing mutual understanding. Such cross-disciplinary interaction also plays a key role in bridging academic and cultural divides (Kiss and Salk 2021).

2.2.2 Multicultural Public Sphere

Addressing challenges such as the climate crisis requires not only initiatives led by governments and corporations but also horizontal linkages among diverse stakeholders. Interaction among people of different backgrounds, generations, and genders fosters mutual respect and tolerance. A multicultural public sphere is defined as an inclusive space—accessible regardless of age, gender, religion, language, occupation, or nationality—where diverse individuals can engage in open debate and reach consensus. Such spaces can, in principle, be established anywhere in the world (Takahashi 2023).

2.2.3 Action Research

Action research is a practice-oriented approach in which researchers and stakeholders collaboratively investigate problems, develop solutions, and iteratively move between practice and research. The objective is to generate knowledge that can be fed back into society to improve existing conditions (Lewin 1946/2017).

In this study, we established a multicultural public sphere where a wide range of actors—including researchers, NPOs, universities, government officials, politicians, corporations, industry associations, citizens, and children with their guardians—could collaborate regardless of demographic or institutional affiliation. Activities included stakeholder interviews, multi-stakeholder workshops, presentations by key actors, group discussions, surveys, and follow-up debates informed by survey analysis. This cyclical process enabled deliberation across multiple dimensions.

Through a series of mobile labs and action research sessions within the multicultural public sphere, we examined local realities in situ and connected them with phenomena in geographically distant regions. This comparative approach revealed both commonalities and differences, enabling CN strategies to be adapted to specific contexts.

Importantly, CN in this research was not treated as an isolated target but as a driver of sustainable transformation aligned with the Sustainable Development Goals (SDGs). Consistent with this orientation, our approach emphasized:

1. Cross-sectoral integration rather than a purely technical focus;
2. Demand-driven solutions in addition to supply-side measures;

3. Recognition of synergies among existing initiatives, rather than treating measures as discrete and unrelated; and
4. Broad stakeholder collaboration, ensuring that strategies reflect social contexts and maintain systemic coherence.

2.3 Research Process

Based on the objectives and methodology described above, the study was conducted in the following phases.

2.3.1 Phase 1: "Sustainable Energy in Utsunomiya—Remarkable Potential of Renewable Energy" (October 2021–March 2022; Published March 2022; Revised August 2022)

One of the key requirements for regional cities to achieve carbon neutrality is the deployment of sustainable energy sources. According to Utsunomiya City's Carbon Neutral Roadmap, total CO_2 emissions amount to 3.59 million tons, over 80% of which are energy related. In this context, the Renewable Energy Division of the Utsunomiya Environmental Action Forum (NPO) approached the Takahashi Laboratory at Utsunomiya University with the idea of producing the city's first *Sustainable Energy* report. The project was launched as part of the 2021 academic courses *Global Environmental Policy* and *Seminar on Environment and International Cooperation.*

Many of the participating students had no prior formal training in energy engineering. They began by studying the fundamentals of renewable energy, then engaged in consultations and field visits with local businesses, researchers, and NPOs. The resulting report adopted a citizen-oriented perspective (Fig. 3).

Using a co-creative approach, the team analyzed public datasets to visualize renewable energy utilization rates and potential at the global, national (Japan), prefectural (Tochigi), and municipal (Utsunomiya) levels. As shown in Fig. 4, global renewable energy statistics distinguish among electricity, heat, and transport sectors; however, in Japan and Tochigi Prefecture, data for non-electric renewable energy are largely unavailable or unreported. In Utsunomiya City, there were no accessible data on renewable energy use outside of electricity, and the renewable share in electricity was only 8.7%.

The study identified renewable energy sources in the region as including photovoltaics, solar thermal, wind, hydro and small hydro, biomass, geothermal, and ground-source heat. As illustrated in Fig. 5, the potential for renewable heat is particularly high, yet remains largely unrecognized and underutilized. The report emphasized that energy conservation measures should be prioritized before large-scale expansion of renewable energy capacity.

The findings were compiled into a booklet at the end of FY2021. In FY2022, the team actively disseminated the results through the establishment of the multicultural

Fig. 3 Cover page of the Report on "Sustainable Energy in Utsunomiya—Remarkable Potential of Renewable Energy: jointly discussed by Utsunomiya University Student and an NPO" *Source* Takahashi et al. (2022)

public sphere. A notable event was the Multicultural Public Sphere Forum held on 19 April 2022, attended by government officials, experts, and members of the public. The report was submitted to the Utsunomiya City Government and the Mayor, and was later incorporated into the *3C Kids College* and *Utsunomiya City Environmental University* programs. The Mayor issued a formal letter of appreciation, and the report received coverage in local media and was cited in deliberations at the Tochigi Prefectural Assembly. Public and political response was strong, with multiple requests from citizens, government officials, and politicians for further investigation into the practical implementation of energy conservation and renewable energy measures.

2.3.2 Phase 2: "Towards a Carbon-Free and Comfortable City—Insights from Swedish Model Cases for Utsunomiya" (September–November 2022; English Report with Japanese Translation Published February 2023)

Phase 1 revealed that sustainable energy deployment in Japan—particularly in Utsunomiya—remains limited and sought to raise awareness among local governments and citizens through the multicultural public sphere. However, awareness alone does not ensure actual implementation.

In August 2022, following the first field research visit to Sweden in three years, the lead author (Takahashi) was struck by the extent to which European regional cities were undergoing substantial transformations toward carbon neutrality (CN). Around the same time, Lund University invited our team to collaborate on a project

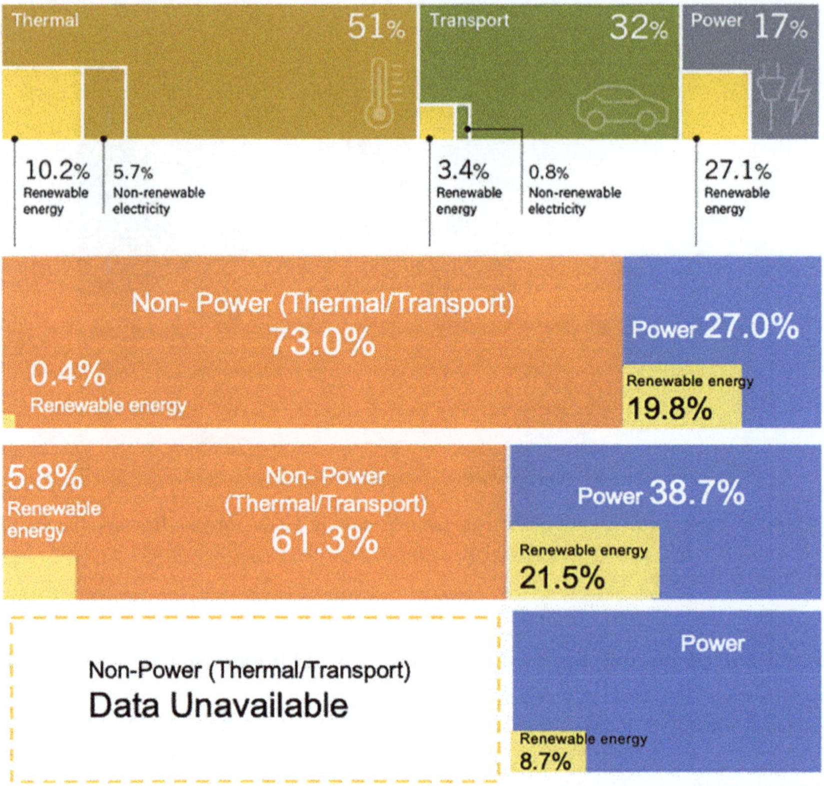

Fig. 4 Percentage of Renewable Energy Installed in the World, Japan, Tochigi Prefecture, and Utsunomiya City (2019). 20212021a *Source* Takahashi et al. (2022), REN21(2021), Agency for Natural Resources and Energy, Government of Japan (). Ministry of the Environment, Government of Japan ()

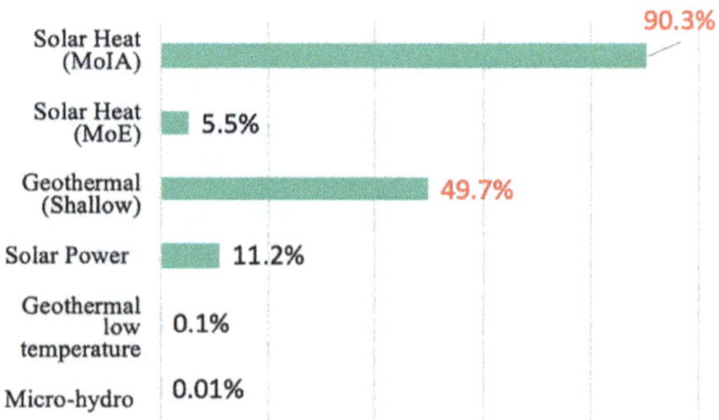

Fig. 5 Renewable Energy Potential of Utsunomiya City. *Source* Ministry of Internal Affairs and Communications (2022) for small hydro and solar thermal (total), Ministry of the Environment (2022) for others * From Utsunomiya City (2002) "Regional New Energy Vision

to develop CN strategies for Japanese regional cities, using Utsunomiya as a pilot case.

In this collaboration, graduate students from Lund University examined Utsunomiya's CN strategy and, applying the mobile lab methodology, identified advanced practices from benchmark municipalities in Sweden that could serve as relevant references. The process comprised two main stages:

1. Case identification and discussion—The Lund University team conducted literature reviews and interviews to understand the characteristics of Utsunomiya's CN strategy, then identified Swedish CN strategies and measures—particularly from benchmark municipalities—that could be adapted to Utsunomiya. These cases were examined in mobile labs and discussed in multicultural public sphere forums involving Utsunomiya City officials, the university, NPOs, civic groups, and residents.
2. Feasibility assessment and recommendations—Based on the forum discussions and additional stakeholder interviews, the team assessed the transferability of Swedish strategies to Utsunomiya and developed specific recommendations.

The findings were consolidated into the report *Carbon-Free Utsunomiya—Building a Cleaner City: Insights from Swedish Model Cases* (Fig. 6). The report presented practical examples in four key areas: mobility, energy efficiency, climate adaptation through nature-based solutions (NbS), and industry–business collaboration. Notably, it proposed innovative measures in mobility and NbS that could be scaled to a city of approximately 300,000 inhabitants. These proposals were evaluated in a co-creative workshop (multicultural public sphere forum) hosted by Utsunomiya University and were generally well received.

The report also identified a critical "blind spot" in Japan's CN efforts: insufficient attention to energy efficiency improvements. Examples included Lund's building insulation projects and Gothenburg's river-water-based district cooling system—both representing highly efficient solutions. The report emphasized the advantages of prioritizing energy efficiency and recommended that Utsunomiya adopt more ambitious measures, noting that the city "appears to have focused on power generation while neglecting energy efficiency" (Fernández et al. 2022, p. 23).

Furthermore, the Swedish cases underscored the importance of governance. Successful projects incorporated resident participation processes that combined top-down and bottom-up approaches, enabling rapid policy deployment supported by data validation. As noted in the report, "Engaging citizens in ongoing dialogue is essential not only for understanding their needs and concerns but also for enabling them to contribute ideas and solutions" (Fernández et al. 2022, p. 30). Particular emphasis was placed on reaching all segments of the population, including older residents—a point of special relevance for aging cities such as Utsunomiya. This finding aligned closely with the concept of the multicultural public sphere employed in the present study.

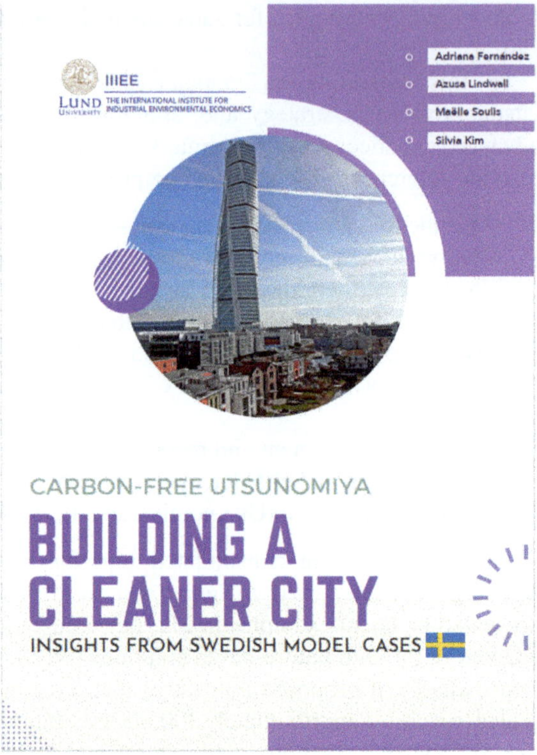

Fig. 6 Cover page of the Report on Towards a Carbon-Free and Comfortable City—Insights from Swedish Model Cases for Utsunomiya. *Source* Fernández et al. (2023)

2.3.3 Phase 3: "Towards a Carbon–Neutral Tochigi Prefecture—Addressing Japan's Blind Spot: Energy Efficiency Improvements"

The joint research with the NPO in Phase 1 revealed the high potential of diverse renewable energy sources—particularly renewable heat—alongside the critical importance of improving energy efficiency. These findings were reinforced by the Swedish recommendations in Phase 2, which identified energy efficiency as a major blind spot in Japan's CN efforts. Accordingly, Phase 3 focused on an in-depth examination of energy efficiency improvements.

While "energy efficiency improvement" encompasses a broad spectrum of measures, this report addressed three thematic areas:

2.3.3.1 Energy Conservation Equipment and Insulation

Tochigi Prefecture experiences a wide annual temperature range, with hot summers and significant nighttime/morning cold in winter—typical of an inland climate. Insulation is therefore particularly important. Many facilities and devices are aging, and replacing them with high-efficiency alternatives during renewal cycles is crucial. Accordingly, this study placed emphasis on building insulation and energy-saving appliances.

2.3.3.2 Local Renewable Heat with High Efficiency

Electricity accounts for less than half of total energy demand; direct use of heat is inherently more efficient than converting it to electricity. Given Tochigi's abundant winter sunshine, solar thermal energy is highly promising, as are locally sourced geothermal and groundwater cooling resources. Renewable heat remains a nationwide blind spot in Japan; thus, this study focused particularly on solar thermal and geothermal energy.

2.3.3.3 Nature-Based Solutions (NbS)

Replacing asphalt or other impermeable surfaces with vegetation can significantly lower surface temperatures. Green roofs and walls, along with strategically planted tall trees, can reduce cooling energy demand. NbS also absorb and gradually release rainwater, making them effective in flood prevention—a growing concern given the increasing frequency of heavy rainfall. Moreover, NbS support biodiversity and foster community engagement, delivering multiple co-benefits.

2.3.3.4 Assessment and Scenario Analysis

Building on these three thematic areas—energy conservation and insulation, renewable heat, and NbS—students from Utsunomiya University who had participated in Phase 2 undertook further training through expert lectures and field surveys. They critically reviewed the CN roadmaps of Tochigi Prefecture, Utsunomiya City, and Nasushiobara City, and estimated the potential CO_2 reductions for 2030 and 2050 for each measure using scenario calculations provided by Utsugawa (National Institute of Advanced Industrial Science and Technology). Specific proposals were developed and presented at a multicultural public sphere forum attended by representatives from Tochigi Prefecture, Utsunomiya City, universities, NPOs, civic groups, and citizens, followed by in-depth discussions.

Refinement and recommendations.

Based on discussions during the forum and further consultations with experts, the study developed recommendations on the transferability of Swedish CN strategies to Utsunomiya, which were compiled into the final report.

The analysis first found that although Tochigi Prefecture, Utsunomiya City, and Nasushiobara City have each formulated CN plans and roadmaps for 2050, the emphasis remains heavily on renewable energy—particularly photovoltaics. While energy conservation and insulation are also cited as priorities, related initiatives are largely confined to public awareness campaigns, and available support schemes are not widely publicized. No subsidies for renewable heat were identified, and NbS was absent from the priority lists of all three municipalities.

Second, the findings indicate that substantial emission reductions can be achieved through systematic deployment of existing technologies without reliance on new energy sources or innovations. Measures such as LED lighting, high-performance insulated housing, high-efficiency heat pumps for renewable heat, and electric vehicles offer significant potential. Scenario calculations suggest that combining energy conservation/insulation with expanded renewable heat utilization could reduce CO_2 emissions by 68.1% in Tochigi Prefecture, 62.2% in Utsunomiya, and 62.8% in Nasushiobara by 2030 compared with 2013 levels, with near-complete CN achievable by 2050 in all three cases. NbS effects were not quantified. In essence, replacing equipment with high-efficiency alternatives during renewal cycles can substantially cut energy consumption.

Finally, the study determined that the additional costs of installing energy-efficient equipment are, in many cases, recovered within a relatively short timeframe through resulting energy savings. It is therefore essential to raise public awareness of these economic benefits and to promote the use of support schemes to encourage widespread adoption.

2.4 Scenario Development and Examination of Social Acceptance

2.4.1 Local Carbon Neutrality Strategies Through Energy Efficiency and Renewable Heat Utilization in Tochigi Prefecture

The preceding analyses underscore the critical importance of improving energy efficiency and diversifying the use of renewable energy sources when formulating local strategies for carbon neutrality (CN). Our findings indicate that even without reliance on novel energy sources or emerging technologies, the systematic and widespread deployment of currently available measures—such as energy efficiency enhancements and diversified renewable heat utilization—could enable Tochigi Prefecture, Utsunomiya City, and Nasushiobara City to exceed their current 2030 CO_2 reduction targets by more than 10% relative to 2013 levels. Under such scenarios, near-complete CN by 2050 would be achievable in all three jurisdictions.

However, this potential remains insufficiently recognized within local communities. To address this, we developed a set of local strategic scenarios for CN and conducted an empirical examination of their social acceptance through a Multicultural Public Sphere Forum. This process forms the core of the present report.

To ensure accessibility for both citizens and students, six distinct scenarios were designed. Each combined measures from three domains—energy-efficient equipment, building insulation, and renewable heat—across three ambition levels: Level 1 (approximately aligned with current national policies), Level 2 (enhanced municipal policies), and Level 3 (further intensified policies incorporating additional options), as shown in Table 1. Through collaborative deliberations among research partners, desirable scenario configurations were identified. Based on these, Utagawa conducted model-based calculations, which were compiled into tabular form.

The modeling procedure proceeded as follows. The study area encompassed Tochigi Prefecture as a whole and Utsunomiya City. The baseline scenario reflected measures consistent with Japan's *Strategic Energy Plan* and *Global Warming Countermeasures Plan.* Additional countermeasures included strengthened equipment replacement policies, accelerated vehicle turnover with improved fuel efficiency, and enhanced building insulation. For renewable energy, estimates of renewable energy potential for Tochigi Prefecture and Utsunomiya City from the Ministry of the Environment—covering both electricity and heat—were compared with actual regional energy consumption. This confirmed that the assumed levels of renewable energy

Table 1 Assumed levels of energy-saving equipment, insulation, and renewable heat

	Energy-saving equipment	Insulation	Alternative energy
Level 1	National Measures and Policies	National Measures and Policies	National target (58% non-fossil by 2030)
Level 2 municipality independent policy	Prefectural Measures Dissemination of technical advice	Prefectural Measures Dissemination of technical advice	58% renewable electricity target; installation of solar power in all new buildings in Tochigi Prefecture; dissemination of technical advice; establishment of a local retail electricity utility to purchase and sell local renewable energy, with expert participation
Level 3 municipality Reinforcement Policy	Reinforced municipal policies; introduction of energy-efficient machinery, vehicles, and EVs at renewal; retrofitting of factories for energy savings; sharing of professional information; creation of local energy advisory centers; conducting energy efficiency audits; collaboration with local retailers to inform consumers about efficient products	Expanded insulation retrofitting (equivalent to new construction annually); sharing of professional information; creation of local energy advisory centers; mandatory insulation standards for new construction; expansion of retrofitting policies	Solar PV and solar thermal in offices and homes; biomass heat where available; promotion of local retail electricity utilities purchasing and selling local renewable energy with expert participation; promotion of district heat supply networks where feasible

deployment under the scenarios were well within available potential. Strategies to prevent environmentally detrimental overdevelopment were also incorporated.

Further measures included the installation of rooftop photovoltaic systems on all new buildings, expansion of corporate and household contracts for 100% renewable electricity, and broader adoption of solar thermal systems for space heating and hot water in the residential and commercial sectors. Greening measures to reduce urban cooling loads were also evaluated. Field surveys were conducted of notable energy efficiency facilities and ground-source heat pump installations within the prefecture.

Multiple simulations were run to compare the effects of these interventions. Energy efficiency measures were categorized into equipment upgrades (including vehicle efficiency improvements) and insulation upgrades, while renewable energy measures were classified as renewable electricity or renewable heat.

The economic impacts of these measures were also assessed. Using the 2015 Tochigi Prefecture input–output table, a bottom-up estimation linked projected demand increases by industry sector to the emission reduction scenarios. The analysis considered three main channels of economic impact: increased capital investment in equipment, savings from reduced utility costs, and revenue from renewable electricity sales. The shares of in-prefecture consumption and out-of-prefecture outflows were held constant at current levels for the purposes of this analysis.

2.4.2 Examination of Social Acceptance

The six scenarios developed through the above process were subsequently evaluated for social acceptance within the Multicultural Public Sphere established at Utsunomiya University, using an action research framework. A dedicated forum was convened on May 27, 2023, in a hybrid format combining in-person and online participation. Outreach began approximately three weeks prior to the event, utilizing the university website, Facebook, Twitter, academic society websites, and the environmental education mailing list within Tochigi Prefecture. This campaign generated 111 pre-registrations.

Participants represented a broad spectrum of stakeholders, including researchers, NGOs, universities, municipal officials, policymakers, business and industry associations, and citizens. Before the scenario presentations, attendees received an overview of domestic and international CN initiatives, relevant technological options and their potential, and the policy measures required to realize this potential. The six scenarios were then presented, after which participants selected their preferred options and engaged in in-depth discussions of the modeling results. The event agenda is shown in Table 2.

3 Results of Analysis

Based on the above, this section presents:

Table 2 Multicultural Public Sphere Forum Progress on the day of the forum

Time	Duration	Session / Content	Presenter(s)
10:00	15 min	Opening and Introduction—Purpose of the meeting and summary of the report	Wakana Takahashi—Professor, Faculty of International Studies, Utsunomiya University
10:15	20	Efforts toward carbon neutrality in Nasushiobara City	Takashi Sagara—Manager, Climate Change Countermeasures Division, Nasushiobara City
10:35	15	Proposals for energy efficiency improvements (examples: insulation, solar thermal, parking lot greening)	Ryota Anno, Chichi Endo, Hane Kusano, Minami Suzuki—3rd-year students, Takahashi Seminar, School of International Studies, Utsunomiya University
10:50	15	CO_2 reductions from energy efficiency improvement measures and local economic benefits	Manabu Utagawa—Senior Researcher, National Institute of Advanced Industrial Science and Technology
11:05	10	Q&A Session	All presenters
11:15	5	Group Selection—"Which scenario do you like? How far can you go?" (introduction with visual materials; participants select scenarios based on own ideas; Scenarios 1–6 include varying 2030 targets, insulation, thermal, and renewable energy measures, economic ripple effects, and policy options)	Facilitators
11:30	2	Workshop Introduction—How to proceed with the workshop	Facilitators
11:30	30	Workshop—Objective: Explore possibilities, challenges, ideas, and actions for the selected scenarios	All participants
12:02	7	Sharing Session—Key points from group discussions	Group representatives
12:09	16	Comment Session—Feedback and reflections	Aya Takahashi—Assistant Director, Policy Deliberation Office, Policy Department, Utsunomiya City; Takashi Sagara—Manager, Climate Change Countermeasures Division, Nasushiobara City; Wakana Takahashi—Professor, Faculty of International Studies, Utsunomiya University
12:25	5	Closing Remarks	Hideki Takashima—Assistant Director, Climate Change Countermeasures Division, Environment and Forest Department, Tochigi Prefecture

Table 3 Combination of energy-saving equipment improvements, insulation, and renewable energy levels in six scenarios

Scenario	Energy-Saving Equipment	Insulation	Renewable Energy / Alternative Energy
1. Large-Scale Power Generation Continuation	Level 1	Level 1	Level 1
2. Enhanced Solar Deployment	Level 1	Level 1	Level 2
3. Strengthened Energy-Efficient Equipment + Solar	Level 2 (Prefectural Plan)	Level 2 (Prefectural Plan)	Level 2
4. Enhanced Energy-Efficient Equipment + Insulation + Solar	Level 3 (Enhanced)	Level 3 (Enhanced)	Level 2
5. Energy-Efficient Equipment + Insulation + Diversified Renewable Energy	Level 2 (Prefectural Plan)	Level 2 (Prefectural Plan)	Level 3 (Municipality-Specific Reinforcement)
6. Enhanced Energy-Efficient Equipment + Insulation + Diversified Renewable Energy	Level 3 (Enhanced)	Level 3 (Enhanced)	Level 3 (Municipality-Specific Reinforcement)

(i) an outline of the six scenarios (Table 3) and
(ii) the results of examining their social acceptability.

3.1 Scenario Presentation

Scenario 1: Continued Large-Scale Power Generation.

Aligned with Japan's Sixth Strategic Energy Plan, this scenario assumes a 36–38% share of renewable electricity by 2030, a 45% reduction in energy-related CO_2 emissions from 2013 levels, and national-level energy efficiency measures applied to Tochigi Prefecture.

Scenario 2: Enhanced Solar Deployment.

Based on the energy efficiency measures of the Sixth Strategic Energy Plan, this scenario assumes that the entire 58% non-fossil share in the power mix is sourced from renewable energy.

Scenario 3: Strengthened Energy-Efficient Equipment + Solar.

Models prefectural-level measures that strengthen energy-efficient equipment compared with Scenarios 1 and 2, while maintaining a focus on solar deployment.

Scenario 4: Enhanced Energy-Efficient Equipment + Insulation + Solar.

In addition to stronger energy-efficient equipment measures, this scenario introduces enhanced insulation. From 2025 onward, new buildings achieve Grade 6 insulation performance, with retrofits of existing buildings to match.

Scenario 5: Energy-Efficient Equipment + Insulation + Diversified Renewable Energy.

Expands renewable energy measures beyond electricity to include heat utilization, targeting a 64% reduction in energy-related CO_2 emissions by 2030 compared to 2013.

Scenario 6: Enhanced Energy-Efficient Equipment + Insulation + Diversified Renewable Energy.

Builds on Scenario 5 by adding enhanced insulation, with the same Grade 6 adoption and retrofitting from 2025. While the measures differ from Scenarios 3/4 and 5/6, overall differences in total sectoral reduction rates remain relatively small.

Table 4 summarizes the above scenario.

3.2 *Social Acceptability and Overall Evaluation*

Figure 7 presents the results of the hybrid co-creation workshop held at Utsunomiya University on May 27, 2023, conducted within the framework of action research using the Multicultural Public Sphere. Outreach for the event began approximately three weeks in advance via the university website, Facebook, Twitter, academic society websites, and the environmental education mailing list in Tochigi Prefecture, resulting in 111 pre-registrations. Ultimately, 86 participants responded to the scenario survey. Approximately 70% were students in their teens or twenties, while the remaining 30% consisted of citizens, municipal officials, NGO representatives, and other stakeholders from diverse backgrounds.

The evaluation results showed that Scenario 1 (Continued Large-Scale Power Generation) and Scenario 2 (Enhanced Solar Deployment) both failed to meet the target of halving CO_2 emissions by 2030. Scenario 2 also exhibited the largest gap between investment costs and savings on utility expenses, and neither scenario was selected by any participants.

In contrast, all scenarios from Scenario 3 onward, which incorporated enhanced energy efficiency measures, exceeded the halving target. Higher levels of enhancement require more investment but also deliver greater savings on utility bills. Scenario 4 (Enhanced Insulation) attracted the highest number of selections overall. When combined with other "enhanced" scenarios, support for insulation measures exceeded 70%.

The following sections summarize participants' comments for each scenario, based on workshop discussions and post-event questionnaires.

Table 4 Six Scenarios (Tochigi Prefecture)

	CO_2 Reduction (Compared to 2013)	Investment Amount (Billion Yen)	Utility Cost Savings (Billion Yen)	Economic Ripple Effect within the Prefecture (Billion Yen)	Increase in Employment (persons)
1. Continued Large-Scale Power Generation	45%	120	80	80	8000
2. Enhanced Solar Deployment	45%	150	80	100	10,000
3. Strengthened Energy-Efficient Equipment + Solar	54%	220	180	150	15,000
4. Enhanced Energy-Efficient Equipment + Insulation + Solar	54%	220	180	150	15,000
5. Energy-Efficient Equipment + Insulation + Diversified Renewable Energy	64%	250	210	170	17,000
6. Enhanced Energy-Efficient Equipment + Insulation + Diversified Renewable Energy All Enhanced Scenario	64%	250	120	170	17,000

3.3 Scenario-Specific Feedback

3.3.1 Scenario 3—Strengthened Energy-Efficient Equipment + Solar

Supporters emphasized that policies promoting the dissemination of high-efficiency home appliances could significantly influence consumer choices, thereby reducing energy consumption and greenhouse gas emissions. They valued the direct household-level reduction in energy demand and the associated alleviation of environmental burdens. Because the adoption of energy-efficient appliances is a familiar

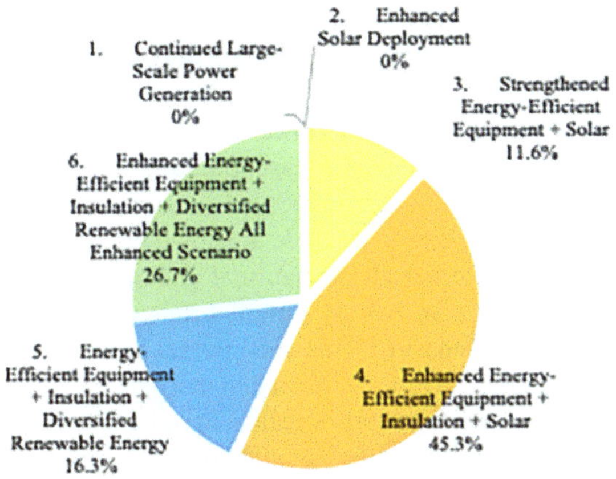

Fig. 7 Social acceptability of scenarios ($n = 86$)

and easily actionable measure, participants considered it an effective way to raise environmental awareness among local citizens.

Student proposals—such as introducing "energy advisers" in electronics retail stores and offering purchase subsidies—were regarded as effective in encouraging environmentally conscious purchases. Several participants noted that, although subsidies already exist, many people are unaware of them, and that providing information could increase uptake. Overall, Scenario 3 was supported primarily for its budgetary feasibility and practicality, with a shared view that measures should avoid excessive idealism and focus on actions individuals can readily implement.

3.3.2 Scenario 4—Enhanced Energy-Efficient Equipment + Insulation + Solar

> This scenario received the highest overall support. Many residents of Utsunomiya, drawing on personal experience of living through all four seasons, expressed an urgent need for improved insulation. A widely shared opinion was that enhancing insulation should take priority over expanding heating and cooling systems, as it directly reduces the need for air conditioning and heating, thereby contributing to energy conservation.

Participants also highlighted that Tochigi Prefecture has the highest winter mortality increase rate in Japan, identifying insulation improvement as an urgent public health issue. The long-term benefits of insulation measures were frequently stressed. While renewable energy adoption was recognized as necessary, many felt it would require more time and investment, and that improving energy efficiency should come first. Overall, insulation improvements and the adoption of energy-efficient appliances in individual homes were seen as the foundation upon which renewable energy expansion should be built.

3.3.3 Scenario 5—Energy-Efficient Equipment + Insulation + Diversified Renewable Energy

Supporters of this scenario concluded, based on student presentations, that a diverse mix of energy sources is essential, and that Scenario 5 could accommodate a wide range of needs. Many expressed surprise at the vast potential of geothermal energy. Although current costs were acknowledged as a challenge, participants argued that, given the scale of recent national Green Transformation (GX) budgets, it is crucial to utilize diverse, energy-efficient renewable sources such as geothermal.

While Scenario 6 was recognized as the ideal long-term objective, supporters argued that, given the financial burden on municipalities, Scenario 5 should be implemented first and expanded gradually. They also believed that strengthening renewable energy policy would simultaneously raise awareness of energy efficiency and insulation. A recurring theme was the importance of national political will, as fundamental transformation would be difficult for municipalities to achieve alone. Participants stressed that both prefectural and national governments must position these measures as policy priorities and foster market development.

3.3.3.1 Scenario 6—Enhanced Energy-Efficient Equipment + Insulation + Diversified Renewable Energy

Supporters of Scenario 6 frequently cited its combined benefits: substantial CO_2 reductions, significant utility cost savings, positive economic ripple effects within the prefecture, and job creation. Although investment requirements are considerable, they argued that Scenario 6 offered the highest overall rationality, given its superior performance in both emissions reductions and economic impacts.

3.3.3.2 Discussion: Implications from the Social Acceptability Analysis

The social acceptability assessment revealed that all scenarios incorporating strengthened measures for energy-efficient equipment, insulation, and renewable heat received high levels of support. Among them, Scenario 4 stood out as the most preferred, primarily due to its ability to deliver multiple benefits simultaneously: sustained reductions in utility costs through improved insulation, reduced health risks during the winter season, and enhanced indoor comfort throughout the year. Participants remarked that "once insulation is installed, the benefits last for many years" and that "in an aging society, reducing winter health hazards is extremely important," underscoring how improvements in health, economic savings, and quality of life resonated strongly with residents.

When including all enhanced scenarios (Scenarios 4 and 6), support for insulation measures exceeded 70% of participants. This indicates that insulation is not merely perceived as an energy-saving measure, but as a foundational policy that safeguards the quality of life and safety of local communities.

In addition, diverse renewable heat sources introduced in Scenarios 5 and 6—such as geothermal and solar thermal energy—were also highly valued for their untapped potential. Participants highlighted that "geothermal and solar thermal systems have higher efficiency than electrification and are easier to utilize as local resources" and that "a diverse energy mix, not dependent solely on electricity, contributes to greater resilience during disasters." These insights emphasize the need to more clearly position renewable heat within regional energy policies and actively promote its deployment.

Overall, the findings demonstrate that the deployment of existing energy-efficient equipment, improved insulation, and renewable heat utilization can generate synergies across three key dimensions:

(1) Environmental benefits—reducing CO_2 emissions and mitigating climate change;
(2) Economic benefits—lowering utility costs and stimulating the local economy through ripple effects;
(3) Social benefits—improving health, comfort, and community well-being, thereby securing resident support.

The analysis also clarified the key conditions that enhance social acceptability:

- Simultaneous realization of multiple benefits—Measures that improve environmental performance, economic savings, health, and quality of life concurrently tend to gain stronger public support.
- Long-term effectiveness and investment certainty—Insulation offers enduring benefits once installed, giving both homeowners and municipalities confidence in the return on investment.
- Utilization of local resources and resilience enhancement—Renewable heat from geothermal or solar thermal sources leverages locally available resources, improves energy self-sufficiency, and strengthens resilience in the face of disasters.
- Immediate deployability and ease of understanding—The fact that the necessary technologies already exist in the market and their benefits are easily understood by residents facilitates acceptance.

These conditions can serve as practical guidelines for designing socially acceptable carbon neutrality policies. In short, the study confirms that the integration of energy-efficient equipment, insulation improvements, and renewable heat utilization should be regarded as core measures in regional CN strategies—ones that simultaneously advance environmental goals, revitalize the local economy, and improve quality of life, while also enjoying broad public support.

4 Conclusion

This study developed and examined a regional carbon neutrality (CN) strategy for Tochigi Prefecture, with particular attention to its social acceptability. Many Japanese regional cities face structural disadvantages compared to large metropolitan areas, including population decline, industrial constraints, and limited access to the financial, human, and technical resources necessary for decarbonization. At the same time, their energy demand is dominated by the residential and commercial sectors, and they possess considerable untapped potential for energy transition through the utilization of local resources. These cities play a decisive role in meeting Japan's CN targets. This makes it essential to design strategies tailored to local contexts and assess their feasibility.

However, domestic CN-related research has tended to focus on technological and institutional evaluations at the national or metropolitan level, with very limited analysis of social acceptability involving diverse stakeholders, including local residents. While numerous scenario studies exist at the national level, relatively few have been localized to the regional scale and examined as concrete policy packages. In Tochigi Prefecture, in particular, no prior studies have assessed decarbonization scenarios incorporating economic and employment impacts, and none have addressed social acceptability. In this respect, the present study is distinctive in the Japanese research context in that it integrates scenario analysis with a social experiment to evaluate environmental, economic, and social dimensions in a unified framework.

The study focused on three categories of measures based on existing technologies: the introduction of energy-efficient appliances, enhancement of building insulation performance, and deployment of renewable heat. Six scenarios were constructed, ranging from measures aligned with national policy to locally tailored, more ambitious interventions (Sect. 1). These scenarios were evaluated not only in terms of quantitative indicators—GHG emission reduction, economic ripple effects, and job creation—but also through empirical testing of social acceptability via the Multicultural Public Sphere Forum (Sect. 2).

Results showed that Scenarios 1 and 2, reflecting current national policy and measures from advanced decarbonization cities, achieved only a 45% reduction from 2013 levels by 2030, falling short of the halving target, and received zero support in the social acceptability assessment. In contrast, the enhanced scenarios (3–6), which included energy-efficient appliances and insulation improvements, achieved 54–64% reductions while outperforming in utility savings, economic ripple effects, and job creation. Scenario 4, which featured enhanced insulation, received the highest support due to its combined economic, health, and comfort benefits, with insulation-related measures garnering over 70% support across all enhanced scenarios. Geothermal and solar thermal technologies were also re-evaluated for their efficiency and local resource attributes, leading to their recognition as high-priority measures for future policy. The analysis identified four key factors linked to high support: (1) multiple co-benefits across environmental, economic, health, and comfort dimensions; (2) long-term effectiveness and investment certainty; (3) improved resilience through local resource utilization; and (4) immediate deployability and intuitive comprehensibility of existing technologies.

Notably, these results came as a major surprise to both the project's co-researchers and many participants. The findings demonstrated that, by making full use of existing energy-efficient and renewable energy technologies, Tochigi Prefecture could reduce CO_2 emissions by 68% by 2030 and achieve CN by 2050. This transformed CN from an "abstract and distant concept" into a tangible and personally relevant goal for many participants when linked to daily life and locally visible technologies. Furthermore, recognizing that the importance of energy efficiency measures remains poorly understood among the general public, numerous participants emphasized the need for active dissemination of such information to residents and communities. Local government officials also pointed out that gaining public understanding and support is essential for advancing ambitious insulation policies in the face of potential resistance from local stakeholders.

These findings align with existing scholarship on socio-technical transitions and social acceptability. Energy transitions are embedded in long-term "lock-in" structures of institutions, infrastructures, and social practices, making technological innovation alone insufficient (Geels 2011; Meadowcroft 2009). This study demonstrates the potential of an integrated approach—iterating between quantitative scenario analysis to visualize impacts and participatory forums to test social feasibility—to help break such lock-in. Viewed through the framework that conceptualizes renewable energy acceptance across "socio-political," "community," and "market" levels (Wüstenhagen et al. 2007), the study simultaneously engaged all three: (a) presenting policy options at the prefectural and municipal levels (socio-political), (b) enabling deliberation and preference expression in community forums (community), and (c) linking these to concrete market actions such as appliance upgrades and building retrofits (market). The high support for insulation, energy-efficient appliances, and renewable heat reflected their fulfillment of core acceptance conditions: co-benefits across multiple dimensions, durable impacts, enhancement of resilience through local resources, and immediate feasibility with clear benefits—thereby empirically supporting the hypothesis that institutionalized information-sharing and dialogue can elevate acceptance.

The scenario content was further informed by international reference cases, including the EU's "100 Climate-Neutral and Smart Cities Mission" and Sweden's advanced insulation programs, with their transferability examined through participatory deliberation. The combination of a mobile lab, action research, and a multicultural public sphere—three methodological pillars of this study—proved adaptable for cross-regional comparison and policy transfer assessment.

Nevertheless, limitations remain. The quantitative analysis focused on 2030, with less attention to mid-term targets for 2035 or 2040. Future work should integrate these with nature-based solutions (NbS), mobility transitions, and deep decarbonization in industrial sectors, while also developing standardized approaches for consensus-building during policy implementation (clarifying who participates, when, with what information, and through which decision-making processes).

In summary, this study empirically demonstrates that, even under the constraints faced by regional cities, policy packages built around existing technologies can deliver concurrent environmental, economic, and social benefits, and that their social acceptability can be enhanced through participatory information-sharing and dialogue. It also underscores that formulating and empirically validating CN strategies in regional contexts is essential to achieving Japan's national decarbonization goals. The integrated "quantitative modeling—participatory governance—implementation evaluation" framework developed here offers a methodological foundation for extending such strategies to other regions and for embedding socially supported CN policies into real-world decision-making at the national scale.

Acknowledgements This research was made possible through funding from the Tochigi Prefecture University–Regional Collaboration Activity Support Program and the Grant-in-Aid for Scientific Research of JSPS (No. 18KT0001). The authors also wish to express their gratitude to Ms. Hiroko Masubuchi (Director, NPO Utsunomiya Environmental Action Forum) and Dr. Noboru Suzuki (Professor Emeritus, Graduate School of Engineering, Utsunomiya University) for their assistance during the workshop.

References

Agency for Natural Resources and Energy, Government of Japan (2021) 2020-nendo enerugī jukyū jisseki (sokuhō) sankō shiryō [FY2020 energy supply and demand results (preliminary) reference materials]. Ministry of Economy, Trade and Industry. https://www.meti.go.jp/press/2021/11/20211126002/20211126002-1.pdf

Bernadett K, Salk CF (2021) Mobile research labs to explore nature-based solutions in Japanese cities: A way to bridge disciplinary divides? Annual Report of the Center for the Multicultural Public Sphere, Utsunomiya University 13:145–154

European Union (2022) EU missions: Climate-neutral and smart cities: List of applicant cities (Updated 31 March 2022). https://research-and-innovation.ec.europa.eu/system/files/2022-03/list_of_public_eligible_applicant_cities-v7.pdf

Fernández A, Lindwall A, Soulis M, Kim S (2023) Building a cleaner city: Insights from Swedish model cases. Lund University, International Institute for Industrial Environmental Economics (IIIEE): CMPS Working Paper, 2022(2) https://uuair.repo.nii.ac.jp/records/2000743

Geels FW (2011) The multi-level perspective on sustainability transitions: Responses to seven criticisms. Environ Innov Soc Trans 1(1):24–40. https://doi.org/10.1016/j.eist.2011.02.002

Intergovernmental Panel on Climate Change (2021) Climate change 2021: The physical science basis. Contribution of Working Group I to the Sixth Assessment Report of the Intergovernmental Panel on Climate Change. In: Masson-Delmotte V et al (eds) Cambridge University Press. https://doi.org/10.1017/9781009157896

Kurt L (2017) Shakai-teki kattō no kaiketsu [Resolving social conflicts] (Original work published 1946 as Action research and minority problems. J Soc (2): 34–46. Chitose Press

Meadowcroft J (2009) What about the politics? Sustainable development, transition management, and long-term energy transitions. Policy Sci 42(4):323–340. https://doi.org/10.1007/s11077-009-9097-z

Ministry of the Environment, Government of Japan (2021a) Jichitai haishutsuryō karute: 09 Tochigi-ken karute isshiki, 09201 Tochigi-ken, Utsunomiya-shi [Municipal emission profile: Tochigi Prefecture and Utsunomiya City]. https://www.env.go.jp/policy/local_keikaku/tools/karte.html (in Japanese)

Ministry of the Environment, Government of Japan (2021b) Chikyuu ondanka taisaku keikaku [Plan for global warming countermeasures] Cabinet Decision, October 22, 2021

Ministry of the Environment, Government of Japan (2022) Datsutanso senkou chiiki (dai 2 kai) sentaku kekka ni tsuite [Results of the 2nd selection of leading decarbonization areas] November 1, 2022. https://policies.env.go.jp/policy/roadmap/assets/preceding-region/2st-DSC-kekka.pdf

Ministry of the Environment, Government of Japan (2023) 2050-nen nisankatanso haishutsu jisshitsu zero hyōmei jichitai (2023-nen 2-gatsu 28-nichi jiten) [List of municipalities declaring net-zero CO_2 emissions by 2050 (as of February 28, 2023)]. https://www.env.go.jp/content/000115115.pdf

REN21 (2021) Renewables 2021 global status report. https://www.ren21.net/wp-content/uploads/2019/05/GSR2021_Full_Report.pdf

Takahashi W (2023) Hajimeni [Introduction]. Annual Report of the Center for the Multicultural Public Sphere, Utsunomiya University, 15.

Takahashi W, Imade Y, Miyake T et al (2022) Udaisei to NPO ga kangaeta Utsunomiya no jizoku kanona enerugii: Odoroki no saiseikanona enerugii no potensharu [Sustainable energy in Utsunomiya: Surprising potential of renewable energy: a report jointly developed by Utsunomiya University students and an NPO] CMPS Working Paper, 2022(1). (in Japanese) https://uuair.repo.nii.ac.jp/search?page=1&size=50&sort=custom_sort&search_type=2&q=962

Takahashi W, Utagawa M, et al (2023) Kābon nyūtoraru na Tochigi-ken o mezashite: Nihon no mōten enerugī kōritsu kaizen o kangaeru [Aiming for a carbon-neutral Tochigi Prefecture: Considering Japan's blind spot—Energy efficiency improvement]. Center for the Multicultural Public Sphere Working Paper Series, 5

United Nations Environment Programme (2022) Emissions gap report 2022: The closing window—Climate crisis calls for rapid transformation of societies. https://www.unep.org/emissions-gap-report-2022

Wüstenhagen R, Wolsink M, Bürer MJ (2007) Social acceptance of renewable energy innovation: An introduction to the concept. Energy Policy 35(5):2683–2691. https://doi.org/10.1016/j.enpol.2006.12.001

System Evaluation of the Optimal Energy Mix

The Fundamental Framework of Green Energy Transition in the Era of Uncertainty

Weisheng Zhou

1 Classification of Energy and the Background of the Green Transition

The relationship between humanity and energy is essentially the history of civilization itself. In agrarian societies, social structures were built upon natural energy sources such as sunlight, water, wind, and biomass. Since the Industrial Revolution, the use of fossil fuels—coal, oil, and natural gas—expanded rapidly, dramatically enhancing productive capacity and living standards, while simultaneously creating new challenges, including environmental degradation and climate change. In the latter half of the twentieth century, nuclear power emerged, diversifying energy supply, yet raising issues related to safety and waste management. Entering the twenty-first century, the development of renewable energy and decarbonization technologies has become urgent, representing an ongoing challenge to realize a sustainable society. Human history can thus be understood as a series of energy transitions, and moving forward, it will be essential to advance to new stages while seeking harmony among environmental, economic, and social dimensions.

Primary energy sources on Earth can be classified according to their origin (geo-physical versus artificially created) and technological dependency, as shown in Table 1.

When categorizing the major primary energy sources on Earth according to frameworks commonly employed in policy, technological, and industrial contexts, they can be broadly divided into three categories: fossil fuels, renewable energy, and nuclear power. This tripartite classification is based on considerations such as energy origin, sustainability, environmental impact, and supply security, with each category exhibiting distinct characteristics and challenges. Consequently, this framework is

W. Zhou (✉)
College of Policy Science, Ritsumeikan University, Osaka, Japan
e-mail: zhou@sps.ritsumei.ac.jp

W. Zhou (ed.), *Decarbonization Policy Engineering*, World Sustainability Series,
https://doi.org/10.1007/978-981-95-5953-4_10

Table 1 Four major energy sources on earth

Category	Representative energy	Origin	Characteristics
Category 1	Solar photovoltaic, solar thermal energy, wind power, hydropower, biomass, and fossil fuels	Solar nuclear fusion	Renewable energies such as solar, wind, hydro, and biomass constitute the main sources, characterized by extremely low greenhouse gas emissions and minimal environmental burden. Conversely, fossil fuels, though belonging to the same origin, excel in supply stability and energy density but present problems such as CO_2 emissions and resource depletion. Renewable energies are practically inexhaustible, and in recent years, their generation costs have been rapidly declining
Category 2	Geothermal energy and nuclear fission (nuclear power)	Radioactive decay and physical phenomena within the Earth	CO_2 emissions during operation are low. Geothermal energy is geographically constrained but has low environmental impact and is sustainable in the long term. Nuclear fission enables high-density and stable electricity supply and is economically advantageous; however, challenges include radioactive waste, accident risks, and the finiteness of fuel resources
Category 3	Tidal energy	Gravitational interactions between the Earth and the Moon (and the Sun)	CO_2 emissions are virtually zero, and because it is based on celestial mechanics, the energy source is effectively inexhaustible. However, the initial investment in generation facilities is large, and potential installation sites are geographically limited
Category 4	Artificial nuclear fusion (including potential future use of antimatter)	Advanced technological creation by humankind	This category includes artificial nuclear fusion and potential future use of antimatter; if realized, these technologies would produce no CO_2 emissions and impose an extremely small environmental burden. Fuels such as deuterium and lithium are abundant and highly inexhaustible, potentially offering long-term economic advantages, but at present, commercialization has not been achieved, and cost and technological uncertainties remain significant

not merely a technical taxonomy; it serves as a fundamental analytical tool for structuring discussions on energy mix composition, decarbonization policy, and energy security. Nonetheless, from an academic perspective, the classification of energy sources is more nuanced, and the tripartite scheme does not fully capture all relevant dimensions. For instance, "zero-emission fuels" such as hydrogen, synthetic fuels (e-fuels), and ammonia are not primary energy sources but rather secondary energy carriers, rendering their placement within the threefold classification challenging, yet they play a pivotal role in decarbonization efforts.

The green energy transition refers to the shift from the energy structures established since the First Industrial Revolution, which have been predominantly dependent on fossil fuels, toward a sustainable energy system underpinned by renewable energy and low-carbon or carbon–neutral technologies. This transition constitutes a core element of climate change mitigation strategies and represents a comprehensive structural transformation spanning economic, social, technological, and institutional dimensions. Notably, the levelized cost of electricity (LCOE) for solar photovoltaics, once considered prohibitively high, fell below that of coal-fired power around 2022, demonstrating the economic feasibility of the green transitional landmark development in energy history (Zhou et al. 2021).

Moreover, as climate change mitigation and decarbonization progress, optimizing the energy mix—i.e., constructing a "best mix"—has become increasingly critical. This approach involves accelerating the deployment of renewable energy while compensating for its inherent variability through short-term reliance on fossil fuels for supply stability, and, in the long term, integrating nuclear power, energy storage technologies, hydrogen energy, and other low-carbon options to achieve a flexible and reliable energy system.

In recent years, in addition to the three primary energy sources, new technologies and approaches such as hydrogen, carbon capture, utilization, and storage (CCUS), energy storage systems, and energy carriers have emerged, rendering energy systems increasingly complex. Consequently, energy policy can no longer rely solely on a single technology or resource; instead, an integrated, multi-faceted approach is indispensable (Fig. 1).

Therefore, the future green transition of energy requires a strategic perspective that moves beyond binary thinking, such as "fossil fuels versus renewable energy," and instead leverages the unique characteristics of each energy source and technology in a context-appropriate manner to construct a sustainable and stable energy system. A flexible and effective configuration based on optimal combinations rather than exclusion is key to facilitating a smooth transition to a decarbonized society.

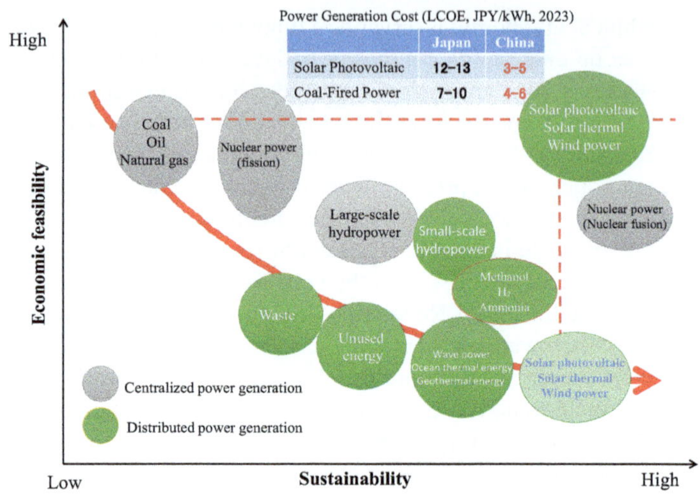

Fig. 1 Schematic Diagram of the Characteristics of Energy Supply (Created by the author)

2 Pathways for the Green Energy Transition Under Uncertainty

2.1 Fundamental Principles of Energy Policy

Simultaneously achieving economic development, environmental protection, and energy security is not merely an ideal; it constitutes a strategic prerequisite for ensuring the long-term stability and prosperity of society. Moreover, in the context of increasing climate change impacts, geopolitical risks, and heightened uncertain-ty, the framework of "3E + 2S", as illustrated in Fig. 2, can be regarded not on-ly as a fundamental principle of domestic energy policy but also as a foundation for international cooperation and trust-building.

The mandate of comprehensive energy policy is to envision the desired future structure of energy supply and demand in response to the policy requirements of each era. To harmonize the elements of 3E + 2S, policymakers must implement various measures in a flexible and proactive manner, ensuring strategic coordination across sectors.

(1) Stability: Secure Energy Supply (Energy Security)

Energy security refers to the uninterrupted provision of energy necessary for both daily life and industrial activity, even in the face of unforeseen events such as geopolitical crises or natural disasters. Achieving this requires securing a diverse mix of energy sources and supply routes, maintaining sufficient reserves, and improving domestic self-sufficiency. To some extent, demand response mechanisms—where households and businesses voluntarily reduce electricity consumption in response

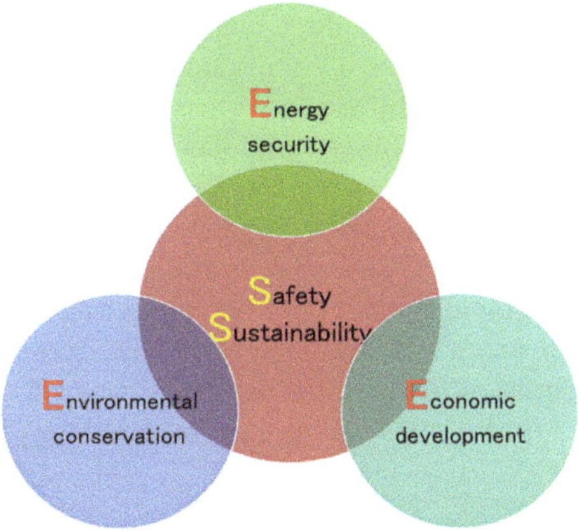

Fig. 2 Fundamental principles of energy policy (3E + 2S)

to utility signals or incentives—are employed. Nevertheless, from the supply-side perspective, ensuring stable energy provision remains the foremost priority.

(2) Economic Development

From the perspective of economic development, it is essential to maintain appropriate energy prices, enhance corporate competitiveness, and alleviate the financial burden on households. In addition, stimulating the economy through efficient energy use, technological innovation, and infrastructure investment is also of critical importance.

(3) Environmental Conservation

In terms of environmental conservation, key objectives include the reduction of greenhouse gas emissions, prevention of air, water, and soil pollution, and protection of ecosystems. Achieving these goals requires the deployment of renewable energy, promotion of energy efficiency, and active utilization of low-carbon tecnologies.

(4) Safety

Safety entails minimizing risks of accidents and disasters across all stages of the energy system, including power generation facilities, fuel transportation, and waste management, as well as mitigating threats such as terrorism. For nuclear and fossil-fuel-based facilities in particular, strict compliance with safety standards and robust risk management are indispensable.

(5) Sustainability

Sustainability refers to the preservation of resources, environmental integrity, and economic stability for future generations. This requires expanding the use of renewable resources, building a circular economy, and transforming the energy system with a long-term perspective.

Balancing the three elements of 3E is crucial, ensuring that the pursuit of one objective does not compromise the others. The two elements of 2S serve as foundational conditions that crosscut all aspects of 3E. In policy design, reconciling short-term stability with long-term sustainability is a key challenge.

The green energy transition is driven precisely by these multifaceted requirements represented by 3E + 2S, encompassing climate change mitigation, finite re-source management, and energy security. However, the pathway is subject to significant uncertainties—not only due to climate variability but also arising from technological, economic, political, and social factors that influence the trajectory of the green transition. This section examines strategic responses that can enable the green energy transition, considering these inherent uncertainties.

2.2 *Types of Uncertainty Affecting the Green Energy Transition and Their Impacts*

2.2.1 Technological Uncertainty: Maturity and Diffusion of Key Technologies

The core technologies supporting the decarbonization of energy systems—such as hydrogen energy, energy storage, CCUS (Carbon Capture, Utilization and Storage), nuclear fusion, and smart grids—face significant uncertainty regarding whether they will achieve large-scale breakthroughs over the next 10 to 20 years. Each of these technologies continues to encounter technical challenges that must be overcome for practical deployment and large-scale adoption. At present, it remains difficult to predict the pace at which these technologies will mature and the extent to which they can diffuse cost-effectively.

For example, if battery costs do not decline as anticipated, it will become challenging to manage the variability of renewable energy output, and the flexibility of the transmission and distribution network may not be ensured. Similarly, hydrogen production, storage, and transportation technologies, as well as CCUS infrastructure, require substantial investment, and any delays in their development could impede the overall green transition strategy.

Conversely, as shown in Fig. 3, the IRENA report Renewable Power Generation Costs 2024 indicates that, on a levelized cost of electricity (LCOE) basis, renewable energy remains the most cost-competitive option for new power generation in 2024. Approximately 91% of newly commissioned utility-scale capacity supplied electricity at lower cost than the cheapest newly installed fossil-fuel alternatives. Importantly, wind power has become more affordable than fossil fuels in all major markets.

Over the past decade, the rapid decline in LCOE for renewable energy has fundamentally altered the rules of the energy economy. Globally, the LCOE of concentrated solar power fell from \$0.46/kWh in 2010 to \$0.044/kWh in 2023, representing

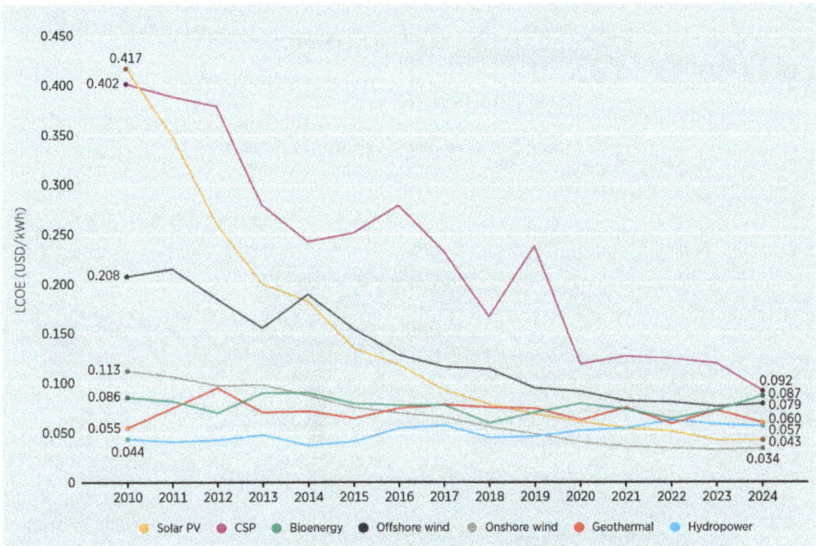

Notes: CSP = concentrated solar power; kWh = kilowatt hour; LCOE = levelised cost of electricity; PV = photovoltaic; USD = United States dollar.

Fig. 3 Decline in LCOE of renewable energy (2010–2024) (IRENA 2024)

a reduction of over 90% (IRENA 2024). In particular, the widespread adoption of wind, solar, and battery technologies has driven prices to historic lows, with the cost of generating 1 MWh of electricity in China being up to 64% lower than in other markets (Stone 2025).

Figure 4 presents the authors' estimated comparison of the levelized cost of electricity (LCOE) between coal-fired power and solar photovoltaic (PV) generation (Zhang et al. 2023). In the baseline scenario, the comparison shows that in 2022, solar PV and coal-fired generation in China reached the same cost level of 0.35 CNY/kWh. By 2030, the cost of solar PV is projected to decrease to 0.1 CNY/kWh, whereas coal-fired power is expected to decline only slightly to 0.29 CNY/kWh. This indicates that, from the perspective of generation costs, solar PV has become more economically advantageous than coal-fired power since 2022.

Furthermore, the actual LCOE values in 2024 are approximately 0.32–0.45 CNY/kWh for coal-fired power (Stuart 2025) and about 0.2 CNY/kWh for solar PV (CREEI 2025), closely matching the estimated results. The LCOE of renewable energy, including solar PV, is expected to decline further due to ongoing technological advancements and market development.

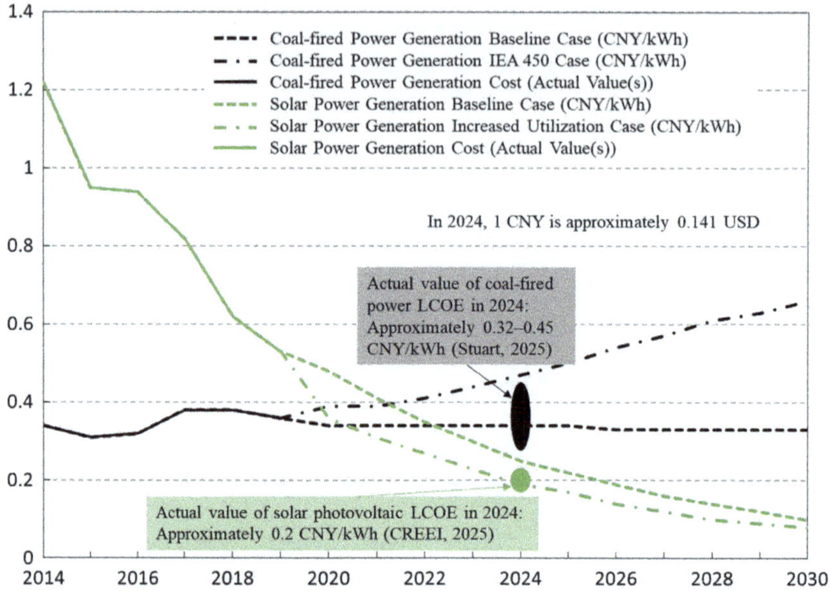

Fig. 4 Actual and estimated LCOE of coal-fired and solar power in China (Zhang et al. 2023)

2.2.2 Economic Uncertainty: Investment Risks Associated With Price and Market Fluctuations

Many investments essential for the green energy transition are made with the expectation of long-term returns. However, economic conditions are constantly changing, creating significant uncertainty in investment decisions. Factors include carbon price trends (e.g., the timing and level of carbon pricing implementation), the profitability of investments in renewable energy infrastructure, and price surg-es or supply instability in critical resources such as rare metals. Notably, as renewable energy revenue models shift from feed-in tariffs (FIT) to feed-in premiums (FIP) or market-linked mechanisms, the predictability of returns for project developers has declined. Furthermore, minerals critical to electrification technologies—such as lithium, cobalt, and nickel—are heavily concentrated in specific countries, exposing them to geopolitical risks. These economic and market fluctuations are thus key determinants of capital flows into the energy transition.

2.2.3 Political and Institutional Uncertainty: Lack of Policy Consistency and Regulatory Stability

Long-term policy consistency and stable regulatory frameworks are indispensable for advancing the energy transition. Nevertheless, political and institutional uncertainties persist, including government transitions, uncertainties in the imple-mentation of international agreements (e.g., the Paris Agreement), sudden cuts or revisions of subsidy programs, and stricter safety regulations. For instance, if a subsidy program promoting renewable energy deployment is abruptly reduced or withdrawn, prior investments may become difficult to recover, undermining the confidence of the private sector. Ensuring the predictability of energy policy and the sustainability of institutional design is therefore essential for a robust green transition.

2.2.4 Uncertainty in Social Acceptance: Local Opposition and Challenges in Trust Building

Even when technical and institutional conditions are met, energy infrastructure development cannot proceed without social acceptance. Notably, opposition from residents—often framed as "Not in My Backyard" (NIMBY) reactions—against the installation of wind or solar power due to landscape or environmental concerns has been observed in various regions. Additionally, public resistance to rising electricity costs and persistent apprehension or distrust toward new technologies such as nuclear power or hydrogen exist. Although these social sentiments and values may appear irrational at first glance, they are deeply rooted in local lifestyles and historical contexts and can only be addressed through careful dialogue and trust-building. Uncertainty in social consensus formation can thus become a bottleneck in the implementation of technologies and policies.

2.2.5 Natural and Environmental Uncertainty: Variability of Renewables and the Bidirectional Impacts of Climate Change

Renewable energy is highly dependent on natural conditions, resulting in significant output variability due to weather, seasonal changes, and regional characteristics. Such "natural variability" poses a major challenge for the stable operation of energy systems. Moreover, climate change itself exerts diverse effects on both energy demand and supply capacity—for example, surges in electricity demand due to heatwaves, reductions in hydropower capacity due to water shortages, and infrastructure damage caused by typhoons or floods. These natural and environmental uncertainties not only influence the pace and direction of the green transition but also necessitate strengthened risk management and redundant system design.

Overall, the green energy transition is continually challenged by uncertainties across five dimensions: technological, economic, institutional, social, and natural. These uncertainties interact in complex ways, affecting policy formation and decision-making. Designing strategies for the green transition under such conditions requires multi-layered, flexible approaches and the sharing of risk across society.

2.3 *Pathways of Energy Transition Under Uncertainty*

The formulation and implementation of energy transition policies are inherently characterized by uncertainty. In particular, energy transitions involving decarbonization must navigate a wide array of intertwined factors, including the pace of climate change, the speed of technological advances in renewable energy, hydrogen, and nuclear power, geopolitical risks, economic growth rates, and shifts in consumer behavior. As a result, the conventional approach of "predicting the future and selecting the optimal solution" has its limitations, making it necessary to develop new frameworks for strategic design and policy formation that explicitly incorporate uncertainty.

Against this backdrop, what becomes essential is the policy engineering approach outlined in Chap. 1. Within this perspective, two methodologies have attracted particular attention: "robust optimization" (Ben-Tal et al. 2009) and "adaptive policymaking" (Walker et al. 2013). The former refers to strategies that remain viable and maintain effectiveness across diverse future scenarios, while the latter emphasizes the flexible adjustment and revision of policies in response to changing circumstances. Both approaches provide conceptual frameworks that directly address uncertainty in energy policy and seek to achieve a balance between stability and adaptability.

2.3.1 Robust Optimization: Strategic Rationality to Minimize Regret

The "robust strategy" is characterized by its ability to withstand a relatively broad range of uncertain future scenarios and not rely on a single "optimal solution." Concrete examples include diversification of the power supply portfolio, parallel development of multiple technological options, and avoidance of dependence on specific resources. In the energy mix, a robust strategy could aim for substantial integration of renewable energy while maintaining a certain level of other low-carbon power sources, such as nuclear and natural gas, to create a structure capable of flexibly responding to changes in technological and societal conditions. Moreover, in setting policy targets, it is more important to adopt a minimax regret approach—focusing on whether catastrophic failure can be avoided even under the worst-case scenario—rather than pursuing an "ideal optimal point."

2.3.2 Adaptive Policymaking: Dynamic and Evolutionary Rationality

On the other hand, adaptive policymaking assumes that future conditions are unpredictable and designs policies with built-in flexibility, implementing them through iterative learning and adjustment. This approach is analogous to biological evolution, emphasizing policies that can be modified or redesigned rather than fixed, immutable decisions. Adaptive policymaking incorporates regular monitoring and evaluation, automated policy adjustments based on pre-determined trigger points, and learning processes through pilot projects or experimental deployment. In energy policy, practical expressions of adaptive policymaking include social implementation of emerging technologies, trial-and-error in institutional design, and adjustments according to regional conditions.

2.3.3 Key Differences and Characteristics of Robust Strategy and Adaptive Policymaking

Table 2 summarizes the main differences and characteristics of "Robust Strategy" and "Adaptive Policymaking." Robust optimization offers rationality through risk minimization by avoiding failures ex ante, whereas adaptive policymaking provides rationality by continuously adapting to changes through ex post learning and adjustments. Both are effective strategies for dealing with uncertainty: robust strategy provides "resilience," while adaptive policymaking provides "adaptiveness.“ These approaches are not mutually exclusive but rather complementary. Robust optimization provides a stable framework that can withstand uncertainty, while adaptive policymaking flexibly reinforces and adjusts that framework in response to changing conditions. This integrated approach to policy design embodies the stance required in contemporary green energy policy: not to avoid uncertainty but to strategically respond to it.

2.3.4 Case Studies of Policy Formation in Different Countries

These concepts are already reflected in the policy formation of several countries. In Japan, particularly after the Great East Japan Earthquake, energy policy has shifted toward a framework that emphasizes "robustness." In the 2024 formulation of the 7th Basic Energy Plan, the policy maintains a renewable energy share of 36–38% in the 2030 power generation mix, while keeping nuclear power at 20–22% and natural gas-fired power in the low 20% range. Additionally, the plan explicitly highlights new priority measures, including the utilization of hydrogen and ammonia-based power generation, grid reinforcement, and accelerated deployment of energy storage systems, thereby facilitating a more flexible and distributed power supply structure. This approach anticipates "scenario-transcending stability," in which alternative low- and zero-carbon energy sources and demand–supply adjustment mechanisms function effectively even if renewable energy deployment does not progress as expected.

The recent introduction of e-methane (synthetic methane), hydrogen, and ammonia in Japan exemplifies adaptive policymaking. Given that these technologies still involve considerable uncertainty in terms of costs and supply infrastructure, the government has established mechanisms that allow gradual deployment alongside technological development support, enabling flexible long-term decisions regarding adoption. This approach demonstrates that the government pro-vides a clear strategic

Table 2 Key differences and characteristics of robust strategy and adaptive policymaking

Comparison items	Robust strategy	Adaptive policymaking
Basic Concept	Building strategies that are "failure-resistant" and can withstand future uncertainties	An approach that assumes uncertainty and "learns and adjusts" policies in response to changes
Objective	To achieve stable outcomes across multiple scenarios	To flexibly respond to future changes while ultimately achieving policy goals
Focus of the Approach	"Minimizing regret": avoiding the worst-case outcomes	"Incremental learning": proceeding by updating decisions step by step
Characteristics of Strategy Design	Diversification of technologies and measures, ensuring redundancy	Trial introduction, phased implementation, and sequential adjustment based on monitoring
Timing of Decision-Making	Decisions often made at the initial stage with a wide range of countermeasures in mind	Ensuring flexibility at the initial stage, with sequential decisions made as needed
Strengths	High stability and reliability; able to withstand sudden changes to some extent	Excellent flexibility and responsiveness; adaptable even to unknown situations
Weaknesses	May become inefficient due to excessive safeguards	Initial policies may lack clarity; frequent policy changes may cause confusion
Situations Suitable for Application	Risks where changes are unpredictable but potential losses are significant (e.g., power mix, critical infrastructure development)	Fields with great uncertainty in technological progress or social trends (e.g., hydrogen, CCUS, and other emerging technologies)
Examples (Japan)	Diversification of power sources through the combined use of renewables, nuclear, and natural gas	Phased support for the introduction of e-methane and hydrogen; policy adjustments through regional demonstrations
Examples (China)	National strategy of the "2060 Carbon Neutrality" target	Phased introduction of the Emissions Trading System (ETS); policy implementation that accounts for regional differences

direction while retaining flexibility in the implementation process, embodying the principles of adaptive policy design.

In contrast, China announced in 2020 the goals of peaking CO_2 emissions by 2030 and achieving carbon neutrality by 2060. Although highly ambitious, the policy implementation strategy emphasizes adaptability. For example, it allows differentiated roadmaps according to regional economic structures and energy re-source availability, rather than enforcing a uniform national approach. Advanced urban and coastal areas are encouraged to accelerate renewable energy deployment, whereas inland and energy-intensive industrial regions are permitted to implement phased reductions.

Furthermore, China launched the national carbon emissions trading system (ETS) in 2021, initially covering only the power generation sector. From March 2025, the scheme expanded to include cement, steel, and aluminum industries, with further expansions planned. This represents an adaptive approach in which the system is implemented on a small scale initially and gradually matured over time, rather than attempting to complete the entire framework at once. On the technological front, the government demonstrates flexibility not only toward established renewable energy sources such as solar and wind but also toward emerging technologies like hydrogen and CCUS, with government-led demonstration projects being actively promoted across the country.

As noted above, Japan has presented a medium- to long-term power generation structure with a strong emphasis on robustness, while adopting a cautious and phased approach to the introduction of new technologies. In contrast, China has established a nationwide direction but seeks policy learning and evolution through flexible, region- and sector-specific responses and experimental institutional de-sign. In both cases, attention should be paid to how each country leverages its strengths through different approaches in addressing the common challenge of uncertainty.

In conclusion, both robust optimization strategies and adaptive policymaking are theoretically and practically advanced methodologies, as they internalize uncertainty and incorporate mechanisms to manage it within institutional design. By combining the two approaches, a "dynamic and resilient green transition strategy" can be constructed, capable of withstanding change while remaining flexible. Energy policy has evolved beyond mere supply issues to encompass complex challenges spanning economic structures, local communities, and environmental jus-tice. Within this context, approaches that manage uncertainty while harnessing it creatively should form the core of strategic design for a sustainable future. As the energy transition becomes an increasingly complex and long-term process, these methodologies are likely to serve as important models not only at the national level but also in international policy formation. Strategic design that combines robustness and adaptability is therefore a crucial methodology for overcoming the uncertainties inherent in the energy transition.

3 The Fate of Fossil Fuels in the Green Transition

3.1 Socioeconomic Structures Dependent on Fossil Fuels

The First Industrial Revolution, which began in the late eighteenth century, was driven by the invention of the steam engine and the widespread use of coal. During this period, coal became the primary source of power, supporting the rapid development of industrial foundations such as the textile industry, steel production, and

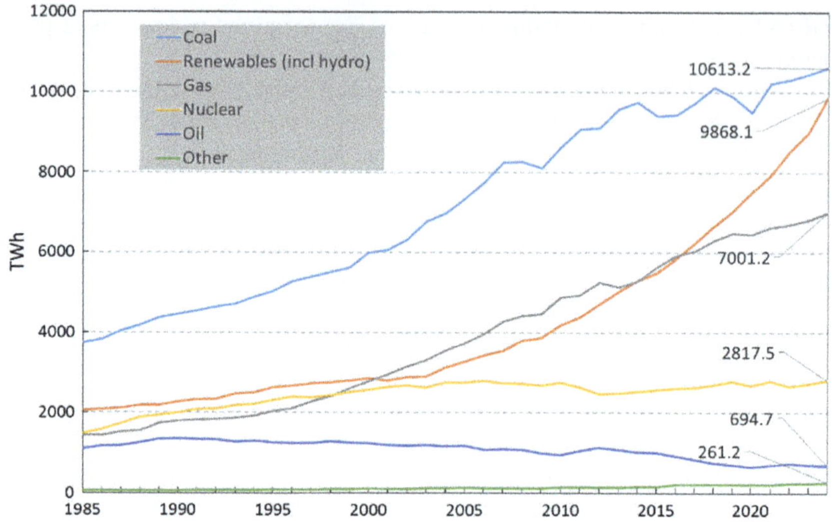

Fig. 5 Changes in the global electricity structure (Energy Institute 2025)

railways. This marked the First Energy Revolution, characterized by a structural shift from traditional energy sources such as wood and hydropower to fossil fuels.

From the late nineteenth century to the early twentieth century, the proliferation of electric-ity and petroleum led to the Second Energy Revolution. Electricity brought inno-vation to urban infrastructure and domestic life, while petroleum, coupled with the development of automobiles and airplanes, transformed transportation systems dramatically. In particular, the mass production of Ford's Model T starting in 1908 became a symbol of 20th-century industrial civilization. As shown in Fig. 5, fossil fuels remain the dominant source of global energy supply (Energy Institute 2025).

This acceleration of industrialization and energy use gave rise to an economic model characterized by mass production, mass consumption, and mass disposal. Companies pursued efficiency and economies of scale, while consumer society enjoyed convenience and comfort; simultaneously, product lifespans shortened, recycling rates remained low, and resource waste became normalized.

At the same time, industrial activities and urbanization exacerbated environmental problems such as air pollution, water contamination, and soil degradation. By the latter half of the twentieth century, both local and global environmental issues—such as acid rain, ozone depletion, and climate change—became apparent.

Thus, while fossil fuels have supported economic growth and technological progress in modern society, they have also caused severe distortions in social, environmental, and resource sustainability. In other words, 20th-century civilization is characterized by an "economic structure dependent on fossil fuels" and a "social structure oriented toward material abundance."

3.2 Global Trends in Fossil Fuel Use

Against the backdrop of intensifying climate change and accelerating interna-tional responses, the "green transition" of energy systems is progressing worldwide. The transition to a low-carbon and decarbonized society based on renewable en-ergy has become an unavoidable challenge for all countries. In this process, the role of fossil fuels, which have long been the mainstay of energy supply, is increasingly being destabilized. The advancement of the green transition does not merely imply a "departure from fossil fuels"; it raises strategic and institutional questions regarding how these resources should be positioned and gradually reduced or transformed.

A major turning point in this context was the 26th Conference of the Parties (COP26) (Table 3). In the final document of the Glasgow Climate Pact, the phrase "phase-down of unabated coal power" was included for the first time in a United Nations document. The explicit mention of coal reduction in an interna-tional agreement was itself groundbreaking and symbolically accelerated the de-carbonization of coal in various countries. Furthermore, coal was clearly positioned as a key pillar alongside measures such as methane reduction and forest conservation, in addition to greenhouse gas mitigation.

In response to this trend, many countries, particularly in Europe, have introduced policies to gradually phase out coal-fired power generation and reduce dependence on oil and natural gas. Notably, at the G7 Climate, Energy, and Environment Ministers' Meeting held in Italy in April 2024, the participating countries agreed to phase out coal-fired power plants without accompanying CO_2 reduction measures by 2035, issuing a joint statement (G7 2024). This agreement aligns with the "transition away from fossil fuels" direction adopted at COP28 in Dubai in 2023 and represents the first time the G7 explicitly specified a timeline for coal phase-out, marking progress in international norm-setting.

However, in the United States, in addition to the Trump administration's withdrawal from the Paris Agreement, the U.S. Department of Energy (DOE) released the Climate Impact Reassessment Report in late July 2025. This report questioned, from a scientific perspective, climate models, the risks of greenhouse gases, and the frequency of extreme weather, while legally supporting the Environmental Protection

Table 3 Major developments surrounding coal-fired power

November 2021	At COP26, Parties agreed to the phase-down of unabated coal power
May 2023	At the G7 Summit, leaders committed to accelerating the phase-down of fossil fuels, including coal
December 2023	At COP28, Parties agreed to transitioning away from fossil fuels in energy systems
April 2024	At the G7 Climate, Energy and Environment Ministers' Meeting, members reaffirmed their commitment to achieving a fully or predominantly decarbonized power sector by 2035

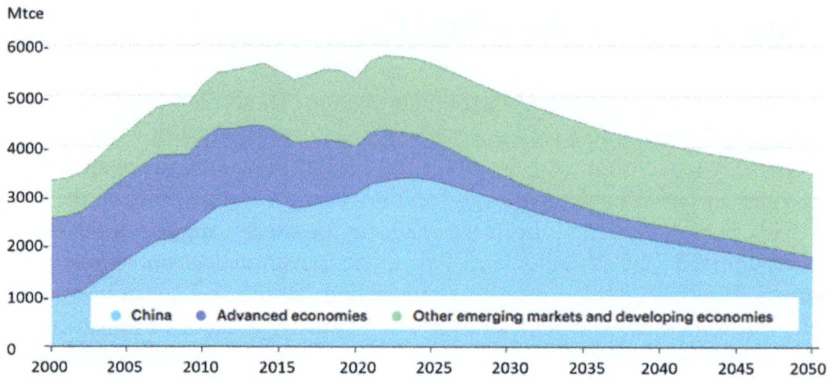

Fig. 6 Coal demand by region in the stated policies scenario, 2000–2050 (IEA 2023)

Agency's (EPA) repeal of greenhouse gas regulations (DOE 2025). As a result, the United States, the world's second-largest emitter, has clearly abandoned a pathway toward carbon neutrality, which is likely to lead to increased fossil fuel consumption.

According to the International Energy Agency (IEA)'s "Net Zero by 2050" roadmap published in 2021, the era of fossil fuels may be approaching its end as one of the consequences of the global energy crisis. Under the IEA's Stated Policies Scenario (STEPS), the transition to clean energy is expected to advance, with demand for coal, oil, and natural gas projected to peak before 2030. Figure 6 illustrates regional coal demand, including that of China, under this policy scenario from 2000 to 2050.

China's trajectory plays a significant role in these projections. The country has set a national target to reach peak carbon emissions by 2030, making reductions in fossil fuel consumption essential for achieving this goal. Indeed, China accounts for approximately 30% of global fossil fuel consumption, reflecting its substantial influence. It still consumes roughly 60% of the world's coal, of which more than 35% is used for power generation (IEA 2023). In this context, the im-pact of the G7 agreement extends beyond the G7 countries themselves and is likely to have considerable spillover effects on China, the world's largest coal consumer.

On the other hand, considering the realities of energy supply and demand as well as economic constraints, the immediate phase-out of fossil fuels remains difficult, and they continue to play a role as "complementary resources during the transition period." As shown in Fig. 7, in both Japan and China, 70–80% of electricity generation still relies on fossil fuels, indicating shared challenges in achieving a green energy transition (IEA 2025a). In this context, the aforementioned G7 agreement is expected to have not only an impact on G7 countries but also a certain ripple effect on China, the world's largest coal consumer. Japan, meanwhile, is advancing the gradual retirement of aging coal-fired power plants while employing "high-efficiency, low-carbon technologies" such as ammonia co-firing, maintaining coal as part of the transitional energy mix for a certain period.

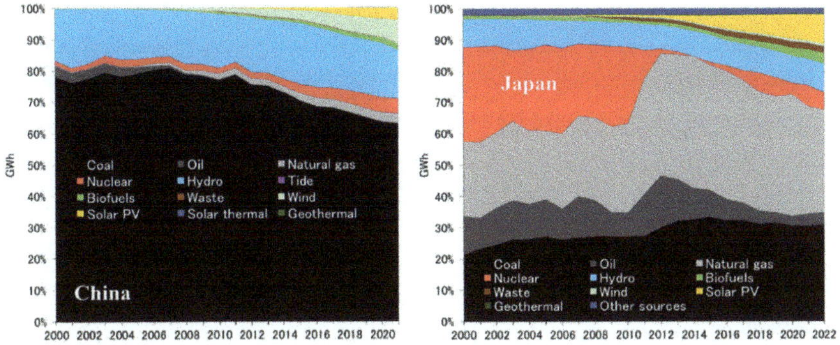

Fig. 7 Electricity generation by source (IEA 2023)

3.3 The Green Transition Is Not the "End of Fossil Fuels"

The green transition in energy should not be understood through a simplistic dichotomy of "fossil fuels = bad." Rather, it represents a technical and institutional challenge aimed at guiding the energy system toward gradual decarbonization while ensuring overall sustainability and flexibility. Therefore, the role of fossil fuels within the green transition must be assessed both in terms of quantitative reduction and qualitative transformation.

A key challenge going forward is the efficient and clean utilization of fossil fuels. Even as decarbonization is globally demanded to mitigate climate change, fossil fuels continue to account for the majority of the world's energy supply. This reali-ty underscores the critical importance of advancing the transition to renewable energy while simultaneously managing the interim use of fossil fuels as efficiently and environmentally responsibly as possible.

Specifically, this involves introducing high-efficiency technologies in power generation and industrial processes, deploying advanced clean coal technologies to reduce CO_2 emissions, and combining these approaches with CCUS technologies, co-firing with ammonia or hydrogen, and the use of synthetic fuels such as e-methane (e-fuel) introduced in Chap. 16. Additionally, mitigating environmental impacts during extraction and transportation—such as preventing methane leaks and reducing flaring—is an urgent priority. In Japan, for example, policies are promoting natural gas as a "transition fuel," gradual phase-out of inefficient coal-fired power, and expansion of co-firing ratios.

One example of clean and efficient coal-fired power generation is the Integrated Gasification Fuel Cell Combined Cycle (IGFC). As illustrated in Fig. 8, IGFC does not combust coal directly but gasifies it to produce a hydrogen-rich gas, which is supplied to a solid oxide fuel cell (SOFC). This is further integrated with a gas turbine and a steam turbine to simultaneously enhance power generation efficiency and CO_2 capture, representing a next-generation thermal power technology (NEDO 2023).

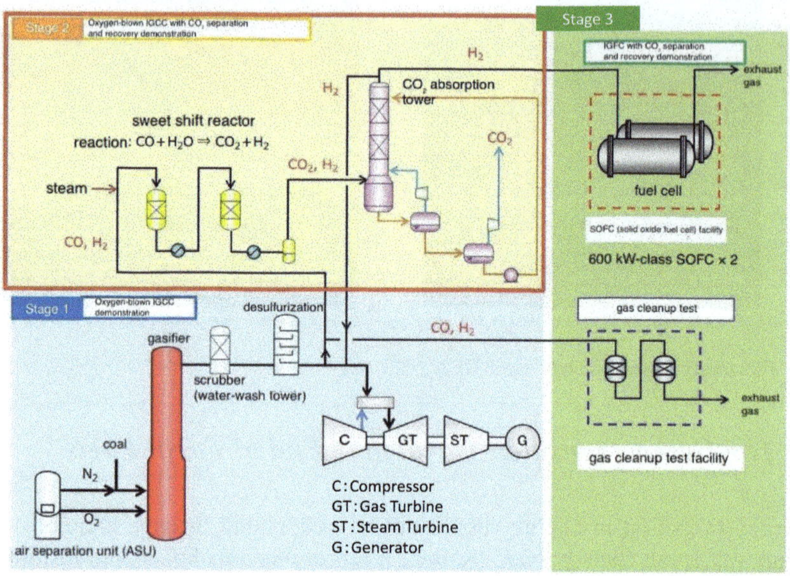

Fig. 8 Overview of the Coal Gasification Fuel Cell Combined Power Generation (IGFC) Demonstration Project (NEDO 2023)

The core of this technology lies in high-temperature gasification of coal to generate a synthesis gas primarily composed of carbon monoxide and hydrogen, with concurrent CO_2 separation and capture to mitigate environmental impact. The high-concentration hydrogen gas is then supplied to the fuel cell for high-efficiency power generation, which, when combined with gas and steam turbines, aims to surpass the efficiency of conventional power generation methods.

IGFC technology is currently under demonstration by Japan's NEDO and Osaki CoolGen Corporation. Results indicate that, when applied to a 500 MW commer-cial-scale system, CO_2 can be captured at a rate of 90%, achieving a grid-side ef-ficiency of 47% on a higher heating value basis. In the third phase of demonstra-tion, a fuel cell was integrated with an oxygen-blown IGCC system equipped with CO_2 separation and capture to verify power generation characteristics and opera-tional feasibility under high-concentration hydrogen supply (NEDO 2023).

Thus, IGFC represents a technology capable of simultaneously achieving significant CO_2 emission reductions and improved power generation efficiency in coal-fired power. With further advancements in CO_2 capture and fuel cell technologies, IGFC is expected to be established as a highly efficient power generation system suitable for practical deployment in the future.

Furthermore, thermal power generation does not compete with renewable energy generation, as discussed later, but in reality functions as a "complementary resource." The expansion of renewable energy depends on the balancing capability of thermal power, while thermal power simultaneously supports the widespread adoption of renewables by incorporating decarbonization technologies. Ultimately, it is desirable to gradually reduce the role of thermal power while transitioning to a stable carbon–neutral electricity system.

Another critical issue is how to avoid the emergence of "stranded assets"—existing fossil fuel infrastructure that loses market value due to the collapse of the carbon bubble, regulatory tightening, technological innovation, or changes in demand. The transition to a decarbonized society and structural industrial changes must be carried out in a socially just and inclusive manner. In particular, it is crucial to design a "Just Transition" that minimizes negative impacts on employment and local economies arising from the contraction of the fossil fuel industry and energy transition, ensuring that no one is left behind. Additionally, considerations of "climate justice" are required, including the alignment of international resource diplomacy and development investments and the accountability associated with cross-border carbon flows. In developing countries, energy access remains a serious challenge, and abrupt fossil fuel phase-out policies could provoke social and economic disruption, necessitating a flexible and phased approach. Such a "green transition of fossil fuels" is increasingly recognized as a practical policy instrument for advancing decarbonization while maintaining economic and geopolitical stability, particularly during the transitional phase of energy transformation.

In summary, the role of fossil fuels in the green transition is not merely one of "termination" but rather of "redefining roles and technologies." Even in a future where renewable energy becomes a major power source, fossil fuels will serve temporarily as complementary resources that support system stability, while also acting as a technological testbed for decarbonization. The key is to strategically design both their utilization and exit strategies; optimizing the "use of fossil fuels" will likely become central to future energy policy.

4 The Prospects of Renewable Energy in the Green Transition

4.1 New Deployment and Investment Trends in Renewable Energy

As previously mentioned, solar and wind power, the core of renewable energy, have seen dramatic cost reductions over the past decade, making them the most economically viable choice for new electricity generation in many regions (Fig. 3). In particular, in developing countries, renewables contribute to expanding energy access as distributed generation in areas without established grids.

As shown in Fig. 9, recent investment trends in the energy sector are driving a major shift toward a more electrified energy system abundant in renewable energy. The adoption of technologies such as solar photovoltaics, electric vehicles, energy

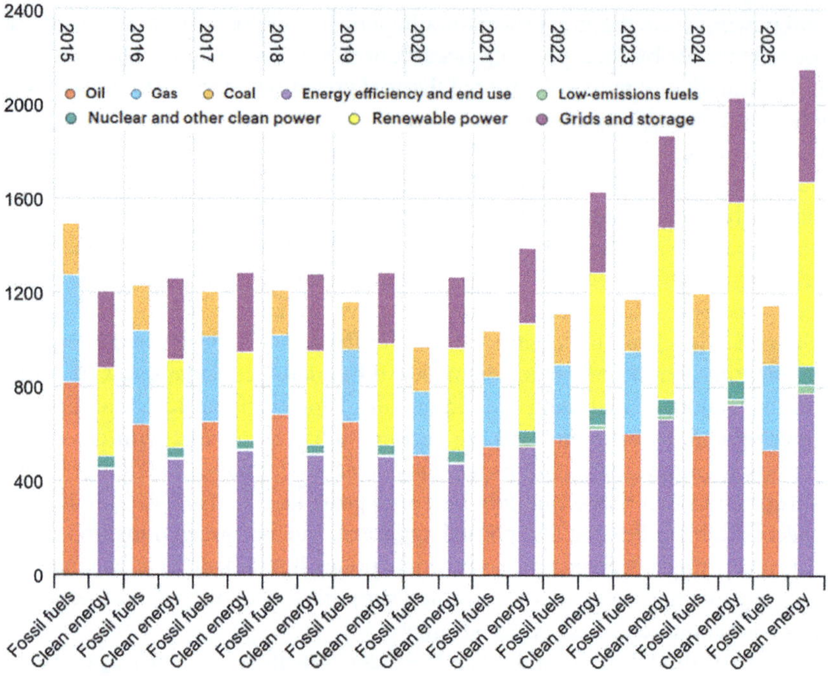

Fig. 9 Global investment in clean energy and fossil fuels, 2015–2025 (IEA 2025b)

storage, and heat pumps is accelerating, and overall investment is rapidly shifting toward clean energy (IEA, 2025).

In 2024, global energy investment exceeded USD 3 trillion, of which approximately USD 2 trillion was directed toward clean energy technologies and related infrastructure. In contrast, investment in fossil fuels amounted to just over USD 1 trillion. In other words, for every dollar invested in fossil fuels, roughly two dollars were invested in clean energy. This ratio was 1:1 five years ago. This trend is par-ticularly pronounced in advanced economies and China. In 2024, clean energy investment reached approximately USD 675 billion in China, USD 370 billion in Europe, and USD 315 billion in the United States, with these three regions accounting for more than two-thirds of global clean energy investment (IEA 2025b).

The main factors driving the recent surge in clean energy investment are as follows:

(1) Strengthened government policy support and fiscal measures

The rapid increase in clean energy investment in recent years has been driven by multiple factors, including policy, economic, and geopolitical considerations. Among these, strengthened government support and fiscal measures have been central to expanding investment. In the United States, the Inflation Reduction Act (IRA); in the European Union, the Green Deal Industrial Plan; and in China, large-scale renewable energy subsidy programs have all been implemented, promoting capital inflows into the clean energy sector through national-level subsidies and tax incentives. Moreover, many countries have set carbon neutrality targets around 2050 and have explicitly positioned the expansion of clean energy as a key policy instrument for achieving these goals.

(2) Declining Costs of Renewable Energy Technologies

From a technological perspective, the declining costs of renewable energy technologies have been a major driving force. In particular, the cost of solar photovol-taic (PV) power has decreased by more than 90% over the past decade, making it cheaper than fossil fuels in some regions and promoting widespread adoption. In addition, related technologies such as batteries (especially lithium-ion), wind pow-er, and heat pumps have also rapidly become more affordable, improving their investment returns.

(3) Energy Security and Geopolitical Risks

Geopolitical risks have further accelerated the transition to renewable energy. Since the Russia–Ukraine war, instability in the international supply of fossil fuels has increased, prompting many countries to re-evaluate renewable energy sources that can be supplied domestically. Political instability in regions such as the Middle East and Africa has similarly highlighted the vulnerabilities of fossil fuel dependence, making the shift to renewables strategically rational from an energy security perspective.

(4) Expansion of Financial Markets and ESG Investments

Financial market trends have also had a significant impact. The expansion of ESG (Environmental, Social, and Governance) investment has made funding for clean energy and decarbonization projects easier than ever. In particular, institutional investors in Europe and the United States are increasingly divesting from fossil fuel-related assets due to climate risk considerations, redirecting capital to-ward the clean energy sector.

(5) Infrastructure Development and Strengthening of Power Systems

Finally, expanding renewable energy requires investment in supporting infrastructure. The development of transmission networks and energy storage systems is a prerequisite for renewable energy deployment, and investment in such infrastructure is accelerating in many countries. Moreover, the widespread adoption of electric vehicles and heat pumps necessitates the reinforcement of regional power grids, which itself represents new investment opportunities.

The IEA points out that "the future success of clean energy investment depends on policy consistency, regulatory stability, supply chain development, and lower capital costs," emphasizing that these structural factors will strongly influence fu-ture investment trends (IEA 2025a).

According to the IEA's 2025 report, investment in low-carbon power sources such as solar, wind, and nuclear reached approximately USD 900 billion, nearly ten times the investment in coal and gas-fired power (about USD 92 billion) in the same year (IEA 2025b). This indicates that global energy capital is rapidly shifting from conventional energy sources to sustainable energy systems. This trend is particularly pronounced in advanced economies and China and will have critical im-plications for addressing global climate change, ensuring energy security, and achieving the Sustainable Development Goals (SDGs).

4.2 Advanced Cases in the Field of Renewable Energy

(1) China

In recent years, China has emerged as a global leader in the renewable energy sector. In solar, wind, and hydro power, China ranks first in terms of installed ca-pacity, deployment scale, and investment, playing a central role in the energy transition.

In solar power, China accounts for more than 80% of global solar panel production, and its cumulative domestic installed capacity reached approximately 609 GW by the end of 2023, representing roughly 40% of the global total (IRENA 2024). China has rapidly expanded solar power bases in high-insolation regions of its western inland areas, and leveraging its manufacturing capacity, it exports large quantities of solar products to Europe, Southeast Asia, and South America.

For example, on September 6, 2021, Dongfang Electric Corporation supplied the core systems and major equipment for China's first concentrating solar power (CSP) project, the 50 MW Hamy Molten Salt Tower CSP Plant in Xinjiang, which successfully achieved full-load operation. This project employs the molten salt tower technology, recognized as one of the most efficient and advanced CSP technologies. It utilizes 14,500 large heliostats to concentrate sunlight onto the tower, heating molten salt (Fig. 10). As a result, stable power generation at a grid-connected capacity of 50 MW has been realized.

Fig. 10 50 MW Molten salt tower Concentrated Solar Power (CSP) Project in Hami, China (DEC 2025)

This achievement represents a significant milestone in securing stable power supply from renewable energy. Particularly, the plant's thermal storage capability enables 24-h continuous operation, addressing the long-standing challenge of output variability in solar power. It also contributes to improved grid flexibility and dispatchability, and is expected to play a central role in future decarbonized energy systems (DEC 2025).

Moreover, the plant integrates power generation and storage, and it is estimated to supply approximately 200 million kWh of clean, stable, and high-quality dispatchable electricity to the grid annually. Consequently, this project functions not only as a baseload power source potentially replacing fossil fuel plants but also as a high-potential peak load balancing resource. In addition, it has spillover effects on the expansion of renewable energy deployment and industrial development in the Xinjiang region, serving as a model for regional energy transition and economic revitalization.

In wind power generation as well, China boasts the largest installed capacity globally, both onshore and offshore. As of the end of 2023, onshore wind capacity reached approximately 402 GW, and offshore wind exceeded 36 GW, ranking first worldwide in both categories (Global Wind Energy Council 2024).

The reason China has achieved such a dominant position in the renewable energy sector lies in its clear national strategic direction. In particular, the "dual-carbon targets"—aiming for carbon peaking by 2030 and carbon neutrality by 2060—serve as a central pillar of the central government–led energy policy. Various support measures, including feed-in tariffs (FIT), auction schemes, and grid connection support, have been implemented to promote deployment.

China has also developed a vertically integrated supply chain for its renewable energy industry. In the solar PV sector, almost all stages from polysilicon production to module assembly are covered domestically. In 2023, China's investment in the renewable energy sector reached approximately USD 890 billion, accounting for more than 55% of the global total (BloombergNEF 2025).

However, the rapid expansion of renewable energy has also revealed several challenges. A representative issue is the "curtailment" of wind and solar power, whereby generated renewable electricity cannot be absorbed by the grid and is thus wasted. In 2018, due to "capacity constraints" in the power system, the combined generation from hydropower, wind, and solar was limited to 102.3 billion kWh—equivalent to the annual output of the Three Gorges Dam and corresponding to approximately 40 million tons of CO_2 emissions (The State Council of the People's Republic of China 2019). Even with mandatory policies such as full purchase guarantees, the situation remains serious. In 2021, total wind power curtailment reached 20.61 billion kWh (average curtailment rate: 3.1%), while solar power curtailment amounted to 6.78 billion kWh (average curtailment rate: 2.1%) (National Energy Administration 2022). These data indicate that a fundamental overhaul of the power system is urgently needed to accommodate large-scale renewable energy deployment.

This issue stems from insufficient transmission infrastructure and regional supply-demand imbalances. Moving forward, integration with storage systems, ther-mal power, smart grids, and hydrogen technologies will be necessary.

Nonetheless, China is now not only the "largest adopter" of renewable energy but is increasingly expected to act as a "rule-maker" on the global stage. For ex-ample, in the Global South, China is providing solar and wind technologies and financial support through the Belt and Road Initiative, indicating an expansion of international cooperation centered on renewable energy (World Bank 2023).

Overall, China leads the world in renewable energy across all dimensions—scale, quality, supply networks, cost, and policy support—and its influence is expected to continue expanding in the coming years.

(2) California, United States

In the United States, renewable energy policies exhibit significant divergence be-tween the federal and state levels. Under the Trump administration, policies fa-vored fossil fuels, as evidenced by the withdrawal from the Paris Agreement and support for the coal industry. Conversely, at the state level, there has been a pro-active push for renewable energy, resulting in a decentralized energy transition. Among the states, California stands out as a leader in U.S. renewable energy poli-cy. According to reports as of November 2024, solar power in California surpassed natural gas-fired generation as the largest electricity source in early 2025 (EMBER 2025). Indeed, as of the end of 2024, the electricity generation mix consisted of 34.4% natural gas and 32.7% solar power, indicating that the transition to a renewable energy–dominated system is becoming a tangible reality (Fig. 11).

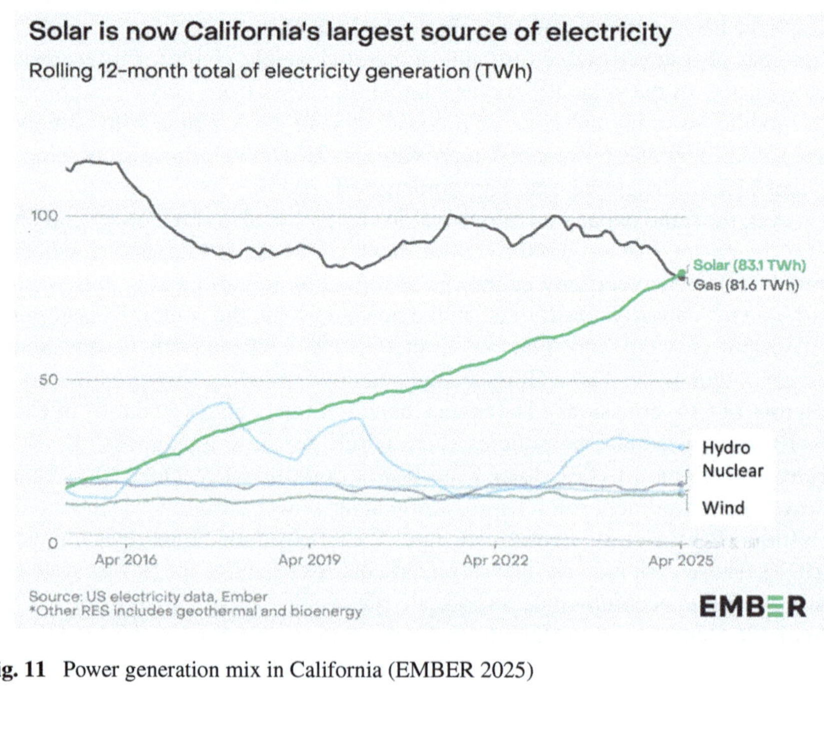

Fig. 11 Power generation mix in California (EMBER 2025)

Table 4 Challenges and countermeasures in renewable energy deployment in California

Challenges	Countermeasures
Fluctuations in power supply (Duck curve phenomenon)	Adjustment through large-scale storage systems, distributed storage, and smart demand–supply control
Transmission network bottlenecks	Energy Imbalance Market (EIM) and reinforcement of transmission infrastructure
Surplus and curtailment of electricity	Demand–supply management and utilization of storage based on coordination with regional consumption
Nighttime power shortages	Strengthening of nighttime-oriented storage systems and introduction of long-duration storage technologies

California's renewable energy policy goes beyond merely increasing generation capacity; it involves transitioning to a smart grid–based power system that precisely controls when, where, and how much electricity is generated. This approach addresses the "instability of power supply" often associated with renewable ener-gy, steadily advancing the shift toward a society underpinned by clean electricity.

Below are the challenges California faces in integrating renewable energy and the specific measures implemented to address them (Table 4).

California has legally mandated the achievement of 100% clean electricity supply by 2045. To realize this target, the state is advancing a three-pronged strategy encompassing the enhancement of energy storage technologies, the development of smart grids, and the reconstruction of transmission infrastructure. These initiatives not only have significant spillover effects for other states but also serve as a model leading the United States' transition to renewable energy.

4.3 Challenges and Prospects of Renewable Energy

However, renewable energy faces several challenges that must be addressed.

(1) Grid Integration and Flexibility Issues

The greatest challenge is the inherent variability of renewable energy, which fluctuates significantly depending on weather, seasons, and regional conditions. Addressing this requires the deployment of energy storage systems, flexible backup generation (e.g., gas-fired power), and demand-side management measures such as demand response. Additionally, large-scale renewable integration has revealed operational challenges for the overall power system, including grid congestion and frequency regulation.

(2) Integration with Decarbonization Strategies

The deployment of renewable energy should not merely replace conventional generation but must be coordinated with comprehensive societal decarbonization strategies (deep decarbonization). For instance, producing green hydrogen using renewable electricity can enable the decarbonization of industrial and transportation sectors, thereby enhancing the overall value of renewable energy.

(3) Social Acceptance and Decentralized Deployment

Widespread adoption of renewable energy requires understanding and cooperation from local communities. Large-scale wind and solar projects often face opposition due to their impacts on landscapes and local environments. In the future, community-based renewable energy projects, such as local consumption systems and energy cooperatives, will be essential to improving social acceptance.

(4) International Collaboration and Supply Chain Resilience

Renewable energy infrastructure is heavily dependent on internationally sourced materials and components, such as rare earth elements and silicon. As dependence on specific countries, particularly China, increases, diversifying supply chains and establishing recycling technologies have become urgent. Furthermore, cross-border electricity trade and coordination with international carbon markets will be key considerations in future institutional design.

(5) Expanding Applications Beyond Electricity

Another important prospect is the expansion of renewable energy applications beyond electricity. These include green hydrogen production, electric vehicles (EVs), and electrified heating and cooling systems. Electrification and sector coupling can maximize the utilization efficiency of renewable energy and reduce the overall carbon footprint.

(6) Institutional and Economic Support

Accelerating renewable energy deployment also requires strong institutional and economic support. Many countries initially promoted renewable adoption through feed-in tariffs (FIT) and priority grid access, but policy frameworks are increasingly shifting toward market-integrated mechanisms such as feed-in premiums (FIP), non-fossil certificates, and carbon pricing. These measures facilitate investment and improve cost competitiveness.

Finally, the future of renewable energy depends on technological innovation, institutional design, local consensus-building, and international cooperation. Next-generation technologies, including advanced energy storage, smart grids, and AI-based predictive optimization, have the potential to significantly enhance the reliability of renewable energy. Moreover, ensuring the stable supply of critical materials (rare earths and metals), promoting recycling of end-of-life equipment, and enhancing supply chain transparency are essential for long-term sustainability.

5 The Future of Nuclear Power in the Context of a Green Transition

Amid the accelerating deployment of renewable energy toward a decarbonized society, the potential and challenges of nuclear power (fission) and future nuclear fusion are once again drawing international attention. In particular, nuclear power—with its characteristics of stable supply, low carbon emissions, and high ener-gy density—is expected to complement highly variable renewable energy and serve as a baseload power source supporting the achievement of carbon neutrality.

5.1 Current Status and Outlook of Fission Nuclear Power

(1) The Decarbonization Contribution of Nuclear Power and Policy Trends

Nuclear power has continuously attracted attention as a zero-emission electricity source in climate change mitigation because it does not emit CO_2 during electricity generation. In some European and Asian countries, nuclear power is being reassessed and reintroduced through measures such as the replacement of aging reactors, the development of advanced reactors, and the deployment of small modular reactors (SMRs).

In Japan, the 7th Strategic Energy Plan explicitly emphasizes the maximized uti-lization of existing nuclear power plants and the development of next-generation innovative reactors. China has also established a policy to actively and orderly expand nuclear power generation while ensuring safety.

As of the end of 2019, China had 48 reactors in operation, including research reactors and reactors that had achieved initial criticality. By the end of 2024, the number of commercially operating reactors had increased to 58, with a total capacity of approximately 60.88 GW. This ranks second in the world after the Unit-ed States (94 reactors) and is comparable to or slightly above France (approximately 56 reactors). Furthermore, with 27 reactors under construction (total out-put ~ 32 GW) and additional reactors approved for construction, the total number of operating, under-construction, and approved reactors reaches 102. If the scheduled commissioning proceeds, China is expected to have more than 100 operating reactors by 2030. Notably, no accidents exceeding Level 2 on the Interna-tional Nuclear Event Scale (INES) have occurred, and no major nuclear accidents have been reported since operations began in 1992.

(2) Global Nuclear Renaissance—"Triple Capacity Declaration" (COP28)

By the end of 2023, 25 countries—including the United States, Japan, the Unit-ed Kingdom, Canada, France, Ghana, South Korea, Sweden, and the United Arab Emirates (excluding China)—jointly adopted at COP28 a declaration to "triple nuclear power capacity by 2050 compared to 2020 levels."

At the end of 2023, the global nuclear power capacity was approximately 416 GW. According to the 2024 edition of the IEA's World Energy Outlook, under the "STEPS" scenario—which assumes continuation of current policies—nuclear capacity is projected to reach 647 GW by 2050. Under the more ambitious "NZE (Net Zero Emissions by 2050)" scenario, nuclear capacity could expand to 916 GW. Meanwhile, the latest IAEA report estimates the operating capacity at the end of 2023 to be approximately 372 GW, projecting a high-case scenario of 950 GW and a low-case scenario of 514 GW by 2 (IEA 2025a). Compared to these projections, the "triple capacity" commitment of COP28 can be considered highly ambitious.

Small Modular Reactors (SMRs) are next-generation nuclear reactors expected to have shorter construction periods and flexible siting compared to conventional large reactors. Their potential for deployment as distributed energy sources has drawn attention, particularly in regions where electricity grids are not yet fully developed. Moreover, integration with non-electric applications, such as heat supply and hydrogen production, is also being envisioned.

(3) Safety, Waste, and Social Acceptance

At the same time, nuclear power continues to face structural challenges, includ-ing accident risks, the final disposal of radioactive waste, counter-terrorism measures, and securing the understanding of local communities. Since the Fuku-shima Daiichi nuclear accident, restoring social trust has required a long-term ef-fort and governance with a high degree of transparency. In particular, achieving consensus on the extended operation of aging reactors or the construction of new ones will significantly influence the feasibility of future nuclear policy implemen-tation (Fig. 12).

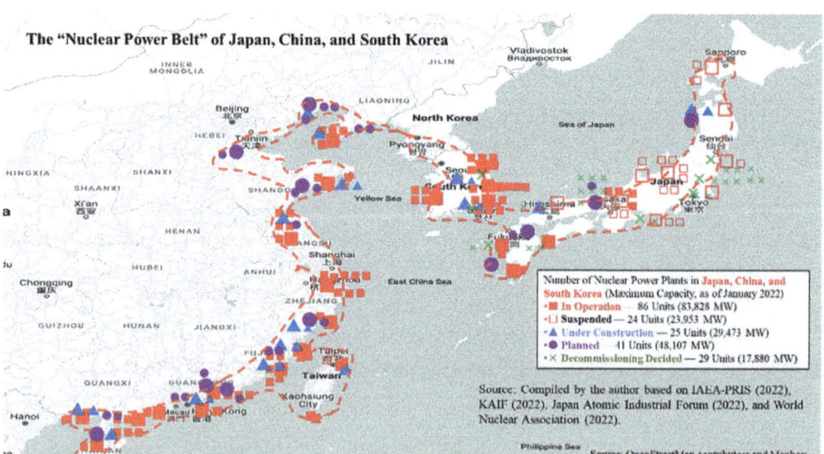

Fig. 12 Distribution of nuclear power plants in China, Japan, and South Korea. Compiled by the author based on IAEA-PRIS (2022), KAIF (2022), Japan Atomic Industrial Forum (2021), and World Nuclear Association (2022)

Fig. 13 The Experimental Advanced Superconducting Tokamak (EAST), a fully superconducting tokamak fusion experimental device under development in China, January 15, 2025. Photo by Xinhua News Agency

As shown in Fig. 13, East Asia (Japan, China, and South Korea) has a high density of nuclear power plants, often referred to as the "nuclear belt." In this re-gion, which shares a narrow sea and atmospheric space, an accident in one country has a high potential to seriously affect neighboring countries, making nuclear safety both a regional and international concern. As evidenced by the Fukushima Daiichi accident and the past Chernobyl disaster, nuclear disasters know no borders. Therefore, building safety systems through international cooperation is in-dispensable. In other words, the risks associated with the "nuclear belt" in Japan, China, and South Korea cannot be addressed by the regulations or responses of a single country alone, necessitating the establishment of a regional nuclear safety system. Harmonizing technical standards, ensuring transparency of information, and establishing mutual assistance systems for emergencies are essential to securing energy security and human safety across East Asia.

5.2 Technological Advances and Long-Term Potential of Fusion Energy

(1) Principle of Fusion and Its Characteristics as Clean Energy

Nuclear fusion is a phenomenon in which light atomic nuclei (e.g., deuterium and tritium) merge to form a heavier nucleus, releasing enormous amounts of energy.

It is also the source of energy emitted by the Sun. In theory, fusion emits no CO_2 and produces little to no long-lived radioactive waste, making it an ultimate clean energy source. As shown in Table 1, fusion is considered an ultimate energy source generated through advanced human technology, capable of utilizing virtually all available resources on Earth.

(2) International Collaborative Projects and Private Sector Innovation

Currently, the International Thermonuclear Experimental Reactor (ITER) under construction in France represents the largest international project aimed at the commercialization of fusion energy. Meanwhile, numerous innovative fusion startups have emerged, particularly in the United States, the United Kingdom, and Japan, pursuing parallel development approaches such as laser heating, magnetic confinement, and superconducting technologies. These companies aim to achieve practical deployment of small-scale fusion technologies by the 2030s.

(3) The Global Fusion Race Is On

On December 5, 2023, at the Atlantic Council Global Energy Forum held during COP28 in Dubai, the United States proposed, for the first time outside of ITER, an international fusion cooperation initiative. John Kerry, U.S. Special Presidential Envoy for Climate, announced that the United States would collaborate with 35 countries, including China, to promote nuclear fusion development through this initiative. The plan encompasses five key areas of cooperation: research and development, supply chains and markets, human resource development, regulation and safety, and public engagement. Its primary objectives are to accelerate the commercialization of fusion energy, facilitate a global clean energy transition, and enhance humanity's capacity to address the climate crisis. The U.S. government aims to construct commercial fusion facilities within ten years to supply electricity to households and businesses. Additionally, the United States and the United Kingdom signed a nuclear fusion cooperation agreement on November 8, 2023.

In Japan, the first national fusion energy development strategy was announced in April 2023, promoting corporate participation in research and development reactors and targeting the realization of fusion power around 2050. Blue Laser Fusion (BLF), a U.S. company founded by Nobel Laureate Shuji Nakamura and others, has established a joint research division with the Laser Science Research Institute at Osaka University. The research focuses on achieving laser-driven fu-sion reactions that do not produce neutrons, with BLF collaborating with Osaka University, which possesses some of the world's most advanced laser technology, to clarify a pathway toward highly clean fusion energy for societal implementation. The joint research division is scheduled to operate until March 2028, with BLF contributing approximately 300 million yen. Faculty, engineers, and re-searchers from Osaka University, Hiroshima University, and Purdue University in the United States are also expected to participate.

According to reports from Xinhua News Agency, on January 20, China's fully superconducting tokamak experimental reactor (EAST), also known as the "Chinese artificial sun," achieved plasma operation in a steady-state, long-pulse, high-confinement mode for 1,066 s at temperatures exceeding 100 million de-grees Celsius, setting a new world record. This achievement is considered a signifi-cant milestone for accelerating the realization of controlled fusion power (Xinhua News Agency 2025). The record was accomplished by a research team at the In-stitute of Plasma Physics, Hefei Institutes of Physical Science, Chinese Academy of Sciences, and is regarded as a major turning point in advancing fusion research from basic science to engineering applications.

Nuclear fusion power is often regarded as a pivotal instrument for decarbonization, with numerous research institutions and emerging enterprises actively en-gaged in its research and development. Global equity investment in nuclear fusion peaked in 2021, during which the United States overwhelmingly dominated the investment landscape. Subsequently, however, both the absolute volume and the proportional share of U.S. investment have declined, while China has demonstrated a notable surge in engagement. Within a mere three-year period, China's equity investment in controlled nuclear fusion expanded from virtually "zero" to an unprecedented level, rapidly surpassing that of the United States. For two consecutive years, equity inflows into Chinese fusion enterprises have exceeded those directed to all other countries. Figure 14 illustrates the global funding trends for fusion enterprises between 2010 and 2023.

Should developments proceed in accordance with current trajectories, the timeline for the commercialization of nuclear fusion may be significantly advanced. It is conceivable that within less than a decade, large-scale deployment of fusion power could become a practical reality.

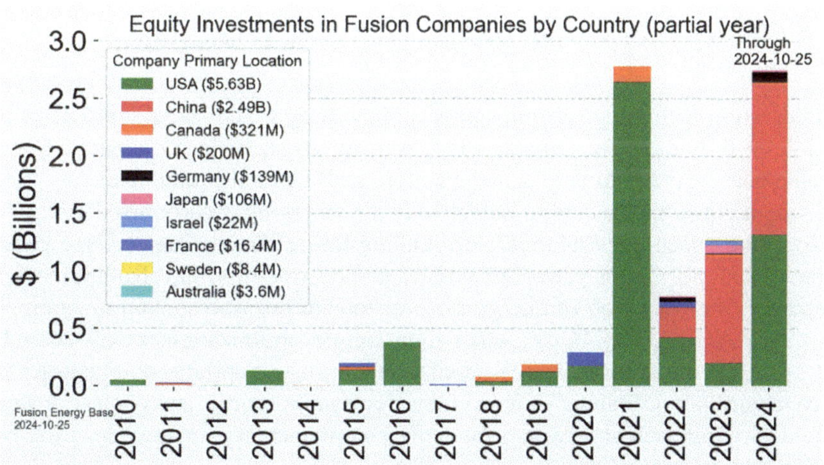

Fig. 14 Funding to fusion companies from 2010 to 2023 globally (Fusion Energy Base 2025)

(4) Technical and Economic Challenges

The greatest barriers to the practical deployment of fusion energy remain the es-tablishment of a positive energy balance and sustained stable operation over long periods. In addition, numerous technical challenges must be overcome, including the high cost of device construction, the radiation resistance of materials, and fuel supply issues such as tritium production. Moreover, the economic viability of commercial fusion reactors remains uncertain, making continued policy support and long-term investment indispensable.

5.3 Green Transition Through the Coexistence of Nuclear Power and Renewable Energy

Even in a future where renewable energy becomes the primary power source, nuclear energy plays a significant role as a "non-fossil baseload" capable of complementing the variability of renewables and ensuring grid stability. In the short term, existing fission reactors and SMRs fulfill this role, while in the medium to long term, the advent of fusion energy offers the prospect of a sustainable and safe energy structure.

The key to a green transition lies not only in technological innovation but also in social consensus building, responsible waste management, and risk diversification through international cooperation. Looking ahead to the future of nuclear power requires more than simple approval or opposition; it demands an empirical and strategic perspective focused on under what conditions and in what ways nuclear energy can coexist and complement other energy sources.

6 A Stepwise Approach to Energy Green Transition

By integrating technological innovation, policy design, and social consensus, it is possible to provide a "sustainable exit" even for fossil fuels. Complementing re-newable energy, the goal is to achieve a more resilient and equitable energy transi-tion, which constitutes a central objective in the emerging decarbonized era.

Given the "deep uncertainties" surrounding future climate impacts, the pace of technological innovation, geopolitical risks, market volatility, and changes in so-cial acceptance, the transition of energy systems toward decarbonization must empha-size robustness and flexibility, rather than simple optimization, as dis-cussed in Sect. 10.2.3. In this context, it is essential to secure an appropriate combination of the "three major energy sources"—fossil fuels, renewable energy, and nuclear power—while deploying each source in a mutually complementary manner, ensuring both stability and sustainability.

The following sections attempt to outline a stepwise roadmap for green transition by structuring it into three layers: technology, policy, and institutional frame-works.

Under the uncertainty of climate change, the overall picture of energy green transition is described as a roadmap constructed around the concepts of the best mix of the three major energy sources (fossil fuels, renewables, and nuclear) and robust optimization, emphasizing technical, policy, and institutional inter-connections.

6.1 Layer 1: Stepwise Transition in the Technological Dimension

From a technological perspective, the expansion and deployment of renewable energy form the central axis of the energy transition, while the use of fossil fuels requires a stepwise promotion of "decarbonization" through emission control technologies (e.g., CCUS) and co-firing techniques. Over the next decade, solar and wind power are expected to be positioned alongside thermal and nuclear power as major energy sources, supported by complementary technologies such as energy storage, smart grids, and hydrogen production—particularly green hydrogen. In parallel, small modular reactors (SMRs) and next-generation nuclear reactors, designed for enhanced safety and flexibility, are likely to be re-evaluated for limited deployment to mitigate the intermittency of renewables.

In this way, constructing a multi-faceted and stepwise technological portfolio is essential as a robust design to address the uncertainties of climate change. Optimization here does not simply mean efficiency; it refers to "a system configuration that remains sustainable even if any given technology becomes temporarily unavailable."

Stage 1: Establishment of Core Technologies and Deployment Expansion (~2030)

In the initial stage, the share of green power is expanded through the thorough introduction of renewable energy—particularly solar and wind—and energy efficiency measures. Simultaneously, efforts to reduce dependence on fossil fuels fo-cus on the decarbonization of coal-fired power plants, the shift toward natural gas, and the demonstration and commercialization of CCUS (Carbon Capture, Utilization, and Storage) technologies. Regarding nuclear energy, the life extension of existing safe reactors and the R&D of new reactor types are advanced to provide a complementary role in supply stability.

Stage 2: System Integration and Enhancement of Flexibility (2030s)

To address supply–demand gaps and grid instability caused by large-scale integration of variable renewable energy, priority is given to strengthening grid interconnections, smart grids, demand-side management (demand response), energy storage, and power interconnections. In addition, the introduction and technical implementation of new energy carriers such as hydrogen, synthetic fuels (e-fuels), and ammonia contribute to cross-sector decarbonization in industry and transport.

Stage 3: Completion of a Highly Reliable, Low-Carbon Energy System (~2050)

By 2050–2060, while renewable energy forms the backbone of the power mix, advanced CCUS-equipped thermal power and next-generation nuclear reactors (including SMRs) are strategically deployed as backup sources to maintain overall grid stability. This approach establishes a robust best mix that balances technical reliability, sustainability, and decarbonization.

6.2 *Layer 2: Stepwise Policy Support and Guidance*

On the policy side, a clear timeline and quantitative targets for decarbonization—such as net-zero deadlines, renewable energy share, and reductions in coal dependence—must be explicitly defined. Effective policy tools should combine economic incentives with regulatory measures. The introduction of carbon pricing mechanisms (carbon taxes or emissions trading) may impose short-term costs on businesses and households but serve as a core driver for market self-transformation in the medium to long term. Direct investments in green innovation and public procurement are also necessary to absorb the initial risks of technological development.

To address energy security and geopolitical risks, policies must avoid excessive reliance on specific energy sources and support diversification of energy sources and supply routes (e.g., cross-border trade of hydrogen or renewable electricity, joint development of wide-area grids). Policy flexibility—that is, a governance structure capable of immediate adjustment and reconfiguration in response to changing conditions—becomes a critical guarantee of robustness in an era of deep uncertainty.

Stage 1: Establishment of Market Incentives and Price Formation

In the initial stage, policies that promote the market-based adoption of low-carbon technologies—such as feed-in tariffs (FIT), carbon taxes, and emissions trading schemes (ETS)—take center stage. What is crucial here is "policy consistency" and "long-term predictability", creating a market environment in which technological investments and corporate strategies are aligned with climate policies.

Stage 2: Risk Reduction and Targeted Support

In the next stage, targeted support for specific technologies—such as hydrogen supply chains, CCUS infrastructure, and high-performance energy storage—is essential. The role of public support is to create initial markets, develop infrastruc-ture, and internalize externalities. Policy must also address risks such as stranded assets and respond to international cooperation, technology transfer, and mecha-nisms like the Carbon Border Adjustment Mechanism (CBAM).

Stage 3: Transition to Collaborative, Integrated, and Distributed Governance

In the final stage, policy implementation moves beyond top-down central gov-ernment measures toward distributed and collaborative governance, involving local governments, industry, and civil society. Policy objectives expand beyond carbon "net-zero" to encompass sustainable, multi-dimensional goals that integrate resilience, just transition, economic development, social inclusion, and environmental protection.

6.3 *Layer 3: Structural Transition in the Institutional Dimension*

Institutionally, reforms in the energy market are a primary requirement, includ-ing priority access for renewables, definitions of non-fossil power, and redesign of electricity market structures. In addition, a framework for participatory energy governance that involves local governments, communities, and private actors is indispensable. Climate action must evolve from national-level policy directives into concrete, actionable initiatives through mechanisms such as local decarbonization plans and regional new energy utilities.

To address the social costs and employment shifts arising from decarbonization, institutional frameworks for a just transition are also required. This includes re-training support for carbon-intensive industries, social protection against energy price spikes, and measures to combat energy poverty.

These institutional responses go beyond simple legal reforms; they aim to establish resilient systems that enable energy transition to be sustainable and socially acceptable even under conditions of uncertainty.

Stage 1: Consolidation and Redesign of Existing Institutions

At this stage, the focus is on creating a market foundation in which renewables, nuclear, and low-carbon thermal power can compete fairly, through electricity market reforms (e.g., unbundling generation and transmission, capacity market development), market transparency, and technology standards. Institutional coherence for balancing energy security and decarbonization is also critical.

Stage 2: Cross-Institutional Coordination Design

It is necessary to ensure policy alignment and interoperability among institutional frameworks. Examples include integrating urban planning with distributed renewable energy deployment, transportation policy with EV infrastructure, and taxation with carbon pricing mechanisms. Furthermore, policy learning and pol-icy experimentation form core components of institutional development, allowing adaptive governance and iterative improvement across sectors.

Stage 3: Rebuilding a Sustainable Social Contract

In the final stage, where decarbonization extends to social structures and lifestyles, it becomes essential to rebuild the social contract surrounding energy. Institutional focus areas include consensus-building for renewable energy facility siting, fairness in pricing structures, support for industrial transformation, and assistance for vulnerable populations during the transition. This stage raises questions of public philosophy, democratic legitimacy, and participatory institutional design.

To address the complex challenges of unpredictable climate impacts, geopolitical risks, and uncertainty regarding the maturity of energy technologies, it is indispensable to strategically and incrementally combine the three major energy sources—fossil fuels, renewables, and nuclear—while maintaining robust optimization that can flexibly adapt to future fluctuations. Rather than excluding any one energy source, complementary utilization ensures technological diversification, policy flexibility, and institutional inclusivity, thereby opening a pathway to simultaneously achieve the threefold goals of decarbonization, energy security, and social sustainability.

The green transition of the three major energy sources requires a phased and complementary approach, as each differs in temporal horizon, technological maturity, and social acceptance. In the short term, rapid expansion of renewables and reduction of fossil fuel dependency are necessary. In the medium to long term, achieving systemic decarbonization depends on the societal deployment of innovative technologies such as CCUS (Carbon Capture, Utilization, and Storage), hydrogen, and SMRs (Small Modular Reactors).

References

Ben-Tal A, El Ghaoui L, Nemirovski A (2009) Robust optimization. Princeton Uni-versity Press

BloombergNEF (2025, February 6) China's clean energy manufacturing capacity drives cost declines. BloombergNEF

China Renewable Energy Engineering Institute (CREEI) (2025, May 28) China renewable energy project cost management annual report 2024

DEC (2025) Xinjiang Hami 50MW molten salt tower CSP project https://www.dbc.com.cn/info/1428/1635.htm

EMBER (2025, February 23) In numbers: solar and batteries are quickly displacing gas in California's power grid. The Progress Playbook. https://theprogressplaybook.com/2025/02/23/batteries-are-displacing-gas-in-californias-power-grid/?utm_source=chatgpt.com

Energy Institute (2025) Statistical review of world energy data 2025. Energy Institute. Retrieved from https://www.energyinst.org/statistical-review-of-world-energy

Fusion Energy Base (2025) The global fusion race is on. Fusion Energy Base. https://www.fusionenergybase.com/articles/the-global-fusion-race-is-on

G7 (2024) G7 Ministers' meeting on climate, energy and environment (2024) Climate, Energy and Environment Ministers' meeting Communiqué, Retrieved from https://www.g7italy.it/wp-content/uploads/G7-Climate-Energy-Environment-Ministerial-Communique_Final.pdf

Global Wind Energy Council (2024) Global wind report 2024. Global Wind Energy Coun-cil

International Atomic Energy Agency (2022) Power Reactor Information System (PRIS). https://pris.iaea.org/PRIS/home.aspx

International Energy Agency (2025a) World Energy Outlook 2025. https://www.iea.org/reports/world-energy-outlook-2025

International Energy Agency (2025b) Global investment in clean energy and fossil fuels, 2015–2025. In world energy investment 2025. International Energy Agency. https://www.iea.org/data-and-statistics/charts/global-investment-in-clean-energy-and-fossil-fuels-2015-2025

International Energy Agency (2023) World energy outlook 2023. OECD Publishing. https://www.iea.org/reports/world-energy-outlook-2023

International Renewable Energy Agency (IRENA) (2024) Renewable power generation costs in 2024. International renewable energy agency. Retrieved from https://www.irena.org/-/media/Files/IRENA/Agency/Publication/2025/Jul/IRENA_TEC_RPGC_in_2024_2025.pdf

International Renewable Energy Agency (IRENA) (2025) Renewable power genera-tion costs in 2024. IRENA. Retrieved from https://www.irena.org/-/media/Files/IRENA/Agency/Publication/2025/Jul/IRENA_TEC_RPGC_in_2024_2025.pdf

Japan Atomic Industrial Forum (2021) Nuclear power reactors in Japan (In operation, under construction, and under preparation). https://www.jaif.or.jp

Korea Atomic Industrial Forum (2022) Korea Atomic Industrial Forum (KAIF). http://www.kaif.or.kr/eng/National

National Energy Administration (2022, April 21) Notice on the completion of the responsi-bility weights for renewable energy power consumption in 2021 (Guonengfa Xineng [2022] No. 47). National Energy Administration. https://zfxxgk.nea.gov.cn/2022-04/21/c_1310587748.htm

NEDO (2023) Interim evaluation report of the demonstration project for integrated coal gas-ification fuel cell combined cycle power generation. New Energy and Industrial Technology Development Organization (NEDO). https://www.nedo.go.jp/content/100976814.pdf. The economics of clean coal power generation with carbon capture and storage technology in China, https://doi.org/10.1007/s10098-023-02531-1

Stuart Stone (2025) Cost of clean power tech expected to fall by up to 11 per cent in 2025, BloombergNEF (BNEF), from https://www.businessgreen.com/news/4409685/bnef-cost-clean-power-tech-expected-fall-cent-2025?utm_source=chatgpt.com. Accessed on 15 Aug 2025

The State Council of the People's Republic of China (2019, July 4) Strive to basically re-solve the problems of curtailment of hydropower, wind, and solar energy by 2020. The State Council. http://www.gov.cn/xinwen/2019-07/04/content_5405844.htm

U.S. Department of Energy (DOE) (2025, July 29) A critical review of impacts of green-house gas emissions on the U.S. climate. U.S. Department of Energy. https://www.energy.gov/topics/climate

Walker WE, Haasnoot M, Kwakkel JH (2013) Adapt or perish: a review of plan-ning approaches for adaptation under deep uncertainty. Sustainability 5(3):955–979. https://doi.org/10.3390/su5030955

World Bank (2023) China's role in green energy transitions in the global south. World Bank. https://doi.org/10.1596/978-1-4648-1984-8

World Nuclear Association (2022) China: nuclear power. https://www.world-nuclear.org/information-library/country-profiles/countries-a-f/china-nuclear-power.aspx

Xinhua News Agency (2025) China's EAST sets new world record with 1,066-second steady-state high-confinement plasma operation. Xinhua. https://english.news.cn/20250120/9f7bff53ece445e182f6513e7b4089fb/c.html?utm_source=chatgpt.com. Accessed on 30 June 2025

Zhang C, Wang J, Zhou W (2023) Study on coal phase-out in China: part 2. Analysis of factors affecting coal consumption and evaluation of power generation costs. Policy Sci 30(2): 60–71

Zhou W, Wang J, Ling Y, Cheon K, Miyawaki N (2021) Study on coal phase-out in China: part 1. Current status of coal consumption and analysis of influencing factors. Policy Science, 29(1), October 2021

Theoretical Optimization Analysis for Maximum Power Generation in Organic Rankine Cycle

Baoju Jia, Yu Lei, Faming Sun, and Weisheng Zhou

1 Introduction

Energy shortage is one critical issue in this century that must be appropriately solved to realize a sustainable society. Utilization of low-temperature heat sources (lower than 100 degrees Celsius) to generate clean power is one promising way since its potential is huge. However, large amounts of the low-temperature heat sources are still not well developed due to their relatively low thermal efficiencies and small power outputs.

Recently, researches on the conversion of the low-temperature heat sources into clean power have received a lot of attention, such as, geothermal energy conversion using Organic Rankine Cycle (ORC) (Hettiarachchi et al. 2007; Zhang et al. 2011; DiPippo 2004; Sauret and Rowlands 2011) and Kalina Cycle (Kalina and Leibowitz 1989), ocean thermal energy conversion using ORC (Uehara and Ikegami 1990; Wu 1990; Uehara et al. 1998a; Ikegami and Bejan 1998; Sun et al. 2012a) using Kalina Cycle (Kalina 1982, 1984) and Uehara Cycle (Uehara et al. 1998b), solar thermal power generation using ORC (Zhang et al. 2005; Wang et al. 2010; Delgado-Torresa and Garcia-Rodriguez 2010; Sun et al. 2013a) and Kalina Cycle (Lolos and Rogdakis 2009; Sun et al. 2012b; Wang et al. 2013; Sun et al. 2013b, 2014), biothermal energy

B. Jia · F. Sun
Research Institute of Global 3E, Kyoto, Japan
e-mail: baojujia@gmail.com

F. Sun
e-mail: sunfamingjia@gmail.com

Y. Lei
Dongfang Electric Engineering & Consulting Co., Ltd., Sichuan, China
e-mail: leiyu@dongfang.com.cn

W. Zhou (✉)
College of Policy Science, Ritsumeikan University, Osaka, Japan
e-mail: zhou@sps.ritsumei.ac.jp

W. Zhou (ed.), *Decarbonization Policy Engineering*, World Sustainability Series,
https://doi.org/10.1007/978-981-95-5953-4_11

conversion using ORC (Liu et al. 2011), and waste heat power generation from various thermal processes using ORC (Hung et al. 1997; Liu et al. 2004; Wei et al. 2007; Dai et al. 2009) and Kalina Cycle (Kalina and Leibowitz 1994; Fallah et al. 2016), etc. In which, the ORC technology is considered as one effective way since it has been installed throughout the world and the Rankine cycle provides most of worldwide electricity production.

ORC uses an organic fluid as the working fluid in Rankine cycle. The organic working fluids play a key role in ORC. Their suitable selection criteria are summarized for the conversion the low-temperature heat sources and depending on the operating conditions (Saleh et al. 2007; Chen et al. 2010; Tchanche et al. 2011; Babatunde and Sunday 2018; Herath et al. 2020; Yang et al. 2024), which is important for a particular application, i.e. solar thermal, geothermal or waste heat recovery, OTEC, etc. And isentropic fluids are considered the most suitable for recovering low-temperature waste heat (Hung et al. 1997). R134a high-density organic working fluid is favored for geothermal power generation from power output viewpoint (Sauret and Rowlands 2011). R245fa is feasible and acceptable in the low-temperature solar power Rankine cycle system (Wang et al. 2010), and R1233zd and R245fa's net power outputs in ORC system are very close found by experimentally investigate (Araya et al. 2020). Ammonia (R717) is considered as one of the suitable working fluids for a closed Rankine cycle OTEC plant (Uehara et al. 1998a; Sun et al. 2012a).

In summary, although many studies described that ORC could be one possible way to convert low-temperature heat sources into electricity effectively. However, theoretical optimum operating parameters for maximum power generation of ORC and its contribution to generate green and zero-carbon energy are not given clearly. This work focuses on the theoretical analysis of some typical or low-GWP organic working fluids, such as R717, R134a, R1234yf, R290, R245fa and R1233zd, in ORC to get their optimization operation conditions directly to develop next generation green and zero-carbon energy generation system to promote the race to zero. Besides, results related to ORC system can also be applied in the marine industry for further high efficiency and lower emission in the future. Thus, this chapter can be organized as follows,

- Modelling and methodology: details of the proposed method for development next generation green and zero-carbon energy generation system to promote the race to zero are presented.
- Results and discussion: results of the proposed method and analysis of the theoretical optimization operation conditions for maximum power generation are described and provided.
- Conclusion and future work: contributions of the chapter, limitations of the proposed method, and potential directions for future research are summarized and suggested.

2 ORC Modelling for Power Generation from Low-Temperature Heat Sources

ORC is named for its use of organic working fluids, which allows Rankine cycle to generate electricity from Low-temperature Heat Sources. Sketch of the ORC is shown in Fig. 1.

Some typical working fluids (Table 1), such as R717, R134a, R1234yf, Propane (R290), R245fa and R1233zd, are selected and their practical issues like environmental-friendly (ozone depletion potential (ODP), global warming potential (GWP)), safety use (toxicity, flammability, corrodibility) and the market availability are also given for reference. From thermodynamics point of view, the selected working fluids can be classified to "wet" (e.g., R717, R134a, R1234yf and R290), and "isentropic" (e.g., R245fa and R1233zd) depending on the slope of the saturation vapor curve on a temperature-entropy diagram. Figs. 2 and 3 show the ORC on the temperature-entropy plane by using "wet" and "isentropic" working fluid, respectively.

Thermodynamic performance of these working fluids in ORC can be evaluated by the following theoretical analysis. State function of every point in ORC are defined in Table 2. And the corresponding coefficient values are shown in Table 3 for these working fluids (R717, R134a, R245fa, R1233zd, R1234yf, and R290).

In ORC (Fig. 1), the net power output can be given as (turbine, pump)

$$\dot{W}_{net} = \dot{W}_t - \dot{W}_p \tag{1}$$

where $\dot{W}_t = \dot{m}_{wf}\ (h_1 - h_2)$ is the gross power output in the turbine, $\dot{W}_p = \dot{m}_{wf}(h_4 - h_3)$ represents the power consumed in the pump, $\dot{m}_{wf}$ is the mass flow rate of working fluid.

In the evaporator of ORC, the heat balance can be given as (evaporator)

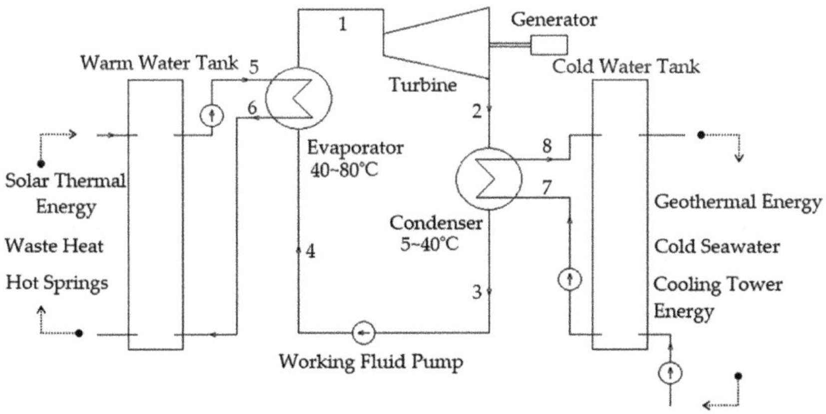

Fig. 1 The sketch of the ORC using Low-temperature Heat Sources

Table 1 Typical organic working fluids in ORC

Working fluids	R717	R134a	R1234yf	R290	R245fa	R1233zd
Fluid type	Wet	Wet	Wet	Wet	Isentropic	Isentropic
Inflammability	Lower	No	Lower	Higher	No	No
Toxicity	Higher	Lower	Lower	Lower	Lower	Lower
Metal corrodibility	Higher	No	No	No	No	No
ODP	0	0	0	0	0	0
GWP (100 years)	<1	1430	4	3	950	1
Market Availability	Easy	Difficult In the future	Easy	Easy	Difficult In the future	Easy

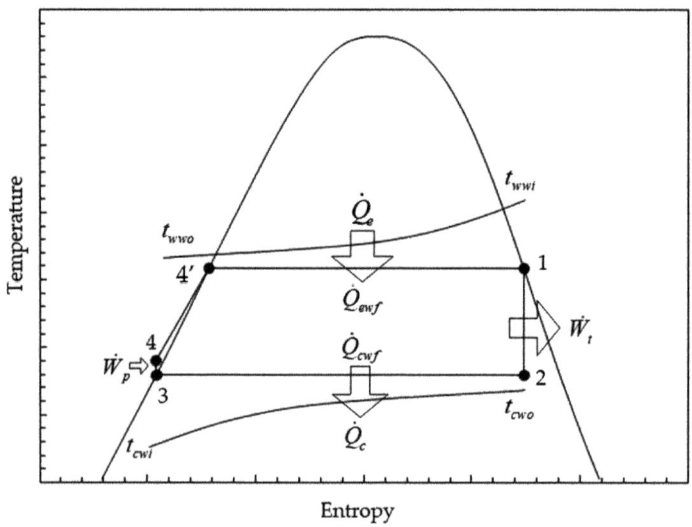

Fig. 2 Temperature-entropy for "wet" working fluids

$$\dot{Q}_e = \dot{Q}_{ewf} \tag{2}$$

where, $\dot{Q}_e = \dot{m}_{ww} c_p \Delta t_{ww}$ is the heat transfer rate absorbed from the warm-water, $\dot{Q}_{ewf} = \dot{m}_{wf}\ (h_1 - h_4)$ is the heat transfer rate supplied to the cycle, $\dot{m}_{ww}$ is the mass flow rate of warm water, c_p represents the specific heat capacity for water in a constant pressure system, $\Delta t_{ww} = t_{wwi} - t_{wwo}$, t_{wwi} and t_{wwo} are respectively the warm-water temperature at the inlet and outlet of the evaporator.

In the condenser of ORC, the heat balance can be given as (condenser)

$$\dot{Q}_c = \dot{Q}_{cwf} \tag{3}$$

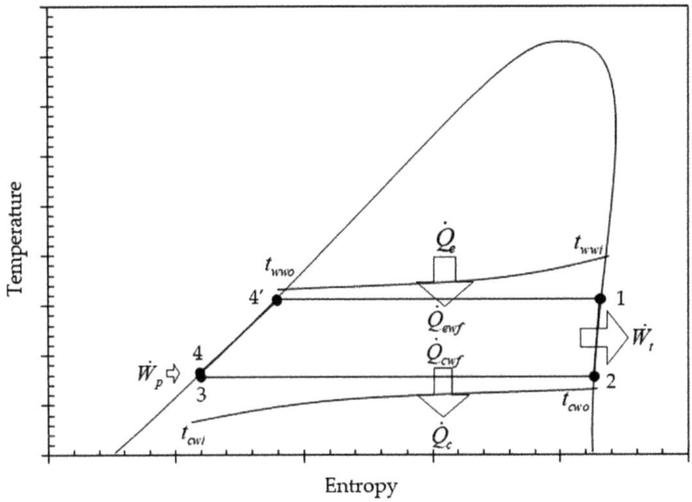

Fig. 3 Temperature-entropy for "isentropic" working fluids

Table 2 State function of every point in ORC

Point	State	Items	Equations
0	Zero Celsius degree	Absolute temperature [K]	$T_0 = 273.15$
1	Saturated vapor	Enthalpy [kJ/kg]	$h_1 = \zeta_{h_{1,1}} t_e^2 + \zeta_{h_1,2} t_e + \zeta_{h_1,3}$
		Pressure [kPa]	$P_1 = \zeta_{P_1,1} t_e^2 + \zeta_{P_1,2} t_e + \zeta_{P_1,3}$
		Entropy [kJ/(kg K)]	$s_1 = \zeta_{s_1,1} t_e^2 + \zeta_{s_1,2} t_e + \zeta_{s_1,3}$
2	Wet vapor	Enthalpy [kJ/kg]	$h_2 = h_3 + (t_c + T_0) \cdot (s_2 - s_3)$
		Pressure [kPa]	$P_2 = P_3$
		Entropy [kJ/(kg K)]	$s_2 = s_1$
3	Saturated liquid	Enthalpy [kJ/kg]	$h_3 = \zeta_{h_3,1} t_c^2 + \zeta_{h_3,2} t_c + \zeta_{h_3,3}$
		Pressure [kPa]	$P_3 = \zeta_{P_3,1} t_c^2 + \zeta_{P_3,2} t_c + \zeta_{P_3,3}$
		Entropy [kJ/(kg K)]	$s_3 = \zeta_{s_3,1} t_c^2 + \zeta_{s_3,2} t_c + \zeta_{s_3,3}$
		Specific volume [m^3/kg]	$v_3 = \zeta_{v_3}$
4	Compressed liquid	Enthalpy [kJ/kg]	$h_4 = h_3 + v_3(P_1 + P_3)$
		Entropy [kJ/(kg K)]	$s_4 = s_3$
		Pressure [kPa]	$P_4 = P_1$

where, $\dot{Q}_c = \dot{m}_{cw} c_p \Delta t_{cw}$ is the heat transfer rate rejected into the cold-water, $\dot{Q}_{cwf} = \dot{m}_{wf}\ (h_2 - h_3)$ is the heat transfer rate rejected from the cycle, $\dot{m}_{cw}$ is the mass flow rate of cold water, $\Delta t_{cw} = t_{cwo} - t_{cwi}$, t_{cwi} and t_{cwo} are the cold-water temperature at the inlet and outlet of the condenser.

In addition, the heat transfer rate between energy resources and working fluid can also be given (evaporator, condenser)

Table 3 Constant coefficients of organic fluids for the state function of every point in ORC

Working fluids	R717	R134a	R1234yf	R290	R245fa	R1233zd
$\zeta_{h_1,1}$	−1.7876e-02	−6.2410e-03	−7.1811e-03	−1.4624e-02	−7.6592e-04	−8.9258e-04
$\zeta_{h_1,2}$	1.7698	9.9516e-01	1.1714	2.1504	7.9390e-01	7.6816e-01
$\zeta_{h_1,3}$	1590.6051	389.4105	351.5388	551.0576	403.1600	403.7864
$\zeta_{h_3,1}$	3.9878e-03	1.9118e-03	2.1823e-03	5.0082e-03	1.0747e-03	4.6113e-04
$\zeta_{h_3,2}$	4.6020	1.3317	1.2838	2.4736	1.2706	1.2224
$\zeta_{h_3,3}$	343.2888	200.0629	200.0424	200.1463	200.0204	200.0223
$\zeta_{s_1,1}$	−2.9320e-06	−1.0923e-05	−1.5180e-05	−2.8389e-05	1.9682e-06	1.5652e-06
$\zeta_{s_1,2}$	−1.0115e-02	6.9016e-04	1.6491e-03	1.9731e-03	3.2470e-04	2.2104e-04
$\zeta_{s_1,3}$	6.0349	1.7004	1.5651	2.3050	1.7412	1.7435
$\zeta_{s_3,1}$	−1.5799e-05	−2.2528e-06	−1.1332e-06	−1.2300e-08	−4.1033e-06	−5.7227e-06
$\zeta_{s_3,2}$	1.6726e-02	4.8447e-03	4.6694e-03	8.9698e-03	4.6347e-03	4.4533–03
$\zeta_{s_3,3}$	1.4722	1.0002	1.0001	1.0005	1.0001	1.0002
$\zeta_{P_1,1}$	5.8267e-01	3.5765e-01	3.1744e-01	3.3461e-01	1.4323e-01	1.1423e-01
$\zeta_{P_1,2}$	−5.5589	−2.7089	−7.4941e-01	3.7371	−3.8044	−2.7190
$\zeta_{P_1,3}$	851.4715	556.6621	543.9449	687.8062	175.1930	142.8210
$\zeta_{P_3,1}$	3.5071e-01	2.1524e-01	1.9500e-01	2.1998e-01	7.6177e-02	6.2164e-02
$\zeta_{P_3,2}$	13.7436	9.2652	9.5793	13.3971	1.7860	1.6218
$\zeta_{P_3,3}$	441.3651	299.7510	321.6670	480.0812	56.2418	50.4481
ζ_{v_3}	1.6510e-03	8.2412e-04	9.1066e-04	2.0200e-03	7.4396e-04	7.8872e-04

$$\dot{Q} = UA\Delta T_m \tag{4}$$

where, U is the overall heat-transfer coefficient, A is the cross-section area normal to the direction of heat transfer, $\Delta T_m = (\Delta t_i - \Delta t_o)/\ ln\ (\Delta t_i/\Delta t_o)$ is called the logarithmic mean temperature difference (LMTD), Δt_i represents the temperature difference of heat exchanger end section (heat source inlet), Δt_o shows the temperature difference of heat exchanger end section (heat source outlet). Thus, according to the energy balance at the evaporation, evaporation temperature of the evaporator can be given as

$$t_e = t_{wwi} - \left(t_{wwo} \cdot \exp\left[\frac{(UA)_e}{\dot{m}_{ww}c_p}\right] \Big/ 1 - \exp\left[\frac{(UA)_e}{\dot{m}_{ww}c_p}\right]\right) \tag{5}$$

Thus, according to Eqs. (1–5), the $\dot{W}_{net}$ can be written as,

$$\dot{W}_{net} = \dot{m}_{ww}c_p \cdot \left(t_{wwi} - \frac{t_{wwi} - t_e(1 - \exp[\alpha])}{\exp[\alpha]}\right)\left(1 - \frac{A(t_c, t_e)}{B(t_c, t_e)}\right) \tag{6}$$

it is used to calculate the net power output of ORC from the heat transfer rate absorbed from the warm-water and ORC's performance (thermal efficiency), where, $t_c = \frac{(t_{cwi}-(t_{cwi}+t_{wwi}-(t_{wwi}-t_e\cdot(1-\exp[\alpha]))/\exp[\alpha])\cdot\exp[\beta])}{(1-\exp[\beta])}$, $40\,^\circ C \le t_e \le 80\,^\circ C, 5\,^\circ C \le t_c \le 40\,^\circ C$, $\alpha = (UA)_e/(\dot{m}_{ww}c_p)$, $\beta = (UA)_c/(\dot{m}_{ww}c_p \cdot A(t_c, t_e)/B(t_c, t_e))$,.$A(t_c, t_e) = (t_c + T_0)\cdot(a_1t_c^2 + a_2t_e^2 + a_3t_e + a_4t_c + a_5), B(t_c, t_e) = b_1t_c^2+b_2t_e^2+b_3t_e+b_4t_c+b_5$.

To get the maximum net power output of ORC theoretically, the theoretical optimization design (Table 4) is used to solve the Eq. (6) as below.

$$\max\{\dot{W}_{net}\}_{orc} = \dot{m}_{ww}c_p\left(t_{wwi} - \frac{t_{wwi} - (t_e)_{opt}(1-\exp[\alpha])}{\exp[\alpha]}\right)\left(1 - \frac{A((t_c)_{opt}, (t_e)_{opt})}{B((t_c)_{opt}, (t_e)_{opt})}\right) \tag{7}$$

In the expression,

$$\begin{aligned} A((t_c)_{opt}, (t_e)_{opt}) =& ((t_c)_{opt} + T_0) \\ &\cdot\left(a_1(t_c)_{opt}^2 + a_2(t_e)_{opt}^2 + a_3(t_e)_{opt} + a_4(t_c)_{opt} + a_5\right) \end{aligned} \tag{8}$$

$$B((t_c)_{opt}, (t_e)_{opt}) = b_1(t_c)_{opt}^2 + b_2(t_e)_{opt}^2 + b_3(t_e)_{opt} + b_4(t_c)_{opt} + b_5 \tag{9}$$

where, $a_i(i = 1, \cdots, 5)$, $b_i(i = 1, \cdots, 5)$ and $c_i(i = 1, \cdots, 6)$ are constant coefficients, varied with working fluid medium. Such as R717, R134a, R245fa, R1233zd, R1234yf, and R290, whose corresponding coefficient values are shown in Table 5.

Thus, it is easy to know that $\max\{\dot{W}_{net}\}_{orc}$ depends on the performance of the evaporator ($(UA)_e$) and heat capacity at constant pressure (c_p) as well as the warm and cold-water temperature at the inlet, t_{wwi}, t_{cwi} and mass flow rate of warm water $\dot{m}_{ww}$.

Meanwhile, the following optimal operation condition values can also be given as follows, in which, Eqs. (11–12, 14) derived from energy balance at the condensation and Eq. (13) derived from the turbine' heat transfer rate supplied to the cycle.

$$t_{wwo} = \frac{t_{wwi} - (t_e)_{opt}\cdot(1-\exp[\alpha])}{\exp[\alpha]} \tag{10}$$

$$t_{cwo} = t_{cwi} + t_{wwi} - t_{wwo} \tag{11}$$

$$\dot{m}_{cw} = \dot{m}_{ww}\cdot\frac{A((t_c)_{opt}, (t_e)_{opt})}{B((t_c)_{opt}, (t_e)_{opt})} \tag{12}$$

$$\dot{m}_{wf} = \frac{\max\{\dot{W}_{net}\}_{orc}}{h_1((t_e)_{opt}) - h_2((t_c)_{opt}, (t_e)_{opt})} \tag{13}$$

$$(UA)_c = (UA)_e\cdot\frac{A((t_c)_{opt}, (t_e)_{opt})}{B((t_c)_{opt}, (t_e)_{opt})} \tag{14}$$

Table 4 Constant coefficients of the state function of every point in ORC for power generation from Low-temperature heat sources

Steps	Optimization details	
Step1	Design Variable	$X = \begin{pmatrix} t_e \\ t_c \end{pmatrix}$
	Find	$\max\limits_{t_e, t_c}\{\dot{W}_{net}(X)\}$
	Satisfy	$g_1(X) = t_e - 40 \geq 0,\ g_2(X) = 80 - t_e \geq 0,$ $g_3(X) = t_c - 5 \geq 0,\ g_4(X) = 40 - t_c \geq 0.$
		$f_1(X) = \frac{t_{cwi} - (t_{cwi} + t_{wwi} - (t_{wwi} - t_e \bullet (1 - \exp[\alpha]))/\exp[\alpha]) \bullet \exp[\beta]}{1 - \exp[\beta]} - t_c = 0$ (energy balance at the condensation)
⇃⇂		
Step2	Penalty Function	$F(X) = \frac{1}{\dot{W}_{net}(X)} + R_k (f_1(X))^2 + \gamma_k \sum_{j=1}^{4} \frac{1}{g_j(X)}$
	Find	$\min\limits_{t_e, t_c, \gamma_k \to 0, R_k \to \infty} \{F(X)\}$
	Assume	$40\ ^\circ C \leq (t_e)_{opt} \leq 80\ ^\circ C,\ 5\ ^\circ C \leq (t_c)_{opt} \leq 40\ ^\circ C$
	Solve	$f_1(X) = 0 \Longleftrightarrow \begin{cases} (t_e)_{opt} = (t_{cwi} + t_{wwi}) - (t_c)_{opt} \\ (UA)_c = (UA)_e \cdot \frac{A((t_c)_{opt}, (t_e)_{opt})}{B((t_c)_{opt}, (t_e)_{opt})} \\ \Delta t_{ww} = \Delta t_{cw} \\ \dot{m}_{cw} = \dot{m}_{ww} \cdot \frac{A((t_c)_{opt}, (t_e)_{opt})}{B((t_c)_{opt}, (t_e)_{opt})} \end{cases}$
⇃⇂		
Step3	Find	$\max\limits_{t_c}\{\dot{W}_{net}(t_c)\}$
	Solve	$(t_c)_{opt} = c_1 t_{cwi}^2 + c_2 t_{wwi} \cdot t_{cwi} + c_3 t_{wwi}^2 + c_4 t_{cwi} + c_5 t_{wwi} + c_6$ $(t_e)_{opt} = -c_1 t_{cwi}^2 - c_2 t_{wwi} \cdot t_{cwi} - c_3 t_{wwi}^2 + (1 - c_4) t_{cwi} + (1 - c_5) t_{wwi} - c_6$
⇃⇂		
Result	$\max\{\dot{W}_{net}\}_{orc} = \dot{m}_{ww} c_p \cdot \left(t_{wwi} - \frac{t_{wwi} - (t_e)_{opt}(1 - \exp[\alpha])}{\exp[\alpha]}\right)\left(1 - \frac{A((t_c)_{opt}, (t_e)_{opt})}{B((t_c)_{opt}, (t_e)_{opt})}\right)$	

Meanwhile, the corresponding ORC thermal efficiency at maximum net power output is

$$\eta_{orc} = 1 - \frac{A\left((t_c)_{opt}, (t_e)_{opt}\right)}{B\left((t_c)_{opt}, (t_e)_{opt}\right)} \tag{15}$$

Table 5 Constant coefficients of organic fluids to get the maximum net power output in ORC

Coefficients		1	2	3	4	5	6
R717	a_i	1.5799e-05	−2.9320e-06	−1.0115e-02	−1.6726e-02	4.5627	–
	b_i	−3.4088e-03	−1.8838e-02	1.7789	−4.5793	1246.6393	–
	c_i	1.7755e-04	3.7940e-04	2.0663e-04	7.4895e-01	2.5004e-01	6.2668e-02
R134a	a_i	2.2528e-06	−1.0923e-05	6.9016e-04	−4.8447e-03	7.0017e-01	–
	b_i	−1.7344e-03	−6.5358e-03	9.9740e-01	−1.3241	189.1358	–
	c_i	1.3316e-04	−6.2219e-04	2.2385e-04	7.6058e-01	2.6402e-01	−7.1341e-01
R − 1234yf	a_i	1.1332e-06	-1.5180e-05	1.6491e-03	−4.6694e-03	5.6500e-01	–
	b_i	−2.005e-03	−7.4702e-03	1.1720	−1.2751	151.2940	–
	c_i	9.6429e-05	−7.2092e-04	1.9362e-04	7.6739e-01	2.7468e-01	−1.1829
R290	a_i	1.2300e-08	−2.8389e-05	1.9731e-03	−8.9698e-03	1.3045	–
	b_i	−4.5639e-03	−1.5300e-02	2.1428	−2.4465	350.4918	–
	c_i	1.1020e-04	−6.6237e-04	2.0319e-04	7.6476e-01	2.6920e-01	−9.6838e-01
R245fa	a_i	4.1033e-06	1.9682e-06	3.2470e-04	−4.6347e-03	7.4107e-01	–
	b_i	−1.0181e-03	−8.7248e-04	7.9673e-01	−1.2693	203.0510	–
	c_i	1.7296e-04	−3.9031e-04	2.5408e-04	7.4516e-01	2.4800e-01	8.1594e-02
R1233zd	a_i	5.7227e-06	1.5652e-06	2.2104e-04	−4.4533e-03	7.4337e-01	–
	b_i	−4.1210e-04	−9.8267e-04	7.7030e-01	−1.2212	203.6913	–
	c_i	1.5612e-04	−3.9490e-04	2.3342e-04	7.4630e-01	2.4951e-01	3.5570e-02

From this equation, ORC thermal efficiency for maximum net power output is only decided by the warm and cold-water temperature at the inlet, t_{wwi} and t_{cwi}, for a given working fluid.

The potential net power output of the ORC or maximum power output of Carnot cycle (Ikegami and Bejan, 1998) is

$$\dot{W}_{carnot} = \frac{\left(\sqrt{t_{wwi} + T_0} - \sqrt{t_{cwi} + T_0}\right)^2}{\left(\dot{m}_{ww} c_p\right)^{-1} + \left(\dot{m}_{cw} c_p\right)^{-1}} \tag{16}$$

The efficiency of ORC in comparison with the Carnot cycle is

$$\eta_{orc,\ carnot} = \frac{\max\left\{\dot{W}_{net}\right\}_{orc}}{\dot{W}_{carnot}} \times 100\% \tag{17}$$

which is defined by comparing the ORC with the Carnot cycle in terms of the maximum net power output. In this way, the parameter can be used to evaluate the designed ORC system is efficient or not. And the rest part $(1 - \eta_{orc,\ carnot})$ implies the energy loss of the ORC.

3 Results and Discussion

It is known that suitable working fluid can make ORC get better performance in terms of more power generation and higher efficiency from low-temperature heat sources. Figure 4 shows the effect and comparison of different working fluids on ORC with the same initial conditions as shown in Table 6. Meanwhile, the parameters of temperature of the warm water at outlet of the evaporator (t_{wwo}), temperature of the cold water at outlet of the condenser (t_{cwo}), mass flow rate of the cold water ($\dot{m}_{cw}$), mass flow rate of the working fluid ($\dot{m}_{wf}$), and performance of the condenser ($(UA)_c$) can be given by using the eqs. (10–14). And corresponding state of every point on ORC is given in Table 7.

Results show that, for the same initial conditions, such as, $t_{wwi} = 80\ ^\circ C$, $t_{cwi} = 15\ ^\circ C$, $\dot{m}_{ww} = 10\ kg/s$, and $(UA)_e = 100\ kW/K$, R717 has the best performance in terms of the maximum power output ($\max\left\{\dot{W}_{net}\right\}_{orc} = 56.0\ kW$) with thermal efficiency of 8.6% (η_{orc}), and the next is the R1233zd, it is 54.4 kW ($\max\left\{\dot{W}_{net}\right\}_{orc}$) with thermal efficiency of 8.3% (η_{orc}). And Fig. 5 display the relationship between t_e and $\dot{W}_{net}$ with same initial conditions of R717 shown in Table 6. The $t_{e,\ opt} = 62.9[^\circ C]$ and corresponding to $\max\left\{\dot{W}_{net}\right\}_{orc,R717} = 56.0[kW]$ match very well with equations of Table 4, which indicates that the solution for theoretical optimization result of maximum power output in ORC is correct. Here, it should be noticed that other optimization conditions are almost the same to get the maximum power output of ORC, such as, $\dot{m}_{cw} = 9.1 \sim 9.2\ kg/s$, $(UA)_c = 91$~$92\ kW/K$. In this case, mass flow rate, volumetric flow rate of the working fluid, saturation pressure of the

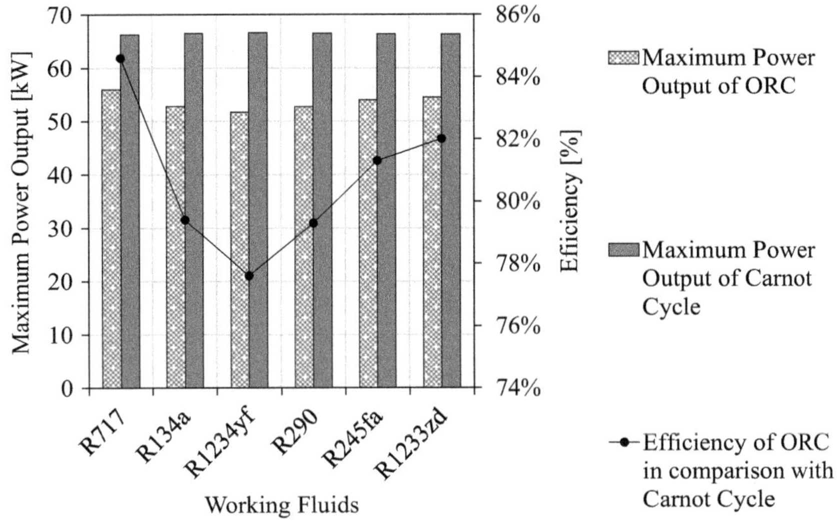

Fig. 4 Comparison of different working fluids on ORC in terms of maximum power output and efficiency in comparison with Carnot cycle

Table 6 Initial conditions and performance comparison of different working fluids on ORC

	R717	R134a	R1234yf	R290	R245fa	R1233zd
t_{wwi} [°C]	80					
t_{wwo} [°C]	64.5	64.1	63.9	64.1	64.3	64.3
t_{cwi} [°C]	15					
t_{cwo} [°C]	30.5	30.9	31.1	30.9	30.7	30.7
$t_{e,\,opt}$ [°C]	62.9	62.5	62.3	62.4	62.7	62.8
$t_{c,\,opt}$ [°C]	32.1	32.5	32.7	32.6	32.3	32.2
$P_{e,\,opt}$ [MPa]	2.808	1.783	1.729	2.225	0.500	0.422
$P_{c,\,opt}$ [MPa]	1.243	0.829	0.843	1.150	0.193	0.167
$\dot{m}_{ww}$ $[kg/s]$	10					
$\dot{m}_{cw}$ $[kg/s]$	9.1	9.2	9.2	9.2	9.2	9.2
$\dot{m}_{wf}$ $[kg/s]$	0.560	3.499	4.164	1.821	3.135	3.119
$Q_{wf,\,l}$ [L/min]	56.8	173	227.5	220.7	139.9	150.3
η_{orc} [%]	8.6%	7.9%	7.7%	7.9%	8.2%	8.3%
$(UA)_e$ [kW/K]	100					
$(UA)_c$ [kW/K]	91	92	92	92	92	92
$\max\{\dot{W}_{net}\}_{orc}$ [kW]	56.0	52.8	51.7	52.7	54.0	54.4
$\dot{W}_{carnot}$[kW]	66.2	66.5	66.6	66.5	66.4	66.4
$\eta_{orc,\,carnot}$ [%]	84.6%	79.4%	77.6%	79.3%	81.3%	82.0%

Table 7 State of every point for different working fluids on ORC

Point	Items	R717	R134a	R1234yf	R290	R245fa	R1233zd
1	Enthalpy [kJ/kg]	1631.3	427.3	396.7	628.4	449.9	448.5
	Pressure [MPa]	2.808	1.783	1.729	2.225	0.500	0.422
	Entropy [kJ/(kg K)]	5.387	1.701	1.609	2.318	1.769	1.764
	Temperature [°C]	62.9	62.5	62.3	62.4	62.7	62.8
2	Enthalpy [kJ/kg]	1531.3	412.1	384.3	599.5	432.7	431.0
	Pressure [MPa]	1.243	0.829	0.843	1.150	0.193	0.167
	Entropy [kJ/(kg K)]	5.387	1.701	1.609	2.318	1.769	1.764
	Temperature [°C]	32.1	32.5	32.7	32.6	32.3	32.2
3	Enthalpy [kJ/kg]	495.1	245.4	244.3	286.1	242.2	239.9
	Pressure [MPa]	1.243	0.829	0.843	1.150	0.193	0.167
	Entropy [kJ/(kg K)]	1.993	1.155	1.152	1.293	1.146	1.138
	Temperature [°C]	32.1	32.5	32.7	32.6	32.3	32.2
4‘	Enthalpy [kJ/kg]	650.3	427.3	289.3	376.0	284.1	448.5
	Pressure [MPa]	2.808	1.783	1.729	2.225	0.500	0.422
	Entropy [kJ/(kg K)]	2.468	1.701	1.289	1.566	1.276	1.764
	Temperature [°C]	62.9	62.5	62.3	62.4	62.7	62.8
4	Enthalpy [kJ/kg]	501.9	247.6	246.8	293.1	242.7	240.3
	Pressure [MPa]	2.808	1.783	1.729	2.225	0.500	0.422
	Entropy [kJ/(kg K)]	1.993	1.155	1.152	1.293	1.146	1.138
	Temperature [°C]	32.5	33.1	33.4	33.5	32.4	32.3

evaporator and saturation pressure of the condenser are noticeable different between them, such as, $\dot{m}_{wf,R717} = 0.560\ kg/s$, $\dot{m}_{wf,R1233zd} = 3.119\ kg/s$; $v_{wf,\,R717} = 56.8\ L/min$, $v_{wf,\,R1233zd} = 150.3\ L/min$; $P_{e,\,opt,\,R717} = 2.808\ MPa$ $P_{e,\,opt,\,R1233zd} = 0.422\ MPa$; $P_{c,\,opt,\,R717} = 1.243\ MPa$ $P_{c,\,opt,\,R1233zd} = 0.167\ MPa$. It means that R1233zd needs a larger piping system in ORC than R717. However, the pressure-resistant standard of R717 ORC system is higher than R1233zd ORC system since the working pressure of R717 is noticeably higher than that of R1233zd.

From the optimized results as shown in Table 6, we also know that R1233zd (GWP: 1) can deserve the best option to replace R245fa (GWP: 950) on the condition of maximum net power output in ORC, because it is not only a non-ozone depleting, low global warming potential (GWP) HFO refrigerant (Table 1), but also a suitable replacement working fluid for R245fa for maximizing net power output with the same initial conditions and almost the same optimization conditions (Table 6), such as, $t_{wwi} = 80\ ^\circ C$, $t_{cwi} = 15\ ^\circ C$, $\dot{m}_{ww} = 10\ kg/s$, $\dot{m}_{cw} = 9.2\ kg/s$, $(UA)_e = 100\ kW/K$, $(UA)_c = 92\ kW/K$, $\dot{m}_{wf} = 3.119 \sim 3.135\ kg/s$, $v_{wf} = 139.9\text{~}150.3\ L/min$, $P_{e,\,opt} = 0.422\text{~}0.500\ MPa$, $P_{c,\,opt} = 0.167\text{~}0.193\ MPa$. In this case, their maximum net power outputs are respectively $\max\{\dot{W}_{net}\}_{orc,R1233zd} = 54.4\ kW$ and $\max\{\dot{W}_{net}\}_{orc,R245fa} = 54.0\ kW$, which shows that R1233zd and R245fa's net power outputs are very close

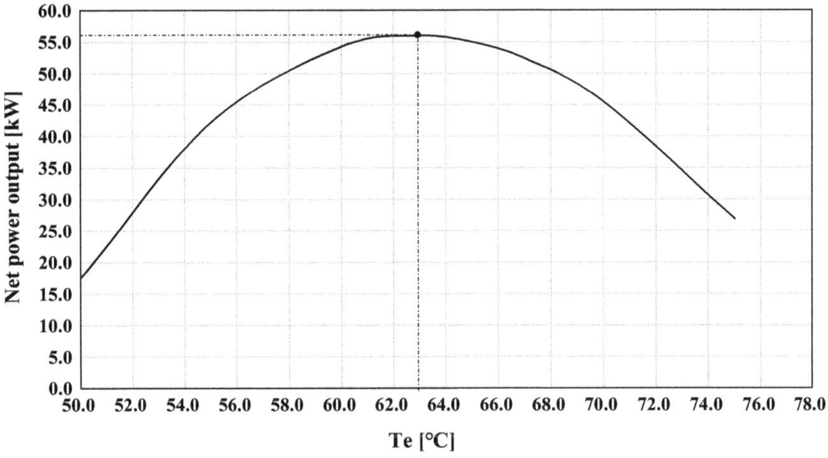

Fig. 5 Relationship between t_e and $\dot{W}_{net}$ with same initial conditions of R717 shown in Table 6

and coincides with Araya's experimentally investigate (Araya et al., 2020). And it also means that the proposed model is correct. In same, R1234yf (GWP: 4) also can replace r134a (GWP: 1430) since their optimization operation conditions are almost same.

For comparison, potential net power output of ORC (maximum power output of the Cannot cycle) and the potential efficiency of the ORC are also shown in Fig. 4 and Table 6. It shows that the maximum power output of the Carnot cycle with these working fluids (R717, R134a, R1234yf, R290, R245fa, R1233zd) are almost the same ($\dot{W}_{carnot} = 66.2 \sim 66.6\ kW$).

Meanwhile, the efficiency of ORC ($\eta_{orc,\ carnot}$) in comparison with the Carnot cycle varies with the working fluid is given here, such as R717 $\eta_{orc,\ carnot,\ R717} = 84.6\%$ is the highest, the next one is R1233zd $\eta_{orc,\ carnot,\ R1233zd} = 82.0\%$ and R1234yf $\eta_{orc,\ carnot,\ R1234yf} = 77.6\%$ is the lowest, etc.

Furthermore, the efficiency of ORC in comparison with Carnot cycle is kept constant for working fluids comparison, such as 85.0% as shown in Fig. 6 and Table 8, by adjusting the mass flow rate of the warm water ($\dot{m}_{ww}$). It shows the same result in comparison with the maximum power output of these working fluids, R717 is the largest, the next is the R1233zd, the following is the R245fa, and the last is the R1234yf. However, the maximum power output of the designed ORC system in R717 is 55.2 kW, R1233zd is 47.7 kW, R245fa is 45.2 kW, and the last is R1234yf (26.0 kW).

In this way, if the R1234yf ORC system is designed to have the same efficiency of ORC and the same maximum power output with the R717 ORC system, the scale of the heat exchangers in R1234yf ORC system ($(UA)_{e,\ c}$) should be enlarged enough, which results in an increase in initial capital investment of the ORC system.

Moreover, R717 ORC is discussed in more detail as below. Effect of different low-temperature heat sources (less than 100 °C), such as the temperature of warm water at

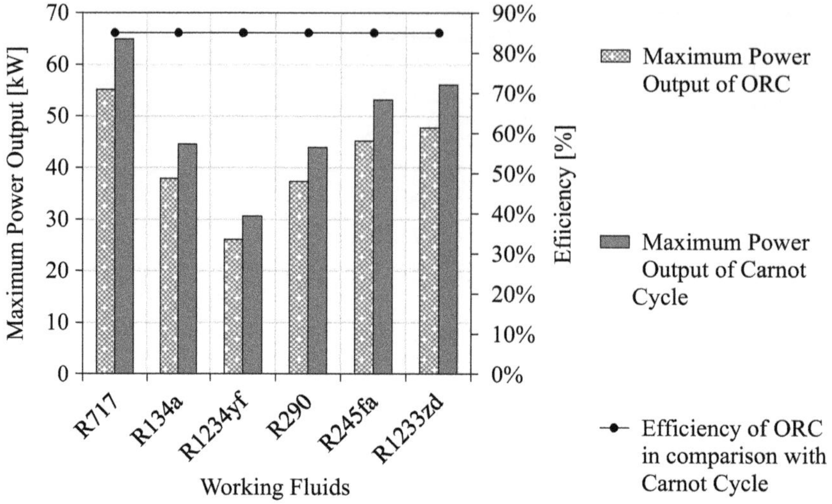

Fig. 6 Comparison of different working fluids on ORC in terms of maximum power output with the same efficiency of ORC in comparison with Carnot cycle

Table 8 Initial conditions and performance comparison of different working fluids on ORC with the same efficiency in comparison with Carnot cycle

	R717	R134a	R1234yf	R290	R245fa	R1233zd
t_{wwi} [°C]	80					
t_{wwo} [°C]	64.4	63.0	62.4	62.9	63.6	63.8
t_{cwi} [°C]	15					
t_{cwo} [°C]	30.6	32.0	32.6	32.1	31.4	31.2
$t_{e,\,opt}$ [°C]	62.9	62.5	62.3	62.4	62.7	62.8
$t_{c,\,opt}$ [°C]	32.1	32.5	32.7	32.6	32.3	32.2
$P_{e,\,opt}$ $[MPa]$	2.808	1.783	1.729	2.225	0.500	0.422
$P_{c,\,opt}$ $[MPa]$	1.243	0.829	0.843	1.150	0.193	0.167
$\dot{m}_{ww}$ $[kg/s]$	9.8	6.7	4.6	6.6	8	8.5
$\dot{m}_{cw}$ $[kg/s]$	9.0	6.2	4.2	6.1	7.3	7.8
$\dot{m}_{wf}$ $[kg/s]$	0.551	2.509	2.099	1.289	2.622	2.728
$Q_{wf,\,l}$ $[L/min]$	55.9	124.1	114.7	156.2	117.0	131.6
η_{orc} [%]	8.6%	7.9%	7.7%	7.9%	8.2%	8.3%
$(UA)_e$ $[kW/K]$	100					
$(UA)_c$ $[kW/K]$	91	92	92	92	92	92
$\max\{\dot{W}_{net}\}_{orc}$ $[kW]$	55.2	37.9	26.0	37.3	45.2	47.6
$\dot{W}_{carnot}[kW]$	64.9	44.6	30.6	43.9	53.1	56.0
$\eta_{orc,\,carnot}$ [%]	85.0%					

the inlet $t_{wwi} = 70{\sim}90\,^{\circ}C$ to the maximum power output of ORC ($\max\{\dot{W}_{net}\}_{orc,\ R717}$) is shown in Fig. 7. Effect of different cold sources, such as the temperature of cold water at the inlet $t_{cwi} = 25{\sim}1\,^{\circ}C$ to $\max\{\dot{W}_{net}\}_{orc,\ R717}$ is shown in Fig. 8. And effect of performance of the evaporator $(UA)_e = 50{\sim}300\ kW/K$ to $\max\{\dot{W}_{net}\}_{orc,\ R717}$ is shown in Fig. 9.

It is clear that $\max\{\dot{W}_{net}\}_{orc,\ R717}$ is nearly linearly increased with increasing of the temperature of warm water at inlet (t_{wwi}) or performance of the evaporator $(UA)_e$ and is linearly decreased with increasing of the temperature of cold water at inlet t_{cwi} with same efficiency of ORC in comparison with Carnot cycle, such as $\eta_{orc,\ carnot} = 85.0\%$, which can be realized by adjusting the operating parameter of the mass flow

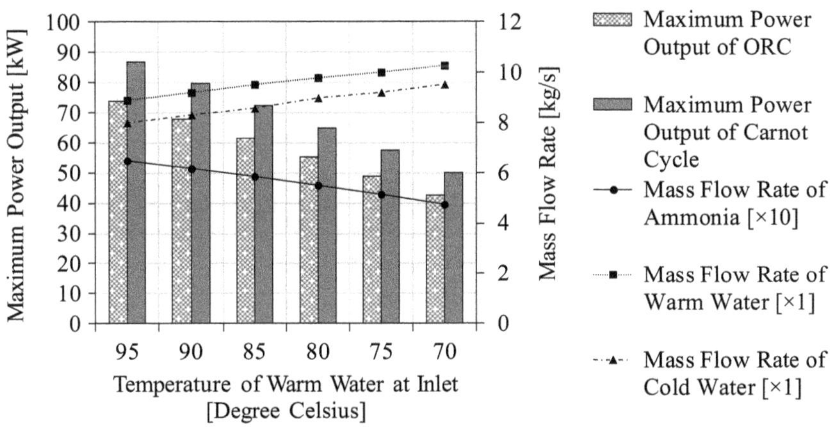

Fig. 7 Influence of t_{wwi} on the R717 ORC in terms of maximum power output with the same efficiency in comparison with Carnot cycle

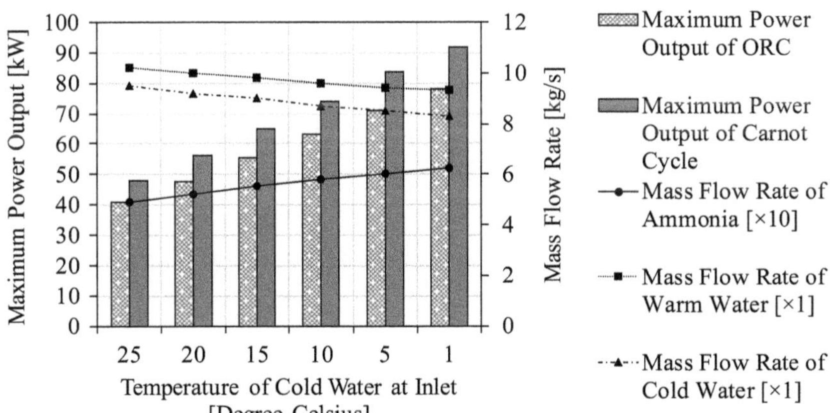

Fig. 8 Influence of t_{cwi} on the R717 ORC in terms of maximum power output with the same efficiency in comparison with Carnot cycle

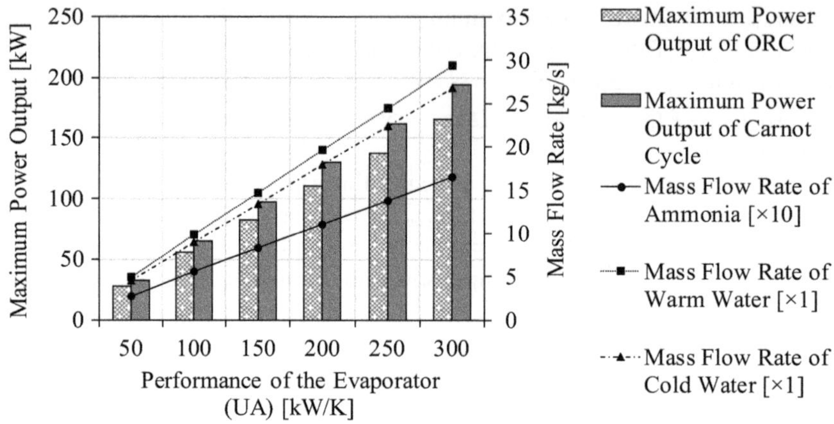

Fig. 9 Influence of $(UA)_e$ on the R717 ORC in terms of maximum power output with the same efficiency in comparison with Carnot cycle

rate of the warm, cold water and the working fluid ($\dot{m}_{ww}, \dot{m}_{cw}, \dot{m}_{wf}$) with the aid of the proposed theoretical optimum model. Meanwhile, it is shown that about 6 kW of the maximum power output can be increased by increasing the t_{wwi} for 5 °C, about 7 ~ 8 kW of the maximum power output can be increased by decreasing 5 °C of t_{cwi}, and about 28 kW of the maximum power output can be increased by increasing performance of the evaporator 50 kW/K for $(UA)_e$. In addition, the others optimal operating conditions, such as the temperature of warm water at the outlet (t_{wwo}), the temperature of cold water at the outlet (t_{cwo}) can be given by using eqs. (10–11), and $(UA)_c$ can be given by eq. (14). And the optimal rate of $(UA)_c$ to $(UA)_e$ is almost the same $(UA)_c/(UA)_e = 0.91$~$0.92[-]$ since the corresponding ORC thermal efficiency is about 8~9%. Thus, from the initial conditions of t_{wwi} and t_{cwi}, for example $t_{wwi} = 85\ ^{\circ}C$ and $t_{cwi} = 15\ ^{\circ}C$ are fixed, the optimal ORC can be designed directly, and its maximum power output is 82.6 kW for $(UA)_e = 150$ kW/K with optimal operation condition of $\dot{m}_{ww} = 14.7[kg/s]$, $\dot{m}_{cw} = 13.4[kg/s]$, $\dot{m}_{wf} = 0.825[kg/s]$ and $(UA)_c = 137$ kW/K. In the way, theoretical optimum model with its optimal operating parameters for maximum power generation of ORC and its contribution to generate green and zero-carbon energy by using some typical or low-GWP organic working fluids can be given clearly.

4 Conclusion

Theoretical analysis of some typical or low-GWP organic working fluids in ORC for maximum power generation is studied for development of green and zero-carbon energy generation system to promote the race to zero. Results show that this study provides a theoretical basis for the design of ORC to optimization with getting their optimization operation conditions, which also will favor the design and optimization for green marine industry (marine waste-heat recovery, ocean thermal energy conversion, etc.). Thus, following results can be given.

1. Temperatures of warm and cold water at inlet, mass flow rate of the warm water and performance of the evaporator play a key role for maximum power generation. And the corresponding thermal efficiency could be evaluated using the temperatures of warm and cold water at inlet.
2. In the case of same initial conditions of temperatures of warm water (85 Celsius degree) and cold water (15 Celsius degree) at inlet, mass flow rate of the warm water (10 kg/s) and performance of the evaporator (100 kW/K), R717 has the best performance in terms of the maximum power output 56.0 kW with thermal efficiency of 8.6%, and the next is the R1233zd (54.4 kW, 8.3%), R245fa (54.0 kW, 8.2%), R134a (52.8 kW, 7.9%), R290 (52.7 kW, 7.9%), and R1234yf (51.7 kW, 7.7%). Here, it should be noticed that other optimization conditions are almost the same (mass flow rate of the cold water 9.1 ~ 9.2 kg/s; performance of the condenser 91 ~ 92 kW/K) to get their maximum power output of ORC.
3. Low-GWP R1233zd (GWP: 1) can deserve the best option to replace R245fa (GWP: 950) and R1234yf (GWP: 4) also can replace r134a (GWP: 1430) since their optimization operation conditions are almost same.
4. Potential net power output of ORC with these working fluids are almost the same. Meanwhile, efficiency of ORC in comparison with the Carnot cycle varies with the working fluid, R717 is the highest, the next one is R1233zd. In addition, in the case of keeping efficiency of ORC as constant for comparison maximum power output of these working fluids in ORC by adjusting mass flow rate of the warm water, it shows the same result, R717 is the largest, the next is the R1233zd.
5. Theoretical Optimum Model with its Operating Parameters for Design Maximum Power Generation of ORC and its Contribution to Generate Green and Zero-Carbon Energy by Using some Typical or Low-GWP Organic Working Fluids Is Given Clearly

It should be noticed that the proposed model is suitable for low-temperature heat sources, in future, we will enhance and enlarge its applicability to medium-temperature heat sources to make it more effective since ORC technology is currently one of the most efficient ways to turn medium and low temperature heat energy into electric power.

Author's note: This chapter is a revised and expanded version of the author's article, "Theoretical analysis of organic Rankine cycle for maximum power generation in optimization operation conditions" published in *Green Energy and Resources*, Vol. 2, Issue 4 (2024).

References

Araya S, Wemhoff A, Jones G, Fleischer A (2020) An experimental study of an organic Rankine cycle utilizing HCFO-1233zd (E) as a drop-in replacement for HFC-245fa for ultra-low-grade waste heat recovery. Appl Therm Eng 180:115757

Babatunde A, Sunday O (2018) A review of working fluids for organic rankine cycle (ORC) applications. *IOP Conf. Series*. Mater Sci Eng 413:012019

Chen HJ, Goswami DY, Stefanakos EK (2010) A review of thermodynamic cycles and working fluids for the conversion of low-temperature heat. Renew Sustain Energy Rev 14(9):3059–3067

Dai YP, Wang JF, Gao L (2009) Parametric optimization and comparative study of organic Rankine cycle (ORC) for low-temperature waste heat recovery. Energy Convers Manag 50(3):576–582

Delgado-Torresa AM, Garcia-Rodriguez L (2010) Analysis and optimization of the low-temperature solar organic Rankine cycle (ORC). Energy Convers Manag 51(12):2846–2856

DiPippo R (2004) Second law assessment of binary plants generating power from low-temperature geothermal fluids. Geothermics 33:565–586

Fallah M, Mohammad S, Mahmoudi S, Yari M, Akbarpour Ghiasi R (2016) Advanced exergy analysis of the Kalina cycle applied for low temperature enhanced geothermal system. Energy Convers Manag 108:190–201

Herath H, Wijewardane M, Ranasinghe R, Jayasekera J (2020) Working fluid selection of organic rankine cycles. Energy Rep 6(9):680–686

Hettiarachchi HDM, Golubovica M, Worek WM, Ikegami Y (2007) Optimum design criteria for an organic Rankine cycle using low-temperature geothermal heat sources. Energy 32(9):1698–1706

Hung TC, Shai TY, Wang SK (1997) A review of organic rankine cycles (ORCs) for the recovery of low-temperature waste heat. Energy 22(7):661–667

Ikegami Y, Bejan A (1998) On the thermodynamic optimization of power plants with heat transfer and fluid flow irreversibilities. J Sol Energy Eng 120:139–144

Jia B, Lei Y, Sun F, Zhou W (2024) Theoretical analysis of organic rankine cycle for maximum power generation in optimization operation conditions. Green Energy Resour 2(4):100101

Kalina AI (1982) Generation of energy by means of a working fluid, and regeneration of a working fluid. N. p, United States

Kalina AI (1984) Combined-cycle system with a novel bottoming cycle. J Eng Gas Turb Power 106:737–742

Kalina AI, Leibowitz HM (1989) Application of the Kalina cycle technology to geothermal power generation. Geotherm Resour Counc Proc 13:605–611

Kalina AI, Leibowitz HM (1994) Applying Kalina cycle technology to high enthalpy geothermal resources. Geotherm Resour Counc Trans 18:531–536

Liu BT, Chien KH, Wang CC (2004) Effect of working fluids on organic Rankine cycle for waste heat recovery. Energy 29(8):1207–1217

Liu H, Shao YJ, Li JX (2011) A biomass-fired micro-scale CHP system with organic Rankine cycle (ORC)-thermodynamic modelling studies. Biomass Bioenergy 35(9):3985–3994

Lolos PA, Rogdakis ED (2009) A Kalina power cycle driven by renewable energy sources. Energy 34(4):457–464

Saleh B, Koglbauer G, Wendland M, Fischer J (2007) Working fluids for low-temperature organic Rankine cycles. Energy 32(7):1210–1221

Sauret E, Rowlands AS (2011) Candidate radial-inflow turbines and high-density working fluids for geothermal power systems. Energy 36(7):4460–4467

Sun FM, Ikegami Y, Arima H, Zhou WS (2013a) Performance analysis of the low temperature solar-boosted power generation system e part I: comparison between Kalina solar system and Rankine solar system. J Sol Energy Eng 135:011006.1-11

Sun FM, Ikegami Y, Arima H, Zhou WS (2013b) Performance analysis of the low temperature solar-boosted power generation system e part II: thermodynamic characteristics of the Kalina solar system. J Sol Energy Eng 135:011007.1-8

Sun FM, Ikegami Y, Jia BJ (2012b) A study on Kalina solar system with an auxiliary superheater. Renew Energy 41:210–219

Sun FM, Ikegami Y, Jia BJ, Arima H (2012a) Optimization design and exergy analysis of organic Rankine cycle in ocean thermal energy conversion. Appl Ocean Res 35:38–46
Sun FM, Zhou WS, Ikegami Y, Nakagami K, Su XM (2014) Energy-exergy analysis and optimization of the solar-boosted Kalina cycle system 11 (KCS-11). Renew Energy 66:268–279
Tchanche BF, Lambrinos G, Frangoudakis A, Papadakis G (2011) Low-temperature heat conversion into power using organic Rankine cycles—a review of various applications. Renew Sust Energ Rev 15(8):3963–3979
Uehara H, Dilao CO, Nakaoka T (1998a) Conceptual design of ocean thermal energy conversion power plants in The Philippines. Sol Energy 41(5):431–441
Uehara H, Ikegami Y (1990) Optimization of a closed-cycle OTEC system. J Sol Energy Eng 112:247–256
Uehara H, Ikegami Y, Nishida T (1998b) Performance analysis of OTEC system using a cycle with absorption and extraction processes. Trans Jpn Soc Mech Eng (Part B) 64:384–389
Wang JF, Yan ZQ, Zhou EM, Dai YP (2013) Parametric analysis and optimization of a Kalina cycle driven by solar energy. Appl Therm Eng 50(1):408–415
Wang XD, Zhao L, Wang JL, Zhang WZ, Zhao XZ, Wu W (2010) Performance evaluation of a low-temperature solar Rankine cycle system utilizing R245fa. Sol Energy 84(3):353–364
Wei DH, Lu XS, Lu Z (2007) Performance analysis and optimization of organic Rankine cycle (ORC) for waste heat recovery. Energy Convers Manag 48(4):1113–1119
Wu C (1990) Specific power optimization of closed-cycle OTEC plants. Ocean Eng 17(3):307–314
Yang M, Liu M, Yeh R (2024) Investigation of low-GWP working fluids as substitutes for R245fa in organic Rankine cycle application. Heliyon 10:e34219
Zhang SJ, Wang HX, Guo T (2011) Performance comparison and parametric optimization of subcritical organic Rankine cycle (ORC) and trans critical power cycle system for low-temperature geothermal power generation. Appl Energy 88(8):2740–2754
Zhang XR, Yamaguchi H, Fujima K, Enomoto M, Sawada N (2005) A feasibility study of CO2-based Rankine cycle powered by solar energy. JSME Int J Ser B Fluids Therm Eng 48:540–547

Optimal Design of Integrated Energy System Based on Supply-Demand Coupling Analysis

Hongbo Ren, Qiong Wu, and Weisheng Zhou

1 Introduction

The issue of global warming due to rising greenhouse gas (GHG) emissions presents a great challenge to the stability of the world's climate, economy and population (Dagoumas and Barker 2010). From the viewpoint of original sources of GHG emissions, it has been recognized that global climate destabilization is primarily due to the combustion of fossil fuels. It may be no exaggeration to say that the only way to achieve a sustainable low-carbon society is to develop a low-carbon energy system. Actually, in most of the medium and long-term scenarios toward a low-carbon society, the innovation of energy system is always paid the main attention. Thus, energy efficiency and distributed energy resources (DER) which are recognized as dominant demand-side and supply-side measures are the orders of the day when it comes to construct a low-carbon society. However, a low-carbon energy system integrating energy efficiency and distributed generation technologies may result in a higher degree of complexity, while determining the optimal combination of energy supply and utilization technologies. To support such a complex decision problem, a procedure based on mathematical optimization has proved its validity. Generally, recent studies on the optimization of DER system mainly focus on the function extension of the optimization model including the consideration of multiple objectives, various uncertainties, etc.

H. Ren (✉) · Q. Wu
College of Energy and Mechanical Engineering, Shanghai University of Electric Power, Shanghai, China
e-mail: tjrhb@163.com

Q. Wu
e-mail: wuqiongrff@163.com

W. Zhou
College of Policy Sciences, Ritsumeikan University, Kyoto, Japan
e-mail: zhou@sps.ritsumei.ac.jp

W. Zhou (ed.), *Decarbonization Policy Engineering*, World Sustainability Series,
https://doi.org/10.1007/978-981-95-5953-4_12

On the other hand, as major priorities in all energy policies worldwide, energy efficiency measures are also paid enough attention while planning local energy systems. Penna developed a multi-objective model for the determination of optimal retrofit solutions in terms of either economic performance or energy benefit (Penna et al. 2015). Karmellos developed a multi-objective mixed-integer non-linear programming (MINLP) model for the prioritization of energy efficiency measures in buildings (Karmellos et al. 2015). Antipova proposed a systematic method for the optimal retrofit of buildings with the consideration of various energy efficiency measures (Antipova et al. 2014). Diakaki proposed a multi-objective optimization model for the improvement of energy efficiency in buildings aiming at reducing the acquisition costs and increasing the energy savings (Diakaki et al. 2010).

According to the above discussions, previous studies generally focused on either the supply side (distributed energy resources) or the demand side (energy efficiency measures) of the local energy system. However, it should be indicated that the introduction of distributed generations is essentially a demand-side action due to the fact that DERs usually mean small-scale energy generators installed at or near the end-users. On the other hand, energy conservation actions may affect the size of the distributed energy system due to the decreased energy consumption. Therefore, as the two pillars of sustainable energy policy, distributed generation and energy efficiency should be analyzed in an integrated framework.

2 Optimization Model for the Design of Integrated Energy Systems

2.1 Problem Definition

As a "no-regrets" option for CO_2 emissions mitigation, energy efficiency is expected to induce more countries to contribute their efforts to take moderate emission reduction target. It is recognized that the use of energy is always directed to satisfy human needs and desires which are usually realized by employing various equipments. Therefore, the potential of energy conservation can be accounted for each component satisfying various forms of final energy demands. Generally, the various measures that may be considered for the improvement of energy efficiency may be distinguished in the following basic categories (Diakaki et al. 2008): measures on architectural design, measures on building materials, measures on energy sources, measures on energy utilization system, measures on management and control, as well as measures on equipment efficiency.

Along with the increased awareness of global warming, besides the aforementioned energy efficiency measures on the demand side, as a "less-regrets" mitigation option, DER technologies are paid more and more attention. Generally, a DER system may include power generation units (renewable ones, typically) and combined heat and power (CHP) units with nominal powers ranging from some kW to tens of MW, and is always designed to deliver production to cover the load demand of a single or a group of customers. For this reason, DER facilities are often located close to the load centre, which makes them convenient to be considered with the energy efficiency measures in an integrated way.

In this study, the local energy system is represented as a network of energy chains, starting from the primary energy supply and ending in the end-use equipments, as shown in Fig. 1. The system is driven by an exogenous energy demand, due to the operation of various end-use equipments. The primary energy sources may include fossil fuels (here, only natural gas which has been recognized as a clean fuel, is considered), and local renewables (wind, solar and biomass energy, etc.). Energy convention given by several generation options comprises: CHP units (micro turbine, gas engine, fuel cell, etc.) and renewable technologies (PV system, wind turbine, etc.). Energy demand sector includes direct electricity use, space heating and cooling loads, as well as hot water consumption. Finally, the energy end-use sector illustrates various types of applications for the aforementioned four energy forms. This sector may include options for fuel and technology substitution, as well as various measures for energy conservation.

Among the various sectors introduced above, as the typical technical options to reduce primary energy requirements, both distributed generation and energy efficiency have been focused in this study. Since a wide variety of appropriate technologies are available in the market, the main issue is to identify those which would be proven to be more effective and reliable based on specific preference. With such a variety of measures, the decision-maker should compare environmental, energetic and financial factors so as to achieve the best possible solution ensuring the optimal combination of energy efficiency measures and distributed generators for the local area. In detail, the decisions regarding local energy system under study concern appropriate choices of system combination, equipment size, and running schedule of selected equipments or measures.

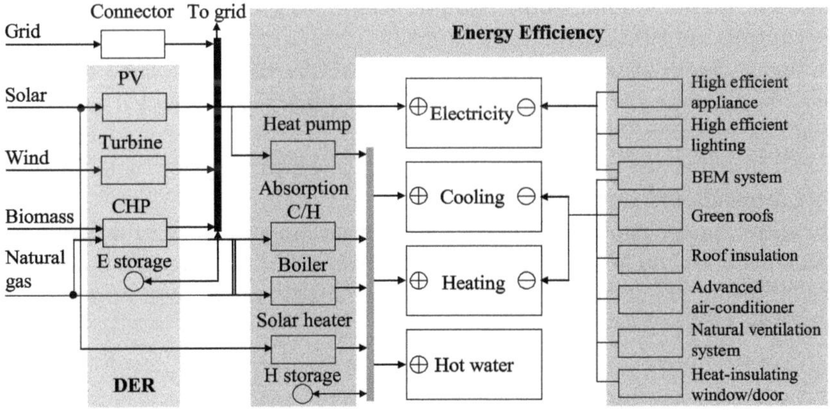

Fig. 1 Sample layout of an integrated local energy system

2.2 Description of the Model

The developed optimization model is a tool for analyzing competition and synergy between different alternatives of the rational use of energy resources. For given time series of fluctuating thermal and electrical requirements, the model calculates the optimal combination of energy efficiency and distributed generation technologies while taking into account local parameters like resource availability, climate condition, energy market, etc.

To this purpose, the model framework proposed in a previous study has been extended and enhanced to include not only distributed generation contributions but also energy efficiency improvements (Ren and Gao 2010). In order to be comparable with the distributed generation under a uniform framework, the energy efficiency measures have been modeled as virtual energy supply equipments, which reduce energy demand rather than provide a more efficient or clean supply. Actually, as is known to all, energy efficiency measures can provide energy and capacity to the end-users in much the same way that a power generator does. Whenever it operates, high-efficient equipments and appliances provide benefits in the form of avoided demand and, in this sense, are always available when needed. Because of this, energy efficiency measures can be designed to deliver the energy equivalent of a conventional generator. The insulation of dwellings and the installation of energy-saving light bulbs are examples of such demand-decreasing measures. In order to compare these measures with the distributed energy supply options and to make them accessible to optimization, they will be modeled as virtual supply processes.

In the pre-developed plan and evaluation model, the objective is to minimize the overall cost of supplying energy services to a specific customer. As shown in Fig. 2, minimization of annual energy cost is based on hourly values of energy demands, local characteristics, and energy tariffs. In addition, technical and economic data of energy efficiency and distributed generation alternatives should be supplied. After optimization, which yields the optimal combination of different technologies considered, the total annual cost, specific energy use, emissions of the resulting energy system are calculated.

In this study, in order to understand the feasibility of developing a low-carbon society, besides the cost optimization, the model has been extended to minimize annual $_{CO}2$ emissions of the local energy system. In this way, the model allows for dynamic emission and cost optimization of the local energy system with fluctuating energy demands and suppliers.

To allow for the application of optimization techniques, an appropriate mathematical model should be developed in accordance to the concepts mentioned above. In the following sub-sections, the mathematical equations of the optimization model are described.

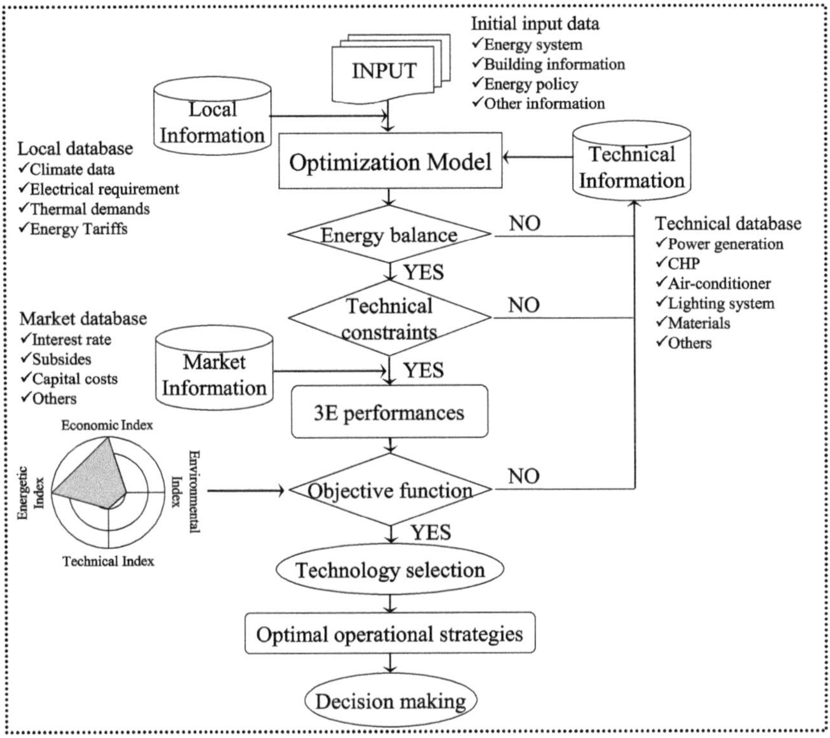

Fig. 2 Flow chart of the optimization model

2.3 Decision Variables

As mentioned earlier, the decisions considered in this study concern three choices. For these three choices, three types of decision variables are defined respectively:

1. Decision variables reflecting the alternative choices regarding system layout (technology combination).
2. Decision variables reflecting the equipment size based on the determined system layout.
3. Decision variables reflecting the operational schedules (energy flows) of all system components.

2.4 Objective Function

In this study, the energy system optimization model aims to determine the optimal use of resources in energy supply and conversion, and to some extent, in final energy demand. Firstly, the system is optimized by using total annual cost of the entire energy system over the whole study period as the objective function to be minimized. In a simplified form the objective function can be written as:

$$[\min]\ \ f_C = C_f + C_v = f(I_i, R_i) + f\left(Q_{i,j}\right) \tag{5.1}$$

where, C_f and C_v indicate the fixed and variable costs, respectively. The fixed cost is usually the initial investment which should be considered from the life-cycle viewpoint. It is a function of the equipment size R_i, as well as an integer variable I_i which indicates the existence or number of each equipment. The variable cost is described as a function of real output ($Q_{i,j}$, the virtual output of energy efficiency measures as well) of the i th technology at the j th time interval.

Alternatively, the environmental objective function is described in Eq. (2). In this study, the environmental performance is assessed based on total CO_2 emissions.

$$[\min]\ \ f_E = CO_e + CO_f = f\left(E_j\right) + f\left(Q_{i,j}\right) \tag{5.2}$$

where, CO_e and CO_f indicate the CO2 emissions from utility electricity and on-site fuel consumption, respectively. E_j indicates the imported amount of utility electricity.

2.5 *Main Constraints*

Generally, equality constraints in terms of energy balances and inequality constraints in terms of technical restrictions should be considered.

Energy generation and consumption balance is formulated following the proposed system model, which includes utility grid, distributed generation, power export, and energy conservation thanks to the introduction of energy efficiency measures. Corresponding equality constraints have the following expressions:

$$\sum_i Q_{i,j} + E_j - S_j = D_j \quad \forall j \tag{5.3}$$

where, S_j is the exported amount of electricity. D_j indicates the energy demands.

For a stable operation, the real power output of each generator is limited in terms of its rated capacity as follows:

$$Q_{i,j} \le R_i \quad \forall i, j \tag{5.4}$$

3 Numerical Study

Energy consumption in the residential sector has been increasing continuously due to the improvements of living standard, such as enlargement of houses, popular use of various types of home electric appliances, and an increase in the number of small families. In this study, the developed optimization model is applied to a residential area in Japan. Table 1 illustrates the general information of the housing complex comprising 19 detached houses, 16 terrace houses and 34 apartments.

Note LDK stands for Living Room, Dining Room, Kitchen.

3.1 Energy Demands

According to the energy consumption intensity (energy consumption per square meter) for the residential buildings in Japan, it is possible to assess the hourly load values for various end-uses (electricity, cooling, heating and hot water) throughout the whole year. Generally, in a residential building, the electrical and thermal demands show considerable hour by hour fluctuation. To illustrate this, for simplicity, 3 days have been selected to represent an extreme day in the heating season, a day in the mid-season, and a summer's day (see Fig. 3). The winter's day, first January, has relatively high thermal requirement, and the daily heat-to-power ratio is 2.3. The thermal demand occurs from morning until night and the coincidence between the thermal and power demand is high — all of the electrical demand occurring coincidentally with the thermal demand. The spring's day, first May, has low thermal demand (hot water only) and on this day the daily heat-to-power ratio is only 0.6. The summer's day, first Sep, has intermediate thermal requirement (cooling and hot water). The thermal demand occurs mainly from 17:00 in the evening until mid-night, which reflects the characteristics of residential buildings.

Table 1 General information of the examined residential area

Type of dwellings		Units	Area of building structures (m)				
		(Units)	Floor	Wall	Roof	Window	Door
Detached house	5LDK	6	150	204	70	20	6
	4LDK	13	132	192	63	15	6
Terraced house	1LDK	8	59			3	2
	2LDK	4	70	1080	532	6	2
	3LDK	4	81			9	2
Apartment	1LDK	10	47	660	96	3	2
	2LDK	24	95	1776	540	6	2

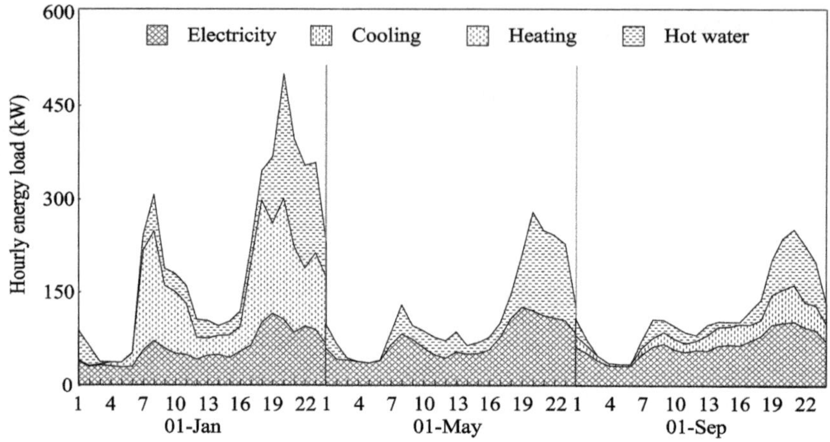

Fig. 3 Hourly load profiles on three typical days

3.2 *Climate Data*

In order to make full use of the renewable energy resources, it is necessary to understand local climate conditions. For example, the strength of solar irradiation is the primary consideration in selecting locations for photovoltaic (PV) system installation. The electricity output of a PV array is directly proportioned to the radiation input. Local climate and environment factors such as ambient temperature and wind speed may also constrain the output of the PV array. In addition, ambient temperature is also an important parameter affecting the effective utilization of solar heating. Wind speed is the key factor determining the output of wind turbines.

In this study, corresponding to the energy demand, three representative days are selected to illustrate local ambient temperature, irradiation and wind speed in different seasons, as shown in Fig. 4. Generally, it can be found that as an area with typical maritime climate, the summer temperature is not very high and the winter temperature is not very low, accompanying with a mild solar radiation and low wind speed. According to the profiles shown in the figure, the examined three climate parameters illustrate different characteristics in various seasons. For example, a winter's day has the lowest ambient temperature and solar radiation, but enjoys the largest wind speed. On the contrary, on a summer's day, although the ambient temperature is the highest, the solar radiation is less than that on a mid-season's day. Looking into the hourly fluctuation of climate data, both ambient temperature and irradiation peak at mid-day, while the wind speed does not illustrate obvious fluctuation throughout a whole day.

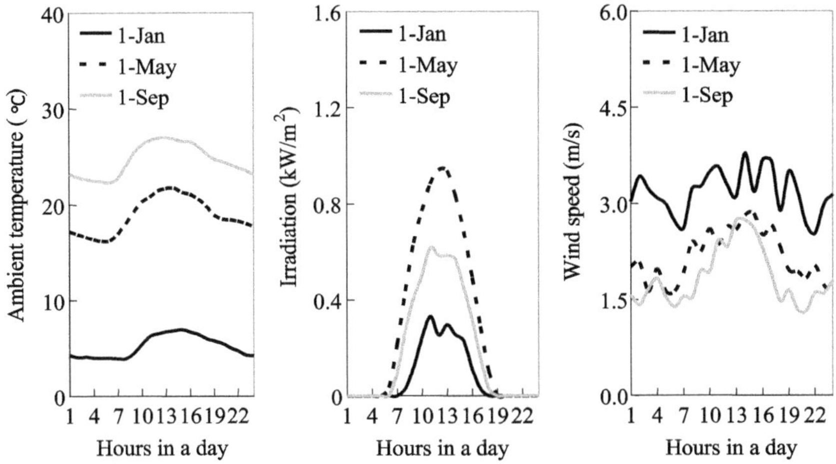

Fig. 4 Hourly ambient temperature, irradiation and wind speed in various seasons

3.3 Technical Options

In this study, the proposed model may select the optimal technical combination from a set of technologies in terms of lowest system cost or lowest emissions. Note that the alternative technologies are exogenously determined by the decision-maker thus may vary from scenario to scenario. While accounting for the characteristics of residential buildings, various DER and energy efficiency technologies available in the market have been selected as the candidates. In detail, the DER technologies considered include: fuel cell (FC), gas engine (GE), microturbine (MT), and PV cells, as shown in Table 2. Except the PV system, all other technologies are fired by natural gas, with heat recovery for heating and absorption cooling. The investigated technical and cost information include rated capacity, lifetime, capital cost, fix operation and maintenance (O&M) cost, variable O&M cost, as well as power and thermal efficiencies.

On the other hand, multiple energy efficiency measures are assumed, as shown in Table 3. In this study, the examined energy efficiency measures can be summarized as high efficiency lighting (HL), high efficiency refrigerator (HR), high efficiency television (HT), high efficiency air conditioner (HA), wall insulation (WI, with materials of polystyrene and plastic fiber for WI-1, WI-2, respectively), heat-insulated door (ID), heat-insulated window (IW), green roof (GR) and solar water heater (SW). The parameters investigated for various energy efficiency measures include: capacity, lifetime, capital cost, reduced load type, and applicable period. Furthermore, as mentioned in the previous section, the energy efficiency measures have been modeled as virtual energy supply equipments. Therefore, the capacity of each energy efficiency measure is assumed to be hourly amount of energy savings due to the introduction of a single unit of the measure. Note that, some measures,

Table 2 Parameters of distributed generators

Type	Capacity (kW)	Lifetime (year)	Capital cost ($/kW)	Fixed O&M cost ($/kW)	Variable O&M cost ($/kWh)	Power efficiency (%)	Thermal efficiency (%)
FC	1	10	12,436	20	0.04	37	50
	5	12	8700	17	0.04	32	36
	10	12	6500	15	0.03	35	35
	125	20	4000	10	0.03	35	27
GE	1	20	10,819	28	0.03	20	65
	30	20	2029	23	0.02	28	64
	60	20	1851	19	0.02	29	62
	75	20	1796	18	0.02	29	61
	100	20	1774	17	0.02	30	62
MT	28	10	3046	23	0.02	23	54
	60	10	2420	20	0.02	25	56
	67	10	2201	16	0.02	25	45
	76	10	2225	17	0.02	24	48
	100	10	2015	14	0.02	26	45
PV	1	20	7450	12	0.00	12	0

for example, the high efficiency lighting is assumed to be available only within the specific period (18:00–24:00).

3.4 *Energy Related Characteristics*

As shown in Table 4, the electricity and city gas tariffs for residential application are determined according to the investigation of local utility companies. It is worth noting that, electricity buy-back is feasible for the surplus power out of both PV system and other DER options, and the prices are also shown in Table 4.

In addition, in order to examine the environmental performance of the local energy system, the carbon intensities of various energy forms have been investigated, as illustrated in Table 4.

3.5 *Scenario Description*

A central part of the model application is the definition of scenarios. In this study, in terms of the energy supply strategy for the residential area, four scenarios have been

Table 3 Parameters of energy efficiency measures

Type	Capacity (kW)	Lifetime (year)	Capital cost ($/Unit)	Reference unit	Reduced load type	Applicable period (hour)
HL	0.096	25	13	Unit	Electricity	18 ~ 24
HR	0.082	15	210	Unit	Electricity	0 ~ 24
HT	0.050	20	174	Unit	Electricity	18 ~ 24
HA	0.190	15	362	Unit	Heating, cooling	0 ~ 24
WI-1	0.003	30	293	m^3	Heating, cooling	0 ~ 24
WI-2	0.004	30	439	m^3	Heating, cooling	0 ~ 24
ID	0.030	30	293	m^2	Heating, cooling	0 ~ 24
IW	0.067	30	37	m^2	Heating, cooling	0 ~ 24
GR	0.030	30	164	m^2	Heating, cooling	0 ~ 24
SW	0.400	15	100	m^2	Hot water	0 ~ 24

Table 4 Energy prices and carbon intensities of various energy forms

Item		Unit	Energy price		Carbon intensity ($kg\text{-}CO_2/kWh$)
			Demand charge ($)	Energy charge ($)	
Electricity tariff	Summer_ On-peak	kWh	3.53	0.41	0.38
	Summer_ Mid-peak			0.26	
	Summer_ Off-peak			0.1	
	Winter_ On-peak			0.34	
	Winter_ Mid-peak			0.26	
	Winter_ Off-peak			0.1	
Buy-back price (PV)			–	0.4	
Buy-back price (other DER)			–	0.05	
City gas tariff (DER system)		m^3	1.14	0.97	0.18

assumed according to the technical combination and different objective functions proposed in Sect. 2.

Scenarios 1–3 focus on the economic performance of the energy system because economic performance always plays the main role in the final decision-making, especially for local energy consumers. Whereas, in Scenario 1, only energy efficiency measures are included; in Scenario 2, only DER technologies are accounted; and in Scenario 3, both of the former two options are considered.

Similar with Scenario 3, Scenario 4 considers energy efficiency and distributed generation measures simultaneously, while taking annual CO_2 emissions as the objective function to be minimized. As a consequence, the categories of technologies with lower release of CO_2 emissions may be preferred. Specially, some renewable technologies are expected to receive the priority in the determined system combination. In addition, the energy efficiency measures without primary energy input are also popular.

On the other hand, besides the aforementioned four scenarios, as a base scenario for reference, a conventional system without DER or energy efficiency adoption is also assumed, in which external grid meets the electrical demand and gas boiler serves the thermal load.

4 Results and Discussion

In order to show the effectiveness of the proposed method, the residential building complex has been examined as a case study while considering various scenarios as introduced above. Besides the discussions on the optimal results corresponding to different technical combinations and objective functions, the influence of possible policy measures is also examined in a detailed way.

4.1 Optimal Adoption Results

Table 5 shows the deduced system combination and corresponding economic, energy and environmental performances for various scenarios. Generally, both energy efficiency (EE) and DER can contribute a lot to the local energy system from either economic, energetic or environmental viewpoint. Especially, as economic objective is considered (Scenarios1 ~ 3), the introduction of proposed local energy system may achieve win-win-win benefits, resulting in reduced total cost, CO_2 emissions and primary energy consumption. For example, compared with the base scenario, Scenario 3 with both DER and EE adoption achieves reasonable reduction ratios of cost, CO_2 emissions and primary energy consumption with values of 17.7%, 18.3% and 21.3%, respectively. In addition, it is interesting to notice that Scenario 2 (with only DER adoption) enjoys better economic performance but worse energetic and environmental performances than Scenario 1 (with only EE introduction).

Table 5 Optimal adoption results of various scenarios

Scenario	DER capacity (kW)	EE capacity (kW)	System combination		Total cost (Thousand $)	Total CO_2 emissions (Ton)	Total primary energy consumption (MJ)
			DER	EE			
Base scenario	0	0	–	–	141.6	314.2	7.9
Scenario 1	0	54	–	HL: 306; HR: 69; SW: 58	130.4	278.1	7.0
Scenario 2	400	0	GE-75: 1; PV: 325	–	126.4	291.5	7.1
Scenario 3	400	23	GE-75:1; PV: 325	HL: 306; HR: 69	116.6	256.7	6.2
Scenario 4	366	815	FC-1: 19; GE-100:2; PV: 147	HL: 306; HR: 69; HT: 69; HA: 237; WI-1: 50; ID: 214; IW: 573; SW: 1316	278.9	127.6	3.4

Moreover, as the environmental issues are paid the most attention (Scenario 4), most of the energy efficiency measures (except WI-2 and GR) are selected, but the total capacity of DER is less than that of Scenario 2 and Scenario 3. Correspondingly, local CO_2 emissions is reduced by 59.4% (compared with the base scenario), while total cost increases dramatically (97% more than the base case).

4.2 Annual Energy Balances

In order to understand the contribution of various DER or EE options on the final energy supply, Figs.5 and 6 show annual electrical and thermal balances of the assumed five scenarios, respectively.

As shown in Fig. 5, the introduction of DER and EE technologies results in reduced power input from the utility grid. In detail, the shares of power supply from DER and EE measures are 18%, 38%, 54% and 76% for scenarios 1 ~ 4, respectively. Generally, DER technologies may serve most of local electrical requirement if necessary, while the energy efficiency measures cannot be neglected with virtual power supply (nearly 18% of total demand). As to Scenario 1, high-efficiency lighting (HL) and high-efficiency refrigerator (HR) are selected to reduce annual electricity load by 8% and 9%, respectively. When DER technology is paid the only attention (Scenario 2), gas engine plant takes a share of about 38% of total annual demand. It is interesting to notice that although PV array is introduced, it supplies no electricity for the local requirements. This is because all the power generation from the PV system is sold back to the grid thanks to the assumed high buy-back price (see Table 4.4). Detailed discussion will be illustrated in the following subsection. Furthermore, as mentioned

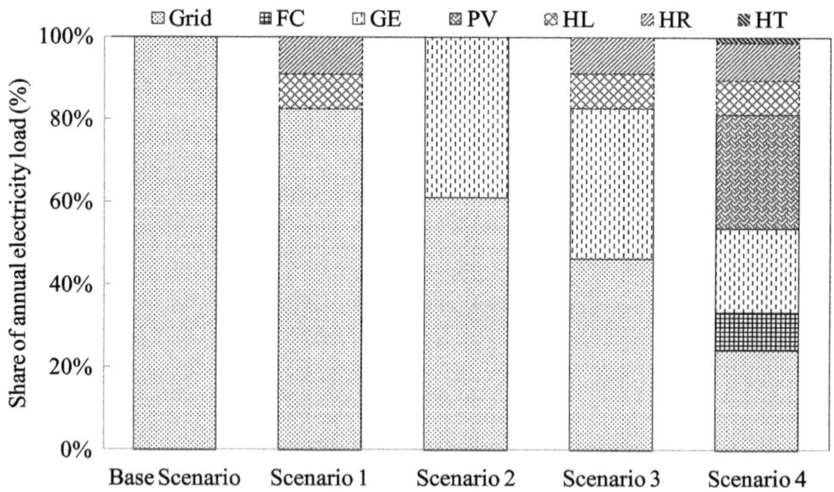

Fig. 5 Annual electrical balance of various scenarios

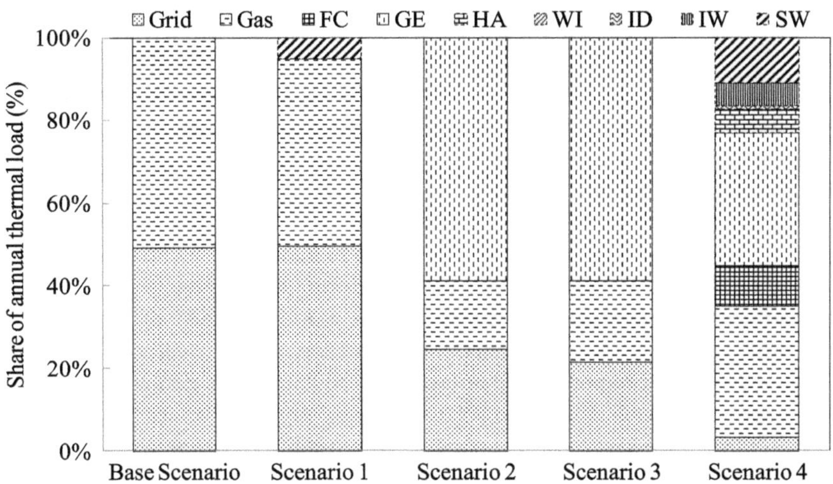

Fig. 6 Annual thermal balance of various scenarios

above, when environmental aspect is focused, most of local electricity demand is served by the on-site generation (including EE based virtual generation). As to the EE measures, HL, HR and HT supplied 8%, 9% and 1% of total electricity load, respectively. On the other hand, the shares of power from FC, GE and PV are 9%, 20% and 28%, respectively. Therefore, it can be found that the potential of PV system for local power supply is considerable if its environmental benefits are paid enough attention.

Figure 6 illustrates the share of various thermal sources for local thermal demands. As to the conventional system (Base Scenario), utility grid servers about half of local thermal demand with the adoption of electric air-conditioner, while city gas is consumed (with gas boiler or direct-fired absorption chiller) for the other half. On the other hand, with the introduction of DER or EE measures, the direct gas and electricity consumption for local thermal demands is reduced more or less.

In Scenario 1, only solar water heater is introduced for the reduction of hot-water load, and it takes up about 12.1% of annual total thermal demand. When the DER technologies are considered in Scenario 2, the recovered heat from gas engine contributed most (about 59%) of local thermal demand, while the shares of electricity and city gas are 24.4% and 16.6%, respectively. In Scenario 3, although both DER and EE measures are considered for optimization, the shares of various thermal sources are similar to the situation in Scenario 2 because only EE measures with electricity reduction are selected (see Table 5). As to Scenario 4, due to the relatively high attention paid to the environmental performance, the EE measures (HA, WI-1, ID, IW, SW) with virtual thermal energy supply about 23.1% of annual total thermal demand, DER technologies (GE and FC) take a share of 42.0%, while the share of utility grid is reduced to a relatively low level (about 3.2%).

4.3 Hourly Operating Strategies

Hourly operating schedules are of vital importance to the understanding of the technical characteristics and corresponding performances of different system alternatives. As an example, Fig. 7 shows the hourly electrical balances of the examined four scenarios on a summer day.

In Scenario 1, the adoption of HR and HL results in reduced electricity consumption during the specific time period as indicated in Table 3.3. As to Scenario 2, gas engine system serves all the electricity demand during the on-peak period, and is out of operation during the off-peak period. This is mainly because of the fluctuating hourly tariff rates according to the TOU tariff structure (see Table 4.4). Moreover, although PV system generates much more electricity than local demand, all of which is sold back to the grid due to the relatively high buy-back price. Scenario 3 is a combination of Scenario 1 and Scenario 2. The running schedule of Scenario 4 is quite different with other three scenarios. In order to reduce local CO_2 emissions, the power out of PV system is preferentially employed for on-site consumption during the daytime period, the residual is sold to the grid. In addition, fuel cell and gas engine operate during the off-peak hours.

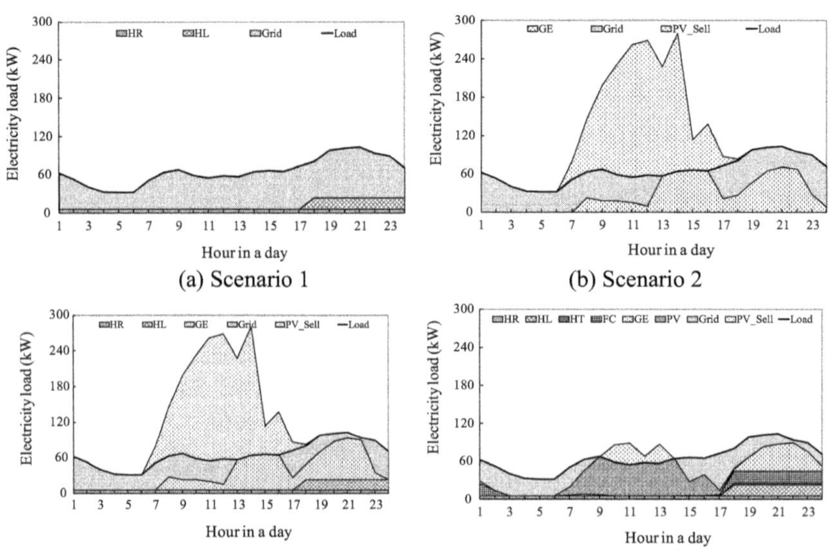

Fig. 7 Hourly electrical balance of various scenarios (Summer). (a) Scenario 1. (b) Scenario 2. (c) Scenario 3. (d) (a) Scenario 4

4.4 Cost-Benefit Analysis

Usually, the economic objective and the environmental objective stand on the opposite sides. In other words, the reduction of environmental emissions (CO_2, in this study) may at the cost of poor economic performance. In order to understand the trade-off relationship between the economic and environmental performances of the examined local energy system, a cost-benefit analysis is executed and the results are shown in Fig. 8. As is expected, along with the increase of reduction ratio of annual CO_2 emissions (from pure economic optimization to pure environmental optimization), annual system cost increases gradually. Looking into the figure, it can be found that the reduced revenue from selling electricity to the grid is the main factor affecting the changing of annual total cost. In addition, a sharp increase of total cost appears when the emissions reduction ratio increases from 55% to the extreme point (Scenario 4).

Moreover, in order to understand the additional cost due to the reduction of CO_2 emissions, the unit reduction cost has been calculated. As shown in the figure, unless the emissions reduction ratio rises to more than 43%, the reduction cost enjoys a minus value. In other words, the proposed local energy system may reduce more than 40% of total CO_2 emissions without increasing the system cost (compared with the base scenario). Generally, the reduction cost increases linearly with the rise of emissions reduction cost; while it rises sharply to a value as high as 0.74 \$/kg-CO_2 when the emissions reduction potential is exhausted.

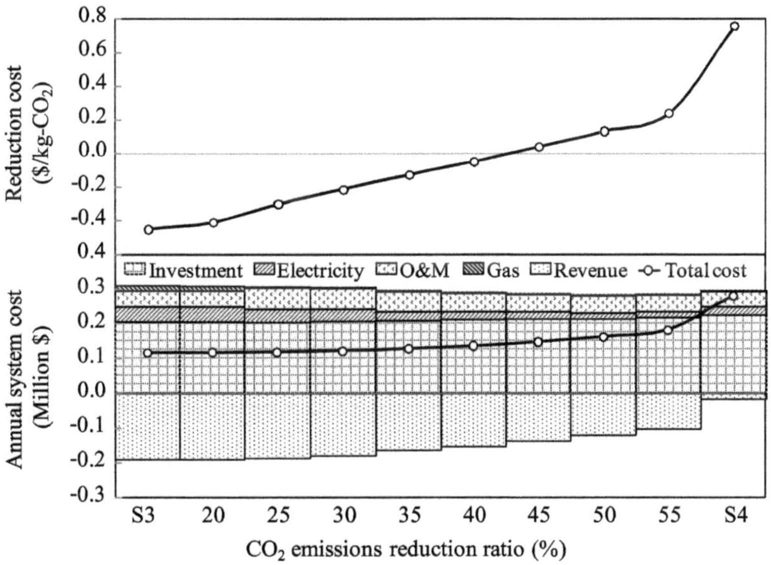

Fig. 8 Trade-off between system cost and CO_2 emissions

5 Conclusions

The introduction of energy efficiency and distributed energy resources is among the main priorities in the energy policy of all countries to cope with the continuing global warming. In this study, an optimization model has been developed to investigate the feasibility of introducing DER and energy efficiency measures in a local energy system. Special attention has been paid to the description of energy efficiency technologies and their contribution to the CO_2 emissions reduction. To this purpose, a residential area located in Kitakyushu, Japan has been considered for analysis. The effectiveness of the proposed methodology has been shown by carrying out simulations on the assumed local energy system. The main results are summarized as follows:

1. The adoption of DER technologies and energy efficiency actions in local energy system may results in satisfied economic, energetic and environmental performances.
2. Local energy prices (including buy-back prices) may have great effect on the design and operation of local energy system.
3. A win-win situation (with both cost saving and emissions reduction) may be achieved with the rational design and operation of local energy system.

References

Antipova E, Boer D, Guillén-Gosálbez G, Cabeza LF, Jiménez L (2014) Multi-objective optimization coupled with life cycle assessment for retrofit-ting buildings. Energ Buildings 82:92–99

Dagoumas AS, Barker TS (2010) Pathways to a low-carbon economy for the UK with the macro-econometric E3MG model. Energy Policy 38(6):3067–3077

Penna P, Prada A, Cappelletti F, Gasparella A (2015) Multi-objectives optimization of energy efficiency measures in existing buildings. Energ Buildings 95:57–69

Karmellos M, Kiprakis A, Mavrotas G (2015) A multi-objective approach for optimal prioritization of energy efficiency measures in buildings: model, software and case studies. Appl Energy 139:131–150

Diakaki C, Grigoroudis E, Kabelis N, Kolokotsa D, Kalaitzakis K, Stavrakakis G (2010) A multi-objective decision model for the improvement of energy efficiency in buildings. Energy 35(12):5483–5496

Ren H, Gao W (2010) A MILP model for integrated plan and evaluation of distributed energy systems. Appl Energy 87(3):1001–1014

Long-Term Design Optimization of Regional Integrated Energy Systems Considering CO2 Emissions Reduction

Hongbo Ren, Qiong Wu, and Weisheng Zhou

1 Introduction

Along with the growing global energy consumption and increasingly serious environmental problems, China has put forward the goal of "carbon peaking and carbon neutrality" and issued a relevant action plan. To realize low-carbon development, the energy sector, which accounts for a large share of carbon emissions, has become the main force for energy conservation and emission reduction (Wang 2020). The development of regional integrated energy systems with the advantages of multi-energy complementarity and step utilization is an important solution for the low-carbon transformation of the energy industry (Li et al. 2020). With the gradual progress of urbanization and modernization, the parks and neighborhoods in the city are constantly expanding, which leads to the problem of rational planning and scientific layout of energy between multiple blocks. Therefore, it is necessary to consider the establishment of efficient energy conversion and collaboration mechanisms among multiple blocks and energy systems at the early planning stage, so as to realize the sustainability and reliability of energy supply and the low-carbon evolution of the region as a whole (Qu et al. 2018). The low-carbon evolution of the integrated regional energy system is complex and diversified, relying on a variety of clean technologies and market regulation, with carbon trading as an effective means to reduce carbon emissions (Liu and Feng 2016).

H. Ren (✉) · Q. Wu
College of Energy and Mechanical Engineering, Shanghai University of Electric Power, Shanghai, China
e-mail: tjrhb@163.com

Q. Wu
e-mail: wuqiongrff@163.com

W. Zhou
College of Policy Sciences, Ritsumeikan University, Kyoto, Japan
e-mail: zhou@sps.ritsumei.ac.jp

W. Zhou (ed.), *Decarbonization Policy Engineering*, World Sustainability Series,
https://doi.org/10.1007/978-981-95-5953-4_13

A lot of research have been conducted on the planning of urban and regional energy systems in the medium- and long-term context, and have obtained many research results (Allegrini et al. 2015). Walmsleyused California as an example for carbon emission analysis, and studied a scenario in which the renewable energy generation in the power system accounted for 33% in a ten-year period, which was found to be more costly and required intervention by the state government to control it (Walmsley et al. 2015). Mirzaesmaeeli presented a new deterministic multi-temporal mixed integer linear programming (MILP) method with many time-varying dynamic parameters for long-term power system generation planning that avoids horizon terminal effects at the end year of the planning period (Mirzaesmaeeli et al. 2010). To determine optimal long-term strategy, Koltsaklis developed a generation expansion planning model that integrates the dynamic electricity market, the evolution of energy policies, and the cost of carbon dioxide emissions (Koltsaklis et al. 2015). Cao et al. proposed a multi-stage planning approach for integrated energy systems that considers construction timing, such as dividing the planning cycle into several phases, and increasing the equipment investment at the beginning of each planning phase to meet the demand of load growth in long time scales (Cao et al. 2015). Liu considered inter-annual variations in technical and economic parameters and updated the supply of electricity, heat and cooling load demand in stages to optimize equipment allocation (Liu et al. 2020).

In this chapter, the long-term multi-stage dynamic planning optimization method is applied to the dynamically expanding integrated energy system, and a multi-regional interconnected integrated energy system planning model is constructed with the medium- and long-term carbon emissions of the multi-regional integrated energy system as the core constraint. The focus is on the impacts of introducing emerging carbon reduction means such as clean energy technologies and stepped carbon trading under the mandatory emission reduction policy. Firstly, the energy flow of the integrated energy system is studied and a mathematical model of equipment is established; secondly, with the minimization of the total cost of the full planning cycle as the objective function, a multi-regional integrated energy system planning model considering the carbon trading mechanism is constructed to find the optimal equipment allocation strategy under different carbon emission constraints; finally, the validity of the planning method is verified under multiple scenarios using a park in Shanghai as an example.

2 Modeling of Regional Integrated Energy Systems

2.1 Structure of Regional Integrated Energy Systems

In the regional integrated energy system analyzed in this chapter, external energy inputs include grid electricity, hydrogen, natural gas, and biomass, which are dispatched and transformed through the system to meet the electricity, heat, and cooling load demands of regional users. Among the local power supply equipment, the power generation output of photovoltaic and wind turbine depends on the local natural conditions, and there are regional spatial constraints in installation, the upper limit of photovoltaic panels laying area, and the power output of the wind turbine is highly fluctuating; the gas combustion engine, as the main power generating equipment, can be utilized in a graded manner in terms of the amount of heat it generates; and the fuel cell is fueled by hydrogen, which realizes the co-generation of heat and power while having zero carbon emission. Heat source equipment includes boilers and heat pumps, and also utilizes waste heat from internal combustion engines and fuel cells. Gas boiler technology is relatively mature and can realize more stable and efficient heat supply; biomass boiler is highly efficient and environmentally friendly, but is limited by biomass energy resources; heat pump converts electricity into heat energy, which comes from the power supply equipment of the integrated energy system or the power grid. Cold source equipment includes absorption refrigerant and electric refrigerator; absorption refrigerator absorbs heat energy for cooling, and its advantage is to consume excess heat energy to realize energy laddering; electric refrigerator consumes electric energy to realize cold supply, and it has high energy supply efficiency. Energy storage equipment such as batteries, heat storage tanks and cold storage tanks can transfer excess energy from off-peak hours to peak hours, while improving system stability and economy. Carbon Capture, Storage and Utilization System (CCUS) converts high-density carbon dioxide into other organic matter and prevents it from being emitted into the atmosphere.

2.2 *Typical Equipment Model*

2.2.1 Energy Conversion Equipment

① Photovoltaic power generation

The photovoltaic (PV) power output mainly depends on the local irradiance and installation area, and its power output model is shown in Eq. 13.1.

$$P^{PV}_{l,n,y,d,h} = G_{l,y,d,h} S_{l,n} \xi_{PV} \varepsilon_{PV} \quad \forall Y \geq y \geq n, \forall l, d, h \tag{13.1}$$

where, $P^{PV}_{l,n,y,d,h}$ is the actual power output of the PV panels installed in year n of block l in year y, day d, hour h. In specific communities or towns where distributed energy systems are really needed, the year of installation of the PV panels, n, depends on the scenario setting, where n takes a value less than or equal to y, and the total duration of the planning cycle is year Y. $S_{l,n}$ refers to the area of PV panels in year n of the park, and the amount of electricity generated by PV panels from absorbing solar energy The amount of electricity generated by the solar energy absorbed by the PV panels depends on their operating characteristics parameters photoelectric conversion ratio ε_{PV}, correction factor ξ_{PV} and hour-by-hour irradiance $G_{y,d,h}$.

② Wind turbine power generation

Wind resources exist in certain areas that can be put into wind turbines (WT) with the specific output shown in Eq. 13.2.

$$P^{WT}_{l,n,y,d,h} = \begin{Bmatrix} 0 & , & H^{speed}_{l,y,d,h} < const1, \\ H^{speed}_{l,y,d,h}\xi_{WT}\varepsilon_{WT}, & & const1 \le H^{speed}_{l,y,d,h} \le const2, \\ H^{const2}_{l,y,d,h}\xi_{WT}\varepsilon_{WT}, & & const2 < H^{speed}_{l,y,d,h}, \end{Bmatrix} \forall Y \ge y \ge n, \forall l, d, h \tag{13.2}$$

where, $P^{WT}_{l,n,y,d,h}$ is the actual power output of the wind turbine installed in year n of the park l in year y, day d, hour h, the value of which depends on the operating characteristics of the parameter kinetic energy conversion rate ε_{WT}, the correction coefficient ξ_{WT} and the hour-by-hour wind speed of the block l of a specific blade height $H^{speed}_{l,y,d,h}$, when the wind speed is less than the value of $const1$, it is not possible to generate electricity; there is an upper limit on the rotational speed, and it can be adapted to the wind speeds greater than the value of $const2$.

③ Alkaline electrolyzer

Alkaline electrolyzer (AE) converts electrical energy into chemical energy level electrolysis of water to produce hydrogen and oxygen, of which hydrogen can be widely used as a chemical raw material, the hydrogen in the model is used to reduce carbon dioxide to achieve the purpose of emission reduction, the model as shown in Eqs. 13.3 and 13.4.

$$P^{AE}_{l,n,y,h} \le P^{WT}_{l,n,y,d,h} \quad \forall Y \ge y \ge n, \forall l, d, h \tag{13.3}$$

$$F^{H2}_{l,n,y,h} = a_{AE}P^{AE}_{l,n,y,h} \quad \forall Y \ge y \ge n, \forall l, d, h \tag{13.4}$$

where, the efficiency of the electrolyzer isa_{AE}, the power consumption is$P^{AE}_{l,n,y,h}$, and the hydrogen production is$F^{H2}_{l,n,y,h}$.

④ Biomass boilers

Biomass boiler (BB) enables a more stable heat supply, the input energy source for this device is wet waste, modeled as shown in Eqs. 13.5 and 13.6.

$$F^{BB}_{l,n,y,d,h} = a_{BB}Q^{BB}_{l,n,y,d,h} \quad \forall Y \ge y \ge n, \forall l, d, h \tag{13.5}$$

$$Q^{BB}_{l,n,y,d,h} <= Q^{nom,BB}_{l,n} \quad \forall Y \ge y \ge n, \forall l, d, h \tag{13.6}$$

where, the fuel consumption power $F^{BB}_{l,n,y,d,h}$ depends on the heat production power $Q^{BB}_{l,n,y,d,h}$ and the coefficient of performance a_{BB}, and the rated installed power is taken as the maximum value of the real time power.

⑤ Fuel Cells

Solid Oxide Fuel Cell Fuel Cell (SOFC) as a new energy device uses hydrogen as the input energy and outputs both electrical and thermal energy, and the energy consumption and supply are shown in Eqs. 13.7, 13.8 and 13.9.

$$F^{SOFC}_{l,n,y,d,h} = a^{SOFC} P^{SOFC}_{l,n,y,d,h} \quad \forall Y \geq y \geq n, \forall l, d, h \tag{13.7}$$

$$P^{SOFC}_{l,n,y,d,h} <= P^{nom,SOFC}_{l,n} \quad \forall Y \geq y \geq n, \forall l, d, h \tag{13.8}$$

$$Q^{SOFC}_{l,n,y,d,h} = \eta^{th,nom,SOFC} F^{SOFC}_{l,n,y,d,h} \quad \forall Y \geq y \geq n, \forall l, d, h \tag{13.9}$$

where, $P^{SOFC}_{l,n,y,d,h}$ is the output electric power,a^{SOFC} is the performance coefficient of SOFC, a as the performance coefficient takes different values for different equipment. $P^{nom,SOFC}_{l,n}$ It refers to the rated power of SOFC equipment installed in the nth year of the park, and takes the maximum value of $P^{SOFC}_{l,n,y,d,h}$. The fuel consumption power $F^{SOFC}_{l,n,y,d,h}$ consists of two parts: electric power and thermal power. The SOFC thermal power $Q^{SOFC}_{l,n,y,d,h}$ depends on the fuel consumption power.

⑥ Gas internal combustion engines

A gas internal combustion engine (ICE) is usually combined with a heat recovery system to generate electrical energy, which is mathematically modeled similarly to SOFC as shown in Eqs. 13.10, 13.11 and 13.12.

$$F^{ICE}_{l,n,y,d,h} = a^{ICE} P^{ICE}_{l,n,y,d,h} \quad \forall Y \geq y \geq n, \forall l, d, h \tag{13.10}$$

$$P^{ICE}_{l,n,y,d,h} <= P^{nom,ICE}_{l,n} \quad \forall Y \geq y \geq n, \forall l, d, h \tag{13.11}$$

$$Q^{ICE}_{l,n,y,d,h} = \eta^{th,nom,ICE} F^{ICE}_{l,n,y,d,h} \quad \forall Y \geq y \geq n, \forall l, d, h \tag{13.12}$$

where, $P^{ICE}_{l,n,y,d,h}$ is the electric power, the introduction of the performance factor a^{ICE} to get the fuel consumption power $F^{ICE}_{l,n,y,d,h}$,$P^{nom,ICE}_{n}$ to take the maximum value of $P^{ICE}_{l,n,y,d,h}$ as the rated installed power in the nth year of the park l, given the value of the thermal efficiency coefficient $\eta^{th,nom,ICE}$ that is, to get the ICE heat power $Q^{ICE}_{n,y,d,h}$.

⑦ Gas boilers

Gas boiler (GB) as a mature technology to achieve more stable heat supply, the input energy source of this equipment is natural gas, the fuel consumption power $F^{GB}_{l,n,y,d,h}$ depends on the heat production power $Q^{GB}_{l,n,y,d,h}$ and performance coefficient a_{GB} as shown in Eq. 13.13, the rated power of the installed GB equipment is taken as the maximum value of the real time power as shown in Eq. 13.14.

$$F_{l,n,y,d,h}^{GB} = a_{GB} Q_{l,n,y,d,h}^{GB} \quad \forall Y \geq y \geq n, \forall l, d, h \tag{13.13}$$

$$Q_{l,n,y,d,h}^{GB} <= Q_{l,n}^{nom,GB} \quad \forall Y \geq y \geq n, \forall l, d, h \tag{13.14}$$

⑧ Carbon Capture, Utilization and Storage (CCUS) Equipment

The amount of carbon dioxide captured is proportional to the amount of hydrogen produced, so this simplified conversion is shown in Eqs. 13.15, 13.16 and 13.17.

$$C_{l,n,y,h}^{CCUS} = a_{CCUS} P_{l,n,y,h}^{CCUS} \quad \forall Y \geq y \geq n, \forall l, d, h \tag{13.15}$$

$$F_{l,n,y,h}^{H2} = \chi_n C_{l,n,y,h}^{CCUS} \quad \forall Y \geq y \geq n, \forall l, d, h \tag{13.16}$$

$$C_{l,n,y,h}^{CCUS} \leq C_{l,n}^{CCUS\max} \quad \forall Y \geq y \geq n, \forall l, d, h \tag{13.17}$$

where, the amount of carbon dioxide captured by the CCUS device $C_{l,n,y,h}^{CCUS}$ is directly proportional to the electricity consumption, the unit hourly electricity consumption is $P_{l,n,y,h}^{CCUS}$,a_{CCUS} is the efficiency, and the scaling factor χ_n is the ratio of hydrogen atoms to carbon atoms in the methanol, where the installed capacity of the CCUS is $C_{l,n}^{CCUS\max}$.

⑨ Heat pumps

A heat pump (EHP) converts electrical energy into thermal energy, and its power can come from all the power supply equipment or the grid of this integrated energy system, and the mathematical relationship between its energy input and output is shown in Eqs. 13.18 and 13.19.

$$P_{l,n,y,d,h}^{EHP} = a_{EHP} Q_{l,n,y,d,h}^{EHP} \quad \forall Y \geq y \geq n, \forall l, d, h \tag{13.18}$$

$$Q_{l,n,y,d,h}^{EHP} <= Q_{l,n}^{nom,EHP} \quad \forall Y \geq y \geq n, \forall l, d, h \tag{13.19}$$

where, $P_{l,n,y,d,h}^{EHP}$ is the power consumption of the heat pump,a_{EHP} is the thermal efficiency,$Q_{l,n,y,d,h}^{EHP}$ is the thermal power, and $Q_{l,n}^{nom,EHP}$ is the rated installed power of the heat pump in year n of block l.

⑩ Absorption chillers

Absorption chiller (ABC) will be converted to cold energy, the advantage is the consumption of excess heat energy to achieve the gradient utilization of energy, energy input and output of the mathematical relationship between the Eqs. 13.20 and13.21 shown.

$$I_{l,n,y,d,h}^{ABC} = a_{\mathrm{ABC}} L_{l,n,y,d,h}^{ABC} \quad \forall Y \geq y \geq n, \forall l, d, h \tag{13.20}$$

$$L_{l,n,y,d,h}^{ABC} <= L_{l,n}^{nom,ABC} \qquad \forall Y \geq y \geq n, \forall l, d, h \tag{13.21}$$

where, $I_{l,n,y,d,h}^{ABC}$ is the heat consumption power of absorption chillers,a_{ABC} is the refrigeration efficiency,$L_{l,n,y,d,h}^{ABC}$ is the refrigeration power, and $L_{l,n}^{nom,ABC}$ is the rated installed power of absorption chillers in year n of block l.

⑪ Electric coolers

Electric cooler (EC) need to consume electricity to realize the cold supply, energy input and output of the mathematical relationship between Eqs. 13.22 and 13.23 shown.

$$I_{l,n,y,d,h}^{EC} = a_{EC} L_{l,n,y,d,h}^{EC} \quad \forall Y \geq y \geq n, \forall l, d, h \tag{13.22}$$

$$L_{l,n,y,d,h}^{EC} <= L_{l,n}^{nom,EC} \qquad \forall Y \geq y \geq n, \forall l, d, h \tag{13.23}$$

where,$I_{l,n,y,d,h}^{EC}$ is the power consumption of the electric chiller,a_{EC} is the refrigeration efficiency,$L_{l,n,y,d,h}^{EC}$ is the real-time refrigeration power, and $L_{l,n}^{nom,EC}$ is the rated installed power of the electric chiller in the nth year of block l.

⑫ Equipment Climbing Constraints

Considering the safety of the technology as well as the level of development, the upper and lower limits of the electrical, thermal and cooling power are set here as shown in Eq. 13.24, and the energy conversion equipment other than PV panels is defined as equipment group e1;

$$l_{e1}^{\min} U_{l,n,e1}^{nom} \leq U_{l,n,e1,y,d,h} - U_{l,n,e1,y,d,h-1} \leq l_{e1}^{\max} U_{l,n,e1}^{nom} \quad \forall Y \geq y \geq n, \forall l, d, \forall h > 1 \tag{13.24}$$

where, $l_e^{\min/\max}$ is the upper and lower limits of the operating boundary of the energy conversion equipment, and $U_{l,n,e1,y,d,h}$ and $U_{l,n,e1}^{nom}$ refer to the actual output power and rated power of the equipment installed in the nth year of the equipment group e1 in block l, respectively, so that the actual output power of the energy conversion equipment is limited to a reliable range without significant fluctuations.

2.2.2 Energy Storage Equipment

Energy storage systems balance the time mismatch between energy production and consumption. The energy storage devices belong to equipment group e2 and their role is to transfer excess energy from off-peak hours to peak hours. The integrated energy system uses lithium batteries (LiB), heat storage tanks (HWT) and cold storage tanks (CWT) to store electrical energy, heat and cold respectively. The following relationship exists between the charging and discharging power and the state of the stored energy as shown by the linear equation of Eq. 13.25.

$$\begin{aligned} E_{l,n,e2,y,d,h} =& E_{l,n,e2,y,d,h-1}(1-\sigma_{e2}) \\ &+ (\gamma_{e2} z^{ch}_{l,n,e2,y,d,h} CH_{l,n,e2,y,d,h} - z^{dis}_{l,n,e2,y,d,h} DIS_{l,n,e2,y,d,h}) \\ & \forall Y \geq y \geq n, \forall l, d, h \end{aligned} \tag{13.25}$$

where, $E_{l,n,e2,y,d,h}$ is the energy storage power in year y, day d, hour h of the equipment installed in year l of block l, the energy storage power in year y, day d, hour h-1 of the equipment $E_{l,n,e2,y,d,h-1}$ installed in year l of block l,σ_{e2} is the percentage of natural decay of the energy,γ_{e2} is the charging loss,$z^{ch}_{l,n,e2,y,d,h}$ and $z^{dis}_{l,n,e2,y,d,h}$ are binary variables of 0–1 that indicate whether or not charging and discharging are turned on, and $CH_{l,n,e2,y,d,h}$ and $DIS_{l,n,e2,y,d,h}$ represent the charging and discharging power of the equipment[10] respectively.

Ensure this continuous day-to-day and year-to-year storage characteristic and achieve hour-by-hour capture of stored energy when describing seasonal and yearly time variations over the lifetime of the energy storage system, as shown in Eqs. 13.26 and 13.27.

$$E_{l,n,e2,y,d-1,H} = E_{l,n,e2,y,d,0} \quad \forall Y \geq y \geq n, \forall l, d, h \tag{13.26}$$

$$E_{l,n,e2,y-1,D,H} = E_{l,n,e2,y,1,0} \quad \forall Y \geq y \geq n, \forall l, d, h \tag{13.27}$$

where, the energy stored at the end of the last hour of each day of the year $E_{l,n,e2,y,d-1,H}$ is equal to the energy stored at the beginning of the first hour of the next day $E_{l,n,e2,y,d,0}$ and the energy stored at the end of the last hour of the last day of the year $E_{l,n,e2,y-1,D,H}$ is equal to the energy stored at the beginning of the first hour of the first day of the next year $E_{l,n,e2,y,1,0}$.

For necessary operational safety, the rated capacity takes the value of the maximum capacity of the real-time energy storage, as shown in Eqs. 13.28 and 13.29.

$$E_{l,n,e2,y,d,h} \leq E^{nom}_{l,n,e2} \tag{13.28}$$

$$CH_{l,n,e2,y,d,h}, DIS_{l,n,e2,y,d,h} \leq m^{\max}_{e2} E^{nom}_{l,n,e2} \tag{13.29}$$

where, $E^{nom}_{l,n,e2}$ is the rated capacity, the charging and discharging energy setting limit parameter $m^{\max}_{e2}$, is less than 1.

For the purpose of achieving a more realistic simulation of the operating conditions, the simultaneous operation of storing and releasing energy is not allowed, as shown in Eq. 13.30.

$$z^{ch}_{l,n,e2,y,d,h} + z^{dis}_{l,n,e2,y,d,h} \leq 1 \tag{13.30}$$

where, $z^{ch}_{l,n,e2,y,d,h}$ and $z^{dis}_{l,n,e2,y,d,h}$ are 0–1 binary variables indicating whether or not charging and discharging are on.

2.3 *Stepped Carbon Trading Model*

Carbon trading market is another driver of carbon emission reduction, as shown in Eqs. 13.31 and 13.32, the region can purchase emission credits to offset carbon emissions in excess of the quota, the carbon trading policy is in the developmental stage in China, and at present, the domestic electric power industry mainly adopts the gratuitous allocation of initial carbon emission allowances, this chapter adopts the baseline method to determine the system's gratuitous carbon emission quotas.

$$C_y^{\sin k} = \begin{Bmatrix} d_y - C_y^{emission}, & 0 < C_y^{emission} \le d_y \\ C_y^{emission} - d_y, & d_y < C_y^{emission} \end{Bmatrix} \quad \forall Y \ge y \tag{13.31}$$

$$K_y = \begin{cases} -\hat{b}_y & 0 < C_y^{emission} \le d_y \\ b_y & d_y < C_y^{emission} \le 1.2d_y \\ b_y(1+c) & 1.3d_y < C_y^{emission} \le 1.4d_y \\ b_y(1+2c) & 1.4d_y < C_y^{emission} \le 1.5d_y \\ b_y(1+3c) & 1.5d_y < C_y^{emission} \le 1.6d_y \\ \ldots & \ldots < C_y^{emission} \le \ldots \end{cases} \quad \forall Y \ge y \tag{13.32}$$

where, $C_y^{\sin k}$ is the carbon allowances purchased or sold in year y, dy is the carbon allowances in year y,κ_y is the price of a ton of carbon credits in year y, c the growth proportionality coefficient,$\hat{b}_y$ can be a fixed value, and the priceby value depends on the year y market.

3 Optimal Planning Model for Integrated Energy Systems

In addition to the introduction of the interconnection mechanism, the model can also assess the impact of carbon emission reduction means and emission reduction intensity on the planning strategy of the regional integrated energy system; the model is able to represent the finite planning scope of any length of time, which can realize the flexible division of the equipment investment stage under the multi-year perspective; and introduce the carbon trading mechanism into the regional integrated energy system, which can assist in realizing the global reduction of carbon dioxide emission; and then optimize the equipment configuration of multiple interconnected integrated energy systems to realize the coordinated operation of multiple regional integrated energy systems. In addition, a reasonable residual value setting can reduce the impact of the end effect of the planning cycle.

Combined with the current development of integrated energy systems, the model is oriented to the demand for cooling, heating and electricity loads in each block, and with the goal of economic optimization, the model will make the best design decisions to achieve decarbonization, taking into account the constraints of candidate technologies, emission reduction methods and other constraints. For candidate energy conversion and storage technologies, the model can determine the need for installation equipment, installation capacity, installation time and location. In addition, it can determine the interconnection of interconnected integrated energy system interconnections between blocks, including whether pipelines are constructed and when they are constructed. This study considers the impact of several dynamic factors on energy planning, notably technological advances and equipment degradation, interannual variations in unit investment costs, daily variations in energy demand and energy prices, as well as variations in carbon dioxide emission factors and carbon quota trading prices.

3.1 Objective Function

The objective function of the model is to minimize the total cost of development and operation of the research object. This is shown in Eq. 13.33.

$$\min \cos\mathrm{t} = C_{AE} + C_{G\&E} + C_{O\&M} + C_{CE} + C_{Net} \tag{13.33}$$

where, Costs include the investment cost of equipment C_{AE}, the cost of purchasing external energy $C_{G\&E}$, the cost of operation and maintenance $C_{O\&M}$ and the cost of purchasing carbon credits C_{CE} and the cost of pipeline construction C_{Net}.

Technology investment cost where salvage value is the recoverable asset cost of equipment that has not yet reached the end of its decommissioning life at the end of the planning cycle, which is linearly proportional to its size, cost of capital, and exchanges for economic value at the end of the planning horizon, as shown in Eqs. 13.34, 13.35 and 13.36.

$$C_{AE} = \sum_{l=1}^{lx} \sum_{e} \sum_{\mathrm{n}}^{y} \phi_{n,e} W_{l,n,e}^{nom} - REV \quad \forall Y \geq y \geq n \tag{13.34}$$

$$REV = \sum_{e} \sum_{l}^{lx} \sum_{n}^{Y} \rho_{n,e} \frac{L_{n,e}}{N_e} \phi_{n,e}^{Y} W_{l,n,e}^{nom} \quad \forall Y \geq y \geq n \tag{13.35}$$

$$L_{n,e} = \begin{cases} 0 & n \leq Y - N_e + 1 \\ N_e - (Y - n + 1) & n > Y - N_e + 1 \end{cases} \tag{13.36}$$

where, the technology investment cost C_{AE} is calculated from the unit investment cost $\phi_{n,e}$, the rated installed power $W_{l,n,e}^{nom}$ and the salvage value REV,$\rho_{n,e}$ is the recovery ratio, which takes a value less than 1, and $\phi_{n,e}^{Y}$ is the unit investment cost of the equipment at the end of the planning year. $L_{n,e}$ is the residual life of equipment e installed in year n. N_e is the life of equipment e. To distinguish it from L, lx is used in the formula to refer to the upper limit of the number of blocks.

The cost of purchasing external energy is obtained by summing the four components over the planning cycle, as shown in Eqs. 13.37, 13.38, 13.39, 13.40 and 13.41.

$$C_{G\&E} = \sum_{l=1}^{lx}\sum_{y}\sum_{d}\sum_{h}(CH2_{l,y,d,h} + CPG_{l,y,d,h} + CNG_{l,y,d,h} + CBIO_{l,y,d,h}) \quad \forall Y \geq y, \forall d, h \tag{13.37}$$

$$CH2_{l,y,d,h} = c_y^{H_2} F_{l,y,d,h}^{H_2} \quad \forall Y \geq y, \forall l, d, h \tag{13.38}$$

$$CNG_{l,y,d,h} = c_y^{NG} F_{l,y,d,h}^{NG} \quad \forall Y \geq y, \forall l, d, h \tag{13.39}$$

$$CPG_{l,y,d,h} = c_y^{PG} F_{l,y,d,h}^{PG} \quad \forall Y \geq y, \forall l, d, h \tag{13.40}$$

$$CBIO_{l,y,d,h} = c_y^{BIO} F_{l,y,d,h}^{BIO} \quad \forall Y \geq y, \forall l, d, h \tag{13.41}$$

where, the cost of purchasing external energy is $C_{G\&E}$, the cost of purchasing hydrogen, natural gas and biomass are $CH2_{y,d,h}$,$CNG_{y,d,h}$ and $CBIO_{y,d,h}$, respectively, and the cost of purchasing electricity from the grid is $CPG_{y,d,h}$, which is the consumption multiplied by the unit price of c_y^{H2},,c_y^{NG} c_y^{BIO} and c_y^{PG}, respectively, and the four components are summed over the planning cycle to obtain $C_{G\&E}$.

The O&M cost contains fixed O&M cost and variable O&M cost, as shown in Eq. 13.42, and the linear function of the installed capacity of the fixed O&M cost equipment can be calculated by Eq. 13.43, which applies to all the equipment of this energy system.

$$C_{O\&M} = C_{FO\&M} + C_{VO\&M} \tag{13.42}$$

$$C_{FO\&M} = \sum_{l=1}^{lx}\sum_{e}\sum_{n}^{y}\psi_{y,e}U_{l,n,e}^{nom} \quad \forall Y \geq y \geq n \tag{13.43}$$

where, CO&M is the total O&M cost, including fixed O&M cost CFO&M and variable O&M cost CVO&M. CFO&M is. $\psi_{y,e}$ is the investment cost per unit capacity, and $U_{l,n,e}^{nom}$ is the rated installed power of the equipment.

Since the mathematical relationship between equipment operation and the corresponding O&M costs is complex and difficult to assess accurately, to simplify the calculation, as shown in Eq. 13.44.

$$\begin{aligned} C_{VO\&M} = &\sum_{l=1}^{lx}\sum_{n=1}^{y}\sum_{d}^{D}\sum_{H}^{H}(\omega_{y,PV}P_{n,y,d,h}^{PV} + \omega_{y,WT}P_{n,y,d,h}^{WT} + \omega_{y,ICE}P_{n,y,d,h}^{ICE} \\ &+ \omega_{y,SOFC}P_{n,y,d,h}^{SOFC} + \omega_{y,GB}Q_{n,y,d,h}^{GB} + \omega_{y,BB}Q_{n,y,d,h}^{BB} + \omega_{y,EHP}Q_{n,y,d,h}^{EHP} \\ &+ \sum_{e2}\omega_{y,e2}(DIS_{n,e2,y,d,h} + CH_{n,e2,y,d,h})) \\ &\forall Y \geq y, \forall d, h \end{aligned} \quad (13.44)$$

where, CVO&M is the variable O&M cost, which is estimated using the real-time operating power of the device; in addition, the cost of charging and discharging energy from the storage device is the same, and CVO&M is equal to the sum of the product of the real-time power of the individual device and the variable O&M cost per unit of the device,$\omega_{y,e}$.

The cost of carbon allowance purchase depends on the unit price of carbon allowance and the quantity of carbon allowance purchased, as shown in Eq. 13.45.

$$C_{CE} = \sum_{1}^{Y} C_y^{\sin k} * \kappa_y \quad \forall Y \geq y \quad (13.45)$$

where, C_{CE} is the cost of carbon allowances,$C_y^{\sin k}$ is the yth degree of purchase of carbon allowances, and κ_y is the unit price of carbon allowances.

The cost of interconnection cost consists of asset cost and O&M cost as shown in Eqs. 13.46, 13.47 and 13.48.

$$C_{Net} = C_{Net}^1 + C_{Net}^2 \quad (13.46)$$

$$C_{Net}^1 = (\max(P_{y,d,h,l,l'}^{exc})\varpi_y^{ele} + \max(Q_{y,d,h,l,l'}^{exc})\varpi_y^{heat})X_{l,l'} \quad \forall Y \geq y, \forall d, h, l \quad (13.47)$$

$$\begin{aligned} C_{Net}^2 = &\sum_{l1}^{l3}\sum_{1}^{Y}\sum_{1}^{D}\sum_{1}^{H}(z_{l,l',y,d,h}^{exc}P_{y,d,h,l,l'}^{exc}\ell_y^{ele} + z_{l,l',y,d,h}^{exc}Q_{y,d,h,l,l'}^{exc}\ell_y^{heat})X_{l,l'} \\ &\forall Y \geq y, \forall d, h, l \end{aligned} \quad (13.48)$$

where,C_{Net} is the interconnection cost,C_{Net}^1 is the asset cost, and C_{Net}^2 is the daily O&M cost for heat and power exchange. The asset cost of interconnection and interaction C_{Net}^1 is the cost of pipeline laying and cable installation, where heat and electricity exchanges are positively correlated with the distance between the integrated energy systems;$X_{l,l'}$ is the distance of heat pipeline laying or cable installation between energy stations;ϖ_y^{ele} and ϖ_y^{heat} are the unit asset cost;ℓ_y^{heat} and ℓ_y^{ele} are the unit O&M cost, respectively.

3.2 Constraints on Energy Supply and Demand Balance

The primary function of the energy system is to satisfy the user's needs, which are met by the equipment in the system and the external grid, as shown in Eqs. 13.49, 13.50 and 13.51.

$$\begin{aligned} TED_{l,y,d,h} =& P_{l,y,d,h}^{pg} + P_{l,n,e1,y,d,h} + DIS_{l,n,y,d,h}^{LiB} - CH_{l,n,y,d,h}^{LiB} - F_{l,n,y,d,h}^{EHP} \\ & - F_{l,n,y,d,h}^{EC} - F_{l,n,y,d,h}^{MGC} - P_{l,n,y,d,h}^{AE} - P_{l,n,y,d,h}^{CCUS} - P_{l,l',y,d,h}^{exc} \\ & + P_{l',l,y,d,h}^{exc} \\ & \quad \forall l, n, y, d, h \end{aligned} \tag{13.49}$$

$$\begin{aligned} THD_{l,y,d,h} \le& Q_{l,n,e1,y,d,h} + DIS_{l,n,y,d,h}^{HWT} - CH_{l,n,y,d,h}^{HWT} \\ & - Q_{l,n,y,d,h}^{ABC} - Q_{l,l',y,d,h}^{exc} + Q_{l',l,y,d,h}^{exc} \\ & \quad \forall l, n, y, d, h \end{aligned} \tag{13.50}$$

$$\begin{aligned} TCD_{l,y,d,h} =& L_{l,n,e1,y,d,h} + DIS_{l,n,y,d,h}^{CWT} - CH_{l,n,y,d,h}^{CWT} \\ & \forall l, n, y, d, h \end{aligned} \tag{13.51}$$

where, $TED_{l,y,d,h}$ is the electrical load demand,$THD_{l,y,d,h}$ is the heat load demand, and $TCD_{l,y,d,h}$ is the cooling load demand.

3.3 Energy Interaction Constraints in Interconnected Regions

In order to ensure the safe and stable operation of the system, there exists an upper limit value for the energy exchange carried out by the blocks, and this value is used to calculate the initial investment cost for the interconnection case, as shown in Eqs. 13.52, 13.53, 13.54 and 13.55.

$$z_{l,l',y,d,h}^{exc,heat} Q_{y,d,h,l,l'}^{exc} \le z_{l,l',y,d,h}^{exc,heat} Q_{y,d,h,l,l'}^{exc,\max} \qquad \forall Y \ge y, \forall d, h, l \tag{13.52}$$

$$z_{l,l',y,d,h}^{exc,ele} P_{y,d,h,l,l'}^{exc} \le z_{l,l',y,d,h}^{exc,ele} P_{y,d,h,l,l'}^{exc,\max} \qquad \forall Y \ge y, \forall d, h, l \tag{13.53}$$

$$z_{l,l',y,d,h}^{exc,heat} + z_{l',l,y,d,h}^{exc,heat} \le 1 \qquad \forall Y \ge y \ge n, \forall l, d, h \tag{13.54}$$

$$z_{l,l',y,d,h}^{exc,ele} + z_{l',l,y,d,h}^{exc,ele} \le 1 \qquad \forall Y \ge y \ge n, \forall l, d, h \tag{13.55}$$

where, $Q_{y,d,h,l,l'}^{exc,\max}$ and $P_{y,d,h,l,l'}^{exc,\max}$ are the upper limit values for thermal and electrical energy exchange respectively,$z_{l,n,e2,y,d,h}^{ch}$ and $z_{l,n,e2,y,d,h}^{dis}$ are 0–1 binary variables, in order to achieve a more realistic simulation of the operating conditions, it is not permitted for block l to introduce the energy of other blocks l' and at the same time transfer energy to l'.

3.4 Carbon Emission Constraints

Carbon dioxide emissions come from energy consumption, which depends on the natural gas consumption of internal combustion engines and gas boilers, as well as the amount of electricity purchased from the grid. In order to find a better environmental policy, the total amount of carbon dioxide emissions from year 1 to year Y should be controlled as shown in Eqs. 13.56 and 13.57, and the total amount of carbon emissions over the whole planning period can be artificially set to a constant value of const, which can be used to achieve mandatory emission reductions. Under this constraint, different values of const will lead to different optimal configurations of equipment and their optimal installation times.

$$\begin{aligned} C_y^{emission} = &\sum_{n=1}^{y}\sum_{l=1}^{lx}\sum_{d}^{D}\sum_{h}\Big((\zeta^{NG}(F_{l,n,y,d,h}^{ICE} + F_{l,n,y,d,h}^{GB}) \\ &+ \zeta^{pg} F_{l,y,d,h}^{pg} - C_{l,y}^{\sin k} - C_{l,n,y,h}^{CCUS}) \end{aligned} \tag{13.56}$$

$$\sum_{y=1}^{Y} C_y^{emission} = const \tag{13.57}$$

where, $C_y^{emission}$ is the CO2 emissions in year y, the natural gas consumption of internal combustion engines and gas boilers are $F_{l,n,y,d,h}^{ICE}$ and $F_{l,n,y,d,h}^{GB}$ respectively,$P_{l,y,d,h}^{pg}$ is the amount of electricity purchased from the grid, which is multiplied by the natural gas carbon factor ζ^{NG} and the grid emission factor ζ^{pg} to calculate the emissions,$Z_{l,y,d,h}^{\sin k}$ is the amount of carbon credits purchased from the carbon market, and $C_{l,n,y,h}^{CCUS}$ is the amount of CO2 captured by CCUS.

CCUS is a technological means from the perspective of energy consumption and cleanliness, emission reduction can also consume new energy, but at present, there are problems of large power consumption and high cost; carbon trading mechanism is also a widely used means of controlling emissions, by giving different carbon quotas to each subject of the market and allowing them to trade carbon in the carbon market to realize the additional incentives for positive carbon emission reduction enterprises and increase carbon cost for negative carbon emission reduction enterprises. The carbon trading mechanism is also widely used as a means of emission control; by giving different carbon quotas to each subject in the market and allowing them to trade in the carbon market, it realizes additional incentives for positive carbon emission reduction enterprises and additional carbon costs for negative carbon emission reduction enterprises, while the quota method of the trading subjects mainly includes the baseline method and the historical intensity method.

4 Numerical Analysis

4.1 Introduction of the Research Object

A local community in Shanghai is selected for analysis. The region is divided into three blocks, each block includes residential and commercial areas, in which the ratio of residential users and commercial users load demand is 4:1, the center of the integrated energy system in the three blocks are 6 km, 8 km, 10 km, respectively, and the same block cannot cross the municipal zoning and control boundaries, the energy system within the block and potential energy exchange between blocks is shown in Fig. 1, the building area of each block and the construction time schedule is shown in Table 1. The same block cannot cross the municipal zoning and control boundaries. The energy system within the blocks and the potential energy exchange between the blocks are shown in Fig. 1, and the floor area and construction schedule of each block are shown in Table 1.

4.2 Input Data

4.2.1 Regional Load Demand

In this study, one day is selected as a typical day from each of the four seasons of the year, and the hour-by-hour energy demand is considered for 24 h per day, and each typical representative day has the duration of the corresponding season, which is 91, 92, 92, and 90 days, respectively. Assuming that the various energy demands of the users grow by 2.0% per year during the planning period, and that the compliance ratio of each block is the ratio of the floor area, the typical day-by-hour loads of the initial year are shown in Fig. 2, taking Block I as an example.

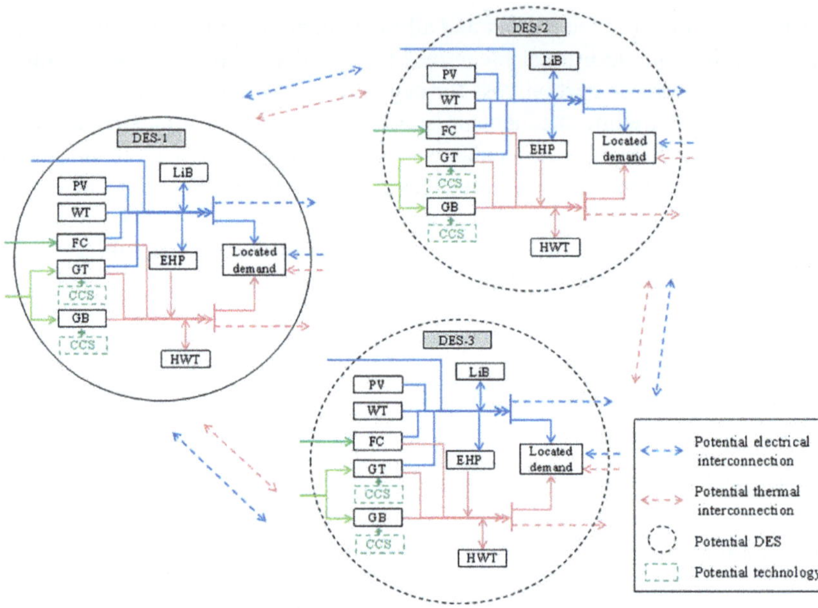

Fig. 1 Schematic diagram of energy exchange between blocks

4.2.2 Equipment Parameters

Input data for equipment efficiency parameters and economic characteristics for the initial year are presented in Table 2 and are consistent across blocks. In order to capture the dynamic characteristics of long-term operation, the trend and magnitude of fluctuations in each set of parameters should be determined. For the installed equipment, the efficiency consistently decreases by 1% due to wear and fatigue, but depreciates by 0.8% per year due to the maturity of boiler O&M technology. As science and technology evolve, the efficiency of new equipment will increase each year. PV efficiency increases every year, but the maximum remains within 4000 m^2 due to the constant size of the existing area. PV, SOFC and LiB have some potential for development, so their unit capital cost decreases more dramatically 1%, while the other technologies are more mature, with a flat decrease in cost, and the unit cost of O&M is assumed to decrease by 0.2–0.8% per year, including fixed O&M and Variable O&M costs. Interconnection costs include $32/m per unit of heat pipeline, and $4.5/m per unit of transmission.

Table 1 Time span of construction phases

Block	Floor area (m^2)	2021–2025	2026–2030	2031–2035	2036–2040	2041–2045	2046–2050	2051–2055	2056–2060
Block I	80,000	–	–	–	–	–	–	–	–
		–	–	–	–	–	–	–	–
		–	–	–	–	–	–	–	–
Block II	100,000	–	–	–	–	–	–	–	–
		–	–	–	–	–	–	–	–
		–	–	–	–	–	–	–	–
Block III	120,000	–	–	–	–	–	–	–	–
		–	–	–	–	–	–	–	–
		–	–	–	–	–	–	–	–

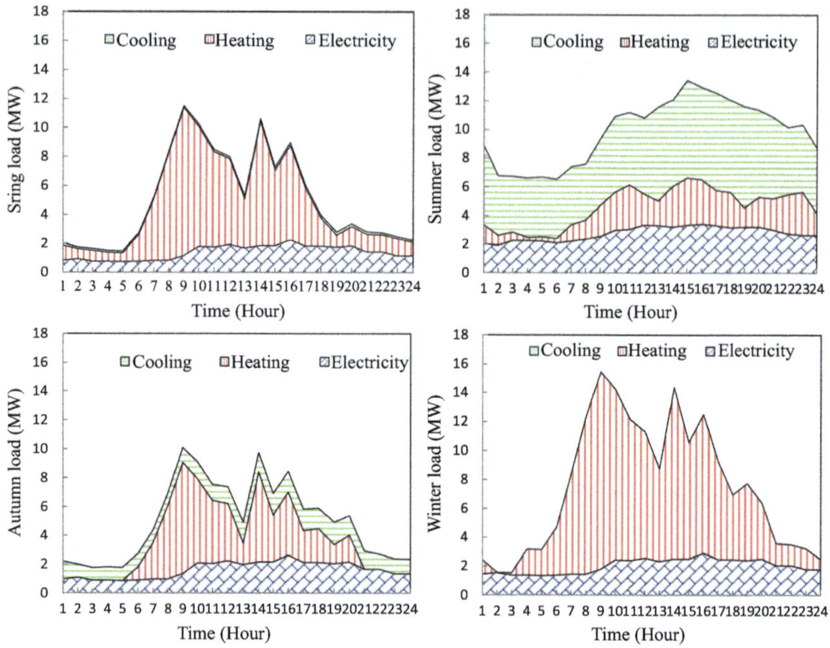

Fig. 2 Typical daily load demand for four seasons in the initial year of block I

4.2.3 Energy Prices and Emission Parameters

The solar irradiance for four typical days in the first year is shown in Fig. 3. The GCS results show that the average solar irradiance fluctuates very little over a 40-year period, so this example assumes that it remains constant throughout the planning period. Since the multi-selected area is inland, the wind speed and direction at 40 m altitude for the four typical days are quoted from the survey data.

In addition to solar energy, external inputs include hydrogen, natural gas, and grid electricity, and energy prices will continue to evolve in an environmentally and policy-driven manner. Natural gas prices in 2021 are \$3.22/m^3, hydrogen prices are \$6.45/m^3, and the grid uses time-of-day electricity pricing, as shown in Fig. 4, and driven by the dual-carbon goal that will continue to promote the use of cleaner energy sources and increase the cost of carbon-containing energy sources in the country. It is assumed that grid electricity and natural gas prices increase at a rate of 1% per year[10], and hydrogen prices decrease at a trend of 2% per year.

Based on the latest carbon emission factor data in Shanghai, this chapter sets the grid emission factor to be 0.42 tCO2 /MWh. The prediction model of the grid electricity emission factor with natural year as the variable is established by collecting publicly available data and performing regression analysis, as well as supplementing the sample observations by interpolation, to obtain the prediction value from 2021 to 2060 that are shown in Fig. 5. Due to the fixed natural gas composition, the carbon

Table 2 Technical and economic parameters in the initial year

Item	Unit cost of capacity	Fixed O&M costs	Variable O&M costs	efficiency	longevity
	Yuan/kW	Yuan/(kW–y)	Yuan/(kWh)	–	surname Nian
PV	3500	100.5	0.207	0.28	20
Ventilator	4600	100.5	0.207	0.23	20
ICE	8280	124.2	0.090	0.39/0.36	20
stove	910.8	138	0.0138	0.89	20
Heat pumps	2277	138	0.0138	3.2	10
Absorption chillers	2656.5	138	0.0138	0.7	15
Electric coolers	1062.6	138	0.0138	4.0	15
Fuel cell	10,070.6	29.67	0.2415	0.42/0.52	10
CCUS equipment	300 yuan/ton	30 yuan/ton	2.1 yuan/ton	0.9	20
Biomass boilers	1200.5	138	0.0138	0.85	20
Battery	5030.1	51.405	0.0157	0.92	20
Heat storage tank	138	13.8	0.0157	0.85	20
Cold storage tank	138	13.8	0.0157	0.85	20

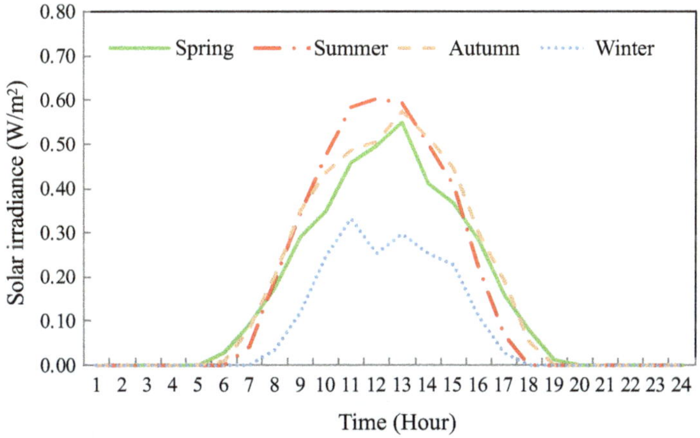

Fig. 3 Hourly solar irradiance on a typical day in the four seasons

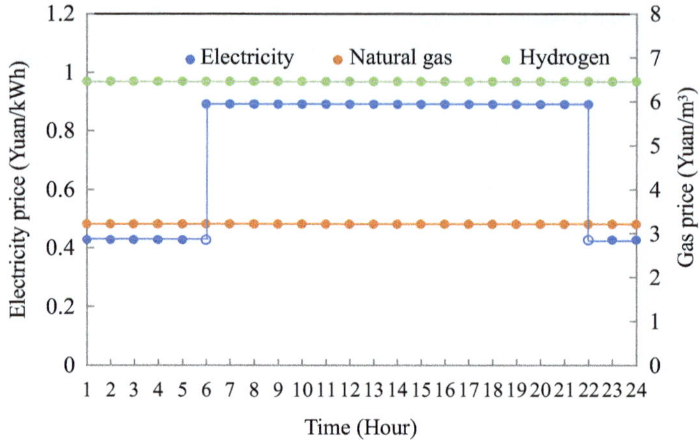

Fig. 4 Initial annual energy prices

emission factor is set to 0.2 tCO2 /MWh and remains unchanged during the planning period. In addition, according to the trading data of Shanghai Environment and Energy Exchange, the average carbon price in 2021 is RMB 42.9/tCO2, and the unit price of carbon trading in the planning period is shown in Fig. 5. In this chapter, according to the "Shanghai Carbon Emission Allowance Allocation Program for 2016", the heat generation of each unit is converted into equivalent power generation capacity, and the carbon emission allowance is allocated according to the total equivalent power generation capacity, and the carbon emission allowance per unit of power generation capacity is determined to be 0.728 t/(MWh).

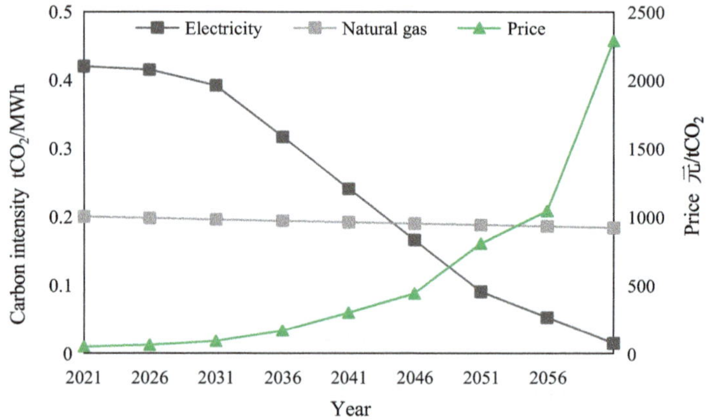

Fig. 5 Natural gas, grid carbon emission factors and carbon price trends

4.3 Scenario Setting

In this chapter, four major scenarios are set up in general to explore the impacts of interconnection mechanisms, carbon emission reduction means and emission reduction intensity on regional integrated energy system planning strategies, as shown in Table 3.

Scenario 1 has a steady development over a 40-year planning cycle, with the three regions being independent of each other and with no emission reduction constraints. Scenario 2 has a steady development over a 40-year planning cycle, with interconnections allowed between the three regions, i.e. exchange of electricity and heat, and no emission reduction requirements. Scenario 3 builds on Scenario 2 and requires a 30% reduction in emissions over the 40-year planning cycle, with participation in the carbon market as a means of reducing emissions. Scenario 4 introduces new energy and higher-cost clean technologies under the trading mechanism of Scenario 3, resulting in the three sub-scenarios set in Scenario 4, which gradually increase the pressure to reduce emissions, respectively, 30% in Scenario 4A, 40% in Scenario 4B, and 50% in Scenario 4C. The technologies introduced include three types of equipment, solid oxide fuel cells, wind turbines, and lower quality wind power used as feedstock for CCUS to capture CO_2.

Table 3 Scenario settings

Scenario setting	Interconnectivity	Reduce emissions or not	Forms of emission reduction	Means to reduce emissions
Scenario 1	×	×	–	–
Scenario 2	√	×	–	–
Scenario 3	√	√	30% full-cycle emission reduction	carbon market
Scenario 4A	√	√	30% full-cycle emission reduction	Carbon trading + clean technology introduction
Scenario 4B	√	√	40% full-cycle emission reduction	Carbon trading + clean technology introduction
Scenario 4C	√	√	50% reduction in full-cycle emissions	Carbon trading + clean technology introduction

5 Results and Analysis

5.1 Optimal Equipment Configuration of the System Under Different Scenarios

As shown in Figs. 6,7 and8, the comparison of Scenario 1 and Scenario 2 reveals that the selection of equipment types is the same, but the regional interconnection in Scenario 2 promotes the matching of energy supply and demand, resulting in a lower cumulative installed capacity of equipment. The power generation equipment and heat generation equipment in Block II in Scenario 2 undergoes a significant downsizing because Block II is 6 km away from Block I, and it is more cost-effective to use the heat and power from Block I. Block III is 10 km away from Block I, and the cost of piping installation is high, so it is economically more favorable to install the equipment independently.

A comparison of Scenario 2, Scenario 3, and Scenario 4A reveals that Scenario 3's emission reductions come from purchased carbon credits, and the installed capacity of the equipment does not change much, remaining the least installed capacity in Block II. Scenario 4A adds fuel cells to the Scenario 3 equipment to alleviate the pressure of expensive purchased emission credits. With increasing emission reduction intensity,

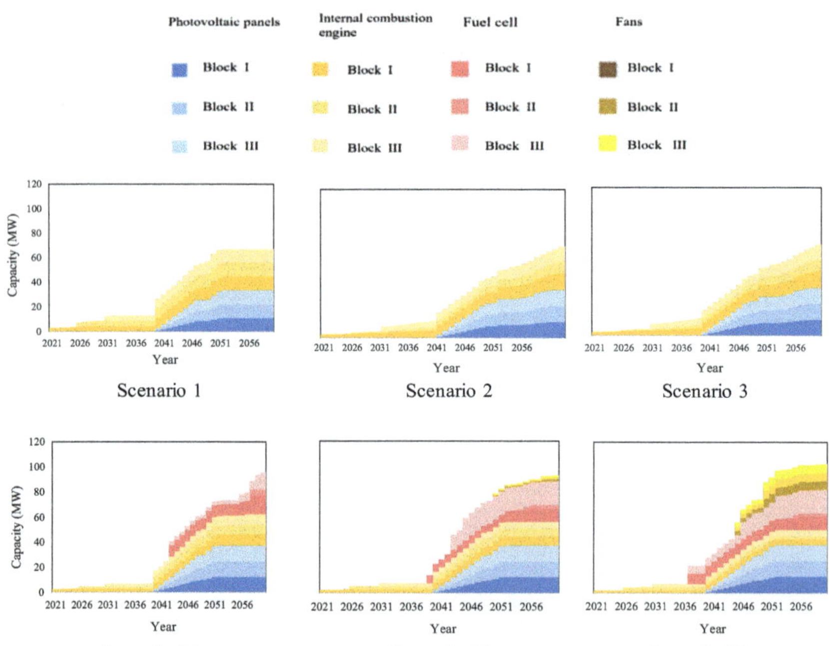

Fig. 6 Evolution of the cumulative installed capacity over the planning period under different scenarios

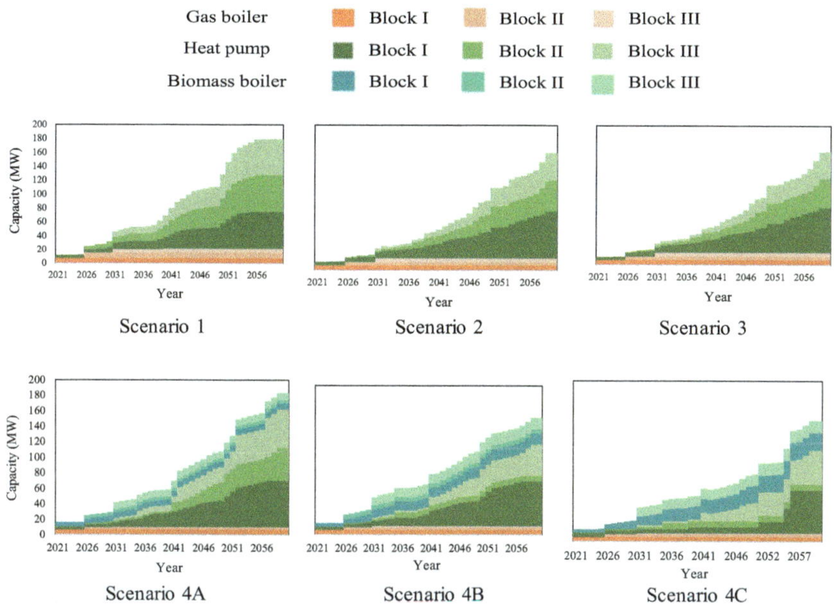

Fig. 7 Evolution of the cumulative installed capacity of heating equipment over the planning period under different scenarios

cleaner, though less efficient, wind turbines are added to Scenario 4B, and CCUS is further put into use in Scenario 4C. In addition, an intrinsic link between equipment and carbon emissions can be found, with a certain decrease in the carbon emission curve occurring from the year in which the cleaner technology is commissioned, as shown in Fig. 9.

Fig. 9Evolution of cumulative installed CCUS capacity over the planning cycle for each block of scenario 4C

The installed capacity of CCUS is shown in Fig. 9, which is only put into use at the end of Scenario 4C to realize high-intensity carbon emission reduction because of the high comprehensive investment cost including capture, transportation and utilization. The CCUS investment in Block II is less than that in the other blocks due to less purchased fossil fuels, and CCUS is suitable for higher density carbon emitters, and there is a possibility of a technological breakthrough in this technology in the long run, which is expected to lead to a significant reduction in the unit cost.

5.2 *Analysis of Interregional Connectivity Results*

Due to the field distance advantage, block II has interconnection and interaction with both block I and III, and the transmission of electric energy is analyzed as an example, as shown in Fig. 10. Overall, all the sub-scenarios of Scenario 4 show an increase

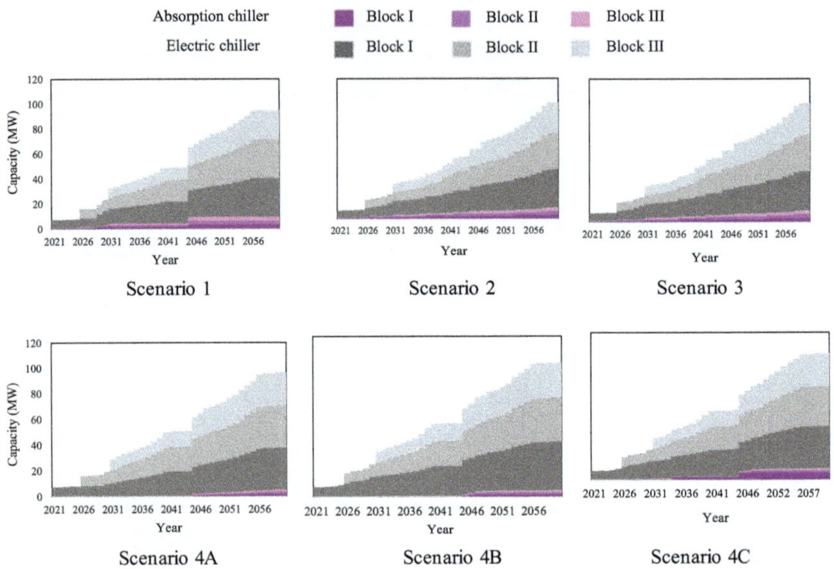

Fig. 8 Evolution of the cumulative installed capacity of cooling equipment over the planning period under different scenarios

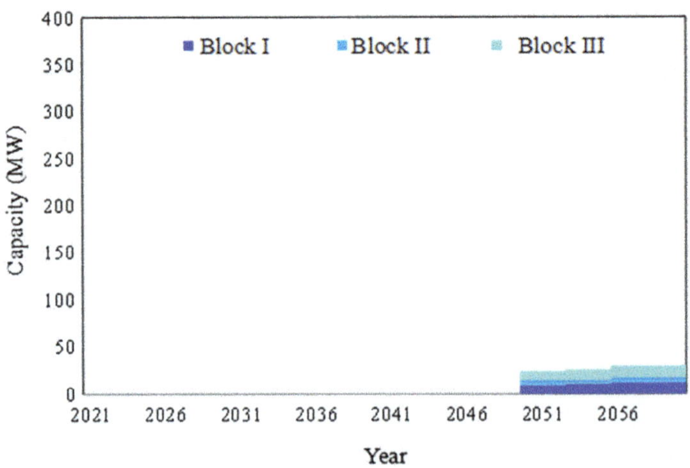

Fig. 9 Evolution of cumulative installed CCUS capacity over the planning cycle for each block of scenario 4C

in fuel cell power generation, but the installed capacity of Block II is small, and the power consumption is mostly supplied from other blocks, with self-production on the low side. Comparison of Scenarios 4B and 4C reveals that when the pressure to reduce emissions is greater, turbines with higher unit costs are selected to generate

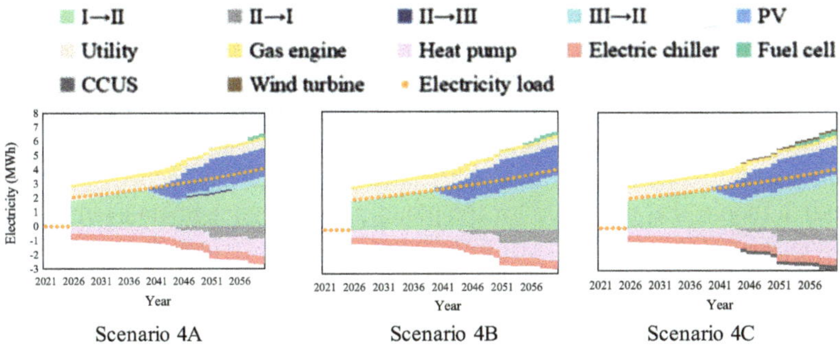

Fig. 10 Electricity supply and demand over the planning cycle (scenario 4 Block II)

electricity, in addition to increasing CCUS, which has less capacity but its power consumption is on the high side.

5.3 Daily Running Strategies on a Typical day in Different years

Due to space constraints, the typical summer day electricity usage of Scenario 4C is taken here to show more detailed hour-by-hour operation details of the energy storage devices, as shown in Fig. 11 for example in 2030, 2045 and 2060 for Block II. The primary role of energy storage devices is to shave peaks and fill valleys within a single day, allowing for a better match between energy supply and demand, reducing costs, and enhancing system stability. With the increase in load demand and the reduction of technology costs, photovoltaic panels are installed and used in the middle and late stages, which makes the power purchased by the grid lower, thus realizing the reduction of carbon emissions, and the photovoltaic equipment has not been selected for use in 2030 due to economic deviations, which can be corresponding to Fig. 11.

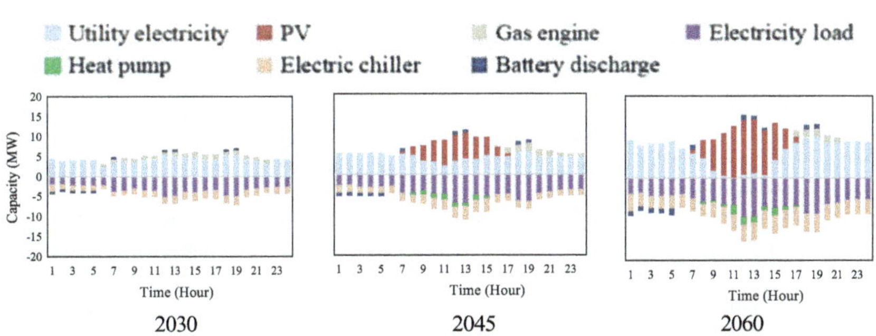

Fig. 11 Hourly electric supply and demand in different years (scenario 4C block II)

6 Conclusions

This study verifies the validity of the model by applying it to an actual case of development in an urban area, demonstrates its advantages in dealing with the task of designing a multi-stage, multi-area interconnected integrated energy system, and can provide a theoretical reference for the planning and management of energy use in urban areas:

① Allowing neighboring regions to interconnect allows for more rational energy allocation, promotes complementary interconnection of multiple blocks, reduces the risk of energy interruptions, and achieves year-over-year reductions in carbon emissions at a lower total cost, and the closer the blocks are geographically, the more cost-effective the energy interconnections will be.
② Carbon emission reduction constraints will have a direct impact on the regional carbon emission curve, the larger the emission reduction, the faster the equipment turnover, which makes the curve fluctuate less, and the value of carbon emissions decreases year by year.
③ The introduction of carbon trading mechanism can make the region as a whole reach a more economical emission reduction effect, but need to be appropriate system design; but continuously improve the operational efficiency of energy equipment to improve the innovation of technology, more beneficial to the healthy development of society and economy.
④ Equipment configuration landing point for the day before the optimization of the operation strategy, a reasonable equipment configuration can enhance the stability and reliability of energy supply at the same time, to achieve the optimization of the way and timing of energy generation, to promote the development and utilization of renewable energy.

Author's note: This chapter is a revised and expanded English version of the author's article, "Optimal Planning of Regional Integrated Energy Systems Considering Long-Term Carbon", originally published in *Policy Science*, Vol. 32, No. 3 (2025).

References

Allegrini J, Orehounig K, Mavromatidis G et al (2015) A review of modelling approaches and tools for the simulation of district-scale energy systems. Renew Sustain Energy Rev 52:1391–1404

Cao Y, Mu Y, Jia H et al (2015) Multi-stage planning of integrated energy system in parks considering construction timing. J Electr Eng 40(21):6815–6828

Koltsaklis NE, Liu P, Georgiadis MC (2015) An integrated stochastic multi-regional long-term energy planning model incorporating autonomous power systems and demand response. Energy 82:865–888

Li C, Wang N, Do X et al (2020) Research review and outlook of multi-energy complementary distributed energy system integration. Chin J Electr Eng 45(8):1–10

Liu L, Feng L (2016) Analysis of influencing factors of synergistic development of carbon market in China. Chin Foreign Energy 16(9):1–7

Liu Z, Zhao Y, Wang X (2020) Long-term economic planning of combined cooling heating and power systems considering energy storage and demand response. Appl Energy 279:115819
Mirzaesmaeeli H, Elkamel A, Douglas PL et al (2010) A multi-period optimization model for energy planning with CO2 emission consideration. J Environ Manage 36:367–378
Qu K, Li H, Yu T et al (2018) Decentralized scheduling of multi-regional integrated energy system under carbon trading mechanism. Chin J Electr Eng 38(3):1–8
Walmsley MRW, Walmsley TG, Atkins MJ (2015) Achieving 33% renewable electricity generation by 2020 in California. Energy 92:260–269
Wang X (2020) China strives to achieve carbon neutrality by 2060. Ecol Econ 36(12):4

Scenario Analysis of Energy Transition Pathways towards Net-Zero Communities

You Li and Weisheng Zhou

1 Background

Addressing climate change and improving environmental health have become global imperatives, prompting countries worldwide to commit to ambitious carbon reduction targets under the Paris Agreement (W. Zhou et al. 2021a). These commitments mark a collective recognition of the urgent need to limit global warming and mitigate its adverse impacts on ecosystems, economies, and human health (Seo 2017). Japan, as a highly industrialized nation with significant energy demand and limited domestic fossil fuel resources, has pledged to achieve net-zero greenhouse gas emissions by 2050. This goal represents a profound challenge, as it requires rethinking and restructuring virtually every aspect of its energy and socio-economic systems over the coming decades (Kii et al. 2023; Ozawa et al. 2022).

Achieving carbon neutrality entails deep, transformative changes across multiple sectors of the economy (Caineng et al. 2024). In Japan, power generation, heating, and transportation alone account for more than 70% of total emissions, making them critical targets for decarbonization efforts (Kiang and Schipper 1996). Analyses suggest that the power sector, in particular, must achieve zero emissions as early as 2040 to make the 2050 economy-wide target feasible (Ozawa et al. 2022). The building sector also warrants urgent attention, as it contributes nearly one-fifth of total emissions through energy use in heating, cooling, lighting, and appliances. In addition, the industrial sector faces unique challenges due to its energy intensity and dependence on high-temperature heat and process emissions.

Y. Li
Asia-Japan Research Institute, Ritsumeikan University, Osaka, Japan
e-mail: liyou@fc.ritsumei.ac.jp

W. Zhou (✉)
College of Policy Science, Ritsumeikan University, Osaka, Japan
e-mail: zhou@sps.ritsumei.ac.jp

W. Zhou (ed.), *Decarbonization Policy Engineering*, World Sustainability Series,
https://doi.org/10.1007/978-981-95-5953-4_14

Transitioning to a low-carbon economy requires more than simply switching fuels; it demands rapid technological innovation, large-scale deployment of renewable energy, improvements in energy efficiency, electrification of end uses, and the development of enabling infrastructure for emerging technologies such as hydrogen and carbon capture (Zhang 2010). At the same time, the transition must be managed carefully to maintain economic stability, ensure energy security, and minimize social disruption (Sylla and Koddenbrock 2025). Policymakers must grapple with significant uncertainties, including the pace of technological development, the evolution of energy markets, demographic shifts, and public acceptance of new technologies and policies.

Against this backdrop, robust scenario analyses and integrated assessment models play a crucial role. They provide a systematic framework for exploring possible pathways, evaluating trade-offs among different strategies, and identifying priorities at various stages of the transition. By simulating how alternative policy and technology combinations affect emissions, costs, and other socio-economic outcomes, these tools inform evidence-based policymaking and guide investment decisions. In particular, models that capture the temporal dynamics and interactions of diverse technologies are essential for designing realistic and resilient pathways toward net-zero communities.

1.1 Integrated Energy Assessment Models

Integrated energy assessment models have become indispensable tools for analyzing the long-term evolution of energy systems under different decarbonization scenarios (Kemfert 2002). By linking energy, economic, and environmental systems, these models provide policymakers and planners with quantitative insights into how various low-carbon technologies and policy instruments could contribute to emission reduction goals. Over the past decades, a variety of integrated models have been developed, each with its own strengths, assumptions, and application scopes.

For example, the GEM-E3 model (Capros et al. 2013) is a widely used multi-regional, multi-sectoral computable general equilibrium model that represents the interaction between goods, services, labor, and capital markets. It allows researchers to assess how emissions can be reduced through measures such as fuel switching, energy efficiency improvements, and demand reduction while accounting for macroeconomic feedback. Similarly, the MARKAL model (Fishbone and Abilock 1981; Tsai and Chang 2015), originally developed by the International Energy Agency (IEA), has been extensively applied to evaluate the role of specific low-carbon technologies—such as hydrogen, renewables, and CCS—in shaping Japan's future energy mix and to inform investment decisions under various regulatory and market conditions in countries like the United States. The GREEN model (Pampanelli et al. 2014), on the other hand, is designed to assess the global and economy-wide costs of climate policies through a dynamic, multi-sectoral equilibrium framework, making it particularly suitable for international comparisons.

Japan has also developed its own integrated models to address national priorities. The DNE21 (Akimoto et al. 2004) and its enhanced version DNE21+ (Akimoto & Yamada) are comprehensive frameworks that combine an energy system model, a macroeconomic model, and a climate change model. These models have been instrumental in identifying cost-effective mitigation strategies for Japan and assessing the trade-offs between nuclear constraints, renewable energy expansion, and energy efficiency measures. The authors proposed the concept of an "East Asia Low-Carbon Community" with the aim of realizing a wide-area low-carbon society from multilayered perspectives—global, glocal, and local. To support the materialization of this vision, they developed the G-CEEP model (Glocal Century Energy-Environment Planning), which enables integrated assessment of economic, environmental, and energy systems (Su et al. 2014; Zhou et al. 2021b). The model has been applied to explore regional cooperation scenarios, analyze nuclear limitations, and evaluate renewable deployment pathways for achieving Japan's 2030 and 2050 targets.

While these models excel at providing long-term, economy-wide assessments, they also have inherent limitations. Most are highly parameter-driven, emphasizing optimal energy shares rather than explicitly detailing the contributions and interactions of different technologies over time. Furthermore, they often assume perfect foresight and smooth transitions, which may overlook critical uncertainties such as technological breakthroughs, resource constraints, and shifts in market or societal preferences. Addressing these gaps calls for more flexible modeling approaches that can capture dynamic interactions and stage-specific priorities in the decarbonization process.

1.2 Regional Energy Planning Models

In contrast to national or global-scale integrated assessment models, regional energy planning models focus on the unique characteristics and constraints of specific local contexts (Feijoo et al. 2024). As energy systems transition toward more decentralized and community-based configurations, these models play an increasingly important role in designing and optimizing low-carbon pathways at the city, district, or microgrid level.

Regional models are particularly well-suited to account for spatial, economic, and social heterogeneity, as well as for incorporating stakeholder preferences and operational constraints that may not be captured at larger scales. For instance, recent research has proposed integrated planning frameworks for electricity-gas-heat systems, which deepen the coupling between sectors to improve overall efficiency and reduce emissions (Massrur et al. 2018). Other studies have demonstrated how 100% renewable energy systems can be achieved at the community level by combining rooftop photovoltaics, wind turbines, biomass, and hybrid energy storage systems (Kroposki et al. 2017). Biomass, in particular, has been shown to play a critical role in providing dispatchable, renewable baseload power in fully renewable scenarios, complementing intermittent resources like solar and wind.

Innovative approaches have also emerged that prioritize user-centric and climate-responsive planning. Examples include occupant-centered models for predicting dynamic heating and cooling loads in urban neighborhoods, which help align energy supply with actual demand patterns, and climate portfolio approaches, which optimize the mix of renewable resources—such as hydropower, wind, and solar—to maximize resilience and minimize costs.

By focusing on feasibility, efficiency, and synergistic interactions between technologies, regional planning models complement the broader insights offered by integrated assessment models. They highlight the importance of tailoring solutions to local resource availability, infrastructure conditions, and social dynamics. Moreover, embedding detailed energy technology modules from regional models into national frameworks can enhance the overall realism and robustness of scenario analyses. This integration enables policymakers to better evaluate how individual technologies interact and how uncertainties in demographics, economics, and technology evolution may shape the pathways toward carbon-neutral communities.

1.3 Policy Relevance and Research Gaps

As Japan and other countries accelerate their efforts toward net-zero emissions, robust evidence from scenario analyses and energy modeling becomes increasingly critical for informing policy design and implementation. Integrated assessment models and regional planning models have already provided valuable insights into the potential of various technologies, the costs and benefits of alternative pathways, and the trade-offs among environmental, economic, and social objectives. These insights help policymakers identify priority sectors, set realistic interim targets, and design supportive regulatory frameworks and market mechanisms. For instance, the recognition of the power sector as a linchpin for early decarbonization has shaped Japan's renewable energy and grid modernization policies, while analyses of building-sector efficiency have guided retrofit standards and incentive programs.

However, significant research gaps remain. Existing models often emphasize optimal end-states—such as the final energy mix in 2050—while paying insufficient attention to the temporal sequence of technology deployment and the evolving interactions among different mitigation measures over time. This limits their ability to provide actionable guidance on which technologies should be prioritized at each stage of the transition, how to coordinate investments in enabling infrastructure, and how to manage uncertainties inherent in long-term planning. Furthermore, many models still adopt deterministic assumptions and idealized market conditions, underestimating the risks posed by technology bottlenecks, cost volatility, and social acceptance issues.

Addressing these gaps calls for developing enhanced modeling frameworks that can capture the dynamic and adaptive nature of the transition to carbon neutrality. Such models should integrate detailed energy technology modules, reflect real-world constraints and behavioral factors, and allow for scenario exploration under different assumptions about technology readiness, policy interventions, and external shocks. By doing so, they can support more resilient and flexible policy design, ensuring that decarbonization pathways remain feasible and cost-effective even in the face of uncertainty. This chapter responds to this need by presenting an improved scenario-based modeling approach that highlights the temporal contributions and interactions of key technologies, providing more granular insights to guide the phased transition toward net-zero communities.

2 Material and Methods

Achieving zero or low carbon emissions in regional energy systems requires not only a technological transition to renewable energy but also the alignment of socio-economic structures to support a low-carbon economy. This approach integrates three key sub-models: a macroeconomic model, an energy balance model, and an emissions constraint model. The overall methodological framework is illustrated in Fig. 1.

2.1 Macro-Economic Constraints

Currently, most energy-related macroeconomic models, or energy models with macroeconomic components, are based on a two-level CES production function. The relationship between production, capital, labor, and energy consumption are described by Eq. (1):

$$Y_t = f(K_t, L_t, E_t, N_t) \quad \text{(Question.14.1)}$$

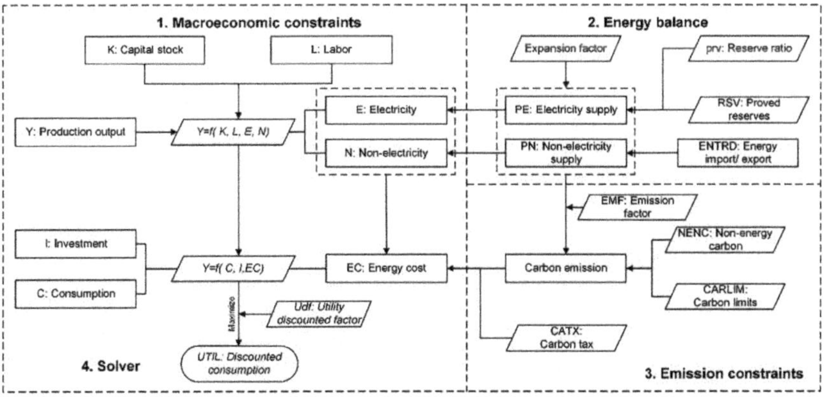

Fig. 1 Overview of the Glocal Century Energy Environment Planning (G-CEEP) model

where Y_t, K_t, L_t, E_t and N_t represent production output, capital, labor, electricity and non-electricity in t year, respectively. It can be detailed as Eq.(2).

$$\Delta Y_t = \gamma \begin{bmatrix} \delta\left(\delta_1 \Delta K_t^{-\rho_1} + (1-\delta_1)\Delta L_t^{-\rho_1}\right)^{\frac{\rho}{\rho_1}} + \\ (1-\delta)\left(\delta_2 \Delta E_t^{-\rho_2} + (1-\delta_2)\Delta N_t^{-\rho_2}\right)^{\frac{\rho}{\rho_2}} \end{bmatrix}^{-\frac{1}{\rho}} \quad \text{(Question.14.2)}$$

where ΔY_t, ΔK_t, ΔL_t, ΔE_t and ΔN_t represent new production output, new capital, new labor, new electricity and non-electricity in t year, respectively. δ, δ_1 and δ_2 are distribution parameters. ρ, ρ_1 and ρ_2 are substitution parameters. The accumulation of production factors can be described by Eq. (3) to (7).

$$Y_t = Y_{t-1} \times (1-\mu)^5 + \Delta Y_t \quad \text{(Question.14.3)}$$

$$L_t = L_{t-1} \times (1-\mu)^5 + \Delta L_t \quad \text{(Question.14.4)}$$

$$E_t = E_{t-1} \times (1-\mu)^5 + \Delta E_t \quad \text{(Question.14.5)}$$

$$N_t = N_{t-1} \times (1-\mu)^5 + \Delta N_t \quad \text{(Question.14.6)}$$

$$\Delta K_t = 5 \times \left[I_t \times (1-\mu)^5 + I_{t+1}\right] \quad \text{(Question.14.7)}$$

where Y_t, K_t, L_t, E_t and N_t represent production output, capital, labor, electricity and non-electricity in t year. I_t is the annual investment. μ is defined as annual depreciate rate. Then, the production output is expressed as the sum of the costs of consumption, investment, and energy system cost.

$$Y_t = C_t + I_t + EC_t \quad \text{(Question.14.8)}$$

The Terminal condition is as follows:

$$K_T \times (\omega + \mu) \leq I_T \quad \text{(Question.14.9)}$$

where T is the last period in this model. ω is annual growth rate.

2.2 Energy Balances

In this model, energy balances primarily take into account energy cost, emission constraints, and energy supply and demand, as follows:

$$[EC_t = \sum \left(EP_{t,e}PE_{t,e} + NP_{t,e}PN_{t,e}\right)] \quad \text{(Question.14.10)}$$

where $EP_{t,e}$ is the electricity energy price coefficient. $PE_{t,e}$ denotes the electricity supply. $NP_{t,e}$ is the non-electricity energy price coefficient. $PN_{t,e}$ denotes the non-electricity supply. EC_t denotes the energy cost.

2.3 Constraints

The energy supply of the energy balance sub-model has to satisfy the energy demand of the macroeconomic sub-model, so that the energy supply and demand constraints can be described as:

$$\sum_{e \in ET} PE_{t,e} + \sum_{e \in NT} PN_{t,e} \geq (E_t + N_t) \times \prod_{\tau=0}^{t-1} (1 - \omega_\tau)^5 \qquad \text{(Question.14.11)}$$

where τ is the Autonomous Energy Efficiency Improvement (AEEI) factor.

2.4 Objective Function

Following the previous approach, for each region, there is a representative producer-consumer. The savings decision is modeled by choosing a consumption sequence for each region that maximizes the sum of the discounted "utilities" of consumption. To optimize the investment and consumption patterns over successive time periods, the utility function is the logarithm of consumption.

$$UTIL = \sum_{t=1}^{T} \left(5 \times \prod_{t=1}^{t-1} (1 - udr)^5 \times \log(C_t) \right) \qquad \text{(Question.14.12)}$$

where $UTIL$ is discounted utility. C is Annual consumption. Udr denotes Utility discount rate.

3 Energy Technology Scenarios

To explore pathways for decarbonizing energy systems in smart communities, this study considers several widely accepted strategies, including phasing out coal-fired power, electrification, improving building energy efficiency, deploying on-site renewables, importing green energy, and advancing emerging technologies such as hydrogen and carbon capture and storage (CCS). Based on these strategies, three scenarios are developed, as shown in Table 1. The All Technologies Accessible (ATA) scenario represents an optimistic pathway where all technologies are assumed to be available by 2050. The Without New Technologies (WNT) scenario reflects a conservative approach that relies only on currently existing technologies, without contributions from emerging options. The Self-Sufficiency (SS) scenario assumes the community does not rely on external energy imports and must meet all demand locally.

4 CO_2 Mitigation Target

In this study, CO_2 emissions are estimated using the carbon emission factors corresponding to each energy source and the amount of energy consumed. Table 2 summarizes the assumed net CO_2 emission limits under the scenarios. According to Japan's revised Nationally Determined Contribution (NDC), total greenhouse gas emissions are to be reduced by 46% from 2013 levels by 2030, with CO_2 emissions from energy use reduced by 45% over the same period. Moreover, previous studies have suggested that achieving carbon neutrality by 2050 would likely require the power sector to fully decarbonize even earlier, potentially by 2040. Building on these insights, this study adopts more stringent targets, assuming a 50% reduction in CO_2 emissions by 2030. To meet these ambitious goals, the scenarios also assume a rapid decarbonization phase between 2025 and 2035.

5 Macroeconomic Scenarios

Projections of economic and energy trends rely on consistent population scenarios. In this study, annual changes in production, capital stock, and labor are derived from Japan's JPNSSP population scenarios, combined with assumptions on total factor productivity (TFP) (Honjo et al., 2021). The JPNSSP provides long-term socio-economic projections for climate impact assessments, with city-level population data available from 2015 to 2100. The statistical assumptions underlying the population scenarios are illustrated in Fig. 2.

TFP reflects the portion of economic output growth not attributable to additional inputs of capital, labor, or energy. This study adopts two TFP pathways—high and low—based on previous literature, as shown in Fig. 2d. Under the high TFP scenario, average productivity in the commercial sector reaches 3.15 in 2030 and 3.35 in 2050. In contrast, the low TFP scenario yields 2.91 in 2030 and 2.98 in 2050. By combining five population scenarios (JPNSSP1–JPNSSP5) with the two TFP assumptions, ten distinct socio-economic scenarios are constructed (Fig. 2c).

To ensure the model's results are broadly applicable, regional socio-economic parameters are aggregated from projections for Japan's 47 prefectures. Figure 3 presents the resulting trends in production, capital stock, labor, and energy demand. Across all scenarios, Japan's population peaks around 2020. JPNSSP3, with the lowest fertility rate, sees the steepest decline—approximately 17% below 2015 levels

Table 1 Energy technical scenarios and technology availability

Scenario	Energy efficiency measures in buildings			On-site renewable energy			Import of green energy		New energy technologies	
	Lighting	Heating/cooling	Other appliances	On-site PV	On-site Wind	Biomass	Imported PV	Imported Wind	Hydrogen	CCS
All technologies-accessible (ATA)	√	√	√	√	√	√	√	√	√	√
Without new technologies (WNT)	√	√	√	√	√	√	√	√	×	×
Self-sufficient (SS)	√	√	√	√	√	√	×	×	√	√

Table 2 Net CO_2 emissions

Year	2010	2015	2020	2025	2030	2035	2040	2045	2050
CO_2 emissions [Kt-CO_2/yr.]	94.15	89.60	80.89	66.42	47.14	27.85	13.38	4.67	0
%	100%	95%	86%	70%	50%	29%	14%	5%	0%
Year	2010	2015	2020	2025	2030	2035	2040	2045	2050

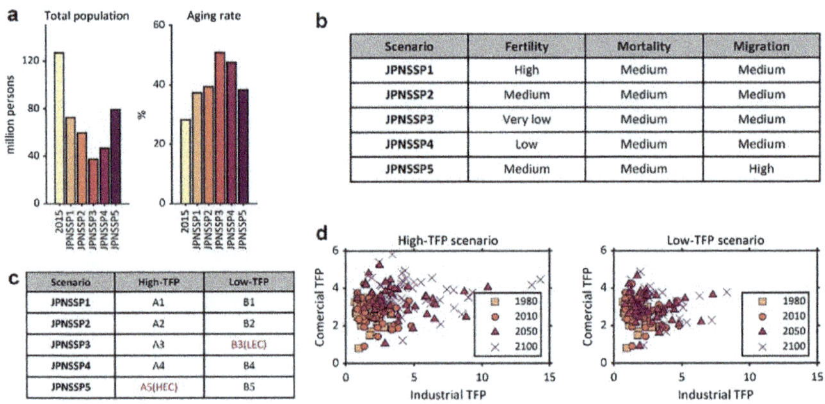

Fig. 2 Macroeconomic scenarios assumptions. a Prediction of Japan's total population and elderly population ratio in 2100 based on JPNSSPs; b the specific statistical assumptions of JPNSSPs; c ten combined scenarios for economic projection; d Growth of industrial and commercial TFP in the 47 prefectures of Japan

by 2050. Conversely, JPNSSP5, the most optimistic scenario, combines moderate fertility and increased immigration, limiting the decline to about 8%.

Japan's economic growth depends heavily on the commercial sector, where TFP is assumed to gradually improve. Nonetheless, population aging and decline constrain labor supply, making sustained production growth challenging and potentially leading to a gradual output decline. Regarding energy demand, although Japan peaked its carbon emissions in 2013, recent business sector growth has kept demand resilient. In the most optimistic scenario (A5), energy demand peaks around 2040 and climbs to 1.25 times the 2015 level by 2050. In the most pessimistic scenario (B3), demand peaks in 2020 and then falls by 18% by 2050. For subsequent analysis, the A5 scenario is designated as the high-energy consumption (HEC) case, while B3 serves as the low-energy consumption (LEC) case, to assess the contributions of current and emerging technologies toward long-term carbon reduction goals.

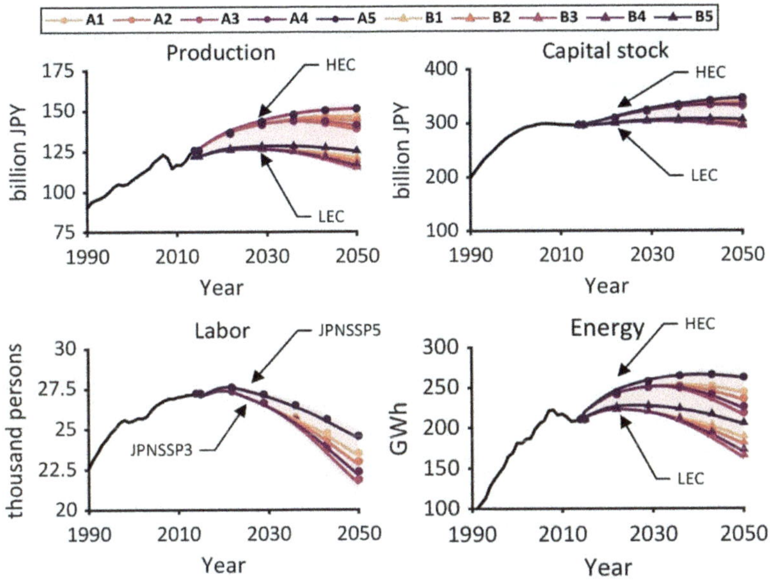

Fig. 3 Production, capital stock, labour population, and energy demand of the select commercial community, 1990–2050. Economic value is measured in constant 2000 JPY (Annual fluctuations are derived from the average of 47 prefectures in Japan)

6 Carbon Emissions

6.1 Early and Middle-Stage Decarbonization: Efficiency and Renewables

Figure 4 illustrates the dynamic contributions of different measures to CO_2 emission reductions under six scenarios. In the early stage of decarbonization (before 2020), building efficiency measures emerge as the primary driver. Thanks to their low retrofit costs and direct impact on energy demand, building efficiency alone meets the initial reduction targets and continues to contribute about 22% of the total abatement potential by 2050 — with heating and cooling accounting for 9.6%, other appliances 8.3%, and lighting 4.1%. On-site renewable energy, mainly rooftop photovoltaics (PV) and small wind turbines, plays a supporting role. However, their deployment is constrained by available space within the community. In the high-energy consumption (HEC) scenarios, on-site PV and wind contribute 5.9% and 1.4%, respectively, by 2050. In low-energy consumption (LEC) scenarios, where total demand is lower, their relative contributions rise to 9.6% and 2.3%, respectively.

In the middle stage, as on-site renewables approach their spatial limits, imported utility-scale PV and wind power become increasingly important in the four scenarios that allow imports (HEC-ATA, HEC-WNT, LEC-ATA, LEC-WNT). Imported PV,

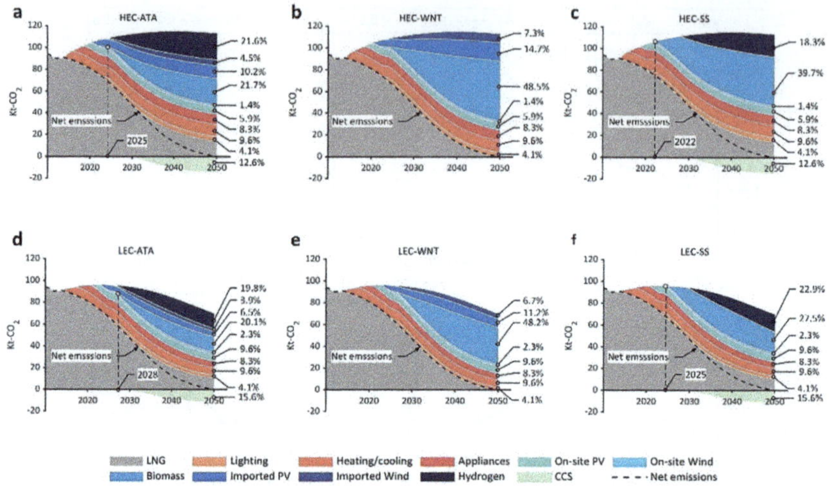

Fig. 4 Energy-related CO_2 emissions and carbon neutrality target realization pathways from 2010 to 2050

benefiting from economies of scale and lower costs, can even replace on-site wind as early as 2021. However, the growth of imported PV and wind remains constrained over time due to intermittency challenges and the added costs of storage and balancing measures, which are addressed in Sect. 3.3. By 2050, imported PV and wind together contribute up to 14.7% and 7.3% of abatement potential in the HEC-WNT scenario, and 11.2% and 6.7% in the LEC-WNT scenario, where no new technologies are assumed.

6.2 *Late-Stage Decarbonization: Dispatchable Low-Carbon Options*

In the late stages of decarbonization, dispatchable low-carbon energy sources — biomass, hydrogen, and fossil fuels with CCS — become the backbone of the power system, providing between 48.2% and 70.5% of the total abatement potential by 2050. Their flexibility is essential to compensate for the intermittency of solar and wind. Among them, hydrogen shows significant promise as a dispatchable and scalable option, although its large-scale use depends on the development of a stable, low-cost supply chain. In the absence of affordable hydrogen and proven CCS technology, biomass becomes even more critical. In extreme scenarios without new technologies (HEC-WNT and LEC-WNT), biomass alone is required to supply nearly half of the total abatement, highlighting its role as a base-load resource. This underscores the urgent need to establish reliable biomass supply chains, which may include importing resources from abroad.

An important finding of this analysis is the transitional role of imported green power from utility-scale PV and wind farms. While not dominant in the long term, imported renewables effectively bridge the gap between low-cost on-site energy and the higher-cost options of biomass and hydrogen. For example, in the HEC-ATA scenario, after on-site PV reaches saturation, imported PV becomes the most cost-effective way to meet additional reductions until biomass ramps up after 2025. Conversely, in the no-import scenario (HEC-SS), biomass must contribute much earlier, even before 2022. This suggests that the period from 2022 to 2025 is a critical window for establishing biomass supply chains and driving down costs to support the final phase of decarbonization.

6.3 Renewable Energy Limits and Energy Mismatch

Our findings reveal a notable divergence from many existing studies: the decarbonization potential of solar and wind energy is significantly lower than often reported. While previous research suggests that renewable energy could contribute over 50% of decarbonized power generation by 2050, our analysis estimates their share to be only around 20–30%. This discrepancy stems from the economic and technical challenges of integrating large shares of variable renewables in the later stages of decarbonization, even under optimistic assumptions about future costs and technology readiness.

As renewable penetration increases, the issue of energy mismatch becomes more pronounced, and the cost advantages of PV and wind diminish. Achieving high levels of grid flexibility requires substantial investment in storage technologies to balance intermittent supply and ensure reliable power. The added costs of these storage solutions offset the initial price benefits of solar and wind, making more flexible options—such as biomass and hydrogen—more competitive in the long run.

7 Insights

Achieving carbon neutrality by 2050 is undoubtedly a formidable challenge, yet our results indicate that it remains within reach. Notably, the optimal deployment of any given technology is highly dependent on the status and progress of others, reflecting the competitive dynamics among lower-cost options. In essence, decarbonization unfolds as a dynamic competition between technologies, with their adoption shaped by regional resource potential, cost trajectories, and governmental policy preferences.

In light of these interdependencies and uncertainties, several robust and actionable insights emerge. First, energy efficiency measures—particularly in buildings—should be prioritized at every stage of the transition. These measures are low-cost, deliver immediate reductions in energy demand, and set the stage for deeper decarbonization. Strengthening public guidance and incentives to foster low-carbon behaviors among businesses and households can further amplify the impact, embedding green practices into everyday life and making them the societal norm.

Between approximately 2022 and 2028, lower-cost renewable energy sources—primarily solar and wind—can play a dominant role in supplying clean energy. During this phase, the additional costs associated with their intermittency remain manageable, and short-term energy storage systems can effectively mitigate mismatches between supply and demand. However, as the transition progresses into later stages, the limitations of intermittent renewables become more pronounced, and the system increasingly relies on flexible, dispatchable energy options. In this context, imported hydrogen, biomass, and fossil fuels with CCS emerge as indispensable complements, compensating for the seasonal and temporal variability of renewables and ensuring reliability.

It is also important to recognize that long-term uncertainties—such as future population trends, economic growth rates, and other macroeconomic factors—introduce significant unpredictability into energy demand projections. These uncertainties can influence not only the optimal technology mix but also the pace and sequencing of the transition. Addressing them requires a two-pronged approach. On the one hand, advancing reliable forecasting techniques can improve our ability to anticipate and adapt to changing conditions, though this remains technically challenging. On the other hand, developing a diverse portfolio of technologies and exploring a broad set of plausible energy scenarios can enhance resilience, ensuring that the energy system remains robust in the face of unforeseen developments.

In sum, while no single pathway can be prescribed with certainty, a flexible, phased approach—grounded in efficiency, leveraging renewables when cost-effective, and ensuring dispatchable capacity through hydrogen, biomass, and CCS—offers a practical and resilient blueprint for achieving carbon neutrality by 2050.

8 Conclusion

This study developed an improved G-CEEP model by embedding detailed energy technology modules into the traditional framework, enabling a more granular and dynamic evaluation of mitigation pathways toward carbon neutrality. The model was applied to assess energy consumption trajectories to 2050, examining the characteristics and economic implications of different technologies over time. This approach offers a unique perspective, highlighting not only the overall feasibility of net-zero targets but also the relative priorities and trade-offs of various measures at each stage of the transition. By explicitly modeling the interactions and temporal dynamics of multiple technologies, the study provides valuable insights into how communities can effectively sequence and coordinate decarbonization efforts.

The findings indicate that while the pathway to carbon neutrality remains highly challenging, it is nonetheless achievable if communities adopt focused and adaptive measures tailored to the evolving context of each phase. In the early stages, energy efficiency improvements—particularly in buildings—along with on-site solar photovoltaics and wind turbines, emerge as cost-effective and impactful actions. During the intermediate phase, between roughly 2022 and 2028, imported utility-scale solar and wind power play a pivotal transitional role, bridging the gap between low-cost on-site generation and more expensive dispatchable options. This period is also identified as a critical window for establishing robust biomass supply chains and driving down costs in preparation for the later stages of the transition.

The results also reveal that absorbing 20–30% penetration of solar and wind in urban communities is feasible, provided sufficient investment is made in energy storage technologies, particularly for long-duration applications. Otherwise, greater reliance on flexible resources such as biomass and hydrogen may be necessary to maintain system balance. Furthermore, leveraging surplus renewable energy for hydrogen production emerges as a promising strategy, as it not only provides a cost-effective long-term storage medium but also facilitates higher levels of renewable deployment by alleviating the constraints of seasonal storage.

In conclusion, achieving carbon neutrality by 2050 requires a phased, adaptive approach that balances immediate, low-cost opportunities with long-term investments in flexible and scalable technologies. The insights derived from this improved modeling framework highlight the importance of prioritizing efficiency, strategically deploying renewables, and preparing enabling infrastructure for biomass, hydrogen, and storage systems. By addressing the dynamic interplay of these elements, communities can navigate the complexities of the transition and realize a sustainable, carbon-neutral future.

Author's note: This chapter is a revised and expanded version of the author's article,"Energy transition roadmap towards net-zero communities: A case study in Japan", originally published in *Sustainable Cities and Society*, Vol. 100, (2024).

References

Akimoto K, Tomoda T, Fujii Y, Yamaji K (2004) Assessment of global warming mitigation options with integrated assessment model DNE21. Energy Econ 26(4):635–653

Akimoto K, Yamada K Research activities in systems analysis group. Feature Acquisition of New Large National Projects and Future Research and Development Prospects 25

Caineng Z, Minjie L, Feng M, Hanlin L, Zhi Y, Yichao Y, Chunxiao G, Ying W, Hao Y, Ping Y (2024) Development, challenges and strategies of natural gas industry under carbon neutral target in China. Pet Explor Dev 51(2):476–497

Capros P, Van Regemorter D, Paroussos L, Karkatsoulis P, Fragkiadakis C, Tsani S, Charalampidis I, Revesz T, Perry M, Abrell J (2013) GEM-E3 model documentation. JRC Scientific and Policy Reports 26034

Feijoo F, Prina MG, Mimica M, Duić N (2024) Multi-scale energy planning for the global transition: local, regional, and global insights. In: In, vol 10. Elsevier, p 100841

Fishbone LG, Abilock H (1981) Markal, a linear-programming model for energy systems analysis: technical description of the bnl version. Int J Energy Res 5(4):353–375

Honjo K, Gomi K, Kanamori Y, Takahashi K, Matsuhashi K (2021) Long-term projections of economic growth in the 47 prefectures of Japan: an application of Japan shared socioeconomic pathways. Heliyon 7(3)

Kemfert C (2002) An integrated assessment model of economy-energy-climate-the model Wiagem. Integr Assess 3(4):281–298

Kiang N, Schipper L (1996) Energy trends in the Japanese transportation sector. Transp Policy 3(1-2):21–35

Kii M, Isikawa R, Kometani Y (2023) Toward a carbon neutral urban transportation system in Japan. IATSS research 47(2):171–178

Kroposki B, Johnson B, Zhang Y, Gevorgian V, Denholm P, Hodge B-M, Hannegan B (2017) Achieving a 100% renewable grid: operating electric power systems with extremely high levels of variable renewable energy. IEEE Power and energy magazine 15(2):61–73

Massrur HR, Niknam T, Aghaei J, Shafie-Khah M, Catalao JP (2018) Fast decomposed energy flow in large-scale integrated electricity–gas–heat energy systems. IEEE Transactions on Sustainable Energy 9(4):1565–1577

Ozawa A, Tsani T, Kudoh Y (2022) Japan's pathways to achieve carbon neutrality by 2050–scenario analysis using an energy modeling methodology. Renew Sust Energ Rev 169:112943

Pampanelli AB, Found P, Bernardes AM (2014) A lean & green model for a production cell. J Clean Prod 85:19–30

Seo SN (2017) Beyond the Paris agreement: climate change policy negotiations and future directions. Reg Sci Policy Pract 9(2):121–141

Su X, Zhou W, Sun F, Nakagami KI (2014) Possible pathways for dealing with Japan's post-Fukushima challenge and achieving CO_2 emission reduction targets in 2030. Energy 66:90–97

Sylla NS, Koddenbrock K (2025) Delinking and global reparations: alternative perspectives on the socio-ecological crisis. transcript Verlag

Tsai M-S, Chang S-L (2015) Taiwan's 2050 low carbon development roadmap: an evaluation with the MARKAL model. Renew Sust Energ Rev 49:178–191

Zhang Z (2010) China in the transition to a low-carbon economy. Energy Policy 38(11):6638–6653

Zhou W, Qian X, Nakagami K (2021a) East Asian low-carbon community: realizing a sustainable decarbonized society from technology and social systems. Springer, Cham, pp 61–81

Zhou W, Qian X, Nakagami K (2021b) East Asian low-carbon community: realizing a sustainable decarbonized society from technology and social systems. Springer, Cham, pp 83–101

Scenario Analysis of Energy Transition Pathways in Japan

You Li and Weisheng Zhou

1 Introduction

1.1 Background and Motivation

The imperative for decarbonization has never been more pressing, as climate change continues to challenge the sustainability, security, and equity of global energy systems. The sixth assessment report of the Intergovernmental Panel on Climate Change (IPCC) underscores that achieving net-zero greenhouse gas emissions by mid-century is essential to limit global temperature rise to within 1.5 °C (Field et al. 2018; Misal et al. 2022). In this context, countries around the world are re-evaluating their energy strategies and accelerating the transition from fossil fuel dependency to clean and renewable energy sources (Kalair et al. 2021).

Japan, as the world's third-largest economy and a long-standing industrial powerhouse, faces a particularly complex challenge. On the one hand, it has pledged to achieve carbon neutrality by 2050 (Ohta 2021) and committed to a 46% reduction in greenhouse gas emissions by 2030 compared to 2013 levels, aligning its goals with the Paris Agreement. On the other hand, Japan's energy system is marked by high dependence on imported fossil fuels—particularly oil and natural gas—due to its limited domestic resources. The 2011 Fukushima Daiichi nuclear accident further reshaped public perception and policy on nuclear energy, significantly constraining one of the country's key low-carbon energy sources (Su et al. 2012). These constraints necessitate a more nuanced and regionally tailored approach to energy transition.

Y. Li
Asia-Japan Research Institute, Ritsumeikan University, Osaka, Japan
e-mail: liyou@fc.ritsumei.ac.jp

W. Zhou (✉)
College of Policy Science, Ritsumeikan University, Osaka, Japan
e-mail: zhou@sps.ritsumei.ac.jp

W. Zhou (ed.), *Decarbonization Policy Engineering*, World Sustainability Series,
https://doi.org/10.1007/978-981-95-5953-4_15

Moreover, the country is undergoing rapid demographic changes, including a declining and aging population, which will significantly influence future energy demand, labor force availability, and regional development dynamics (Kumagai 2014). Regional disparities in renewable energy potential—such as abundant wind resources in Hokkaido and strong solar radiation in Kyushu—suggest that Japan's energy transition must be planned not only nationally but also regionally, with attention to local resource endowments, technological capacities, and social acceptance.

Against this backdrop, scenario-based long-term energy planning emerges as a powerful tool to explore future transition pathways under multiple assumptions. Such planning allows policymakers, researchers, and stakeholders to examine the implications of different choices, trade-offs among technologies, and the timing of interventions. In particular, integrated modeling approaches can simulate how different sectors interact over time, highlight key bottlenecks, and support the design of balanced strategies that combine technological innovation with policy coherence.

This chapter responds to these challenges by developing a comprehensive scenario analysis for Japan's energy system transformation toward 2050. It builds on recent advances in integrated energy-environment-economic modeling and introduces a regionally disaggregated perspective that accounts for spatial heterogeneity and inter-regional energy flows. By capturing both national goals and local realities, this work aims to contribute to more informed, inclusive, and effective pathways toward Japan's carbon-neutral future.

1.2 National Energy and Climate Strategies

Japan's national energy strategy has undergone significant transformation in recent decades, shaped by a combination of global climate commitments, domestic energy security challenges, and public sentiment toward different energy sources (Sergeevna 2021). At the core of this strategy is the Basic Energy Plan, a policy document periodically revised by the Ministry of Economy, Trade and Industry (METI) to reflect evolving technological, geopolitical, and socio-economic conditions (Nakai 2024). The latest version, published in 2021, explicitly aligns Japan's long-term energy direction with its commitment to achieving carbon neutrality by 2050 and reducing greenhouse gas emissions by 46% by 2030 compared to 2013 levels.

The Basic Energy Plan articulates the "3E + S" principle—Energy security, Economic efficiency, Environmental protection, and Safety—as the foundation for Japan's energy policy (An et al. 2025). This principle seeks to balance reliable energy supply, cost competitiveness, climate responsibility, and public safety, particularly in light of the Fukushima Daiichi nuclear accident's enduring impact on public trust in nuclear power. As a result, Japan's nuclear future remains highly uncertain, placing additional pressure on other low-carbon options to fill the gap.

To meet its ambitious climate targets, Japan is now pursuing a multi-pronged strategy that involves a significant expansion of renewable energy deployment (Whittaker 2024). Solar photovoltaics and offshore wind power are prioritized as key contributors, with offshore wind seen as particularly promising in regions such as Tohoku and Hokkaido. Simultaneously, hydrogen and ammonia are positioned as next-generation energy carriers, especially for sectors where electrification is difficult. Japan has launched large-scale demonstration projects for hydrogen production and import, with the vision of establishing an international hydrogen supply chain involving partners such as Australia, Brunei, and Saudi Arabia (Wen and Aziz 2024).

Electrification is another strategic pillar, aiming to decarbonize energy demand in buildings, transportation (Rinawati et al. 2025), and industry. This entails not only replacing fossil fuel-based appliances and vehicles with electric alternatives but also ensuring that the electricity powering them is increasingly renewable. Smart grid technologies (Ida et al. 2024), digitalization, and decentralized energy systems are being promoted to enhance the flexibility and resilience of the power system, which is essential for accommodating the variable output of renewable sources.

Moreover, Japan's energy strategy now places growing emphasis on regional heterogeneity and local innovation (Onodera et al. 2024). Recognizing that renewable resource potential, industrial composition, and population dynamics differ markedly across the country's regions, METI and local governments are working to develop "regional decarbonization roadmaps." These roadmaps integrate local energy resource assessments, demographic trends, and industrial strengths to design tailored transition pathways. For instance, while Kyushu and Shikoku have abundant solar capacity, northern regions like Hokkaido are more suited to wind power and biomass. This spatial differentiation is seen as vital not only for technical feasibility but also for ensuring equitable development and community engagement.

In sum, Japan's national strategy reflects a growing sophistication in integrating climate targets with energy security, economic imperatives, and local realities. However, the effectiveness of this strategy depends heavily on the successful implementation of enabling technologies, timely investments, and continuous policy coordination across scales and sectors. Scenario-based analysis, therefore, plays a critical role in testing the robustness of these strategies under various assumptions, guiding adaptive policymaking in an uncertain future.

1.3 Global Landscape of Energy Transitions

The energy transition is a global imperative shaped by the dual challenges of climate change mitigation and sustainable development. Since the adoption of the Paris Agreement in 2015, countries around the world have intensified their efforts to reduce greenhouse gas emissions and restructure their energy systems toward net-zero targets (Le et al.). While the pathways and timelines differ by national context, common themes include increasing the share of renewable energy, electrifying end-use sectors, phasing out coal, improving energy efficiency, and investing in enabling infrastructure such as energy storage, smart grids, and green hydrogen.

European countries have been at the forefront of this transition. The European Union's "Fit for 55" package and the European Green Deal outline a legally binding framework to reduce emissions by 55% by 2030 and reach carbon neutrality by 2050 (Rodrigues 2024). Germany, in particular, has pursued the *Energiewende* (energy transition) for over two decades, aiming to eliminate nuclear power while rapidly expanding wind and solar power (Sacco et al. 2024). However, this model has also faced challenges such as grid congestion, regional imbalances, and the intermittency of renewables, prompting discussions around sector coupling, cross-border electricity trade, and long-duration storage.

The United States has re-entered the climate policy arena with renewed vigor under the Inflation Reduction Act (IRA) of 2022 (Bistline and Wolfram 2025), which allocates over $370 billion for clean energy and climate programs. The U.S. approach emphasizes market-based incentives, technological innovation, and domestic manufacturing of clean energy components. States such as California and New York have independently set aggressive renewable energy and net-zero targets, demonstrating the potential of sub-national actors in driving decarbonization.

In the Asia-Pacific region, China has emerged as both the largest emitter and the largest investor in clean energy (Li et al. 2024). Its "Dual Carbon" goals—peaking CO_2 emissions before 2030 and achieving carbon neutrality by 2060–are supported by massive investments in solar, wind, electric vehicles, and battery manufacturing. However, the country also faces significant challenges, including reliance on coal, regional disparities, and the need to manage industrial restructuring in carbon-intensive sectors. South Korea and Singapore have similarly committed to net-zero targets and are exploring hydrogen and digital energy technologies as core pillars of their transition.

Developing countries present a more complex landscape, where energy access and economic development often take precedence over emission reductions. Nevertheless, many countries in Latin America and Africa are leapfrogging to cleaner systems by harnessing abundant renewable resources such as hydropower, solar, and geothermal energy. The challenge here is less about technology and more about financing, institutional capacity, and aligning international support with local priorities (Kariuki 2025).

This global diversity of energy transition approaches highlights the importance of context-specific strategies. While high-income countries can often afford early and aggressive decarbonization, developing countries require pathways that align with development goals and financial realities. Furthermore, the geopolitical implications of the energy transition—such as new dependencies on critical minerals, shifts in trade flows, and competition over technological leadership—are reshaping international relations.

For Japan, the global energy transition offers both opportunities and lessons. The rise of international hydrogen markets, the declining costs of renewable energy technologies, and the emergence of flexible grid solutions all provide new avenues for energy security and innovation. At the same time, Japan's aging population, high urban density, and vulnerability to natural disasters present unique constraints. By situating its national strategies within the broader global context, Japan can better benchmark its progress, anticipate future risks, and identify areas for international collaboration—particularly in technology deployment, supply chain development, and climate finance.

1.4 Research Objectives of this Chapter

To support evidence-based policymaking and contribute to the academic understanding of Japan's energy transition, this chapter sets out to examine plausible mitigation pathways toward a carbon-neutral society by 2050. While national-level strategies and global benchmarks provide a directional vision, regional and temporal heterogeneities in resource availability, demographic trends, and technological feasibility necessitate a more granular approach. In this context, we employ an enhanced Glocal Century Energy Environment Planning (G-CEEP) model that integrates socioeconomic projections, renewable energy resource assessments, and energy technology modules to evaluate the interactions and sequencing of various decarbonization strategies across time and space.

The core objectives of this chapter are threefold. First, we aim to project Japan's future energy demand and supply scenarios under different socioeconomic and technological assumptions. This includes evaluating the impacts of population decline, aging demographics, and productivity shifts on long-term energy consumption patterns. Second, we assess the regional renewable energy potential across ten distinct geographic areas in Japan, highlighting spatial disparities and opportunities for local energy autonomy. Third, we examine the evolution of the electricity mix, energy mismatch challenges, and the role of flexible technologies such as storage, hydrogen, and carbon capture in achieving a balanced and resilient energy system.

2 Scenario Assumptions

2.1 Demographic Transitions and Economic Growth Assumptions

Japan's long-term energy transition cannot be separated from its evolving demographic and economic foundations. As shown in Fig. 1a, Japan's total population peaked around 2020 and is projected to decline steadily, reaching just above 100 million by 2050. At the same time, the aging rate continues to climb, with the share of the elderly population rising from approximately 28% in 2020 to over 35% by mid-century. This demographic shift signifies not only a reduction in total energy

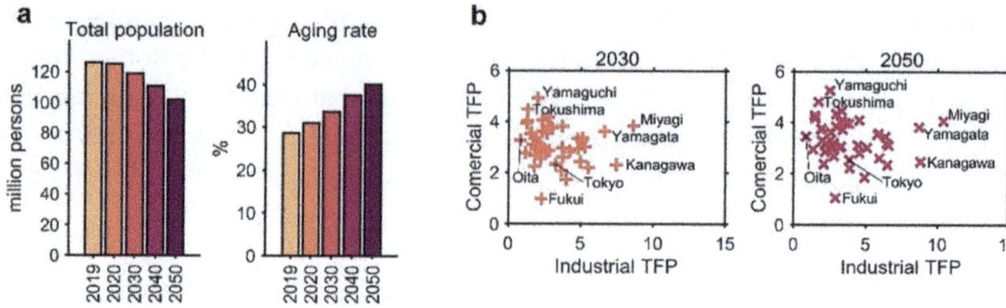

Fig. 1 Socioeconomic Projections of Japan and Regional Economic Productivity Distribution a Projected trends in total population (left) and aging rate (right) in Japan from 2019 to 2050, based on the JPN-SSP scenarios. b Spatial distribution of total factor productivity (TFP) across prefectures in 2030 (left) and 2050 (right), highlighting differences in industrial and commercial sector development

demand but also a transformation in sectoral energy use, especially in residential and commercial buildings where cooling, heating, and health-related demands become more dominant.

The analysis incorporates socioeconomic projections derived from the Japan Shared Socioeconomic Pathways (JPN-SSP), a set of long-term national scenarios aligned with global SSPs but tailored to Japan's demographic and policy contexts. The five JPN-SSP narratives differ in their assumptions on fertility, mortality, and immigration, leading to divergent population trajectories and structural changes. These variations serve as a critical input for downstream energy modeling, particularly in shaping future demand (Honjo et al., 2021).

In parallel, assumptions regarding Total Factor Productivity (TFP) are integrated to simulate differentiated economic growth potentials. As illustrated in Fig. 1b, considerable regional heterogeneity is observed in both industrial and commercial TFP, with Tokyo, Kanagawa, and Fukui showing relatively lower values compared to high-performing prefectures such as Miyagi, Yamagata, and Yamaguchi. Between 2030 and 2050, modest shifts in productivity are expected, but the divergence across regions persists. TFP encapsulates not only labor and capital efficiency but also structural economic transformation, technological innovation, and institutional capacity—all of which play pivotal roles in determining energy demand elasticity.

These socioeconomic assumptions—shrinking and aging population, and spatially varied economic productivity—highlight the importance of embedding regional differentiation and demographic-economic interactions in energy transition modeling. Without such granularity, policy prescriptions risk being overly centralized and insensitive to local conditions, particularly in regions facing acute depopulation or economic stagnation.

2.2 *Macroeconomic and Demographic Projections*

Understanding the future trajectory of Japan's energy system requires a thorough examination of its macroeconomic and demographic evolution. Based on the national scenarios outlined in the Japanese Shared Socioeconomic Pathways (JPN-SSPs), we integrated population, economic, and energy demand assumptions into the G-CEEP modeling framework to generate consistent projections through 2050. These projections provide the foundational assumptions for evaluating decarbonization pathways.

Japan's total population is projected to decline significantly—from approximately 126 million in 2020 to just over 100 million by 2050—while the aging rate continues to rise sharply, reaching over 40% by mid-century (Fig. 2a). This demographic shift not only reduces the size of the working-age population but also poses structural challenges to energy demand and productivity. The declining population is closely associated with a gradual contraction in residential and transportation energy use, especially in rural prefectures with accelerated depopulation and limited industrial renewal.

Complementary to this demographic trend, regional economic performance also exhibits notable shifts. Figure 2b compares prefecture-level total factor productivity (TFP) in commercial and industrial sectors in 2030 and 2050, highlighting a growing disparity between urban and rural areas. High-performing regions such as Tokyo,

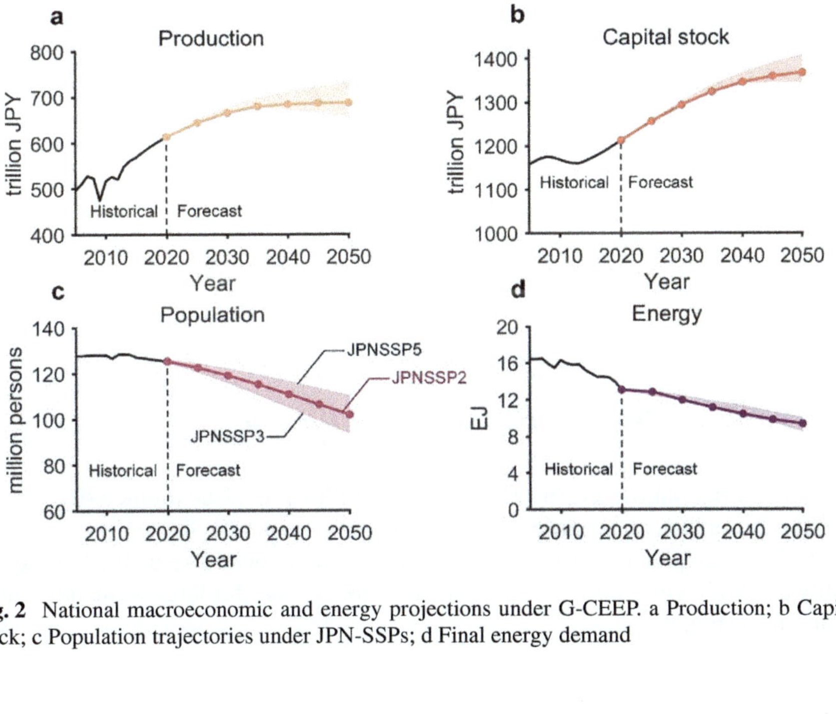

Fig. 2 National macroeconomic and energy projections under G-CEEP. a Production; b Capital stock; c Population trajectories under JPN-SSPs; d Final energy demand

Kanagawa, and Miyagi remain relatively stable, while others may experience stagnation, reflecting diverging regional capacities to invest in and absorb energy-related technologies.

To account for broader macroeconomic trends, the G-CEEP model integrates forecasted values for key indicators including national production, capital stock, population, and final energy demand under various scenario settings (Fig. 2). While capital and production are expected to continue moderate growth until 2050, they are accompanied by a significant contraction in population. These combined factors lead to a gradual decline in aggregate final energy demand from approximately 16 EJ in 2020 to below 10 EJ by 2050 under the median scenario. The narrowing gap between production and capital stock growth rates also reflects a transition toward more capital-efficient but demographically constrained economic structures.

Figure. 2 depict these interlinked trends. The solid lines indicate baseline trajectories produced by the G-CEEP model, while shaded bands represent uncertainties across different JPN-SSP pathways—particularly JPN-SSP2 (slow growth, rapid aging) and JPN-SSP5 (high growth, delayed demographic decline). These results underscore the complex interplay between demographic contraction, economic restructuring, and energy use, providing a consistent and regionally adaptable foundation for subsequent scenario analysis.

3 Regional Renewable Energy Potential in Japan

The geographical diversity of Japan offers a wide range of renewable energy potentials across its regions, from abundant wind on the northern coasts to high solar irradiance in the southern islands. Evaluating regional resource endowments is essential for designing spatially differentiated transition strategies, optimizing interregional energy flows, and guiding technology deployment. To this end, Japan is divided into ten energy planning regions, each with distinct profiles in terms of solar photovoltaic (PV), onshore and offshore wind, biomass, geothermal, and hydropower potential. These regional disparities directly affect the feasibility, cost-effectiveness, and sequencing of decarbonization efforts.

3.1 Northern Regions: Hokkaido and Tohoku

In Japan's northernmost regions, Hokkaido and Tohoku stand out for their rich renewable energy endowments, particularly in wind and solar power. These regions exhibit favorable geographical and meteorological conditions, with expansive land availability and high wind speeds, making them strategic areas for large-scale renewable deployment (Fig. 3).

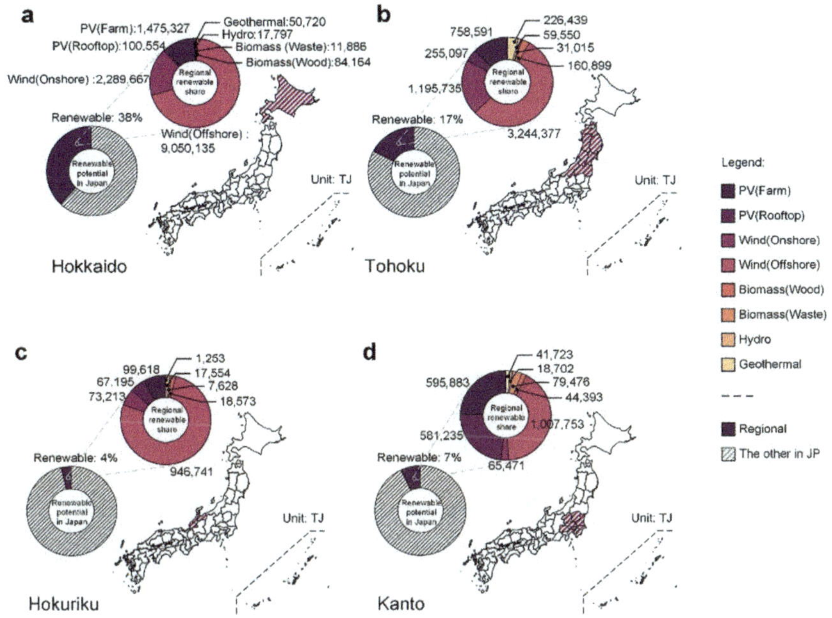

Fig. 3 Renewable energy potentials by source and regional share in a Hokkaido, b Tohoku, c Hokuriku, and d Kanto

Hokkaido possesses the highest regional share of renewable potential in the country, accounting for approximately 38% of the national total. The region is particularly endowed with offshore wind resources, totaling over 9050 PJ, which dominates its renewable portfolio. Onshore wind also contributes significantly, with an estimated 2290 PJ. In addition, Hokkaido exhibits strong potential in solar photovoltaic deployment (both farm-based and rooftop), geothermal sources, and biomass from wood and waste. Such a diverse resource mix positions Hokkaido as a cornerstone of Japan's long-term decarbonization strategy, although grid interconnection and seasonal demand variability remain critical challenges for full-scale integration.

Tohoku, adjacent to Hokkaido, also demonstrates notable renewable potential, though its share of the national total is lower, at approximately 17%. Like Hokkaido, offshore wind plays a leading role (3244 PJ), supported by onshore wind and solar PV. Biomass and geothermal resources are also present, though in smaller quantities. Despite its lower share, Tohoku's renewable output remains substantial in absolute terms and complements Hokkaido in supplying clean energy to the eastern part of Japan. These northern regions' robust wind capacity, especially offshore, aligns with Japan's national energy plans that emphasize offshore wind expansion. However, the full realization of this potential depends heavily on infrastructure investment, regional coordination, and technological advancement in long-distance power transmission.

3.2 Central Regions: Kansai and Chubu

Located in the heart of Honshu, Kansai and Chubu are Japan's major population and industrial centers, yet they exhibit relatively limited renewable resource endowments compared to northern and southern regions (Fig. 4). Despite their economic prominence, the constrained land availability and dense urban development in these regions restrict large-scale deployment of certain renewable technologies, particularly wind and utility-scale photovoltaics.

Kansai holds approximately 4% of Japan's total renewable energy potential. The region's renewable resource mix is dominated by onshore wind (approximately 669 PJ), followed by PV farm installations (around 275 PJ). Rooftop PV systems also contribute moderately, though biomass, hydro, and geothermal potentials are minimal. Kansai's highly urbanized landscape and relatively low natural resource productivity present challenges to achieving high levels of energy self-sufficiency through local renewables alone. Nonetheless, its central role in electricity demand and grid operation makes it a vital node in the national energy transition.

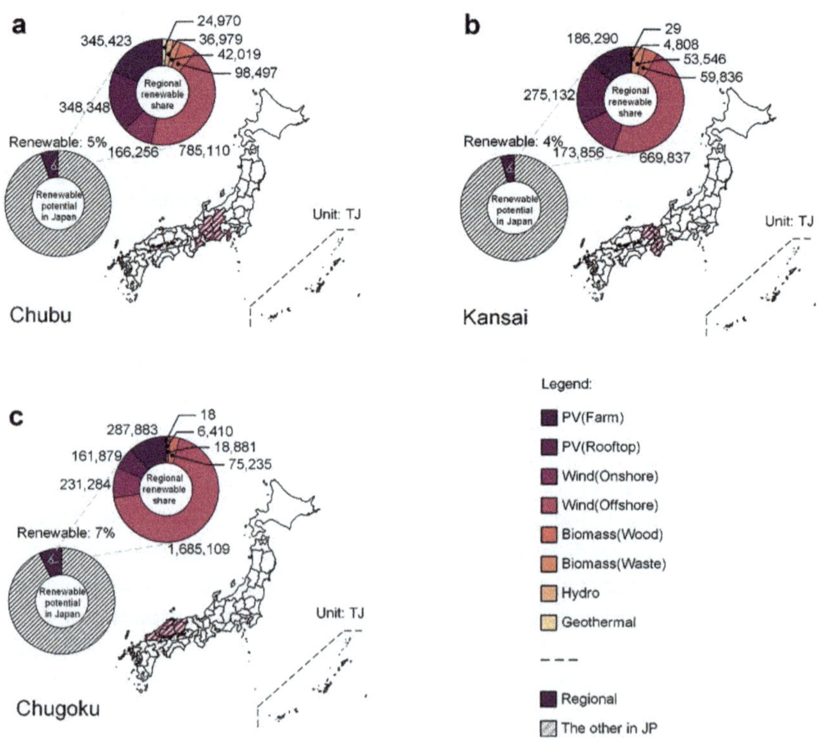

Fig. 4 Renewable energy potentials by source and regional share in a Chubu, b Kansai, and c Chugoku

Chubu, while slightly better endowed than Kansai, still represents a modest 5% share of the national renewable potential. The total estimated renewable resource in the region is around 1685 PJ, primarily composed of onshore wind (785 PJ) and PV farm capacity (345 PJ). Chubu also demonstrates non-negligible contributions from rooftop PV and biomass, while hydro and geothermal resources remain limited. As a major manufacturing and transportation hub, Chubu's decarbonization efforts will likely rely on a combination of localized efficiency improvements, rooftop solar expansion, and integration with energy-surplus regions such as Hokkaido and Tohoku.

Given their limited local renewable capacity, both Kansai and Chubu are expected to play pivotal roles in national-scale power balancing, interregional grid interconnection, and flexible demand-side management strategies. Their future contributions may focus less on energy production and more on innovation in system integration, smart grid deployment, and efficient electrification technologies.

3.3 Southern Islands: Kyushu, Shikoku, and Okinawa

The southern island regions of Japan—Kyushu, Shikoku, and Okinawa—demonstrate distinctive renewable energy potentials shaped by their geography and climate. These areas benefit from high solar irradiance, expansive coastlines suitable for wind development, and localized biomass and geothermal resources. While Kyushu holds substantial strategic importance due to its high share in the national renewable resource pool, Shikoku and Okinawa also contribute meaningfully within their geographic and demographic constraints (Fig. 5).

Kyushu stands out with a total renewable energy potential of approximately 2522 PJ, representing about 14% of the national total. Among the sources, onshore wind power (614 PJ) and PV farms (254 PJ) dominate, with rooftop PV (242 PJ) and biomass (totaling around 204 PJ) also contributing significantly. The region's potential for geothermal (70 PJ) and hydropower (33 PJ) is notable, offering a balanced mix of dispatchable and variable renewable sources. Its renewable profile positions Kyushu as a key region for energy transition, particularly due to its geographic proximity to East Asian trade and resource flows.

In contrast, Shikoku has a more moderate renewable potential of 950 PJ, accounting for about 4% of the national share. The majority of its resources come from onshore wind (183 PJ) and PV farms (93 PJ), while biomass, hydro, and geothermal potentials are relatively small. The compact size of Shikoku limits its development area but allows for localized grid flexibility and distributed PV applications.

Okinawa, Japan's southernmost prefecture, presents a more limited renewable energy profile with a total potential of 435 PJ, comprising just 1.5% of the national potential. Despite its high solar resource availability, the island's small land area and remoteness constrain large-scale deployment. PV farms (52 PJ) and onshore wind (34 PJ) form the core of its potential, with marginal contributions from biomass and

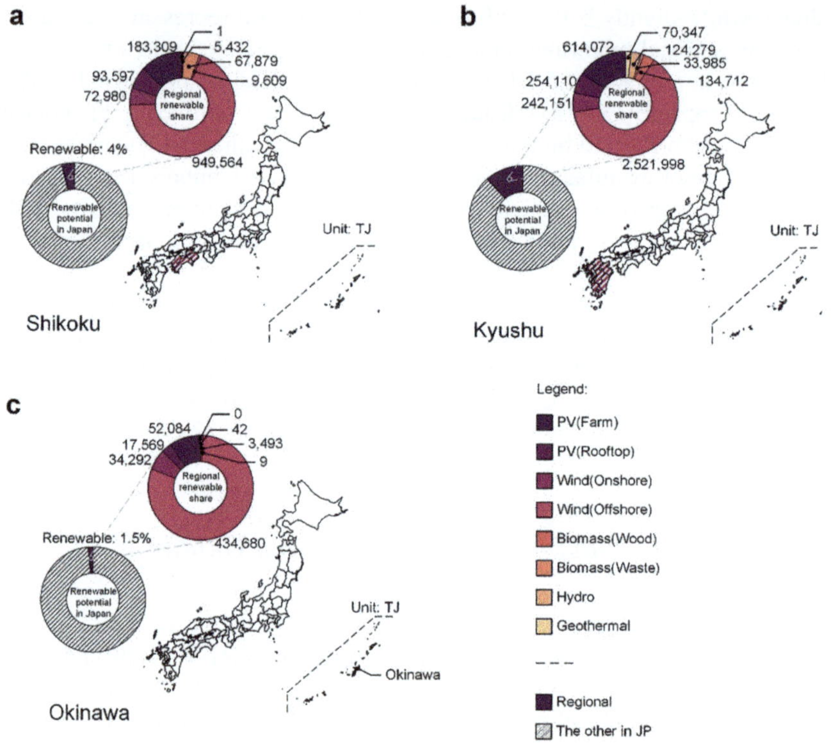

Fig. 5 Renewable energy potentials by source and regional share in a Shikoku, b Kyushu, and c Okinawa

rooftop PV. Nonetheless, Okinawa's case highlights the importance of tailored energy strategies for remote islands, including microgrids, battery storage, and marine renewables.

4 Energy Supply and Consumption Projections toward 2050

4.1 Energy Supply Projections

Japan's energy supply landscape is expected to undergo a profound transformation over the next three decades, driven by decarbonization targets and technological advancement. In 2019, the primary energy supply was dominated by fossil fuels, with oil, coal, and gas together accounting for more than 75% of the total energy input. However, by 2050, this structure is projected to shift significantly toward low-carbon and renewable sources.

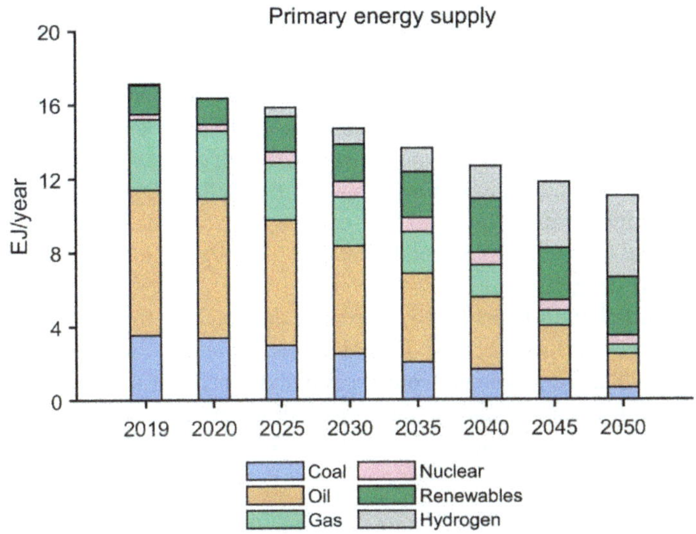

Fig. 6 Projected changes in Japan's primary energy supply by source (2019–2050), highlighting the decline in fossil fuels and the rise of hydrogen and renewables

The contribution of coal and oil is expected to decrease sharply. Coal supply, which stood at around 3.8 EJ/year in 2019, will be nearly phased out by 2050, while oil consumption will decline from over 7 EJ/year to less than 2 EJ/year. Natural gas will also see a substantial decline, although to a lesser extent, making room for cleaner alternatives (Fig. 6).

Conversely, renewable energy and hydrogen are projected to play increasingly central roles. Renewable sources such as solar, wind, and biomass are projected to surpass natural gas and oil by the early 2040s, reaching around 4 EJ/year by 2050. Even more notably, hydrogen will become a cornerstone of Japan's future energy system, expanding from a marginal presence in 2020 to over 4 EJ/year by 2050, thereby equaling or surpassing renewables in primary supply. This is consistent with Japan's strategic roadmap for a "hydrogen society."

The transition also reflects a growing reliance on nuclear energy, which maintains a relatively stable but modest contribution throughout the period. Overall, the energy supply system in 2050 is expected to be characterized by diversification, decentralization, and deep decarbonization, laying the groundwork for achieving net-zero CO_2 emissions.

4.2 Energy Demand Projections

In parallel with the anticipated transformation on the supply side, Japan's final energy demand is projected to decline steadily over the coming decades, reflecting improvements in energy efficiency, electrification, and structural economic shifts. In 2019, the country's final energy consumption was approximately 13 EJ/year, with oil products dominating the energy mix, followed by electricity and coal-derived fuels. However, by 2050, the total demand is expected to fall below 9 EJ/year, indicating a substantial contraction of over 30% (Fig. 7).

This downward trend is primarily driven by aggressive efficiency gains and the decarbonization of end-use sectors. The most significant reductions are observed in fossil-derived fuels. Oil products, which accounted for the largest share in 2019, are projected to diminish significantly, with residual usage limited to hard-to-electrify sectors such as long-haul transport and heavy industry. Similarly, coal products and town gas will see near-total phaseout by 2050, aligning with Japan's climate ambitions.

Meanwhile, electricity is expected to take on an increasingly prominent role, becoming the backbone of end-use energy services by mid-century. Its share in final consumption rises steadily as sectors like buildings and transport undergo electrification. Notably, hydrogen also emerges as a vital carrier in the demand structure. From virtually zero in 2019, hydrogen consumption is projected to exceed 2 EJ/year by 2050, representing a crucial complement to electricity in decarbonizing sectors that are less amenable to direct electrification.

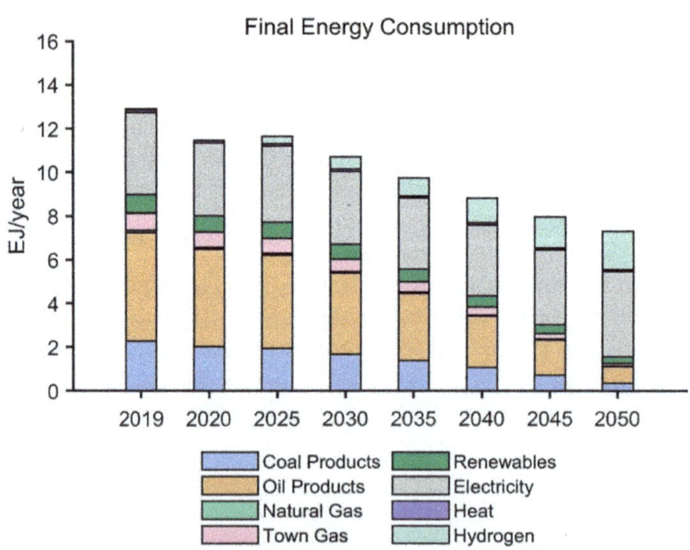

Fig. 7 Projected changes in Japan's final energy consumption by energy type (2019–2050), showing declining fossil fuels and the rise of electricity and hydrogen

Renewables in final energy form (e.g., biofuels and solar thermal) maintain a modest yet stable role, while direct heat usage declines due to the electrification of heating systems. Overall, the projections point to a leaner, cleaner, and more diversified energy demand system in Japan, characterized by a sharp reduction in fossil fuel reliance and a strategic shift toward electricity and hydrogen.

4.3 Electricity Mix Forecasting

Japan's electricity generation mix is projected to undergo a fundamental transformation over the next three decades, reflecting national decarbonization goals and the scaling-up of emerging technologies. In 2019, the total electricity output was under 900 TWh/year, primarily reliant on fossil-based generation, including oil, gas (without CCS), and a substantial share from nuclear and coal (Fig. 8).

By 2050, total generation is forecasted to exceed 1100 TWh/year, supported by increased electrification across sectors and the deployment of low-carbon generation technologies. Notably, the share of renewable energy, particularly solar PV and wind, expands dramatically. Solar PV becomes the dominant contributor by mid-century, with wind power (both onshore and offshore) rising steadily as a complementary source.

Hydrogen-based power generation, nearly negligible in 2020, surges after 2040 and becomes a key dispatchable source by 2050, occupying a significant fraction of the mix. This reflects Japan's strategic push for hydrogen as a firm power solution

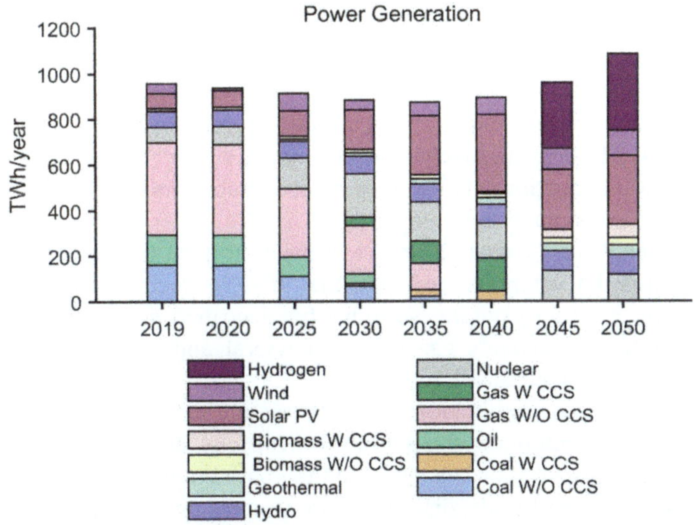

Fig. 8 Projected evolution of Japan's electricity generation mix by technology (2019–2050), highlighting the decline of fossil fuels and the emergence of solar, wind, and hydrogen

to balance variable renewables. Meanwhile, the use of coal and oil without carbon capture and storage (CCS) is phased out almost entirely by 2040, replaced in part by CCS-equipped thermal generation, including gas and limited coal CCS installations.

Nuclear power experiences a moderate rebound in the 2030s and stabilizes thereafter, contributing to both energy security and emissions reduction. Hydro and geothermal remain relatively constant in absolute terms but decline in proportional contribution due to overall expansion in total generation.

Overall, the Japanese power mix in 2050 is characterized by deep decarbonization, high penetration of renewables, firm low-carbon backstops like hydrogen and nuclear, and limited reliance on fossil-based CCS technologies. This diversified structure enhances both system resilience and climate alignment.

5 Policy Recommendations

In light of the projected transformations in Japan's energy demand, supply, and regional renewable potentials, several policy recommendations emerge to support an effective and equitable transition toward carbon neutrality:

First, regional differentiation and decentralization of energy policy should be emphasized. The analysis reveals that renewable energy potential varies significantly across Japan's ten regions. Areas such as Hokkaido and Kyushu possess abundant wind and solar resources, whereas metropolitan regions like Kansai and Chubu rely more on distributed and imported clean energy. Therefore, national energy strategies should be complemented by region-specific planning, supported by localized incentive structures and inter-regional grid infrastructure investments.

Second, hydrogen and renewable integration should be prioritized as core pillars of Japan's long-term energy security and decarbonization. As the role of hydrogen grows in both supply and end-use projections, early investment in production, storage, and transportation infrastructure—particularly for green hydrogen using regional renewables—will be critical. Likewise, updating market mechanisms (e.g., capacity markets, renewable auctions) is essential to encourage large-scale deployment of solar PV, wind, and storage technologies.

Third, demand-side innovation and electrification must be accelerated through building retrofits, EV adoption, and industrial process electrification. Policy tools such as dynamic electricity pricing, subsidies for electrified equipment, and building energy performance standards can drive behavioral and structural changes that reinforce the supply-side transition.

Fourth, cross-sector coordination and infrastructure readiness are indispensable. As final energy consumption shifts toward electricity and hydrogen, integrated planning across transportation, industry, and the built environment is needed. This includes deploying smart grids, district energy systems, and multi-vector hubs that cooperate electricity, heat, and fuel distribution networks.

Lastly, policy certainty and long-term signaling are vital to unlock private investment and innovation. A clear, enforceable roadmap—backed by climate legislation, carbon pricing, and periodic review mechanisms—will help align government, industry, and local communities toward common targets.

6 Conclusion

This chapter presented a comprehensive scenario analysis of Japan's energy transition pathways toward carbon neutrality, integrating national-level projections and regionally disaggregated data. Through the development and application of an enhanced G-CEEP modeling framework, the study reveals that Japan's decarbonization is not only technically feasible but also highly dependent on spatial, technological, and structural considerations.

Population dynamics and economic assumptions set the foundation for long-term projections of energy demand and supply, which highlight a substantial decline in fossil fuel dependence and a parallel rise in electricity and hydrogen as clean energy carriers. Meanwhile, regional assessments underscore the unequal distribution of renewable resources, emphasizing the need for tailored, location-specific strategies and infrastructure development.

The findings suggest that while technological solutions such as renewable energy, hydrogen, and CCS offer crucial leverage points, achieving net-zero emissions by 2050 will require coordinated efforts across multiple levels of governance, sectors, and regions. Importantly, the transition must be underpinned by proactive policy frameworks that incentivize innovation, ensure social equity, and enable regional empowerment.

As Japan navigates the complexities of deep decarbonization, this integrated scenario analysis provides critical insights for long-term energy system planning. It serves as a valuable reference for policymakers, planners, and researchers seeking to align national ambitions with practical implementation, ensuring that the pursuit of carbon neutrality remains both strategic and inclusive.

References

An M, Xu W, Wang X, He W, Fang X, Song M, Wang B (2025) Synergetic evolution of energy-economy-environment-society system: a case study of Chengdu-Chongqing urban agglomeration. China Energy Strat Rev 59:101755

Bistline JE, Wolfram C (2025) Inflation reduction act: origins, policy implications, and research gaps. Rev Environ Econ Policy 19(2):000–000

Field CB, Barros V, Stocker TF, Dahe Q, Dokken DJ, Ebi KL, Mastrandrea MD, Mach KJ, Plattner G-K, Allen SK (2018) IPCC, 2012: summary for policymakers: managing the risks of extreme events and disasters to advance climate change adaptation. In: Planning for climate change. Routledge, pp 111–128

Honjo K, Gomi K, Kanamori Y, Takahashi K, Matsuhashi K (2021) Long-term projections of economic growth in the 47 prefectures of Japan: An application of Japan shared socioeconomic pathways. Heliyon 7(3)

Ida T, Tanaka M, Ito K (2024) Smart grid economics. Adv Jpn Bus Econ

Kalair A, Abas N, Saleem MS, Kalair AR, Khan N (2021) Role of energy storage systems in energy transition from fossil fuels to renewables. Energy Storage 3(1):e135

Kariuki JM (2025) Nuevo Clima: energy in transition—the role of smart conventional energy systems, hydrogen, and Lithium-ion batteries in Latin America and the Caribbean's (LAC) energy future. In: Energia Progresiva: An intertemporal analysis of Latin America and the Caribbean's energy landscape in the industrial age. Springer, pp 575–611

Kumagai F (2014) Demographic changes in Japan. In: Family issues on marriage, divorce, and older adults in Japan: with special attention to regional variations. Springer, pp 11–38

Le VLT, Gupta R, Pham KD, Nguyen LH (n.d.) Financial inclusion, ICT development, and CO2 emissions: An ARDL approach. Sustain Dev

Li R, Fang D, Xu J (2024) Does China's carbon inclusion policy promote household carbon emissions reduction? Theoretical mechanisms and empirical evidence. Energy Econ 132:107462

Misal H, Hoare VH, Miles V (2022) Responding to the climate crisis–taking action on the IPCC 6th assessment report. Weather 00431656), 77(4

Nakai T (2024) Japan's triple sustainability challenge. Asia Pac Bus Rev 30(3):499–513

Ohta H (2021) Japan's policy on net carbon neutrality by 2050. East Asian Policy 13(01):19–32

Onodera H, Delage R, Nakata T (2024) The role of regional renewable energy integration in electricity decarbonization—a case study of Japan. Appl Energy 363:123118

Rinawati DI, Keeley AR, Itsubo N, Managi S (2025) Beyond carbon: environmental and social impacts of Japan's vehicle electrification transition. J Environ Manag 383:125509

Rodrigues AA (2024) "Fit for 55": the EU plan for a green transition. In: Sustainable finances and the law: between public and private solutions. Springer, pp 333–344

Sacco RL, Megre M, de Medeiros Costa HK, Brito TLF, dos Santos EM (2024) Energy transition policies in Germany and the United Kingdom. Energy Res Soc Sci 110:103460

Sergeevna PZ (2021) Energy strategy and transition to green energy in Japan. Russ JpnOlogy Rev 4(2):29–53

Su X, Zhou W, Nakagami KI, Ren H, Mu H (2012) Capital stock-labor-energy substitution and production efficiency study for China. Energy Econ 34(4):1208–1213

Wen D, Aziz M (2024) Perspective of staged hydrogen economy in Japan: a case study based on the data-driven method. Renew Sust Energ Rev 189:113907

Whittaker DH (2024) Japan's quest for a sustainable, virtuous circle of growth and innovation. Asia Pac Bus Rev 30(3):451–466

Low-Carbon Transition of City Gas Companies in Japan and China: Evaluation and Cooperation Opportunities

Ze Ran and Weisheng Zhou

1 Introduction

Natural gas is considered a cleaner energy than coal and oil, but natural gas is still a type of fossil fuel. Burning natural gas emits carbon dioxide, but if methane—the main component of natural gas—leaks into the atmosphere, it has a much higher Global Warming Potential (GWP) than carbon dioxide, resulting in an even greater impact on global warming. It is important to decarbonize the natural gas industry to achieve carbon neutrality. City gas companies are an important part of the natural gas value chain. City gas companies purchase natural gas (and LPG, Liquefied Petroleum Gas, in some companies' case) from upstream natural gas producers, typically big oil and gas companies; distribute natural gas through pipelines or tank trucks; and sell natural gas to end users, including residential, commercial and industrial customers. City gas companies own infrastructure such as distribution pipelines in general, and transmission pipelines and LNG terminals in some cases. Since more and more city gas companies have become integrated energy companies, some of them also own power plants, including gas-fired power plants and renewable power plants such as solar power stations and wind farms. To prevent the assets owned by city gas

Z. Ran (✉)
Graduate School of Policy Science, Ritsumeikan University, Osaka, Japan
e-mail: ps0697px@ed.ritsumei.ac.jp

W. Zhou
College of Policy Science, Ritsumeikan University, Osaka, Japan
e-mail: zhou@sps.ritsumei.ac.jp

W. Zhou (ed.), *Decarbonization Policy Engineering*, World Sustainability Series,
https://doi.org/10.1007/978-981-95-5953-4_16

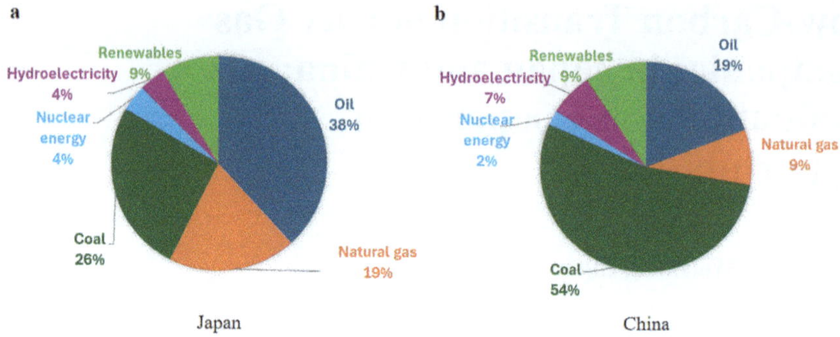

Fig. 1 The energy consumption mix of Japan and China in 2023. **a** Japan; **b** China

companies from becoming stranded assets,[1] it is essential for city gas companies to develop low-carbon transition strategies.

Japan and China are major consumers and importers of natural gas. China is the world's fourth largest natural gas producers, while Japan imports almost all of its natural gas from overseas. Both Japan and China have a large city gas business, which provides a basis for comparison. Based on the calculation of energy data of Japan and China (Energy Institute, 2024), Japan has a larger share of natural gas than China in the energy consumption mix, and more than half of China's energy consumption still comes from coal, as shown in Fig. 1.

Japan consumed 99.2 billion cubic meters (bcm) of natural gas in 2024, and the natural gas consumption breakdown by sector is as follows: 65% for power plants, 11% for industry, 20% for residential, services and agriculture, and 4% for other sectors (Enerdata, 2024). In Japan, natural gas is mainly used to generate power and gas-fired power plants are mainly owned by power companies, but city gas companies also own a small share of power plants. In comparison, China consumed 394.5 bcm of natural gas in 2023, and the natural gas consumption breakdown by sector is as follows: 33% for city gas,[2] 42% for industrial fuel, 17% for power plants, 8% for chemicals and fertilizers (NEA, 2024). Due to China is abundant in coal and the technology of gas turbines in China is not as advanced as Japan, natural gas is not mainly used to generate power in China. Whether in Japan or China, city gas companies primarily provide services to residential and commercial end users, but they may also own other businesses, such as power plants.

[1] Stranded assets refer to assets that were once considered valuable but may suffer significant loss in economic value due to changes in external factors such as market conditions or social and environmental circumstances. In particular, fossil fuel-related assets are frequently at risk of becoming stranded due to environmental regulations, technological innovation, and shifts in economic conditions.

[2] In China, consumption in residential, commercial services, etc., is generally categorized as "city gas" in the natural gas consumption mix. However, large city gas companies in China may also own other businesses such as power plants.

Three major city gas companies in Japan and five major city gas companies in China will be studied and compared in this paper, hoping that the result will shed light on the future development of both Japanese and Chinese city gas companies. Although there are around 210 city gas companies in Japan, Tokyo Gas, Osaka Gas and Toho Gas are the three largest, accounting for approximately 80% of Japan's nationwide city gas sales in terms of volume (*Tokyo Gas,* 2022).

There are more than 3600 city gas companies in China, but there are five major city gas companies, namely Kunlun Energy Company Limited (Kunlun Energy), the Hong Kong and China Gas Company Limited (Towngas), China Gas Holdings Limited (China Gas), ENN Energy Holdings Limited (ENN Energy), and China Resources Gas Group Limited (CR Gas). In 2022, these five companies accounted for 53% of China's total natural gas consumption (CGA, 2023). Since the supply areas of these Chinese city gas companies are scattered nationally, the supply areas are not listed here.

There are different names for the low-carbon transition plans, such as energy transition plans, carbon neutrality roadmaps, etc. In this paper, low-carbon transition strategies refer to any kind of these plans. Up to now, there are no comparative studies, especially quantitative comparison, between major Japanese city gas companies and major Chinese city gas companies in the literature. It should be noted that the strategies of the group companies will be compared among these companies, because generally it is the group company that releases the transitional strategies.

This paper designed a methodology to evaluate the low-carbon transition strategies of city gas companies for quantitative comparison. This paper first examines the overall low-carbon transition strategies of major city gas companies in Japan and China, and then examines specific strategies on hydrogen, e-methane, renewable energy, and methane emissions control, while the "S + 3E" principle is used when necessary. A comparison between Japanese and Chinese companies is made in each part.

2 Methodology

2.1 Existing Methodologies to Assess the Low-Carbon Transition Progress of Companies

There are many existing methodologies to disclose or assess the low-carbon transition progress of companies, such as the Task Force on Climate-related Financial Disclosure (TCFD), Transition Pathway Initiative (TPI), the Low Carbon Transition Rating (LCTR), and the Net Zero Assessment (NZA).

The Financial Stability Board (FSB) of G20 created the TCFD in 2015, but it disbanded TCFD in 2023 and has asked the IFRS Foundation to take over the monitoring of the progress of companies' climate-related disclosures. The framework of

TCFD are widely followed by companies to disclose their climate-related information. There are four core elements of climate-related financial disclosures, which are metrics and targets, risk management, strategy, and governance. The TCFD has three disclosure recommendations for metrics and targets (TCFD, 2021):

(a) "Disclose the metrics used by the organization to assess climate related risks and opportunities in line with its strategy and risk management process."
(b) "Disclose Scope 1, Scope 2, and, if appropriate, Scope 3 greenhouse gas (GHG) emissions, and the related risks."
(c) "Describe the targets used by the organization to manage climate related risks and opportunities and performance against targets."

For recommended disclosure (c), it suggested "in describing their targets, organizations should consider including the following: whether the target is absolute or intensity based; time frames over which the target applies; base year from which progress is measured; and key performance indicators used to assess progress against targets. (TCFD, 2021)".

The TPI is a global initiative established in 2017, led by asset owners and supported by asset managers. "As of September 2023, 143 investors globally, representing around US$60 trillion combined Assets Under Management and Advice, had pledged support for TPI (Jahn et al., 2023)". The TPI has two ways to assess the low-carbon transition of companies, Management Quality (MQ) and Carbon Performance (CP). The indicators of MQ are not sector specific, while the indicators of CP are sector specific. There are 23 indicators or questions for MQ. According to the answers to the indicators, the companies evaluated are ranked from Level 0 to Level 5. The CP assessments are primarily based on the Sectoral Decarbonization Approach (SDA) developed by Science-Based Targets Initiative (SBTi). But the SDA does not include the oil and gas sector. For MQ, Tokyo Gas is ranked Level 4 as of January 5, 2024 (*TPI,* 2025b), and Osaka Gas is ranked Level 5 as of January 5, 2024 (*TPI,* 2025a). But the TPI did not assess the CP of Tokyo Gas or Osaka Gas.

The LCTR of Morningstar Sustainalytics "assesses the degree to which a company's projected Greenhouse Gas Emissions (GHG) differ from its fair-share budget for GHG emissions (Barr et al., 2024)". Morning Sustainalytics is an independent ESG research, ratings and analytics firm. The gap between the baseline GHG emissions and the expected GHG emissions can be achieved through management. The Management Score consists of more than 80 management indicators, including both carbon performance indicators and other indicators such as financial indicators. The relative difference between the baseline GHG emissions projection and the GHG emissions budget projection is called the Baseline GHG Emissions Gap Percentage, which is known as Exposure when it is translated to an implied temperature rise. The relative difference between the expected GHG emissions projection and the GHG emissions budget projection is called the Expected GHG Emissions Gap Percentage, which is known as the Low Carbon Transition Rating when it is translated to an implied temperature rise. For example, if the implied temperature rise is 2.4 degrees Celsius, the score is Significantly Misaligned. However, Morningstar Sustainalytics does not release the LCTR results of any city gas companies publicly.

The NZA is a framework developed by Moody's Ratings. Moody's Ratings is previously known as Moody's Investors Service, and it is the credit rating business of Moody's Corporation. The NZA "considers the entity's ambitions, the implementation of its plan and its governance around greenhouse gas emissions reductions (Moody's Ratings, 2023)". The NZA adopts a five-point scale, which is from NZ-1 ("Leading") to NZ-5 ("Limited"). For ambition, it adopts a six-point scale, from 1.5°C to Above 2.5°C degrees Celsius. For implementation, it adopts a five-point scale, from Strong to Undeveloped. For GHG governance, it adopts a four-point scale, from Tier 1 (best) to Tier 4 (worst). Even though Moody' Ratings releases the credits ratings of some companies, it does not release the NZA results of any city gas companies publicly.

These existing methodologies cannot be used directly for evaluating the of the low-carbon transition strategies of city gas companies, especially the specific approaches used to reduce GHG emissions.

2.2 *Methodology of This Paper*

This paper focuses on evaluating the GHG reduction commitment and approaches to achieve the reductions of the low-carbon transition strategies of city gas companies. The existing methodologies discussed in Sect. 2.1 usually contain two parts, the type of goals and the implementation plans or the approaches. Based on this observation, this paper also designs an evaluation framework with these two parts, but this paper does not evaluate the operational management factors or financial factors. The evaluation framework used in this paper includes two parts and 6 core indicators, i.e., accounting boundary, timeframe, reduction type, clean energy action, methane emissions control, and carbon capture, utilization and storage (CCUS), as shown in Fig. 2. The total points are 7 points.

For accounting boundary, since it's much more difficult to set up a Scope 3 target, when a company set up a Scope 3 target together with Scope 1 and Scope 2 targets, 1 point is assigned. Without a Scope 3 target, only 0.4 points are assigned.

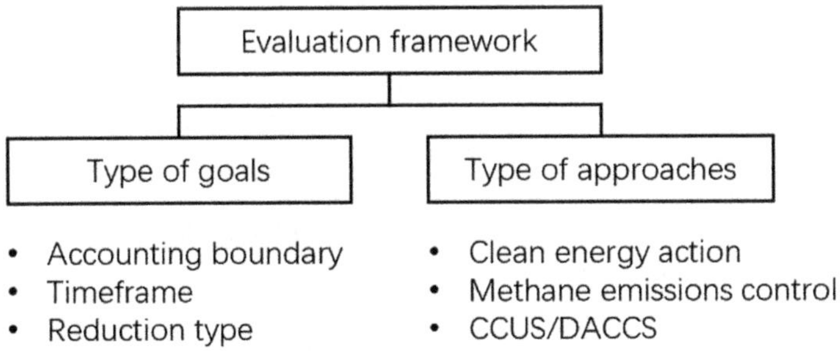

Fig. 2 Evaluation framework of the low carbon transition strategies

For timeframe, 0.5 points are assigned to net zero by 2050. If a company has interim targets from 2035 to 2045, 0.5 points are added. If a company has a net zero target later than 2050 or without setting a net zero target, 0 points are assigned even if the company has interim targets, and 0 points will be assigned to indicators like accounting boundary and reduction type.

For reduction type, since it is more meaningful to set up an absolute reduction target than an intensity reduction target, 1 point is assigned to absolute reduction target, and 0.4 points are assigned to intensity reduction target. If 2050 target is absolute target but the main interim targets are intensity targets before 2050, 0.4 points will be assigned, too.

For clean energy action, considering the different development stages of technologies, 0.4 points are assigned to action to develop renewable energies, which include solar and wind energy, and biogas; 0.6 points are assigned to clean hydrogen, which includes green, blue and turquoise hydrogen (0.4 points will be assigned if the type of hydrogen is not mentioned); and 1 point is assigned to e-methane. The technological and economic thresholds to develop e-methane are much higher than hydrogen and renewables.

Since methane is the world's second largest greenhouse gas, it is necessary for city gas companies to control and reduce the emissions of methane. For methane emissions control, 0.4 points are assigned if a company only has corporate strategies, and 1 point is assigned if a company has corporate strategy and joins international initiative like the Oil & Gas Methane Partnership 2.0 (OGMP2.0), the Oil and Gas Climate Initiative (OGCI) and/or the Oil & Gas Decarbonization Charter (OGDC). Only these three international initiatives are recognized in this methodology since these initiatives have either strict monitoring, reporting and verification (MRV) requirements or quantitative reduction targets.

For CCUS, 0.4 points are assigned to temporary utilization of CO_2 captured, and 1 point is assigned to permanent storage of CO_2 with or without considering the utilization of CO_2, since it is more difficult to storage CO_2 permanently. Direct air carbon capture and storage (DACCS) or bioenergy with carbon capture and storage (BECCS) will be assigned 1 point, the same as CCS.

The methodology to evaluate low-carbon transition strategies is shown in Table 1.

In addition, this paper uses the "S + 3E" (Safety + Energy Security, Economic Efficiency, Environment) principle to evaluate specific strategies, such as hydrogen, e-methane, and renewables. Safety is the premise of developing any kind of energy. Energy security, i.e., stable energy supplies, is valued by each government. Economic efficiency, or economic affordability, is key for the scale development of any kind of energy. Environmental sustainability is indispensable to achieve net-zero society. In Japan, the "S + 3E" principle is introduced by the 4th Basic Energy Plan since 2014 which is still maintained by the 7th Basic Energy Plan of Japan. In China, the 3E is also well discussed and sometimes known as the "impossible triangle", because energy security, economic efficiency and environmental sustainability are difficult to achieve together. The "S + 3E" principle is a preferable framework to evaluate the low-carbon strategies based on different situations of Japan and China.

Table 1 Low-carbon transition strategy evaluation methodology

Indicators	Score Cap	Sub-indicators	Scores
Accounting boundary	1	Scope 1 & 2	0.4
		Scope 1 & 2 and Scope 3	1
Timeframe	1	Net zero by 2050	0.5
		Interim targets from 2035 to 2045	0.5
		Net zero by a year later than 2050 (In this case, the interim targets will also be assigned 0 points.)	0
Reduction type	1	Intensity reduction	0.4
		Absolute reduction (with or without intensity reduction targets)	1
Clean energy action	2	Renewable energy (including biogas)	0.4
		Hydrogen	0.4
		E-methane	1
Methane emissions control	1	Having corporate strategy	0.4
		Having corporate strategy and joining initiatives like OGMP 2.0, OGCI, and/or OGDP	1
CCUS/DACCS	1	Temporary utilization of CO_2 captured	0.4
		Permanent storage of CO_2 (with or without considering CO_2 utilization)	1
Total	7		

3 Comparison of the Low-Carbon Transition Strategies of City Gas Companies in Japan and China

3.1 The Low-Carbon Transition Strategies of City Gas Companies in Japan

Japan has a much higher natural gas consumption rate in the energy mix than China as introduced in the introduction section. After the 2011 Fukushima nuclear disaster, the role of natural gas-fired power plants is strengthened in Japan. In addition, natural gas is a source of heat, especially industrial heat, which is difficult to decarbonize by electrification. Therefore, for Japan to achieve net zero by 2050, the decarbonization of the natural gas system itself is among the top priorities of Japan's total low-carbon transition strategy.

The Japan Gas Association released the *Carbon Neutral Challenge 2050 Action Plan* in June 2021. According to the Plan, the gas industry in Japan is aiming for achieving carbon neutralization rate of 5% or more for gas (injection of carbon neutral methane into city gas pipeline 1% or more) by 2030 and realizing carbon neutralization of gas by 2050. In 2050, carbon neutral methane will account for 90% of total gas, direct use of hydrogen accounting for 5%, biogas and other decarbonization measures accounting for 5%. There are three approaches to realize these targets: (1) Thorough shift to natural gas and advanced use of natural gas (demand side). Fuel conversion from oil and coal, widespread use of cogeneration and fuel cells, higher efficiency of equipment, etc. (2) Decarbonization of the gas itself (supply side). Decarbonization of the gas itself for innovation on the supply side such as methanation and hydrogen utilization. (3) CCU / CCS and overseas contribution, etc. CCU / CCS, overseas expansion of innovative gas appliances and engineering capabilities developed in Japan, utilization of carbon–neutral LNG, etc. (*The Japan Gas Association*, 2021).

As explained in the introduction section, this paper examines the low-carbon transition strategies of the group companies of the city gas companies, i.e., Tokyo Gas Group, Daigas Group (formerly known as Osaka Gas Group), and Toho Gas Group.

3.1.1 Tokyo Gas Group

Tokyo Gas Group released *Compass 2030* in November 2019, *Compass Action* in November 2021 and the Tokyo Gas Group Carbon Neutrality Roadmap 2050 in March 2024. In *Compass 2030*, Tokyo Gas Group has already stated that it will work to achieve Net-Zero CO_2 emissions.

According to the Tokyo Gas Group Carbon Neutrality Roadmap 2050, the accounting boundary of Tokyo Gas Group includes Scope 1, Scope 2 and Scope 3. Tokyo Gas Group aims to achieve net zero by 2050 and has interim targets in 2030 (20% reduction compared to FY2022) and 2040 (60% reduction compared to FY2022), which are absolute reduction targets.

According to the *Compass Action*, Tokyo Gas Group will responsibly lead the transition to Net-Zero CO_2, establish a value co-creation ecosystem, and transform the LNG value chain. To responsibly lead the transition with gaseous and renewable energy, there are three focuses: (1) transforming into the top leader at the forefront of the transition by thoroughly using natural gas and developing practical CCUS; (2) creating value chain for carbon–neutral methane by forming alliance for achieving low-cost, practical deployment of hydrogen and carbon–neutral methane; (3) creating renewable energy value chain by leveraging the Group's unique strengths by establishing revenue model that covers from power source development, operation and maintenance (O&M) to renewable energy sales (*Tokyo Gas*, 2021).

The Tokyo Gas Group Carbon Neutrality Roadmap 2050 listed three approaches that build upon the responsible transition to Net-Zero CO_2: (1) Decarbonize both gas and electricity. In addition to making sophisticated use of natural gas, decarbonize both gas and electricity; (2) Partner with customers. Advance decarbonization on both the demand side and the supply side; (3) Optimize social benefits. Pursue real-world deployment that flexibly adapts to S + 3E (Safety + Energy Security, Economic

Table 2 Low-carbon transition strategy evaluation for Tokyo Gas Group

Indicators	Scores	Sub-indicators	Scores
Accounting boundary	1	Scope 1 & 2	
		Scope 1 & 2 and Scope 3	1
Timeframe	1	Net zero by 2050	0.5
		Interim targets from 2035 to 2045	0.5
		Net zero by a year later than 2050 (In this case, the interim targets will also be assigned 0 points.)	
Reduction type	1	Intensity reduction	
		Absolute reduction (with or without intensity reduction targets)	1
Clean energy action	2	Renewable energy (including biogas)	0.4
		Hydrogen	0.6
		E-methane	1
Methane emissions control	0.4	Having corporate strategy	0.4
		Having corporate strategy and joining initiatives like OGMP 2.0, OGCI, and/or OGDP	
CCUS/DACCS	1	Temporary utilization of CO_2 captured	
		Permanent storage of CO_2 (with or without considering CO_2 utilization)	1
Total	6.4		

Efficiency, Environment) needs, while maintaining different choices. To decarbonize gas and electricity, there are four approaches: (1) Thoroughly reduce greenhouse gas (GHG) emissions across the LNG supply chain (shift to clean energy); (2) Adopt e-methane, biomass, etc. (3) Expand renewable power source transaction volume; (4) Achieve net-zero CO_2 emissions from gas-fired power plants (e-methane, hydrogen, CCS, etc.) (Tokyo Gas, 2024).

The total score for the low-carbon transition strategy of Tokyo Gas Group is 6.4 points, as shown in Table 2.

3.1.2 Daigas Group (Osaka Gas Group)

Daigas Group announced its plan to achieve carbon neutrality by 2050 in 2021. Daigas Group released the *Energy Transition 2030* in March 2023 and the Daigas Group Energy Transition 2050 in February 2025.

According to the Daigas Group Energy Transition 2050, the accounting boundary includes Scope 1, 2 and 3. It aims to reduce 5 million tons of CO_2 emissions from 2017 levels by 2030, 10 million tons of CO_2 emissions by 2040, and achieve carbon neutrality by 2050, which are absolute targets.

Daigas Group's *Energy Transition 2030* has already included the 2050 targets. The plan introduced four strategies to achieve carbon neutrality: (1) Low carbon transition with the advanced utilization of natural gas. (2) Initiative for the social implementation of e-methane. (3) Initiatives in the electricity business to achieve carbon neutrality. (4) Initiatives concerning the capture, utilization, and storage of CO_2 (Daigas, 2023). The first strategy is using natural gas to substitute coal and oil, because natural gas emits less greenhouse gases. The efficiency of the gas utilization should also be improved through cogeneration of cooling, heating and cooling, also known as CCHP (combined cooling heating and power). The second strategy is to develop e-methane, which is generated from renewable hydrogen and captured CO_2, to decarbonize the city gas system. The supply chain of e-methane does not only consider Japan but also consider overseas. The third strategy is to develop solar, wind, biomass, together with battery storage and Virtual Power Plant (VPP). The fourth strategy is to develop carbon capture and utilization (CCU) and carbon capture and storage (CCS). The captured CO_2 can be used to produce e-methane.

Daigas Group's *Energy Transition 2050* added a milestone year of 2040. It stated that "Driving the energy transition by reducing carbon emissions through energy savings and existing technologies until 2030, shifting to carbon neutrality with CN energy from 2030, and accelerating the growth of CN energy through innovation from 2040" (Daigas, 2025). From 2030 to 2040, the strategy for carbon neutrality in thermal energy is the transition to e-methane and bio-methane. From 2040 onward, the strategy is the accelerated rollout of e-methane and bio-methane, and SOEC methanation will be launched. The *Energy Transition 2050* used the term of bio-methane instead of biogas used in the *Energy Transition 2030*.

In Daigas Group's *Energy Transition 2050*, Daigas Group did not explicitly mention reducing methane emissions from its own value chain.

The total score for the low-carbon transition strategy of Daigas Group is 6 points, as shown in Table 3.

3.1.3 Toho Gas Group

Toho Gas Group released the *Toho Gas Group 2050 Carbon Neutrality Initiative* in July 2021. The *Toho Gas Group 2050 Carbon Neutrality Initiative* did not mention Scope 1, 2 and 3 explicitly, but it mentioned to achieve carbon neutrality in their entire value chain, including customer locations, in 2050 (Toho Gas, 2021). Therefore, the accounting boundary includes Scope 1, 2 and 3. Toho Gas Group aims to reduce CO2 emissions by 3 million tons from FY2020 by 2030, which is an absolute reduction target, but it does not have interim targets between 2035 and 2045.

For gas, there are three initiatives: (1) Reductions of carbon and decarbonization at customer locations. This initiative includes two approaches, switching fuel from coal and petroleum to city gas, and advanced utilization of energy through energy conservation, CO2 reductions, high-efficiency gas equipment, etc. (2) Carbon recycling. This initiative includes the social implementation of CO2 separation, capture,

Table 3 Low-carbon transition strategy evaluation for Daigas Group

Indicators	Scores	Sub-indicators	Scores
Accounting boundary	1	Scope 1 & 2	
		Scope 1 & 2 and Scope 3	1
Timeframe	1	Net zero by 2050	0.5
		Interim targets from 2035 to 2045	0.5
		Net zero by a year later than 2050 (In this case, the interim targets will also be assigned 0 points.)	
Reduction type	1	Intensity reduction	
		Absolute reduction (with or without intensity reduction targets)	1
Clean energy action	2	Renewable energy (including biogas)	0.4
		Hydrogen	0.6
		E-methane	1
Methane emissions control		Having corporate strategy	
		Having corporate strategy and joining initiatives like OGMP 2.0, OGCI, and/or OGDP	
CCUS/DACCS	1	Temporary utilization of CO_2 captured	
		Permanent storage of CO_2 (with or without considering CO_2 utilization)	1
Total	6		

and utilization that can be carried out right now, and the expansion of usage applications for CO2 in the future. (3) Decarbonization of gas itself. This initiative includes two approaches. One approach is the development and verification of technology for methanation, etc. in the near term and the large-quantity introduction of clean synthetic methane, etc. The other approach is the expansion of the introduction of carbon neutral LNG and biogas.

For hydrogen, there are three initiatives: (1) Demand creation: mobility applications. Expand the demand of mobility such as the preparation of hydrogen stations and the support for expansion of vehicle types and applications. (2) Demand creation: further expansion of applications, to heating and other fields. (3) Supply chain creation. This initiative includes building a hydrogen supply terminal at the Chita-Midorihama Works.

For electricity, there are two initiatives: (1) Reductions of carbon and decarbonization at customer locations. Encourage the popularization of diverse distributed energy sources, including solar power generation, storage batteries, electric automobiles, etc. (2) Decarbonization of electrical power sources. This initiative includes the development of renewable energy power sources and the procurement of renewable energy.

Table 4 Low-carbon transition strategy evaluation for Toho Gas Group

Indicators	Scores	Sub-indicators	Scores
Accounting boundary	1	Scope 1 & 2	
		Scope 1 & 2 and Scope 3	1
Timeframe	0.5	Net zero by 2050	0.5
		Interim targets from 2035 to 2045	
		Net zero by a year later than 2050 (In this case, the interim targets will also be assigned 0 points.)	
Reduction type	1	Intensity reduction	
		Absolute reduction (with or without intensity reduction targets)	1
Clean energy action	2	Renewable energy (including biogas)	0.4
		Hydrogen	0.6
		E-methane	1
Methane emissions control		Having corporate strategy	
		Having corporate strategy and joining initiatives like OGMP 2.0, OGCI, and/or OGDP	
CCUS/DACCS	1	Temporary utilization of CO_2 captured	
		Permanent storage of CO_2 (with or without considering CO_2 utilization)	1
Total	5.5		

The Toho Gas Group 2050 Carbon Neutrality Initiative did not explicitly mention reducing methane emissions from its own value chain.

The total score for the low-carbon transition strategy of Toho Gas Group is 5.5 points, as shown in Table 4.

3.1.4 Summary of Low-Carbon Transition Strategies of Major City Gas Companies in Japan

The three major city gas companies, Tokyo Gas, Daigas, and Toho Gas, have all announced plans for carbon neutrality, and all aim to achieve the goal of carbon neutrality by 2050, in line with Japan's national goal. The carbon neutrality targets of the three companies all include Scope 1, Scope 2, and Scope 3, i.e., the entire value chain. Tokyo Gas Group has the most detailed plan compared to the other two companies. The targets are compared as follows in Table 5.

The approaches to achieving carbon neutrality of these three companies are very similar. In the near term, coal and oil should be substituted by natural gas. In the long term, e-methane is seen as the main approach to decarbonizing the gas system. To decarbonize electricity, renewable energy should be expanded, high-efficiency

Table 5 Carbon neutrality targets of major city gas companies in Japan

Companies	Items	2030	2040	2050	Accounting boundary
Tokyo Gas Group	Emissions reduction (vs. FY2022)	20% reduction	60% reduction	Net-Zero	Scope 1, 2 and 3
	Carbon neutrality rate in gas supplied domestically	Begin	50%		
	Carbon neutrality rate in electricity supplied domestically	30%	50%		
Daigas Group	Emissions reduction (vs. 2017)	5 million tons reduction	Max 10 million tons reduction	Carbon Neutrality	Scope 1, 2 and 3
	Gas carbon neutrality rate		40%		
Toho Gas Group	Emissions reduction (vs. FY2020)	3 million tons reduction		Carbon Neutrality	Scope 1, 2 and 3
	Gas carbon neutrality rate	at least 5%			
	Renewable energy power sources handled	500,000 kW			

natural gas-fired power plants should be introduced, and e-methane, hydrogen, biogas, CCS, etc. should be used or applied for thermal power plants. The main approaches to achieving carbon neutrality of major city gas companies in Japan are compared as follows in Table 6.

3.2 *The Low-Carbon Transition Strategies of City Gas Companies in China*

The China Gas Association released the Research on Low-Carbon Transition and Development of China's City Gas Industry 2023 in 2024 (CGA, 2023). The research proposed four low-carbon transition pathways for the city gas industry: (1) Traditional gas business: Enhancing methane emissions control, achieve the three targets of "increasing supply, increasing revenue, and delivering climate benefits". (2) Traditional gas business: Developing a new gas industry energy system integrating "natural

Table 6 Main approaches to achieving carbon neutrality of major city gas companies in Japan

Companies	Items	2030	2040	2050
Tokyo Gas Group	Gas	Begin adopting e-methane	Expand use of e-methane and utilize hydrogen and offsetting methods	Achieve by fully harnessing innovation in DACCS/BECCS, turquoise hydrogen, etc
	Electricity	Expand renewable energy transaction volume	Work to achieve net-zero CO_2 emissions from thermal power generation	Fully achieve net-zero emissions from thermal power
Daigas Group	Items	2020–2030	2030–2040	2040–2050
	Gas	Energy saving, rollout of e-methane	Increase of e-methane use	Accelerated increase of e-methane produced by SOEC methanation
	Electricity	Lauch of high-efficiency natural gas-fired power plants		Zero emissions from thermal power plants
Toho Gas Group	Items	current–2030	2030–2050	
	Gas	Development and verification of technology for methanation, etc	Large-quantity introduction of clean synthetic methane, etc	
		Expansion of introduction of carbon neutral LNG and biogas		
	Hydrogen	Demand creation; Creation of a domestic supply chain; Importation of hydrogen from abroad		
	Electricity	Acceleration of development and procurement of renewable energy resources; Measures for thermal power plants (CO_2 separation and capture and hydrogen utilization)		

gas + renewable gas (including hydrogen)". (3) Traditional gas business: Developing the new energy storage and transportation infrastructure system consisting of natural gas pipelines, storage reservoirs, and LNG terminals. (4) New low-carbon business: Developing the new electricity system and heating system with the interaction of natural gas and new energy (it should be noted that solar and wind energy are called new energy in China). The research also proposed five low-carbon transition pathways for city gas companies: (1) Lean management of the main business: Strengthening the capacity building for methane emissions control. (2) Strategic transition: energy sales → energy production service → energy community service. (3) Business transition: city gas business → city gas integrated energy business (distributed energy, clean heating, etc.) → coordination between renewable energy and gas business. (4) Technology application: low-carbon technology → zero-carbon technology

→ negative-carbon technology. (5) Carbon management system: the "Dual Carbon (carbon peaking and carbon neutrality)" management platform of city gas industry; carbon emission data statistics and accounting, carbon asset management tools.

There are more than 3600 city gas companies in China, but there are five major city gas companies, namely Kunlun Energy Company Limited (Kunlun Energy), the Hong Kong and China Gas Company Limited (Towngas), China Gas Holdings Limited (China Gas), ENN Energy Holdings Limited (ENN Energy), and China Resources Gas Group Limited (CR Gas). In 2022, these five companies accounted for 53% of China's total natural gas consumption (CGA, 2023). This paper will examine the transition strategies of these five major companies.

3.2.1 Kunlun Energy

Kunlun Energy released the *Kunlun Energy Company Limited Action Plan Towards Carbon Peaking and Carbon Neutrality 2024 Edition* in May 2024. The action plan stated that Kunlun Energy "will improve the basic capacity of carbon emission statistics and accounting to gradually account for Scope 3 carbon emissions and carbon footprint for key products (Kunlun Energy, 2024)", therefore Scope 3 emissions are not included in the accounting boundary yet. Kunlun Energy aims to peak carbon emissions by 2030 and achieve net-zero emissions by 2050. It also committed to achieving net-zero emissions across its LPG, urban gas distribution network, and branch pipeline operations by 2035 and many other interim targets by 2035 and by 2040. The action plan consists of four approaches:

(1) Promoting energy-saving, efficiency improvement, and emission control.
   - Advancing industrial energy-sabing and carbon reduction.
   - Adopting green design and expanding green buildings.
   - Enhancing recycling and advancing methane emissions control.

(2) Accelerating the transition to low-carbon energy and expanding new energy business.
   - Phasing out coal and oil to advance cleaner energy consumption.
   - Developing green power to achieve low-carbon energy consumption.
   - Facilitating the fusion of natural gas and new energy business to develop green and low-carbon integrated energy services. The new energy business includes solar, wind, hydrogen and energy storage.

(3) Fostering green technological innovation to cultivate zero-carbon and carbon-negative drivers.
   - Enhancing R&D and reserves in green technologies to drive a low-carbon transition.
   - Joining in the development of zero-carbon and carbon-negative projects. These technologies include biogas, hydrogen, CCUS, BECCS (Bioenergy with CCS), DAC (Direct Air Capture), etc.

Table 7 Low-carbon transition strategy evaluation for Kunlun energy

Indicators	Scores	Sub-indicators	Scores
Accounting boundary	0.4	Scope 1 & 2	0.4
		Scope 1 & 2 and Scope 3	
Timeframe	1	Net zero by 2050	0.5
		Interim targets from 2035 to 2045	0.5
		Net zero by a year later than 2050 (In this case, the interim targets will also be assigned 0 points.)	
Reduction type	1	Intensity reduction	
		Absolute reduction (with or without intensity reduction targets)	1
Clean energy action	1	Renewable energy (including biogas)	0.4
		Hydrogen	0.6
		E-methane	
Methane emissions control	0.4	Having corporate strategy	0.4
		Having corporate strategy and joining initiatives like OGMP 2.0, OGCI, and/or OGDP	
CCUS/DACCS	1	Temporary utilization of CO_2 captured	
		Permanent storage of CO_2 (with or without considering CO_2 utilization)	1
Total	4.8		

- Advancing breakthroughs in key technologies for digital and intelligence control.

(4) Improving the green low-carbon system and building a green brand.

- Establishing a framework of green and low-carbon systems and standards.
- Strengthening digital and intelligent carbon footprint management to support the building of an industry background database.
- Driving upstream and downstream players to build green and low-carbon industrial and supply chains together.
- Playing the demonstration and leading role to boost its green driving force.

The total score for the low-carbon transition strategy of Kunlun Energy is 4.8 points, as shown in Table 7.

3.2.2 ENN Energy

ENN Energy released its Decarbonization Action 2030—The Journey to Net Zero 2024 Edition in 2024. ENN Energy aims to achieve carbon neutrality in scope 1 and scope 2 by 2050 without relying on the purchase of green certificates and other offsets (*ENN*, 2024). ENN Energy aims to reduce GHG emission intensity of the city-gas business by 50% compared to 2019 and reduce the GHG emission intensity per unit of energy generated from the energy generating facilities of Integrated Energy (IE) business by 48% compared to 2019, which are intensity targets.

Its future plans by 2030 include four parts.

1 City-gas business: Continue to expand the application of unmanned stations, reducing GHG emissions generated by the operation of city-gas business; Continue to explore innovative technologies to enhance the efficiency of natural gas and organically integrate hydrogen with city-gas business.
2 IE business: Continue to increase the share of renewable energy supply, optimize the energy supply structure; Explore models of integrating natural gas and renewable energy; Continue to explore the application prospects of new types of renewable energy such as hydrogen energy and geothermal energy.
3 Value added business: Continue to explore energy efficiency improvement technologies for household energy consumption devices based on digital intelligence technologies; Continue to enrich value added product models, activate the value of existing customers, and commit to upgrading to more intelligent and lower-carbon products.
(4) Low-carbon office: In addition to steadily increasing photovoltaic power consumption, further expand the application of geothermal energy in office building scenarios; Improve energy efficiency management in office buildings by applying digital intelligence technologies.

From 2030 to 2050, ENN Energy will take intensive actions to scale up renewable energy consumption and supply, increase the ratio of zero-carbon energy consumption, and expand the application of green and carbon-negative technologies.

The total score for the low-carbon transition strategy of ENN Energy is 3.5 points, as shown in Table 8.

3.2.3 Towngas

Towngas released the Towngas' Carbon Neutrality Roadmap by 2050 in 2024. According to the roadmap, Towngas aims to reduce group operational GHG emissions by 10% by 2025 compared with 2020 levels and achieve carbon neutrality by 2050 (Towngas, 2024).

To mitigate climate change, it has been reducing its direct and indirect GHG emissions by upgrading its infrastructure with energy efficient equipment, strengthening its control measures to reduce methane leakage, and increasing the proportion of low-carbon, renewable energy sources such as natural gas and landfill gas for gas production processes. For downstream operations, efforts include generating renewable energy, implementing Combined Heat and Power (CHP) solutions, improving the efficiency of gas appliances, and developing innovative waste-to-energy projects. These initiatives can be summarized as follows: Develop low-carbon fuel; Expand

Table 8 Low-carbon transition strategy evaluation for ENN Energy

Indicators	Scores	Sub-indicators	Scores
Accounting boundary	0.4	Scope 1 & 2	0.4
		Scope 1 & 2 and Scope 3	
Timeframe	0.5	Net zero by 2050	0.5
		Interim targets from 2035 to 2045	
		Net zero by a year later than 2050 (In this case, the interim targets will also be assigned 0 points.)	
Reduction type	0.4	Intensity reduction	0.4
		Absolute reduction (with or without intensity reduction targets)	
Clean energy action	0.8	Renewable energy (including biogas)	0.4
		Hydrogen	0.4
		E-methane	
Methane emissions control	0.4	Having corporate strategy	0.4
		Having corporate strategy and joining initiatives like OGMP 2.0, OGCI, and/or OGDP	
CCUS/DACCS	1	Temporary utilization of CO_2 captured	
		Permanent storage of CO_2 (with or without considering CO_2 utilization)	1
Total	3.5		

renewable energy; Facilitate hydrogen distribution and blending; Enable fuel switch; Enhance energy efficiency.

The total score for the low-carbon transition strategy of Towngas is 3.1 points, as shown in Table 9.

3.2.4 China Gas

China Gas released the 2022 China Gas Carbon Neutrality Action Plan Report in 2022. According to the action plan, China Gas aims to achieve carbon peaking by 2030 and carbon neutrality in its operation by 2050 (China Gas, 2022). It also aims to reduce CO2 emission intensity by 50% compared to the 2021 level by 2030. China's Nationally Determined Contributions in 2022 aims to lower its carbon dioxide emissions per unit of GDP by over 65 percent from the 2005 level by 2030 (MEE, 2022). Compared with the national target, the target of China Gas reducing CO2 emission intensity by 50% in 10 years is ambitious.

There are seven approaches to reduce carbon emissions: green energy; energy conservation and innovation; low-carbon logistics; smart operation; methane emissions control; net-zero pathway; green ecosystem. The so-called net-zero pathway

Table 9 Low-carbon transition strategy evaluation for Towngas

Indicators	Scores	Sub-indicators	Scores
Accounting boundary	0.4	Scope 1 & 2	0.4
		Scope 1 & 2 and Scope 3	
Timeframe	0.5	Net zero by 2050	0.5
		Interim targets from 2035 to 2045	
		Net zero by a year later than 2050 (In this case, the interim targets will also be assigned 0 points.)	
Reduction type	1	Intensity reduction	
		Absolute reduction (with or without intensity reduction targets)	1
Clean energy action	0.8	Renewable energy (including biogas)	0.4
		Hydrogen	0.4
		E-methane	
Methane emissions control	0.4	Having corporate strategy	0.4
		Having corporate strategy and joining initiatives like OGMP 2.0, OGCI, and/or OGDP	
CCUS/DACCS		Temporary utilization of CO_2 captured	
		Permanent storage of CO_2 (with or without considering CO_2 utilization)	
Total	3.1		

includes three parts: Developing CCUS pilot projects; Exploring China Certified Emission Reduction (CCER) self-developed projects; Establishing a Zero-Carbon Environment & Energy Research Center and an Urban Carbon Service Center. The so-called green ecosystem also includes three parts: Low-carbon life (shifting from coal to gas, hydrogen blending into natural gas, etc.); Supplier emission reduction (developing green procurement standards); Urban–rural low-carbon renovation and upgrading (building regional "Dual Carbon" digital platform systems, etc.). For methane emissions control, China Gas is the only Chinese company joining OGMP 2.0.

The total score for the low-carbon transition strategy of China Gas is 4.1 points, as shown in Table 10.

Table 10 Low-carbon transition strategy evaluation for China gas

Indicators	Scores	Sub-indicators	Scores
Accounting boundary	0.4	Scope 1 & 2	0.4
		Scope 1 & 2 and Scope 3	
Timeframe	0.5	Net zero by 2050	0.5
		Interim targets from 2035 to 2045	
		Net zero by a year later than 2050 (In this case, the interim targets will also be assigned 0 points.)	
Reduction type	0.4	Intensity reduction	0.4
		Absolute reduction (with or without intensity reduction targets)	
Clean energy action	0.8	Renewable energy (including biogas)	0.4
		Hydrogen	0.4
		E-methane	
Methane emissions control	1	Having corporate strategy	
		Having corporate strategy and joining initiatives like OGMP 2.0, OGCI, and/or OGDP	1
CCUS/DACCS	1	Temporary utilization of CO_2 captured	
		Permanent storage of CO_2 (with or without considering CO_2 utilization)	1
Total	4.1		

3.2.5 CR Gas

CR Gas has not yet announced its carbon neutrality target. It aims to reduce CO2 emission intensity by 10% compared with 2020 levels by 2025, and it expects that the carbon emissions will peak by 2035 (CR Gas, 2025a). However, since it did not announce its carbon neutrality target, relevant indicators will not be assigned points.

It stated that it will reduce its own carbon emissions through methane emission management and clean energy utilization, and will engage in in-depth cooperation and R & D with external institutions on CCS technologies, thereby contributing to the realization of the national "Dual Carbon" targets (CR Gas, 2025b).

The total score for the low-carbon transition strategy of CR Gas is 2.2 points, as shown in Table 11.

3.2.6 Summary of Low-Carbon Transition Strategies of Major City Gas Companies in China

The year to achieve carbon neutrality announced by major Chinese city gas companies are almost all 2050, but the scope is Scope 1 & 2. The scope is quite different

Table 11 Low-carbon transition strategy evaluation for CR Gas

Indicators	Scores	Sub-indicators	Scores
Accounting boundary		Scope 1 & 2	
		Scope 1 & 2 and Scope 3	
Timeframe		Net zero by 2050	
		Interim targets from 2035 to 2045	
		Net zero by a year later than 2050 (In this case, the interim targets will also be assigned 0 points.)	
Reduction type		Intensity reduction	
		Absolute reduction (with or without intensity reduction targets)	
Clean energy action	0.8	Renewable energy (including biogas)	0.4
		Hydrogen	0.4
		E-methane	
Methane emissions control	0.4	Having corporate strategy	0.4
		Having corporate strategy and joining initiatives like OGMP 2.0, OGCI, and/or OGDP	
CCUS/DACCS	1	Temporary utilization of CO_2 captured	
		Permanent storage of CO_2 (with or without considering CO_2 utilization)	1
Total	2.2		

from their Japanese counterparts. It should be noted that China's national target to achieve carbon neutrality is 2060, so these companies set their targets 10 years earlier than the national target. Among the major Chinese companies, Kunlun Energy has the most detailed plan for carbon neutrality, with interim targets for every five years. CR Gas has not announced its carbon neutral target. The carbon neutrality targets of major Chinese city gas companies are compared as follows in Table 12.

The major Chinese city gas companies all see methane emissions control as a way to decarbonize their business. No major Chinese city gas companies see e-methane as a way to achieve carbon neutrality. In addition, no major Chinese companies has paid enough attention to how to achieve carbon neutrality for the thermal power plant. This is probably because the major Chinese companies only considered Scope 1 & 2 in their carbon neutrality strategies. The main approaches of major Chinese city gas companies to achieving carbon neutrality are compared as follows in Table 13.

Table 12 Carbon neutrality targets of major city gas companies in China

Companies	2025	2030	2035	2050	Scope
Kunlun Energy	Achieve an installed capacity of 200 MW of green power, etc	Carbon Peaking	Achieve net-zero emissions across its LPG, urban gas distribution network, and branch pipeline operations, etc	Carbon Neutrality	Scope 1 & 2
ENN Energy		Reduce GHG emission intensity of the city-gas business by 50% compared to 2019		Net-zero emission	Scope 1 & 2
Towngas	Reduce group operational GHG emissions by 10% compared with 2020 levels			Carbon Neutrality	Scope 1 & 2
China Gas		Carbon Peaking		Carbon neutrality	Scope 1 & 2
CR Gas	Reduce CO2 emission intensity by 10% compared with 2020 levels		Carbon Peaking		

Note Kunlun Energy also has targets for 2040 and 2045, which are not shown in this table

3.3 Quantitative Comparison Between Major City Gas Companies in Japan and China

From the above analysis, the scores of the low-carbon transition strategies of major city gas companies in Japan and China are listed and compared in Table 14. The three major Japanese city gas companies rank higher than their Chinese counterparts. This is mainly because Japanese companies consider Scope 3 emission reductions and aim to develop e-methane to decarbonize the natural gas system and the thermal power plant system. The reason behind this is that share of natural gas in Japan's energy mix is much higher than that of China, and Japan pays more attention to the decarbonization of the natural gas system.

Table 13 Main approaches to achieving carbon neutrality of major city gas companies in China

<table>
<tr><th>Companies</th><th colspan="2">Approaches</th></tr>
<tr><td>Kunlun Energy</td><td colspan="2">Promoting energy-saving, efficiency improvement, and emission control (both CO_2 and methane)
Accelerating the transition to low-carbon energy and expanding new energy business (phasing out coal and oil, expanding solar and wind power, etc.)
Fostering green technological innovation (biogas, hydrogen, CCUS, BECCS, DAC, etc.)</td></tr>
<tr><td rowspan="2">ENN Energy</td><td>Current—2030</td><td>2030–2050</td></tr>
<tr><td>Continue to expand the application of unmanned stations, reducing GHG emissions (both CO_2 and methane)
Continue to explore the application prospects of new types of renewable energy such as hydrogen energy and geothermal energy
Continue to enrich value added product models</td><td>Scale up renewable energy consumption and supply
Increase the ratio of zero-carbon energy consumption
Expand the application of green and carbon-negative technologies</td></tr>
<tr><td>Towngas</td><td colspan="2">Strengthening its control measures to reduce methane leakage
Expanding renewable energy
Facilitating hydrogen distribution and blending</td></tr>
<tr><td>China Gas</td><td colspan="2">Methane emissions control
Green energy (solar and wind power)
Blending hydrogen into natural gas</td></tr>
<tr><td>CR Gas</td><td colspan="2">Methane emission management
Clean energy utilization
CCS</td></tr>
</table>

Table 14 Scores of the low-carbon transition strategies of major city gas companies in Japan and China

Countries	Companies	Scores
Japan	Tokyo Gas Group	6.4
	Daigas Group	6.0
	Toho Gas Group	5.5
China	Kunlun Energy	4.8
	China Gas	4.1
	ENN Energy	3.5
	Towngas	3.1
	CR Gas	2.2

4 Comparative Studies of Specific Low-Carbon Transition Strategies

This chapter will analyze specific low-carbon transition strategies, such as hydrogen, e-methane, renewables, methane emissions reduction, etc.

4.1 Comparative Study of Hydrogen Strategies

All major Japanese and Chinese city gas companies discussed in this paper included hydrogen in their low-carbon transition plans, but the major city gas companies in Japan have more sophisticated hydrogen strategies than Chinese companies. One of the reasons for this is that Japan has led the world in developing national hydrogen strategy. In 2017, Japan released the Basic Hydrogen Strategy, which is the world's first national hydrogen strategy. China released its national hydrogen strategy, Hydrogen Energy Industry Development Medium- and Long-Term Plan (2021–2035), in 2022, five years later than Japan. Japan is one of the leaders in hydrogen technologies, but China is catching up. Details will be discussed in Sect. 4.1.3.

Hydrogen can be used to decarbonize both the gas system and the electricity system. For the gas system, hydrogen can be blended into the gas system to reduce carbon emissions. For the electricity system, hydrogen can be burned together with methane to reduce the carbon emissions of gas-fired power plants. Hydrogen can also be used to produce e-methane and other hydrogen derivatives such as ammonia and methanol.

4.1.1 Hydrogen Production and Supply

Japanese city gas companies are considering the production of turquoise hydrogen (hydrogen produced by methane pyrolysis, which produces hydrogen and solid carbon), and considering importing hydrogen from abroad, which are not options for major Chinese city gas companies.

Tokyo Gas Group has made it clear that it will promote "onsite hydrogen production through renewable energy water electrolysis and methane pyrolysis (turquoise hydrogen)" (Tokyo Gas, 2024). Tokyo Gas Group is developing technologies for the mass production of key components of water electrolyzers that help lower the cost of green hydrogen. Tokyo Gas Group has co-developed water electrolysis cell stack for low-cost hydrogen production and has co-developed a novel low-cost catalyst that doesn't use rare metals. For turquoise hydrogen, even though Tokyo Gas Group mentioned turquoise hydrogen in its Carbon Neutrality Roadmap 2050, it did not elaborate its plan on developing turquoise hydrogen. The advantage of turquoise hydrogen is that the process does not emit CO2. However, the disadvantages of turquoise hydrogen are that the technology is not widely used, the cost is linked to the price of natural gas, which is highly volatile, the market for solid carbon may not be available locally, etc. In short, there are some technological and economic challenges for turquoise hydrogen.

Daigas Group considers producing hydrogen from biomass, which can simultaneously produce low-cost hydrogen, CO2 and electricity.

Toho Gas Group considers importing hydrogen from abroad and producing hydrogen by reforming natural gas and carbon recycling technology (blue hydrogen),

Table 15 Hydrogen production and supply pathways considered by major city gas companies in Japan and China

Major city gas companies	in Japan	in China
Green hydrogen	Yes	Yes
Blue hydrogen	Yes	Without mentioning
Turquoise hydrogen	Yes	Without mentioning
Importing hydrogen	Yes	Without mentioning

Note It should be noted that if a major city gas company has considered a certain pathway, it will be marked as "Yes", which does not mean that all companies have considered that pathway. The same applies hereinafter

and by utilizing the surplus electricity. It plans to change the LNG terminal to a hydrogen import terminal at the Chita-Midorihama Works in the long term and create a hydrogen supply chain.

Hydrogen production and supply pathways considered by major city gas companies in Japan and China are as follows in Table 15.

The reason why some Japanese city gas companies are considering importing hydrogen from abroad is that hydrogen produced overseas is more economical than that produced in Japan. Mizuho's study shows that "In terms of supply cost estimates, Middle Eastern blue ammonia/hydrogen will be the most cost-competitive by 2050, followed by U.S. blue ammonia and Australian blue hydrogen (Mizuho, 2023)". The Mizuho study estimates that in 2050, the supply cost of blue hydrogen using ammonia as the carrier from Middle East is below 20 yen/m3, but green hydrogen[3] produced in Japan is above 30 yen/m3. Based on Mizuho's study, around 80% of Japan's total hydrogen demand in 2050 (about 24 million tons of hydrogen) will be met by imported hydrogen (about 20 million tons of hydrogen). According to Hydrogen Factsheet—Japan, "Japan's capacity for domestic hydrogen production, especially for green hydrogen, is strongly limited due to scarcity of suitable land (Sieler et al., 2021)". Japan has already begun importing hydrogen since 2022. There is one hydrogen import terminal called Hy Touch Kobe, owned and operated by Kawasaki Heavy Industries in collaboration with other partners, located at the Port of Kobe. The hydrogen handled at the terminal is imported from Australia and is produced from brown coal gasification with pilot CCS. Although no new plans have been announced, it is possible that Japan will build more hydrogen import terminals.

The major Chinese city gas companies generally do not specify the source of hydrogen production. Only Kunlun Energy mentioned the development of green hydrogen in its Action Plan Towards Carbon Peaking and Carbon Neutrality 2024 Edition. Since China has the largest producing capacity in solar and wind energy and is developing the green hydrogen fast, it's easier for Chinese city gas companies to produce hydrogen from renewables at a lower cost in the future. The comparison of the cost of hydrogen produced in China and in Japan can be referred to Sect. 4.1.3.

[3] Since there are limited CO2 storage sites in Japan, hydrogen produced in Japan in the future will be mainly green hydrogen.

In addition, Japan enacted the Hydrogen Society Promotion Act in May 2024, which provides subsidies to address the price gap between low-carbon hydrogen and fossil fuels. The development of hydrogen in Japan is boosted by this law. In China, there is no such national subsidies for low-carbon hydrogen production.

4.1.2 Hydrogen Utilization

In Japan, it is considered better to use hydrogen in coastal areas, because the distance between hydrogen utilization and hydrogen import terminals or potential hydrogen production sites is short. According to Japan's Basic Hydrogen Strategy revised in 2023, "Ports and coastal areas have the potential to create large-scale demand for hydrogen through the concentration of existing industries" (The Ministerial Council on Renewable Energy, Hydrogen and Related Issues, 2023).

Tokyo Gas Group has several strategies for hydrogen utilization. (1) Using hydrogen for power plants, such as co-firing of hydrogen to reduce the carbon emissions from gas-fired power plants. Tokyo Gas has already adopted power generation equipment that supports hydrogen co-firing for its Chiba-Sodegaura Power (CSP) project. The CSP is located near Tokyo Bay. (2) Using hydrogen for residential use. Tokyo Gas is also leading Japan's first hydrogen supply project at Harumi Flag, a residential compound in Tokyo. The compound will be powered by hydrogen through hydrogen pipelines and fuel cells. Tokyo Gas is also promoting the ENE-FARM system, which uses hydrogen produced from natural gas or LPG to generate electricity and heat. (3) Using hydrogen to meet transportation needs. Tokyo Gas has begun selling hydrogen produced by AEM water electrolyzer at hydrogen refueling station. (4) Developing hydrogen burners for specific industrial needs. Tokyo Gas has developed the world's first hydrogen-only burner for use at asphalt plants and Japan's first hydrogen burner with a built-in device for recovering waste heat from aluminum manufacturing. (5) Using hydrogen to produce e-methane. Hydrogen produced from renewables can also be used to produce e-methane.

Daigas Group considers using hydrogen carriers to decarbonize thermal energy and electric energy. Hydrogen carriers are defined as hydrogen and its compounds, including ammonia and carbon-recycling fuel, such as e-methane, SAF, and e-fuel. In short, hydrogen carriers include hydrogen and hydrogen derivatives. Hydrogen, ammonia and e-methane can be used to decarbonize thermal power plants and hard-to-abate industries. Ammonia can be used in co-firing with coal.

Toho Gas Group views hydrogen as one of the three types of energy that will make up the future energy system, the other two being gas (city gas and LPG) and electricity. Toho Gas Group has the following strategy for hydrogen utilization. (1) Expanding the use of hydrogen in the mobility sector by expanding vehicle types and refueling stations. (2) Expanding the use of hydrogen in heating and other fields. Toho Gas Group is developing hydrogen burners that can be used in aluminum-melting furnaces, etc. It is also researching the technology of mixed combustion of city gas and hydrogen for cogeneration.

Table 16 Hydrogen application directions considered by major city gas companies in Japan and China

Major city gas companies	In Japan	In China
Decarbonize thermal power plant	Yes	Without mentioning
Decarbonize thermal energy / Specific industry use (hydrogen burners)	Yes	Without mentioning
Residential use (fuel cells)	Yes	Without mentioning
Hydrogen blending into natural gas	Yes	Yes
Hydrogen refueling stations	Yes	Yes

Kunlun Energy stated that it will develop comprehensive wind-PV-gas-power-hydrogen utilization demonstrations and will expand the applications of green hydrogen without specifying detailed application directions.

ENN Energy stated that it is exploring the application prospects of natural-gas-based hydrogen energy projects, which mainly means hydrogen blending into natural gas. ENN Energy has already started its hydrogen blending projects in the Taixing Economic and Development Zone, Jiangsu Province. It has also carried out a hydrogen-rich tail gas combustion project by collecting hydrogen from the tail gas of a photoelectronic technology company. The collected hydrogen is used to drive a steam boiler with the addition of natural gas.

Towngas stated that it will carry out the distribution of hydrogen and hydrogen blending. China Gas stated that it will carry out hydrogen blending. CR Gas is developing hydrogen blending and hydrogen refueling stations.

Hydrogen application directions considered by major city gas companies in Japan and China are as follows in Table 16.

For hydrogen utilization, Chinese city gas companies mainly consider blending hydrogen into the natural gas system and developing hydrogen refueling stations when feasible. However, Japanese city gas companies also consider using hydrogen to decarbonize the thermal power plant and thermal energy. It is because that Japan's technologies of gas turbine, co-firing of hydrogen and fuel cell are more advanced than China. Japan's Hydrogen Society Promotion Act also support the public utilization of hydrogen by lowering the cost of hydrogen and funding hydrogen refueling stations, ammonia terminals, and pipeline networks.

4.1.3 Evaluate the Hydrogen Strategy by the "S + 3E" Principle

Japan's revised Basic Hydrogen Strategy in 2023 has evaluated the hydrogen strategy by the "S + 3E" principle. This paper will review some of the evaluations in detail and add the evaluation of China's hydrogen strategy from the "S + 3E" perspective.

For safety, both Japan and China have regulations to make sure the value chain of hydrogen is safe to prevent accidents since hydrogen is easy to burn and explode. For environmental sustainability, green hydrogen and blue hydrogen are clean energy with low GHG emissions, which applies to both Japan and China. The differences between Japan and China on hydrogen strategy are mainly on energy security and economic efficiency.

For energy security, Japan prioritizes producing hydrogen domestically, but Japan has to import hydrogen because the domestically produced hydrogen is not enough to meet Japan's demand (METI, 2024). The current hydrogen production of Japan is mainly from natural gas, which is mainly imported from overseas. Since hydrogen can be produced from renewables, it can be produced and stored domestically in Japan. In addition, hydrogen can be imported from some Asian countries and Australia. Japan consumed 1.7 million tons of hydrogen in 2020 (Intralink Limited, 2022). But Japan has set ambitious targets for the future, e.g., 3 million tons annually by 2030, 12 million tons by 2040, and 20 million tons by 2050 (The Ministerial Council on Renewable Energy, Hydrogen and Related Issues, 2023). These ambitious targets cannot be met only by domestic production. To secure stable supply, green hydrogen, blue hydrogen and other types of clean hydrogen are all considered by Japan. In contrast, China is the world's largest hydrogen producer, and China will not consider importing hydrogen from overseas. In 2024, the hydrogen production of China surpassed 36.5 million ton, while the world's hydrogen production and consumption scale is around 105 million ton (NEA & Guoneng Qingchuang Technology, 2025). The majority of the hydrogen production in China is from coal, the production mix being 63% from coal, 19% from natural gas, 18% from industrial byproduct, and 1% from renewables in 2023 (Diao, 2025). Since China is abundant in coal, the supply of hydrogen can be easily secured. The problem for China is to increase the share of clean hydrogen. Since China is the world's largest renewable energy producer, China values green hydrogen more than blue hydrogen. China's national hydrogen plan only set a quantitative target for 2025, which is for green hydrogen. By 2025, hydrogen produced from renewables, i.e., green hydrogen, in China reaches 0.1–0.2 million tons per year (NDRC, 2022). Even though China's green hydrogen target is conservative, the potential for China to develop green hydrogen is huge since China is the world's largest renewable energy producer. The lack of quantitative targets beyond 2025 and lack of detailed implementation pathways in China's plan shows that China does not value hydrogen as much as Japan.

For economic efficiency, Japan enacted the Hydrogen Society Promotion Act in 2024, aiming to bridge the price gap between clean hydrogen and fossil fuels through holistic policies. Japan aims to reduce the cost of hydrogen to 20 yen/Nm3, a level that the cost of power generated by hydrogen is below that of power generated by natural gas by 2050. In China, there is no holistic incentive policies for hydrogen, except for some incentive policies for fuel cell vehicles in certain pilot regions. In addition, there is no official cost reduction target in China. The difference in incentive policies between Japan and China also shows that China does not prioritize hydrogen energy as much as Japan does. Although China does not have holistic incentive policies for hydrogen, it should be noted that green hydrogen produced in China is cheaper than green hydrogen produced in Japan. It is estimated that in 2030, the levelized cost of hydrogen (LCOH) produced by proton exchange membrane (PEM) electrolysis is 0.22 USD/m3 and 0.11 USD/m3 in China, 0.55 USD/m3 and 0.18 USD/m3 in

Japan (Ran and Zhou, 2025). This is also true for hydrogen produced from offshore wind. For example, the levelized cost of hydrogen (LCOH) produced from offshore wind in China and delivered to Japan in the form of methylcyclohexane (MCH) is 2 USD/kg and 1.8 USD/kg in 2030 and 2050, which is lower than the prices targeted for these years by the Japanese government, 3 USD/kg and 2 USD/kg (Song et al., 2021). Therefore, from the perspective of economic efficiency, there is a potential for Japan and China to cooperate in green hydrogen production.

One of the reasons that Japan prioritizes the development of hydrogen is that Japan is one of the leaders in hydrogen related technologies such as hydrogen fuel cell and gas turbines. In recent years, China has begun to catch up in hydrogen technologies. For example, China has surpassed Japan in hydrogen-related patents between 2013 and 2022 and ranked No.1 in overall patent competitiveness and in four areas (production, storage, transportation and supply, and safety management), but China is not No.1 in patent competitiveness in the area of utilization (United Daily News, 2025). Advanced technology in gas turbines is a foundation to develop the co-firing of hydrogen and natural gas or even to develop pure hydrogen-fired turbines. China is a leader in electric vehicles, which squeezed the development potential of hydrogen for road transportation in China. Different technological edges highly influenced the focus of low-carbon transition strategies of the two countries. Each country will leverage its technological edges to the maximum and compete through differentiation.

The hydrogen strategy evaluated by the "S + 3E" principle is illustrated as in Fig. 3. China has the strength of producing hydrogen at a lower cost. Japan has the strength in fuel cell technologies and a huge importing demand. These are the basis of future cooperations between these two countries. Some Japanese companies have begun to cooperate with Chinese partners in exporting green ammonia which is produced from green hydrogen from China to Japan. For example, Marubeni has signed a long-term offtake agreement with China's Envision Energy for green ammonia in Inner Mongolia, China in 2025 (Marubeni, 2025). Jilin Electric Power, a subsidiary of China's State Power Investment Corporation (SPIC) has signed an agreement to export green ammonia with many international customers including Japan's Itochu in 2025 (Yicai, 2025).

4.2 *Comparative Study of e-Methane Strategies*

4.2.1 E-methane Strategies of City Gas Companies

The use of e-methane does not result in a net increase of CO2 in the atmosphere because the CO2 used to produce e-methane is captured from the atmosphere or from point sources such as power plants or industrial facilities. One of the biggest advantages of e-methane is that it can use the existing infrastructure and end-use equipment.

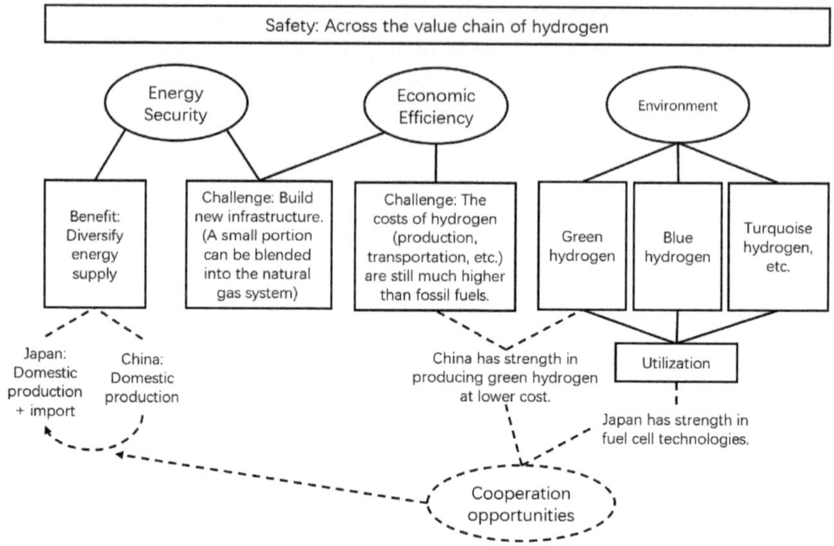

Fig. 3 The hydrogen strategy evaluated by the "S + 3E" principle

Developing e-methane is considered as a way to decarbonize both the gas system and the gas-fired power plants by major Japanese companies. Both Tokyo Gas Group and Daigas Group have set a target of 1% of e-methane in the gas grid by 2030, in line with Japan's national target. Japan's national target for 2050 is that e-methane is injected 90% into the gas infrastructure (METI, 2021).

All three companies have started demonstration projects in Japan to verify the methanation technologies. Tokyo Gas Group and Daigas Group are developing innovative methanation technologies. Tokyo Gas Group is developing PEMCO2 reduction method, which employs a Polymer Electrolyte Membrane (PEM) to simultaneously cause electrolysis of water and reduction of CO2 to form methane (Synthetic methane could smooth the path to net zero). Daigas is developing Solid Oxide Electrolyzer Cell (SOEC) methanation method. These innovative technologies can improve the efficiency of methanation and thus reduce the cost of e-methane production. Tokyo Gas aims to commercialize both Sabatier methanation and PEMCO2 technology by 2030 (Tokyo Gas, 2025). Daigas Group aims to achieve commercialization of Sabatier methanation around 2030 and to accelerate SOEC methanation around 2040 (Daigas, 2025). Toho Gas Group aims to start city gas production utilizing methanation technology by 2030 (Toho Gas, 2021).

In the meantime, Tokyo Gas Group, Daigas Group and Toho Gas Group are seeking opportunities to produce e-methane overseas with international partners. Target countries include the United States, Australia, Malaysia, UAE, Peru, etc. For example, Tokyo Gas, Daigas, Toho Gas, Mitsubishi Corporation, and Sempra Infrastructure are considering building e-methane projects in North America. The reason for Japanese city gas companies considering importing e-methane from abroad is the same as hydrogen, because e-methane produced in target countries is more economical than e-methane produced in Japan. In addition to e-methane, Daigas and Toho Gas also value bio-methane or biogas.

All three companies have joined the e-NG Coalition, an international organization promoting the international development of electric natural gas, another name for e-methane, members including companies from Japan, South Korea, the United States and Europe.

Unlike their Japanese counterparts, no major Chinese city gas companies have started to integrate e-methane into their low-carbon transition strategies. Chinese city gas companies mainly focus on reducing operational GHG emissions and have not included Scope 3 emissions in their low-carbon transition strategies. E-methane is a method to decarbonize Scope 3 emissions. Therefore, Chinese city gas companies have not included e-methane in their strategies. In addition, the cost of e-methane is much higher than natural gas at the current stage. Comparison of the cost of e-methane and natural gas is to be introduced in Sect. 4.2.2.

A comparison of the e-methane strategies of major city gas companies in Japan and China is shown in Table 17.

As for the reasons for Japan to develop e-methane, the Agency for Natural Resources and Energy (ANRE) of Ministry of Economy, Trade and Industry (METI) has made it clear that there are four reasons to develop e-methane in Japan (METI, 2023). Firstly, the existing city gas infrastructure can be used. Secondly, there is higher potential for e-methane to realize higher mass production than biomethane. Thirdly, e-methane produced in Japan contributes energy self-sufficiency. Fourthly, Japanese companies work vigorously and have strength in e-methane production technology, which can also contribute to Japan's economic development.

In addition, Japan has incorporated e-methane into the 6th and the 7th Strategic Energy Plans. Therefore, Japanese city gas companies put e-methane at a very high level in their low-carbon strategies.

Table 17 E-methane strategies of major city gas companies in Japan and China

Major city gas companies	in Japan	in China
Have an e-methane strategy	Yes	Without mentioning
Consider producing e-methane domestically and has started pilot projects	Yes	Without mentioning
Consider importing e-methane from abroad	Yes	Without mentioning

4.2.2 Evaluate the e-Methane Strategy by the "S + 3E" Principle

E-methane is one of the derivatives of hydrogen. There are some similarities between e-methane and hydrogen when they are evaluated by the "S + 3E" principle, but there are also some differences.

The safety of e-methane is easier to secure than hydrogen. Since methane is the main content of natural gas and can use the existing natural gas infrastructure, the safety of e-methane is the same as the safety of natural gas. Both countries have a long history of developing the natural gas system, so that the safety issue of e-methane is not an issue difficult to address for either Japan or China.

Since e-methane can be produced domestically in Japan, it is a type of energy that can diversify Japan's energy sources, which will enhance Japan's energy security. But similar to hydrogen, Japan has to produce e-methane domestically and import e-methane from overseas to meet its demand in the future. Japan aims to achieve mass production of e-methane by 2040 (METI, 2025), which means there are still challenges to provide a stable supply of e-methane. China produces about 60% of natural gas it consumes. China is importing natural gas, but it is hard to imagine that China will import e-methane in the foreseeable future. Since the existing infrastructure can be used, whether it is domestic infrastructure or infrastructure in other countries, the supply system does not need to be built, which contributes both securing energy supply and improving economic efficiency.

For economic efficiency, e-methane poses a greater challenge than hydrogen, since e-methane is produced by hydrogen and CO2, which is captured either from point sources like factories and power plants or from the atmosphere directly. In addition, the methanation technologies have not been deployed at large scale. The cost of e-methane is much higher than natural gas. Previous research finds that "In the base case, the levelized cost of e-methane in Japan is 5.71–10.18 times the JKM monthly prices in 2024, which is 3.12–5.56 times in China's case" (Ran and Zhou, 2025). This result also shows that it is more cost-effective to produce e-methane in China and export it to Japan. The fact that existing infrastructure can be used will improve economic efficiency to a certain extent. Similar to hydrogen, the cooperation opportunity between these two countries is also very large, but it is necessary to build a carbon accounting mechanism and a carbon pricing mechanism that are mutually recognized by both Japan and China (Ran and Zhou, 2025).

For environmental sustainability, e-methane is considered a good way to decarbonize the natural gas system. The environmental benefit of e-methane is obvious: "While e-methane does create CO_2 emissions when burned, it is made by recycling captured CO_2. This means the amount of CO_2 in the atmosphere does not increase in real terms" (Elliott, 2025).

The e-methane strategy evaluated by the "S + 3E" principle is illustrated as in Fig. 4. Japan is a promotor of e-methane and is one of the leaders in e-methane technologies. From the perspective of production capacity, it is considered that the technology development level of e-methane in Europe is the same as Japan at present, which is tens to hundreds m3/h (Nagahama, 2024). But the challenge to develop e-methane is still huge, especially from the perspective of economic efficiency. Like

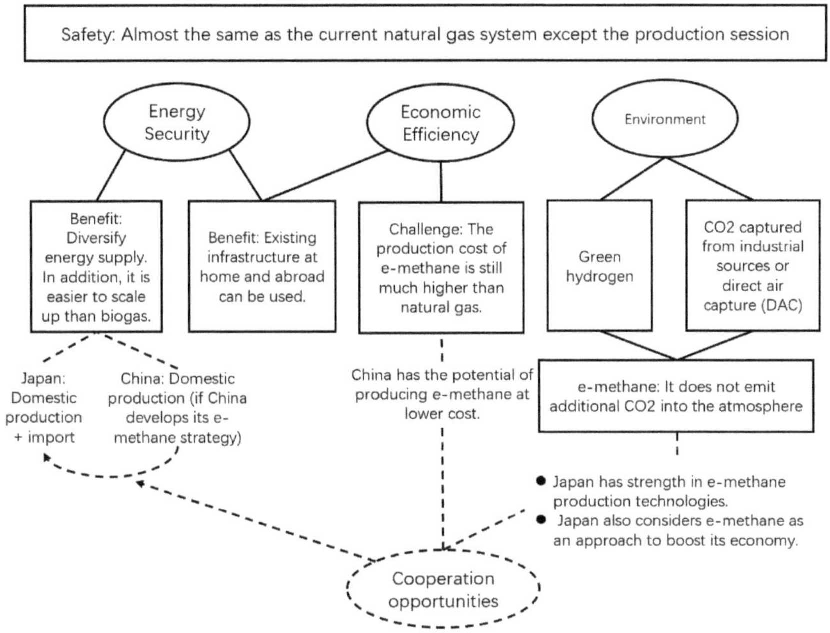

Fig. 4 The e-methane strategy evaluated by the "S + 3E" principle

hydrogen, there is an opportunity for Japan and China to cooperate on e-methane. If the cost of e-methane can be reduced considerably, China can also be a potential market for e-methane, which also provides business opportunities for Japanese companies.

4.3 Comparative Study of Renewable Energy and Related Strategies

All major Japanese and Chinese city gas companies consider developing renewable energy to decarbonize the electricity system. All major Japanese companies consider using natural gas-fired power plants to balance renewable power. Some major Chinese companies also consider using natural gas-fired power plants to balance renewable power, but generally use the term of integrating natural gas and renewables.

Tokyo Gas is developing floating offshore wind power in deep waters because of Japan's lack of shallow waters. For energy storage, Tokyo Gas is developing grid storage batteries to stabilize the grid and storage batteries at renewable energy power plants to balance the output of renewable power. Tokyo Gas is also developing virtual power plants (VPPs).

Daigas promotes the development of renewable energy power sources, including solar, onshore and offshore wind, and biomass. Daigas aims to develop 5 GW of renewable energy by 2030 and 10 GW of renewable energy by 2040 (Daigas, 2025). Daigas aims to advance grid storage battery technologies to enhance grid stability. Daigas is developing grid battery management/degradation diagnosis technologies that enable the use of low-cost second-life batteries. Daigas is also developing EV charging with renewable power. Daigas is also developing virtual power plants, like Tokyo Gas.

Toho Gas's FY2030 target is 500,000 kW of renewable energy, including the developing and purchasing renewable power at home and abroad. It is also developing storage batteries.

As mentioned above, Kunlun Energy aims to achieve the goal of having 100% of its electricity come from green power (Kunlun Energy, 2024). Kunlun Energy has projects that integrate gas-fired power plants and renewable energy. It is also developing comprehensive wind-PV-gas-power-hydrogen utilization demonstration projects and PV-storage-charging projects.

ENN Energy aims to increase the total installed capacity of PV projects to 9,900 MW by 2030 and to increase the proportion of renewable energy (including photovoltaic, biomass, geothermal, air, and hydrogen energy) and zero-carbon energy to 36% by 2030 (ENN, 2024). ENN Energy is considering exploring business models for integrating natural gas and renewable energy. It is also developing PV-energy storage projects.

China Gas owns BIPV projects, distributed PV projects and distributed wind power projects. It also has an electricity sales business which deals with green electricity transactions. Towngas is developing distributed PV projects and virtual power plants. CR Gas is developing distributed PV projects and EV charging stations.

It can be concluded that the renewable energy and related strategies of major Japanese and Chinese city gas companies are quite similar, with only minor differences.

A comparison of the renewable energy and related strategies of major city gas companies in Japan and China is shown in Table 18.

Table 18 Renewable energy and related strategies of major city gas companies in Japan and China

Major city gas companies	in Japan	in China
Develop solar, wind and biomass energy	Yes	Yes
Develop geothermal energy	Without mentioning	Yes
Purchase green electricity	Yes	Yes
Use gas-fired power plants to balance renewable power	Yes	Yes
Develop energy storage	Yes	Yes
Develop EV charging stations	Yes	Yes
Develop virtual power plants	Yes	Yes

China leads the world in developing solar and wind energies. According to the data of Ember Energy, "China accounted for more than half of the global increase in wind and solar power in 2024" and the share of wind and solar in electricity generation is 18% in China in 2024, which surpassed the global average of 15% and Japan's 11% (Ember, 2025). Due to the strength China has in this sector, it is comparatively easier for Chinese companies to deploy renewable energy to decarbonize their system than their Japanese counterparts.

4.4 Comparative Studies of Methane Emissions Control Strategies

Almost all major city gas companies take measures to control methane emissions from their operations, but major Chinese city gas companies see methane emissions control as an important way to achieve carbon neutrality and are more active in joining international methane-related initiatives. Japan as a country has joined the Global Methane Pledge, an international initiative to reduce global methane emissions by 30% from 2020 levels by 2030, but Japan has not released its national plan on controlling methane emissions. It should be noted that the target of the Global Methane Pledge is a global target, not a national target. China has not joined the Global Methane Pledge, but has released the Methane Emissions Control Action Plan, which is a national policy.

Tokyo Gas has included methane emissions control in its Carbon Neutrality Roadmap 2050. For example, Tokyo Gas is promoting thorough methane leak detection at LNG terminals and pipelines through daily patrols and other inspections. It is also working with LNG producers to improve visibility of upstream leaks.

Daigas did not include methane emissions control in its Energy Transition 2050, even though it stated that it strives to reduce GHG emissions from its business activities, including CO2 and methane in the Daigas Group Sustainability Report 2024 (Daigas, 2024). Toho Gas did not include methane emissions control in its 2050 Carbon Neutrality Initiative.

All five major Chinese city gas companies have included methane emissions control in their low-carbon transition plans and all of them have joined the China Oil and Gas Methane Alliance. The China Oil and Gas Methane Alliance aims to reduce the average methane emission intensity of the natural gas production process to less than 0.25% by 2025, and to reach the world-class level by 2035.

Kunlun Energy aims to reduce methane emission intensity by 20% from the 2020 level by 2030 and by 40% from the 2020 level by 2040 (Kunlun Energy, 2024). Kunlun Energy is advancing methane leak detection and repair (LDAR) by utilizing methods such as Pan-Tilt-Zoom (PTZ) lasers, drones, mobile vehicles, handheld devices and detection dogs. Kunlun Energy also manages Boil Off Gas (BOG) recycling.

ENN is also a member of the Methane Guiding Principles (MGP). ENN Energy regularly inspects the integrity of the pipeline network to prevent methane leaks. ENN Energy is optimizing the identification of methane emission sources to achieve full coverage of PTZ laser methane monitoring system in all city gas stations. Towngas is a member of the MGP.

China Gas has joined the OGMP 2.0 initiated by the United Nations Environment Programme (UNEP). China Gas is the only Chinese company joining the OGMP 2.0 and has achieved OGMP 2.0 Gold Standard recognition. China Gas has also conducted LDAR and BOG recycling. Japan's INPEX has joined OGMP 2.0, but INPEX is not a city gas company.

The methane emissions control strategies of major city gas companies in Japan and China are compared as follows in Table 19.

Japanese companies begun to pay attention to reduce methane emissions, but Japanese stakeholders mainly focus on promoting the methane emissions reduction from the LNG producers or exporters. In 2023, JERA Co., Inc. and Korea Gas Corporation jointly launched the Coalition for LNG Emission Abatement towards Net-zero (CLEAN) initiative, supported by the Japanese Ministry of Economy, Trade and Industry (METI), the Japan Organization for Metals and Energy Security (JOGMEC), the governments of the United States of America, the Republic of Korea, and Australia, and the European Commission. In 2024, 22 more Japanese companies has joined the CLEAN initiative (JOGMEC, 2024). The CLEAN initiative aims to promote the LNG producers to increase data transparency and take action to reduce methane emissions. In the future, there is possibility that Japanese and Chinese city gas companies working together to ask the LNG producers to take action to reduce upstream emissions.

Table 19 Methane emissions control strategies of major city gas companies in Japan and China

Major city gas companies	in Japan	in China
Take measures to control methane emissions by using LDAR, etc	Yes	Yes
Include methane emissions control into their low-carbon transition plans	Yes (some)	Yes (all)
Join international initiatives	Not yet	Yes
Methane Guiding Principles (MGP)		ENN Energy, Towngas, (and Beijing Gas)
Oil and Gas Methane Partnership 2.0 (OGMP 2.0)		China Gas

5 Discussion

Currently, Japan is mainly considering importing hydrogen and e-methane from Australia, Middle East, Southeast Asia, and North America. But as shown in Sect. 4.1 on hydrogen and Sect. 4.2 on e-methane, there are opportunities for Japan and China to cooperate in these two areas. This part will dive deeper into the opportunities of cooperation in these areas. Japan is one of the leaders in hydrogen and e-methane technologies although China is also catching up in hydrogen technologies. China has a huge domestic market for the deployment and scaling up of these technologies. The strengths of Japan and China are complementary in these areas, just like the cooperation between Japan's Mitsubishi Heavy Industries and China's Dongfang Electric Corporation on gas turbines. Therefore, Japan and China can build a "Technology × Market" linkage promotion mechanism between these two countries, as shown below in Table 20.

As introduced in Sect. 4.1.3, some Japanese companies like Marubeni and Itochu have signed contracts with Chinese companies to import hydrogen derivatives such as green ammonia from China to Japan, but these are only efforts from companies. To accelerate and expand the cooperation between Japan and China in these areas, governmental dialogues are also necessary. This paper recommends that in future bilateral governmental dialogues on economy or climate change, the mechanism introduced in this section could be included as a topic to discuss. Other topics such as cooperation in methane emissions control and reduction discussed in this paper could also be included in future bilateral talks.

Table 20 "Technology × Market" linkage promotion mechanism between Japan and China

Area	Promotion strategies
Standards and certification	Establish a standard and certification system on key issues such as carbon footprint accounting and quality requirements of hydrogen and e-methane that is mutually recognized by both Japan and China
Demonstration investment	Set up a mutual fund which can be named as "H_2 and e-methane Tech Bridge" by public and private consortium of both countries
Business cooperation	Develop and scale up products that best meet the market demand through Original Equipment Manufacturer (OEM), licensing and/or Mergers and Acquisitions (M&A)
Talent educational system cooperation	Design educational programs that foster talents on energy transition, carbon management, etc., which are carried out by universities and/or companies

6 Conclusion

This paper uses quantitative method to compare the low-carbon transition strategies of major city gas companies between Japan and China. In addition to the comparison of the overall strategies, this paper also compares the specific approaches to achieve the low-carbon transition strategies in detail, and this paper evaluates the hydrogen and e-methane strategies by the "S + 3E" principle.

Comparing the low-carbon transition plans of major Japanese city gas companies such as Tokyo Gas Group, Daigas Group, and Toho Gas Group, and major Chinese city gas companies such as Kunlun Energy Company Limited (Kunlun Energy), the Hong Kong and China Gas Company Limited (Towngas), China Gas Holdings Limited (China Gas), ENN Energy Holdings Limited (ENN Energy), and China Resources Gas Group Limited (CR Gas), the following conclusions can be drawn and related suggestions are provided.

(1) Major Japanese city gas companies have more robust low-carbon transition strategies and therefore rank higher than major Chinese city gas companies using the evaluation methodology designed in this paper. It is recommended that Chinese city gas companies should raise their ambitions to include the Scope 3 emissions reduction into their strategies to achieve net-zero.
(2) Major Japanese city gas companies have more comprehensive hydrogen strategies than their Chinese counterparts. But Japan faces more challenges to develop hydrogen than China, in terms of energy security (stable hydrogen supply) and economic efficiency. These challenges Japan is facing are also opportunities for Japan and China to cooperate, because China has advantage in mass production of clean hydrogen and China is already a huge market for hydrogen, though the share of clean hydrogen should increase.
(3) Japan is a promotor of e-methane and is one of the leaders the world in e-methane technologies, but the challenge to develop e-methane is still huge, especially from the perspective of economic efficiency. Similar to hydrogen, there are also opportunities for Japan and China to cooperate on e-methane. To achieve these opportunities, this paper proposed a "Technology × Market" linkage promotion mechanism between these two countries.
(4) Both Chinese city gas companies and Japanese city gas companies begin to transition into integrated energy companies by developing renewable projects, but Chinese companies have an edge in deploying renewables.
(5) Major Chinese city gas companies place a higher priority on methane emissions control than their Japanese counterparts in the low-carbon transition strategies. There are opportunities for Japanese and Chinese city gas companies working together to ask upstream LNG producers to increase data transparency on methane emissions and strengthen emission reduction actions.

Looking to the future, cooperation between Japan and China in the areas of hydrogen, e-methane, and methane emissions control could be a win–win situation for the city gas companies and other industrial stakeholders as they achieve their low-carbon transition strategies.

References

Barr C, Osborne-Saponja A, Ng A (2024) Low carbon transition rating methodology abstract.
CGA (2023) Research on low-carbon transition and development of China's city gas industry 2023. Urban and Rural Development 3:68–73
CR Gas (2025a) China Resources Gas Group Limited 2024 Environmental, Social and Governance Report. https://www.crcgas.com/English/2025esg/2025-04-28/b9ba07c4-e554-462a-9058-4ba0657b2c9d.pdf
CR Gas (2025b) CR Gas Special Column. Accessed April 5, 2025 from https://www.crcgas.com/jnjt/index.html
China Gas (2022) Read the China Gas Carbon Neutrality Action Report in One Chart. Accessed April 5, 2025 from http://www.jjckb.cn/2022-06/25/c_1310632540.htm
Daigas (2023) Daigas Group Energy Transition 2030. Accessed March 25, 2025 from https://www.daigasgroup.com/en/files/data/sustainability/reportpolicy/integrated_report/report2023_33-38.pdf
Daigas (2025) Daigas Group Energy Transition 2050. Accessed March 29, 2025 from https://www.daigasgroup.com/en/files/data/sustainability/reportpolicy/sustainability_report/energy_transition.pdf
Daigas (2024) Daigas Group Sustainability Report 2024. Accessed April 8, 2025 from https://www.daigasgroup.com/en/files/data/sustainability/reportpolicy/sustainability_report/report2024_all_interactive.pdf
Diao J (2025) How to make green hydrogen good to use and affordable? China City News, 08. https://paper.people.com.cn/zgcsb/pc/content/202503/24/content_30063825.html
ENN (2024) Decarbonisation Action 2030 - The Journey to Net Zero 2024 Edition. Accessed April 5, 2025 from http://ir.ennenergy.com/en/ir/sustainability/decarbonisation2030_netzero_2024.pdf
Elliott D (2025) E-methane's potential role in reducing emissions explained. Accessed August 13, 2025 from https://spectra.mhi.com/energy-transition/e-methanes-potential-role-in-reducing-emissions-explained
Ember (2025) China's electricity data in 2024. Accessed June 25, 2025 from https://ember-energy.org/countries-and-regions/china/#data
Enegy Institute (2024) Statistical Review of World Energy 2024. https://www.energyinst.org/__data/assets/pdf_file/0006/1542714/684_EI_Stat_Review_V16_DIGITAL.pdf
Enerdata (2024) Japan Energy Information. Enerdata. Accessed July 22, 2025 from https://www.enerdata.net/estore/energy-market/japan/
Intralink Limited (2022) The Hydrogen Economy Japan Market Report. https://exporttojapan.co.uk/wp-content/uploads/2023/03/Hydrogen-Report-2022-06-07-Final-amended_public.pdf
JOGMEC (2024) JOGMEC published the Results of the CLEAN initiative, "CLEAN Annual Report" at the LNG Producer-Consumer Conference 2024. Accessed June 6, 2025 from https://www.jogmec.go.jp/english/news/release/news_08_00019.html
Jahn V, Ingham R, Dietz S (2023) TPI's methodology report: management quality and carbon performance. https://www.transitionpathwayinitiative.org/publications/uploads/2023-methodology-report-management-quality-and-carbon-performance-version-5-0
Kunlun Energy (2024) Kunlun Energy Company Limited Action Plan Towards Carbon Peaking and Carbon Neutrality 2024 Edition. Accessed April 2025 from https://www.kunlun.com.hk/s2/n217/c7404/content.html
MEE (2022) Progress on the Implementation of China's Nationally Determined Contributions (2022). Accessed from https://www.mee.gov.cn/ywgz/ydqhbh/qhbhlf/202211/W020221111760730462299.pdf
METI (2023) About e-methane. Accessed from https://www.meti.go.jp/shingikai/enecho/denryoku_gas/denryoku_gas/gas_jigyo_wg/pdf/027_03_01.pdf
METI (2024) About the Hydrogen Society Promotion Act. Accessed from https://www.meti.go.jp/shingikai/enecho/shoene_shinene/suiso_seisaku/pdf/014_01_00.pdf

METI (2021) The 6th strategic energy plan. Ministry of Economy, Trade and Industry
METI (2025) The 7th strategic energy plan. Accessed from https://www.enecho.meti.go.jp/category/others/basic_plan/pdf/20250218_01.pdf
Marubeini (2025) Marubeni Signs a Long-Term Offtake Agreement for Green Ammonia Produced in Inner Mongolia, China. Accessed August 12, 2025 from https://www.marubeni.com/en/news/2025/info/00028.html
Mizuho (2023) How Japan Can Win the Global Competition for Hydrogen. https://www.mizuhogroup.com/binaries/content/assets/pdf/mizuho-bank/insights/industry/mif_237.pdf
Moody's Ratings (2023) Assessment Framework Net Zero Assessments. https://www.moodys.com/research/doc--PBC_1384532
NDRC (2022) Hydrogen Energy Industry Development Medium- and Long-Term Plan (2021–2035). Accessed from https://www.ndrc.gov.cn/xxgk/zcfb/ghwb/202203/P020220323314396580505.pdf
NEA & Guoneng Qingchuang Technology (2025) China Hydrogen Energy Development Report 2025. https://www.nea.gov.cn/20250430/96022785b3a747248288ad1c57d3a025/2025043096022785b3a747248288ad1c57d3a025_35aeee443346424eb4a3da029cb007003c.pdf
NEA (2024) China Natural Gas Development Report (2024). Petroleum Industry Press. https://www.nea.gov.cn/1310782456_17217223060611n.pdf
Nagahama Y (2024) Initiatives of e-methane (Synthetic Methane) development and social implementation. J Jpn Boiler Assoc (445):11–16. https://cir.nii.ac.jp/crid/1520300788745968128
Ran Z, Zhou W (2025) Greenhouse gas emissions and economic analysis of e-methane in Japan and China. Sustainability 17(8)
Sieler RE, Cames L, Schuster B, Borghardt S, Trobe BL (2021) Hydrogen factsheet - Japan. https://adelphi.de/en/system/files/mediathek/bilder/H2%20Factsheet%20Japan.pdf
Song S, Lin H, Sherman P, Yang X, Nielsen CP, Chen X, McElroy MB (2021) Production of hydrogen from offshore wind in China and cost-competitive supply to Japan. Nat Commun 12(1):6953. https://doi.org/10.1038/s41467-021-27214-7
TCFD (2021) Implementing the Recommendations of the Task Force on Climate-related Financial Disclosures. https://assets.bbhub.io/company/sites/60/2021/07/2021-TCFD-Implementing_Guidance.pdf
TPI (2025a) Management Quality of Osaka Gas. Accessed May 5, 2025 from https://www.transitionpathwayinitiative.org/companies/osaka-gas
TPI (2025b) Management Quality of Tokyo Gas. Accessed May 5, 2025 from https://www.transitionpathwayinitiative.org/companies/tokyo-gas
The Ministerial Council on Renewable Energy, Hydrogen and Related Issues (2023) Basic Hydrogen Strategy. Accessed from https://www.meti.go.jp/shingikai/enecho/shoene_shinene/suiso_seisaku/pdf/20230606_5.pdf
The Japan Gas Associaton (2021) Carbon Neutral Challenge 2050 Action Plan. https://www.gas.or.jp/pdf/en/newsrelease/newsrelease_20211101.pdf
Toho Gas (2021) Toho Gas Group 2050 Carbon Neutrality Initiative. Accessed March 29, 2025 from https://www.tohogas.co.jp/corporate/ir/pdf/carbon-neutrality-initiative.pdf
Tokyo Gas (2025) Synthetic methane could smooth the path to net zero. Accessed April 7, 2025 from https://www.nature.com/articles/d42473-022-00166-2
Tokyo Gas (2022) Tokyo Gas Frequently Asked Questions. Accessed March 30, 2025 from https://www.tokyo-gas.co.jp/en/IR/library/pdf/anual/07e24.pdf
Tokyo Gas (2024) Tokyo Gas Group Carbon Neutrality Roadmap 2050. https://www.tokyo-gas.co.jp/en/IR/support/pdf/20240322-03e.pdf
Tokyo Gas (2021) Tokyo Gas Group Compass Action. Accessed March 29, 2025 from https://www.tokyo-gas.co.jp/Press_e/20211126-02e.pdf
Towngas (2024) Towngas' Carbon Neutrality Roadmap by 2050. Accessed April 5, 2025 from https://www.towngas.com/media/qrllfo1v/towngas-carbon-neutrality-roadmap-by-2050.pdf

United Daily News (2025) China Surpasses Japan in Hydrogen Energy Patent Competitiveness, Ranks First Globally. United Daily News. https://uniteddaily.my/en/6dd98321-9925-40d5-939d-65d27b0b7842

Yicai (2025) Jilin Electric Power's world's largest single green ammonia project has been commissioned, with multiple green hydrogen-based energy off-takers and partners already secured in advance. Accessed August 12, 2025 from https://finance.sina.com.cn/jjxw/2025-07-26/doc-infhvfqm0525858.shtml?froms=ggmp

Impact of Urban Land-Sea Distribution on Future Building Energy Consumption in Chinese Cities Under Climate Change: A Comprehensive Study

Wanxiang Yao, Shuangshuang Li, Teng Yang, Ting Lei, and Bart Julien Dewancker

1 The Impact of Land-Sea Distribution on Future Building Energy Consumption Under Climate Change

Building operations constitute a major contributor to global greenhouse gas emissions. It is estimated that the construction sector consumes between 30 and 40% of total global energy, with fossil fuels remaining the dominant source. Such dependence on fossil energy intensifies climate change while simultaneously creating profound environmental and economic challenges (Lu and Lai 2019). Nevertheless, the building sector possesses considerable capacity for energy conservation (Yao et al. 2024). Projections suggest that by 2050, global carbon dioxide emissions could be reduced by no less than 820 million tons, equivalent to nearly 61% of the baseline scenario (Mirasgedis et al. 2024). Existing research addressing the implications of future climate change on building energy consumption can be broadly categorized into four domains: (1) the effect of climate change on building energy demand; (2) the adaptation of building design and energy efficiency measures in response to climate

W. Yao (✉) · S. Li
iSMART, Qingdao University of Technology, Qingdao, China
e-mail: yaowanxiang@126.com

S. Li
e-mail: lishuangshuangJZ@163.com

T. Yang · T. Lei · B. J. Dewancker
Faculty of Environmental Engineering, The University of Kitakyushu, Kitakyushu, Japan
e-mail: ytseven77@163.com

T. Lei
e-mail: Lei19931129@163.com

B. J. Dewancker
e-mail: bart@kitakyu-u.ac.jp

W. Zhou (ed.), *Decarbonization Policy Engineering*, World Sustainability Series,
https://doi.org/10.1007/978-981-95-5953-4_17

change; (3) the conversion and application of climate datasets, particularly the role of typical meteorological year (TMY) data under global warming scenarios; and (4) the influence of climate change on construction-related policies and regulatory frameworks.

Firstly, a growing body of research has examined how climate change influences building energy requirements across different regions. For example, Hietaharju et al. (2021) constructed a stochastic dynamic model of building stock to assess the combined impacts of climate change and renovation strategies on regional heating demand. Similarly, Jalali et al. (2023) utilized statistical downscaling techniques to generate future weather datasets and evaluated their effects on residential building performance in New Zealand's diverse climate zones. Their analysis demonstrated that climate change will significantly reshape the thermal performance of housing in the region. In Japan, Yuan et al. (2024) investigated building energy consumption across four cities using representative meteorological data for the 2020s and 2086s, recommending design strategies to improve resilience against climate impacts. Extending this approach to China, Xiong et al. (2023) applied global climate models under varying Shared Socioeconomic Pathways (SSP) to generate future weather inputs and forecast heating and cooling loads for cities across multiple climate zones. On a broader European scale, Yang et al. (2021) incorporated future climate projections from 2010 to 2099 using the RCA regional climate model to simulate the energy performance of urban residential buildings in 38 cities spanning five climate regions. Finally, Kamal et al. (2023) addressed the combined influence of climate change and urban heat islands on the present and future cooling loads of high-rise residential buildings in Qatar's hot-humid environment, emphasizing the necessity of accounting for both drivers in building energy simulations.

Secondly, in recent years, the exploration of building design strategies and energy efficiency under projected climate change scenarios has become an important direction in the field of building energy conservation. A number of representative studies illustrate this trend. For instance, Rai et al. (2024) evaluated the implications of climate change for heating and cooling (H/C) load demands in typical residential rooms in India, proposing a novel approach to quantify load contributions from components such as walls, windows, and infiltration, thereby enhancing the energy efficiency of future building design. Wehrli et al. (2024) constructed a reference dataset for building designers and engineers by compiling one year of hourly weather data from the Swiss CH2018 climate scenarios, grounded in records from 45 monitoring stations across Switzerland. In China, Cao et al. (2017) studied five representative cities across major climate zones, identifying significant increases in outdoor design temperatures for both heating and cooling, thus underscoring the necessity of differentiated energy-saving strategies tailored to regional climatic variations. In addition, Ashrafian (2023) examined the interplay between climate change and building performance under cost-optimal and nearly zero-energy scenarios, highlighting the regional importance of climate-responsive design in balancing energy consumption, costs, and occupant comfort. Expanding to a systems perspective, Ran et al. (2024) developed a coordinated design framework for integrating buildings with Regional Integrated Energy Systems (RIES) under future climate conditions, validated through both Typical Meteorological Years (TMYs) and SSP2-4.5 and SSP5-8.5 scenarios. At the urban scale, Chen et al. (2023) investigated the combined effects of climate change and energy conservation measures (ECMs) on more than 50,000 buildings in Changsha, offering important insights into urban energy planning and carbon mitigation. Finally, Rostam et al. (2024) introduced a dynamic building energy design framework (CABED), which enables adaptive responses to evolving climate conditions, including extreme weather events.

Thirdly, in projecting building energy consumption under prospective climate change scenarios, the transformation of baseline meteorological records and the construction of Typical Meteorological Year (TMY) datasets play a decisive role in securing the reliability of simulation outcomes. Reveshti et al. (2023) examined the implications of employing both updated and conventional TMY datasets on future energy demand estimations, concluding that the revised dataset yielded more accurate forecasts, particularly indicating reduced heating loads and increased cooling requirements in cities such as Tabriz and Tehran. Similarly, Cui et al. (2017) employed a 55-year meteorological record from China to compare building simulation results derived from typical-year data against those from multi-year datasets, revealing substantial discrepancies between the two approaches. To further incorporate climate change attributes into performance evaluation, Farah et al. (2019) introduced a method for embedding climate change signals into historical weather datasets, thus producing more representative weather inputs for building simulations. In parallel, Seyed Salehi et al. (2024) developed a TMY generation approach grounded entirely in long-term energy simulations, thereby enhancing predictive robustness. More recently, Rodrigues et al. (2023) advanced an open-source, cross-platform weather morphing tool that integrates Coupled Model Intercomparison Project Phase 6 (CMIP6) outputs to deliver high-resolution future hourly datasets for building performance assessments.

Lastly, numerous researchers emphasize that climate change is poised to reshape regulatory frameworks and policy orientations within the construction sector, prompting urgent intervention by governments to curb global warming and lower carbon emissions. Delia D'Agostino et al. (D'Agostino et al., 2024) analyzed the implications of climate change for the energy performance of European buildings, undertaking a detailed assessment of energy demand within the context of existing supply-side policies. Their findings provide critical guidance for adapting current policies to ensure buildings remain resilient under evolving climatic conditions. Similarly, Fernandes et al. (2024) examined future climate scenarios for office buildings in Portugal, proposing performance-enhancement strategies that strengthen both energy efficiency and environmental sustainability, thereby offering a conceptual foundation for policy and industry application. In Italy, Baglivo et al. (2023) explored how improvements in building envelope performance can reinforce climate adaptation strategies within the framework of national energy policies, offering direct recommendations for policy design. Meanwhile, Andric et al. (2020) investigated the residential sector in Qatar by selecting representative buildings from the national stock and modeling energy demand under current and projected climate data. They advocated for the integration of regional environmental policies with energy-efficient building technologies to alleviate climate-induced impacts. Beyond national perspectives, Mirasgedis et al. (2024) evaluated the broader contribution of buildings to climate mitigation and sustainable development, highlighting their synergies with the United Nations Sustainable Development Goals (SDGs) and emphasizing their pivotal role in advancing global sustainability agendas.

A growing body of literature has examined the nexus between climate change and building energy use, with most studies focusing on the four major dimensions highlighted above. In the context of accelerating global warming, the building sector is confronted with increasingly severe challenges. Existing research consistently shows that climate change not only reshapes global temperature regimes and meteorological conditions but also exerts significant influence on building energy demand and operational performance. Anticipated future climatic shifts are projected to substantially modify regional requirements for heating and cooling. Nevertheless, investigations linking urban land-sea distribution with building energy consumption remain absent, and the trajectory of this relationship under future climate scenarios is poorly understood. The spatial configuration of cities relative to coastlines plays a crucial role: coastal areas often benefit from oceanic moderation, whereas inland regions tend to experience larger thermal fluctuations and more frequent extreme events. Current scholarship is limited in several respects: (1) insufficient clarity on how urban meteorological parameters affected by land-sea distribution will evolve relative to baseline conditions; (2) the dynamic interaction between land-sea distribution and building energy consumption under projected climate change has not been systematically addressed; and (3) quantitative frameworks capable of capturing this relationship in future urban contexts are still lacking, which constrains theoretical support for urban planning, energy-efficient building design, and policy formulation.

To address the research questions, this study adopts a three-step methodological framework: (1) future meteorological data for sixty representative cities in China, characterized by diverse land–sea distributions and climate zones, were projected using the latest CMIP6 climate model; (2) the control variable approach was then applied to systematically examine the correlations and evolutionary trends between land–sea distribution and urban building energy consumption, under both present and projected global warming conditions; (3) finally, a dimensionless analysis was employed to quantitatively characterize the relative variation patterns of energy demand in relation to land–sea distribution, with projections specifically targeted for the mid- and late-twenty-first century (2050s and 2080s).

2 Methodology for Assessing Land-Sea Distribution's Impact on Building Energy Consumption

2.1 Research Framework

The research was designed in accordance with the analytical framework illustrated in Fig. 1. To explore how urban land–sea distribution influences building energy consumption, sixty representative cities located within the same climatic zone and latitude band in China were selected using a control variable approach and classified into six groups. For these cities, typical meteorological year (TMY) datasets were processed, and future TMY projections for the 2050s and 2080s were generated based on CMIP6 outputs under the SSP3-7.0 scenario. These datasets were subsequently applied in building energy simulation software to evaluate annual heating and cooling demands under both present and future conditions. Finally, by combining the cities' land–sea distribution characteristics with the simulation outcomes, the study analyzed the correlation between spatial distribution and building energy consumption across different time horizons. The quantified results offer a scientific basis for formulating energy-saving strategies and policies for future urban development.

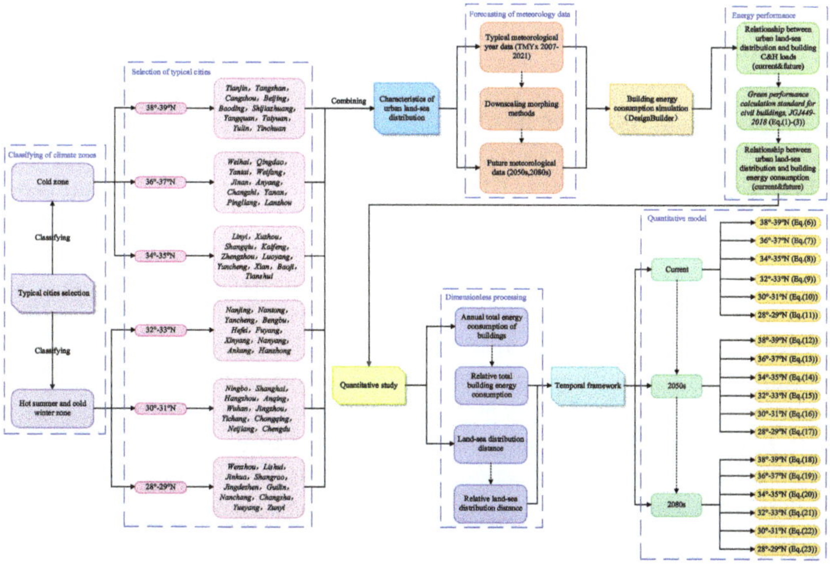

Fig. 1 Research framework

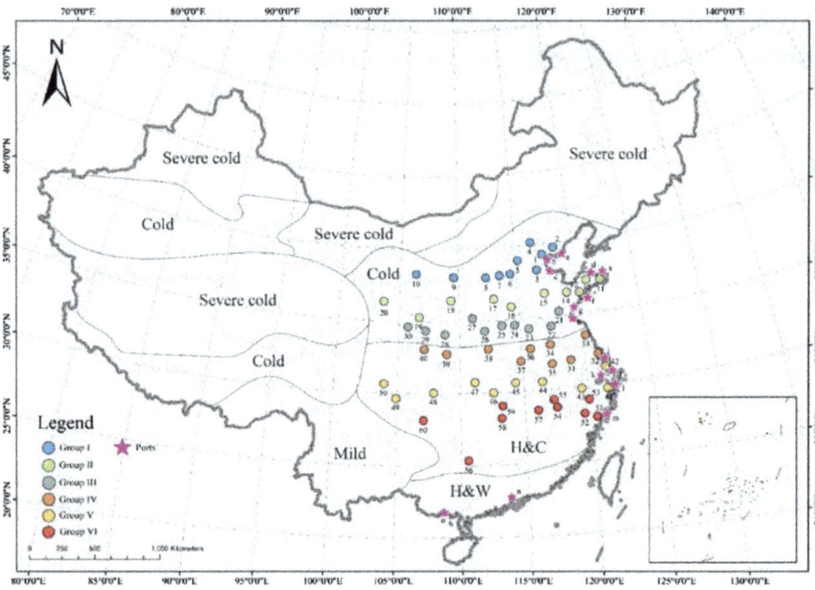

Fig. 2 Distribution of target cities and ports in China

2.2 *Data Processing*

2.2.1 Selection of Typical Cities

Due to its vast geographical span, China exhibits pronounced climatic heterogeneity. Accordingly, the nation is conventionally divided into five representative thermal zones: the Severe Cold region, the Cold region, the Hot Summer–Cold Winter (H&C) region, the Hot Summer–Warm Winter (H&W) region, and the Mild region. For the purpose of this study, the H&C and Cold zones were selected as the focus due to their geographical distribution across China, with cities in these zones being evenly spread throughout the country and a substantial number of cities located within them.

Using the control variable method, sixty representative cities from these two climate zones were chosen as target cities for the study, as illustrated in Fig. 2 To ensure that the selected cities experience similar levels of solar radiation, these cities were further categorized into six groups based on their latitude ranges, as outlined in Table 1.

2.2.2 Determination of Land-Sea Distribution Distances

In defining the characteristics of the target cities' land-sea distribution, this study takes into account the dynamic nature of coastline changes over time, which complicates precise quantification (Cai et al. 2022). Consequently, the shortest distance

Table 1 Target cities grouping information

City number	Climate zone	Latitude range	Group number
1–10	Cold	38°-39°N	I
11–20	Cold	36°-37°N	II
21–30	Cold	34°-35°N	III
31–40	H&C	32°-33°N	IV
41–50	H&C	30°-31°N	V
51–60	H&C	28°-29°N	VI

from the centroid of each Chinese city to the nearest port was selected as a reference factor (Sun and Wang 2024). By utilizing these varying shortest distances as variables, the study investigates the relationship between land-sea distribution and building energy consumption. The data on the shortest distances from the centroids of the target cities to the nearest ports were obtained from the Ministry of Transport of the People's Republic of China (MOT 2006). The distribution of the ports involved in this study is depicted in Fig. 2, with specific latitude and longitude coordinates for each port provided in Table 2.

Table 2 Information on the geographic location of ports

Port Number	Province	Port name	North latitude (degree)	Longitude (degree)
a	Hebei	Tangshan port	39.22234	119.0078
b	Tianjin	Tianjin port	38.9822	17462
c	Hebei	Huanghua port	38.31365	19119
d	Shandong	Yantai port	37.55524	121.3919
e	Shandong	Weihai port	37.44626	122.1982
f	Shandong	Qingdao port	36.09743	120.3286
g	Shandong	Rizhao port	35.38067	119.5557
h	Jiangsu	Lianyungang port	34.74736	119.435
i	Jiangsu	Nantong port	32.0072	120.8203
j	Shanghai	Shanghai port	30.84651	121.8349
k	Zhejiang	Jiaxing port	30.59819	121.0855
l	Zhejiang	Ningbo port	29.9206	121.8684
m	Zhejiang	Wenzhou port	27.96256	120.7997
n	Guangdong	Guangzhou port	23.09421	113.452
o	Guangxi	Qinzhou port	21.72705	108.5884

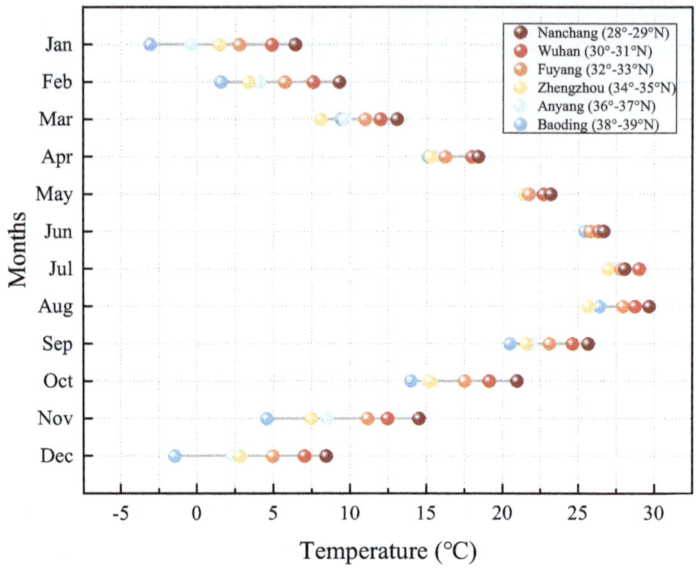

Fig. 3 TMY mean monthly temperatures for the six median cities

2.3 *Analysis of the Measured Meteorological Data*

2.3.1 Analysis of the Current Meteorological Data

The TMY data (TMYx, 2007–2021) for sixty cities across various climate zones in China were gathered and organized for analysis (Climate. OneBuilding. Org 2024). Among the six groups of target cities, the monthly average dry bulb temperatures for typical years in six cities, whose centroid distance to the nearest port lies at the median, are presented in Fig. 3. As illustrated in Fig. 3, notable variations in the typical monthly average temperatures exist among cities located in different latitude ranges. The temperature differences are more pronounced in winter and relatively smaller in summer. Therefore, it is essential to categorize cities in different climate zones according to their latitude to minimize experimental errors.

2.3.2 Prediction of Future Meteorological Data

In climate science research, CMIP6 serves as a crucial framework that offers detailed predictions of future climate change scenarios. This initiative consolidates leading climate models from across the globe to enhance the understanding of climate system changes and the impact of human activities on the climate (Eyring et al. 2016).

Through the CMIP6 framework, scientists are able to assess the potential consequences of different greenhouse gas emission pathways, thereby providing a scientific foundation for policy development.

The Coupled Model Intercomparison Project Phase 6 (CMIP6) constitutes the most recent stage of the internationally coordinated climate modeling initiative led by the Working Group on Coupled Modeling (WGCM) under the World Climate Research Programme (WCRP). Officially initiated in 2016, CMIP6 represents the most extensive and systematic comparison of global climate models to date. Its scientific design is structured around a hierarchical framework of "Grand Challenges" and associated scientific questions, focusing on three central issues: the response of the Earth system to external forcing, persistent model biases, and modes of climate variability. Within this framework, CMIP6 incorporates 23 specialized Model Intercomparison Projects (MIPs), which collectively address a broad spectrum of topics ranging from reconstructions of paleoclimate conditions to projections of future climate scenarios (Zhou et al. 2019).

The CMIP6 modeling framework is constructed on an advanced system of coupled Earth components, integrating atmospheric models (AGCMs), ocean models (OGCMs), land surface modules, and sea ice dynamics. The atmospheric component has been upgraded to resolutions between 25 and 100 km, comprising 60–90 vertical layers, while the ocean counterpart attains horizontal resolutions of 0.25–1 degree with 50–100 vertical levels. Major refinements have been introduced in physical parameterizations, covering cloud microphysics, convective processes, planetary boundary layer dynamics, and land–atmosphere interactions. Additionally, the inclusion of sophisticated biogeochemical processes enables explicit representation of carbon and nitrogen cycles, alongside other critical feedback mechanisms. From a scenario perspective, CMIP6 adopted the Shared Socioeconomic Pathways (SSPs), which provide a spectrum of future emission trajectories ranging from SSP1-1.9 to SSP5-8.5, thereby strengthening the integration of socioeconomic development narratives with climate projections. Technically, CMIP6 advances are characterized by finer spatial resolution, enhanced physical process schemes, and stronger coupling across the Earth system. The program has mobilized contributions from around 50 modeling centers worldwide, generating more than 100 climate model versions and producing an estimated 100 petabytes of data. (O'Neill et al. 2016).

This study employs the BCC-CSM2-MR model from the CMIP6 climate models (World Climate Research Program, WCRP) to project future climate scenarios. Developed by the Beijing Climate Center in 2017, this model encompasses 51,200 data grid points. The SSP3-7.0 medium–high greenhouse gas emission scenario is selected, as it aligns most closely with the current shared socio-economic pathways. Consequently, using the collected current TMY data, future TMY climate data for sixty cities across various climate zones in China were predicted under the SSP3-7.0 emission scenario for the temporal periods of the 2050s (2036–2065) and 2080s (2066–2095). This was done through downscaling to the general circulation model (GCM) global grid (Cui et al. 2017), with reference to the CMIP6 baseline climate model years from 1985 to 2014. The prediction results are presented in Sect. 3.1.

Table 3 Meteorological data morphing reference factors

Meteorological Factors	Typical meteorological year data	Future meteorological data	DesignBuilder simulation
Dry Bulb Temperature (°C)	√	√	√
Dew Point Temperature (°C)	√	√	√
Relative Humidity (%)	√	√	×
Atmospheric Pressure (Pa)	√	√	√
Horizontal Infrared Radiation from the Sky (W/m^2)	√	√	×
Direct Normal Radiation (W/m^2)	√	√	√
Diffuse Horizontal Radiation (W/m^2)	√	√	√
Direct Normal Illuminance (lux)	√	√	×
Wind direction (°)	√	×	√
Wind Speed (m/s)	√	√	√
Total Sky Cover	√	√	×
Opaque Sky Cover	√	√	×

Note √ means the changes are included, × means the changes are not included

Under the projected global warming scenario, several meteorological factors, including dry bulb temperature, dew point temperature, relative humidity, solar radiation, wind direction, and wind speed, are expected to change. This study examines these changes in meteorological factors in the context of future climate change and compares them with energy consumption simulation data, as detailed in Table 3.

2.4 Typical Building Models

The prototypical building adopted in this study is a ten-story office facility with a total floor area of 24,643 m^2. The building geometry was developed using Building Information Modeling (BIM) software and subsequently imported into DesignBuilder for the purpose of energy demand simulation, as illustrated in Fig. 4. The reliability of the simulation framework was confirmed in a previously published study, where the model was benchmarked against an established case reference. For the present analysis, the model was further refined and adjusted to facilitate extensive computational experiments involving large-scale simulation tasks. (Lei et al. 2024).

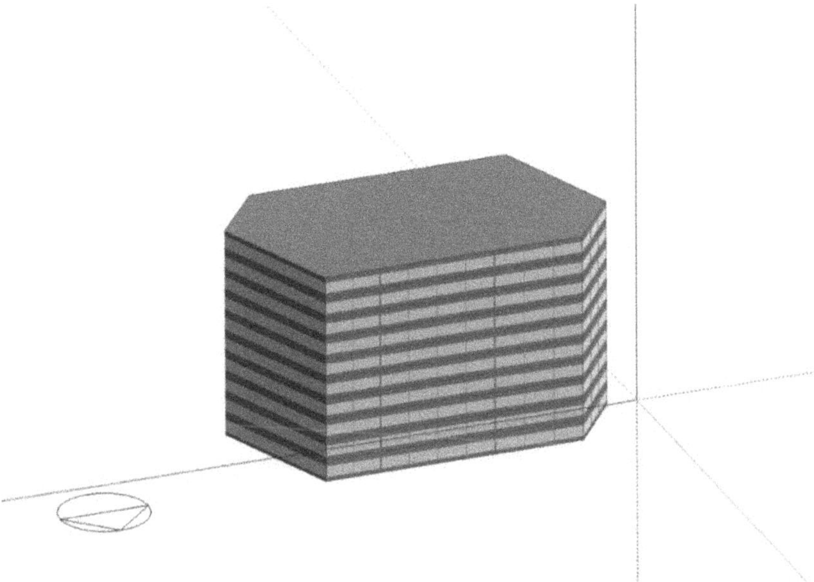

Fig. 4 Building energy simulation model

The current and future TMY weather data in EPW format were input into DesignBuilder to simulate the building's annual load demand, with a time step of 1/8 h. The building's energy model information, including details on orientation, materials, window-to-wall ratio, and fence structures, is provided in Table 4. The settings for different functional areas, including density, operating schedules, and set temperatures based on usage, are outlined in Table 5 (Qian et al. 2020). Previous studies have indicated that the measured annual energy consumption per unit area for office buildings in China's hot summer and cold winter zone ranges from 36.19 to 38.44 kWh/m^2 (Gao et al. 2023). In this study, the simulated average annual energy consumption per unit area for the target building in the same climate zone is approximately 33.42 kWh/m^2. This value shows a small deviation from the measured values, suggesting that the simulation model is of practical relevance in engineering research.

Table 4 Integral parameter setting

Items	Unit	Settings
Orientation	-	South
Roof heat transfer coefficient	W/(m^2·K)	0.42
External wall heat transfer coefficient	W/(m^2·K)	0.45
Outside or elevated floor slabs transfer coefficient	W/(m^2·K)	0.46
External window transfer coefficient	W/(m^2·K)	2.0
Window-wall ratio (E,S,W,N)	%	0.55, 0.56, 0.54, 0.54

Table 5 Functional area parameter setting

Function	Persons per unit area (Person/m^2)	Daily working time	Equipment heat dissipation (W/m^2)	Setting temperature (Cold,°C)	Setting temperature (Heat,°C)
Office	0.1	7:00–19:00	15	26	22
Corridors and others	0.01	7:00–22:00	0	26	22

3 Analysis of Future Trends and Impacts

This section first contrasts the baseline meteorological conditions of sixty representative Chinese cities across diverse climate zones with projected climate scenarios, emphasizing the disparities linked to varying land–sea distributions. It then examines, through the control variable method, the correlations and evolving trends between urban land–sea configurations and building energy demand under both present and anticipated climatic contexts. Finally, it quantifies the projected relative variations in urban land–sea distribution and associated building energy consumption, providing a basis for future assessments.

3.1 Prediction of Meteorological Data Under Future Scenarios

Based on simulations with CMIP6 climate models, this study generates hourly meteorological datasets for sixty representative cities spanning diverse climate regions in China under future climate scenarios. Figure 5 illustrates the projected variations in annual mean dry-bulb temperature relative to the baseline period. Under the SSP3-7.0 scenario, these cities are expected to undergo discernible warming by the 2050s and 2080s, with magnitudes differing across climatic zones (see Fig. 5). Both future periods exhibit pronounced increases in air temperature when compared to the reference year. Specifically, by the 2050s, temperatures in cold regions are projected to rise by approximately 2.0–3.5°C, while those in hot summer and cold winter (H&C) zones may increase by 2.4–3.9°C, corresponding to mean changes of 2.69°C and 2.74°C, respectively. By the 2080s, these values are projected to range from 3.4–5.4 °C in the cold zone and 2.7–4.5 °C in the H&C zone, with average increases of 4.4 °C and 3.4 °C, respectively. The temperature rise in the cold zone is projected to be greater than that in the H&C zone by the 2080s, while the temperature increase in the two zones is relatively similar by the 2050s. Notably, when transitioning from the 2050s to the 2080s, the cold zone exhibits a higher warming trend compared to the H&C zone. In particular, cities in groups V (30°–31°N) and VI (28°–29°N) demonstrate a smaller increase in annual average temperature from the 2050s to the

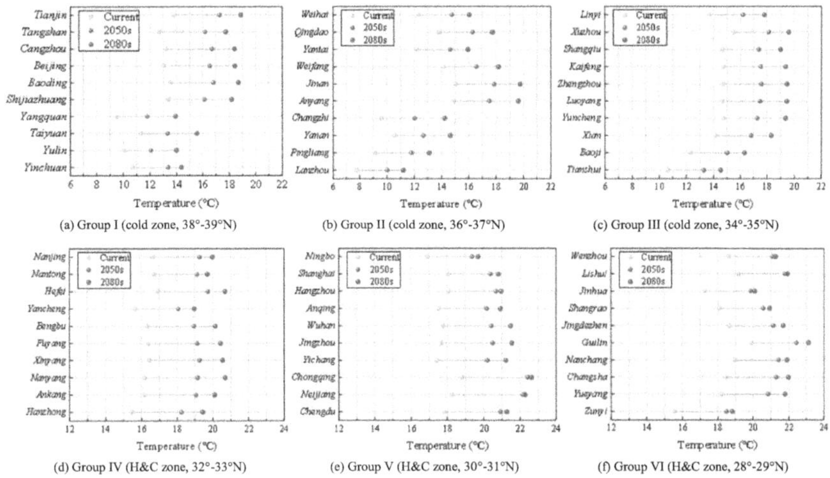

(a) Group I (cold zone, 38°-39°N) (b) Group II (cold zone, 36°-37°N) (c) Group III (cold zone, 34°-35°N)
(d) Group IV (H&C zone, 32°-33°N) (e) Group V (H&C zone, 30°-31°N) (f) Group VI (H&C zone, 28°-29°N)

Fig. 5 Increase in the annual mean dry-bulb temperature by the 2050s and 2080s in major cities in China under the SSP3-7.0 scenario

2080s. This phenomenon may be attributed to the latitudinal position of these cities, as those at lower latitudes receive higher annual average solar radiation. However, this also means they are closer to the maximum solar heat load, so, despite continued global warming, the additional solar radiation absorbed remains relatively lower.

In the context of projected global warming, this study investigates the evolution of solar radiation and relative humidity (RH) across sixty representative cities, as illustrated in Fig. 6. The analysis reveals a consistent increase in the annual mean diffuse horizontal radiation (DHR) and a concurrent decline in RH in both the cold region and the hot summer–cold winter (H&C) climatic zone. Compared with the baseline period, DHR demonstrates a pronounced upward trend in all cities. Under the SSP3-7.0 scenario, the cold zone is expected to experience an average rise of approximately 4 W/m^2 by the 2080s, whereas the H&C zone shows a larger increment of about 24.3 W/m^2. The magnitude of this increase is more significant in lower-latitude H&C cities. For instance, in Groups IV–VI, cities such as Xinyang, Jingzhou, and Wenzhou are projected to exceed 90 W/m^2 in DHR by the 2080s, a level not anticipated in the higher-latitude Groups I–III. Regarding humidity, coastal cities generally maintain higher annual mean RH compared with their inland counterparts, though local topographic influences create exceptions. For example, Weihai, Qingdao, and Yantai (Group II) consistently display markedly higher RH than inland cities such as Weifang and Jinan, while Yinchuan (Group I) is projected to show a slight increase in RH owing to its unique geographical features. Overall, RH values are higher in the H&C zone than in the cold zone, reflecting the greater solar input at lower latitudes, which enhances evaporation and atmospheric moisture content. Nevertheless, as climate warming intensifies, RH is expected to decline. By the 2080s, the cold zone is projected to experience an average reduction of about 3.1%, while the H&C

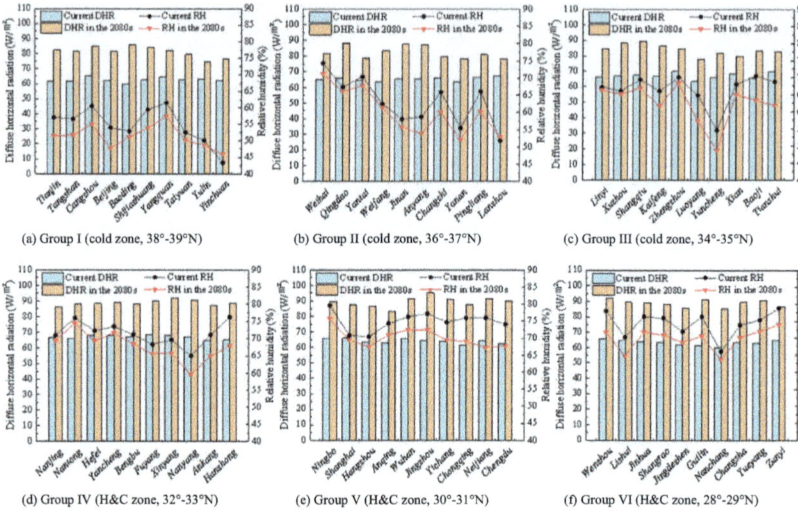

(a) Group I (cold zone, 38°-39°N)
(b) Group II (cold zone, 36°-37°N)
(c) Group III (cold zone, 34°-35°N)
(d) Group IV (H&C zone, 32°-33°N)
(e) Group V (H&C zone, 30°-31°N)
(f) Group VI (H&C zone, 28°-29°N)

Fig. 6 Trends in the annual DHR and RH by the 2080s in major cities in China under the SSP3-7.0 scenario

zone may face a sharper decrease of roughly 4.2%, suggesting a more pronounced drying trend in the latter.

3.2 *Effects of Urban Land and Sea Distribution on Future Building Energy Consumption*

The sixty representative cities located in China's cold and heating–cooling (H&C) zones were systematically classified based on their geographic coordinates (latitude and longitude). For each city, simulations were conducted using building energy consumption software, which incorporated baseline-year meteorological records as well as climate projections for the 2050s and 2080s under the SSP3-7.0 scenario. The analysis computed annual heating and cooling demands, together with total energy consumption, at a high temporal resolution of 0.125 h.

3.2.1 Cooling and Heating Load

The annual cooling load for buildings in the sixty target cities, compared between the baseline year, the 2050s, and the 2080s, is presented in Fig. 7, with the annual heating load depicted in Fig. 10 The distribution maps showing the percentage change (either increase or decrease) in cooling load for the 2050s and 2080s relative to the baseline year are shown in Figs. 8 and 9, respectively, while the heating load changes are illustrated in Figs. 11 and 12. As depicted in Fig. 7, the cities within each of the six groups are ordered from left to right based on the distance from the city centroid to the nearest port, reflecting the variations in land-sea distribution for each city. It is apparent from the figures that, under the influence of future global warming, the annual cooling load in cities across both the cold zone and H&C zone is expected to rise in the 2050s and 2080s. As shown in Figs. 8 and 9, the growth in the cold zone is relatively uniform, with similar rates of increase projected for both the 2050s and 2080s, resulting in a higher demand for cooling. Under the SSP3-7.0 scenario, compared to the baseline year, the annual cooling load in cities within the cold zone is predicted to increase by an average of 36.3% by the 2050s, and by approximately 85.5% by the 2080s. In contrast, cities in the H&C zone are expected to experience a more pronounced increase from the baseline year to the 2050s, followed by a smaller increase from the 2050s to the 2080s, in line with temperature changes. By the 2050s, the annual cooling load in cities within the H&C zone is projected to increase by an average of 38.8%, with a further growth of approximately 55.1% by the 2080s. These results demonstrate that the growth rate of annual cooling load varies significantly among cities with different land-sea distributions. Typically, as the distance of a city from the coastline increases, the annual cooling load initially increases, then subsequently decreases.

In the context of global warming, not only is the annual cooling load of buildings expected to undergo significant changes, but the annual heating load will also be influenced to some extent. As depicted in Fig. 10, across the six experimental groups, the annual heating load of buildings is projected to decrease over the future temporal framework as climate change progresses. By examining Figs. 11 and 12, it is evident that, compared to the baseline year, the annual heating load in cities within the cold zone and H&C zone will experience a marked reduction by the 2050s, followed by a relatively smaller decrease by the 2080s. This trend is more pronounced in cities

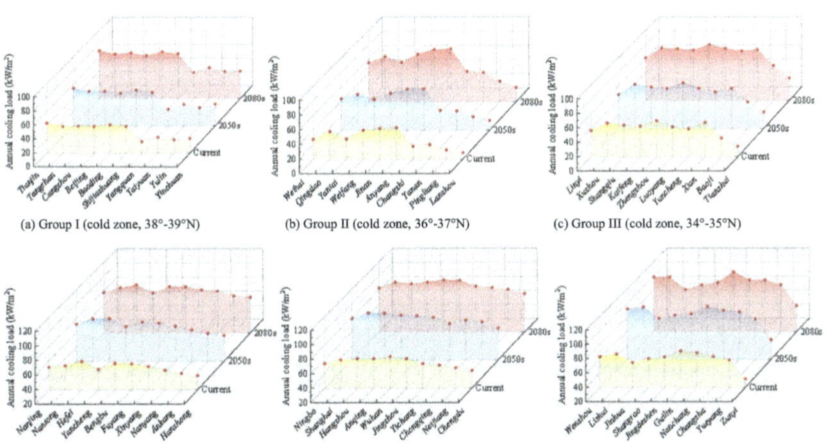

Fig. 7 Trends in the annual cooling load by the 2050s and 2080s in major cities in China under the SSP3-7.0 scenario

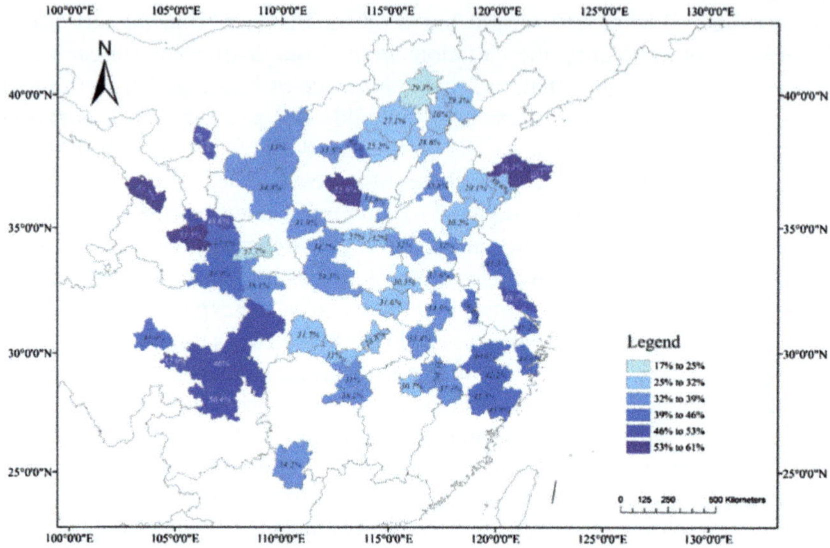

Fig. 8 The distribution maps of the percentage change in cooling load in 2050s compared to the baseline year

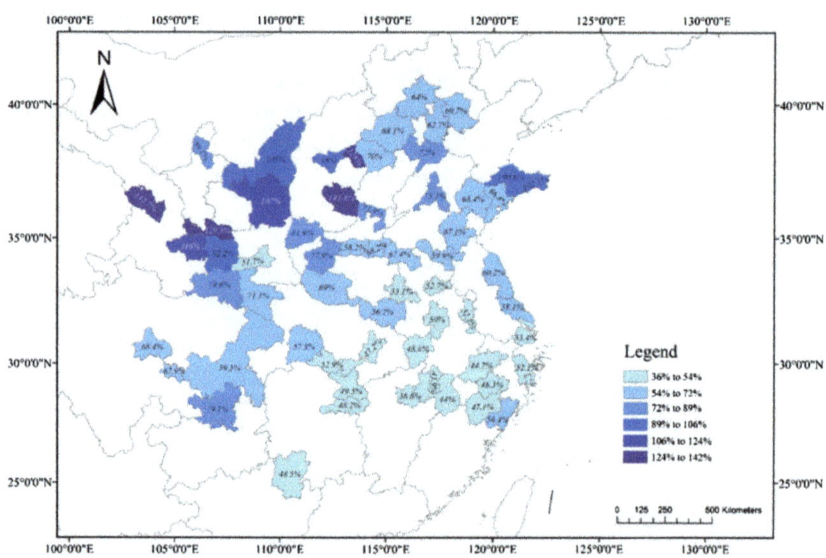

Fig. 9 The distribution maps of the percentage change in cooling load in 2080s compared to the baseline year

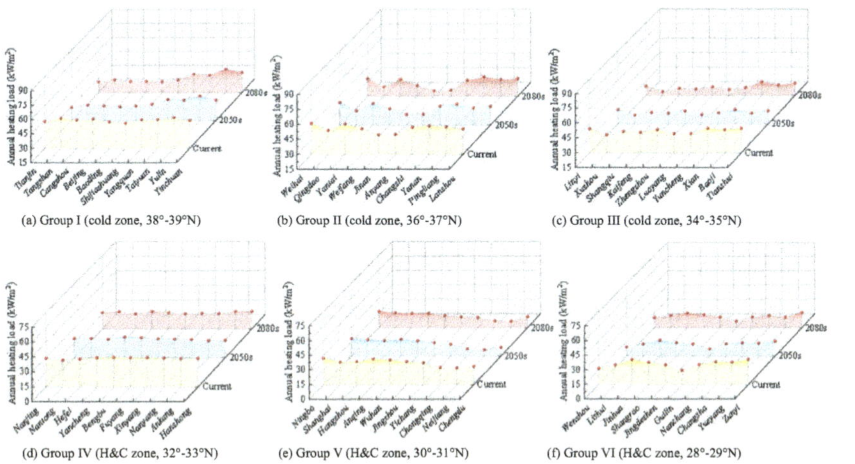

Fig. 10 Trends in the annual heating load by the 2050s and 2080s in major cities in China under the SSP3-7.0 scenario

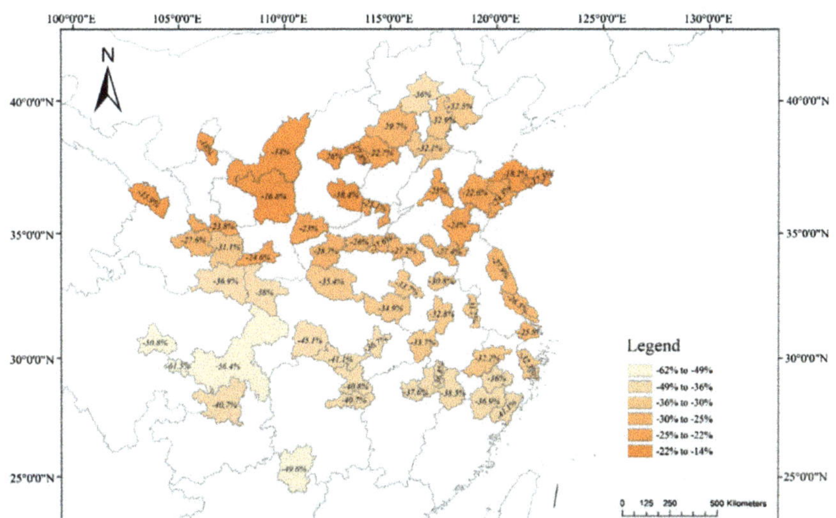

Fig. 11 The distribution maps of the percentage change in heating load in 2050s compared to the baseline year

located at lower latitudes within the H&C zone, which also exhibit lower annual heating load demands. Under the SSP3-7.0 scenario, compared to the baseline year, the annual heating load in cities within the cold zone is expected to decrease by an average of 24.3% by the 2050s, with a further reduction of approximately 31.8% by the 2080s. For cities in the H&C zone, the annual heating load is projected to

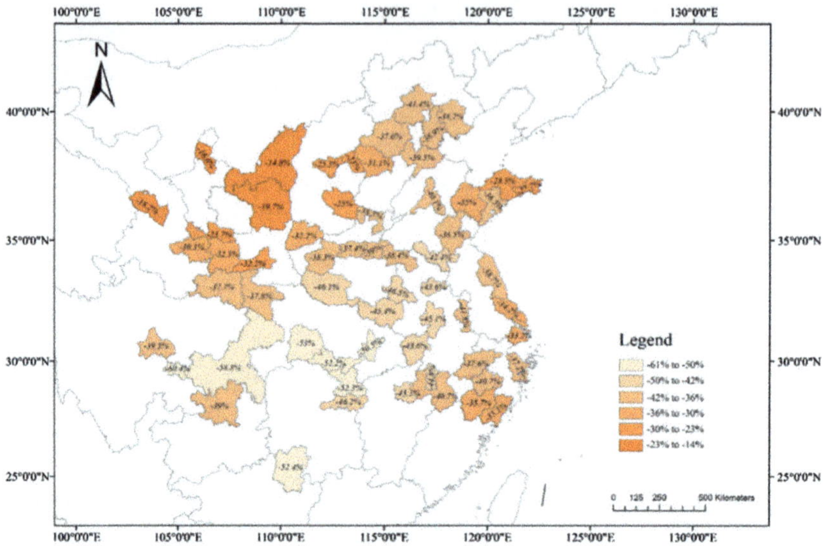

Fig. 12 The distribution maps of the percentage change in heating load in 2080s compared to the baseline year

decrease by around 38.1% on average by the 2050s, and by 43.8% by the 2080s. The land-sea distribution of cities continues to influence changes in the annual heating load of urban buildings both presently and in the future. As temperatures rise, the demand for building cooling loads during the summer is expected to persistently increase, while the heating load in winter will decrease to a certain extent before stabilizing.

3.2.2 Annual Energy Consumption

To evaluate the annual building energy demand of the target cities, this study accounts for regional variations in the annual coefficients of performance (COP) of heating and cooling systems across China's diverse climate zones. Following the *Green Performance Calculation Standard for Civil Buildings* (JGJ449-2018) (MOHURD 2018), the annual integrated heating and cooling energy consumption for buildings in sixty representative cities is calculated, as expressed in Eqs. (1)–(3).

$$E_{bld} = E_{H,bld} + E_{C,bld} \tag{1}$$

$$E_{H,bld} = \frac{Q_{H,bld}}{\theta_1} \tag{2}$$

$$E_{C,bld} = \frac{Q_{C,bld}}{\theta_2} \tag{3}$$

E_{bld} is the amount of combined heating and cooling energy consumed by the building throughout the year, (MWh). $E_{H,bld}$ is the annual heating energy consumption of the building, (MWh). $E_{C,bld}$ is the annual cooling energy consumption of the building, (MWh). $Q_{H,bld}$ is the building's cumulative annual heating load, (MWh), determined by simulation. $Q_{C,bld}$ is the building's cumulative annual cooling load, (MWh), determined by simulation. θ_1 is the comprehensive efficiency conversion weight of the heating system, which is 1.6 for public buildings in cold zone and 2.2 for public buildings in H&C zone. θ_2 is the comprehensive efficiency conversion weight of the cooling system, this study taking 2.5 for public buildings in cold and H&C zones.

Based on the simulated annual heating and cooling loads and in accordance with energy consumption calculation standards, Fig. 13 presents the comparative results of total annual building energy use for sixty representative cities across different climate zones, benchmarked against the baseline year under the SSP3-7.0 scenario for the 2050s and 2080s. Furthermore, Figs. 14 and 15 provide spatial distribution maps depicting the percentage variations (increase or decrease) in energy consumption relative to the baseline year. As shown in Fig. 13, climate change leads to divergent energy consumption trajectories between the cold zone and the H&C (heating and cooling) zone. Specifically, Figs. 14 and 15 indicate that in the cold zone, annual energy consumption is projected to experience a modest decline by the 2050s, a trend consistent across three latitude-based experimental groups (Group I–III), particularly evident in higher-latitude cities where reduced winter heating demand is likely the main driver. By contrast, by the 2080s, most cities are projected to exhibit a significant increase in annual building energy use compared to the baseline. Under SSP3-7.0, the cold zone is expected to show an average decline of approximately 4.8% in the 2050s, followed by an increase of about 5.4% by the 2080s. Conversely, cities in the H&C zone are projected to undergo a consistent rise in energy use across both future periods. This growth can be attributed to the dominance of cooling loads in the total energy balance of the H&C zone, which are expected to escalate with ongoing global warming, while reductions in heating demand gradually diminish or stabilize. Relative to the baseline year, the H&C zone is anticipated to record an average increase of about 13% in the 2050s and 21.5% in the 2080s under the SSP3-7.0 scenario.

3.3 *Influence Patterns of Urban Land and Sea Distribution on Future Building Energy Consumption*

This study investigates the relative patterns of change between urban land-sea distribution and building energy consumption using regression analysis, and forecasts the future trends of these changes for the 2050s and 2080s. Sixty representative cities from various climate zones were categorized into six groups based on latitude, and a function representing the annual energy consumption of buildings was developed for

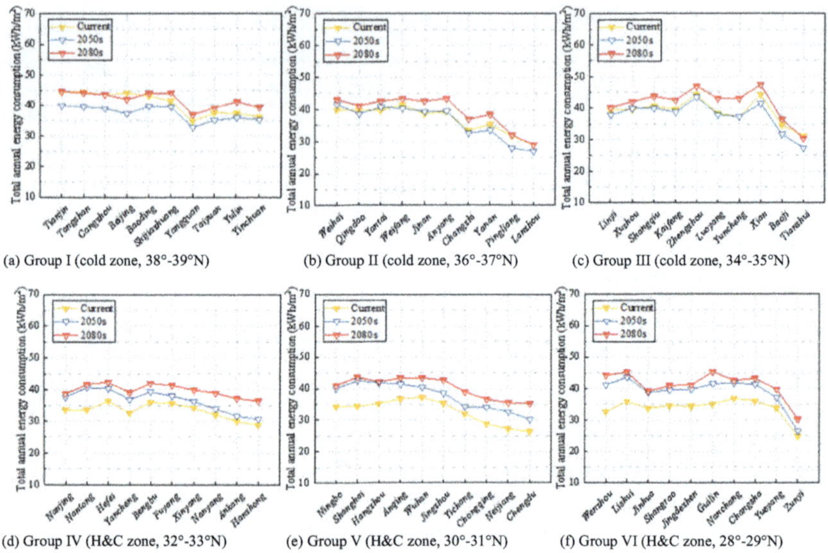

Fig. 13 Trends in the total annual energy consumption by the 2050s and 2080s in major cities in China under the SSP3-7.0 scenario

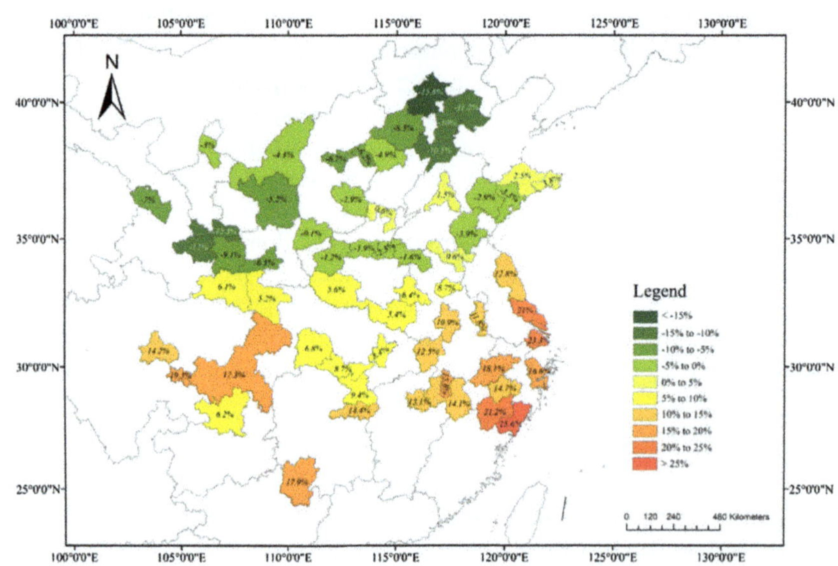

Fig. 14 The distribution maps of the percentage change in energy consumption in 2050s compared to the baseline year

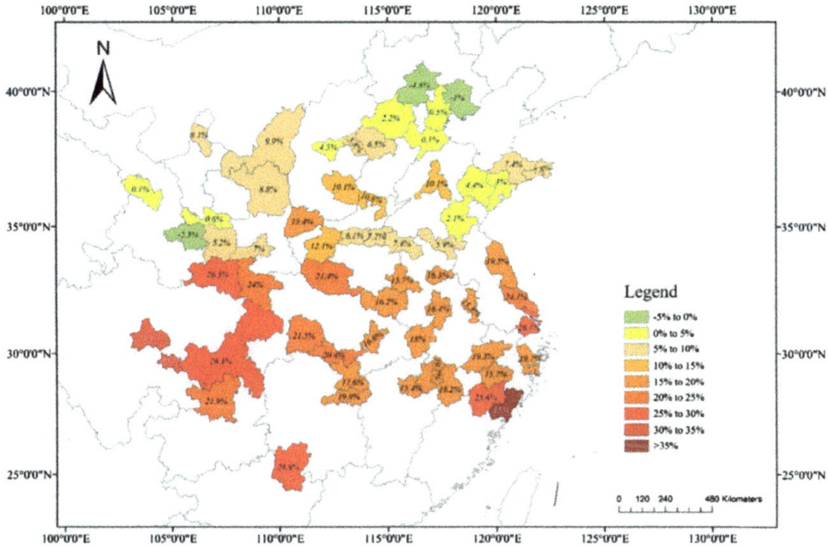

Fig. 15 The distribution maps of the percentage change in energy consumption in 2080s compared to the baseline year

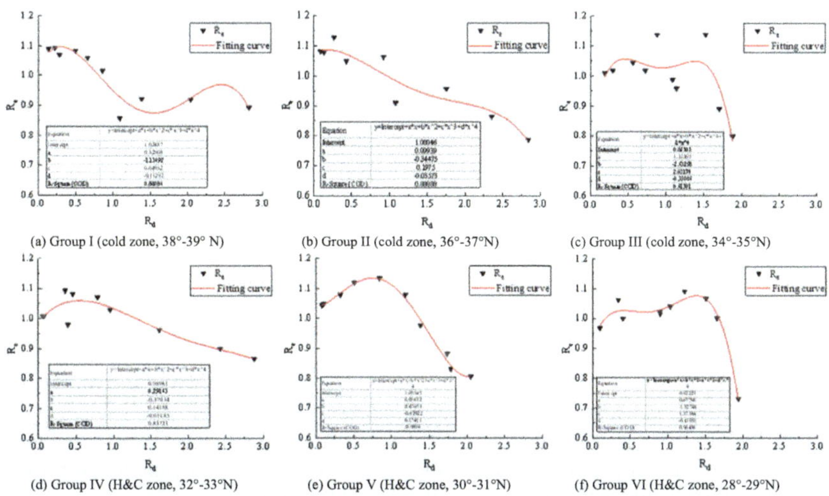

Fig. 16 Patterns of changes in total annual building energy consumption with distance from the urban sea-land distribution under the baseline year

each group, with respect to the distance from land to sea. The goal of this study is to provide valuable insights for governmental bodies regarding urban planning, building energy efficiency design, and policy development in both current and future contexts, while addressing existing research gaps. The land-sea distribution distance and the

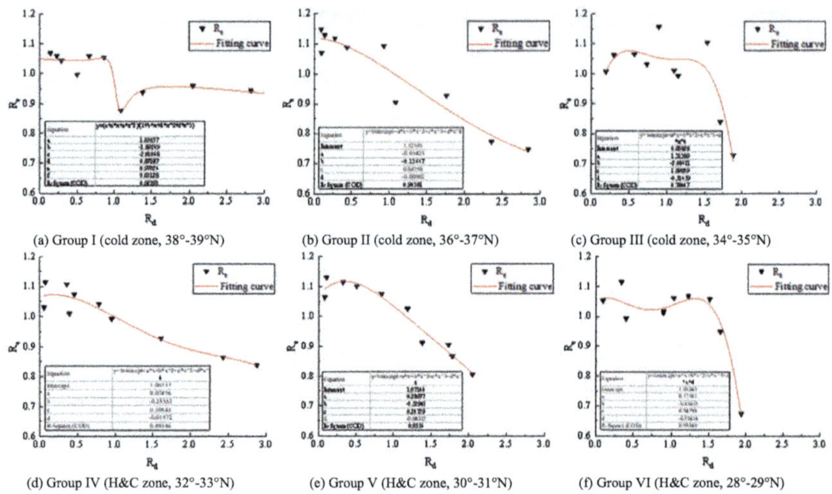

Fig. 17 Patterns of changes in total annual building energy consumption with distance from the urban sea-land distribution in the 2050s under the SSP3-7.0 scenario

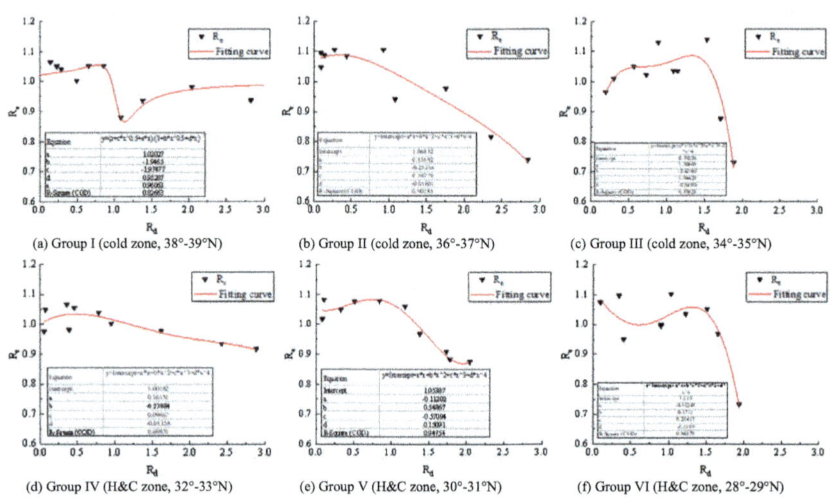

Fig. 18 Patterns of changes in total annual building energy consumption with distance from the urban sea-land distribution in the 2080s under the SSP3-7.0 scenario

total annual energy consumption of buildings were processed in a dimensionless form using the equations presented in Eqs. (4) and (5).

$$R_d = \frac{d}{d_{average}} \tag{4}$$

$$R_e = \frac{e}{e_{average}} \quad (5)$$

d is the land-sea distribution distance of the cities, (km). $d_{average}$ is the average urban land-sea distribution distance, (km). R_d is the relative urban sea-land distribution distance, dimensionless. e is the total annual energy consumption of the buildings, (MWh). $e_{average}$ is average annual total building energy, (MWh). R_e is the relative annual total energy consumption of buildings corresponding to different urban sea-land distribution distances, dimensionless.

Based on the results of the study, a quantitative model of the effect pattern of urban sea and land distribution was established on the current and future annual total energy consumption of buildings. As shown in Eqs. (6)–(23). In the baseline year, future 2050s and 2080s temporal frameworks, the change patterns of annual total building energy consumption with the distance of urban sea-land distribution are shown in Figs. 16, 17 and 18, respectively.

For baseline years (2007s–2021s),

$$R_e = 1.02457 + 0.52568R_d - 1.13692R_d^2 + 0.64952R_d^3 - 0.11292R_d^4, \quad (38° - 39°\text{N}) \quad (6)$$

$$R_e = 1.08046 + 0.09939R_d - 0.34475R_d^2 + 0.1975R_d^3 - 0.03573R_d^4, \quad (36° - 37°\text{N}) \quad (7)$$

$$R_e = 0.82505 + 1.33367R_d - 2.63293R_d^2 + 2.05174R_d^3 - 0.55006R_d^4, \quad (34° - 35°\text{N}) \quad (8)$$

$$R_e = 0.98981 + 0.29163R_d - 0.37034R_d^2 + 0.14158R_d^3 - 0.011851R_d^4, \quad (32° - 33°\text{N}) \quad (9)$$

$$R_e = 1.03543 + 0.05652R_d + 0.47655R_d^2 - 0.62922R_d^3 + 0.17411R_d^4, \quad (30° - 31°\text{N}) \quad (10)$$

$$R_e = 0.92229 + 0.67766R_d - 1.52748R_d^2 + 1.37584R_d^3 - 0.41051R_d^4, \quad (28° - 29°\text{N}) \quad (11)$$

For the future 2050s (2036s-2065s) in the SSP3-7.0 scenario,

$$R_e = 1.12101 - 0.01423R_d - 0.13667R_d^2 + 0.04758R_d^3 - 0.00502R_d^4, \quad (38° - 39°\text{N}) \quad (12)$$

$$R_e = 0.83828 + 1.31389R_d - 2.48611R_d^2 + 1.89959R_d^3 - 0.51459R_d^4, \quad (36° - 37°\text{N}) \quad (13)$$

$$R_e = 0.83828 + 1.31389R_d - 2.48611R_d^2 + 1.899959R_d^3 - 0.51459R_d^4, \quad (34° - 35°\text{N}) \quad (14)$$

$$R_e = 1.06717 + 0.07496R_d - 0.23553R_d^2 + 0.10845R_d^3 - 0.01572R_d^4, \quad (32^\circ - 33^\circ\text{N}) \tag{15}$$

$$R_e = 1.07144 + 0.28977R_d - 0.52943R_d^2 + 0.24759R_d^3 - 0.04327R_d^4, \quad (30^\circ - 31^\circ\text{N}) \tag{16}$$

$$R_e = 1.05243 + 0.17581R_d - 0.81453R_d^2 + 0.94798R_d^3 - 0.32414R_d^4, \quad (28^\circ - 29^\circ\text{N}) \tag{17}$$

For the future 2080s (2066s–2095s) in the SSP3-7.0 scenario,

$$R_e = (1.02027 - 1.97477R_d^{0.5} + 0.96063R_d)/(1 - 1.9465R_d^{0.5} + 0.95287R_d), \quad (38^\circ - 39^\circ\text{N}) \tag{18}$$

$$R_e = 1.06832 + 0.13592R_d - 0.25216R_d^2 + 0.10279R_d^3 - 0.01601R_d^4, \quad (36^\circ - 37^\circ\text{N}) \tag{19}$$

$$R_e = 0.79228 + 1.30609R_d - 2.42965R_d^2 + 1.94429R_d^3 - 0.54918R_d^4, \quad (34^\circ - 35^\circ\text{N}) \tag{20}$$

$$R_e = 1.00182 + 0.16151R_d - 0.23084R_d^2 + 0.09617R_d^3 - 0.01358R_d^4, \quad (32^\circ - 33^\circ\text{N}) \tag{21}$$

$$R_e = 1.05387 - 0.11202R_d + 0.54867R_d^2 - 0.57094R_d^3 + 0.15091R_d^4, \quad (30^\circ - 31^\circ\text{N}) \tag{22}$$

$$R_e = 1.115 - 0.35245R_d + 0.1711R_d^2 + 0.25417R_d^3 - 0.1559R_d^4, \quad (28^\circ - 29^\circ\text{N}) \tag{23}$$

As shown in Figs. 16, 17 and 18, under both present and projected global warming scenarios, the annual energy consumption of the examined cities generally follows a trend of initial increase followed by decline as the relative distance from the coastline expands. With the exception of Group VI, the building energy demand in the 2050s and 2080s, when compared with the baseline period, presents a fluctuating trajectory characterized by a decrease at first, then a subsequent increase, and another decline thereafter. Relative to the H&C zone, cities located in the cold zone reveal a more moderate growth in energy use with increasing distance from the port. This differentiated pattern can be attributed to multiple interrelated urban factors, including climatic variations, levels of urbanization and building density, economic development trajectories, technological progress, as well as energy supply and management approaches.

Climate is the primary factor influencing energy consumption. Coastal regions typically experience milder climate conditions, characterized by cooler summers and warmer winters, which result in lower cooling and heating demands for buildings, thereby leading to relatively lower energy consumption. As the distance from the coastline increases, inland areas tend to face more extreme climate conditions. For instance, higher summer temperatures and lower winter temperatures necessitate greater energy usage to maintain comfortable indoor conditions, consequently increasing energy consumption. In more remote inland areas, the implementation of adaptive construction measures, such as advanced insulation and thermal technologies, may mitigate the rise in energy consumption for buildings.

Urbanization and building density constitute the second critical determinant. Coastal cities, serving as major centers of population and economic activity, generally exhibit high-density development patterns. Such compact urban forms facilitate greater energy efficiency, particularly through centralized heating and cooling systems, which lower energy use per unit of floor area. By contrast, cities located at intermediate distances from the coast—often in phases of rapid expansion and infrastructure transition—tend to experience elevated energy demand due to dispersed spatial layouts and underdeveloped service networks. Further inland, however, long-term urban planning and construction optimization have typically fostered more rational building configurations, comprehensive infrastructure systems, and higher energy efficiency, thereby mitigating overall consumption levels.

Economic and technological levels constitute the third factor. Coastal areas typically exhibit higher levels of economic development, enabling the widespread adoption of advanced energy-saving technologies and materials, such as efficient cooling and heating systems, intelligent control systems, and energy-efficient building designs, all of which contribute to reduced building energy consumption. Regions located at moderate distances from the coastline may be undergoing rapid economic growth, accompanied by extensive building construction. However, the adoption of energy-saving technologies has not yet become widespread, resulting in higher energy consumption. In contrast, inland areas farther from the coastline, which may experience more stable economic development, are gradually seeing the broader implementation of energy-saving technologies, leading to a gradual reduction in building energy consumption.

Energy supply and management constitute the final factor. Coastal areas generally benefit from a reliable and diverse energy supply, which facilitates the optimization of energy management and utilization, thereby reducing building energy consumption. Regions located at moderate distances from the coastline may face challenges such as unstable energy supply and inefficient management, which can result in higher building energy consumption. In contrast, inland areas, with advances in technology, may increasingly adopt more localized and renewable energy solutions, gradually improving energy management and leading to a reduction in building energy consumption.

Based on the findings presented, it is recommended that government policymakers focus on the following aspects when planning future urban development, coordinating building energy efficiency design strategies, and formulating related policies:

(1) Establishment of a Zoning Management System: By incorporating the characteristics of land-sea distribution, cities should be divided into distinct building energy efficiency management zones. Differentiated building energy efficiency standards and design guidelines should be established for each zone.
(2) Optimization of Urban Form: Urban ventilation should be improved, and building energy consumption reduced, by rationalizing building density and height to leverage the sea-land wind effect.
(3) Enhancement of the Monitoring and Assessment System: A comprehensive meteorological and building energy consumption monitoring network should be developed across the region to provide data-driven support for decision-making.
(4) Development of a Stage Goal Plan: Short-term, medium-term, and long-term goals for building energy efficiency should be set, with gradual implementation of measures in each stage.

However, the model does have certain limitations. For instance, in Groups III and VI, the model demonstrates a rapid decline toward the end, which may be attributed to geographical factors unique to these cities, resulting in significant climatic variations.

4 Key Findings and Future Directions

In this research, meteorological datasets for sixty Chinese cities—distributed across distinct climate zones and characterized by varying land-sea configurations—were projected for the mid-twenty-first century (2050s) and late-twenty-first century (2080s) under the SSP3-7.0 scenario, utilizing simulations from the CMIP6 climate model. The analysis focused on examining the interplay between urban land-sea distribution and building energy consumption under both present and future climate conditions, with outcomes expressed through a dimensionless evaluation framework. The principal findings can be summarized as follows:

(1) Future meteorological projections for the selected cities, generated using the CMIP6 model under SSP3-7.0, indicate significant warming trends. By the 2050s, average urban temperatures in cold and hot-summer-cold-winter (H&C) zones are expected to rise by approximately 2.69 °C and 2.74 °C, respectively. These increases are projected to reach 4.4 °C and 3.4 °C by the 2080s. Concurrently, diffuse horizontal radiation is projected to increase by 4 W/m^2 in the 2050s and 24.3 W/m^2 in the 2080s, whereas relative humidity is anticipated to decline by 3.1% and 4.2%. The extent of these climatic shifts exhibits spatial variability that is strongly influenced by the land-sea distribution of each city.
(2) Employing a control-variable approach, the study further explored the impacts of land-sea distribution on building energy demand under both baseline and future warming scenarios. Results suggest that annual cooling and heating loads undergo significant shifts with global warming. Specifically, by the 2050s, cooling loads are projected to increase by 36.3% in cold-zone cities and 38.8% in H&C-zone cities, while heating loads are expected to decrease by 24.3%

and 38.1%, respectively. By the 2080s, cooling demand may rise by as much as 85.5% in cold-zone cities and 55.1% in H&C-zone cities, while heating demand is anticipated to decline by 31.8% and 43.8%. The magnitude and direction of these changes vary according to local land-sea configurations.

(3) In terms of overall annual building energy consumption, divergent patterns emerge across different climate zones. Under SSP3-7.0, compared with the baseline, cold-zone cities are expected to experience a 4.8% reduction in total energy use by the 2050s, whereas H&C-zone cities are projected to see an increase of 13%. By the 2080s, total consumption in cold-zone cities may increase slightly by 5.4%, while in H&C-zone cities it could rise substantially by 21.5%.

(4) Finally, by employing a dimensionless analytical framework (Eqs. (6)–(11)), the study quantified relative change patterns between land-sea distribution and building energy consumption, extending these predictions to the 2050s and 2080s (Eqs. (12)–(23)). These results provide a valuable theoretical basis to inform future urban planning, energy-efficient building design, and climate-responsive policy development.

Although the analysis primarily centers on sixty Chinese cities across multiple climate zones, the methodological framework is broadly transferable, offering applicability to comparative studies in other regions worldwide.

Author's note: This chapter is based on the author's article, "Study on the impact of land-sea distribution on future building energy consumption in Chinese cities under the background of climate changes", published in *Journal of Energy & Buildings*, Vol. 332, No. 115434 (2025).

References

Andric I, Al-Ghamdi SG (2020) Climate change implications for environmental performance of residential building energy use: the case of Qatar. Energy Rep 6:587–592. https://doi.org/10.1016/j.egyr.2019.09.030

Ashrafian T (2023) Enhancing school buildings energy efficiency under climate change: A comprehensive analysis of energy, cost, and comfort factors. J Build Engin 80:107969. https://doi.org/10.1016/j.jobe.2023.107969

Babu ADCA, Srivastava RS, Rai AC (2024) Impact of climate change on the heating and cooling load components of an archetypical residential room in major Indian cities. Build Environ 250:111181. https://doi.org/10.1016/j.buildenv.2024.111181

Baglivo C, Congedo PM, Malatesta NA (2023) Building envelope resilience to climate change under Italian energy policies. J Clean Prod 411:137345. https://doi.org/10.1016/j.jclepro.2023.137345

Cai H, Li C, Luan X, Ai B, Yan L, Wen Z (2022) Analysis of the spatiotemporal evolution of the coastline of Jiaozhou Bay and its driving factors. Ocean Coast Manag 226:106246. https://doi.org/10.1016/j.ocecoaman.2022.106246

Cao J, Li M, Wang M, Xiong M, Meng F (2017) Effects of climate change on outdoor meteorological parameters for building energy-saving design in the different climate zones of China. Energy Build 146:65–72. https://doi.org/10.1016/j.enbuild.2004.045

Chen Y, Ren Z, Peng Z, Yang J, Chen Z, Deng Z (2023) Impacts of climate change and building energy efficiency improvement on city-scale building energy consumption. J Build Engin 78:107646. https://doi.org/10.1016/j.jobe.2023.107646

Cui Y, Yan D, Hong T, Xiao C, Luo X, Zhang Q (2017) Comparison of typical year and multiyear building simulations using a 55-year actual weather data set from China. Appl Energy 195:890–904. https://doi.org/10.1016/j.apenergy.2003.113

D'Agostino D, Congedo PM, Albanese PM, Rubino A, Baglivo C (2024) Impact of climate change on the energy performance of building envelopes and implications on energy regulations across Europe. Energy 288:129886. https://doi.org/10.1016/j.energy.2023.129886

Eyring V, Bony S, Meehl GA, Senior CA, Stevens B, Stouffer RJ, Taylor KE (2016) Overview of the Coupled Model Intercomparison Project Phase 6 (CMIP6) experimental design and organization. Geosci Model Develop 9(5):1937–1958. https://doi.org/10.5194/gmd-9-1937-2016

Farah S, Whaley D, Saman W, Boland J (2019) Integrating climate change into meteorological weather data for building energy simulation. Energy Build 183:749–760. https://doi.org/10.1016/j.enbuild.2018.11.045

Fernandes MS, Coutinho B, Rodrigues E (2024) The impact of climate change on an office building in Portugal: measures for a higher energy performance. J Clean Prod 445:141255. https://doi.org/10.1016/j.jclepro.2024.141255

Gao B, Zhu X, Ren J, Ran J, Kim MK, Liu J (2023) Multi-objective optimization of energy-saving measures and operation parameters for a newly retrofitted building in future climate conditions: a case study of an office building in Chengdu. Energy Rep 9:2269–2285. https://doi.org/10.1016/j.egyr.2023.01.049

Hietaharju P, Pulkkinen J, Ruusunen M, Louis J-N (2021) A stochastic dynamic building stock model for determining long-term district heating demand under future climate change. Appl Energy 295:116962. https://doi.org/10.1016/j.apenergy.2021.116962

Jalali Z, Shamseldin AY, Ghaffarianhoseini A (2023) Impact assessment of climate change on energy performance and thermal load of residential buildings in New Zealand. Build Environ 243:110627. https://doi.org/10.1016/j.buildenv.2023.110627

Kamal A, Mahfouz A, Sezer N, Hassan IG, Wang LL, Rahman MA (2023) Investigation of urban heat island and climate change and their combined impact on building cooling demand in the hot and humid climate of Qatar. Urban Clim 52:101704. https://doi.org/10.1016/j.uclim.2023.101704

Lei T, Yang T, Yao W, Cao J, Gao W, Li Y (2024) Mechanism analysis of climate change impacts on the performance of ice storage systems. J Energy Storage 99:113184. https://doi.org/10.1016/j.est.2024.113184

Lu M, Lai JHK (2019) Building energy: a review on consumptions, policies, rating schemes and standards. Energy Procedia 158:3633–3638. https://doi.org/10.1016/j.egypro.2019.01.899

Ministry of Housing and Urban-Rural Development of China (2018) Green Performance Calculation Standards for Civil Buildings (JGJ 449–2018). China Building Industry Press, Beijing, China

Ministry of Transport of the People's Republic of China (2006) Ministry of Transport of the People's Republic of China. China. https://www.mot.gov.cn/

Mirasgedis S, Cabeza LF, Vérez D (2024) Contribution of buildings climate change mitigation options to sustainable development. Sustain Cities Soc 106:105355. https://doi.org/10.1016/j.scs.2024.105355

Climate. OneBuilding. Org. (2024). WMO Region 2 - Asia / CHN _ China. https://climate.onebuilding.org/WMO_Region_2_Asia/CHN_China/

O'Neill BC, Tebaldi C, van Vuuren DP, Eyring V, Friedlingstein P, Hurtt G, Knutti R, Kriegler E, Lamarque J-F, Lowe J, Meehl GA, Moss R, Riahi K, Sanderson BM (2016) The scenario model intercomparison project (ScenarioMIP) for CMIP6. Geosci Model Develop 9(9):3461–3482. https://doi.org/10.5194/gmd-9-3461-2016

Qian F, Gao W, Yang Y, Yu D (2020) Potential analysis of the transfer learning model in short and medium-term forecasting of building HVAC energy consumption. Energy 193:116724. https://doi.org/10.1016/j.energy.2019.116724

Ran J, Qiu Y, Liu J, Zhu X, Liu J, Tian Z (2024) Coordinated optimization design of buildings and regional integrated energy systems based on load prediction in future climate conditions. Appl Therm Eng 241:122338. https://doi.org/10.1016/j.applthermaleng.2024.122338

Reveshti AM, Ebrahimpour A, Razmara J (2023) Investigating the effect of new and old weather data on the energy consumption of buildings affected by global warming in different climates. Int J Thermofluids 19:100377. https://doi.org/10.1016/j.ijft.2023.100377

Rodrigues E, Fernandes MS, Carvalho D (2023) Future weather generator for building performance research: an open-source morphing tool and an application. Build Environ 233:110104. https://doi.org/10.1016/j.buildenv.2023.110104

Rostam MG, Abbasi A (2024) Enhancing building energy efficiency through a climate adaptive design: considering upcoming extreme climate conditions. J Clean Prod 450:141747. https://doi.org/10.1016/j.jclepro.2024.141747

Salata F, Falasca S, Ciancio V, Curci G, de Wilde P (2023) Climate-change related evolution of future building cooling energy demand in a Mediterranean Country. Energy and Buildings 290:113112. https://doi.org/10.1016/j.enbuild.2023.113112

Salehi SSS, Kalamees T, Kurnitski J, Thalfeldt M (2024) New typical meteorological year generation method based on long-term building energy simulations. Build Environ 256:111504. https://doi.org/10.1016/j.buildenv.2024.111504

Sun G, Wang M (2024) Pythagorean fuzzy information processing based on centroid distance measure and its applications. Expert Syst Appl 236:121295. https://doi.org/10.1016/j.eswa.2023.121295

Wehrli K, Sidler F, Gubler S, Settembrini G, Koschenz M, Irigoyen SD, Kotlarski S, Fischer AM, Zweifel G (2024) Building design in a changing climate – Future Swiss reference years for building simulations. Climate Services 34:100448. https://doi.org/10.1016/j.cliser.2024.100448

Xiong J, Guo S, Wu Y, Yan D, Xiao C, Lu X (2023) Predicting the response of heating and cooling demands of residential buildings with various thermal performances in China to climate change. Energy 269:126789. https://doi.org/10.1016/j.energy.2023.126789

Yang Y, Javanroodi K, Nik VM (2021) Climate change and energy performance of European residential building stocks – A comprehensive impact assessment using climate big data from the coordinated regional climate downscaling experiment. Appl Energy 298:117246. https://doi.org/10.1016/j.apenergy.2021.117246

Yao W, Xu A, Kong X, Wang Y, Li X, Gao W (2024) Analysis of dust deposition law at the micro level and its impact on the annual performance of photovoltaic modules. Energy 306:132448. https://doi.org/10.1016/j.energy.2024.132448

Yuan J, Jiao Z, Xiao X, Emura K, Farnham C (2024) Impact of future climate change on energy consumption in residential buildings: a case study for representative cities in Japan. Energy Rep 11:1675–1692. https://doi.org/10.1016/j.egyr.2024.01.042

Zhou T, Zou L, Chen X (2019) Overview of the coupled model intercomparison project phase 6 (CMIP6). Clim Change Res 15(5):445–456. https://www.climatechange.cn/CN/10.12006/j.issn.1673-1719.2019.193

Innovations for a Decarbonized Society

Evolution of Thermodynamic Systems and Technological Breakthroughs in Carbon Dioxide Cycle

Xinrong Zhang, Yang Nie, Yiyi Wang, and Zhaorui Peng

1 Introduction

The historical development of thermodynamic systems represents a remarkable journey of scientific discovery and engineering innovation that has fundamentally shaped modern civilization. From the early steam engines that powered the Industrial Revolution to today's sophisticated energy conversion technologies, this evolution reflects humanity's persistent quest for more efficient and sustainable ways to harness thermal energy. The transformation of thermodynamic systems over the past three centuries has not only revolutionized energy production and utilization but also profoundly influenced economic development, environmental considerations, and technological progress across multiple industries. This historical narrative reveals how theoretical breakthroughs in thermodynamics have consistently informed practical engineering solutions, while real-world challenges have in turn stimulated new scientific inquiries. The interplay between fundamental scientific principles and their engineering applications has created a dynamic field that continues to address some of society's most pressing energy challenges while pushing the boundaries of what is thermodynamically possible.

X. Zhang (✉) · Y. Nie · Y. Wang · Z. Peng
Department of Energy and Resources Engineering, School of Mechanics and Engineering Science, Peking University, Beijing, China
e-mail: xrzhang@pku.edu.cn

Y. Nie
e-mail: 2101111999@pku.edu.cn

Y. Wang
e-mail: ajax.yywang@stu.pku.edu.cn

Z. Peng
e-mail: zhaorui_peng@pku.edu.cn

W. Zhou (ed.), *Decarbonization Policy Engineering*, World Sustainability Series,
https://doi.org/10.1007/978-981-95-5953-4_18

The development process of thermodynamic systems is a historical microcosm of human understanding of energy conversion and improvement of energy utilization efficiency. From the early steam engines to modern complex refrigeration, power generation and heat pump systems, thermodynamic engineering has undergone several centuries of evolution. This section will systematically review this development process and explore its technological breakthroughs and social impacts.

1.1 The Germination of Early Thermal Machinery (Seventeenth -Eighteenth Centuries)

The prototype of thermodynamic systems can be traced back to steam-powered devices in the late seventeenth century. In 1698, British engineer Thomas Savery invented the first practical steam pump for mine drainage. Although inefficient, this device marked humanity's first attempt to convert thermal energy into mechanical work (Thurston 1972). In 1712, Thomas Newcomen improved the steam engine design, creating the "atmospheric engine" with a thermal efficiency of approximately 0.5%. This machine was widely adopted in coal mines and became a crucial power source in the early Industrial Revolution (Review et al. 1992). British historian of science D.S.L. Cardwell noted that the Newcomen engine signified the beginning of humanity's deliberate utilization of thermal energy to perform work, moving beyond reliance solely on wind and water power (Mayr 1972).

1.2 The Establishment of Thermodynamic Theory and the Improvement of the Steam Engine (Nineteenth Century)

The nineteenth century marked a golden age of parallel development in thermodynamic theory and engineering practice. In 1824, French engineer Sadi Carnot published *Réflexions sur la puissance motrice du feu*, introducing the Carnot cycle theory which established the theoretical efficiency limit for heat engines. This seminal work laid the foundation for the second law of thermodynamics.

Concurrently, James Watt implemented revolutionary improvements to steam engines, including inventions such as the separate condenser, double-acting cylinder, and centrifugal governor. By 1800, Watt's steam engine achieved an efficiency of 3% -a sixfold increase over Newcomen's design (Rolt and Dickinson 2011). During the 1850s, Rudolf Clausius and Lord William Thomson (later Kelvin) independently formulated precise statements of the first and second laws of thermodynamics, providing the theoretical framework for heat engine design (Smith 1998). Figure 1 shows the major breakthroughs in thermodynamics from the 17th to the nineteenth century.

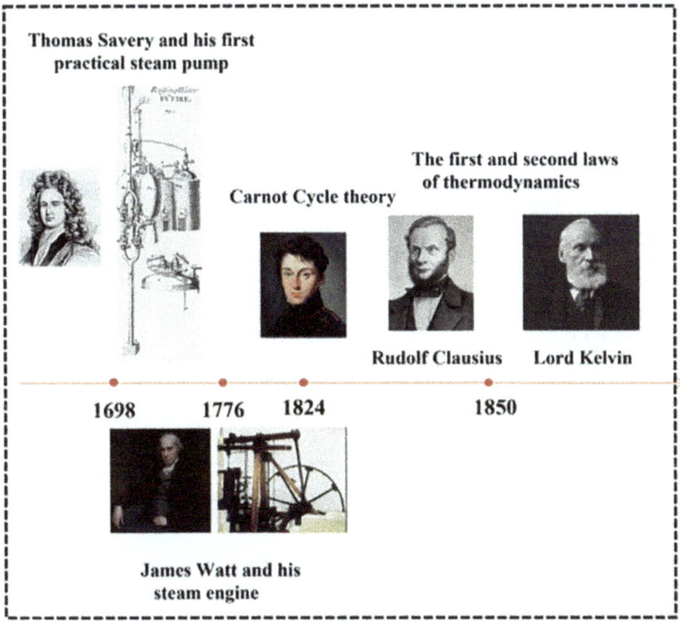

Fig. 1 Major Breakthroughs in Thermodynamics from the 17th to the 19th Century

1.3 The Development of Diversified Thermal Systems (Late 19th Century—Early Twentieth Century)

The late 19th to early twentieth century marked a pivotal era in the evolution of thermal systems, driven by the maturation of thermodynamics as a scientific discipline. This period witnessed the transition from steam-dominated industrial applications to a broader spectrum of thermal systems, including internal combustion engines, refrigeration cycles, gas turbines, and early heat pump concepts. The diversification of thermal energy conversion was further accelerated by advancements in materials science, precision manufacturing, and a deepening understanding of heat transfer mechanisms. As a result, thermal systems expanded beyond traditional power generation, finding applications in transportation, industrial processes, and even domestic environments—laying the groundwork for modern energy technologies.

1.3.1 The Origin and Development of Refrigeration Systems

The practice of food preservation using ice cellars traces back to ancient times, but the advent of mechanical refrigeration systems marked the true beginning of the modern refrigeration industry. Figure 2 shows the historical evolution of refrigerants. In 1755, William Cullen at the University of Edinburgh conducted the first documented

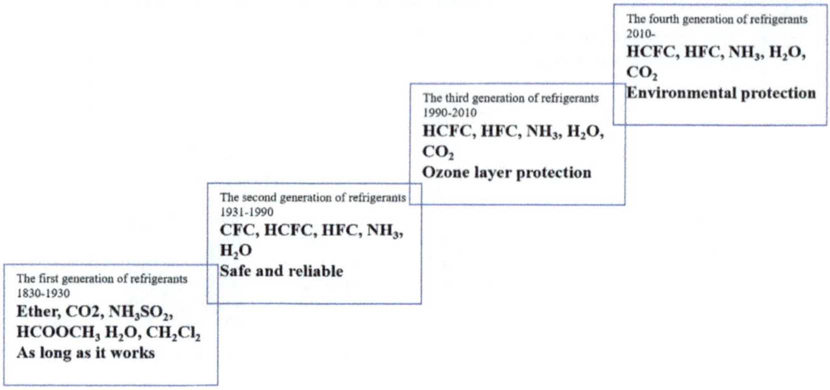

Fig. 2 The historical evolution of refrigerants

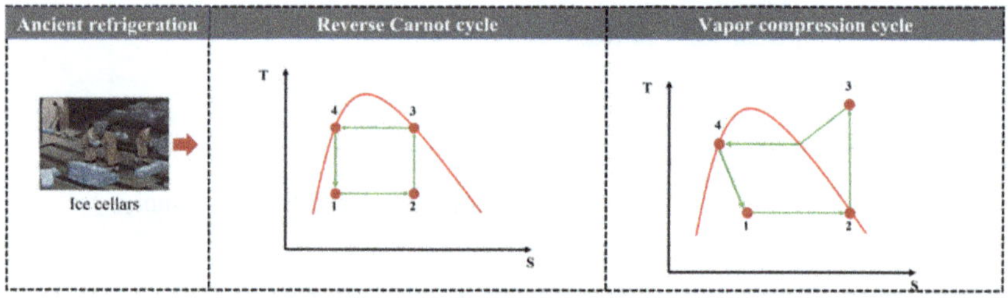

Fig. 3 Construction of vapor compression cycle

artificial refrigeration experiment, demonstrating cooling effects through ether evaporation. Although this experiment did not yield immediate practical applications, it established the fundamental principle of vapor-compression refrigeration, which is shown in Fig. 3.

In 1805, American inventor Oliver Evans conceived the first theoretical design for a vapor-compression refrigeration machine, though he never constructed a working prototype (Thévenot 1979). Building upon Evans' concept, Jacob Perkins developed the first functional vapor-compression refrigerator in 1834, employing ether as the refrigerant. While commercially unsuccessful, this machine established the basic architecture of modern refrigeration systems. The medical field saw its first application of refrigeration technology in 1851 when Dr. John Gorrie developed a hospital air-conditioning system utilizing air expansion cooling to treat malaria patients. By 1856, Alexander Twining implemented the first commercial refrigeration system in the United States.

Early refrigerants, including ether, ammonia, and sulfur dioxide, were predominantly flammable or toxic, exhibiting poor thermodynamic performance and safety characteristics. A significant advancement occurred in 1876 when German engineer Carl von Linde introduced a continuous refrigeration machine using ammonia as the refrigerant, dramatically improving both efficiency and operational safety. Due to its superior thermodynamic properties, ammonia (NH_3) emerged as the dominant refrigerant from the late 19th through the early twentieth century.

1.3.2 Preliminary Development of the Heat Dissipation System

As thermal mechanical power increased significantly during the Industrial Revolution, heat dissipation emerged as a critical engineering challenge. Early steam engines relied on rudimentary air or water cooling methods. A major breakthrough in radiator technology occurred during the mid-nineteenth century alongside railway locomotive development. In 1853, French engineer Alphonse Beaudry pioneered the tubular radiator design, dramatically improving the thermal efficiency of steam locomotives. This innovation was further refined in 1897 when Wilhelm Maybach developed the honeycomb radiator for Daimler automobiles, establishing the prototype for modern automotive cooling systems.

The advent of electronic devices introduced new thermal management requirements. In the field of electronic devices, after Lee De Forest invented the vacuum tube in 1906, passive heat dissipation technology began to be applied to early electronic devices. In the 1920s, with the development of radio stations, forced air cooling technology began to emerge (Kraus and Bar-Cohen 1983).

1.3.3 Development of Thermal Energy to Other Forms of Energy Conversion Systems

From the end of the nineteenth century to the beginning of the twentieth century marked a transformative era in the development of thermal energy conversion systems, witnessing groundbreaking innovations that reshaped industrial capabilities and energy utilization worldwide. This period saw the maturation of thermodynamics as a scientific discipline alongside remarkable engineering achievements, enabling more efficient and diverse methods of converting heat into mechanical work and electrical power. From the refinement of steam turbine technology to the revolutionary emergence of internal combustion engines, these advancements not only enhanced energy conversion efficiency but also expanded the practical applications of thermal systems across transportation, manufacturing, and electricity generation.

This critical phase in energy history was characterized by the interplay between theoretical insights and practical engineering solutions, as researchers and inventors sought to overcome the limitations of existing technologies. The development of new thermodynamic cycles, improved materials, and innovative system designs during this era laid the foundation for modern power generation and established fundamental principles that continue to guide thermal energy conversion today. Examining this pivotal period provides valuable insights into how thermal energy systems evolved to meet growing industrial demands while addressing emerging challenges in energy efficiency and scalability.

It was a period of rapid progress in thermal energy conversion technology. A variety of new thermal systems emerged one after another, which greatly promoted the Industrial Revolution and the process of electrification.

2 Maturity and Innovation of Modern Thermal Systems

The period spanning the latter half of the twentieth century through the early decades of the twenty-first century represents a remarkable epoch in the evolution of thermal energy systems, characterized by revolutionary breakthroughs that fundamentally transformed energy conversion paradigms. During this critical phase, unprecedented progress was achieved across three key dimensions: dramatic enhancements in thermodynamic conversion efficiency, sophisticated integration of complex system components, and extraordinary expansion of practical applications. These technological leaps emerged in response to a confluence of powerful driving forces-escalating global energy requirements, growing environmental consciousness, and rapid advancements in materials science and computational capabilities. The following discussion will systematically examine the principal technological milestones that defined this transformative era, from revolutionary combined-cycle configurations to cutting-edge thermal management solutions, each contributing to the remarkable evolution of modern thermal energy systems. This historical progression not only reflects humanity's enduring quest for more efficient energy utilization but also demonstrates the increasingly sophisticated interplay between thermodynamic theory and practical engineering implementation.

2.1 Breakthroughs in Combined Cycle Power Generation Technology

The Combined Cycle Gas Turbine (CCGT) technology represents the high efficiency level of modern thermal systems. By effectively integrating the Brayton cycle (gas turbine) with the Rankine cycle (steam turbine), this system achieves cascaded energy utilization, as shown in Fig. 4. The first experimental gas-steam combined system was built by BBC in 1949 with 28% efficiency, while the first commercial CCGT plant reached 34% efficiency in 1960. By the early twenty-first century, micro combined heat and power (CHP) systems combining electricity generation with heat supply reached theoretical LHV efficiencies exceeding 90% (80–85% in practical operation), making them suitable for distributed energy applications (Alanne and Saari 2006). General Electric's H-class gas turbine combined cycle systems have surpassed 60% efficiency (Horlock and Bathie 2004) with the Kawagoe Thermal Power Plant of Chubu Electric Power Company in Japan achieving a record net efficiency of 61.5% for its combined cycle units (Ishikawa et al. 2008).

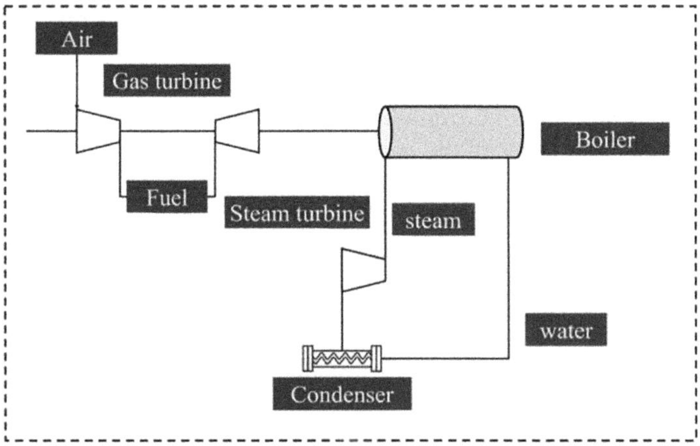

Fig. 4 The gas-steam combined cycle

Despite the significant advancements in efficiency achieved by CCGT technology, several limitations remain. This technology is highly reliant on fossil fuels, primarily natural gas, leading to continued carbon emissions, which conflicts with global decarbonization goals. Secondly, the efficiency improvements of gas turbines are constrained by material limitations—high-temperature operation, which is critical for performance, accelerates component wear and shortens service life, thereby increasing maintenance costs. Additionally, under dynamic load conditions, the suboptimal flexibility in the heat-to-power ratio limits its adaptability to fluctuating energy demands. Potential feasible improvement measures may include further integrating carbon capture and storage (CCS) technologies to reduce emissions, and exploring hybrid configurations with renewable energy sources (such as solar-assisted gas-steam combined cycles) to decrease reliance on fossil fuels.

2.2 *Evolution and Application Expansion of Heat Pump Technology*

The latter half of the twentieth century witnessed significant advancements in heat pump technology, with its development trajectory being closely intertwined with energy crises and environmental concerns. The 1970s oil crisis marked a pivotal turning point, as this efficient heating technology -which utilizes reversed thermodynamic cycles to upgrade low-temperature heat sources to usable temperatures—gained unprecedented development opportunities. Modern heat pump systems have achieved remarkable breakthroughs in coefficient of performance (COP), with modern air-source heat pumps achieving COP of 3–4 at 7 °C ambient temperature, while ground-source systems reach 4–5 COP by utilizing stable 10 °C underground heat (BoroumandJazi et al. 2013).

The technological evolution of heat pumps can be traced back to early engineering practices. In 1928, British engineer T.G.N. Haldane accomplished a landmark engineering application by successfully implementing an air-source heat pump for residential heating for the first time. This pioneering achievement laid the foundation for subsequent civilian applications of heat pump technology. During the 1940s, Robert C. Webber made significant breakthroughs in the field by developing the first direct-expansion ground-source heat pump system, establishing a novel approach for utilizing stable underground heat sources.

However, the performance of air-source heat pumps is highly dependent on ambient temperatures. In cold climates (below 0 °C), their coefficient of performance (COP) decreases significantly, reducing their energy efficiency and practicality in regions with severe winters. On the other hand, most traditional heat pumps rely on hydrofluorocarbon (HFC) refrigerants. Although these refrigerants have low ozone depletion potential, they still possess high global warming potential (GWP), posing indirect environmental risks in cases of refrigerant leakage during operation or maintenance. In addition, while ground-source heat pumps offer higher stability, they face challenges such as high upfront installation costs (especially for ground loop systems) and geographical constraints. Their effectiveness is limited by subsurface thermal properties and available land space. Furthermore, the efficiency of existing systems is constrained by the thermodynamic limitations of traditional refrigerants and cycle designs, making it difficult to efficiently utilize ultra-low-temperature heat sources (e.g., industrial waste heat below 50 °C) and operate stably under highly fluctuating load conditions. These limitations create opportunities for transitioning to CO_2-based thermal conversion cycles and heat pump systems. Leveraging the excellent thermophysical properties of CO_2 (such as high heat transfer coefficients, non-flammability, and negligible global warming potential), such systems can overcome refrigerant-related environmental issues and enhance performance in extreme temperature conditions and low-grade heat utilization scenarios.

2.3 Revolutionary Progress in Refrigeration Systems

In 1928, Thomas Midgley's invention of Freon (R12) initiated the era of chlorofluorocarbon (CFC) refrigerants (Midgley and Henne 1930). This non-toxic, nonflammable synthetic refrigerant enabled the widespread adoption of household refrigerators, marking a revolutionary advancement in refrigeration technology. General Electric introduced the first CFC-based domestic refrigerator, the "Monitor-Top," in 1930 (Mcconnell 2000). A critical turning point came in 1974 when Mario Molina and Sherwood Rowland discovered CFCs' ozone-depleting potential (Mario et al. 1974), leading to the 1987 Montreal Protocol that mandated the phase-out of CFCs.

During the 1990s, hydrofluorocarbons (HFCs) like R134a served as transitional alternatives. Recent developments have focused on natural refrigerants (CO_2, ammonia, hydrocarbons) and new synthetic refrigerants (HFOs) as sustainable solutions.

The technological evolution progressed through several milestones: Willis Carrier's invention of modern air conditioning in 1902; the transformation of refrigerators from luxury items in the 1920s to household essentials; the development of cascade refrigeration systems for ultra-low temperature applications in the 1950s; advancements in absorption refrigeration utilizing industrial waste heat during the 1960s; and the significant improvement in part-load efficiency through variable-speed technology after the 1990s (Wang 1993). These innovations collectively expanded refrigeration applications across diverse sectors.

2.4 Technological Breakthroughs in the Heat Dissipation System

The latter half of the twentieth century witnessed a paradigm shift in thermal management technologies, primarily catalyzed by the exponential growth in power density of electronic systems (Chapman 1974; Tuckerman and Pease 1981). This transformative period saw aluminum heat sinks emerge as the dominant passive cooling solution during the 1940s, marking a substantial leap in heat dissipation capabilities. A pivotal advancement occurred when IBM introduced forced convection cooling architectures for mainframe computers in the 1960s, fundamentally altering the landscape of active thermal regulation. This innovation paved the way for axial fan configurations to become ubiquitous in personal computing platforms by the 1980s. Concurrently, the domain of liquid cooling experienced groundbreaking developments, commencing with Cray Research's implementation of phase-change fluorocarbon-based systems for supercomputers during the 1960s (Thornton 1964). The technological progression continued with water-cooled workstations gaining prominence in the 1980s, ultimately culminating in the widespread deployment of advanced liquid cooling infrastructure within data center environments post-2000. These collective advancements not only addressed immediate thermal challenges but also enabled unprecedented computational performance across various technological domains.

Breakthroughs in new cooling technologies were particularly notable: In 1963, George Grover of Los Alamos National Laboratory invented heat pipe technology, which has become a core solution for cooling electronic devices and spacecraft. The microchannel cooling concept proposed by Tuckerman and Pease in 1981 is now widely used in high-power chip cooling. Phase change materials researched since the 1990s have provided innovative solutions for transient thermal management, while jet impingement cooling technology developed after 2000 meets the extreme cooling demands of high-performance computing.

The theoretical framework also developed significantly. In the 1970s, electronic device cooling formally developed into a systematic discipline. J.P. Holman's

1985 book "Heat Transfer" systematically summarized cooling engineering theories. The development of computational fluid dynamics (CFD) transformed cooling design from experience-based to simulation-based optimization, marking the entry of cooling technology into a new scientific and precise stage.

While the aforementioned systems and technologies have demonstrated significant application value in advancing thermal energy conversion, their inherent limitations underscore the necessity for targeted improvements. Addressing these drawbacks is not merely a matter of incremental optimization; rather, it constitutes a crucial measure to align thermal energy systems with increasingly stringent sustainability standards and diverse operational requirements.

3 The Construction and Development of Transcritical CO_2 Thermodynamic Cycle System

The construction of CO_2 thermodynamic systems aims to meet the more diversified and higher-standard demands for energy systems put forward by people with the development of the times. On one hand, over the past few centuries, the extensive use of traditional fossil fuels has given rise to severe issues such as global climate change and environmental pollution, posing significant threats to human survival and development. CO_2, as a natural working fluid, has been re-evaluated to develop viable thermodynamic cycle systems. On the other hand, CO_2 has attracted widespread attention due to its excellent thermodynamic and transport properties in the supercritical state, which are conducive to the establishment of new and efficient thermodynamic cycles. Here, the applications of CO_2 thermodynamic systems in the energy field will be discussed.

3.1 Trans-Critical CO_2-based Power Generation System

In the global environment of energy saving and emission reduction, the effective use of low-grade renewable thermal energy (such as environmental thermal energy, solar heat, low-temperature waste heat, etc.) has attracted attention. But the construction of reasonable and efficient low-grade thermal power cycle has always been a difficult scientific problem due to the shortage of effective measure and methods. Studies (Yin et al. 2020) have shown that in the Rankine cycle with a low-temperature side of 10–30°C and a high-temperature side of 100°C, CO_2 has a higher thermal efficiency than other working fluids. The relatively low critical temperature of CO_2 (31.1°C) makes it suitable for recovering low-grade thermal energy in the transcritical CO_2 Rankine power cycle, which was first proposed by Zhang (2005). In addition, unlike other heat transfer processes, the transcritical process can effectively absorb heat under certain supercritical conditions. The temperature slip in the supercritical single-phase

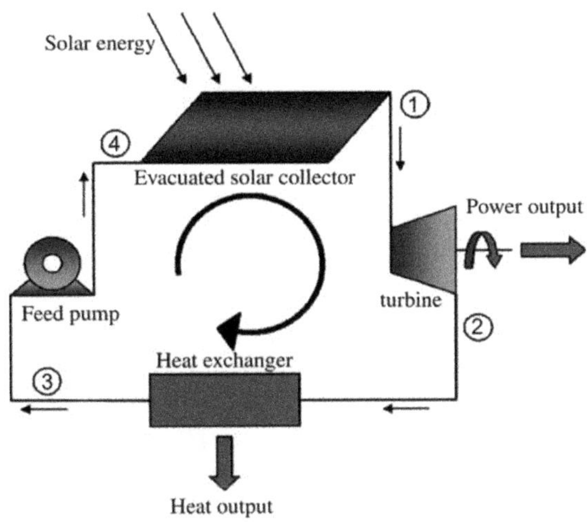

Fig. 5 Schematic diagram of a solar energy powered Rankine cycle using supercritical CO_2 for combined power output and heat collection (Zhang et al. 2007)

state makes it more efficient than the traditional isothermal endothermic evaporation process, which can not only reduce the thermodynamic loss of the process but also avoid the instability of flow and heat transfer during the process. Moreover, under specific conditions, supercritical carbon dioxide (SCO_2) can carry more heat and enable the heat exchanger to reach a higher outlet temperature. The supercritical CO_2 absorption refrigeration cycle and the transcritical CO_2 Rankine cogeneration cycle proposed on this basis provide support for the application of low-grade thermal energy power generation, as shown in Fig. 5 (Zhang et al. 2007).

On this basis, a novel experimental solar collector utilizing supercritical CO_2 natural convective flow is proposed and tested, as shown in Fig. 6. The study shows that this natural convection system can achieve stable supercritical flow and accordingly provide a steady supply of hot water. During the test season, the collector achieves a relatively high efficiency, generally above 60%. This efficiency is significantly higher than that of previous collectors based on natural water convection and transcritical CO_2 flow, as reported in the literature (Chen and Zhang 2014a).

3.2 *Trans-Critical CO_2-based Refrigeration System*

3.2.1 A Novel Type of CO_2 Refrigeration Thermodynamic Cycle Across the Triple Point

To protect the ozone layer and mitigate global warming, the replacement of hydrofluorocarbons (HFCs) and hydrochlorofluorocarbons (HCFCs) has become an urgent task. Therefore, current research on natural refrigerants represented by CO_2 and their energy conversion systems has become a hotspot. CO_2 is non-toxic, non-flammable, and does not damage the ozone layer. It is currently the substance with the greatest

Fig. 6 A novel solar collector system achieved by supercritical CO_2(Chen and Zhang 2014a)

potential for large-scale replacement of HFCs and HCFCs, and its excellent characteristics in refrigeration thermodynamic cycles have also attracted widespread attention. With the popularization of heat pump and refrigeration systems for civil and industrial use, the demand for low-temperature refrigeration and heat control for long-term continuous operation is increasing. In fact, in the CO_2 solid–gas two-phase sublimation heat transfer system, there are not only challenging scientific issues related to dynamic phase change, but also great significance for the research and development of new CO_2 low-temperature heat pumps and refrigeration thermodynamic cycle systems. To address the above issues, it is proposed to generate CO_2 particle solid–gas two-phase flow through the action of CO_2 low-temperature throttling, thereby extending the refrigeration temperature of the CO_2 liquid evaporation cycle to a lower level. The actual refrigeration temperature has dropped from –30°C to below the triple point temperature of CO_2 (–56.6°C), as shown in Fig. 7 (Chen and Zhang 2014b; Zhang and Yamaguchi 2009). Through theoretical analysis of the changes in the CO_2 phase diagram, the research direction for achieving CO_2 low-temperature refrigeration has been further clarified.

3.2.2 Trans-Critical CO_2 Vapor Compression Thermodynamic Cycle Refrigeration and Heat Pump

In China, the transcritical CO_2 vapor compression thermal system faces complex hydraulic and thermal issues rarely encountered abroad, due to factors such as high return water temperature in heating, low ambient temperature, coexisting cooling and

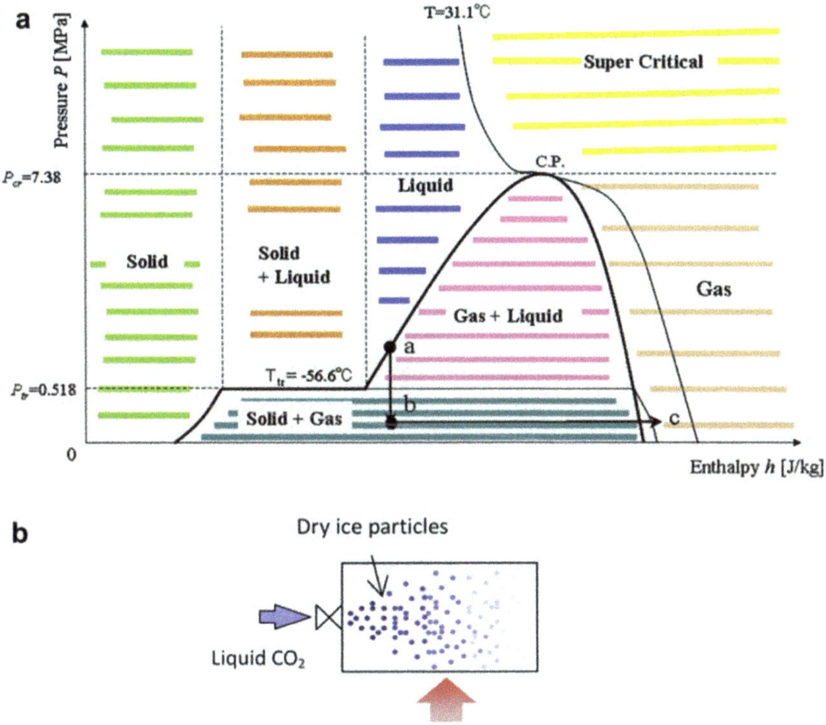

Fig. 7 Principle schematic of the refrigeration method using CO_2 solid–gas two phase fluid flow. **a** CO_2 P–h diagram. a-b process represents liquid CO_2 expanding into solid–gas two phase fluid flow, which goes below the CO_2 triple point of -56.6 °C. **b-c** is the sublimation process of the CO_2 solid–gas fluid. **b** image of the evaporator using CO_2 solid–gas two phase fluid flow, in which dry ice particles sublimate, absorb heat quantity from the refrigeration objective and therefore achieve a refrigeration below –56.6°C(Zhang and Yamaguchi 2009)

heating demands, increasingly severe haze pollution in recent years, and the maturity of conventional technologies for HCFC vapor compression cooling and heating systems. In response, researchers started from the physical essence of the hydraulic and thermal problems in the unsteady vapor compression thermodynamic cycle of transcritical CO_2, and established a universal mathematical model for unsteady hydraulic-thermal coupling and efficient flow and heat transfer. A prototype test platform adaptable to different working conditions was built. Figures 8 and 9 shows the CO2 transcritical heat pump laboratory prototype and the air source CO2 transcritical heat pump laboratory prototype. After correction with test data, its hydraulic and thermal models can accurately describe the hydraulic and thermal behaviors of CO_2 in the transcritical state, overcoming the flaws of traditional models. The establishment of the calculation model for bidirectional unsteady flow and heat transfer of transcritical CO_2 has solved the problem that traditional models can only perform one-way thermal calculation and cannot accurately solve the bidirectional critical heat transfer process. This model can also be used to calculate supercritical cooling problems under small temperature differences.

Variable diameter structures such as sudden expansion and sudden contraction often exist in transcritical CO_2 thermodynamic cycle systems like condensers. When transcritical CO_2 fluid flows through sudden expansion/contraction structures, the pressure changes drastically, which significantly affects various physical properties of transcritical CO_2, thereby making the flow and heat transfer processes complex and difficult to predict and control. Therefore, it is necessary to study the fluid dynamics and heat transfer characteristics of transcritical CO_2 in sudden expansion and contraction structures. The research mainly adopted a combination of experimental visualization and theoretical analysis to reveal the flow bifurcation phenomenon of supercritical CO_2 in sudden expansion and sudden contraction structures, as well as its impact on flow and heat transfer characteristics. It discovered the relationship between asymmetric unsteady bifurcation flow and the variation law of the critical

Fig. 8 CO_2 transcritical heat pump laboratory prototype (60kW)

Fig. 9 Air source CO_2 transcritical heat pump laboratory prototype (90kW)

Reynolds number. Additionally, it identified the possible multi-peak phenomenon of heat transfer of supercritical CO_2 in microchannels and its intrinsic physical mechanism, and further proposed flow stabilization measures, which provide guidance for the design and operation of various heat exchangers using supercritical CO_2 and transcritical CO_2 thermodynamic cycles.

3.2.3 Natural Refrigerant CO_2 Trans-Critical Cooling and Heating Technology and Application

The transcritical CO_2 direct expansion refrigeration thermodynamic cycle with full heat recovery and its technical application were employed in the ice rinks of the 2022 Winter Olympics. The National Speed Skating Oval, also known as the "Ice Ribbon", as shown in Fig. 10, is located at the northwest corner of the Bird's Nest. As a landmark venue of the 2022 Beijing Winter Olympics, it hosted 6 Winter Olympic events including speed skating and figure skating. Operational data from the ice hockey arena and short track speed skating arena shows that the application of the transcritical CO_2 direct expansion ice-making technology with full heat recovery can meet the ice-making needs of the Winter Olympics. It replaces hydrofluorocarbons (HFCs) and hydrochlorofluorocarbons (HCFCs), and also replaces traditional boiler systems in terms of venue heating, dehumidification, preheating, and anti-freezing, achieving energy savings of over 60%. This technology is listed as the biggest technical highlight of the 2022 Beijing Winter Olympics. By adopting transcritical CO_2 ice-making technology to replace HFCs and HCFCs, China has fulfilled its commitments under the Paris Agreement. This technology also marks the first time in the history of the Winter Olympics that natural refrigerant CO_2 is used instead of HFCs and HCFCs for ice-making. Moreover, the temperature difference on the ice surface is small, realizing the fastest ice-making method in human history. In addition, this achievement is not only applied to the 2022 Winter Olympics but also to the construction of ice rinks in schools and various other private sectors.

Fig. 10 Transcritical CO_2 direct expansion refrigeration technology developed and used in the National Speed Skating Oval— "the Ice Ribbon" for 2022 Winter Olympic Game Beijing, which was first proposed by Prof. Xin-Rong Zhang in Peking University

Fig. 11 Prototype completed in Shijinglong Ski Resort by using CO_2 for Winter Olympic (The red banner in the photograph reads: "Peking University College of Engineering – Artificial Snowmaking for the Winter Olympics.")

To ensure the smooth running of the 2022 Beijing Winter Olympics and Paralympics in case of insufficient natural snowfall, a new type of snowmaking equipment has been researched and designed, as shown in Fig. 11. This equipment, integrating the accumulated experience in CO_2 transcritical vapor compression refrigeration thermodynamic cycle technology and through the efficient coordination of the two, has enabled the development of an integrated artificial snowmaking and snow storage technology that can realize artificial snowmaking, snow storage and heating at 0–20 °C. This technology and equipment have been tested and demonstrated on the Winter Olympic tracks, serving as a solution for above-zero artificial snowmaking at the Beijing Winter Olympics and pioneering the innovative emerging industry of above-zero artificial snowmaking. It has achieved an international first in this field, laying a solid foundation for the development and promotion of this emerging industry in China in the post-Winter Olympics era.

The research results can not only realize efficient, environmentally friendly and energy-saving artificial snowmaking and snow storage in environments above the freezing point but also generate hot water and hot air to meet heating needs. It not only provides a safe, environmentally friendly, low-cost and high-quality snow guarantee plan for the 2022 Winter Olympics under above-zero temperature conditions but also offers a way to carry out snow sports in snow-scarce and snow-deficient areas in China. The continuous development of this technology and equipment has greatly expanded the geographical scope of snow sports, promoted the vigorous development of snow sports, and driven the snow sports industry towards energy conservation, greenness and low carbon.

The developed trans-critical CO_2 vapor compression cold and heat source supply technology has achieved significant social and economic benefits in fields such as hotels, office buildings, coal-to-electricity conversion, power plants, and coal mines.

3.3 *Diversified CO_2Thermodynamic System*

3.3.1 CO_2 Energy Storage System

CO_2 energy storage technology utilizes **supercritical carbon dioxide** (sCO_2) to realize energy storage and release through thermodynamic cycles, with its core process consisting of three stages. In the charging phase, excess electricity (usually from renewable energy sources) drives compressors to pressurize CO_2 beyond its critical point (7.38 MPa, 31.1°C), while capturing waste heat generated during compression through **Thermal Energy Storage** (TES) systems. In the storage phase, high-pressure sCO_2 is stored in geological structures such as salt caverns or artificial containers. In the discharging phase, the stored sCO_2 expands through turbines to drive generators for electricity generation, and the pre-stored heat in TES is used to reheat the sCO_2 to improve efficiency. This technology is applied in multiple fields: 1)in grid-scale energy storage, sCO_2 Brayton cycles can achieve 10–100 MW of energy storage.2) in industrial waste heat recovery, transcritical cycles can increase net power by 33.6%-192.3%0.3) in renewable energy integration, the capacity utilization rate in regions with a high proportion of renewable energy reaches 60%, salt caverns can achieve century-scale storage, and the use of nickel alloys to mitigate corrosion allows equipment life to exceed 30 years.

CO_2 energy storage, leveraging the unique thermodynamic properties of sCO_2 and the advantages of geological storage, demonstrates enormous dual potential in enhancing grid resilience and carbon management. The competitiveness of salt cavern projects is highlighted by their investment payback period of less than 10 years, gigaton-level storage potential, and efficiency comparable to mature technologies such as pumped hydro storage. By integrating renewable energy, alleviating intermittency issues, and recovering industrial waste heat, this technology accelerates the process of transitioning to net-zero emissions. Further technological advancements such as modular turbine designs and digital twin optimization are expected to reduce costs and consolidate its position as a core component of sustainable energy infrastructure. With continuous demonstration applications and large-scale promotion, CO_2 energy storage is well-prepared to address key challenges in the global transition to low-carbon energy systems.

3.3.2 Carbon Capture & Utilization (CCU)

Carbon Capture & Utilization (CCU) represents a critical technological pathway toward decarbonization, focusing on intercepting anthropogenic CO_2 emissions from industrial point sources—such as fossil-fuel power plants, cement factories, iron/steel mills, and chemical refineries—and converting this captured carbon into economically valuable products. The capture phase employs mature techniques including post-combustion capture, pre-combustion capture, and oxy-fuel combustion, selectively isolating high-purity CO_2 streams. Crucially, CCU diverges from

Carbon Capture & Storage (CCS) by emphasizing the productive valorization of captured CO_2 through chemical, biological, or electrochemical transformation pathways. Major utilization routes encompass: (i) Enhanced Oil Recovery (EOR), where supercritical CO_2 is injected into depleted reservoirs to mobilize residual hydrocarbons. (ii) chemical synthesis of platform chemicals via catalytic hydrogenation or electrochemical reduction. (iii) mineral carbonation, binding CO_2 into stable carbonates for construction materials. (iv) biological conversion via microalgae cultivation for biofuels. This technological paradigm not only reduces atmospheric CO_2 loading but also creates circular carbon economies where waste emissions become industrial feedstocks.

Despite its transformative potential, CCU deployment faces significant scientific, economic, and scalability challenges, necessitating integrated innovation across materials science, process engineering, and systems optimization. The kinetic limitations and high energy penalties associated with catalytic CO_2 activation remain substantial barriers. Ultimately, scaling CCU requires techno-policy synergies, including carbon pricing mechanisms, standardized CO_2 purity specifications for cross-sectoral utilization, and incentives for early-mover industrial clusters, positioning CCU as an indispensable pillar in the global net-zero transition.

3.3.3 CO_2 Jet Technology

CO_2 phase change jet technology leverages the rapid phase transition characteristics of CO_2 under specific pressure and temperature conditions to achieve energy release or material processing. In the process, high-pressure liquid CO_2 is injected into a low-pressure environment through a nozzle, undergoing an instantaneous phase change from liquid to gas (or supercritical state), accompanied by a sharp volume expansion and a significant drop in temperature. This sudden phase transition generates a high-speed jet flow, which can be applied in fields such as industrial cleaning, material cutting, and geothermal energy development. For instance, in precision cleaning, the low-temperature and high-velocity CO_2 jet effectively removes contaminants without damaging the surface of materials, while avoiding secondary pollution due to its natural vaporization property. In energy extraction, the phase change jet can assist in breaking rock formations, improving the efficiency of oil and gas or geothermal resource exploitation. The technology's core lies in regulating parameters such as pressure (typically 5–30 MPa), temperature, and nozzle structure to control the intensity and duration of the phase change, thereby adapting to diverse application scenarios.

CO_2 phase change jet technology stands out as an environmentally friendly and efficient solution in industrial and energy sectors, with unique advantages in sustainability and precision. Compared to traditional technologies relying on chemical solvents or mechanical tools, it reduces reliance on non-renewable resources and minimizes waste emissions, aligning with global low-carbon development goals. Its applicability across multiple fields—from microelectronics cleaning to large-scale resource extraction—highlights its versatility. However, challenges remain, such as the high energy consumption required for pressurizing CO_2 and the need for advanced materials to withstand extreme pressure and temperature fluctuations in equipment. Future advancements in energy recovery systems and high-performance nozzle designs could further enhance its cost-effectiveness. Overall, as industries increasingly prioritize green production, CO_2 phase change jet technology is poised to play a pivotal role in driving cleaner and more efficient industrial processes.

4 Conclusion

In summary, the historical evolution of energy systems has made significant progress in meeting the growing global energy demand, but it has also been accompanied by some long-standing challenges, such as low thermal efficiency, reliance on environmentally harmful working fluids, and inherent thermodynamic limitations. These drawbacks, ranging from ozone layer depletion and high global warming potential to inefficient utilization of thermal energy in low-grade energy recovery, highlight the urgency of seeking innovative solutions. Against this backdrop, the emergence of new energy systems using CO_2 as a working fluid represents a promising paradigm shift. CO_2 possesses unique thermophysical properties, including a relatively low critical temperature, excellent heat transfer characteristics in transcritical and supercritical states, and inherent environmental friendliness (zero ozone depletion potential and low global warming potential). These properties address many limitations of traditional systems. It can enhance thermal efficiency in Rankine cycles operating within specific temperature ranges, facilitate effective low-grade heat recovery, and reduce thermodynamic losses through temperature slip in supercritical heat absorption processes, all of which demonstrate its versatility and potential in sustainable energy applications.

However, it must be acknowledged that further research is needed to optimize system configurations, address operational challenges, and scale up these technologies for widespread industrial application. By fully leveraging the advantages of CO_2, these new energy systems not only align with global environmental protection goals but also provide a pathway for more efficient, reliable, and sustainable energy conversion and utilization. Since the initial proposal of using carbon dioxide for power generation in the solar energy sector in 2005(Zhang et al. 2005), the development of a thermodynamic energy system centered on CO_2—encompassing CO_2 refrigeration, combined cooling and heating technologies, and CO_2 energy storage—has continuously advanced, serving as a key driver of the global energy revolution and opening up broad prospects for the future of clean energy.

References

Alanne K, Saari A (2006) Distributed energy generation and sustainable development. Renew Sustain Energy Rev 10:539–558
Chapman (1974) Heat transfer. Magnet Resonan Mater Biol Phys Med 9(3):146–151
Chen L, Zhang XR (2014a) Experimental analysis on a novel solar collector system achieved by supercritical CO2 natural convection. Energy Convers Manage 77(jan.):173–182
Chen L, Zhang XR (2014b) A review study of solid–gas sublimation flow for refrigeration: from basic mechanism to applications. Int J Refrig 40:61–83
BoroumandJazi, G, Rismanchi B, Saidur R (2013) A review on exergy analysis of industrial sector. Renew Sustaina Energy Rev
Horlock JH, Bathie WW (2004) Advanced gas turbine cycles. J Eng Gas Turbines Power 126(4):924
Ishikawa M, Terauchi M, Komori T, Yasuraoka J (2008) Development of high efficiency gas turbine combined cycle power plant. Tech Rev - Mitsubishi Heavy Indust 45(1)
Kraus AD, Bar-Cohen A (1983) Thermal analysis and control of electronic equipment
Mario J, Molina FS, Rowland (1974) Stratospheric sink for chlorofluoromethanes: chlorine atom-catalysed destruction of ozone. Nature 249(5460):810–812
Mayr RBO (1972) From watt to Clausius: the rise of thermodynamics in the early industrial age by D. S. l. Cardwell. Isis 63(3):451–452
Mcconnell EH (2000) More work for mother: the ironies of household technology from the open hearth to the microwave. by ruth Schwartz Cowan. home and work: housework, wages, and the ideology of labor in the early republic. By Jeanne Boydston. The Female Economy: the millinery. J Museum Educat 20(1):12–14.
Midgley T, Henne AL (1930) Organic fluorides as refrigerants. ind.eng.chem
Review, by:, John, Bowditch (1992) Power from steam: a history of the stationary steam engine by Richard L. Hills. Technology & Culture
Rolt LTC, Dickinson HW (2011) A short history of the steam engine. Augustus M. Kelley
Smith C (1998) The science of energy: a cultural history of energy physics in Victorian Britain. The science of energy: a cultural history of energy physics in Victorian Britain
Thévenot R (1979) A history of refrigeration throughout the world. Int J Refrig 2(6):249–250
Thornton JE (1964) Parallel operation in the control data 6600. ACM
Thurston RH (1972) A history of the growth of the steam-engine. Nature 19(487):381–382
Tuckerman DB, Pease RFW (1981) High-performance heat sinking for VLSI. IEEE Electron Device Lett 2(5):126–129
Wang SK (1993) Handbook of air conditioning and refrigeration. McGraw-Hill
Yin JM, Zheng QY, Peng ZR, Zhang XR (2020) Review of supercritical CO2 power cycles integrated with CSP. Int J Energy Res 44(3)
Zhang XR, Yamaguchi H (2009) Experimental investigation on heat transfer of CO2 solid-gas two phase flow with dry ice sublimation. In: ASME 2009 heat transfer summer conference collocated with the InterPACK09 and 3rd energy sustainability conferences
Zhang XR, Yamaguchi H, Fujima K, Enomoto M, Sawada N (2005) A feasibility study of CO2 based rankine cycle powered by solar energy. JSME Int J, Ser B 48(3):540–547
Zhang XR, Yamaguchi H, Uneno D (2007) Experimental study on the performance of solar Rankine system using supercritical CO2. Renew Energy 32(15):2617–2628

Analysis on the Spatial Correlation Network and Driving Factors of Carbon Emission Intensity in OECD Countries & China

Chenyang Lee and Seiichi Ogata

1 Background

With rapid economic development and urbanization, CO2 emissions now account for approximately 70% of global greenhouse gas (GHG) emissions. To address climate change and reduce CO2 emissions, the United Nations adopted the Kyoto Protocol in 1997, initiating a collaborative framework for global climate governance. The Protocol not only facilitated a transition in energy structures but also significantly influenced the energy security policies of various countries. Building on this foundation, the United Nations adopted the Paris Agreement in 2015 and introduced the Sustainable Development Goals (SDGs) to further strengthen global efforts toward emissions reduction and sustainable development. According to the International Energy Agency (IEA), approximately 80% of the OECD's regional commercial energy demand remains reliant on fossil fuel combustion (IEA 2019). This dependency has heightened global concerns that rapid economic growth, coupled with high energy consumption, could exacerbate greenhouse gas emissions and accelerate climate change. However, since carbon emissions cross national borders, tackling them effectively requires international cooperation. By combining different efforts across countries, such as technology sharing, joint regulations, and aligned incentives, it is possible to achieve greater emission reductions than by acting alone. This kind of effect is often referred to as "synergistic emission reductions".

In recent years, the issue of national-level carbon emissions has attracted significant attention in academic research. Existing studies on carbon emissions primarily focus on four key areas: measurement, spatio-temporal distribution, influencing factors, and policy regulation. Measurement research typically involves indicators

C. Lee · S. Ogata (✉)
Graduate School of Energy Science, Kyoto University, Kyoto, Japan
e-mail: ogata.seiichi.5x@kyoto-u.ac.jp

C. Lee
e-mail: li.chenyang.48u@st.kyoto-u.ac.jp

W. Zhou (ed.), *Decarbonization Policy Engineering*, World Sustainability Series,
https://doi.org/10.1007/978-981-95-5953-4_19

such as total emissions, per capita emissions, and carbon intensity (emissions per unit of GDP), offering different perspectives on carbon performance. Analyses of spatio-temporal distribution examine regional disparities, temporal trends, and sectoral variations, such as differences between urban and rural areas or among industrial, residential, and transportation sectors. Studies on influencing factors investigate how economic development, population growth, industrial structure, and energy use affect emission levels. Ultimately, policy-related research examines the effectiveness of instruments such as carbon pricing, emissions trading, and renewable energy subsidies in facilitating low-carbon transitions. With regard to emission measurement, methodologies such as input–output analysis (IOA) and life cycle assessment (LCA) have been widely adopted to evaluate carbon emissions at the national scale (Baz and Zhu 2025; Wiedmann 2009). These studies reveal a pronounced global imbalance in carbon emissions (Sun et al. 2022), with some countries achieving "carbon leakage" through industrial outsourcing and international trade—effectively reducing domestic direct emissions by importing high-carbon products, while shifting the emission burden to producing countries (Haga and Zhou 2021). Moreover, the evolving structure of the global value chain (GVC) has complicated countries' roles in global emissions. The spatial mismatch between high-value-added and high-carbon industries has further intensified the redistribution of carbon emissions across nations (Wei et al. 2024). In terms of spatial distribution, research has identified a clear trend of spatial convergence in carbon emissions among OECD countries (Romero-Ávila 2008). Emissions are influenced not only by domestic factors—such as economic development, energy consumption, and industrial structure—but also by the emissions levels of other countries, suggesting a strong spatial interdependence. Some scholars have applied spatial econometric methods to analyze the spatial evolution of carbon emissions in OECD countries, finding evidence of spatial path dependence. These studies suggest that the externalities of emission reduction policies enhance the importance of regional cooperation in achieving effective carbon mitigation (Zhou 2021).

In terms of the factors influencing carbon emissions, existing studies have primarily focused on economic growth, energy structure, green technology innovation, industrial restructuring, trade openness, and environmental policies. On the one hand, some research suggests that the promotion of renewable energy has played a positive role in reducing carbon emissions, with increased shares of wind and solar energy effectively lowering carbon emission intensity. Green technological innovation is widely regarded as a key driver of emission reductions, as it enhances energy efficiency and fosters the development of low-carbon industries. On the other hand, several studies have raised concerns that the substitution effect of renewable energy has not yet been fully realized. As a result, the total consumption of fossil fuels has not declined significantly, leading to the so-called "green paradox", where the deployment of renewable energy fails to suppress the overall growth of carbon emissions (Van Der Ploeg 2011). Building upon this, other scholars have highlighted the possibility of a "rebound effect" (Wang and Wei 2020), whereby technological improvements reduce the per-unit cost of energy consumption, thereby stimulating increased energy demand and, paradoxically, causing a rise in carbon emissions (Croutzet and Dabbous 2021). In addition, there has been a growing body of research examining the nonlinear relationships between carbon emissions and their drivers, such as the threshold effects of renewable energy use and economic development (Nan et al. 2022). The Environmental Kuznets Curve (EKC) hypothesis has been widely applied to explain the relationship between economic growth and carbon emissions, positing that emissions initially rise with economic development but eventually decline as technological progress and industrial upgrading take effect (Umar and Safi 2023). However, whether countries can truly reach the declining stage of the EKC remains debated. Some scholars argue that the observed emissions reductions in OECD countries are largely attributable to the "carbon transfer effect", whereby these nations effectively outsource carbon-intensive production through the import of high-emission goods, sometimes with implicit subsidies (Lin and Zhao 2024). Finally, the effectiveness of carbon reduction policies varies significantly across countries. While measures such as carbon taxes, emissions trading systems, and subsidies for renewable energy have been shown to reduce emissions in many contexts, their success often depends on institutional quality, economic structure, and public acceptance. Consequently, policy implementation and its impacts are highly heterogeneous across national contexts (Dasgupta and De Cian 2018; Knill and Lenschow 1998; Spandagos et al. 2022).

In summary, although existing studies have made notable progress in examining carbon emissions at the national level, several limitations remain. First, traditional spatial analyses often rely on geographic adjacency or attribute similarity, which makes them insufficient for capturing non-proximate but functionally important cooperative emission reduction networks across regions. Second, there is still a scarcity of research focusing on the spatial network structures of carbon emissions among OECD countries and China. This gap is particularly critical, as OECD countries and China together account for the majority of global carbon emissions and play a pivotal role in shaping the global climate governance landscape. The OECD bloc represents highly industrialized economies with advanced low-carbon technologies and mature regulatory frameworks, while China, as the largest developing emitter and a key player in global supply chains, is central to any meaningful global decarbonization pathway. Understanding the spatial interdependencies and network structures among these actors is essential for fostering cross-border coordination, technology transfer, and policy alignment. While social network analysis (SNA) has recently been applied in studies of carbon emissions at the provincial level or within specific sectors in China, such as industry (Liu and Xiao 2021) and transportation (Bai et al. 2020), there remains a lack of comprehensive global analysis of the spatial network structure, central nodes, and influence mechanisms within OECD countries. This gap constrains both theoretical understanding and practical advancements in facilitating coordinated emission reduction efforts across borders.

To address these issues, the present study adopts the social network analysis (SNA) framework to construct a spatial correlation network of carbon emission intensity

(CEI) for OECD countries and China from 1995 to 2022, aiming to explore the formation mechanisms underlying this network. Although traditional spatial econometric methods can reveal spatial autocorrelation patterns of carbon emissions, they are insufficient for fully depicting the complex and dynamic inter-country interactions necessary for synergistic transnational emission reduction. In contrast, social network analysis offers a promising alternative by comprehensively capturing interregional relational data and uncovering the structural features and temporal evolution of spatial linkages within a complex system.

The main contributions of this study are threefold: (1) it employs social network analysis to reveal the structural characteristics of the spatial CEI network across OECD countries and China, thereby enriching the analytical perspective of carbon emission spatial effects; (2) it identifies the core nodes and key transmission pathways within the network, providing valuable references for the design and implementation of targeted emission reduction policies in OECD countries; and (3) it advances the understanding of the driving mechanisms behind carbon emissions, offering empirical evidence to support more effective global cooperation on emission reduction.

The remainder of this chapter is structured as follows: Sect. 2 outlines the research methodology and data sources; Sect. 3 analyzes the spatial network structure and influencing factors of CEI in OECD countries and China; and Sect. 4 summarizes the findings and presents policy recommendations.

2 Materials and Methods

This study focuses on the carbon emission intensity of 39 countries (OECD and China), and the methodological process is as follows: First, the per capita carbon emission index is selected as the primary measure of carbon emission intensity, allowing for an assessment of the relative differences in carbon emissions across countries. Second, based on a modified gravity model, the gravitational strength of the spatial correlation of carbon emissions is calculated, from which the spatial correlation network matrix of carbon emission intensity for the 39 countries is constructed, providing the data foundation for subsequent network analysis. Next, social network analysis is applied to systematically uncover the structural characteristics of the spatial correlation network of carbon emission intensity, focusing on overall network structure, node centrality, and clustering patterns.

2.1 Carbon Emission Intensity Measurement

This study selects carbon emissions per unit of GDP as a proxy variable for carbon emission intensity. The specific calculation formula is as follows:

Question

$$CEI_{it} = \frac{CO_{2\,it}}{GDP_{it}} \tag{1}$$

where CEI_{it} represents the carbon intensity of country i in year t, measured as the ratio of carbon emissions to population; $CO2_{it}$ represents the carbon emissions of country i in year t, in megatonnes (Mt); and GDP_{it} represents the GDP of country i at the end of year t, in billions of US dollars (current price).

2.2 Construction of Spatial Correlation Network for CEI

To more accurately reveal the spatial correlation network structure of carbon emission intensity in OECD countries, this study adopts the following modified gravity model, with reference to the research of Huo et al. (2022):

Question

$$F_{ij} = (\frac{C_i}{C_i + C_j}) * \frac{\sqrt[3]{P_i G_i C_i} * \sqrt[3]{P_j G_j C_j}}{\left(\frac{d_{ij}}{g_i - g_j}\right)^2} \tag{2}$$

where F_{ij} denotes the strength of the spatial linkage of CEIs between countries i and j; C_i and C_j epresent the CEIs of countries i and j; P_i and P_j are the year-end populations (in millions) of countries i and j; G_i and G_j are the gross domestic products (GDP) (in millions of current US dollars) of countries i and j, treated logarithmically; d_{ij} is the spherical distance (1000 km) between countries i and j; g_i and g_j represent the per capita GDPs of countries i and j, also treated logarithmically. The gravity matrix $(F_{ij})_{39\times39}$ for the 39 countries is calculated from Eq. (2), and the matrix is binarized using the mean value of each row as the threshold. If the gravitational value is greater than or equal to the threshold, a correlation is considered to exist between the two countries, assigning a value of 1; otherwise, the value is assigned as 0, thus yielding the spatial correlation matrix of international CEIs.

2.3 Spatial Correlation Network Characteristics of CEI

In conducting an in-depth analysis of the spatially interconnected network structure of carbon emission intensities (CEIs) across countries, the application of social network analysis (SNA) proves particularly valuable. This method not only facilitates a comprehensive understanding of the interrelations and mutual influences of CEIs among nations but also enables the identification of key nodes and the underlying mechanisms driving influence within the network. Drawing on the framework established by Kang et al. (2024), Table 1 presents the selected indicators employed in this study to evaluate both overall and individual network characteristics of the CEI spatial correlation network, along with their corresponding measurement formulas.

2.4 *Data Source*

The objective of this study is to analyze the spatial correlation networks of CEI across countries, along with their underlying driving factors. The study focuses on the period from 1995 to 2022, encompassing data from 39 countries. All data were sourced from the World Bank database to ensure the credibility, consistency, and reliability of the information used in the analysis.

3 Results and Discussion

3.1 *The Trends in Carbon Emission Intensity (CEI) Across Different Nations*

From 1995 to 2022, the carbon emission intensity across major economies worldwide has generally exhibited a declining trend, indicating progress in enhancing energy efficiency, transitioning to cleaner energy sources, and enacting relevant policy measures. Nevertheless, substantial heterogeneity exists among countries regarding the pace and trajectories of emissions reduction.

Taking China as a case study, carbon emission intensity decreased markedly from 45.75 in 1995 to 7.08 in 2022, representing a reduction of over 80%. This substantial decline positions China among the most successful countries globally in terms of emissions mitigation. The rapid decrease is closely associated with the systematic emissions reduction policies implemented post-2006, including the 11th Five-Year Plan, which emphasized the regulation of energy-intensive industries and the large-scale deployment of renewable energy. It is noteworthy that the rate of decline in China's carbon emission intensity slowed after 2013, likely reflecting the country's transition into a more complex phase of economic development and energy structure adjustment. As the world's second-largest carbon emitter, the United States experienced a steady decline in carbon emission intensity from 6.90 in 1995 to 1.89 in 2022, corresponding to an approximate 73% reduction. The U.S. emissions reduction trajectory has been relatively smooth, driven primarily by energy structure optimization resulting from the shale gas revolution and market-driven improvements in energy efficiency. However, the pace of reduction has decelerated in recent years, indicating challenges faced by emissions reduction policies. European Union member states have demonstrated overall strong performance, with Germany reducing its carbon emission intensity from 3.50 to 1.65 and France from

Table 1 Indicators of overall and individual network characteristics

	Indicators	Abbreviation	Formulas	Description
Overall Network	Network Density	ND	$N_d = \frac{T}{N-1}$	T: existing correlation number; N: total correlation number between nations
	Network Connectivity	NC	$N_c = 1 - \frac{M}{(N-1)*N/2}$	M: number of unreachable point pairs in the network
	Network Hierarchy	NH	$N_H = 1 - \frac{H}{\max(H)}$	H: number of points that are symmetrically reachable in the network
	Network Efficiency	NE	$N_e = 1 - \frac{E}{\max(E)}$	E: number of redundant lines in the network
Individual Network	Degree Centrality	DC	$DC_j = \frac{k_j}{N-1}$	k_j:number of directly related nations
	Betweenness centrality	BC	$BC_j = \frac{1}{(N-1)*(N-2)/2} \sum_{s \neq j \neq t} \frac{n_{st}^j}{g_{st}}$	N:number ofall connections; n_{st}^j:number of connections through nation j in all connections from nation s to nation t; g_{st}:number of connections from nation s to nation t
	Closeness Centrality	CC	$CC_j = \frac{1}{\sum_{i=1}^{N} d_{ji}}$	d_{ji}:the shortcut length between nation j and nation i

(continued)

Table 1 (continued)

	Indicators	Abbreviation	Formulas	Description
Block Model	Bidirectional spillover plate (≈0); net benefit plate (>0)	BSP/NBP	$\geq (g_k - 1)/(g - 1)$	g_k:number of nations within the plate; g:number of nations in the network relationship
	Net spillover plate (≈0); agent plate (>0)	NSP/AG	$< (g_k - 1)/(g - 1)$	Position and function of the modules in the spatial network and linkages and interactions between internal and external modules

2.38 to 1.13. These countries have achieved sustained and stable emissions reductions through mechanisms such as carbon markets, renewable energy subsidies, and stringent energy efficiency standards. Notably, Nordic countries like Sweden have achieved exceptionally low carbon emission intensity levels, reaching 0.64, exemplifying the feasibility of deep decarbonization. In contrast, some emerging economies, such as Turkey, have exhibited slower progress, with carbon intensity fluctuating around 5 over the past decade, highlighting the challenges of balancing economic growth with emissions reduction objectives. Resource-rich countries such as Australia and Canada have recorded some reductions; however, their carbon intensities remain higher than those of EU countries, underscoring their continued economic dependence on fossil fuels.

3.2 Analysis of Spatial Correlation Network Structure of CEI

To provide an intuitive understanding of the spatial linkage structure of carbon emissions, this study employs a modified gravity model to map the carbon emission linkage intensities among 39 countries for the years 1995, 2010, and 2020 (see Appendix A, Figs. 2, 3, 4 for details). The results reveal that, over time, the spatial linkage intensity of carbon emission intensity (CEI) has been progressively strengthened. The CEI network among countries has evolved into a more complex and multi-layered structure, indicating increasingly intricate interdependencies. As a result, the spatial configuration of international carbon emission linkages has become progressively more defined and transparent.

Figure 1 illustrates the structural evolution of the spatial correlation network of carbon emission intensity (CEI) among 39 countries—including OECD member states and China, during the period 1995–2022, based on Ucinet analysis. The figure focuses on three core network metrics: network connectivity, network density, and network efficiency. The value of network connectivity (NC) consistently remains at 1 throughout the study period, indicating that a fully connected spatial network has been maintained and that there is a strong and persistent spatial correlation in CEI across these countries. In terms of overall trends, network density increases markedly over time, with a particularly sharp rise observed around the year 2000. This suggests a continuous intensification of spatial linkages in carbon emission intensity, reflecting closer convergence in CEI performance among countries. In other words, countries with similar CEI patterns have become more tightly interconnected over time. The strengthening of this network structure may be attributed to a combination of regional policy diffusion, technological spillovers, and the broader influence of global climate governance mechanisms. These factors have likely contributed to greater interaction, mutual influence, and convergence in national carbon intensity trajectories.

However, in contrast to the rising trend in network density, network efficiency has shown a steady decline since 2000. This indicates that while the number of inter-country connections within the CEI network has increased, the effectiveness of information transfer and the marginal synergistic benefits per connection has diminished. In other words, although countries with similar carbon emission intensities are becoming more interconnected, the network is also exhibiting greater structural redundancy. This redundancy may hinder the effective diffusion of best practices in emission reduction and the dissemination of energy efficiency technologies, leading to potential "structural bottlenecks" within the network. This trend suggests that

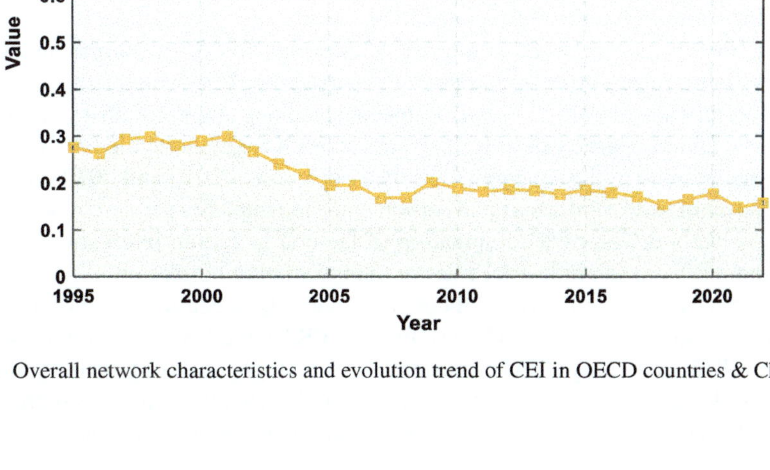

Fig. 1 Overall network characteristics and evolution trend of CEI in OECD countries & China

global efforts toward coordinated carbon mitigation are entering a pivotal phase—one that requires a shift from merely establishing structural connectivity to actively pursuing structural optimization. Enhancing the quality, rather than simply the quantity, of network linkages will be essential to overcoming inefficiencies and unlocking the full potential of transnational collaboration in addressing climate change.

From a comprehensive perspective, the CEI spatial network over the past three decades has exhibited a structural pattern characterized by "high density but low efficiency." While nations are increasingly interconnected in terms of emission intensity, the current network structure may not sufficiently support the effective diffusion of low-carbon technologies or the efficient coordination of emission reduction strategies. At the policy level, greater emphasis should be placed on optimizing the structure of transnational low-carbon technology diffusion networks. Enhancing the efficiency of information transmission and improving marginal synergistic effects are crucial steps toward overcoming structural redundancies. Such efforts would facilitate a shift from "surface-level synergy" to "substantive synergy", thereby promoting a more systematic, inclusive, and equitable global pathway toward low-carbon transition.

3.3 Individual Network Characteristics and Evolution Trend

This section aims to uncover the individual characteristics within the spatial network structure of carbon emissions by analyzing the degree of centrality of all nodes for the years 1995, 2010, and 2020. The results are presented in Table 2. Degree centrality quantifies the number of direct connections a node has with other nodes, serving as a key metric to assess the extent of a node's immediate linkage within the network. In the context of the spatial correlation network of carbon emission intensity, degree centrality is a crucial indicator reflecting a country's direct connectivity and influence on the trends of carbon emission changes across the network. As shown in Table 3, distinct dynamic evolutionary patterns and structural differences emerge both temporally and at the individual country level.

First, the overall network connectivity between OECD countries and China exhibits an increasing trend across the three periods analyzed, indicating a growing spatial correlation of carbon emission intensity among these countries. While the out-degree and in-degree values for most countries were at moderate levels in 1995, the substantial rise in these degree measures observed in 2010 and 2020 reflects an enhanced similarity or interaction in carbon emission intensity across countries. This pattern is likely driven by a combination of factors, including international carbon reduction mechanisms, industrial transfer, and technology diffusion.

From a horizontal comparison of countries with high centrality, traditional European nations such as Germany (DEU), France (FRA), the United Kingdom (GBR), Belgium (BEL), and Italy (ITA) consistently exhibit high levels of both outgoing and incoming degree centrality across all three periods. This pattern underscores their sustained central role within the carbon emission intensity network. These countries

Table 2 Density Centrality Analysis of CEI in OECD countries & China

Country	OutDegree1995	InDegree1995	OutDegree2010	InDegree2010	OutDegree2020	InDegree2020
AUS	11	11	27	26	26	22
AUT	30	31	30	30	32	33
BEL	33	32	32	32	33	33
CAN	29	29	29	28	32	28
CHE	34	35	36	37	35	37
CHL	16	12	21	22	24	23
CHN	32	28	36	35	36	35
COL	19	19	27	28	31	30
CRI	13	10	19	19	19	23
CZE	32	30	33	32	33	33
DEU	36	36	36	36	34	35
DNK	33	34	36	36	34	35
ESP	31	32	36	36	34	34
EST	25	17	30	24	27	26
FIN	29	29	29	29	32	32
FRA	33	34	35	36	35	36
GBR	34	35	36	36	34	36
GRC	27	28	33	32	26	28
HUN	27	28	30	30	30	30
IRL	29	29	30	30	35	37
ISL	6	13	18	20	24	27
ISR	23	26	32	33	29	28

(continued)

Table 2 (continued)

Country	OutDegree1995	InDegree1995	OutDegree2010	InDegree2010	OutDegree2020	InDegree2020
ITA	32	32	36	36	36	36
JPN	35	36	32	32	32	31
KOR	21	27	30	26	35	29
LTU	28	23	28	29	29	29
LUX	31	32	37	37	37	37
LVA	26	23	28	30	27	29
MEX	25	21	29	29	32	31
NLD	32	32	34	35	34	34
NOR	32	34	37	38	35	36
NZL	3	4	10	10	9	9
POL	32	31	33	31	35	33
PRT	24	28	29	32	27	28
SVK	28	25	31	30	29	29
SVN	26	26	30	31	28	28
SWE	29	32	33	36	33	34
TUR	27	30	31	32	34	34
USA	32	31	34	31	37	36

Table 3 Spillover effect of the correlation network

Year	Plate	Issued connection	Received connection	Expected internal relationship ratio(%)	Actual internal relationship ratio(%)	Actual internal relations
1995	I	47	53	7.89	4.08	Proxy
	II	299	310	39.47	41.49	Mainstream In
	III	277	246	28.95	7.05	Mainstream Out
	IV	168	182	18.42	10.16	Proxy
2010	I	261	249	23.68	6.79	Mainstream Out
	II	282	281	26.32	28.06	Mainstream Out
	III	226	248	21.05	4.64	Proxy
	IV	235	226	21.05	17.25	Mainstream Out
2021	I	227	220	21.05	10.28	Mainstream Out
	II	287	298	28.95	26.97	Proxy
	III	177	174	15.79	10.61	Mainstream Out
	IV	261	260	26.32	22.78	Mainstream Out

not only achieved early industrialization but also demonstrated strong leadership in energy transition and the application of low-carbon technologies, enabling them to establish stable linkages within the spatial correlation network. Notably, China (CHN) already exhibited a relatively high degree of centrality in 1995 (OutDegree: 32, InDegree: 28), which further increased in 2010 and 2020 to 36 and 35, respectively. This significant rise reflects the strengthened linkage between China's carbon emission intensity and that of other countries. This increase highlights China's critical influence on global carbon intensity dynamics and underscores its growing importance in international efforts for carbon emission governance.

Conversely, certain countries exhibit clear signs of "marginalization" within the network, such as New Zealand (NZL) and Iceland (ISL). New Zealand's degree centrality remains consistently low—only 3/4 in 1995 and 9/9 in 2020—indicating that its carbon intensity changes relatively independently from other countries. This may be attributed to its high renewable energy share in the energy mix and the relatively small scale of its economy. In contrast, Iceland shows a rapid increase in centrality, rising from a low to moderate level (6/13) in 1995 to 24/27 in 2020, suggesting a significant strengthening of its carbon intensity linkages with other countries. This change could be influenced by sample expansion or by the effects of energy transition policies within the period under study.

From a temporal perspective, the degree centrality of most countries exhibits a synchronous upward trend, particularly among mid-level countries such as South Korea (KOR), Ireland (IRL), Israel (ISR), and Colombia (COL), whose centrality has increased markedly. This pattern reflects a "homogenization" trend in the global correlation of carbon emission intensity. In other words, changes in carbon intensity are no longer predominantly driven by a few traditional industrialized nations; instead, a broader range of countries are increasingly converging in their energy consumption structures and climate policy responses. This convergence has contributed to a denser and more cohesive spatial correlation network.

Furthermore, the heterogeneity in degree centrality reveals a multipolar pattern in the structure of the carbon emission intensity network. For instance, the United States (USA) consistently maintains a high degree of centrality (1995: 32/31; 2020: 37/36), with relatively minor fluctuations in its in-degree and out-degree over time, underscoring its stable dominant position within the global carbon emission system. In contrast, Central European countries such as Portugal (PRT) and Slovenia (SVN) exhibit more volatile degree centrality patterns. Although their centrality values increase intermittently, the lack of continuity may reflect constraints imposed by the nonlinear nature of their domestic economic restructuring processes, which affect their carbon emission intensity linkages.

In summary, the degree centrality analysis of carbon emission intensity reveals that the global carbon emission network is gradually evolving from dominance by a few core countries toward a more complex, multipolar linkage structure. Within the framework of international cooperation on climate change, understanding the evolution of this network is crucial for identifying key countries, enhancing regional policy coordination, and facilitating the diffusion of low-carbon technologies.

4 Block Model Analysis

Referring to the prior study by Gao and Gao (2024), this chapter employs block model analysis to examine the roles of regional and locational relationships within spatial networks. Specifically, countries are classified into four groups (Blocks I–IV) based on the block model, enabling the identification of carbon emission linkages and their evolutionary trends through the direction and intensity of spillover effects. Block modeling serves as a clustering technique to detect similarities and interactions among nodes—in this case, countries—within the network. Countries grouped within the same block exhibit similar carbon emission intensity characteristics, while the connections between blocks represent the direction and magnitude of spillover flows.

The four blocks (Blocks I–IV) are defined as follows: (1) The "Mainstream In" block, which receives more connections from other blocks than it sends out, with the actual proportion of internal connections exceeding the expected proportion; (2) The "Mainstream Out" block, which sends more connections to other blocks than it receives, characterized by an actual internal connection ratio lower than the expected ratio; (3) The "Bidirectional" block, which both sends and receives numerous connections to and from other blocks while maintaining a relatively high level of internal connectivity; and (4) The "Proxy" block, which receives connections from other blocks and sends connections outward, serving as a "bridge" within the network. Based on these classification criteria and the results of the block model analysis, this study employs the block model to examine the locational relationships and roles of 39 countries within the spatial correlation network of carbon emission intensity. The analysis is conducted using the CONCOR program, with the maximum partition depth and convergence criteria set at 2 and 0.2, respectively.

Table 3 systematically presents the internal structural characteristics of each block within the spatial correlation network of carbon emission intensity and their associated spillover effects at the three-time points of 1995, 2010, and 2020. This analysis fully captures the dynamic evolution and complexity of the network structure, as well as the interregional carbon emission flow patterns. In the 1995 network, Block II, classified as a "mainstream inlet" block, exhibited an actual proportion of internal connections (41.49%) slightly higher than the expected proportion (39.47%). This indicates that Block II primarily functioned as a concentrated receiver of carbon emission impacts from other blocks, demonstrating a clear "convergence" effect in absorbing external carbon emission influences. In contrast, Block III functioned as a "mainstream outflow" block, with a notably lower actual internal connection ratio (7.05%) compared to the expected (28.95%), suggesting a stronger tendency to export carbon emission impacts outward, thereby acting as a "diffuser" within the network. Blocks I and IV, meanwhile, served predominantly as "proxy" blocks, characterized by significantly lower actual internal connection ratios than expected. This pattern implies that these two blocks primarily functioned as bridges linking other blocks, with relatively weak internal cohesion but critical importance for overall network connectivity in 1995.

In 2010, the network structure underwent significant changes. Blocks I, II, and IV all transitioned to the role of "mainstream outflow," indicating that the carbon intensity within these blocks not only exhibited active internal interactions but also demonstrated enhanced capacity to radiate and diffuse impacts externally, thereby becoming primary sources driving the spread of carbon emission effects. For instance, although the actual proportion of internal connections in Block II decreased to 28.06% from its 1995 level, it still maintained a relatively high level, reflecting stable internal linkages alongside its role in disseminating carbon emission impacts. Conversely, Block III experienced a sharp decline in its internal connection ratio, dropping to 4.64%, and continued to function as a "proxy". This suggests a weakened ability to radiate and absorb carbon emission influences, serving more as a bridge for the transmission of information or effects rather than as a core driver within the network. Overall, the 2010 network exhibited a trend toward a greater number of blocks actively spreading carbon emission impacts, with the network structure becoming more polarized and the interregional interactions and dependencies of carbon emissions growing increasingly complex.

By 2020, the network structure underwent further adjustments. Block II shifted to the role of a "proxy", with the proportion of its actual internal connections (26.97%)

closely aligning with the expected proportion (28.95%). This indicates that Block II assumed a more stable intermediary function, connecting other blocks within the network through a balanced and decentralized internal structure, and no longer serving as a dominant source of carbon emission diffusion. Meanwhile, Blocks I, III, and IV continued to fulfill the role of "mainstream outflow". Notably, the actual internal connection ratio of Block IV (22.78%) increased significantly, approaching or even exceeding its expected proportion, highlighting its enhanced diffusion influence and establishing it as a key node in promoting carbon emission spread. The continued "mainstream outflow" roles of Blocks I and III further suggest that carbon emission activities in these regions remain active, exerting significant influence on other regions through spatial correlation. Overall, the 2020 spatial correlation network of carbon emission intensity demonstrates a dynamic rotation of roles and an evolving division of labor among blocks, with some regions transitioning from diffusion sources to intermediaries, while others strengthen their radiative capacity, reflecting the spatial transfer and evolution of carbon emission influences.

In summary, the structural changes observed at the three-time points highlight the spatial and temporal heterogeneity of carbon emission linkages and their complex interaction mechanisms. Variations in the ratio of internal connections among blocks, along with the evolution of their roles as "mainstream out", "mainstream in", or "proxy," not only reflect shifts in the direction and intensity of interregional carbon emission flows but also illustrate the processes of functional differentiation and structural adjustment within each region's carbon emission network. This dynamic evolution carries significant implications for the design of regional carbon reduction strategies. It underscores the necessity for policymakers to consider the spatial patterns of carbon emission diffusion and aggregation and to foster coordinated interregional governance to promote the systematic optimization of the low-carbon transition.

Table 4 presents the density matrix and image matrix of the spatial correlation network of carbon emission intensity among blocks for the years 1995, 2010, and 2020, revealing both the strength and the significance of relationships between different blocks over time. The density matrix quantifies the magnitude of spatial correlation strength among blocks, while the image matrix binarizes these values based on a predefined threshold to indicate whether the relationships are statistically significant. The combined comparison of these matrices offers a more comprehensive understanding of the interaction patterns and their temporal dynamics within the network.

Specifically, in 1995, the density matrix indicates a strong relationship between Block II and Blocks III and IV, with values of 0.927 and 0.844, respectively. The corresponding image matrix confirms these as significant connections (indicated by a value of 1), suggesting a robust spatial correlation of carbon emissions between Block II and these two blocks. Additionally, the relationship strengths between Block III and Blocks I and II are also relatively high—0.75 and 0.969, respectively—and are likewise identified as significant, reflecting frequent and intensive spatial interactions of carbon emissions among these blocks in 1995. In contrast, the relationships between Block I and the other blocks did not reach the significance threshold, with the

Table 4 Density and image matrices of the correlation network

Year	Plate	Density matrix				Image matrix			
		I	II	III	IV	I	II	III	IV
1995	I	0.333	0.333	0.583	0.417	0	0	0	0
	II	0.271	0.883	0.927	0.844	0	0	1	1
	III	0.75	0.969	0.159	0.667	1	1	0	0
	IV	0.542	0.844	0.49	0.339	0	1	0	0
2010	I	0.211	0.845	0.922	0.844	0	1	1	1
	II	0.773	1	1	0.99	0	1	1	1
	III	0.933	1	0.153	0.642	1	1	0	0
	IV	0.889	0.99	0.704	0.681	1	1	0	0
2020	I	0.361	0.769	0.857	0.909	0	0	1	1
	II	0.667	0.97	0.988	1	0	1	1	1
	III	0.889	0.988	0.5	0.494	1	1	0	0
	IV	0.929	1	0.481	0.7	1	1	0	0

image matrix showing a value of 0, indicating weak or insignificant spatial correlation within the network at that time.

By 2010, the overall network density had increased significantly. Several relationships among Blocks I, II, III, and IV exhibit high-density values; for example, the connections from Block I to Blocks II, III, and IV are 0.845, 0.922, and 0.844, respectively, all of which are binarized as significant. This reflects the diffusion and intensification of spatial correlations in carbon emissions. Notably, Block II demonstrates relationship densities close to or reaching 1 with all other blocks, indicating its position as a central hub within the network, where carbon emission impacts are transmitted more frequently and substantially across blocks. Connections between Blocks III and IV and Blocks I and II are also generally significant, revealing a more cohesive network structure and stronger inter-block interactions, which reflect high-frequency exchanges of carbon emission impacts across regions.

The 2020 density matrix reveals that, despite slight fluctuations in some inter-block densities, the overall levels remain high. Block II continues to maintain a strong association with other blocks, with densities approaching 1. Block I also sustains high-density values with Blocks III and IV (0.857 and 0.909, respectively), both of which are identified as significant in the image matrix, indicating that spatial linkages of carbon emissions remain active. The association density between Block III and Blocks I and II remains high (0.889 and 0.988), whereas its density with Block IV decreases to 0.494, which the image matrix classifies as not significant, reflecting a weakening and adjustment of certain spatial relationships. Overall, the 2020 network structure demonstrates that core linkages between blocks remain robust, while some structural optimization and differentiation occur, illustrating the nuanced evolution of the spatial pattern of regional carbon emissions.

In summary, the comparative analysis of the density and image matrices in Table 3 demonstrates that the spatial correlation network of carbon emission intensity has evolved from a loosely connected structure to a more cohesive and tightly linked system. The spatial linkages of carbon emissions among the blocks have gradually strengthened and become more significant over time. Block II consistently functions as the core hub within the network, playing a pivotal role in connecting and transmitting carbon emission impacts. Meanwhile, the relationships among Blocks I, III, and IV exhibit dynamic adjustments. This evolution of the spatial correlation structure elucidates the spatial and temporal diffusion mechanisms of regional carbon emissions and provides a scientific foundation for the formulation and enhancement of coordinated regional emission reduction policies.

5 Toward Cooperative Emission Reduction: Conclusions and Policy Implications

This chapter systematically investigates the structural characteristics of the spatial linkage network of carbon emission intensity among OECD countries and China from 1995 to 2022 by applying social network analysis methods. The findings reveal a significant strengthening of spatial linkages in carbon emission intensity among OECD countries, as evidenced by the steady increase in network density over time, indicating a growing regional synergy in carbon emissions. Concurrently, the observed decline in network efficiency suggests that despite the increasing connectivity, there is a risk of redundancy in information transmission, potentially diminishing the marginal effectiveness of policy diffusion and technology sharing. At the individual node level, core countries—such as traditional European nations and China—maintain a central position within the carbon network, whereas smaller countries tend to experience structural marginalization. Further, block model analysis demonstrates that the roles of countries and regions within the carbon emission network have dynamically evolved, shifting from "mainstream in" to "mainstream out" or "proxy" roles, reflecting the evolving spatial diffusion pathways of carbon emissions. This evolution highlights the ongoing adjustment and redistribution of spatial diffusion channels and the functional roles of regions within the carbon emission network.

Based on the above findings, this chapter proposes the following policy recommendations: First, emission reduction cooperation mechanisms among OECD countries should be strengthened by establishing more efficient information-sharing platforms to optimize cross-border diffusion pathways of carbon reduction experiences and green technologies, thereby alleviating synergistic bottlenecks caused by network redundancy. Second, policymakers should prioritize "key node" countries with high centrality in the network, such as Germany, the United Kingdom, and China, to leverage their leadership roles in global carbon emission reduction and enhance spillover effects, fostering the development of regional "carbon governance alliances". Third, the international community should increase investments in green technologies and provide targeted policy support to countries that act as "proxies" or are marginalized within the network, enhancing their agency and adaptability, reducing the risk of carbon leakage, and building a more inclusive and resilient emission reduction system. Finally, attention should be given to the optimization and dynamic management of the network structure, taking into account differences in economic development stages and energy profiles across countries to promote stratified and classified synergistic emission reduction strategies, thereby advancing a new paradigm of "multilateral linkage, division of labor and cooperation, and systemic optimization" for low-carbon transformation.

However, this study has several limitations. Primarily, it does not delve deeply into the specific mechanisms underlying the implementation of carbon emission reduction policies or the cross-country heterogeneity in their effects. Future research could conduct comparative policy analyses across countries to evaluate the effectiveness of interventions such as carbon pricing, carbon trading, and technology subsidies within different institutional contexts, and investigate how these policies influence the evolution of network structures. Additionally, given the increasing role of non-OECD countries in global carbon emissions, subsequent studies should incorporate a broader set of emerging economies, expand the analytical framework, and explore the evolution and cooperation potential of the global carbon emission network. Such efforts would provide more forward-looking theoretical insights and practical pathways to achieving fair, effective, and synergistic global climate governance.

Appendix

See (Figs. 2, 3, and 4).

Network analysis based on closeness centrality provides a clear perspective on the evolution of China's position within the spatial correlation network of carbon emission intensity. This study constructs the networks for OECD countries and China at three time points (1995, 2010, 2020), and employs closeness centrality to examine the structural positions of each country. Closeness centrality measures the reciprocal of the average shortest path length from a given node to all other nodes, reflecting a country's accessibility and information propagation efficiency within the network. In the graphical representation, node size is inversely proportional to closeness centrality: smaller nodes indicate more central positions and stronger influence, whereas larger nodes represent more peripheral positions with relatively weaker influence.

The 1995 network results indicate that developed European economies such as France, Germany, and Italy occupied central positions within the network. At that time, these countries had largely completed industrial modernization and were closely interconnected through coordinated EU energy and environmental policies, allowing fluctuations in their carbon emission intensity to propagate rapidly among member states. In contrast, China's node was noticeably larger, reflecting a relatively

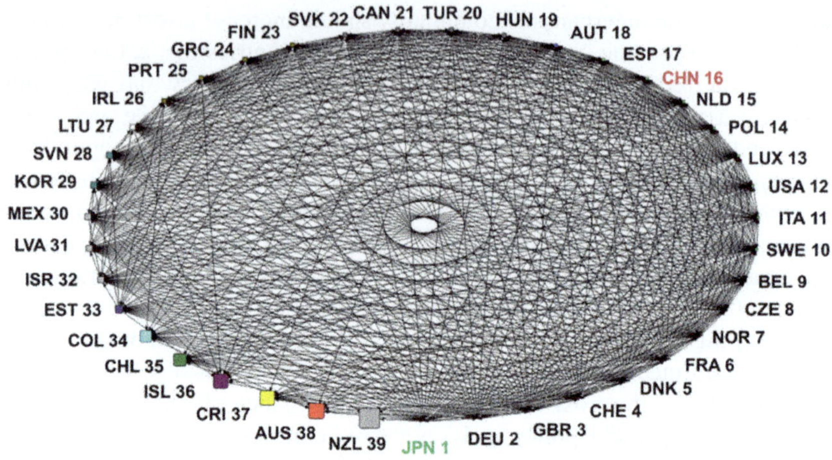

Fig. 2 Spatial correlation networks of CEI in OECD countries & China (1995)

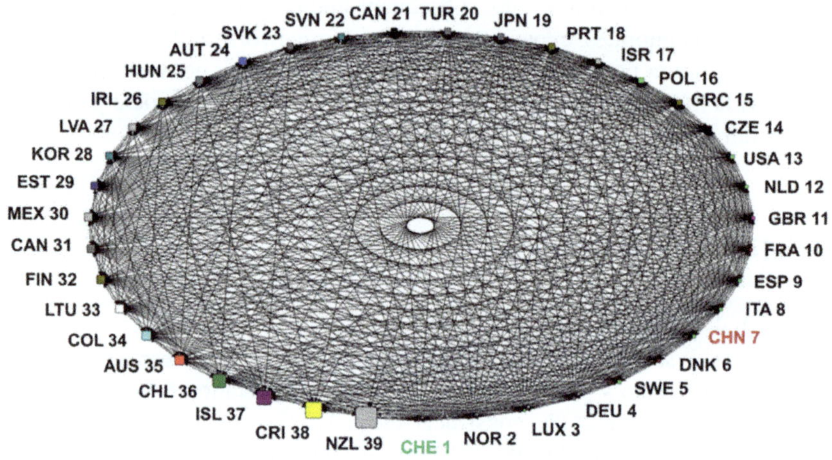

Fig. 3 Spatial correlation networks of CEI in OECD countries & China (2010)

low closeness centrality. Consequently, variations in China's carbon emission intensity had limited influence on other countries. This pattern is consistent with China's economic structure at the time, which remained dominated by labor-intensive, low-value-added manufacturing. In the 1990s, China's energy consumption was heavily reliant on coal, and technological levels were limited; as a result, changes in carbon emissions were primarily driven by domestic demand and exhibited minimal transmission effects within the international network.

By 2010, the network structure had undergone substantial changes, with China's node shrinking significantly and rising to a central position. This shift is closely linked to industrial and policy transformations in China during the 2000s. Between

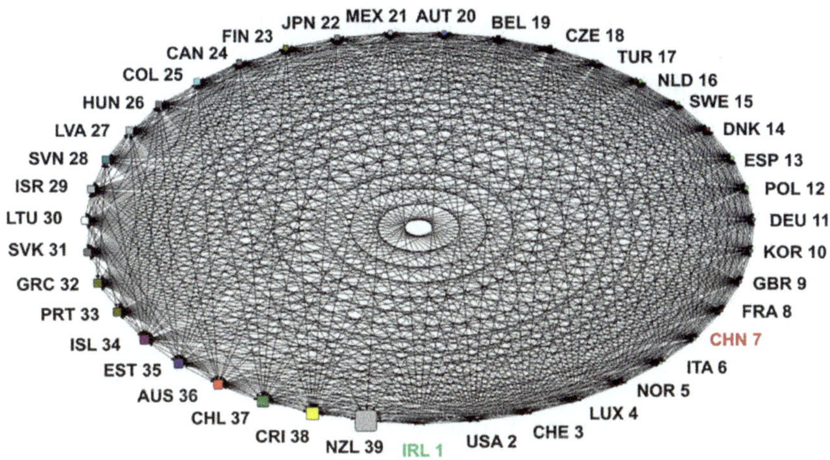

Fig. 4 Spatial correlation networks of CEI in OECD countries & China (2020)

2000 and 2010, China's carbon emissions surged from 3.65 billion tons to 8.62 billion tons, surpassing the United States to become the world's largest emitter.[1] During this period, China's export-oriented industries expanded rapidly, with large volumes of energy-intensive products such as steel, cement, and electrolytic aluminum exported to Europe, North America, and Asia. Consequently, fluctuations in China's carbon emission intensity became increasingly synchronized with the industrial demand of major OECD economies. In particular, during the 2008 global financial crisis, China implemented a 4-trillion-yuan economic stimulus plan, focusing on infrastructure investment, which spurred a large-scale recovery in the steel and cement sectors.

By 2020, China had firmly established its central position within the global carbon emission intensity network, with its node size reaching a level comparable to that of traditional core countries such as France and Italy. A key factor behind this development is China's increasing share of global energy consumption: in 2020, China's primary energy consumption reached 4.98 billion tons of standard coal, accounting for 26% of the global total (Dai et al. 2021). This energy consumption pattern not only enhanced the explanatory power of China's carbon emission intensity on global trends but also influenced other economies through industrial supply chain linkages. For example, between 2015 and 2020, the rapid expansion of China's renewable energy manufacturing sector enabled the export of photovoltaic panels and wind turbines, providing crucial support for Europe's energy transition. During the same period, some smaller European countries, such as Luxembourg, experienced relative increases in node size, indicating a decline in their connectivity within the global emission intensity network and greater dependence on the emission fluctuations of larger economies.

[1] https://ourworldindata.org/co2-emissions.

Overall, between 1995 and 2020, China gradually transitioned from a peripheral to a central position within the carbon emission intensity network. This shift was not merely reflected in abstract network metrics, but was driven by a series of concrete economic and policy developments. In the early period, emissions were largely self-driven and coal-based; during the 2000s, export-oriented industrial expansion and macroeconomic stimulus policies led to more synchronized emission patterns with major economies; by around 2020, China's global energy consumption share and its industrial supply chain connections enabled significant transmission effects.

References

Bai C, Zhou L, Xia M, Feng C (2020) Analysis of the spatial association network structure of China's transportation carbon emissions and its driving factors. J Environ Manage 253:109765

Baz K, Zhu Z (2025) Life cycle analysis of green technologies: Assessing the impact of environmental policies on carbon emissions and energy efficiency. Geosci Front 16(2):102004

Croutzet A, Dabbous A (2021) Do FinTech trigger renewable energy use? Evidence from OECD countries. Renewable Energy 179:1608–1617

Dai H, Su Y, Kuang L, Liu J, Gu D, Zou C (2021) Contemplation on China's Energy-Development Strategies and Initiatives in the Context of Its Carbon Neutrality Goal. Engineering 7(12):1684–1687

Dasgupta S, De Cian E (2018) The influence of institutions, governance, and public opinion on the environment: synthesized findings from applied econometrics studies. Energy Res Soc Sci 43:77–95

Gao Y, Gao M (2024) Spatial correlation network of municipal solid waste carbon emissions and its influencing factors in China. Environ Impact Assess Rev 106:107490

Haga H, Zhou W (2021) An empirical analysis of international carbon transfer. In: Zhou W, Qian X, Nakagami K (eds) East Asian low-carbon community. Springer, pp 289–305

Huo T, Cao R, Xia N, Hu X, Cai W, Liu B (2022) Spatial correlation network structure of China's building carbon emissions and its driving factors: a social network analysis method. J Environ Manage 320:115808

Kang X, Chen L, Wang Y, Liu W (2024) Analysis on the spatial correlation network and driving factors of carbon emissions in China's logistics industry. J Environ Manage 366:121916

Knill C, Lenschow A (1998) Coping with Europe: the impact of British and German administrations on the implementation of EU environmental policy. J Eur Publ Policy 5(4):595–614

Lin B, Zhao H (2024) Asymmetric trade barriers and CO2 emissions in carbon-intensive industry. J Environ Manage 349:119547

Liu S, Xiao Q (2021) An empirical analysis on spatial correlation investigation of industrial carbon emissions using SNA-ICE model. Energy 224:120183

Nan S, Huang J, Wu J, Li C (2022) Does globalization change the renewable energy consumption and CO2 emissions nexus for OECD countries? New evidence based on the nonlinear PSTR model. Energ Strat Rev 44:100995

International Energy Agency (2019) Renewable and non-renewable electricity generation in OECD countries. International Energy Agency. https://www.iea.org/data-and-statistics/charts/renewable-and-non-renewable-electricity-generation-in-oecd-countries-2008-2018

Romero-Ávila D (2008) Convergence in carbon dioxide emissions among industrialised countries revisited. Energy Economics 30(5):2265–2282

Spandagos C, Tovar Reaños MA, Lynch MÁ (2022) Public acceptance of sustainable energy innovations in the European Union: A multidimensional comparative framework for national policy. J Clean Prod 340:130721

Sun X, Dong Y, Wang Y, Ren J (2022) Sources of greenhouse gas emission reductions in OECD countries: composition or technique effects. Ecol Econ 193:107288

Umar M, Safi A (2023) Do green finance and innovation matter for environmental protection? a case of OECD economies. Energy Economics 119:106560

Van Der Ploeg F (2011) Macroeconomics of sustainability transitions: second-best climate policy, Green Paradox, and renewables subsidies. Environ Innov Soc Trans 1(1):130–134

Wang H, Wei W (2020) Coordinating technological progress and environmental regulation in CO2 mitigation: the optimal levels for OECD countries & emerging economies. Energy Economics 87:104510

Wei L, Li W, Jin Z (2024) Global value chains participation and trade-induced carbon inequality: a comparative analysis of developed and developing economies. Ecol Econ 220:108186

Wiedmann T (2009) A review of recent multi-region input–output models used for consumption-based emission and resource accounting. Ecol Econ 69(2):211–222

Zhou W (2021) Concept and framework of an East Asian Low-Carbon Community. In: Zhou W, Qian X, Nakagami K (eds) East Asian low-carbon community. Springer, pp 61–81

Translating Consumer Behavior into Policy Signal Acceptance: A Behavioral Chain Analysis of Carbon Tax and EV Willingness to Pay in Japan

Liu Cao and Seiichi Ogata

1 Introduction: Connecting Policy and Consumer Behavior to Achieve Low-Carbon Transition

Global climate change poses a severe and long-lasting threat to human society, requiring urgent and coordinated decarbonization actions across all sectors by all countries. International agreements, such as the Paris Agreement, emphasize the shared responsibility of countries in reducing greenhouse gas (GHG) emissions (OECD 2020). As a major industrialized country, Japan has committed to ambitious climate goals, aiming to achieve carbon neutrality by 2050 and to reduce greenhouse gas emissions by 46% from 2013 levels by 2030, while further striving for a 50% reduction. These goals imply a profound transformation of Japan's energy system and economy, particularly in the transportation sector (OECD 2023). Although Japan has set a clear decarbonization direction and ambitious targets, there exists a significant gap between its policy aspirations and the current trajectory. According to the assessment by Climate Action Tracker (CAT), Japan's current policies are projected to reduce emissions by only 31% to 38% from 2013 levels by 2030, which falls far short of the 46% target set in its Nationally Determined Contribution (NDC). This gap indicates that while Japan has clearly defined its development direction, existing mechanisms or their modes of implementation may be insufficient to achieve the expected behavioral and industrial transitions. This disconnection between policy and actual outcomes highlights the urgency of analyzing how policy signals influence consumer behavior. It implies that the "behavioral chain" connecting policy signals to consumer actions may contain weak links, misalignments, or substantial

L. Cao · S. Ogata (✉)
Graduate School of Energy Science, Kyoto University, Kyoto, Japan
e-mail: ogata@energy.kyoto-u.ac.jp

L. Cao
e-mail: cao.liu.66a@st.kyoto-u.ac.jp

W. Zhou (ed.), *Decarbonization Policy Engineering*, World Sustainability Series,
https://doi.org/10.1007/978-981-95-5953-4_20

resistance, thus requiring a more in-depth investigation into its dynamics (OECD 2021).

Recent assessments further highlight challenges in specific sectors. In the transportation domain—responsible for approximately 18% of Japan's total CO_2 emissions in 2021 —the shift toward electrification remains modest (IEA 2024). While the Japanese government has promoted battery electric vehicles (BEVs) through subsidies and tax incentives, BEVs accounted for less than 3% of new passenger vehicle sales in 2023, significantly lagging behind the European Union (over 15%) and China (over 20%) (IEA 2024). Hybrid electric vehicles (HEVs), particularly those produced by domestic manufacturers, continue to dominate the market due to consumer familiarity, perceived convenience, and infrastructure readiness.

This limited progress suggests that economic incentives alone are insufficient to drive large-scale behavioral change. Carbon pricing, another critical tool, remains weak in Japan. Japan's current carbon tax rate—JPY 289 per ton of CO_2—formally known as the Tax for Climate Change Mitigation, has remained unchanged since its introduction in 2012.This level is significantly lower than benchmarks recommended by international bodies such as the International Energy Agency, which suggests a price of USD 130 per ton by 2030 for advanced economies to remain aligned with the Paris Agreement.

Furthermore, a growing body of behavioral economics research suggests that consumers interpret and respond to policy instruments through complex cognitive filters. Factors such as policy trust, perceived fairness, environmental awareness, and risk perception significantly mediate the effect of carbon pricing or electric vehicle (EV) subsidies on actual decisions. Thus, understanding the psychological and social determinants of willingness to pay (WTP) for both carbon taxes and electric vehicles—is essential to align consumer behavior with policy targets.

This study aims to deconstruct the behavioral chain that links policy signals to consumer actions. Special attention is given to whether and how visible low-carbon behaviors, such as EV adoption, can reinforce public acceptance of latent carbon pricing mechanisms. Identifying the shared or divergent determinants of WTP for EVs and carbon pricing may provide critical insights for designing mutually reinforcing policy strategies that close the ambition-implementation gap and accelerate Japan's transition to a low-carbon society.

2 Low-Carbon Policy Design and Public Acceptance in Japan

Japan has implemented a variety of low-carbon policies to support its national decarbonization goals, including regulatory measures, market-based instruments, and technology promotion strategies. However, the effectiveness of these policies largely depends on public acceptance and behavioral responses. Among these tools, carbon pricing, especially carbon taxation, and the promotion of electric vehicles have emerged as central components of Japan's climate strategy. This section reviews the design and implementation of these two key policy instruments, with a focus on analyzing consumer acceptance in Japan and the underlying behavioral mechanisms.

2.1 Definition of Carbon Pricing: Instrument and Economic Signal

Carbon pricing is an economic policy instrument aimed at curbing greenhouse gas emissions by charging for emission activities or providing incentives for emission reductions. Its fundamental purpose is to internalize the "external costs" of greenhouse gas emissions, such as damage to crops, medical expenses caused by heatwaves and droughts, and property losses resulting from flooding and sea-level rise, by linking them directly to emission sources.

The main forms of explicit carbon pricing include emissions trading systems (ETS) and carbon taxes. ETS, such as cap-and-trade systems, set an upper limit on total emissions and allow regulated entities to trade emission units, thereby establishing a market price for greenhouse gases. This mechanism offers certainty in terms of environmental impact while maintaining price flexibility. In contrast, carbon taxes set a fixed price per ton of carbon dioxide equivalent (tCO2e), providing price certainty but allowing flexibility in the number of emissions reduced. Other mechanisms include carbon credit schemes and result-based climate financing (RBCF). Japan currently maintains a relatively low explicit carbon tax. This rate is significantly lower than those under many international carbon pricing frameworks. The government has introduced the Basic Policy for Green Transformation (GX), a comprehensive strategy aimed at mobilizing approximately 150 trillion yen (USD 1 trillion) in public and private investment over the next decade to accelerate decarbonization. A key component of the Green Transformation (GX) initiative is the GX Emissions Trading System (GX-ETS), which was launched in April 2023 as a voluntary carbon trading program. This system is scheduled to transition into a mandatory compliance scheme in 2026, at which point large enterprises, with annual carbon emissions of 100,000 tons or more, accounting for about 60% of Japan's total greenhouse gas emissions, will be required to participate. In addition, Japan plans to introduce a carbon tax targeting fossil fuel importers in 2028, aimed at providing a more direct price signal for fossil fuel consumption and complementing the GX-ETS. An emission quota auction mechanism is expected to be introduced by 2033 (Government of Japan 2021; METI 2022).

Carbon pricing serves a dual purpose, which holds significant implications for its behavioral impact. It is designed both to reduce emissions and to generate revenue for the government. In the Japanese context, revenues from carbon pricing mechanisms, namely, the Green Transformation Emissions Trading System auctions and the fossil fuel tax, are explicitly allocated to fund Green Transformation (GX) bonds, with a total amount of 20 trillion yen. This dual-purpose highlight potential trade-offs in policy design. If revenue generation becomes the primary focus, the carbon price may be set too low to deliver a sufficiently strong "price signal" to meaningfully alter consumption and investment patterns—precisely the original intent of carbon pricing.

Therefore, the effectiveness of carbon pricing in driving behavioral change depends not only on its existence but also on the strength and salience of the price signal.

A low price, even if it generates substantial revenue overall, may be insufficient to overcome consumer inertia or perceived upfront costs of sustainable alternatives.

This suggests that the behavioral impact of carbon pricing may be indirect (through subsidizing green technologies) rather than direct (by increasing the cost of carbon-intensive goods).

Table 1 provides a summary of Japan's current and planned carbon pricing instruments.

Compared to Western countries, Japan has lagged behind in researching public acceptance of carbon taxes, especially in terms of systematic empirical analyses focusing on individual WTP. Internationally, a growing body of literature highlights the crucial role of public WTP in shaping and implementing effective carbon tax policies. Although Japan has long implemented a carbon levy under the name of the Tax for Climate Change Mitigation, the scheme continues to face structural challenges such as limited institutional transparency and insufficient carbon price levels. As a result, the policy fails to demonstrate functions beyond merely raising prices, particularly in terms of incentivizing behavioral change or establishing social acceptability. In practice, the tax is commonly perceived by the public as simply a fuel price increase rather than as a climate policy instrument. This narrow understanding is compounded by inadequate public communication regarding the use of tax revenue and the environmental effectiveness of the policy, which weakens public trust and legitimacy. While a carbon tax is, in principle, designed to adjust prices and suppress emissions, Japan's current system falls short of fulfilling these objectives and remains a symbolic measure. Addressing these limitations requires fundamental policy reform, enhanced transparency, and improved communication with the public to restore the tax's credibility and effectiveness. Existing studies indicate that Japanese consumers do exhibit a certain level of WTP under specific conditions—particularly when the use of tax revenue is clearly earmarked for environmental improvement or clean technology investments. Among different revenue use options, allocating funds to renewable energy subsidies tends to garner the highest support. This finding aligns with European experiences that emphasize "green redistribution", suggesting that transparency and targeted allocation of carbon tax revenues are key drivers in enhancing public acceptance (IMF 2023).

Moreover, WTP for carbon tax in Japan exhibits clear socio-economic stratification. High-income individuals, corporate managers, and those with better understanding of carbon tax policy demonstrate greater willingness and capacity to pay. This underscores the importance of bridging informational and cognitive gaps, to prevent carbon taxation from being perceived merely as an economic burden by the general public.

Table 1 Overview of Japan's carbon pricing instruments (current and planned)

Instrument	Status	Coverage	Price / Rate	Key Dates	Main Objectives
Tax for Climate Change Mitigation	Implemented (since 2012)	Levy on fossil fuels, covers approx. 70% of national CO_2 emissions (explicit carbon pricing, Ministry of Environment)	JPY 289/ tCO_2 (≈ USD 2.16, based on 2023 avg. exchange rate)	Introduced in October 2012	Reduce GHG emissions, generate revenue for climate measures (MOE)
GX-ETS	Voluntary phase (2023–2025), Mandatory phase (from 2026)	Approx. 700 major emitters (energy, manufacturing, steel), cover ~ 50% of emissions (GX Basic Policy, METI)	First 2023 transaction price: JPY 2400/tCO_2; auction system to start in 2033 (GX-ETS documents)	Trial in 2023, mandatory from 2026, auction system from 2033	Establish a carbon market, encourage industrial investment in decarbonization (GX Basic Policy)
GX Surcharge (Carbon Levy)	Planned (to be introduced in FY2028, not yet legislated)	Fossil fuel importers and suppliers, applied to end-use prices (based on GX Strategy)	To be introduced in 2028; revenue to repay ¥20 trillion GX Transition Bonds (GX Strategy)	Expected introduction from FY2028 (per GX Strategy and MOF plan)	Subsidize industrial decarbonization, repay GX Bonds (Cabinet Office, MOF)
Fuel Excise Tax (Implicit Carbon Pricing)	Long-standing tax, serves as implicit carbon pricing	Tax on gasoline, diesel, etc., indirectly covers ~ 73.1% of CO_2 emissions (OECD estimate)	OECD estimated average implicit carbon price: EUR 21.44/tCO_2 (2021 data)	Long implemented, remains major fiscal source	Provide price signals to curb fossil fuel use, indirectly drive decarbonization (OECD)

Regional variation also plays an important role in shaping public acceptance of carbon taxation. For example, Tokyo, as a policy front-runner, has higher environmental awareness and more extensive experience in climate policy. The level of public willingness to pay in Tokyo is exemplary. As shown in Fig. 1, Cao et al. (2024) found that the level of public understanding of carbon taxation among Tokyo residents is broadly comparable to the national average. However, whether such regional successes can be extended to other parts of Japan still depends on further institutional refinement and enhanced public engagement.

In sum, Japanese consumers are not inherently resistant to carbon taxation. Rather, their WTP is highly sensitive to policy design, particularly in contexts where revenue

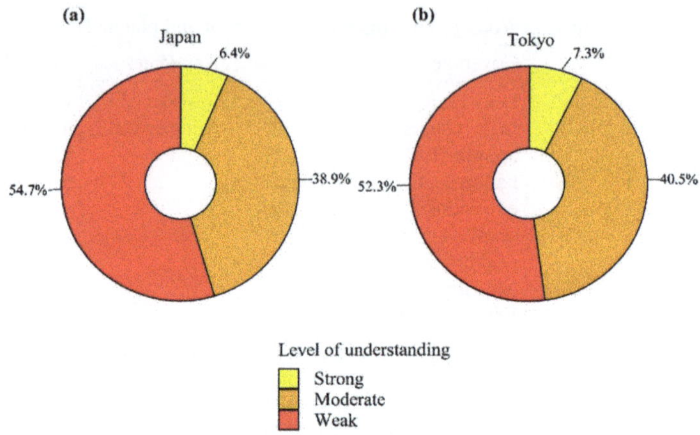

Fig. 1 (a) understanding of Carbon Tax in Japan; (b) understanding of Carbon Tax in Tokyo. *Source* Cao et al. (2024)

use is transparent, goals are well-communicated, and environmental benefits are clear. These findings suggest that there exists a solid social foundation for advancing carbon pricing reforms in Japan (Cao et al. 2024).

2.2 *Electric Vehicle Promotion Policies in Japan*

Electric vehicles are broadly defined as vehicles that are powered by electric motors, obtain electricity from batteries, and can be charged from external power sources. This definition includes battery electric vehicles, which operate entirely on electricity, and plug-in hybrid electric vehicles (PHEVs), which combine electric motors with internal combustion engines. Fuel cell vehicles (FCVs) are also regarded as clean energy vehicles under Japan's policy framework (METI 2023).

Electric vehicles play a key role in global decarbonization efforts, particularly in the transportation sector, by reducing dependence on fossil fuels and mitigating local air pollution and noise. Japan has set a target that by 2035, all new passenger vehicle sales will be "electrified vehicles," including electric vehicles, plug-in hybrid electric vehicles, fuel cell vehicles, and hybrid electric vehicles.

2.2.1 Incentives and Subsidies

Japan has set a national target that by 2035, all new passenger vehicle sales will be "electrified vehicles" (including electric vehicles, plug-in hybrid electric vehicles, fuel cell vehicles, and hybrid electric vehicles). The Ministry of Economy, Trade and Industry (METI) and the Ministry of the Environment (MOE) provide the "Subsidy for Promoting the Introduction of Clean Energy Vehicles" (CEV subsidy) (Reuters 2025).

For fiscal year 2024, the EV subsidy budget is 110 billion yen (approximately USD 710 million), and the charging infrastructure subsidy is 36 billion yen (approximately USD 232 million). These subsidies are allocated for the purchase of new electric vehicles, plug-in hybrid electric vehicles, fuel cell vehicles, and V2H (Vehicle-to-Home) charging equipment. It is noteworthy, however, that hybrid electric vehicles are not included in these purchase subsidies. The maximum subsidy amounts are substantial: 850,000 yen (USD 5,500) for electric vehicles, 550,000 yen (USD 3,500) for kei-class electric vehicles and plug-in hybrid electric vehicles, and 2,550,000 yen (USD 16,600) for fuel cell vehicles. Starting from fiscal year 2024, the evaluation method for subsidies has been revised to more comprehensively assess automobile manufacturers' efforts in green transformation (including GX values such as life-cycle emissions and charging infrastructure), rather than focusing solely on vehicle performance indicators.

Research has confirmed that purchase subsidies are an effective policy to promote the adoption of battery electric vehicles in Japan. If continued, they may help EVs (including BEVs and PHEVs) reach a market penetration rate of 20–30% by 2030 and even exceed 50% by 2050 (Kakishima et al. 2023).

Japan also provides tax incentives, such as the "Eco-Car Tax Reduction," which includes a 100% tax exemption for electric and hybrid vehicles. These direct and easily understandable tax exemptions have been found to have a significantly greater impact on the sales of eco-friendly vehicles than gasoline prices (Japan Automobile Manufacturers Association 2024).

Considering the coexistence of multiple technological pathways in today's electric vehicle market, different types of policy incentives may have complex effects on technology choice. Figure 2 is drawn from a simulation study by Jung et al. (2022), which focuses on the Japanese market. The study sets up three incentive scenarios aimed at promoting fuel cell vehicles: purchase subsidies, free hydrogen refueling, and a combination of the two. It then projects the market share changes of various vehicle types, including internal combustion engine vehicles, hybrid vehicles, plug-in hybrid electric vehicles, battery electric vehicles and Fuel Cell Electric Vehicle (FCEVs), under each policy condition. Although the incentives were originally designed for FCEVs, the simulation results show that other types of electric vehicles, including BEVs, are also affected. Providing support for one specific technology may unintentionally reduce the market appeal of another. This suggests that in promoting electric vehicles, it is not enough to simply offer subsidies. Equal attention must be paid to how the incentives are structured and who they are targeted at. Policies are more likely to be effective in a multi-technology competitive market when they directly address the actual concerns of consumers, such as driving range, cost or charging convenience.

The above-mentioned judgment regarding the influence of incentive structure on actual consumer choices is also supported by a stated choice experiment conducted by Khan et al. (2020) in Japan. The study found that although consumers expressed a certain level of interest in BEVs and Hydrogen Fuel Cell Vehicle (HFCVs), their

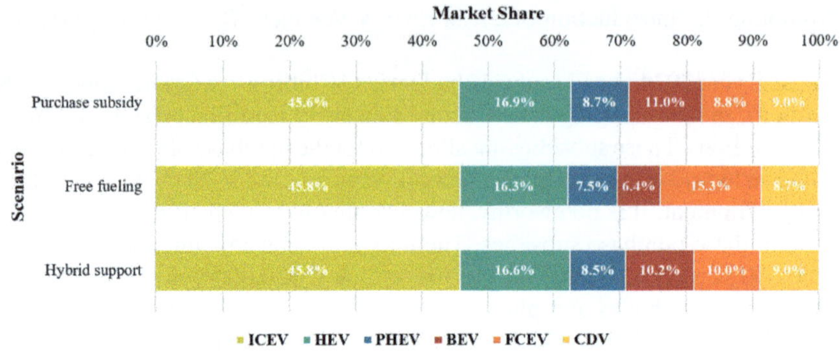

Fig. 2 Projected market shares of major vehicle types in Japan under different incentive scenarios. *Source* Jung et al. (2022)

actual behavior in simulated decision-making scenarios did not fully reflect these preferences. As shown on Fig. 3, 21.2% of respondents indicated interest in BEVs, but only 9.5% chose a BEV in the experiment. In contrast, only 6.4% expressed interest in HFCVs, yet 16.6% ultimately selected this technology. These results suggest that consumer behavior is influenced not only by attitudes but also by their perceived concerns about technology cost, supporting infrastructure, and ease of use. Therefore, to effectively promote market adoption, policy incentives must reduce both financial barriers and psychological resistance, facilitating a meaningful transition from interest to action.

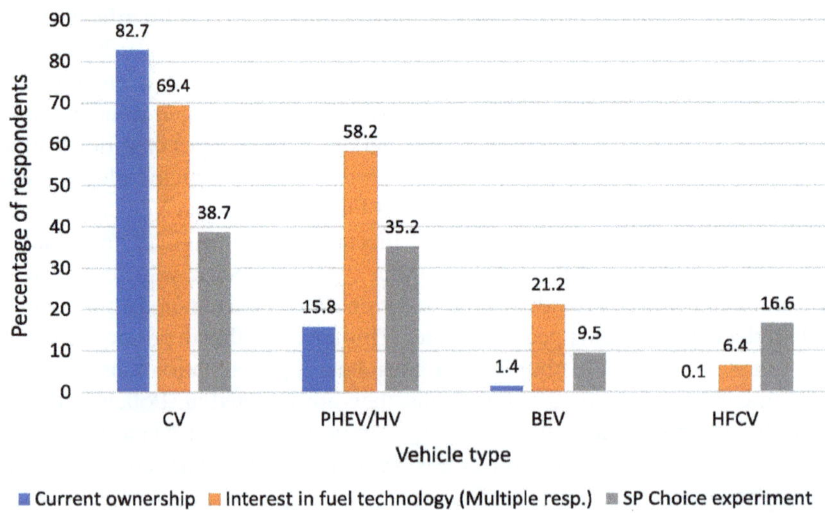

Fig. 3 Comparison between stated interest and actual choice of vehicle fuel types in Japan. *Source* Khan et al. (2020)

2.2.2 Development of Charging Infrastructure

Japan's Green Growth Strategy has set ambitious targets aimed at significantly expanding electric vehicle charging infrastructure. The plan envisions the deployment of 150,000 charging stations by 2030, including 30,000 fast chargers, with the objective of achieving a level of convenience comparable to that of traditional gasoline refueling stations (Government of Japan 2023).

However, actual deployment progress remains notably sluggish. As of end-2024, Japan had only about 33,000 public EV charging points, indicating a substantial gap between the current status and the 2030 target (IEA 2025).

Japan faces multiple critical challenges in accelerating the deployment of charging infrastructure. Chief among them is the high installation cost and technical complexity of DC fast chargers, which are essential for meeting rapid charging demand but poses a significant barrier. In addition, grid capacity limitations, supply chain constraints for key materials such as lithium and semiconductors, and the need for a unified roaming network across different charging operators further hinder the efficient expansion of the charging network and the enhancement of user experience (Deloitte 2025).

Apart from the aforementioned institutional and technical bottlenecks, consumers' acceptance of usage-related conditions also affects the market expansion of EVs in Japan to a certain extent. As shown in Fig. 4, Tanaka et al. (2014) found through a stated choice experiment that Japanese respondents exhibited relatively low WTP for "public charging facility availability." This attribute refers to whether drivers can easily find available charging stations or hydrogen refueling stations in places such as highways, commercial areas, or public parking lots when using electric vehicles away from home. A low WTP does not indicate that consumers do not value this factor. Instead, it reflects that, given the current state of underdeveloped infrastructure, the slow rollout of fast-charging networks, and the lack of interoperability among charging operators, consumers do not have sufficient confidence in this convenience and are therefore unwilling to pay extra for it. This uncertainty in expectations constitutes a psychological barrier in itself. In addition, Japanese consumers also demonstrate a negative WTP (−16.9 USD) for "home plug-in installation costs," indicating that they are generally unwilling to bear the upfront expenses for necessary equipment. This sensitivity to charging-related conditions further suppresses users' willingness to adopt EVs and reduces the momentum for expanding the charging network through market demand.

Despite years of policy efforts and infrastructure development, consumer confidence in the convenience of EV charging has shown no significant improvement. Given the persistent gap between targets and actual outcomes, this structural barrier remains unresolved at the current stage.

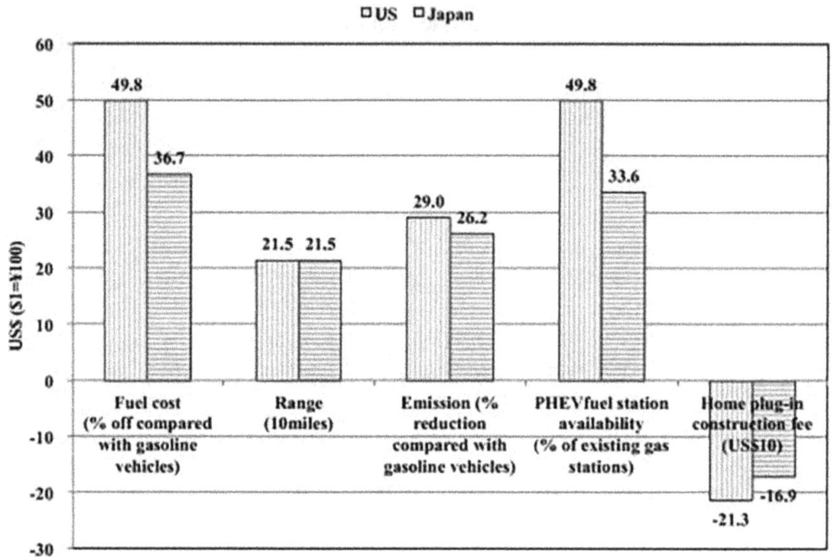

Fig. 4 Willingness to pay for key electric vehicle attributes among Japanese consumers. *Source* Tanaka et al. (2014)

2.2.3 Japan's Electrification Strategy: Emphasis on Hybrid and Hydrogen Vehicles

Japan's unique approach to "electrified vehicles" includes hybrid electric vehicles in its 2035 target, which markedly differs from many countries that focus primarily on battery electric vehicles. Japanese consumers have shown a strong preference for hybrid vehicles, which accounted for 55% of new car sales in 2023 (Mayo Komiyama 2024). This strong preference is partly attributable to the long-standing leadership and success of domestic manufacturers such as Toyota in hybrid technology.

Major Japanese automakers—Toyota, Honda, Nissan, and Mazda—have historically prioritized the development of hybrid technologies over battery electric vehicles. Japan has also made significant investments in hydrogen technologies and fuel cell vehicles, with a target of selling 800,000 FCVs by 2030 and receiving substantial R&D subsidies. However, due to the limited availability of refueling stations and the high cost of operation, FCV sales have fallen far short of the target.

Path dependency and strategic diversification are a double-edged sword. The Japanese auto industry has deep historical advantages and substantial investments in hybrid technologies. This has shaped both policy (including hybrids in the "electrification" goal) and consumer preferences (where hybrids dominate). At the same time, Japan is also actively promoting hydrogen. This "diversified" approach means that, compared to countries focusing exclusively on the adoption of battery electric vehicles, Japan's efforts to address BEV-specific challenges (such as charging infrastructure) may be less concentrated.

Such path dependency creates significant behavioral barriers to the rapid diffusion of BEVs. Consumers are satisfied with the perceived "greenness" of hybrid vehicles, while manufacturers, concerned about existing supply chains and profitability, are reluctant to fully transition. While diversification offers strategic flexibility, it may dilute the focus and resources necessary to overcome the specific economic and infrastructural barriers to BEVs, thus slowing the overall transition to zero-emission vehicles.

Table 2 summarizes Japan's EV and carbon pricing incentives for FY2024.

The accelerating demands of global decarbonization and Japan's ambitious 2030–2050 climate targets make it critically urgent to understand how consumers' visible investments in electric vehicles translate into their tacit acceptance of carbon-pricing policies. Despite generous fiscal incentives for EV purchase and aggressive charging-infrastructure rollouts, a behavioral disconnect persists: individuals who freely choose battery electric or plug-in hybrid vehicles do not necessarily embrace nascent carbon taxes, undermining the virtuous feedback loop that sustains both technology diffusion and policy legitimacy. By empirically mapping the pathways through which demographic profiles, technology trust, and policy perceptions channel overt "green" spending into implicit support for carbon levies, this article addresses a pressing knowledge gap. Its purpose is to furnish policymakers and industry leaders with a rigorous behavioral framework that aligns monetary incentives with cognitive drivers—thereby hastening the dual transition to zero-emission mobility and broad-based public endorsement essential for Japan's net-zero trajectory.

Therefore, this work makes contributions in the following areas:

- For the first time, this study constructs and empirically validates a quantitative model that links willingness-to-pay for EVs (BEVs & PHEVs) to willingness-to-pay for carbon taxes, employing partial-correlation and structural-equation techniques to capture their interdependencies.
- Moving beyond traditional economic metrics, this study incorporates psychological constructs, technology trust, safety perception, and policy comprehension into the analysis of payment behaviors, enriching explanatory power and offering deeper insight into consumer decision-making processes.
- Contrary to one-way causal assumptions, this study proposes and tests a dynamic, two-way feedback model in which overt decarbonization actions and covert policy acceptance mutually reinforce over time, providing a sophisticated theoretical tool for designing more cohesive and sustainable low-carbon strategies.

3 Theoretical Framework

This study proposes an integrated behavioral framework to examine the relationship between overt consumer actions, such as WTP for electric vehicles, and implicit acceptance of decarbonization-related policy signals, such as carbon taxation. By

Table 2 Japan EV and carbon pricing incentives (FY2024, USD)

Incentive Type	Applicable Vehicle Types	Incentive/Tax Reduction	FY2024 Budget (USD billion)	Governing Authority	Main Policy Objective
Car Purchase Subsidy (CEV Subsidy)	EV; Light EV; PHEV; FCV	EV: \$5,612; Light EV: \$3,632; PHEV: \$3,632; FCV: \$16,837	0.726	METI	Promote uptake of clean-energy vehicles
Charging/ Refueling Infrastructure Subsidy	Public chargers (fast/ slow); Hydrogen refueling stations; V2H devices	Total: \$3,301,419,610.432 billion (≈\$2,377,022,119.511 billion for charging infrastructure); equipment cost subsidy 50–100%; special rates for H_2/V2H	3.301	METI (Next-Gen Vehicle Promotion Center)	Expand charging & hydrogen infrastructure
Eco-car Tax Incentives	EV; PHEV; FCV (HEV for fuel-efficiency reductions)	EV/PHEV/FCV: 100% exemption of acquisition tax & vehicle weight tax (new regs until Apr 30, 2026); HEV: 25–75% weight tax reduction		MLIT & JAMA	Reduce ownership costs, promote electrification
Long-term Electrification Target	All passenger vehicles (EV, PHEV, FCV, HEV)	100% of new-vehicle sales electrified by 2035		METI	Achieve full new-car electrification by 2035

Note USD figures based on 2024 average exchange rate (1 USD = 151.45 JPY); sources aggregated from METI press release, METI subsidy overview, MLIT & JAMA guidelines, and METI Green Growth Strategy comparing the EV pathway and the carbon tax pathway, the model illustrates how individual characteristics, perception of technology or policy, and psychological mechanisms jointly shape both market and policy acceptance. The framework highlights potential transmission and feedback mechanisms between consumer-side actions and policy signal reception.

3.1 Economic and Behavioral Foundations of Willingness to Pay

The willingness to pay is one of the fundamental concepts in economics and consumer behavior, representing the maximum price a consumer is willing to pay for one unit of a product or service. Although traditionally viewed as a single reservation price, some researchers conceptualize WTP as a range.

WTP is not fixed; it can vary significantly depending on a variety of factors. These factors include:

- **External differences**: observable characteristics such as age, gender, income, education level, and geographic location.
- **Internal differences**: unobservable individual traits such as risk tolerance, desire for social conformity, and enthusiasm for a particular issue.
- **Contextual factors**: urgent need or perceived scarcity can increase WTP.

Willingness to pay can be estimated using various methods, including surveys, focus groups, and conjoint analysis, the latter of which requires respondents to rank different bundles of attributes to determine the value of each one.

As a dynamic and multifaceted construction, WTP goes beyond pure rationality. The definition of WTP recognizes its susceptibility to both observable (external) and unobservable (internal) influences. This immediately suggests that a purely rational economic calculation of WTP is incomplete. For environmental goods, where benefits are often diffuse or long-term, internal factors such as environmental values, social identity, and risk perception become especially important.

Therefore, in order to effectively stimulate WTP for carbon pricing or electric vehicles, policies must go beyond simple price adjustments. They need to address the underlying psychological and social drivers that shape perceived consumer value. This means understanding the emotional, social, and cognitive dimensions of WTP, not just the financial one. For example, if the internal desire for "belonging" or "environmental contribution" is high, policies can make use of social norms.

3.2 Deconstructing the Concept of Willingness to Pay in the Context of Carbon Pricing and Electric Vehicles

This section will explore in depth the characteristics of WTP for carbon pricing and WTP for electric vehicles, identifying their shared psychological foundations as well as key differences in terms of specificity and perceived focus.

3.2.1 Core Psychological Drivers of Willingness to Pay for Carbon Pricing and Electric Vehicles

The WTP for both carbon pricing and electric vehicles is driven by three fundamental psychological mechanisms. These mechanisms are core components of behavioral economic models, enabling comparison between the two and providing a rationale for a unified analytical framework.

- Environmental Motivation (Sense of Environmental Responsibility): An individual's intrinsic desire or perceived responsibility to protect the environment and a willingness to contribute to a low-carbon society. This reflects a prosocial or altruistic motivation.
- Trust (in Government or Product/Manufacturer): Confidence in the policy-implementing institution (government) or the entity producing low-carbon technologies (product/manufacturer). This includes belief in the effectiveness and fairness of policies as well as the reliability of products.
- Risk Perception (Economic Burden, Technical/Implementation Uncertainty): The perceived economic costs that policies or technologies may entail and the uncertainty regarding their performance, safety, or practical implementation. This reflects a desire to avoid potential negative outcomes.

These psychological drivers do not exist in isolation but form a complex, interrelated system. For example, low trust in government may amplify the perceived economic burden of carbon pricing, even when environmental motivation is high. Similarly, strong trust in manufacturers may alleviate concerns over technological risks associated with electric vehicles, thereby increasing WTP. This interaction suggests that the intensity of one factor can compensate for or weaken another, making overall WTP a function of their dynamic interplay. This complexity calls for a holistic approach in policy design.

Moreover, trust, as a cross-domain mediating factor, plays a particularly critical role. It applies to both "government" and "product/manufacturer." This observation is essential. Trust in the legitimacy and transparency of government can influence WTP for carbon pricing, while trust in product reliability and manufacturer reputation directly affects WTP for electric vehicles. The shared function of "trust" in both domains indicates its role as a general mediating factor in WTP, meaning that a lack of trust in one domain (e.g., government) can spill over and negatively impact WTP in another domain (e.g., electric vehicles), even if the product itself is satisfactory.

This highlights the systemic importance of building and maintaining public trust across different institutional levels.

3.2.2 The Distinction Between Latent and Explicit Willingness to Pay: Perceptual Differences and Shifts in Focus

There is a significant difference in nature between WTP for carbon pricing and WTP for electric vehicles: the former is more "latent," while the latter is more "explicit." This distinction is mainly reflected in the focus of perception and the nature of the expected returns.

The WTP for carbon pricing and electric vehicles reveals distinct perceptual and motivational mechanisms, differentiated as "latent" and "explicit" respectively. The WTP for carbon pricing (latent) is primarily shaped by perceptual focuses such as government reliability, tax transparency, and collective fairness. This reflects an abstract and collective willingness to contribute, where the nature of return is unclear and consists of indirect societal benefits that are difficult for consumers to perceive. Consequently, the reward structure is delayed, diffused, and less tangible. In contrast, the WTP for electric vehicles (explicit) centers on product quality (e.g., driving range, safety), personal convenience, direct day-to-day benefits, and specific, tangible product experience. These factors reflect a concrete, individualized willingness to invest in a product, where the nature of return is direct, immediate, and concrete benefits that are easily perceived by consumers and involve personal utility and instant gratification. This contrast highlights the concept of psychological distance in perceived benefits. According to Construal Level Theory (CLT), greater psychological distance leads to abstract representations of events or objects, while shorter psychological distance results in more concrete representations. The benefits of carbon pricing are distant (in the future, collective, and indirect), making them less salient and less motivating for individual action. In contrast, the benefits of electric vehicles are immediate, personal, and concrete—psychologically "closer" and more directly motivating. This distinction helps explain why a high WTP for an abstract concept does not necessarily translate into concrete action. These perceptual differences carry implications for communication strategies. For carbon pricing, emphasis should be placed on transparency, accountability, and collective benefit, potentially illustrated through vivid examples of future societal gains. For electric vehicles, the focus should be on immediate personal benefits, convenience, and tangible user experience. Blurring these communication strategies may confuse consumers and thereby reduce their WTP for either.

3.2.3 Electric Vehicle Adoption as Behavioral Validation: Enhancing the Salience of Carbon Pricing Signals

This section aims to elaborate on how explicit low-carbon consumption behaviors, such as electric vehicle adoption, can reinforce the acceptance of abstract policy signals, such as carbon pricing, by shaping consumer psychology and value orientation. It conceptualizes a feedback loop in which visible green behaviors elevate the cognitive and normative basis for supporting systemic climate policy.

The adoption of electric vehicles, as a concrete and individualized low-carbon action, not only contributes directly to emissions reduction but also has psychological spillover effects that may enhance public acceptance of broader decarbonization policies. Behavioral economics and cognitive psychology suggest that such tangible pro-environmental actions strengthen personal identity as a "green" consumer, reduce cognitive dissonance, and lead to attitudinal alignment with broader policy goals.

Specifically, individuals who have invested in EVs are more likely to support carbon pricing because they have already accepted the underlying premise that high-carbon behaviors should be discouraged. The personal commitment represented by EV adoption helps bridge the psychological distance associated with carbon pricing, which is often perceived as abstract, future-oriented, and collective in nature.

According to Construal Level Theory (CLT), behaviors that are psychologically close, such as those that are immediate and experiential, can make related abstract concepts more concrete and relatable. Ownership of electric vehicles, by providing tangible and direct benefits while reinforcing low-carbon values, helps reduce the perceived distance between individual actions and broader climate policy. This, in turn, enhances the salience and perceived fairness of carbon pricing mechanisms.

Moreover, adopting EVs creates a sense of behavioral consistency. Once individuals take visible low-carbon actions, they are more likely to align their beliefs and policy preferences accordingly. This behavioral confirmation effect has been documented in domains such as household energy savings, diet change, and waste reduction. It indicates that actions can precede attitude formation, especially when the action involves effort or cost.

The perception of EVs as an environmentally meaningful choice also amplifies social norms related to climate responsibility. As EVs become more normalized, early adopters serve as influencers whose actions can redefine what is considered socially desirable. This normative shift can extend to acceptance of pricing instruments like carbon taxes, which might otherwise face resistance due to their cost implications.

However, for this reinforcement mechanism to operate effectively, consumers must perceive their behavior as connected to systemic change. If EV adoption is viewed merely as a personal convenience or economic decision, the reinforcing effect on carbon pricing support may weaken. Therefore, public communication and policy design should explicitly link consumer actions to broader societal benefits, emphasizing that individual choices are part of a collective decarbonization effort.

In this context, EV adoption is not only a market trend but also a psychological and normative anchor that fosters a receptive environment for carbon pricing. A strategically framed narrative—highlighting synergies between consumer action and policy outcomes—can help construct a behavioral chain in which the explicit act of purchasing EVs retroactively validates and strengthens support for carbon pricing policies.

3.2.4 Behavioral Transformation and Consumer Response: From Action to Policy Recognition

The adoption of electric vehicles is not only a low-carbon consumer choice but also a behavioral entry point that may shape how individuals perceive and respond to climate policy. After engaging in pro-environmental behaviors such as purchasing an electric vehicle, consumers often develop two key psychological responses:

- **Macro-level civic reinforcement**: A stronger sense of social responsibility and alignment with collective low-carbon goals.
- **Micro-level risk sensitivity**: Increased awareness of potential future costs associated with high-carbon lifestyles, such as carbon taxes or reputational burdens.

These responses suggest that concrete low-carbon actions, once taken, can influence how consumers interpret and evaluate policy instruments like carbon pricing. As individuals experience the practical benefits of clean technologies, they tend to become more accepting of supportive policy frameworks, particularly when those policies are perceived as fair, necessary, and consistent with their values and actions.

Motivations such as environmental concern and economic self-interest that drive EV adoption can also enhance the perceived legitimacy of carbon pricing. Social responsibility adds moral significance, while risk aversion offers tangible incentives. Together, they reinforce policy acceptance and contribute to narrowing the gap between abstract climate goals and personal behavior.

Furthermore, when consumers begin to view EV ownership as a hedge against future carbon-related costs, they are more likely to internalize the logic of carbon pricing. This is especially relevant for individuals who are less driven by environmental ideals but more responsive to financial implications. Clearly communicating how EVs reduce long-term carbon cost exposure can increase support for carbon policies across a broader segment of the population.

In this context, EV adoption becomes a behavioral catalyst that enhances openness to broader climate policy engagement. It may also generate spillover effects, encouraging peers to adopt similar behaviors and reinforcing societal momentum toward decarbonization.

This underscores the importance of linking personal low-carbon behaviors to broader systemic transformations. By highlighting how individual actions align with collective goals, policymakers can strengthen public support for carbon pricing mechanisms and accelerate the transition toward a low-carbon society.

Figure 5 presents the theoretical model of this study.

Building upon the dual-path theoretical framework, this study formulates the following hypotheses, focusing on the potential behavioral transmission from overt consumer behaviors to implicit policy acceptance in Japan:

- **H1:** Certain consumer characteristics, such as gender and income, may be associated with similar tendencies in both the WTP for electric vehicles and the acceptance of carbon taxes. This suggests that the attitudes individuals exhibit toward

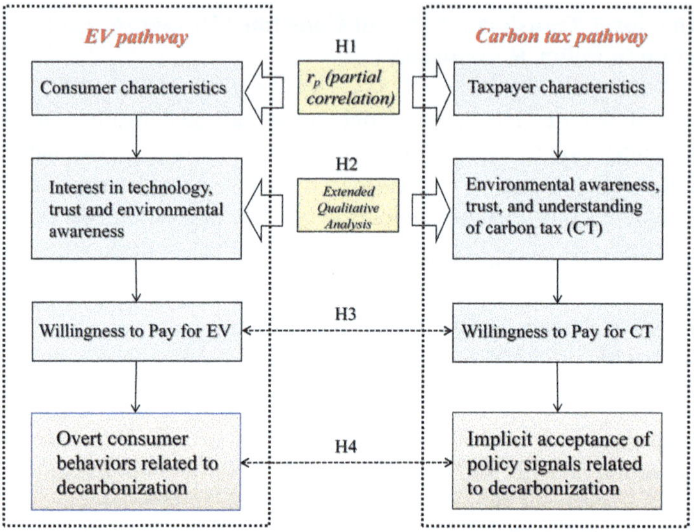

Fig. 5 Theoretical model. *Note* illustrates the influence of the pathway of consumer characteristics on WTP for electric vehicles and taxpayer characteristics on WTP for carbon tax. Four hypotheses are proposed to explain the structure: H1 tests the consistency of demographic transmission across both pathways, H2 explores how variables such as environmental awareness and interest in technology and policy function as key factors in the psychological transfer pathway between green technologies and institutional policies, H3 analyzes the behavioral coupling between green product adoption and policy payment intention, and H4 verifies whether overt green consumption behaviors can be transformed into implicit acceptance of decarbonization policy signals

green products may, to some extent, carry over to their responses to environmental policies, forming a demographic-based linkage between personal consumption behavior and institutional payment preferences.

- **H2:** Environmental awareness and interest in technology and policy exhibit similar influences on both electric vehicle adoption and carbon tax acceptance, suggesting that cognitive factors supporting engagement with green technologies can be psychologically transformed into support for institutional policies. This indicates a transferable pathway from product-level environmental concern to policy-level fiscal compliance—and vice versa.
- **H3:** The WTP for green products (such as electric vehicles) and for institutional policies (such as carbon taxes) is linked through a behavioral chain. However, the stability and transferability of this chain may vary depending on specific factors.
- **H4:** There exists a potential behavioral transformation mechanism between overt green consumption behaviors and the implicit acceptance of institutional decarbonization policies. Under specific cognitive and social conditions, visible pro-environmental actions may facilitate psychological adaptation and fiscal alignment with abstract policy signals, thereby laying the foundation for the societal diffusion of green fiscal instruments. The reverse pathway may also hold true.

4 Materials and Methodology

To examine whether consumers' WTP for carbon pricing and electric vehicles share a common psychological foundation, this study conducts a meta-analytic synthesis of existing empirical findings in Japan. Specifically, it integrates studies on carbon tax WTP (latent policy acceptance) and EV WTP (explicit product preference) and evaluates the degree to which demographic and psychological predictors form a behavioral chain linkage between the two domains. The overarching goal is to identify psychological continuity between abstract environmental policy acceptance and concrete sustainable product adoption.

4.1 Data Sources and Literature Collection Strategy

To construct a robust cross-domain synthesis of WTP determinants between carbon pricing and electric vehicles, a structured and rigorous literature screening process was undertaken. Studies were selected from peer-reviewed academic journals, government-commissioned research reports, and reputable academic databases, including Web of Science, Scopus, JSTOR, and CiNii, ensuring both international and Japan-specific coverage.

Only studies meeting all of the following inclusion criteria were retained for quantitative synthesis:

- The research must focus on Japanese population samples, ensuring cultural and policy contextual consistency.
- It must present a quantitative evaluation of WTP either for carbon pricing mechanisms (e.g., carbon tax or carbon surcharge) or for EV-related low-carbon technologies.
- Studies must report regression-based outputs, including t-statistics, standard errors, or confidence intervals, enabling the derivation of standardized partial correlation coefficients $r_{partial}$.
- Key explanatory variables (e.g., income, age, environmental concern, technological trust) must be consistently defined across multiple studies, allowing for valid cross-study comparison and meta-analytic synthesis.

Regarding the temporal distribution of the data, studies on EV WTP span a broad period from 2014 to 2023, covering technological shifts, market maturation, and subsidy program changes. In contrast, carbon pricing WTP studies remain limited, with eligible quantitative studies identified only in three temporal clusters: 2008, 2012, and 2024. Among these, the 2024 study represents the only fully modern survey with detailed regression outputs compatible with meta-analytic techniques. Earlier studies (2008, 2012) were included only for exploratory comparison where partial statistical data were available, though not all were used for the chain modeling due to incomplete standardization of variables.

This focused sampling strategy ensures high internal validity within the Japanese policy context while maximizing the comparability of latent (carbon pricing) and explicit (EV adoption) WTP determinants. The resulting dataset provides a rare opportunity to construct a cross-temporal and cross-domain behavioral map linking economic and psychological variables to pro-environmental financial preferences.

4.2 Calculation of Partial Correlation Coefficients

To unify effect sizes across studies and enable comparison, all regression coefficients were converted into partial correlation coefficients ($r_{partial}$). This was done using the following transformation based on the t-distribution:

$$r_{partial} = \frac{t}{\sqrt{t^2 + df}} \tag{1}$$

Specifically, the t-value was derived using the following equation:

$$t = \frac{\beta}{SE} \tag{2}$$

where β represents the standardized regression coefficient, SE is the standard error of β, df denotes the degrees of freedom, calculated as $n - k - 1$, with n being the sample size and k the number of predictors.

When studies only reported standardized regression coefficients (β) and their standard errors (SE), the following alternative formula was used:

$$r_{partial} = \frac{\beta}{\sqrt{\beta^2 + SE^2 \times (n - k - 1)}} \tag{3}$$

where: β represents standardized regression coefficient, SE represents standard error of β, n represents sample size, k represents number of predictors, $df = n - k - 1$ represents residual degrees of freedom.

This methodology follows the procedures recommended by Cohen (2003), which are widely used in behavioral and psychological meta-analyses.

4.3 Fisher's Z Transformation and Aggregation

Each value was normalized using Fisher's z-transformation:

$$z = \frac{1}{2} \ln\left(\frac{1 + r}{1 - r}\right) \tag{4}$$

The transformed z-values were then aggregated using inverse-variance weighting:

$$z = \frac{\sum w_i z_i}{\sum w_i}, \text{ where } w_i = n_i - 3 \text{ or } df_i \tag{5}$$

Finally, the aggregated z-value was converted back into $r_{summary}$ using the hyperbolic tangent function:

$$r_{summary} = \tanh(z) \tag{6}$$

This transformation corrects for the non-normal distribution of correlation coefficients in small samples and stabilizes variance (Fisher 1921; Hedges and Olkin 1985).

4.4 Behavioral Chain Judgment Criteria

To assess whether a variable constitutes a "behavioral chain linkage" between carbon tax WTP and EV WTP, the following thresholds were used:

- $r_{partial} > 0.3$ in both domains: strong linkage,
- $0.1 < r_{partial} \leq 0.3$ in both domains: moderate linkage,
- $r_{partial}$ significant in only one domain or < 3: weak or no linkage.

These thresholds were derived from Jacob Cohen (1988) classification of effect sizes (small: 0.1, medium: 0.3, large: 0.5) and are adapted for use in partial correlation meta-analysis (Cohen 2003).

This methodological structure enables a theory-grounded yet empirically rigorous identification of shared determinants between latent policy support and explicit product behavior. By applying a consistent framework to synthesize findings across domains, the study contributes to understanding the extent to which pro-environmental financial preferences, whether for public taxation or private purchases-stem from psychologically continuous motivations.

5 Results and Discussion

Building on the analytical framework introduced earlier, this section presents the calculated standardized partial correlation coefficients $r_{partial}$ for key predictors across carbon tax and EV WTP studies. These coefficients serve as the core empirical outputs used to assess whether common psychological and demographic factors exhibit consistent behavioral influence across both domains.

By aligning variables and standardizing outputs across studies, the analysis enables a direct comparison of effect magnitudes and directions. The resulting coefficients reveal the extent to which behavioral linkages are observable in the data and form the basis for identifying potential "behavioral chains."

The results are summarized in Table 3, which reports the coefficients of variables used in the final empirical analyses across studies, including standard errors, degrees of freedom, partial correlation coefficients, and their mean values calculated using Fisher-z aggregation.

This study attempts to construct a behavioral chain between WTP for electric vehicles and carbon taxes, primarily based on the high degree of consistency between the two research domains in terms of policy context and analytical framework. From a macro-level perspective, Japan, as a developed economy actively promoting low-carbon transition, has long focused its policy priorities on the coordinated development of green technology consumption (such as electric vehicles) and institutional carbon pricing (such as carbon taxes). The diffusion of electric vehicles reflects consumers' acceptance of green products and technologies, while the implementation of carbon taxes demonstrates the public's capacity and awareness to bear institutional environmental costs. Although situated at different stages—consumption behavior and institutional design—both are closely connected under the overarching research theme of "economic acceptability in building a low-carbon society. "Furthermore, from a methodological standpoint, a systematic review of the literature reveals that existing Japanese studies on WTP for electric vehicles commonly adopt the contingent valuation method (CVM) to design questionnaires, employ Logit or Tobit models for econometric analysis, and use demographic variables such as gender, age, and income as primary explanatory factors. The 2024 carbon tax WTP data utilized in this study was likewise designed under the CVM framework, with a questionnaire structure and modeling approach that are highly consistent. The core explanatory variables included in both domains are well-aligned in terms of definition, measurement dimension, and data type. This structural consistency in variables and modeling provides an empirical basis for integrated cross-study analysis, making the exploration of the relationship between green consumption behavior and policy payment acceptance not only theoretically plausible but also methodologically feasible.

Specifically, the selection criteria in this study include consistency in policy context, similarity in methodological frameworks, and overlap in key demographic and attitudinal variables. Based on the resulting highly comparable sources, a behavioral chain linking green consumption and institutional payment behavior was constructed using partial correlation coefficients $r_{partial}$, and the transferability and stability of this chain across population groups were further examined.

Among the many influencing factors, six variables ultimately identified as having highly consistent definitions and quantitative methods across the studies. These variables are gender, income, age, education level, environmental awareness, and interest in or understanding of technology or policy. Such consistency meets the unified standards for cross-study comparison and establishes them as the core factors for behavioral chain analysis.

Table 3 Summary of fisher z-transformed partial correlation coefficients and cross-year comparison between electric vehicle and carbon tax WTP studies

Author and sample	Variable	*Coeff*	*SE*	*df*	$r_{partial}$
Cao et al. (2024)					
Subsidies for low-income households	Gender (Female = 1)	−1471.480	489.210	1616	−0.075
	Age	11.553	21.750		0.013
	Educational level	−383.340	154.510		−0.062
	Annual household income	120.180	792.470		0.004
	Perceived impact of climate change	455.220	247.540		0.046
	Understanding of carbon tax	999.510	347.880		0.071
Subsidies for clean energy	Gender (Female = 1)	−925.990	496.920		−0.046
	Age	−6.740	22.090		−0.080
	Educational level	−379.160	157.060		−0.060
	Company employees (relative to executives)	−2487.190	983.600		−0.063
	Civil servants (relative to executives)	−2632.060	1264.900		−0.052
	Part-time/temporary workers (relative to executives)	−2828.670	1138.390		−0.062
	Annual household income	2775.210	805.070		0.085
	Perceived impact of climate change	344.818	251.580		0.034
	Understanding of carbon tax	1143.160	354.010		0.080
General national budget allocation	Gender (Female = 1)	−1142.030	476.380		−0.060
	Age	−5.000	21.210		−0.006
	Educational level	−369.670	150.530		−0.061
	Annual household income	2258.450	772.140		0.073
	Perceived impact of climate change	32.560	241.050		0.003
	Understanding of carbon tax	1091.290	338.930		0.080

(continued)

Table 3 (continued)

Author and sample	Variable	*Coeff*	*SE*	*df*	$r_{partial}$
Kajiwara and Muromachi (2023)					
Common sample	Gender (Female = 1)	−1.068	0.392	240	−0.173
	Household income	0.001	0.000		0.156
	High tech interest	0.991	0.249		0.248
BEV purchasers	Gender (Female = 1)	−1.606	0.498		−0.204
	Household income	0.001	0.000		0.136
	High tech interest	1.621	0.413		0.245
Non-BEV purchasers	Gender (Female = 1)	0.065	0.631		0.007
	High tech interest	0.565	0.312		0.116
Jung et al. (2022)					
HEV	Gender (Female = 1)	−0.025	0.074	8346	−0.004
	Age	0.012	0.079		0.002
	Annual mileage	−0.036	0.105		−0.004
	Housing type	0.230	0.077		0.033
	Driving frequency	0.037	0.085		0.005
PHEV	Gender (Female = 1)	−0.036	0.063		−0.006
	Age	0.010	0.067		0.002
	Annual mileage	−0.040	0.086		−0.005
	Housing type	0.050	0.065		0.008
	Driving frequency	−0.227	0.070		−0.035
BEV	Gender (Female = 1)	−0.304	0.064		−0.052
	Age	0.042	0.067		0.007
	Annual mileage	0.088	0.086		0.011
	Housing type	0.081	0.066		0.013
	Driving frequency	−0.430	0.070		−0.067
FCEV	Gender (Female = 1)	−0.073	0.060		−0.013
	Age	0.104	0.063		0.018
	Annual mileage	0.120	0.081		0.016
	Housing type	−0.072	0.037		−0.022
	Driving frequency	−0.142	0.062		−0.025

(continued)

Table 3 (continued)

Author and sample	Variable	*Coeff*	*SE*	*df*	$r_{partial}$
CDV	Gender (Female = 1)	−0.303	0.067		−0.050
	Age	0.025	0.074		0.004
	Annual mileage	0.070	0.096		0.008
	Housing type	−0.119	0.058		−0.022
	Driving frequency	−0.507	0.071		−0.078
Khan et al. (2020)					
PHEV/HV	Gender (Male = 1)	−0.999	0.212	1971	−0.106
	Homemaker	−0.947	0.352		−0.060
	Apartment parking	−0.772	0.212		−0.082
	Current vehicle price	0.185	0.065		0.064
	Current AFV owner	3.014	0.340		0.196
	No vehicle currently	1.565	0.445		0.079
HFCV	Gender (Male = 1)	−2.403	0.370		−0.145
	Homemaker	−1.974	0.566		−0.078
	Apartment parking	0.352	1.029		0.008
	Current AFV owner	0.352	1.029		0.008
	Graduate school	2.113	0.507		0.093
	Junior college & undergrad	0.833	0.315		0.059
BEV	Current AFV owner	1.867	0.396		0.106
Okada et al. (2019)					
Purchase intention among non-EV users	Evaluation of EVs regressed on number of household vehicles	0.071	0.003	106,439	0.072
	Evaluation of EVs regressed on daily driving duration	0.002	0.003		0.002
	Evaluation of EVs regressed on age	0.063	0.003		0.064
	Evaluation of EVs regressed on gender (Male = 1)	0.093	0.004		0.071
	Evaluation of EVs regressed on annual household income	−0.008	0.003		−0.008

(continued)

Table 3 (continued)

Author and sample	Variable	*Coeff*	*SE*	*df*	$r_{partial}$
	Evaluation of EVs regressed on environmental awareness	0.287	0.004		0.215
	Purchase intention regressed on evaluation of EVs	0.087	0.004	106,443	0.067
	Purchase intention regressed on environmental awareness	0.038	0.006		0.019
	Environmental awareness regressed on number of household vehicles	0.022	0.003	106,440	0.022
	Environmental awareness regressed on daily driving duration	0.013	0.003		0.013
	Environmental awareness regressed on age	0.230	0.003		0.229
	Environmental awareness regressed on gender (Male = 1)	−0.137	0.003		−0.139
	Environmental awareness regressed on annual household income	−0.001	0.003		−0.001
Tanaka et al. (2014)					
EV	Household income	0.114	0.018	31,990	0.036
	Age	0.000	0.006		0.000
	Married dummy for EV	0.304	0.147		0.012
	Gender (Female = 1)	−0.082	0.139		−0.003
	University degree	0.034	0.136		0.001
	Intent to purchase a car for EV	0.102	0.101		0.006
	Price range for EV	0.344	0.100		0.019

(continued)

Table 3 (continued)

Author and sample	Variable	*Coeff*	*SE*	*df*	$r_{partial}$
	Interest in alternative fuel vehicles for EV	2.426	0.070		0.189
PHEV	Household income	0.141	0.015		0.053
	Age	−0.011	0.005		−0.012
	Married dummy for PHEV	0.589	0.127		0.026
	Gender (Female = 1)	−0.378	0.119		−0.018
	University degree	−0.127	0.117		−0.006
	Intent to purchase a car for PHEV	0.117	0.083		0.008
	Price range for PHEV	0.572	0.082		0.039
	Interest in alternative fuel vehicles for PHEV	2.254	0.063		0.196
Gas vehicles	Age	−0.062	0.006		−0.056
	Household income	0.062	0.018		0.019
	Gender (Female = 1)	−0.982	0.139		−0.039
	Married dummy for gas vehicles	1.786	0.154		0.065
	University degree	0.140	0.131		0.006
	Intent to purchase a car for gasoline vehicles	−0.846	0.102		−0.046
	Price range for gasoline vehicles	0.348	0.102		0.019
	Interest in alternative fuel vehicles for gasoline vehicles	0.493	0.054		0.051

Note Fisher-z aggregated $r_{partial}$ values by study: Cao et al. (2024): Gender: −0.060; Age: 0.000; Income: 0.054; Educational: −0.061; Policy understanding: 0.077; Environmental awareness: 0.028. Jung et al. (2022): Gender: −0.025; Age: 0.006. Khan et al. (2020): Gender: −0.126. Okada et al. (2019): Gender: 0.001; Age: 0.004; Income: 0.000; Environmental awareness: 0.019. Tanaka et al. (2014): Gender: −0.011; Age: −0.006; Educational: −0.002; Income: 0.044; Interest: 0.193

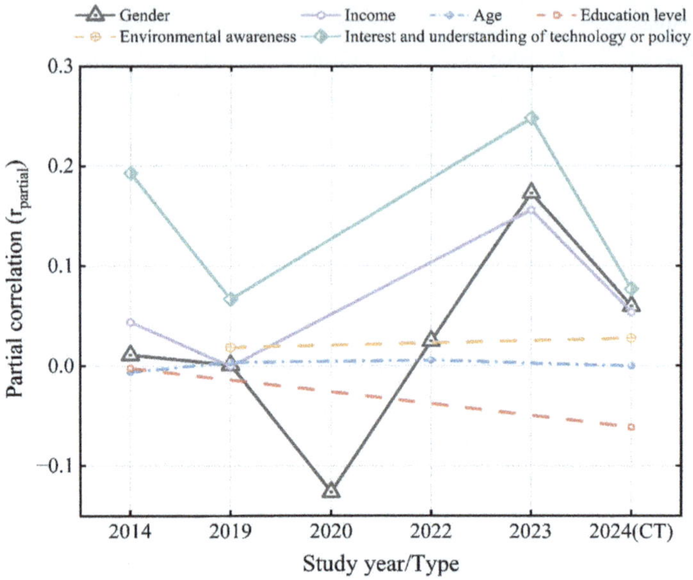

Fig. 6 Partial correlation of demographics in EV and carbon tax WTP studies over time

Figure 6 visualizes the trends of partial correlation coefficients. This study integrates the variables "interest in technology," "interest in policy," and "policy understanding" into a unified category: "interest and understanding of technology or policy." From a semantic and cognitive perspective, these indicators all reflect respondents' levels of concern for, and comprehension of issues related to the green transition, thereby exhibiting a high degree of theoretical consistency.

As shown in Fig. 6, in terms of gender, the continuity of the behavioral chain exhibits a typical structural characteristic. In studies on WTP for electric vehicles, the partial correlation coefficient $r_{partial}$ for gender slightly declined from 0.011 in 2014 to 0.001 in 2019. A reversal occurred in 2020 with a coefficient of −0.126, indicating a temporary shift in the direction of gender influence under the specific contextual background. However, this shift did not persist; the coefficient returned to a positive value in 2022 (0.025) and further increased to 0.173 in 2023, showing a trajectory of recovery from short-term fluctuation to structural consistency.

Before 2020, data showed that male consumers generally had a higher WTP for electric vehicles than their female counterparts. However, in the early stage of the COVID-19 outbreak in 2020, a temporary reversal occurred in which female WTP surpassed that of males. Since 2021, male consumers have once again assumed the dominant position, and this trend has continued through 2024. This fluctuation suggests that complex social, economic, cultural, and policy factors may have interacted to shape gendered consumer behaviors.

In early 2020, as COVID-19 swept across the globe, Japan issued its first-ever state of emergency, which drastically altered the country's daily social and operational structures. The Japanese government called on businesses to reduce commuting by 70%, thereby promoting the widespread adoption of remote work and significantly increasing the frequency and duration of intra-household interactions (TAKAMI 2021).

Japanese women are generally more attentive and sensitive to environmental risks and family health management (Davidson and Freudenburg 1996). This "domestication" effect may have intensified women's leading role in indoor environmental monitoring and health risk management, thereby elevating their sensitivity toward indoor air quality, energy consumption, and health-related travel risks.

In contrast, men tend to focus more on macroeconomic conditions and disaster preparedness under such circumstances, which may have contributed to a typical delay in making large-scale household expenditures such as EV purchases. This temporary weakening of purchasing intent is consistent with the observed decline in demand for durable goods under the combined effects of the pandemic and policy uncertainty. However, as containment measures eased and economic activity gradually normalized, the dominant purchasing trend among male consumers began to re-emerge.

In 2024, the partial correlation coefficient for gender in the carbon tax WTP study was 0.060, aligning directionally with the main trend found in EV WTP research. This suggests that gender, as a structural demographic variable, possesses a certain degree of transferability across different types of green consumption and policy payment contexts. Despite heterogeneity in products and institutional mechanisms, the transmission structure of the gender variable remained unbroken in both research paths, preserving the continuity of the behavioral chain. Overall, the chain is characterized by "directional consistency of the dominant group and unbroken structural transmission," providing a basis for cross-domain structural transfer.

In terms of income, this variable demonstrates a relatively consistent transmission structure between green consumption and institutional payment. In studies on WTP for electric vehicles, the partial correlation coefficient for annual income decreased from 0.044 in 2014 to 0.000 in 2019 but rebounded to 0.156 in 2023, indicating an upward trend in WTP for green consumption among high-income groups in the post-pandemic period. In the 2024 carbon tax WTP study, the partial correlation coefficient for income was 0.054, which aligns with the positive trend observed in later EV studies.

Overall, the income variable did not experience any directional reversal throughout the behavioral chain, consistently maintaining a positive influence. This suggests that high-income groups exhibit a relatively stable psychological recognition pathway from green products to institutional payments. The synergy between value consistency and payment capacity provides a strong basis for their leading role in the green transition.

From a behavioral economics perspective, it can be further inferred that high-income groups possess stronger risk tolerance in bearing the uncertain costs associated with future transitions, making them more receptive to the structural costs of green transformation. In addition, they are more likely to access information related to environmental and climate policies, leading to stronger policy cognition and institutional trust, thereby showing a higher acceptance of institutional payments such as carbon taxes. Furthermore, high-income consumers often prefer to express social responsibility or symbolic value through their consumption, with green products and carbon tax payments potentially internalized as part of their identity or value expression.

Given that the time frame covered by this study spans both the pre- and post-pandemic periods, the findings of LIN (2020) further support this inference. LIN's research pointed out that during the COVID-19 pandemic, the proportion of people who prioritized "spiritual richness" over "material richness" significantly increased, and the income threshold dividing these two value orientations rose from USD 8,000 to USD 15,000. Although EVs possess both material and symbolic attributes, their promotion is fundamentally driven by frontier environmental technologies and sustainability ideologies. Therefore, EVs can be conceptually categorized as a form of spiritual consumption, aligning well with the value orientation identified in Lin's study.

This shift lends strong support to the hypothesis that high-income groups are more inclined to derive life satisfaction from non-material dimensions and may view green consumption and institutional payments as viable pathways toward achieving "spiritual richness."

As shown in Fig. 6, the consistency of age variable within the green payment behavior chain is extremely weak. In studies on WTP for electric vehicles, the partial correlation coefficient for age increased from –0.006 in 2014 to 0.004 in 2019, and further slightly rose to 0.006 in 2022, all of which remained at very low levels and exhibited unstable directions, lacking any significant trend. In the 2024 study on WTP for carbon tax, the partial correlation coefficient for age was 0.000, indicating that its impact on institutional payment intention remains very limited.

Overall, the transmission path structure represented by the age dimension is unstable. The partial correlation coefficients consistently remained close to zero and lacked directional consistency, suggesting the absence of a continuous psychological mechanism or behavioral logic linking green consumption (such as EV adoption) with institutional acceptance (such as support for carbon taxes). This structural instability constitutes a "weak link" in behavioral terms, indicating that age does not serve as an effective anchor variable in constructing the green payment behavior chain.

A large body of literature has pointed out that the age variable often lacks statistical significance in green payment studies or exhibits diverse directions of influence, implying that the weak impact of age is a consistent finding across different studies in the green domain. This characteristic is also supported by the data in the present study.

It is further hypothesized that this phenomenon may stem from significant changes in Japan's demographic structure. In recent years, Japanese society has shown an accelerating trend of population aging. The elderly population remains active in the labor market and possesses strong information-receiving capabilities and social adaptability. Moreover, a relatively well-established social security system provides them with stable economic support. Therefore, although the conventional view that "younger individuals are more supportive of green technologies and policies" is widely accepted, the above structural factors are likely weakening the actual influence of generational differences. In terms of WTP for electric vehicles and carbon taxes, the significant differences among age groups appear to be narrowing, which helps explain the absence of a continuous or directionally significant age effect observed in this study.

Currently, only one-dimensional comparative analysis can be conducted for the variables of education and environmental awareness. According to the data presented, the partial correlation coefficients for education are −0.002 in the EV study and −0.061 in the carbon tax study, while those for environmental awareness are 0.019 and 0.028, respectively. Although the values are limited, the directional consistency suggests a potential for stable behavioral linkage across green payment domains.

The observation that higher education groups exhibit slightly lower WTP is not uncommon in behavioral economics. Several explanations are plausible: First, from the perspective of cognitive complexity and cost–benefit sensitivity, individuals with higher education levels tend to possess stronger analytical capabilities and are more inclined to assess the expected returns of green technologies or institutional fiscal burdens, such as carbon taxes. When the anticipated environmental benefits do not appear to justify the personal financial costs, WTP may decline. Second, in the Japanese sociopolitical context, highly educated individuals often pay closer attention to the transparency and fairness of policymaking processes. If the dissemination of a policy lacks explicit explanations of redistribution mechanisms or empirical validation of its effects, trust in the system may diminish, leading to institutional resistance. Additionally, the phenomenon known as the "cold cognition of the highly educated" warrants attention—while these groups may exhibit pro-environmental attitudes, their behavioral commitment may falter, especially when financial expenditures are involved. As demonstrated in the nationwide empirical study by Cao et al. (2024), correlation tests confirm that individuals with higher education levels in Japan exhibit notable caution toward carbon tax payments. This finding underscores the need for improved policy transparency, fairness, and procedural trust to enhance public acceptance, resonating with arguments by Klenert et al. (2018) and Rafaty (2018), who identify lack of procedural trust as a critical barrier to environmental policy implementation. On the surface, the finding that higher educational attainment is associated with lower WTP appears to contradict the intuitive linkage between education, income, and payment capacity. However, this is not necessarily the case. It is important to emphasize that the reported partial correlation coefficient reflects the net marginal effect of education on WTP after statistically controlling covariates such as income. In other words, this coefficient isolates the independent contribution of education to WTP by removing the confounding influence caused by collinearity between education and income, thus capturing the direction and magnitude of education's effect at a constant income level.

In contrast, environmental awareness functions as an internalized value orientation. Individuals with strong environmental consciousness tend to view payment behavior as a moral responsibility and are more inclined to act accordingly. They also exhibit higher cognitive-behavioral consistency to avoid psychological dissonance.

In Japan's collectivist culture, individuals with high environmental awareness are more likely to be driven by social identity and group affiliation, leading to consistent WTP behaviors.

Nevertheless, whether these two behavioral pathways can be solidified into stable chains remains an open question, and further empirical studies are needed to verify the robustness and persistence of these linkages.

Understanding and interest in technology and policy reflect individuals' attention to environmental technologies and their deep understanding of relevant systems. It serves as an important psychological predictor of green behavior and policy acceptance. According to the partial correlation coefficients, this variable shows consistently positive effects across both policy domains, with values of 0.193, 0.248, and 0.067 in the electric vehicle study and 0.077 in the carbon tax study. These results indicate a stable and positive behavioral linkage between green consumption and institutional payment willingness.

From the perspective of "information-based trust formation," individuals with a deeper understanding of technology and policy are more likely to comprehend the rationale and goals behind policy design, thereby increasing their trust in the legitimacy of such systems and improving acceptance. Furthermore, interest in technology and policy tends to enhance individuals perceived controllability, mitigating psychological resistance when facing complex institutions or uncertainties and thus facilitating greater willingness to make economic contributions. Accordingly, this behavioral chain demonstrates a certain degree of stability.

The four proposed hypotheses in this study are partially supported overall. H1 is supported. Partial correlation analysis indicates that demographic characteristics such as gender and income exhibit relatively similar directional influences on both WTP for electric vehicles and WTP for carbon taxes, thereby validating the cross-domain behavioral linkage hypothesis based on demographic attributes. H2 is partially supported. Environmental awareness and interest in technology and policy show stable and significant positive effects in both the EV and carbon tax dimensions. This suggests that cognitive factors supporting engagement with green technologies can be psychologically translated into acceptance of institutional policies, confirming the transferability pathway from product-level environmental concern to policy-level fiscal compliance. However, the role of environmental awareness still requires further research exploration. H3 is partially supported. The data reveals the existence of a behavioral chain between WTP for green products and WTP for institutional policies. Nevertheless, the strength and stability of this chain vary, and currently only a unidirectional mechanism—from product behavior to policy acceptance—can be confirmed. Further research is needed to verify the reverse pathway from policy acceptance to product behavior. H4 is partially supported. Empirical results suggest that a transformation mechanism may exist between overt green consumption behaviors and the implicit acceptance of institutional decarbonization policies. However, the validity of this mechanism depends on specific cognitive and social conditions, and its effect is characterized by selectivity and contextual dependency. Further investigation is needed to examine the reverse behavioral chain from policy to consumption.

6 Strategic Implications of the Behavioral Chain and Future Outlook

This study, from the perspective of Japanese micro-level consumers, investigates the behavioral linkage mechanism between explicit green consumption behaviors, particularly WTP for electric vehicles, and acceptance of institutional decarbonization policies such as carbon taxes. The results partially confirm that demographic attributes (such as gender and income) and cognitive factors (such as environmental awareness, and interest and understanding in technology or policies) have consistent directions of influence in both types of behaviors, thereby preliminarily establishing the basis for the existence of a behavioral chain. However, the empirical results indicate that this behavioral chain exhibits only a unidirectional nature. That is, the intention to adopt green products like EVs helps to enhance the acceptance of institutional policies such as carbon taxes, while the reverse influence path from policy signals to consumer green behavior still requires further research.

Further analysis shows that under specific social and cognitive conditions, explicit pro-environmental behaviors can strengthen the public's psychological readiness and fiscal acceptance of abstract policy instruments. However, the establishment of this mechanism depends on key psychological support factors such as institutional trust and perceived fairness. When the cognitive distance between individual behavior and policy outcome is too great, the stability of behavioral conversion significantly declines.

From the perspective of behavioral economics, this study provides preliminary evidence on how consumer behavior can feed back into the acceptance of institutional policies. Therefore, this study suggests that the design of low-carbon policies should place consumers at the core and treat them as interlocutors of climate institutions rather than passive recipients. At the same time, this perspective also offers a new approach for green product promotion enterprises to collaborate with policy. By enhancing consumer awareness and engagement, it can indirectly promote the diffusion and legitimacy-building of low-carbon policies.

Groups with high WTP for EVs (such as males, high-income individuals, and those with a certain level of understanding and interest in technology or policies) show stronger behavioral linkage in terms of institutional carbon tax acceptance. Policy design should treat such groups as core drivers of the behavioral chain, encouraging them to shift from green consumption to institutional participation, for example by providing personalized tax reduction incentives or policy co-creation mechanisms to enhance their sense of institutional belonging and responsibility. In contrast, for low-income groups and those with low policy interest, relying solely on green product adoption is insufficient to translate into support for carbon taxes. It is necessary to enhance the fiscal visibility of the policy (such as carbon tax revenue returns and transparent designation of uses) to shorten the cognition and behavior distance. Additionally, although women show weaker influence in EV payment willingness, they may hold a hidden dominant position in household energy decision-making. Therefore, embedded communication in everyday life settings and participation mechanisms should be used to activate their potential for institutional participation.

In conclusion, it is recommended that future policies adopt a three-pronged strategy of layered design, behavioral chain guidance, and institutional transparency enhancement, in order to stabilize and expand the public behavioral pathway from green consumption to institutional acceptance. This will build a more solid public opinion foundation for the social implementation of institutional tools such as carbon taxes. By focusing on how to facilitate systemic alignment between consumer behaviors and institutional legitimacy, policymakers should strive to move from fragmented interventions toward a system-oriented climate governance paradigm anchored in the evolving environmental psychology of citizens.

References

Cao L, Toyohara A, Li Y, Zhou W (2024) Willingness to pay for carbon tax in Japan. Sustain Prod Consum 52:427–444. https://doi.org/10.1016/j.spc.2024.11.004
Cohen J (1988) Statistical power analysis for the behavioral sciences
Cohen (2003) Applied multiple regression/correlation analysis for the behavioral sciences. Routledge. https://doi.org/10.4324/9780203774441
Davidson DJ, Freudenburg WR (1996) Gender and Environmental Risk Concerns. Environ Behav 28(3):302–339. https://doi.org/10.1177/0013916596283003
Deloitte (2025) Driving the global EV battery supply chain
Fisher RA (1921) On the "probable error" of a coefficient of correlation deduced from a small sample
Government of Japan (2021) The long-term strategy under the Paris agreement
Government of Japan (2023) Japan climate transition bond framework
Hedges, LV, Olkin I (1985) Statistical methods for meta-analysis
IEA (2024) Global EV outlook 2024 moving towards increased affordability. www.iea.org
IEA (2025) Global EV Outlook 2025 Expanding sales in diverse markets. www.iea.org
IMF (2023) Public perceptions of climate mitigation policies : evidence from cross-country surveys. International Monetary Fund
Japan Automobile Manufacturers Association (2024) Taxes on automobiles
Jung J, Lee DJ, Yoshida K (2022) Comparison between Korean and Japanese consumers' preferences for fuel cell electric vehicles. Transportation research part D: Transport and environment 113. https://doi.org/10.1016/j.trd.2022.103511
Kajiwara N, Muromachi Y (2023) Factors affecting the purchase of battery electric vehicles (BEV) of the BEV purchasers
Kakishima R, Kikuchi H, Fukuda A (2023) Evaluation of tax and subsidies for BEVs diffusion in the Japanese automobile market
Khan U, Yamamoto T, Sato H (2020) Consumer preferences for hydrogen fuel cell vehicles in Japan. Transportation research part D: transport and environment, 87. https://doi.org/10.1016/j.trd.2020.102542
Klenert D, Mattauch L, Combet E et al (2018) Making carbon pricing work for citizens. Nature Clim Change 8:669–677. https://doi.org/10.1038/s41558-018-0201-2
Komiyama M (2024) Green drive: patents fuel Japan's EV and hybrid technology surge
LIN Xiangwei (2020) A comparative study of the relationship between life satisfaction and economic growth and their influencing factors in China and Japan

METI (2022) Clean energy strategy interim report (outline) chapter 1 ensuring energy security future directions of energy policy
METI (2023) Guidelines for promoting development of EV charging infrastructure (in Japanese)
OECD (2020) Aligning short-term climate action with long-term climate goals opportunities and options for enhancing alignment between NDCs and long-term strategies. www.oecd.org/environment/cc/ccxg.htm
OECD (2021) Exploring the impact of shared mobility services on CO2. OECD Environment Working Papers, Vol 175. https://doi.org/10.1787/9d20da6c-en
OECD (2023) Financing solutions to foster industrial decarbonisation in emerging and developing economies. OECD Environment Working Papers, Vol 226. https://doi.org/10.1787/24a155ab-en
Okada T, Tamaki T, Managi S (2019) Effect of environmental awareness on purchase intention and satisfaction pertaining to electric vehicles in Japan, Transportation Research Part D: Transport and Environment, Vol 67, pp 503–513, ISSN 1361–9209, https://doi.org/10.1016/j.trd.2019.01.012
Rafaty Ryan (2018) Perceptions of corruption, political distrust, and the weakening of climate policy. Glob Environ Polit 18(3):106–129. https://doi.org/10.1162/glep_a_00471
Reuters (2025) Green steel needs incentives to work and Japan has a plan
TAKAMI T (2021) Working from home and work-life balance during COVID-19: The latest changes and challenges in Japan
Tanaka M, Ida T, Murakami K, Friedman L (2014) Consumers' willingness to pay for alternative fuel vehicles: A comparative discrete choice analysis between the US and Japan. Transp Res Part a: Policy Prac-Tice 70:194–209. https://doi.org/10.1016/j.tra.2014.10.019

Climate Finance Empowering Green Recovery and Green Low-Carbon Economic Development

Rong Zhu, Xuepeng Qian, and Weisheng Zhou

1 Introduction

1.1 The Concept and Evolution of Climate Finance

1.1.1 Definition and Background of Climate Finance

Climate change has emerged as one of the most pressing challenges confronting the global political, economic, and social order (Tan et al. 2025). The Intergovernmental Panel on Climate Change (IPCC), in its Sixth Assessment Report released in 2023, unequivocally identified climate change not merely as a long-term, systemic issue but as an imminent and intensifying crisis that poses a profound threat to the survival and development security of humanity. In response, a growing international consensus has coalesced around the need to drastically reduce greenhouse gas emissions as the cornerstone of effective climate action (Röck et al. 2020). However, achieving this objective hinges on the mobilization of substantial financial resources, particularly in areas such as technology innovation, energy transition, infrastructure modernization, and climate adaptation (Bhandary et al. 2021; Omukuti 2024). The vast scale of these financial requirements has given rise to the development of climate finance.

The origins of climate finance can be traced back to the 1992 United Nations Framework Convention on Climate Change (UNFCCC), which laid out the first

R. Zhu · X. Qian (✉)
Graduate School of Global Environmental Studies, Sophia University, Tokyo, Japan
e-mail: qianxp@sophia.ac.jp

R. Zhu
e-mail: r-zhu-9o7@eagle.sophia.ac.jp

W. Zhou
College of Policy Science, Ritsumeikan University, Osaka, Japan
e-mail: zhou@sps.ritsumei.ac.jp

W. Zhou (ed.), *Decarbonization Policy Engineering*, World Sustainability Series,
https://doi.org/10.1007/978-981-95-5953-4_21

international framework for coordinated climate action and emphasized the vital role of financial support in addressing climate challenges. The Convention defines climate finance as "local, national or transnational financing—drawn from public, private and alternative sources of financing—that seeks to support mitigation and adaptation actions to address climate change." (UNFCCC 1992).

The 2009 Copenhagen Accord (UNFCCC 2009) marked a significant milestone by concretizing the concept of climate finance and assigning explicit responsibility to developed countries. The Accord committed developed nations to collectively mobilize $100 billion annually by 2020 to support mitigation and adaptation efforts in developing countries. The following year, the Cancun Agreements (UNFCCC 2010) further elaborated the institutional framework for climate finance, reaffirming the obligations of developed countries and establishing the Green Climate Fund (GCF) as the principal multilateral vehicle for climate finance disbursement.

Since Copenhagen and Cancun, the concept of climate finance has been widely embraced and progressively refined through successive international negotiations. The 2015 Paris Agreement represented a turning point in global climate governance by introducing a novel "bottom-up" architecture centered on nationally determined contributions (NDCs). Under the Agreement, all Parties are required to submit, update, and implement their mitigation pledges while developed countries are called upon to provide concrete support to developing nations through finance, technology transfer, and capacity countries. Climate finance flows must align with the goals of low-carbon development and climate resilience. In doing so, the Paris Agreement elevated climate finance to a legal status on par with mitigation and adaptation, enshrining it as one of the three foundational pillars for achieving global climate goals.

1.1.2 Evolving Financial Needs: From Climate Response to Low-Carbon Transition

The evolution of climate finance is inseparable from the trajectory of global climate action. Shifts in financial demand reflect a broader transformation in strategy, from responding to short-term climate risks to advancing long-term structural changes toward a low-carbon economy. This transition underscores both the deepening understanding of climate change and the instrumental role of finance in enabling systemic economic and social transformations.

In its early phase, climate finance primarily focused on the immediate impacts of climate change, such as funding mitigation projects to reduce greenhouse gas emissions and adaptation initiatives to address risks posed by extreme weather events. At this stage, financial needs were relatively modest and largely dependent on public-sector allocations, international aid, and support from development finance institutions. As scientific consensus around climate change deepened, it became increasingly clear that climate change is not merely a transient environmental issue but a catalyst for profound change across the energy, industrial, agricultural, economic, and social spheres. At the 2023 United Nations Climate Change Conference (COP28), the first Global Stocktake sent a powerful message to the international community: the era of fossil fuels is drawing to a close. This conclusion was strongly endorsed by the World Resources Institute (WRI) and the UNFCCC Secretariat (WRI 2023). Consequently, climate finance is now poised to shift increasingly toward investments in green technologies, clean energy systems, and low-carbon infrastructure.

1.2 Global Trends in Climate Finance

1.2.1 Scale and Growth Trajectory of Climate Finance

Between 2015 and 2023, the world experienced its nine warmest years on record. According to the World Meteorological Organization, the frequency of extreme weather events has increased fivefold over the past half-century (WMO 2023). At the same time, global greenhouse gas emissions have reached unprecedented levels, further intensifying climate-related impacts. To meet the Paris Agreement's 1.5 °C target, countries must not only maintain their current mitigation efforts but also significantly raise the ambition of their nationally determined contributions. Achieving this target will depend heavily on the availability of large-scale climate finance.

According to the Climate Policy Initiative (CPI), its Global Landscape of Climate Finance 2024 projects that meeting the 1.5 °C target will require global annual investments of between USD 6.7 trillion and USD 10 trillion through 2050. The report notes that although global climate finance grew from USD 674 billion in 2018 to USD 1.46 trillion in 2022, this level remains far below what is needed. Under the 1.5 °C scenario, global climate finance must reach USD 7.4 trillion per year by 2030, nearly five times the current level (Climate Policy Initiative 2024).

From 2018 to 2022, financing for climate change mitigation grew at an average annual compound rate of 20%, reaching a cumulative total of USD 1.3 trillion. At the same time, climate adaptation finance also expanded significantly, more than doubling from 2018 levels to reach USD 76 billion in 2022. However, global climate finance still accounts for only 1% of global GDP, falling short of actual needs. For emerging markets and developing economies, climate action could require national spending equivalent to approximately 6.5% of GDP by 2030 (Climate Policy Initiative 2024). The global effort to scale up climate finance remains a major challenge.

1.2.2 Evolution of Climate Finance Instruments

Traditional climate finance instruments include loans, grants, equity investments, and guarantees. In the early stages, concessional loans played a central role, particularly in financing infrastructure projects in developing countries aimed at addressing climate change. For example, since the 1990s, the European Investment Bank (EIB) has supported renewable energy initiatives through low-interest loans, with cumulative investments exceeding one trillion euros. The Green Climate Fund has provided grant-based financing for adaptation projects in the most vulnerable countries and communities, such as drought-resilient agricultural infrastructure in Africa (Garschagen and Doshi 2022). Multilateral development banks have supported private-sector participation in climate projects through equity investments. The World Bank Group, for instance, has deployed equity capital to assist renewable energy enterprises in emerging markets. Guarantee mechanisms have been used to de-risk investments and attract additional capital for climate-related initiatives. The Asian Development Bank (ADB) has provided risk guarantees for renewable energy projects in Southeast Asia.

A new generation of climate finance instruments has emerged in recent years. These include results-based financing mechanisms, carbon credit support tools, secured lending against renewable energy subsidies, green and transition bonds, climate insurance products, and derivative instruments.

1.3 The Role of Climate Finance in in Green Recovery and Green Low-Carbon Economy

1.3.1 Financial Support for Energy Structure Optimization and Clean Energy Development

In recent years, climate finance has played an increasingly important role in accelerating the transition to clean and low-carbon energy systems through technological innovation and industrial upgrading (Lee et al. 2024). In Belgium, for instance, government subsidies for solar panel installations have significantly increased the deployment of renewable energy technologies (De Groote et al. 2024). Research also demonstrates that climate finance has a strong positive effect on the renewable energy transition in sub-Saharan African developing countries (Amoah et al. 2025). These cases highlight the crucial function of climate finance in supporting global energy transformation.

1.3.2 Promoting Green Recovery Through Low-Carbon Economic Development

Since 2021, developed countries in Europe and North America have promoted multilateral cooperation through the Just Energy Transition Partnership (JETP), an intergovernmental initiative that provides financial support to low- and middle-income developing countries. The program particularly targets emerging economies with coal-dominated energy systems, offering funding to facilitate the phaseout of fossil fuels, promote renewable energy adoption, and provide assistance to populations most affected by the transition. These efforts aim to enable a balanced and inclusive shift toward a low-carbon economy at the national level. To date, countries such as South Africa, Indonesia, Vietnam, and Senegal have established JETP agreements with developed country coalitions. These partnerships have mobilized nearly USD 50 billion in initial funding through a combination of grants, concessional loans, commercial lending, and guarantees (Karg, Gupta, and Chen 2025).

As an innovative climate finance mechanism, the JETP has advanced the expansion of renewable energy in developing countries, reduced dependence on fossil fuels, improved energy security, and created new employment opportunities. It has also contributed to green economic growth. By linking financial support with low-carbon technology deployment, the JETP offers a powerful model for sustainable development and has created new growth drivers in participating countries. These efforts have stimulated the application of low-carbon technologies and the expansion of green industries.

1.3.3 Challenges and Opportunities in Financing Green Low-Carbon Transitions

Climate finance serves as a critical tool for promoting green and low-carbon development by directing capital toward sustainable sectors and providing essential financial support for systemic transformation. However, the low-carbon transition presents a range of financing challenges and opportunities. At the CAIJING Annual Conference 2025: Forecasts and Strategies and the 2024 Global Asset Management Forum, Wang Xin, Director-General of the Research Bureau of the People's Bank of China, noted that China faces a significant financing gap in its efforts to transition toward a green and low-carbon economy, enhance climate resilience, and protect ecosystems and biodiversity. Several structural issues underlie this financing shortfall. First, the absence of unified sustainable finance standards hampers the credibility and consistency of financial support mechanisms. Second, the lack of a comprehensive carbon accounting framework makes it difficult to assess emissions and quantify mitigation outcomes. Third, the incentive and regulatory frameworks remain underdeveloped, limiting the effective redirection of capital toward green and low-carbon sectors. Lastly, institutional capacity building has not kept pace with the rapid evolution of green finance, making it difficult to meet the demands of a fast-expanding sustainable finance ecosystem.

2 The Policy Framework for Climate Finance

2.1 International and Domestic Climate Policy Frameworks

2.1.1 Overview of Climate Finance Policy Systems Across Nations

(1) United States

In recent years, the United States has advanced the development of its climate finance system through legislation, financial innovation, and public–private collaboration. The overarching goal is to accelerate the clean energy transition, reduce carbon emissions, and simultaneously stimulate economic growth and job creation. In 2022, the United States enacted the Inflation Reduction Act (IRA), which commits USD 369 billion over the next decade to support climate and clean energy initiatives. The IRA aims to reduce national greenhouse gas emissions by 40% by 2030 compared to 2005 levels (Wang et al. 2024a, b). Key provisions include tax credits and subsidies for clean technology manufacturing, such as expanded incentives for renewable energy, including wind, solar, and energy storage, as well as electric vehicles and hydrogen production. It also supports domestic supply chains for batteries, photovoltaic components, and green hydrogen to reduce dependence on imports, particularly from China. The IRA allocates USD 60 billion to support renewable energy and climate resilience projects in low-income communities.

At the state level, California has developed a comprehensive climate finance framework. First, it operates a statewide Cap-and-Trade program that covers approximately 85% of the state's emissions, using auction revenues to fund clean transportation and community-based emissions reductions. Second, the state has set renewable energy targets, aiming to source 60% of its electricity from renewables by 2030 and achieve 100% clean electricity by 2045. Third, California has enacted a zero-emission vehicle (ZEV) mandate that bans the sale of new internal combustion engine vehicles beginning in 2035 and invests in supporting charging infrastructure (CEC 2024). New York State, under the Climate Leadership and Community Protection Act (CLCPA), has established targets to achieve 100% clean electricity by 2040 and carbon neutrality by 2050 (Zhou et al. 2023).

(2) Germany

Germany has promoted climate finance through its Climate Action Plan 2050 and Renewable Energy Sources Act (EEG), primarily by issuing government-backed green bonds to fund renewable energy, energy efficiency, and sustainable transportation projects. As of 2023, Germany had issued over EUR 50 billion in green bonds, with funds directed toward renewable energy projects, including wind and solar, and electrified transport systems, helping to increase the share of renewables to over 50%. The state-owned development bank KfW has provided approximately EUR 46 billion in climate-related financing, supporting energy efficiency upgrades for more than 500,000 households and the green transition of thousands of small and medium-sized enterprises (KfW Development Bank 2021).

(3) France

France has advanced its climate finance system under the Energy Transition for Green Growth Act and the National Low Carbon Strategy. France is the world's second-largest issuer of green bonds, with cumulative issuance exceeding EUR 40 billion. The proceeds support renewable energy, energy efficiency, and low-carbon transportation projects. France has also introduced a green finance label to encourage financial institutions to design products aligned with climate goals. More than 300 institutions have joined the program, with total green fund assets reaching EUR 300 billion by 2023. The French public investment bank, Bpifrance, provides loans and equity capital to support green innovation and corporate transition initiatives.

(4) Japan

Japan's central government has advanced green transformation by consolidating public funds, encouraging private investment, improving regulatory frameworks, and fostering technological innovation. In 2021, Japan released the Green Growth Strategy toward Carbon Neutrality by 2050, identifying 14 priority sectors, including renewables, hydrogen, electric vehicles, and carbon recycling, to guide investment and financing efforts. To support this strategy, Japan established a JPY 2 trillion fund, equivalent to approximately USD 18 billion, to finance the research and development of low-carbon technologies such as hydrogen, carbon capture and storage, and advanced battery systems.

2.1.2 Bilateral and Multilateral Climate Finance Institutions

According to the Global Landscape of Climate Finance 2024 published by the Climate Policy Initiative (CPI) in October 2024, climate finance demonstrated strong resilience and growth amid global crises, reaching USD 1.46 trillion in 2022. Key findings include the following: First, preliminary estimates suggest that global climate finance exceeded USD 1.5 trillion in 2023. Renewable energy and low-carbon transportation continued to be the primary growth drivers. However, under the 1.5 °C scenario of the Paris Agreement, climate finance must increase fivefold by 2030 to reach the required USD 7.4 trillion annually. Second, between 2018 and 2022, mitigation finance grew at a compound annual growth rate of 20%, reaching USD 1.3 trillion in 2022. During the same period, adaptation finance more than doubled, reaching USD 76 billion in 2022. Third, despite growth in adaptation finance, current climate finance levels in emerging markets and developing economies remain only one-third of the projected need by 2030. Outside of the energy, building, and transport sectors, progress in mitigation finance has been slow. Since 2020, investment in fossil fuels has continued to rise annually, reaching USD 1.1 trillion in 2023.

Multilateral development finance institutions have made significant contributions to the expansion of climate finance. Organizations such as the Multilateral Development Banks (MDBs), Asian Infrastructure Investment Bank (AIIB), Inter-American Development Bank (IDB), and Asian Development Bank (ADB) have played critical roles in supporting climate finance.

In 2018, ADB announced a climate finance target of USD 80 billion in its own resources for the 2019 to 2030 period, with the aim of ensuring that at least 75% of its operations support climate action. In October 2021, just ahead of COP26, ADB increased this commitment to USD 100 billion. In the same year, the World Bank Group (WBG) announced a new set of climate goals for 2021 to 2025, pledging to double its five-year climate investment to USD 200 billion. In fiscal year 2023, WBG delivered a record USD 38.6 billion in climate financing. WBG also published its first climate-related disclosures in line with the recommendations of the Task Force on Climate-related Financial Disclosures (TCFD). In 2023, AIIB launched its first Climate Action Plan (CAP), outlining its climate goals for the 2024 to 2030 period. The plan defines key investment areas and climate financing principles, with a commitment that by 2025, at least 50% of AIIB's approved financing will support climate-related investments. In 2022, climate finance accounted for 56% of AIIB's total approved investments (MDB 2015).

2.1.3 Establishing and Improving China's Domestic Climate Finance System

In recent years, under the strong guidance of government agencies, the active engagement of financial institutions, and broad societal support, China has taken significant steps toward building a national climate finance policy framework.

In October 2020, the Ministry of Ecology and Environment, along with four other agencies, issued the "Guiding Opinions on Promoting Investment and Financing in Response to Climate Change" (MEE General Office Climate Document No. 57 [2020], hereinafter referred to as the Guiding Opinions), which established a system-wide framework for advancing climate finance. This document defined clear goals, tasks, and timelines, marking the first top-level policy design in China for climate finance and the first climate policy document linked to the country's dual carbon targets. Based on the Guiding Opinions, a series of supporting policy documents were issued to institutionalize climate finance, including the Notice of the Pilot Work on Climate Investment and Financing, the National Climate Change Adaptation Strategy 2035, the Notice on the List of Pilot Cities for Climate Investment and Financing, and the Notice on Issuing the Reference Standards for the Inclusion of Local Climate Investment and Financing Projects in Pilot Project Libraries. These documents collectively shaped a policy system centered on pilot programs, using pilot project libraries as a key lever, and advancing with a dual focus on mitigation and adaptation.

In October 2021, the Climate Investment and Finance Association of the Chinese Society for Environmental Sciences published the Guidance on Taxonomy of Climate Investment and Financing Projects (T/CSTE0061–2021, hereinafter referred to as the Guidance), which established China's first national standard for identifying climate investment projects. This was followed by the 2022 release of the Evaluating Guide for Climate Friendly Financial Institutions (TCSTE0289–2022) and the 2023 release of the International Guide on the Taxonomy of Climate Investment and Financing

Projects. Together, these frameworks have accelerated the development of climate finance in China.

In August 2022, the Ministry of Ecology and Environment and eight other agencies jointly announced the first list of 23 national climate finance pilot cities. These pilot cities have actively explored innovative approaches tailored to local conditions and resource endowments. By the end of 2023, the pilot program had yielded significant results. First, policies became increasingly concrete and actionable. For example, Taiyuan issued an Action Plan to Strictly Curb the Blind Expansion of High-Energy-Consumption and High-Emission Projects, Baotou released a Green Transportation Plan for New Energy Heavy-Duty Trucks (2023–2025), and Lanzhou issued a Green Low-Carbon Development Plan for the Transport Sector. Second, pilot regions developed supportive incentive mechanisms. To strengthen the synergy between fiscal and financial policies and to increase funding for low-carbon development, the Nansha New Area in Guangzhou introduced China's first dedicated financial support policy for climate finance, titled Measures for Promoting Climate Investment and Financing in Nansha, followed by its detailed implementation guidelines. By the end of 2023, the national project library contained more than 3000 registered or reserve climate finance projects, representing over CNY 3 trillion in total investment and nearly CNY 170 billion in approved credit.

2.2 Impact of International Standards and Agreements on Climate Finance

2.2.1 The Role and Influence of International Standards on Climate Finance

In the domain of standard-setting and evaluation, a coalition of eight financial institutions including the International Development Finance Club (IDFC), African Development Bank (AFDB), and Asian Development Bank (ADB) jointly released A Framework and Principles for Climate Resilience Metrics in Financing Operations. This framework emphasizes the importance of applying consistent metrics for climate adaptation across financial operations. It serves as a common reference for multilateral financial institutions when assessing and reporting on the effectiveness of adaptation-related climate finance. To ensure methodological alignment in financing, the Organization for Economic Co-operation and Development (OECD) has released key policy documents, including the G20 Transition Finance Framework and the Climate Adaptation Investment Framework, aiming to establish globally applicable standards and indicators for assessing and reporting the impact and effectiveness of sustainable finance. They also provide detailed implementation guidance and technical support to national governments and financial institutions, enabling regular progress assessments and enhancing alignment and cooperation among global financial actors in advancing climate action.

2.2.2 The Influence of International Agreements on Climate Finance

Climate finance has remained a core agenda item in successive United Nations climate conferences. The establishment of the United Nations Framework Convention on Climate Change (UNFCCC) in 1992 laid a solid foundation for the global climate finance architecture. Under the UNFCCC framework, the evolution of global climate finance has progressed from foundational principles to the formation of structured financing mechanisms. These developments span funding mobilization, allocation, transparency in usage, and reporting systems. A financing mechanism was established to provide support to developing country parties, with increasing emphasis placed on the role of the private sector in climate finance. Subsequent international agreements such as the 1997 Kyoto Protocol and the 2015 Paris Agreement adopted numerous resolutions urging developed countries with greater financial capacity to extend both policy and financial support to developing countries with fewer resources and higher vulnerability. During this period, several multilateral climate funds were established, including the Global Environment Facility (GEF), Green Climate Fund (GCF), Special Climate Change Fund (SCCF), and Least Developed Countries Fund (LDCF) (as shown in Table 1).

These funds have provided practical financial support for global climate governance, particularly in developing countries, thereby strengthening their capacity to respond to climate change and mitigating the adverse effects of climate-related losses and damage.

3 Climate Finance Instruments and Products

Climate finance encompasses a wide array of instruments, including grants, donations, green bonds, equity financing, debt-for-nature swaps, guarantees, and concessional loans.

3.1 Climate Credit

3.1.1 Development of Climate Credit

Climate credit refers to specialized loan products provided by financial institutions for projects that meet defined environmental criteria. These loans are typically characterized by lower interest rates and flexible repayment terms and are designed to support climate mitigation and adaptation efforts. In recent years, international financial institutions have made significant strides in expanding climate-related credit facilities. For example, the Asian Development Bank (ADB) has committed to mobilizing $100 billion in climate finance between 2019 and 2030. In 2023 alone, ADB disbursed $9.8 billion in climate-related loans. Similarly, the World Bank Group has emerged

Table 1 Climate Change Conventions, Agreements, and Key Topics

Year	Agreement	Key Topics
1992	UNFCCC	Established foundational principles for climate finance, including common but differentiated responsibilities, the obligation of developed countries to provide new and additional financial resources to developing countries, and the provision of technology transfer. However, due to the absence of detailed mechanisms for the mobilization, allocation, and use of climate funds, progress in implementation has been limited. Following the entry into force of the Convention in 1994, the Global Environment Facility (GEF) became the first operational international financial mechanism under the UNFCCC framework
1997	Kyoto Protocol	A legally binding instrument under the UNFCCC, the Kyoto Protocol, set out enforceable measures to achieve the Convention's long-term goals. It introduced three market-based mechanisms to help developed countries meet their emission reduction targets cost-effectively: Clean Development Mechanism (CDM), International Emissions Trading (IET), and Joint Implementation (JI)
2001	Marrakesh Accords	Established the implementation rules for the Kyoto Protocol, including operational details for the three market mechanisms
2009	Copenhagen Accord	This outcome of COP15 proposed short- and long-term finance commitments by developed countries, including USD 30 billion for the period 2010–2012 and USD 100 billion per year by 2020. It also introduced the plan to establish the Green Climate Fund (GCF) as a financial mechanism under the UNFCCC to support developing countries in climate governance
2010	Cancun Agreements	Officially established the GCF to provide long-term financial support for developing countries in their climate response efforts. The GCF became a major multilateral fund for mobilizing and allocating climate-related finance

(continued)

Table 1 (continued)

Year	Agreement	Key Topics
2015	Paris Agreement	As the second legally binding climate accord under the UNFCCC after the Kyoto Protocol, the Paris Agreement marked the first time both developed and developing countries committed to collective action on climate change. It required developed countries to provide at least USD 100 billion annually by 2020 to support climate actions in developing countries. The Agreement also included provisions for international carbon markets and established rules for carbon emissions cooperation
2022	Sharm El-Sheikh Implementation Plan	Highlighted that the global transition to a low-carbon economy would require annual investments of USD 4 trillion to USD 6 trillion. The conference emphasized the central role of adaptation funds in the climate finance system and recognized the importance of the LDCF and SCCF in supporting climate action in the least developed countries
2023	UAE Consensus	At COP28, parties conducted the first Global Stocktake under the Paris Agreement and reached a consensus to transition away from fossil fuels, known as the UAE Consensus. On the finance front, the conference achieved major milestones, including operationalizing the Loss and Damage Fund and identifying initial funding sources, securing USD 3.5 billion in additional resources for GCF's second replenishment, and facilitating further contributions to the LDCF and SCCF. With support from BlackRock, the UAE announced an initial USD 30 billion commitment to ALTÉRRA, a newly established climate fund. ALTÉRRA is poised to become the world's largest private climate investment vehicle, with ambitions to mobilize an additional USD 250 billion in global capital by 2030 for climate action

as a leader in climate credit. In fiscal year 2022, it allocated $31.7 billion—36% of its total lending portfolio—to climate-related investments, surpassing the 35% climate finance target set in its 2021–2025 Climate Change Action Plan. At COP28, the World Bank Group pledged to raise this share to 45% by fiscal year 2025 (July 1, 2024, to June 30, 2025).

Despite this progress, climate credit faces several systemic challenges, with sovereign debt distress in developing countries being among the most pressing.

According to the International Monetary Fund (IMF), roughly half of all low-income countries are already in or at high risk of debt distress, significantly limiting their capacity to finance climate action. Lidy Nacpil, a representative of the Asian Peoples' Movement on Debt and Development, has argued that climate finance should not be loan-based, as this exacerbates debt burdens (United Nations 2024). This underscores the urgent need for the international community to explore more diversified mechanisms, such as grants, technical assistance, and innovative financial instruments, to ensure the effectiveness and equity of global climate action.

3.1.2 Operational Models and Case Studies of Climate Credit

China's domestic experience with climate credit began with the 2021 release of the Climate Investment and Financing Pilot Program jointly issued by the Ministry of Ecology and Environment and eight other national agencies. This policy directive encouraged local governments to explore innovative financial tools for climate action. Following the selection of the first cohort of 23 pilot cities in 2022, provinces such as Zhejiang, Guangdong, and Yunnan introduced dedicated climate credit products tailored to regional needs.

The operational model for climate credit typically follows a four-step framework: project identification → environmental risk assessment → preferential financing → ongoing monitoring and supervision. This model integrates both policy incentives and market-based mechanisms. First, projects are screened for compliance with environmental standards, focusing on areas such as clean energy (e.g., solar and wind) and energy efficiency (e.g., green buildings, industrial upgrades). During the selection process, projects undergo environmental benefit assessments (e.g., expected carbon reduction levels) to prevent greenwashing. Second, banks conduct risk assessments, examining potential environmental liabilities such as pollution or ecological degradation. High-risk projects may be rejected or subject to remediation requirements. Based on these evaluations, banks apply differentiated pricing strategies, offering discounted interest rates (typically 10%–20% below the benchmark rate) to projects with high environmental ratings. Third, once approved, climate credit funds are disbursed through dedicated accounts to ensure that loans are strictly allocated to eligible projects. Loan terms are also aligned with project lifecycles. For instance, wind energy projects may receive financing with tenors of 10 to 15 years. Finally, banks conduct post-lending monitoring to ensure continued environmental compliance. This includes periodic reviews of energy consumption and emission reduction data. Breaches of environmental commitments may trigger penalty clauses, such as interest rate increases or early loan recall.

The Egypt Suez Canal Green Hydrogen/Ammonia Facility (SCZONE Green Hydrogen/Ammonia Facility), credit-backed by the European Bank for Reconstruction and Development (EBRD), the Green Climate Fund (GCF), and the European Union (EU), stands as the world's largest climate-finance-supported green hydrogen project (European Bank 2024). The facility achieves carbon–neutral production through renewable energy-powered electrolysis: solar/wind power plants within the

Suez Canal Economic Zone (SCZONE) directly supply electricity to electrolyzers, splitting water to produce green hydrogen, which is then synthesized into green ammonia.

The project's climate finance mechanism operates through three core approaches. First, the project employs low-cost financing to catalyze technology investment: the Green Climate Fund (GCF) provides a USD 25 million concession loan (at rates 1.5% below market) to cover procurement of critical equipment like electrolyzers, while EU grants fund technical design to mitigate early-stage risks. Second, it establishes multilateral risk-sharing: the European Bank for Reconstruction and Development (EBRD) leads commercial lending, creating a layered financing structure with GCF capital—where GCF absorbs first losses if the project incurs deficits due to technological or policy volatility, thereby attracting private capital participation. Third, the mechanism integrates climate finance incentives, green hydrogen production costs, and carbon market revenues through tailored financial instruments (results-based lending + carbon price linkages), overcoming green hydrogen's economic viability challenges and providing a replicable model for decarbonizing emerging industrial projects.

3.1.3 Application of Credit Financing in Energy Transition

Credit financing serves as a critical enabling of the energy transition by mobilizing long-term capital for clean energy deployment, energy efficiency improvements, and grid modernization. Its key application areas include financing for renewable energy infrastructure, transition funding for high-emission industries, and the digitalization and smartification of power systems.

On one hand, renewable energy infrastructure projects often require high upfront capital expenditures and have extended payback periods. These conditions deter traditional commercial capital due to tenor mismatches and elevated project risk. Climate credit, with its long-duration and low-cost structure, is well suited to bridge this funding gap and catalyze initial investment. On the other hand, credit financing acts as a strategic bridge in decarbonizing high-emission sectors. These sectors frequently encounter substantial transition costs, complex technological pathways, asset stranding risks, and uncertain short-term returns—factors that discourage private investment. By offering tailored financing tools, climate credit mechanisms help de-risk these transitions and enhance the bankability of decarbonization projects, thereby enabling large-scale capital flows into industrial transformation.

Moreover, political risk, credit risk guarantees, and related concerns remain key considerations for enterprises and investors in the energy transition process. Credit financing serves as an effective instrument for risk mitigation and allocation, enhancing investor confidence and significantly reducing the perceived risk for both investors and lenders. This is particularly critical in regions with political and economic instability or weak rule-of-law environments. As such, credit financing plays a pivotal role in mobilizing private capital for clean energy projects in emerging markets.

3.2 Climate Bonds

3.2.1 Distinctions Between Green Bonds and Climate Bonds

Green bonds and climate bonds are both categorized as sustainable finance instruments designed to fund environmentally beneficial projects. However, they differ in definition, scope, standard frameworks, and areas of emphasis.

First, in terms of definition and scope, green bonds are specialized debt instruments used to finance or refine new and existing projects that yield environmental benefits. According to the 2021 revised edition of the Green Bond Principles (GBP) published by the International Capital Market Association (ICMA), proceeds from green bonds must be allocated exclusively to projects that meet stringent environmental eligibility criteria (ICMA 2021). The GBP outlines that qualifying projects must contribute to both the achievement of net-zero emissions and the protection of ecological integrity (ICMA 2021). Eligible investment areas include, but are not limited to, renewable energy development (e.g., wind and solar), pollution prevention and control (e.g., industrial wastewater treatment), sustainable water resource management (e.g., water-efficient irrigation), biodiversity conservation (e.g., establishment of protected areas), and energy-efficient technologies in green buildings (e.g., LEED-certified infrastructure) (ICMA 2021; Climate Bonds Initiative 2022).

In contrast, the Climate Bonds Standard developed by the Climate Bonds Initiative (CBI) focuses exclusively on projects related to climate change mitigation and adaptation (CBI 2022). This renders green bonds applicable to a broader set of environmental objectives than climate bonds. According to the World Bank's 2022 Green Bond Market Development Report, green bonds finance not only climate mitigation and adaptation projects but also initiatives aimed at general environmental protection (World Bank 2022). This broader scope provides issuers and investors with greater flexibility in project selection and better alignment with the Paris Agreement and the United Nations Sustainable Development Goals (SDGs) (UNEP 2021). Data from Bloomberg New Energy Finance (BNEF) in 2023 indicate that approximately 65% of global green bond proceeds were allocated to renewable energy projects, 15% to green buildings, 10% to clean transportation, and the remaining 10% to pollution control, water management, and other environmental sectors (BNEF 2023).

3.2.2 Role of Climate Bonds in Financing Green Projects

First, climate bonds help reduce the cost of capital for green projects. As a cornerstone instrument in climate finance, climate bonds offer targeted funding and embedded risk mitigation mechanisms (Edeminam and Akpasoh 2023; Ngwenya and Simatele 2020), making them attractive to institutional investors (Fatica and Panzica 2021). Through mechanisms that spread costs across generations, they facilitate sustained climate action (Heine et al. 2019). The climate bond market has expanded concurrently in developed and emerging economies (Alonso-Conde and Rojo-Suárez 2020), serving as a pivotal funding channel for sustainable infrastructure in countries such as China and Brazil (Deus et al. 2021). In several cases, the capital raised via climate bonds has surpassed that of traditional climate finance (Long 2021), underscoring their strategic value in building climate resilience (Nabil 2023).

Second, the use of standardized frameworks and enhanced transparency fosters investor confidence. Climate bonds require issuers to ensure that proceeds are ring-fenced and allocated exclusively to eligible green projects. They also mandate detailed disclosure of use of proceeds, supported by independent third-party verification of both project eligibility and environmental outcomes, in alignment with international standards such as the ICMA GBP and the CBI Climate Bonds Standard. These measures significantly mitigate the risk of greenwashing. Furthermore, a harmonized project assessment and classification framework that clearly defines qualified climate projects is instrumental in avoiding market fragmentation and misalignment.

Third, climate bonds help diversify risk and enhance market liquidity. Through asset securitization, distributed green assets such as rooftop solar panels or EV charging infrastructure can be bundled into green asset-backed securities (ABS), thereby enabling small-scale investors to participate in sustainable infrastructure finance. This structure allows project developers to recover capital earlier, accelerating reinvestment in new initiatives. Moreover, the growth of secondary markets for climate bonds, supported by green bond indices and exchange-traded funds (ETFs), has improved market liquidity and encouraged greater trading activity, further reinforcing the financing pipeline for green projects.

3.3 Climate Funds

3.3.1 Types and Functions of Climate Funds: GEF, GCF, and Others

Since their inception in 1992, climate funds have undergone three major phases of development: an initial stage (1992–2001), an expansion stage (2001–2009), and a transition phase (2009–2012). From 2012 onward, climate finance entered a new phase centered on the Green Climate Fund (GCF). Due to differences in financial market development across regions, green investment funds demonstrate varying characteristics. For example, in the United States and Western Europe, green funds are mainly issued by non-governmental organizations and institutional investors, while in Japan, corporations play the leading role. Table 2 below summarizes the major global climate funds and their respective features:

Table 2 Major global climate funds

Climate Fund	Institution	Investment Scope	Key Achievements
Global Environment Facility (GEF)	Founded in 1991 by UNDP, UNEP, and the World Bank	Provides grant support for mitigation projects such as renewable energy, energy efficiency, sustainable transport, and climate-smart agriculture, as well as some adaptation projects	As of 2022, GEF had disbursed over USD 24 billion in grants across more than 170 countries, leveraging over USD 90 billion in co-financing
Least Developed Countries Fund (LDCF)	Established in 2001 under the Marrakesh Accords	Provides financial support for climate adaptation projects in the world's least developed countries	As of March 2022, the LDCF had raised USD 1.829 billion and leveraged more than USD 50 billion in co-financing, supporting projects in over 50 countries
Special Climate Change Fund (SCCF)	Launched at COP7 under the Marrakesh Accords	Supports a wide range of climate-related activities, particularly adaptation	As of March 2022, the SCCF had raised USD 355 million in funding
Adaptation Fund (AF)	Established at COP3 of the Kyoto Protocol in 2007 in Bali	Addresses gaps in climate finance for adaptation, particularly in improving access to funds for vulnerable countries	As of 2022, the fund had mobilized USD 878 million and financed 127 adaptation projects
Green Climate Fund (GCF)	Established under the Cancun Agreements at COP16 in 2010	Supports the implementation of climate policies and actions in developing countries, with a focus on providing predictable and adequate financial resources	By 2022, the GCF had secured USD 8.9 billion in pledges and financed a total of 190 projects

3.4 *Climate Insurance*

3.4.1 Types and Characteristics of Climate Insurance

Climate insurance is a financial instrument designed to manage risks associated with climate change and extreme weather events. Its primary function is to provide financial support and protection for individuals, businesses, and governments affected by climate-related disasters. The development of climate insurance is rooted in the urgent need for effective adaptation mechanisms, especially in developing countries that remain highly vulnerable to climate extremes. The United Nations Framework Convention on Climate Change (UNFCCC) aims to stabilize greenhouse gas concentrations to prevent dangerous interference with the climate system (UNFCCC 1992, Article 2). While mitigation remains a central objective, many developing countries have called for international support to help them adapt to the impacts of climate change (Jarzabkowski et al. 2019). Market-based insurance and other financial risk transfer mechanisms can serve as integral components of adaptation strategies, helping to reduce vulnerability to the direct impacts of climate change. Despite their potential, weather-related insurance markets remain underdeveloped in many low-income economies, posing risks to poverty alleviation and sustainable development goals (UNFCCC 2008). In response, the Intergovernmental Panel on Climate Change (IPCC) highlighted in its 2012 special report the urgent need for new and sustainable risk transfer solutions. Building on this, the Group of Seven (G7) launched the Climate Risk Insurance Initiative in 2015, aiming to extend insurance coverage to 400 million poor and vulnerable people worldwide (BMZ 2015).

3.4.2 Application of Climate Insurance in Agriculture and Natural Disaster Management

Agriculture is one of the sectors most sensitive to climate change (Mendelsohn et al. 1994). Farmers globally are expected to face increasingly severe challenges, including rising temperatures, erratic rainfall, and more frequent extreme weather events (IPCC 2007; 2013). Adaptation measures are thus essential and may include changes in crop varieties, soil conservation practices, and water harvesting techniques (Deressa et al. 2009; Mendelsohn and Dinar 2009; Di Falco et al. 2014). Agricultural insurance plays a critical role in enabling climate adaptation. It helps reduce financial risks associated with crop losses due to droughts, floods, hail, and other extreme events (Dolan et al. 2001; Mcleman and Smit 2006; Garrido and Zilberman 2008). As climate variability increases, so too does the demand for tailored insurance products (Mills 2009).

In the United States, the 1994 Federal Crop Insurance Reform Act marked a significant expansion of agricultural insurance. The Act introduced a mandatory participation scheme and created a "catastrophic insurance" (CAT) program, providing full government-subsidized coverage for yield losses exceeding 50% of average production. This policy led to a sharp increase in participation, with insured acreage rising to over 180 million acres in 1998, triple the figure from 1988 (Froot 2001). In France, crop insurance plays a significant role in the financial portfolio of insurance companies and has become central to agricultural risk management (Enjolras et al. 2012). The Uruguay Round Agreement on Agriculture (URAA) further supported the development of climate-related insurance by allowing government subsidies for income protection and disaster assistance programs. For instance, it permits up to 70% compensation for losses exceeding 30% of a farmer's three-year average income (URAA, Annex II, Articles 7 and 8). Italy's National Agricultural Solidarity Fund

(FSN) offers another example of climate insurance in practice. The FSN supports farmers through premium subsidies and post-disaster payments. Although premium subsidies averaged 40%, insurance coverage remained low, accounting for only 14% of total crop output value between 2001 and 2012 (Capitanio et al. 2011; 2013). Following reforms in 2003, the Italian government increased support for multi-risk insurance, raising premium subsidies to 80% and promoting direct contracts between farmers and insurers through innovative schemes (Legislative Decree No. 102, 2004). In China, climate insurance has seen notable progress. In Xinjiang, China United Property Insurance introduced a weather index insurance product that determines payouts based on specific weather parameters, streamlining the claims process and improving compensation efficiency. As of June 2024, this scheme had provided RMB 7.33 million in risk protection for 40 farming households, with nearly RMB 1.5 million in payouts, effectively mitigating the economic losses farmers faced due to natural disasters. In Sichuan's Renshou County, China Life Property and Casualty Insurance launched a catastrophic weather index insurance for agriculture, offering fast, data-driven compensation. By quantifying climate risk, the product has enabled local farmers to receive timely economic relief. From 2021 to 2023, total payouts reached RMB 6.54 million (China United Property Insurance 2024; China Life Property and Casualty Insurance 2023).

3.5 Climate Finance Operational Models: Dedicated Institutions and PPP

3.5.1 Dedicated Institutions

Dedicated institutions play a pivotal role in climate finance by integrating specialized services, mobilizing capital, mitigating risks, implementing policy frameworks, and setting technical standards. These institutions, including dedicated climate finance branches and business divisions within financial institutions, serve as critical enablers of the green and low-carbon transition and broader climate action efforts.

Dedicated institutions, such as climate-focused bank branches and specialized business units, leverage the technical expertise and project sourcing capabilities of financial institutions to deliver targeted financing support. For example, Liuzhou has established a dedicated climate finance branch that trains managers with dual expertise in environmental and financial fields to channel loans toward low-carbon projects (Han 2023). These institutions also support the creation of climate project databases to enhance information transparency between supply and demand. In Lishui, a "climate investment and financing smart platform" has been developed to improve project matchmaking efficiency through digital tools.

These institutions also help leverage public funding to crowd in private investment. In Qingdao's West Coast New Area, banks were encouraged to set up green finance divisions using fiscal funds as leverage to attract private capital into climate-related projects. New financial instruments, such as carbon credit-backed loans and carbon sink mortgage loans, have been piloted to unlock the financing potential of climate assets. A notable example includes a carbon credit pledge loan program in Pudong, Shanghai, which enabled enterprises to monetize unused carbon allowances (Mao 2024).

In addition to financing functions, these institutions often serve as policy pilots. Sichuan Province has encouraged financial institutions to establish dedicated climate units and explore mechanisms for foreign climate investment. Liuzhou has also issued Interim Guidelines for Managing the Establishment and Operation of Climate Finance Entities to support the low-arbon transition of local industries.

3.5.2 Public–Private Partnership (PPP) Models

The public–private partnership (PPP) model, originally developed for infrastructure development, has evolved into an optimized financing and implementation mechanism characterized by the principle of mutual benefit. In the context of climate-related initiatives, PPPs offer several distinctive features, including risk sharing and long-term collaboration, the introduction of advanced technologies and management expertise, and the capacity to amplify financing through leverage. These attributes make PPPs a powerful instrument for advancing climate action, though their effective implementation depends on institutional support and cross-sector collaboration.

By distributing risks between public and private actors, PPPs help reduce the overall project risk. In the energy sector, for example, private entities can participate in renewable energy infrastructure investments under PPP arrangements, while the government provides supportive policies—such as feed-in tariffs—allowing firms to assume technical risks (Ray 2025). In the Philippines, the New Clark City PPP project utilizes nature-based solutions (NbS) to manage flood risks, with both public and private partners sharing long-term maintenance responsibilities (Lu 2022). PPP arrangements also attract private-sector expertise and innovation. For instance, the European Investment Bank leveraged a PPP framework to convert landfill methane into electricity, delivering both emission reductions and energy generation benefits (UNECE 2018). In China, China National Nuclear Corporation's financial leasing arm collaborated with banks to launch a clean energy PPP project that employed market-based approaches to improve capital efficiency (Mao, 2024). Furthermore, PPPs can expand the scale of financing through structured financial instruments such as green bonds and real estate investment trusts (REITs). As of 2022, China had launched 3,135 climate-related PPP projects with a total investment of RMB 3.67 trillion (Xie, 2022). Chongqing has made significant progress in advancing climate investment and financing pilot initiatives, with intended financing totaling RMB 407.3 billion, of which PPPs account for a significant portion (Chen, 2024).

Despite their advantages, PPP models continue to face institutional and financial constraints in many developing countries. Addressing these challenges requires targeted capacity building, enabling policy frameworks, and enhanced international cooperation (Bisbey et al. 2022).

3.5.3 Innovation Models: Synergies Between Dedicated Institutions and PPP

Combining dedicated climate finance institutions with PPP mechanisms can unlock new operational models that transcend the limitations of traditional green finance. This integrated approach enables a "policy-guided, professionally serviced, and market-driven" cycle, fostering full-lifecycle project management, regional coordination, and cross-jurisdictional standard alignment.

One approach is to establish specialized PPP climate funds under the management of dedicated institutions to attract private capital. In Lishui, the Climate Finance Joint Innovation Lab designed financing products such as "hydropower value conversion loans," which apply PPP structures to support state reserve forest projects. Internationally, the UK Green Investment Bank (GIB) utilized PPP frameworks to develop tailored financing schemes for low-carbon initiatives, enhancing scalability and replicability.

Dedicated institutions can also provide end-to-end project services, from early-stage screening to exit. For example, the Lishui branch of the Agricultural Bank of China introduced a streamlined "one survey, one review, one approval" lending process for PPP projects, effectively shortening approval timelines. These institutions may also offer climate-smart solutions to private partners, enhancing resilience and ensuring long-term adaptation of PPP investments to climate risks.

Lastly, establishing cross-regional alliances among dedicated institutions and co-developing PPP project pipelines can improve coordination and mutual recognition. In China, Chongqing's Liangjiang New Area partnered with other pilot cities to launch a Climate Change Response Alliance, promoting interoperability of climate projects (Li 2024). At the supranational level, the EU's Climate Bank Roadmap promotes the harmonization of PPP standards across member states to facilitate cross-border investment in climate initiatives.

4 Successful Cases of Climate Finance: Advancing Energy Transition and Green Low-Carbon Economy

4.1 International Climate Finance Success Cases

As innovative practices in global climate governance, international climate finance initiatives are accelerating the large-scale deployment of low-carbon infrastructure and deep decarbonization of energy systems through multi-level market development and cross-regional coordination mechanisms. This section highlights representative case clusters that showcase localized financing adaptation, phased capital injections, and coordinated responses from international institutions. These cases reveal breakthrough practices in risk mitigation and demand alignment, illustrating diverse pathways in international climate finance and providing replicable financial solutions for the global green economic transition.

4.1.1 New York Issues Climate Municipal Bonds to Support Urban Resilience Infrastructure

In the United States, it is estimated that $2 trillion per year is required to maintain and repair aging infrastructure, with an additional $2 trillion annually needed to cope with accelerated climate-related disasters. Amid limited public budgets, green municipal bonds with their high credit quality and unique tax advantages have unlocked significant opportunities for green investment, sparking strong interest from city and state governments. New York's green municipal bond program exhibits the following features:

(1) Tailored investment to support medium- and long-term climate-resilient infrastructure at the state and municipal levels.

The New York State Department of Financial Services has specified the eligible investment scope for climate-resilient infrastructure under its Green Municipal Bond Financing Program. Priority areas include clean water supply, building energy efficiency, renewable energy deployment, solar installation, urban flood control systems, sustainable forestry, agriculture, and land use projects. The program's focus areas are closely aligned with New York's climate objectives and infrastructure financing needs, ensuring targeted resource allocation. Amid fiscal constraints at the state and municipal levels, the green municipal bond program has played an effective role in addressing underinvestment in climate-resilient infrastructure.

(2) Broad investor participation supported by tax exemptions and other incentives.

New York's municipal green bonds have consistently been oversubscribed by more than 20%. This strong investor response can be attributed to two primary factors. First, as a global financial hub, New York hosts a large pool of ESG-focused investors. The program requires issuers to provide annual updates on project progress and fund allocation, improving transparency, reinforcing investor confidence, and further enhancing the attractiveness of green municipal bonds. Second, the federal tax exemption policies applicable to municipal bonds also apply to green municipal bonds. In addition to enjoying the same tax benefits as traditional municipal bonds, green bonds direct capital toward environmentally beneficial projects, helping to improve local living conditions.

(3) Long maturities aligned with medium- and long-term climate infrastructure.

A defining feature of New York's municipal green bonds is their relatively long tenors. The average maturity ranges from 10 to 20 years, reflecting the alignment of long-term capital with the nature of climate-resilient infrastructure investment.

Currently, some Chinese cities have initiated the issuance of green local government bonds. However, on the whole, China's municipal bonds remain at an early stage in supporting urban climate project investments. The scope and standards for such investments require further clarification to strengthen their alignment with urban climate and environmental action goals. Concurrently, drawing on relevant experiences, there is a need to enhance investor subscription mechanisms and tax-exemption policies by developing localized incentive measures tailored to domestic contexts.

4.1.2 Indian Renewable Energy Developer Secures Phased Debt Financing to Support Sustainable Development Goals

In February 2025, Indian renewable energy developer Juniper Green Energy secured $1 billion in phased debt financing from leading financial institutions including Power Finance Corporation Limited (PFC), DBS Bank, HSBC, and the Indian Renewable Energy Development Agency (IREDA). This funding will support the company's growth and the development of its wind-solar hybrid and firm dispatchable renewable energy (FDRE) projects. The CEO of Juniper Green Energy emphasized that the financing will enable business scale-up, accelerate renewable energy deployment, and contribute significantly to India's clean energy transition.

This case exemplifies the use of staged capital disbursement to balance funding needs and risk control, providing sustainable capital support for large-scale renewable projects and highlighting the vital role of enterprises and financial institutions in national energy transition.

4.1.3 EBRD Financing Advances Bulgaria's Solar Photovoltaic Project

The European Bank for Reconstruction and Development (EBRD), in collaboration with Raiffeisen Bank International (RBI), announced in 2024 a loan package of up to €103 million to Tenevo Solar Technologies EAD for the construction and operation of a fully commercial solar photovoltaic power plant in southeastern Bulgaria. The project is expected to have a significant scale, generating over 300 GWh of electricity annually, equivalent to reducing carbon emissions by approximately 250,000 tons per year.

As a leader in climate investment and financing, the EBRD will align this project with the InvestEU Sustainable Transition Framework, ensuring sustainable investment practices comply with EU standards to guarantee low-carbon operations. Additionally, the EBRD will provide first-loss risk coverage for the Tenevo project, with senior loans during the construction phase and subordinated loans from RBI during the operational phase. The EBRD's first-loss risk guarantee mechanism will mitigate early-stage cash flow risks, providing robust financial assurance for the project's successful implementation.

The successful execution of this solar project will significantly contribute to Bulgaria's target of achieving a 34.7% share of renewable energy in electricity consumption by 2030. The substantial investment will not only drive economic growth in Bulgaria but also lay a solid foundation for the country's green energy transition.

This case demonstrates the integration of EU standards with project financing through the Invest EU Sustainable Transition Framework, ensuring both environmental benefits and financial returns. Furthermore, the EBRD's first-loss risk coverage reduces banking risk exposure, addressing the challenge of financing green projects in emerging markets.

4.2 Representative Climate Finance Cases in China

As a key practical example in global climate action, China's climate finance efforts are accelerating low-carbon transformation and green economic development in key sectors through innovative financial instruments and coordinated policy mechanisms. This section highlights typical cases featuring financial liquidity support for low-carbon transition, multi-stakeholder collaboration among government, banks and enterprises, sustainable development-linked loans, and deep empowerment by digital technologies. Together, they demonstrate China's comprehensive climate finance model and contribute valuable insights to global climate governance.

4.2.1 Medium- and Long-Term Loans Supporting Low-Carbon Transition Projects in Key Industries with Core Performance Indicators

Jiuquan Iron and Steel (Group) Co., Ltd. (hereinafter "Jiuquan Steel Group") is the earliest established, largest-scale, and most diversified modern enterprise group in Northwest China, engaged in both ferrous and non-ferrous metallurgy. Jiayuguan Hongdian Ferroalloy Company, a wholly-owned subsidiary of Jiuquan Steel Group, specializes in the production and sale of ferroalloys and non-ferrous metals. To enhance energy efficiency in ferroalloy production and achieve energy savings and carbon emission reductions, Jiayuguan Hongdian Ferroalloy Company initiated a retrofit of its outdated submerged arc furnace. The project is expected to reduce annual energy consumption by more than 1300 tons of standard coal equivalent and cut carbon dioxide emissions by over 3000 tons annually. According to the feasibility study, the project requires a total investment of approximately RMB 58 million, with bank loans of over RMB 30 million needed to fully support its construction.

To facilitate the implementation of these transition measures, in June 2023, the Jiayuguan branch of Lanzhou Bank extended a five-year, RMB 34 million medium- and long-term loan to support the environmentally friendly, fully enclosed submerged arc furnace retrofit project. This loan, the first sustainability-linked loan in Gansu Province, also qualifies for the People's Bank of China's carbon emission reduction support tool policy. This tailored loan designates the enterprise's comprehensive energy consumption per unit of product as the key performance indicator (KPI), with quantified sustainability performance targets (SPTs) linked to the loan's interest rate. A one-way benefit mechanism was established, which not only effectively reduces environmental pollution during production but also lowers the enterprise's financing costs. The loan's interest rate is approximately 110 basis points lower than the weighted average interest rate for similar loans offered by the bank. If the enterprise meets the predetermined targets during the loan term, the effective interest rate will be further reduced by 10 basis points. According to calculations by Jiuquan Steel Group's budget and finance department, this loan is expected to save approximately RMB 500,000 in interest expenses. Upon project completion and achievement of performance targets, the loan's interest rate in subsequent periods could be further reduced, potentially lowering financing costs even more.

On November 26, 2023, the environmentally friendly, fully enclosed submerged arc furnace retrofit project of Jiayuguan Hongdian Ferroalloy Company passed final acceptance and entered the thermal load production phase. The retrofitted No. 7 submerged arc furnace is projected to reduce total energy consumption by 1394 tons of standard coal equivalent annually and achieve carbon dioxide emission reductions of 3080.74 tons per year, meeting the project's designed emission reduction targets.

This project exemplifies three innovative highlights in the application of climate investment and financing to support the transition of traditionally high-carbon enterprises. First, through medium- and long-term loans integrated with key performance indicators and quantified sustainability performance targets, the project ensures sustained transition outcomes. Second, the project design combines transition finance products with the People's Bank of China's monetary policy tools, reducing the cost of transition activities and achieving a "win–win" for financial institutions and enterprises. Third, continuous monitoring of enterprise transition outcomes is conducted through regular supervision and evaluation of key performance indicators across the project's lifecycle, enabling financial institutions to track and assess the enterprise transition progress to ensure the quality and effectiveness of the transition while preventing transition-washing.

4.2.2 Government-Bank-Enterprise Collaboration Advances Shenzhen Xiaomeisha's "Internet + " Smart Energy Project

The Xiaomeisha community, located in Shenzhen's Yantian District, has consistently received accolades since 2006, including designations as a "Green Community," "Model Village for Comprehensive Environmental Improvement," "Top 100 Community," "Beautiful Natural Community," and Guangdong Province's "Livable Community." As a pilot for Shenzhen's near-zero carbon emission zone, Xiaomeisha is focusing on projects such as the "Internet + " Smart Energy initiative to explore pathways to near-zero carbon emissions, further strengthening its low-carbon advantages.

The project centers on the development of a centralized cooling and heating system (including cold storage, air conditioning and heating, and domestic hot water systems) and a comprehensive energy management system. Upon completion, the project will employ regional centralized cooling and heating combined with cold and heat storage technologies. Through operational oversight of a "24-h" energy supply system and scientifically adjusted real-time operational strategies, it will provide safe, reliable, and high-quality energy services to users, ultimately achieving the goals of multi-energy complementarity, integrated optimization, and green, efficient smart energy utilization.

The project enhances energy utilization efficiency across various sectors in the Xiaomeisha region, offering a replicable model for smart energy management and unlocking commercial value. To support this, the Shenzhen Municipal Ecology and Environment Bureau and the Shenzhen Branch of the People's Bank of China have repeatedly organized government-bank-enterprise matchmaking events to promote green low-carbon projects, such as climate investment and financing initiatives, to commercial banks. In 2023, the project secured a loan from a Shenzhen branch of a bank, with a 15-year term and an interest rate lower than the five-year Loan Prime Rate (LPR) for the same period.

Following project implementation, compared to conventional energy supply models, the project achieves energy savings of over 12%, with an estimated annual reduction in carbon dioxide emissions exceeding 907 tons, delivering significant economic value and environmental benefits.

This case demonstrates that government-bank-enterprise collaboration is key to addressing challenges in climate investment and financing. Through policy guidance from government and financial institutions, coupled with the establishment of bank-enterprise matchmaking platforms, resources can be precisely aligned with green low-carbon projects.

4.2.3 Dalian Jinpu New Area Implements Dual-Linked Sustainability Loan for Distributed Photovoltaic Project

Dalian Jinpu New Area, one of the 23 national climate investment and financing pilot regions in China, has consistently pursued a goal-oriented approach since its designation as a pilot. It has actively integrated resources, advanced innovation in climate finance products and services, and continuously explored differentiated and distinctive pathways for climate investment and financing. Against this backdrop, the Huicheng Aluminum Distributed Photovoltaic Construction Project in Dalian Jinpu New Area, under the guidance of the Jinpu New Area (Jinzhou) Ecology and Environment Bureau and the Jinpu New Area branch of the People's Bank of China, secured the nation's first climate investment and financing pilot dual-linked sustainability loan from the Dalian Branch of Industrial Bank. This loan provides RMB 16.3 million in medium- and long-term funding support for the project.

As an innovative green financial instrument, the sustainability-linked loan ties the loan's interest rate to the borrower's cumulative installed capacity of distributed photovoltaic projects and the number of projects certified with the Renewable Energy Green Power Certificate issued by the National Energy Administration. Additionally, it is linked to multiple key performance indicators, including the enterprise's pollution and carbon emission reduction outcomes, energy consumption per unit of output, and ESG ratings. Through a dynamic pricing adjustment mechanism, the loan incentivizes the borrowing enterprise to optimize its industrial structure and achieve sustainable development.

Upon project completion and target achievement, annual reductions are estimated at 10,296.78 tons of CO_2, 3560.49 tons of fossil energy saved, 1.25 tons of SO_2, 1.88 tons of NOx, and 0.27 tons of particulate matter emissions, demonstrating substantial climate finance impact.

This dual-linked sustainable development loan marks a significant innovative practice within the climate investment and financing pilot framework. It not only supports enterprises in scaling up distributed photovoltaic industry development and optimizing financing costs but also provides valuable practical experience for other regions to implement climate investment and financing pilots and contribute to the national dua-carbon goals.

4.2.4 Carbon-Finance Digital Intelligence Platform Empowers Green Transition Finance

Data and information are foundational to the development of a transition finance system and to the effective operation of transition markets. In climate investment and financing practices, the processes of carbon emissions accounting, green financial product administration, and carbon asset management require substantial technical expertise and supporting management systems. These requirements not only increase operational costs but also present challenges for financial institutions seeking to scale up green financial deployment. In response to emerging development needs, China is advancing a new pathway that integrates digital and green transitions. Leveraging digital technologies to support green transition finance has become a prevailing trend.

iGreenBank (Guangzhou) Information Technology Co., Ltd. (hereinafter "iGreenBank"), a subsidiary of Green and Sustainable Development Co., Ltd., has pioneered the development of the "Carbon-Chain-Finance" Smart Carbon-Finance Service Platform (hereinafter "Carbon-Chain-Finance"), a data-driven digital platform that integrates carbon projects, carbon assets, and green financial services through the use of big data, artificial intelligence, and blockchain technologies.

The Carbon-Chain-Finance platform has already been deployed and scaled in multiple application scenarios involving green and low-carbon transitions and sustainable development, with notable results. For instance, leveraging the Carbon-Chain-Finance system, iGreenBank developed the GD Power Carbon-Chain-Finance Platform for GD Power Development Co., Ltd. through a Software-as-a-Service (SaaS) model. This platform encompasses four core functionalities: First, carbon emission reduction project (library) management and certification provide enterprises with services such as carbon emission reduction project information management, operational data management (including carbon emission reduction accounting, environmental benefit assessments, Monitoring, Reporting, and Verification [MRV]), carbon account management, and intelligent green certification. Second, green credit management and certification offers enterprises online green credit management, intelligent certification, online report generation, loan-term monitoring and verification, and related data analytics. Third, green bond management and certification provide enterprises online green credit management, intelligent certification, online report generation, loan-term monitoring and verification, and related data analytics. Fourth, carbon asset management supports enterprises with online green bond management, intelligent certification, monitoring and verification during the bond term, and related data analytics.

This platform has addressed a series of challenges for China Datang, including low efficiency and high costs in managing carbon emission reduction projects and carbon data, unclear carbon asset valuation, and complex green finance certification processes. Through the platform, China Datang manages renewable energy projects online, with green bond financing exceeding RMB 6 billion, green credit financing surpassing RMB 5 billion, annual carbon emission reductions of approximately 15 million tons, and an annual carbon asset value exceeding RMB 800 million.

This case illustrates the establishment of a carbon and green finance integrated management platform, leveraging the combined strengths of state-owned financial investment platforms and leading green fintech and carbon neutrality digital service providers. It mitigates information barriers for financial institutions, ensures the accuracy of accounting data, and significantly enhances the quality and efficiency of green financial services.

4.3 The Role of Climate Finance in Promoting Green Energy Projects

4.3.1 Reducing Financing Costs and Ensuring Capital Availability

Climate finance plays a crucial role in directing capital flows towards green and low-carbon industries and projects, supporting the development of renewable energy, energy efficiency, and emission reduction initiatives. This facilitates economic structural upgrading to effectively address climate change and mitigate climate risks (Chen 2025). Through innovative financing methods and financial instruments, enterprises gain smoother access to funding with reduced financing costs. In 2024, various pilot regions in China promoted the deployment of innovative financial tools and financing models to channel more funds into climate mitigation and adaptation sectors, advancing the achievement of the dual carbon targets. Examples include: Shanghai Pudong Development Bank's first CCER-linked loan in Shenzhen; Hubei carbon market's inaugural trilateral carbon quota financing among banks, securities, and enterprises; two distributed photovoltaic sustainable development dual-linked loans in Jinpu New Area, Liaoning Province; Industrial Bank's Taiyuan branch issuing Shanxi's first carbon footprint-linked loan; Nansha New Area, Guangdong Province issuing the nation's first climate-friendly corporate green bond; and Bank of Communications' Shanghai branch successfully issuing Shanghai's first transition bond in the interbank market. These measures have substantially lowered enterprise financing costs and secured funding for green low-carbon projects.

4.3.2 Establishing Risk-Sharing Mechanisms to Reduce Project Risks

Climate finance mitigates project risks by implementing risk-sharing mechanisms, involving diverse financial institutions, enhancing corporate climate disclosures, and providing policy guarantees. These efforts reduce investment risks and attract capital for green low-carbon projects. Chinese pilot regions design flexible financial contracts and policies that translate abstract emission reduction goals into quantifiable, tradable cash flows while safeguarding the interests of all participants, thereby lowering project investment risks and encouraging social capital participation. For instance, the Shenzhen government explicitly allows CCER project developers to secure loans against future emission reduction revenues. Financial institutions manage risks through dynamic interest rate adjustments—if project outcomes fall short, higher rates compensate capital providers. Moreover, recent policies, standards, and guidelines issued by government, investors, and other entities have promoted enhanced corporate climate disclosure, thereby increasing transparency and mitigating investor risk.

4.3.3 Incentivizing Corporate Green Transformation and Advancing Green Technology

Climate finance incentivizes enterprises to pursue green transformation by means of policy guidance and financial rewards, promoting investment in green technology research and development. For example, Lishui in Zhejiang Province, a climate finance pilot, developed the Lishui Climate Finance Project Inclusion Guidelines to steer corporate green transformation. Zhejiang Yuxin Industrial Co., Ltd. upgraded products and optimized processes by replacing restricted equipment with mature green technology, achieving capacity-equivalent substitution. Following project inclusion, certain local commercial banks moderately lowered credit access thresholds for the steel sector, providing over RMB 10 million in credit support. Simultaneously, enterprises were actively encouraged to apply for ultra-long-term national bond projects, receiving RMB 67.53 million in backing.

4.4 Application of Climate Finance in Adaptation Projects

The Climate Investment and Financing Pilot Program clearly defines the support scope of climate finance to include both mitigation and adaptation. Adaptation projects primarily cover critical sectors such as agriculture, water resources, forestry, and ecosystems. These sectors face challenges including unstable funding sources and underdeveloped adaptive technologies. Climate finance, leveraging integrated financial service advantages, offers diversified financial support for low-carbon transition in key adaptation sectors such as agriculture and water resources, thus addressing sustainability challenges and injecting vital capital.

4.4.1 Water Resources Sector

Climate finance supports the water resources sector mainly through innovative financial products and the establishment of project databases to meet capital needs and promote green technology transformation. First, innovative products such as water rights loans and water resource-specific bonds address funding demands for green transformation. Water rights loans are a novel green finance instrument allowing enterprises to pledge water withdrawal rights or water use rights as collateral for bank loans. This approach has been piloted in locations like Yangzhou, where multiple companies received loans under the water rights loan scheme to support the replacement of water-saving and energy-efficient equipment. Some financial institutions raise funds via special bonds; for example, the Agricultural Development Bank of China issued carbon–neutral bonds adhering to the Guidance on Taxonomy of Climate Investment and Financing Projects (T/CSTE0061–2021) with proceeds allocated to forestry greening, regional water conservation, and soil preservation. Second, establishing climate finance project databases broadens enterprise financing channels and alleviates financing difficulties. In 2025, Changchun's municipal-level Climate Change Finance Project Reserve Database approved a RMB 364 million, 20-year loan with an interest rate of 3.9% per annum for the "Phase I comprehensive river management of Xinlicheng and Shitoukoumen reservoirs" by Changfa Water Resources Company, backed by the Agricultural Development Bank of China. This key municipal infrastructure project aims to dredge and excavate rivers, reinforce slopes, and restore ecosystems to reduce pollutant impact on water sources, improve ecological environment, maintain social stability, and secure sustainable regional economic growth.

4.4.2 Agriculture Sector

Although China has begun promoting climate-resilient agriculture, the overall development remains nascent, with climate risk hedging and adaptive effects still insufficient. In this context, financial institutions innovate comprehensive service models focusing on enhancing agricultural resilience by flexibly providing long-term, stable, large-scale, and low-cost financing alongside technical assistance and capacity building to support sustainable agricultural development.

The Agricultural Bank of China has researched and developed an integrated "commercial bank + investment bank" service that flexibly connects direct and indirect financing markets to meet diversified financing needs in agricultural sustainability. This encompasses credit, bonds, equity, funds, advisory services, and more, enriching climate finance capital supply for agriculture. Specifically: First, credit-dominant financing supports frequent and large-scale capital demands during agricultural transformation. Second, bond issuance and other direct financing provide more flexible and market-driven capital, broadening financing channels and increasing visibility in public markets. Third, equity-type financing tools such as industrial and sci-tech innovation funds address capital shortages, supporting agricultural project construction and agricultural technology R&D, promotion, and application. Fourth, asset securitization mobilizes existing assets to enhance liquidity and profitability, promoting green transformation of agricultural enterprises.

5 Challenges and Future Development of Climate Finance

5.1 Current Challenges in Climate Finance

5.1.1 The Climate Finance Gap Remains a Critical Constraint

According to the Global Landscape of Climate Finance 2025 released by the Climate Policy Initiative, the world will require at least USD 6.3 trillion in annual climate investment between 2024 and 2030. To meet the 1.5 °C global warming target, annual needs will rise to USD 7.1 trillion between 2024 and 2030, and further escalate to USD 9.2 trillion annually from 2031 to 2050. While recent years have witnessed a marked acceleration in global climate finance flows, the current volume remains insufficient to achieve internationally agreed climate objectives (Wang et al. 2025).

At present, approximately half of global climate finance originates from public sources, including national fiscal budgets and development finance institutions. Private sector participation remains limited, resulting in a narrow funding base (Lee and Fang 2025). Moreover, the dominance of debt-based instruments, such as green bonds and green loans, is evident, with debt financing accounting for 70% of total climate finance. In contrast, equity-based financing comprises just 30%. The application of other financial instruments, such as derivatives or insurance products, remains minimal. The lack of diversified instruments continues to hinder the ability of capital markets to respond effectively to the full spectrum of climate investment needs.

Globally, climate financing for emerging markets and developing economies primarily originates from international channels, yet significant funding gaps persist (Zhang et al. 2025). Between 2021 and 2022, only USD 30 billion was directed to the least developed countries and regions, accounting for less than 3% of the global total. These vulnerable economies require substantial capital to address climate change, which must be mobilized through large-scale investment and financing activities (Briera and Lefèvre 2024).

5.1.2 Inadequate Standards and Measurement Frameworks

Robust standards for climate finance accounting, benefit measurement, and disclosure are essential to enabling effective climate investment. However, current national and international frameworks remain underdeveloped, impeding accurate assessment of climate finance flows and their outcomes (Watson 2022). First, most countries have yet to establish dedicated statistical and verification systems for climate finance. Existing data collection practices primarily focus on green finance in general, without disaggregating climate-specific components. As a result, critical baseline data for climate finance remains fragmented or incomplete. Second, there is no unified national standard for climate impact accounting or climate-related information disclosure. While taxonomies for green industries and projects exist, they are not fully aligned with the scope and criteria of climate investment, complicating the identification and evaluation of eligible projects. Enhancing the granularity of project taxonomies is therefore a pressing need (Weikmans and Roberts 2019). Some international organizations have proposed methodologies for measuring climate benefits, but these approaches vary widely. There is a pressing need for harmonization to improve the consistency and operability of climate impact assessments. In addition, the absence of standardized disclosure and performance evaluation frameworks makes it difficult to monitor the implementation and effectiveness of climate-financed projects over time.

5.1.3 Underdeveloped Investment and Regulatory Mechanisms

A sound enabling environment is fundamental to accelerating climate-aligned investment. Legal, regulatory, and institutional frameworks are essential to ensure the scalability and credibility of climate finance. Yet, in most countries, climate investment and financing mechanisms remain in their infancy. Many national systems are still at the pilot stage, and there is a lack of clear, uniform legislation or regulatory guidelines specific to climate finance (Fan et al. 2025). Currently, regulatory oversight of climate investment is often subsumed under general green finance regulations. Detailed rules and enforcement mechanisms tailored specifically to climate finance are lacking. Moreover, companies involved in climate projects frequently disclose insufficient information, making it difficult to verify whether capital is actually directed toward activities aligned with mitigation or adaptation goals. This opacity raises the risk of greenwashing, where environmental benefits are overstated or fabricated, thereby undermining the credibility of climate finance initiatives (Jiang and Ma 2025). The lack of verifiable information about the climate finance projects deters both institutional and private investors, limiting the inflow of social capital into climate finance markets.

5.2 *Innovation in Climate Finance*

5.2.1 Innovation in Financial Products and Services

Innovation in climate finance is increasingly reflected in the diversification and specialization of financial instruments and service models. It encompasses three primary dimensions: the development of diversified financial instruments, breakthroughs in climate insurance, and the expansion of green bonds and asset securitization mechanisms.

One notable example is Nansha District in Guangzhou, which has introduced a series of pioneering climate finance products, including China's first climate guarantee-backed loan, the EU carbon tariff index insurance, carbon–neutral asset-backed commercial papers, and the green climate loan. These instruments are designed to lower financing thresholds, mitigate investment risks, and offer differentiated interest rates tailored to the specific needs of low-carbon projects. Emerging carbon-related financial derivatives, such as carbon quota pledge financing and forest carbon sink–backed loans, are also gaining traction. In Beijing's Miyun District, for instance, carbon sink assets have been monetized to unlock the latent ecological value of forestry resources.

Climate insurance products such as extreme weather index insurance and climate disaster compensation insurance are also being rolled out to help enterprises and local governments hedge against physical climate risks. The EU carbon tariff index insurance launched in Nansha offers a novel risk-transfer mechanism that protects businesses from volatility stemming from international carbon border adjustment measures.

The green bond market continues to expand, with proceeds earmarked for renewable energy, low-carbon transportation, and other mitigation-related sectors. In 2024, Nansha issued China's first batch of climate-friendly corporate green bonds through the public market, attracting social capital for climate-aligned investments. Asset securitization schemes, such as ABS products backed by wastewater treatment revenue, enable early monetization of future cash flows, accelerating capital recovery and enabling reinvestment. Faced with escalating climate risks, financial institutions can adapt by transferring such risks to specialized securitization entities—such as mortgage-backed securities issuers—that have the expertise to structure diversified asset pools. (Kahn et al., 2024)

5.2.2 Innovation in Technologies and Decision-Support Tools

Technological innovation in climate finance comprises three critical areas: the further application of blockchain technology, artificial intelligence and big data–driven decision-making, and the establishment of digital carbon accounting systems.

Blockchain technology has emerged as a key enabler for enhancing the credibility and transparency of climate finance. By ensuring the uniqueness and immutability of carbon credits, blockchain addresses longstanding challenges in traditional carbon markets such as double-counting and fraud. Digital carbon credit platforms employing smart contracts can automatically verify emission reductions, thereby reducing transaction costs (Miller 2023). Moreover, blockchain can record carbon emissions throughout the full product lifecycle, supporting corporate carbon–neutral supply chain management and enabling energy traceability. For example, tracking the source and consumption of renewable electricity (Patel et al 2023). Distributed ledger technology also facilitates decentralized climate financing platforms, allowing micro-investors to participate in climate projects and thereby broadening the capital base (Akgiray 2019).

The use of big data enables the integration of climate, economic, and social variables into dynamic risk modeling. Platforms such as the carbon account system developed by the Beijing Green Exchange leverage AI to monitor enterprise-level carbon emissions in real time, producing high-precision carbon accounting reports that help financial institutions improve credit risk assessments. AI can also forecast the impact of extreme weather events on asset portfolios, guiding insurance pricing strategies and the provisioning of catastrophe reserves (RMI 2024).

Lastly, the development of digital carbon accounting systems helps break down data silos and provides financial institutions with reliable green certification. For example, Taiyuan has launched China's first municipal-level carbon account platform, which uses IoT sensors to collect real-time energy consumption data from enterprises. When integrated with the government's environmental protection databases, this system enables dynamic carbon emissions accounting and regulatory oversight.

5.2.3 Innovation in Models and Mechanisms

Innovations in models and mechanisms for climate investment and financing encompass three key dimensions: exploration of distinctive pilot city models, establishment of a government-bank-enterprise collaborative ecosystem, and breakthroughs in blended financing approaches.

In the realm of pilot city model exploration, Miyun has developed an innovative Water Revitalization-Forest Utilization-Industry Promotion-Carbon Reduction framework. This model activates forestry carbon sink assets, establishes dedicated funds to channel social capital, and drives the green transformation of industries, fostering a virtuous cycle between ecological resources and financial capital. Similarly, Lishui has leveraged a digital intelligence platform to integrate climate investment and financing, creating a comprehensive digital platform and a carbon account supervision system. This initiative connects data chains among government, enterprises, and financial institutions, enabling fully online processes for project screening, risk assessment, and capital allocation. Additionally, Pudong New Area has advanced project library standardization by formulating evaluation criteria for climate investment and financing project inclusion, streamlining project selection processes, and enhancing capital allocation efficiency.

In terms of building a collaborative government-bank-enterprise ecosystem, Nansha has institutionalized regular climate investment and financing branding events to create a cooperative network characterized by "government guidance + financial institution services + enterprise participation." This framework facilitates the alignment of climate financial products with tangible projects. The Beijing Green Exchange, in collaboration with institutions such as the Pudong Development Bank, offers integrated services, including carbon asset management and green project certification, providing a one-stop solution for stakeholders.

Breakthroughs in blended financing models involve combining public and private capital to mitigate private investment risks through risk-sharing mechanisms. For instance, policy banks provide initial risk mitigation funds to attract private equity funds to invest in high-risk technology research and development projects (Nani 2024). In Zhejiang Province, policies such as increasing tolerance for non-performing loans and implementing fiscal incentive mechanisms have encouraged banks to expand the scale of climate-related credit.

5.3 *Emerging Trends in Climate Investment and Finance*

The trajectory of climate finance is being shaped by actions across government, industry, enterprise, and the financial sector. At the policy level, countries are expected to further clarify the strategic role of climate finance and establish medium- to long-term capital planning frameworks. Global mechanisms such as the Paris Agreement, climate-related corporate disclosure regulations, the advancement of green taxonomies, and other climate-aligned policies have created benchmarks for reform, sending clear and consistent signals to capital markets. In China, the Climate Investment and Financing Pilot Program has launched localized experiments and is developing standardized frameworks for project classification, information disclosure, and performance evaluation. The European Union, through its green taxonomy, has further refined capital allocation criteria, thereby enhancing market transparency. These initiatives influence investor perception of climate-related opportunities and inform how climate factors are integrated into investment decision-making.

Corporations are increasingly recognizing the importance of reducing greenhouse gas emissions across their operations, not only direct (Scope 1) and purchased energy (Scope 2) emissions but also value chain emissions (Scope 3). Although voluntary carbon markets have underperformed relative to expectations, the growing number of net-zero corporate commitments has stimulated investment in innovative emissions reduction strategies. Many of these strategies offer co-benefits such as biodiversity preservation, which in turn improves the commercial viability of climate-aligned projects.

Blended finance is emerging as a key topic despite still accounting for a small share of total climate finance. As of 2024, the global volume of climate finance stands at approximately USD 470.5 billion, significantly short of the USD 9 trillion needed annually by 2030. The shortfall is particularly acute in developing economies.

For example, sub-Saharan Africa currently receives just 1% of total global climate finance. To bridge this gap, expanding the use of blended finance, particularly by leveraging public funds to crowd in private capital, will be essential.

As climate risk awareness continues to grow, investors are increasingly aligning portfolios with climate impact considerations. This marks a broader shift in the global economy toward sustainable development pathways.

References

Akgiray V (2019) The potential for blockchain technology in corporate governance (OECD Corporate Governance Working Papers No. 21). OECD Publishing. https://doi.org/10.1787/ef4eba4c-en

Amoah A, Amoah B, Kwablah E, Asiama RK (2025) Renewable energy transition and climate finance nexus in sub-Saharan Africa. Global Environmental Change Advances 4:100013. https://doi.org/10.1016/j.gecadv.2025.100013

Bhandary RR, Gallagher KS, Zhang F (2021) Climate finance policy in practice: A review of the evidence. Climate Policy 21(4):529–545. https://doi.org/10.1080/14693062.2020.1871313

Bisbey J, Lee CB, Ryan P (2022) Integrating climate risk management in infrastructure PPPs in developing countries: key concepts, best practices, and broader considerations. WAPPP J. https://doi.org/10.55898/wappp2022_climate_risks_PPPs

Briera T, Lefèvre J (2024) Reducing the cost of capital through international climate finance to accelerate the renewable energy transition in developing countries. Energy Policy 188:114104. https://doi.org/10.1016/j.enpol.2024.114104

California Energy Commission (CEC) (2024) Electric vehicle charging infrastructure assessment—AB 2127 Report. https://www.energy.ca.gov/data-reports/reports/electric-vehicle-charging-infrastructure-assessment-ab-2127

Capitanio F, Bielza MDC, Cafiero C, Adinolfi F (2011) Does market competitiveness significantly affect public intervention in agricultural insurance: the case in Italy. Appl Econ 43(27):4149–4159.

Chen K (2025) Research on implementation pathways and strategies for promoting green low-carbon development through climate investment and financing in China. Strait Science and Technology and Industry 38(1):33–37 In Chinese

Chen J (2024) Quantifying climate risk: catastrophe insurance explores "new solutions". China Bus J. http://www.cb.com.cn/index/show/bzyc/cv/cv135229121642. Accessed 27 Oct 27 2024

Chen Y (2024) Chongqing advances climate investment and financing pilot work with intended financing reaching RMB 407.3 billion. Xinhua News. https://cq.ifeng.com/c/8ZbcvHMzpBv

Climate Policy Initiative (2024) Global landscape of climate finance 2024: insights for COP29. https://www.climatepolicyinitiative.org/publication/global-landscape-of-climate-finance-2024/

Climate Policy Initiative (2025) Global landscape of climate finance 2025. https://www.climatepolicyinitiative.org/publication/global-landscape-of-climate-finance-2025/

De Groote O, Gautier A, Verboven F (2024) The political economy of financing climate policy—evidence from the solar PV subsidy programs. Resource and Energy Economics 77:101436. https://doi.org/10.1016/j.reseneeco.2024.101436

Deressa TT, Hassan RM, Ringler C, Alemu T, Yesuf M (2009) Determinants of farmers' choice of adaptation methods to climate change in the Nile Basin of Ethiopia. Glob Environ Chang 19(2):248–255. https://doi.org/10.1016/j.gloenvcha.2009.01.002

Di Falco S, Adinolfi F, Bozzola M, Capitanio F (2014) Crop insurance as a strategy to adapt to climate change. J Agric Econ 65(2):485–504. https://doi.org/10.1111/1477-9552.12053

Dolan AH, Smit B, Skinner MW, Bradshaw B, Bryant CR (2001) Adaptation to climate change in agriculture: evaluation of options (Occasional Paper No. 26). Department of Geography, University of Guelph

Enjolras G, Capitanio F, Adinolfi F (2012) The demand for crop insurance: combined approaches for France and Italy. Agric Econ Rev 13(1):5

European Bank (2024) SCZONE green hydrogen. https://www.ebrd.com/home/what-we-do/projects.html

Fan M, Shira RK, Hunjra AI, Zhao S (2025) Climate governance actions, financing constraints, and green transformation of energy-intensive firms. Res Int Bus Financ 79:103041. https://doi.org/10.1016/j.ribaf.2025.103041

Froot KA (2001) The market for catastrophe risk: a clinical examination. J Financ Econ 60(2–3):529–571. https://doi.org/10.1016/S0304-405X(01)00052-6

Garrido A, Zilberman D (2008) Revisiting the demand for agricultural insurance: the case of Spain. In: Agricultural finance review, vol. 68, no. 1. Emerald Group Publishing Limited, pp 43–66

Garschagen M, Doshi D (2022) Does funds-based adaptation finance reach the most vulnerable countries? Glob Environ Chang 73:102450. https://doi.org/10.1016/j.gloenvcha.2021.102450

Han D (2023) Industrial Bank Liuzhou Branch awarded the title of Liuzhou climate investment and financing professional institution. Guangxi News. Accessed 27 Oct 2024 https://www.gxnews.com.cn/staticpages/20230630/newgx649eb8c6-21204009.shtml

IPCC (2007) Climate change 2007: impacts, adaptation and vulnerability. Contribution of working group ii to the fourth assessment report of the IPCC. Cambridge University Press. https://www.ipcc.ch/report/ar4/wg2/

Jarzabkowski P, Chalkias DK, Clarke DD et al (2019) Insurance for climate adaptation: opportunities and limitations. Birkbeck institutional research online

Jiang M, Ma J (2025) Policy, investment, and growth: How green finance drives corporate sustainable development? Int Rev Econ Financ 102:104326. https://doi.org/10.1016/j.iref.2025.104326

Kahn ME, Ouazad A, Yönder E (2024) Adaptation using financial markets: climate risk diversification through securitization (Working Paper No. 32244). National Bureau of Economic Research. http://www.nber.org/papers/w32244

Karg A, Gupta J, Chen Y (2025) Just energy transition partnerships: An inclusive climate finance approach? Energy Res Soc Sci 125:104103. https://doi.org/10.1016/j.erss.2025.104103

KfW Development Bank—Facts and Figures 2021, Retrieved from https://www.kfw-entwicklungsbank.de/PDF/Download-Center/PDF-Dokumente-Selbstdarstellung/2021_Sebstdarstellung_EN.pdf

Lee C-C, Fang Y (2025) Climate finance for energy security: An empirical analysis from a global perspective. Economic Analysis and Policy 85:963–978. https://doi.org/10.1016/j.eap.2025.01.006

Lee C-C, Xuan C, Wang F, Wang K (2024) Path analysis of green finance on energy transition under climate change. Energy Economics 139:107891. https://doi.org/10.1016/j.eneco.2024.107891

Lu X (2022) Accelerating private sector engagement in adaptation in Asia and the Pacific (ADB Sustainable Development Working Paper Series No. 84). Asian Development Bank

Mao GJ (2024) Pudong advances the development of a global climate investment and financing center with 12 green finance projects signed, totaling nearly RMB 10 billion. Shanghai Observer. https://finance.sina.com.cn/jjxw/2024-12-05/doc-incymitn9630771.shtml

McLeman R, Smit B (2006) Vulnerability to climate change hazards and risks: Crop and flood insurance. The Canadian Geographer / Le Géographe Canadien 50(2):217–226. https://doi.org/10.1111/j.0008-3658.2006.00136.x

MDB, IDFC (2015) Common principles for climate change adaptation finance tracking [EB/OL]. (2015–04–03). https://www.worldbank.org/en/news/feature/2015/04/03/common-principles-for-tracking-climate-finance

Mendelsohn R, Nordhaus, WD, Shaw D (1994). The impact of global warming on agriculture: a ricardian analysis. Am Econ Rev 753–771

Mendelsohn RO, Dinar A (2009) Land use and climate change interactions. Annu Rev Resour Econ 1(1):309–332

Mills E (2009) A global review of insurance industry responses to climate change. The Geneva Papers on Risk and Insurance—Issues and Practice 34(3): 323–359. https://doi.org/10.1057/gpp.2009.14

Omukuti J (2024) The need for a climate-resilient development-aligned framing of innovative climate finance. Current Opinion in Environmental Sustainability 66:101400. https://doi.org/10.1016/j.cosust.2023.101400

Patel D, Britto B, Sharma S, Gaikwad K, Dusing Y, Gupta, M (2020) Carbon credits on blockchain. In: 2020 International Conference on Innovative Trends in information Technology (ICITIIT), IEEE, pp 1–5, February 13-14, Kottayam, India

Ray SN (2025) Public–private partnerships in the development of clean energy technologies. Federal University of Agriculture

Rocky Mountain Institute (RMI) (2024) Clean energy: growing globally [Annual report]. https://rmi.org/wp-content/uploads/dlm_uploads/2024/10/rmi_annual_report_2024.pdf

Röck M, Saade MRM, Balouktsi M, Rasmussen FN, Birgisdottir H, Frischknecht R, Habert G, Lützkendorf T, Passer A (2020) Embodied GHG emissions of buildings—The hidden challenge for effective climate change mitigation. Appl Energy 258:114107. https://doi.org/10.1016/j.apenergy.2019.114107

Tan X, Peng Z, Cheng Y, Wang Y, Chao Q, Huang X, Yan H, Chen D (2025) Leveraging artificial intelligence for research and action on climate change: Opportunities, challenges, and future directions. Science Bulletin. https://doi.org/10.1016/j.scib.2025.06.035

United Nations Economic Commission for Europe (UNECE) (2018) Guiding principles on people-first public-private partnerships in support of the United Nations Sustainable Development Goals (ECE/CECI/WP/PPP/2018/3). United Nations

United Nations Framework Convention on Climate Change (2009) Draft decision -/CP.15: copenhagen accord. Conference of the parties, fifteenth session, Copenhagen, 7–18 December 2009. https://unfccc.int/resource/docs/2009/cop15/eng/l07.pdf

United Nations Framework Convention on Climate Change (2010) Decision 1/CP.16: The Cancun agreements. https://unfccc.int/resource/docs/2010/cop16/eng/07a01.pdf

United Nations (1992) United nations framework convention on climate change. https://unfccc.int/resource/docs/convkp/conveng.pdf

United Nations (2024) COP29 draft deal proposes wealthy nations give $250 billion in climate finance, November 22. https://www.globalissues.org/news/2024/11/22/38386

Wang FF, Cao Y, Wang C (2024a) Challenges and strategies for China's climate investment and financing pilot programs. Environmental Management in China 16(6):52–60 In Chinese

Wang W, Zhang P, He T (2025) Climate investment and financing, economic growth target constraints and urban green development efficiency. J Environ Manage 382:125369. https://doi.org/10.1016/j.jenvman.2025.125369

Wang Y, Liu M, Feng Y, et al (2024) The global impact of the U.S. Inflation Reduction Act. North China Finance (10): 29–34, 69. In Chinese

Watson C, ODI, Schalatek L, HBS, Evequoz A (2022) The global climate finance architecture. Climate Funds Update

Weikmans R, Roberts JT (2019) The international climate finance accounting Muddle: Is there hope on the Horizon? Climate Dev 11:97–111

World Meteorological Organization (2023) Climate change indicators reached record levels 2023. WMO. https://public.wmo.int/news/media-centre/climate-change-indicators-reached-record-levels-2023-wmo

World Resources Institute (2023) Unpacking COP28: Key outcomes from the Dubai climate talks, and what comes next. https://www.wri.org/insights/cop28-outcomes-next-steps#:~:text=The%20decision%20to%20transition%20away%20from%20%22fossil%20fuels%22

Xie F (2022) COP27 side event on climate investment and finance: "anchoring a strategic highland of dual-carbon goals and leading innovative climate investment and finance practices." Sina Finance. https://finance.sina.cn/esg/2022-11-23/detail-imqmmthc5672401.d.html

Zhang J, Li Y, Du R, Hua X (2025) Importing innovation or indigenous innovation: Evaluating the effect of climate finance on promoting environmental sustainability in developing countries. Energy Economics 149:108726. https://doi.org/10.1016/j.eneco.2025.108726

Zhou JY, Cui Y, Liu HX (2023) The impact of new climate finance trends in the US and EU on China's participation in international climate development cooperation. Environmental Economy 19:42–47 In Chinese

Monitoring, Reporting and Verification of Emission Data in China

Rong Zhu, Xuepeng Qian, and Weisheng Zhou

1 Introduction

Presently, climate change stands as the most pressing environmental challenge threatening human society (Wang et al. 2023). Over the past two centuries, rising global temperatures have shifted global warming from a scientific concern to an urgent issue at the forefront of public and policy agendas. Emissions trading, initially proposed by the US, is a promising and cost-efficient policy framework for combating global climate change. This market-based mechanism not only enforces emissions limits but also creates financial incentives for compliance, encouraging both developing and transitioning economies to adopt more sustainable, energy-efficient pathways (Matsuo 1998).

China has made substantial strides in developing its carbon market as part of its commitment to mitigating climate change. In June 2013, China launched emissions trading scheme (ETS) pilot projects across seven regions. Building on the successes and lessons from these pilots, China formally inaugurated its national carbon market on July 16, 2021, initially covering the power generation sector. With the introduction of administrative measures, this market is anticipated to expand to encompass eight major industries, including power generation, petrochemicals, chemicals, building materials, steel, non-ferrous metals, papermaking, and domestic civil aviation (Liu et al. 2022).

R. Zhu (✉) · X. Qian
Graduate school of Global Environmental Studies, Sophia University, Tokyo, Japan
e-mail: r-zhu-9o7@eagle.sophia.ac.jp

X. Qian
e-mail: qianxp@sophia.ac.jp

W. Zhou
College of Policy Science, Ritsumeikan University, Osaka, Japan
e-mail: zhou@sps.ritsumei.ac.jp

W. Zhou (ed.), *Decarbonization Policy Engineering*, World Sustainability Series,
https://doi.org/10.1007/978-981-95-5953-4_22

Despite these advancements, the first compliance period of the national carbon market has exposed critical issues in carbon emissions data integrity, drawing significant attention from regulatory authorities. To address data authenticity concerns, the authorities have implemented a top-down regulatory framework focused on rigorous monitoring, reporting, and verification (MRV) protocols. However, this centralized approach has notable limitations, including local resistance and a potential disconnect from industry-specific contexts and regional needs (Međugorac and Schuitema 2023). Uniform standards may inadequately address the diverse conditions across industries, while stringent regulations can stifle innovation, redirecting corporate focus toward compliance over the pursuit of novel emissions reduction strategies. Additionally, limited stakeholder engagement in policymaking may constrain the diversity of perspectives necessary for effective policy adaptation and implementation.

As technological advancements continue to reshape MRV systems, new models for ensuring carbon data quality are emerging. These models reflect a transition towards a more flexible and robust framework for carbon emissions data assurance. This paper employs the Multi-Level Perspective (MLP) theory, a widely adopted framework for analyzing socio-technical transitions, to construct a comprehensive model aimed at enhancing carbon emission data quality. This model is designed to ensure data integrity and outline a pathway for transitioning towards an adaptive and sustainable carbon market framework that addresses the unique challenges and opportunities within China's carbon market system.

2 Research design

2.1 Research methods

This qualitative study employs a multi-faceted research approach, combining literature review, field research, interviews, and case analysis to investigate the mechanisms of carbon data quality assurance and the transition to digital MRV systems. A comprehensive literature review of academic sources, government reports, international agency documents, and news analyses sheds light on key aspects of carbon market activities, data distortion, the MRV system, and governmental responses. To ground this study in practical insight, the author conducted field visits to four Chinese enterprises, observing emission reporting practices, including boundary verification and coal sample preparation, and analyzing documentation to capture firsthand data on enterprise compliance practices.

Semi-structured interviews with government officials, project managers, and verification agency staff revealed the quality of carbon data, challenges and incentives in emissions reporting, and regulatory practices by government bodies. These insights inform the study's case analysis of the transition from traditional MRV to digital MRV, which considers how niche, regime, and landscape factors accelerate this shift. The analysis suggests that a new multi-dimensional data quality assurance framework may replace existing approaches.

Using deductive reasoning and inferential analysis, the study applies multi-level perspective theory to examine potential advancements in policy, technology, and market dynamics within the carbon emissions data quality sociotechnical system.

2.2 Interview result

This study conducted interviews across three key sectors: government agencies, enterprises, and third-party entities. Each sector was assigned three teams, with each team interviewing two distinct groups, resulting in a comprehensive dataset derived from six unique sources (as shown in Table 1).

The findings indicate the presence of robust guidelines within the Monitoring, Reporting, and Verification (MRV) systems, which offer a clear operational framework for enterprises and third-party entities. However, the implementation of these guidelines faces significant barriers, impeding the optimal functionality of MRV systems.

In the governmental sector, authorities responsible for oversight face considerable pressures due to their supervisory roles, which are compounded by frequent understaffing. This limitation hinders effective oversight of carbon emission reporting and verification processes, affecting data reliability and adherence to guidelines.

The enterprise sector demonstrates a notable gap in both understanding and implementing the MRV guidelines for carbon emission reporting and verification. This gap, coupled with constraints in management capacity and resources, often results

Table 1 Barriers and incentives: insights from interview results

	Barrier	Incentive
Government	High pressure in supervision	Innovating regulatory tactics and methods
	Inadequate staff	Collaborate with third-party
	Imperfect guidelines	Constantly updated
Enterprise	Lack of understanding of the guidelines	Government and third-party agency training
	Objective constraints and limited capabilities	Enhance management, technology innovation
	Insufficient awareness	Active engagement and learning
Third-party	Mismatch between workload and cost	High volume of business
	Intense competition	Provide quality service
	Communication with enterprises	Support from competent authority

in incomplete or erroneous documentation, compromising the accuracy of emissions monitoring data.

In the third-party sector, entities face a competitive market with numerous similar organizations contending for market share. This competition imposes pressures on these entities, further strained by discrepancies between service fees and the substantial workloads required. Communication issues between third-party organizations and enterprises also create challenges, complicating efficient MRV system operations.

Despite these barriers, the interviews revealed positive dynamics across the sectors. The government is innovating regulatory strategies and enhancing collaboration with third-party providers, aiming to develop clearer, more practical monitoring and verification guidelines. Enterprises are progressively improving their carbon emission monitoring and management capabilities through continuous adaptation and learning initiatives. Third-party entities, under governmental supervision, maintain operational independence, ensuring objectivity in their functions. Collectively, these factors suggest a developing system for enhanced carbon emission data quality management, emerging from the interview findings.

3 Transition to a multi-dimensional carbon emission data quality assurance scheme: multi-level perspective (MLP) theory deduction

3.1 Definition of the interconnected components in the socio-technical framework of the carbon emission data quality assurance scheme

According to the socio-technical perspective, technology cannot work effectively on its own without interacting with human activity, organizations, and social structures (Geels 2002). This viewpoint highlights how social and technological variables interact to influence how technology is developed and used. As a result, technology's success cannot be only ascribed to its intrinsic qualities; rather, it also depends on how well it integrates with societal structures. As such, effective arrangements or successful technological integrations into society are difficult to disentangle from the broader social context. They are intricately woven into the intricate web of society rather than existing as separate entities (Geels 2002). This shows that comprehending the function and effects of technology requires a more comprehensive perspective that takes into account the dynamic interactions between technology and society.

The data quality assurance schemes' function is partially accomplished by linking heterogeneous parts. In Fig. 1, sociotechnical configuration elements in carbon emission data quality assurance schemes are identified.

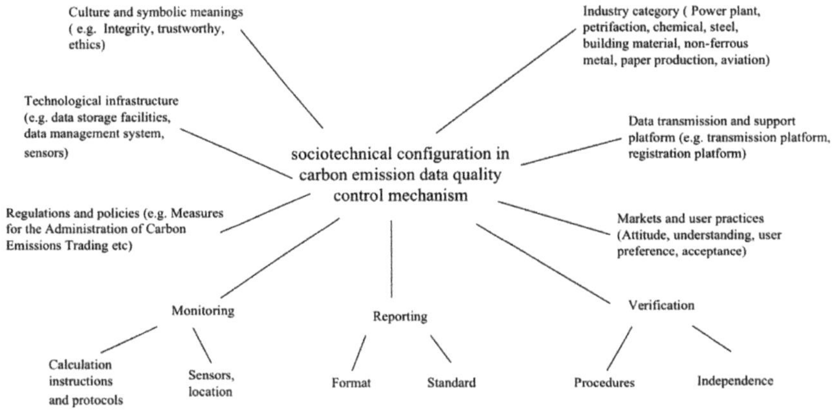

Fig. 1 Elements of the sociotechnical configuration in carbon emission data quality assurance schemes

Industry Category: Eight major industries included in national carbon market, Power generation, petrochemicals, chemicals, building materials, steel, nonferrous metals, paper, and civil aviation companies. Culture and symbolism: The authenticity of emission data influences policy decisions, scientific research, and public understanding of climate change. The Federal Data Strategy in 2019 and the Data Ethics Framework in 2020 demonstrate the importance of trustworthiness in data collection and management. These actions convey a deep commitment to respect and protect personal information, emphasizing fairness, transparency, and morality in data handling. The values upheld in these actions embody the government's resolve to advocate data privacy and information security.

The IT infrastructure: Technology infrastructure includes smart sensors, data gathering technologies, advanced storage facilities, and data management systems. These technologies ensure data integrity, efficient management, and reliability in complex, large datasets.

Data transmission and support platform: The carbon emission data quality assurance schemes utilizes platforms for seamless data transfer and access. Transmission platforms bridge data sources and storage systems, ensuring real-time data availability for analysis and decision-making. Registration platforms organize data entry and retrieval, improving system efficiency and effectiveness.

Rules and policies: The Chinese Ministry of Ecology and Environment has released policy documents to standardize and guide carbon emission monitoring, reporting and verification.

User behavior and markets: Enterprises dominate the carbon emission data quality assurance process, adhering to strict compliance requirements and understanding the importance of accurate reporting in the emission trading market. Attitude towards the assurance scheme significantly impacts report quality, with proactive participants seeking reliable data. Individual preferences and strategic aims also influence assurance scheme dynamics, with some prioritizing comprehensive data verification for corporate environmental responsibility.

Monitoring: The MRV system for carbon emissions starts with monitoring, which is crucial for measuring and collecting data on emission-generating operations. The main monitoring methods include calculation and continuous monitoring, with real-time tracking providing more precise data but potentially limited by cost and technical feasibility. Monitoring data is used for accurate reporting and verification, strengthening the carbon data quality assurance schemes. Reporting: Reporting is the second assurance scheme component, transforming raw data into structured information. It requires a consistent format for transparency and uniformity across companies.

Verification: The MRV approach relies on verification to ensure the accuracy and reliability of carbon emission data quality assurance. This process involves a thorough study of data and monitoring systems, followed by government agencies or third parties. Independent and unbiased verification institutions prevent conflicts of interest and maintain credibility, enhancing the trustworthiness of carbon data.

The carbon emission data quality assurance schemes operates in a complex socio-technical network, involving technological infrastructure, industry categories, cultural meanings, data transmission platforms, regulations, policies, markets, and the MRV framework. The Multiple-Level Perspective emphasizes co-evolution and interaction between levels, emphasizing the need for a comprehensive understanding of the system's interconnection and dynamics.

3.2 Regimes in the carbon emission data quality assurance scheme

In MLP theory, the term "regime" denotes the prevailing set of rules, practices, and technologies that govern a specific sector. Each social group possesses unique characteristics, and the actors within these groups adhere to a certain set of rules. Recognizing that these different groups abide by diverse rules allows us to distinguish among various regimes, e.g., technological regimes, policy regimes, science regimes, user and market regimes and socio-cultural regimes. Table 2 tries to present the regime under the carbon emission data quality assurance scheme.

The effectiveness of carbon emission data quality assurance depends on the interaction of various regimes. The technological regime, which includes advanced monitoring systems and artificial intelligence, could improve data accuracy. However, proper regulation and guidance are crucial to prevent misuse and compromises in data quality. The science regime provides basic methodologies for emission calculations, but if not widely accepted, they could lead to inconsistencies. The user and market regime reflects participant behavior in the emission trading market, with compliance responsibility influencing accurate emission reporting. The socio-cultural regime

Table 2 Regimes under the carbon emission data quality assurance scheme

Regime type	Description	Example
Technological regime	Accepted standards, practices and technologies used in the measurement, reporting and verification of carbon emissions	Continuous emission monitoring system, Integrated satellite-ground-air observation technology
Science regime	Scientific norms, methodologies and principles that guide how carbon emissions are calculated	2006 IPCC guidelines for national greenhouse gas inventories, 2019 refinement to the 2006 IPCC guidelines for national greenhouse gas inventories
Policy regime	Rules, regulations, and policies dictating how carbon emission should be measured, reported and verified	Guidelines for the establishment of a greenhouse gas emission inventory management system, Guidelines for the Accreditation of institutions engaged in the verification of greenhouse gas emission
User and market regime	User preference and practices, market competition rules and growth strategy of the participant involved in emission trading	Enterprises included in the first compliance cycle of the national carbon market, Competition among third parties
Socio-cultural regime	Societal and cultural norms and values that reflected in high carbon emission data quality	The principles of integrity and reliability, fairness, and justice trading standards

sets the tone for all other regimes, influencing individuals and companies to report authentic and accurate emissions.

3.3 Stability of existing socio-technical systems

The socio-technical system, with five regimes united by laws, is rigid and difficult to change. Cognitive, normative, regulative, and formal principles that affect perceptions and actions (Geels 2004) and complex interdependencies among persons and organizations in the system provide stability. Core talents can become core rigidities (Leonard-Barton 1995) and serve as invested "cognitive capital". Established firms struggle to adopt disruptive technologies (Tushman and Anderson 1986; Christensen 1997).

Cognitive capital from enterprises and government standards is needed to calculate carbon emissions. Carbon emission data quality assurance is possible with current strong verification requirements. Regulation and oversight are crucial in the early stages of the national emissions trading market, with legally enforceable contracts and technological standards ensuring system stability. Companies may not implement new technologies and schemes if their management systems can monitor and report carbon data. Current knowledge and practices have created "cognitive capital," making the switch to new ways difficult. Path dependence depends on cumulative learning and prior knowledge. If they think the current system will fix problems, they may not invest in bold solutions. Social understanding takes time to develop, demonstrating the complexity and time-intensive nature of change within socio-technical systems.

3.4 Align innovations and interconnected systems

Sociotechnical transitions can be accelerated by combining multiple innovations, such as advanced analytics, AI-based algorithms, and blockchain technologies. This integration of technologies allows for effective monitoring, reporting, and verification of carbon emissions, fostering transparency and accountability. Linkages between systems can also improve carbon emission data quality. Research by Luo et al. (2023) shows that companies implementing carbon assurance measures typically have higher quality of carbon disclosure compared to those without, particularly in countries with a strong stakeholder orientation. Carbon assurance can potentially substitute for national-level carbon regulation and societal trust. The quality of carbon disclosure also increases with the assured percentage of reported emissions and the level of carbon assurance. Therefore, changes in a single regime can potentially influence the overall system, impacting the assurance of high-quality carbon emission data. Therefore, the integration of these technologies can lead to faster transitions towards sustainable carbon reporting.

3.5 Tensions and misalignment

System transitions can be understood through tensions and misalignment. These systems' regimes fluctuate due to political and business cycles, technical trajectories, cultural movements, and industry lifecycles (Freeman and Louça 2001). Connections with other regimes generally mitigate these variations, facilitating coordination. Sometimes, these oscillations can cause misalignments and tensions, undermining sociotechnical (ST) systems when social group activities diverge. Thus, multiple social groups interact to stabilize and modify ST systems. These tensions and mismatches in sociotechnical regimes can give players freedom, especially when regulations aren't matched. ETS pilot companies, large, and stated-owned companies are better at monitoring and statistical management than non-ETS pilot companies, which are weaker at carbon management (Gao and Li 2022).

3.6 *A multi-level perspective on system innovations*

The three-level transition model suggests that stable sociotechnical regimes may limit radical novelties' breakthrough opportunities, but tensions within social groups or within ST regimes can create opportunities for successful introduction and acceptance. (Geels 2004).

(a) Increasing niche innovation momentum

Niche innovations offer unique opportunities for new technologies and methods, allowing deviations from mainstream rules. They provide learning spaces, enabling insights into technical specifications, user preferences, and public policies. (Garud and Karnøe 2001) The momentum of niche innovations impacts system stability and drives system transitions.

The Mandatory Greenhouse Gas Reporting Program (GHGRP) was initiated in 2009 by the U.S. Environmental Protection Agency to track CO2 emissions in power companies. The China Electricity Council issued a guide in 2019 mentioning online monitoring methods. However, standardized regulations and pilot test sites are not yet established in China. Despite this, CO2 online monitoring in thermal power and waste incineration industries has good application prospects and costs are comparable to calculation methods. Further pilot evaluation is needed to further develop continuous emission monitoring technology.

(b) weakening of existing ST system

Phasing out entrenched technologies and supply chains can accelerate sociotechnical transitions by allowing room for niche innovations like blockchain technology and Continuous Emission Monitoring Systems. Phase-out strategies can include standardized carbon reporting formats, digital data adoption, rewards for accurate reporting, and punitive measures for third-party verification. Transitional strategies, such as phased regulation tightening, financial penalties, and talent cultivation, are crucial for ensuring the smooth transition and promoting innovation.

(c) powerful external pressure

On the landscape level, changes usually take place slowly, e.g. cultural changes, demographic trends, broad political changes. (Geels 2002). The rapid changes in the environment are forcing established regimes to adapt and evolve, necessitating the creation of a new carbon emission data quality assurance scheme.

4 Case study: Transition from Conventional MRV to Digital MRV

Digital technologies have significantly transformed numerous industries, particularly in the area of carbon emission data monitoring, reporting, and verification. Various digital technologies, including satellite remote sensing, continuous emission monitoring systems, big data, the Internet of Things (IoT), and blockchain technology, offer substantial support for the measurement, reporting, and verification (MRV) of carbon emission data. The integration of advanced technologies facilitates precise measurement and tracking of carbon emission data, establishing a foundation for the specialized innovation of digital MRV development.

4.1 Limitations of Traditional Manual MRV Calculation Approaches

Ballassen (2015) and Boyd (2011) highlight the limitations of conventional MRV methods, which rely on bottom-up calculations and self-reported data. These methods introduce biases and inaccuracies, compromising the reliability and integrity of the collected information. Boyd (2011) emphasizes the need for measurement data availability, making MRV impossible in cases where no measurement has been conducted. Manual data entry and aggregation also contribute to potential human errors, causing delays in obtaining reliable datasets. Schaltegger and Csutora (2012) highlights the challenge of differentiation among companies in carbon management accounting research. The bottom-up approach faces difficulties in scaling up to monitor large-scale emissions across diverse sectors and geographical regions. As a result, there is a growing recognition of the need for more advanced and automated MRV systems.

4.2 Digital technology facilitates the emergence of digital MRV

In recent years, there has been a growing recognition of the potential of digital technologies to enhance the accuracy, efficiency and transparency of MRV processes.

An internet of things (IoT) based monitoring system allows for continuous monitoring of emissions from smokestacks, collecting data on various gases. This data assists companies in meeting environmental standards and making informed decisions to enhance emission output. (Martillano et al. 2017). Then, combined with near-real-time dataset, the satellite observation supports the independent, low-cost, spatially distributed and directly observed data, which are especially beneficial for areas that lack bottom-up data (Liu et al. 2023). Furthermore, a designed model of carbon emissions trading, leveraging blockchain technology, not only ensures a secure platform for carbon emissions trading but also offers a wealth of comprehensive trading reference information (Zhang et al. 2022). The utilization of big data enhances performance monitoring capabilities, eliminates human error, and fosters greater interdependencies among components in the maritime industry (Zaman et al. 2017). Therefore, with the support of these digital technologies, the digital MRV is gradually taking shape.

Table 3 reveals that several countries or regions have started implementing digital technologies for monitoring and reporting carbon emission data. Nonetheless, D-MRV are still in the phase of niche level, due to the existence of certain barriers, such as immature technology, regulatory integration, market acceptance, and implementation challenges. Immature technologies like blockchain, big data, and satellite detection require time to mature, while regulatory frameworks and standards require substantial changes. Market acceptance may be reluctance due to lack of understanding, concerns about risks, or resistance to change. Implementation challenges in remote mountainous areas and regions with limited information and communication networks can also pose challenges.

4.3 Regimes transformation

D-MRV systems have many obstacles to development and deployment. These include the lack of technical competence needed to build and implement digital solutions (Bernd 2017), the lack of relevant technology in certain jurisdictions (Vishwanath 2010), and the need for strict data security protocols (Kushilevitz et al. 2009). D-MRV development demands time and money. Traditional methods for determining emission reduction volume work well, which slows D-MRV progress (Belenky et al. 2022). Different regimes must change to remove these hurdles.

Science and technology regime.

Advances in technology cause change.Science and technology hasten the switch to digital MRV. Improved data science and environmental science algorithms offer more accurate and broad carbon measurement and analysis. The rapid expansion of blockchain, smart sensors, and big data has enabled these scientific discoveries. Technology can speed up data collecting, processing, and storage/transmission security. Thus, scientific and technical synergy is expediting the move to a more reliable, efficient, and advanced digital MRV framework.

Policy regime.

MLP theory defines policy regime as a comprehensive system of regulations, procedures, and programs that govern a legislative process or sector. Administrative norms, formal technical regulations including safety standards, subsidy and procurement programs, and policy goals are involved. Interactions between industry and government, institutional adherence to current systems, perceived government role, and policy ideals and ideas also influence the regime. Since conventional MRV will become D-MRV, the policy regime will evolve and improve. Through its Green Economy Transition Knowledge Hub, the European Bank for Reconstruction and

Table 3 Global overview of current application of digital MRV in various countries

NAME	COUNTRY	STAGE	DIGITAL TECHNOLOGY APPLIED	TECHNOLOGY APPLICATION
Rooftop Solar	India	Pilot	Smart sensors	Sensor technology measured energy usage, replacing paper processes
Electronic database and Information management systems	Uganda	Operation	Smart electric meters	Smart meters report on prepaid electricity usage used to calculate consumption for an adapted methodology
Waste to Energy	Chile	Pilot	Smart sensors, distributed ledger technology	Sensor data secured on the blockchain is provided to Chile's MRV system
Open surface, mangrove trust, wood tracking protocol	Peru, Chile, Singapore	Various	Blockchain, satellite and drone data, AI	Digital imagery of the forests is analyzed by Al algorithms to identify changes to the forest and conduct remote monitoring
Sylvera	Global	Operation	Satellite, AI, drones, LiDAR	Using satellite and LiDAR data in combination with AI to rate nature-based carbon credit projects
European bank for reconstruction and development D-MRV system for renewable energy	Kazakhstan, Jordan	Pilot	Smart sensors, cloud computing	Near real-time digital data acquisition and processing in the cloud, immutable storage of raw data, automation of verification process

Source Belenky et al. 2022.

Development (EBRD) launched the Protocol for Digitalized MRV (D-MRV Protocol) with support from the Spanish Office of Climate Change and the Joint Multilateral Development Banks (MDB) Working Group on Article 6. This Protocol promotes climate community trust in D-MRV solutions (Autenrieth 2020).

Market regime.

An effective MRV system is crucial for the success of the developing international climate regime (Aldy 2014). For success, the MRV system must be structured to maximize benefits while minimizing costs associated with this information (Barrett et al. 2015). It is important to recognize that information can incur significant costs, both in terms of direct production expenses and the potential limitations on personal autonomy and decision-making that may arise from its disclosure (Frederick 2011). The implementation of digital information technology enhances the accuracy and comprehensiveness of MRV, although it may also increase associated costs (Barrett et al. 2015). Consequently, profit-oriented organizations may postpone or decline the implementation of D-MRV owing to financial factors. It is plausible that a segment of risk-tolerant investors may acknowledge the potential of D-MRV and be inclined to explore it. The success of this subset of early adopters may result in a transformation of the market regime. The adoption of D-MRV may become more prevalent due to its demonstrated success and inherent advantages, notwithstanding initial concerns regarding cost implications.

The sociocultural regime.

Due to initiatives like the Gold Standard, Verra, and the forthcoming Article 6.4 mechanism based on the Clean Development Mechanism (CDM), D-MRV is showing promise in pilot programs worldwide (Oberpriller et al. 2022). With its ability to reduce on-site inspections and manual data verification, D-MRV can improve data integrity and accuracy (South Pole 2020).

Standards and automated data pipelines are being connected for a fully automated carbon credit issuance system (South Pole 2020; Bürgi 2019). Recently, the CDM has replaced on-site visits with remote digital monitoring, demonstrating D-MRV's potential even when it differs from traditional methods (UNFCCC CDM 2021). The certification process for more comprehensive D-MRV approaches is still underway, and the adoption of more integrated digital monitoring and quantification platforms is still early. The Gold Standard and Verra are forming D-MRV working groups and expert networks to help them transition to digital methods. Despite some drawbacks, such as lengthy and unpredictable feedback processes to methodology changes, D-MRV has great potential. To reduce approval times and costs, review and feedback processes must be streamlined. Thus, D-MRV pilots' continued development and success show that these systems will be used in more sectors. D-MRV technologies, supported by key standards bodies, and ongoing efforts to overcome implementation barriers point to a future where D-MRV systems will be essential to fighting climate change.

4.4 Landscape level

(a) NDC requirement

The 2015 Paris Agreement broke new ground in the global fight against climate change. Each participant gave their climate change strategies and pledged to improve them every 5 years. Nationally Determined Contributions (NDCs) include many pledges. They include emission-reduction strategies, adaptation strategies, and financial and technology support. "MRV is the key to unlocking climate finance and showing progress on climate goals" (World Bank 2022). This system is crucial for tracking countries' revised climate commitments. They ensure transparency, allow evaluation of implemented measures, and hold nations accountable for their promises. MRV systems also help identify gaps and improvements in climate action strategies, guiding policy changes and improving climate change effectiveness. Thus, upgrading NDCs increases demand for reliable MRV systems.

(b) Improving emission trading liquidity

Lin and Huang (2022) found limited liquidity in China's emission market. Lack of liquidity hinders trade and price signals, which businesses need to adopt emission-reduction methods. Thus, national carbon market expansion is hampered (Lin and Jia 2019). Credit generation efficiency and rate can improve market liquidity during D-MRV emergencies. Digital platforms could revolutionize emission trading. They simplify emission trading for market participants by speeding up these processes. Efficiency boosts trade volume, market liquidity, and price signal accuracy (Belenky et al. 2022). Digital data reduces costs and improves information dependability, making carbon trading more efficient and dependable (Kengo 2023).

The two external influences are pressuring conventional MRV systems to evolve and upgrade.

4.5 Transition to a D-MRV system

Figure 2 shows how standard MRV systems become D-MRV systems. Self-reported data, human errors, lack of transparency, and scale issues are downsides of manual MRV approaches. Digital MRV, enabled by IoT, satellite remote sensing, big data, and blockchain, improves carbon emission monitoring and tracking. Although immature technology, regulatory integration, market acceptance, and accessibility barriers remain, several regimes are undergoing transition to promote digital MRV adoption. Science and technology advances data science and digital technologies. The D-MRV Protocol introduces digital MRV standards and frameworks to the policy regime. The market regime may initially be hesitant due to cost, but early adopters and intrinsic benefits can propel adoption. Standards bodies are establishing digital methodology and certification processes for digital MRV as the socio-cultural regime recognizes its potential.

At the landscape level, the Paris Agreement's need for NDCs underlines the need for robust MRV systems to measure progress and hold governments responsible. MRV systems allow transparency, effectiveness evaluation, gap detection, and policy modifications. Upgraded NDCs and climate action increase demand for reliable and robust MRV systems.

Digital technology in MRV operations can improve carbon emission monitoring accuracy, efficiency, and transparency. Despite obstacles, continuous regime reforms and the global commitment to tackling climate change promise a future where digital MRV systems play a vital role in combatting climate change and attaining climate goals.

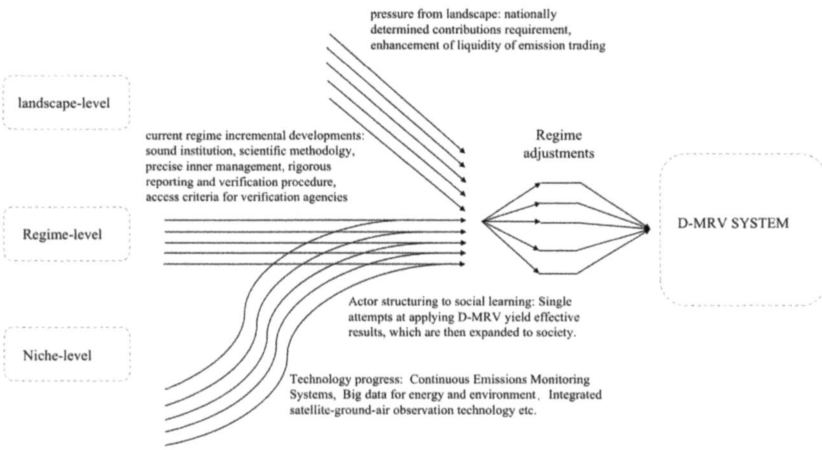

Fig. 2 Transition pathway from a conventional MRV system to a D-MRV system

5 The multi-dimensional carbon emission data quality assurance Scheme

5.1 The foundation of this scheme

The introduction of this design indicates a significant sociotechnical shift that includes technological, economic, social, and political dimensions. Sociotechnical changes possess both academic intrigue and societal significance (Geels 2007), particularly concerning the transfer to new carbon emission data quality assurance frameworks. The significance of this transition is apparent as the responsible authority convened a special press conference about carbon emission data assurance, and there is a growing interest among corporations, scientific institutions, and third parties. A comprehensive carbon emission data quality assurance mechanism would significantly influence the national emission market and the attainment of carbon neutrality. Nonetheless, "transitions are exceedingly intricate processes" (Geels 2007) that necessitate an appreciative theory. The multi-level perspective on transition is such an appreciative theory. Consequently, the basis of this framework is anchored in the synergistic interactions among the five regimes outlined in MLP theory, as illustrated in Fig. 3. The breakthroughs in technology and research facilitate the development of innovative methods for measuring emissions volumes with increased accuracy. In response to these improvements, the competent authority establishes niches for the assessment and refinement of breakthrough technologies via pilot market testing. In these pilot markets, specific industries or companies implement the technology and offer feedback. This input is essential; it informs the scientific and technology frameworks, directing the optimization of technical

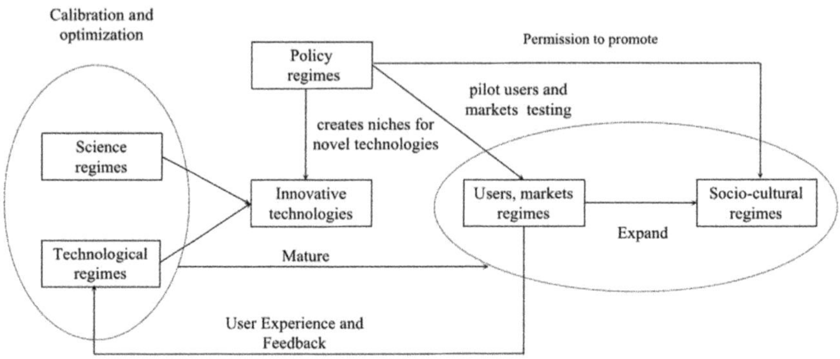

Fig. 3 Interconnected relationships between the five regimes outlined in MLP theory

parameters through iterative development. The repetitive process illustrates the intricacy and cyclical characteristics of sociotechnical transformations, emphasizing the critical importance of inter-regime communication. Upon maturation, the technology achieves widespread societal adoption, illustrating how the amalgamation of concepts from diverse regimes facilitates transitions that are both technologically sound and socially advantageous.

5.2 *Characteristic of this scheme*

The case studies and analysis show that niche, regime, and landscape characteristics influence the evolution of carbon emission data quality assurance schemes. These dynamics should modify existing assurance methods, creating a more robust and effective carbon emission data quality assurance framework. Such a strategy must have some critical properties to be effective and reliable.

(a) Complete. China's national ETS objectives—determining coverage, setting caps, assigning allowances, and verifying compliance—require the data quality assurance scheme to efficiently acquire and integrate data from all relevant carbon emission sources. This comprehensiveness requires focusing on data type and level. Data type includes six categories: production data, which includes product types and output; emissions data, which includes emission volumes and relevant calculation parameters like energy consumption, calorific value, and carbon content; technology data, which covers manufacturing and emissions-reduction technologies, process flows, and capacities; and management data, which covers internal enterprise aspects like ow. Data level means "fineness" (Zeng et al. 2018).

(b) Trackability. Verifying references to prove a data production system's integrity and origins is crucial (Lee 2019). It involves tracking and validating the data processing route. References—standardized procedures, data processing software, or data verification models—are crucial to data traceability (Lee 2019). Traceability guarantees data is supported by records of raw material purchases and consumption, timestamps and sample production people, and working papers for verification. This documented trail proves the data's legitimacy and credibility, making it essential for accurate data analytics.

(c) Transparency. National initiatives to improve emissions control and transparency have grown (Hahn et al. 2015). The US Environmental Protection Agency (2012) requires all facilities emitting at least 25,000 metric tons of CO_2 equivalent to disclose their emissions, while the UK requires stock-listed companies to report their GHG emissions (U.K. Government 2013). Carbon disclosure increases transparency and improves carbon management (Matisoff 2013). Transparency and data quality are integrated in a Carbon Data Quality Assurance System. Bertino et al. (2019) define data transparency as the ability of persons to access all data utilized in process and decisions that affect them. Transparency about data collection, processing, and use is essential for quality assessment. The assurance system's methods must be transparent to evaluate carbon data credibility and dependability. Transparency underpins data quality evaluation, ensuring informed decision-making and carbon data confidence.

5.3 Four dimensions of this scheme

The schemes include four essential dimensions, each defined by specific stakeholders: government, enterprise, scientific institutions, and third-party companies. In their respective dimensions, these stakeholders act as key players, each bearing distinct responsibilities and obligations. The stakeholders, by fulfilling their roles, collectively enhance the system and ensure data quality preservation. This structure parallels the five regimes identified in MLP theory. In a coordinated interaction, similar to the interconnected components in MLP theory, each dimension within the schemes maintains its distinct role while remaining closely linked to the others. This establishes a synergistic relationship among the dimensions, whereby changes in one dimension induce alterations in the others. This synergy and interdependency create a stabilizing force within the systems, allowing them to effectively serve as protectors of data quality.

5.3.1 Government dimension

Governments ensure data quality in multi-dimensional carbon emission data quality assurance schemes, which are crucial to national emission trading markets. An ETS requires substantial government effort because carbon credits derive value from government-enforced constraints, according to Brian (2008). Government engagement includes regulation, standardization, supervision, and providing incentives.

Establishing a robust legal framework and enhancing enforcement mechanisms: The government must adopt and improve carbon data quality laws and regulations. A solid method to precisely measure and maintain legal standards ensures accurate counting (ton is always a ton). Combining this with penalties for noncompliance makes the system reliable. Trustworthiness justifies allowance pricing (ICAP 2021). Thus, the MEE urges local ecological and environmental departments to strictly implement the policies, strengthen supervision and regulation throughout the national emissions market operations, and effectively prevent management and market risks from data falsification. Such enforcement improves governance and reduces data tampering management and market risks. The development of these rules and regulations provides a solid legal foundation for carbon emission data authenticity, accuracy, and integrity. Similarity, the Vietnamese government established a complete legal framework for the national MRV system, including a high-level decision defining stakeholder involvement and a well-defined structure (Long et al. 2019). In addition to legislative frameworks, the government must establish and fund a greenhouse gas credit check department and law enforcement. To reduce deceptive behaviour, the market should act openly and successfully (Brian 2008).

Standardizing and optimizing carbon accounting and verification: Optimizing and standardizing carbon emission accounting and verification is another important government duty. In response to data accounting and oversight difficulties in 2021, the MEE revised the "Guidelines for Accounting Methods and Reporting of Greenhouse Gas Emissions in Enterprises, Power Generation Facilities (revised in 2022)". This revision clarified accounting borders, improved coal sample inspection methods, and standardized raw data handling. Carbon accounting and reporting are unsystematic and inconsistent (Klaaßen and Stoll 2021). Distinct sectors have distinct data accuracy rules. Power plants must measure fuel energy (net calorific value), while aluminum manufacturers can use preset numbers. These fixed numbers vary among industries, thus companies using the same fuel, such anthracite, report very different emissions. This mix-up makes emissions statistics incorrect and unjust (Tang et al. 2018). Given this, the government should standardize carbon accounting and verification. This would make emissions data more reliable and comparable and promote industry accountability. A unified strategy helps enterprises understand their carbon footprints and compare their progress. Harmonized systems also enable cross-sector collaboration and knowledge sharing. Industries can collaborate on emission reduction initiatives using the same methods, resulting in a more holistic environmental sustainability strategy. Standardized criteria give stakeholders and investors precise and comparable data, which can boost confidence. Investments in cleaner technologies and sustainable practises may result.

Innovative supervision and support: Governments are realizing that carbon emission oversight must be strengthened and innovated. They can improve oversight and cooperation by combining strengthened regulatory measures with training and assistance. Advanced technology like D-MRV can improve supervision. This provides precise carbon emission data monitoring and prompt discrepancy detection. However, strict supervision isn't enough. Businesses may struggle to comply with regulations, but governments are giving training and support to help. This support-oriented approach is essential to supervisor methodology innovation. Examples include China's Ministry of Ecology and Environment's campaign to enhance carbon emission reports. MEE taught coal sample collection and data verification under close supervision. This helped industries comply and showed the government's commitment to good change. This multimodal approach, combining technological innovation with supportive measures, protects carbon emission data.

Transparent Information Disclosure Improves Carbon Data quality: To increase carbon data quality, governments must adopt policies and organizations that assure transparency. According to a 2018 CDP questionnaire survey, corporations with higher greenhouse gas emissions employ extended, low-quality climate statements to hide their emissions (Park et al. 2023). An entity dedicated to carbon information disclosure requires companies to publish their carbon emissions data. This institution should verify data accuracy and quality. Governments must also encourage transparent and high-quality carbon emission data reporting. Transparency ensures data veracity and encourages public interaction by disclosing company emissions profiles. Public inspection deters data falsification and allows stakeholders, including the public, to hold firms accountable for their carbon emissions.

5.3.2 Enterprise dimension

Enterprises are crucial to the emission trading system and must accurately declare their carbon emissions. They handle compliance, internal validation, and management.

Active Carbon Asset Management and Planning: Businesses must move beyond carbon allowances to manage carbon assets. This covers carbon audits, compliance systems, and carbon finance and trading. Zhang et al. (2018) suggested five carbon asset management metrics, including carbon fund turnover ratio, carbon emission rate, and carbon technology transformation efficiency, to stress a holistic approach. By using such a strong framework, companies may efficiently manage their carbon assets and stay current on legislation and trends.

Internal Validation and Integrity Assurance: Carbon emissions data must be validated and verified. This is possible with internal audits and validation. Enterprises should assign carbon emissions management staff to record, preserve, and validate data. Carbon data integrity requires this degree of examination and rigor, akin to environmental management systems.

Compliance with Regulations and Standards: Government rules and international standards are a crucial requirement for businesses. This entails following the regulations and actively participating by comprehending policy implementation and forecasting development patterns. Staying compliant and proactive with regulatory requirements promotes company responsibility and sustainability.

Transparency and Public Engagement: In addition to internal validation and compliance, companies should publicly disclose their carbon emissions statistics. This disclosure should include raw statistics and carbon emission reduction and sustainability measures. Transparency builds confidence among stakeholders, especially the public, and shows environmental responsibility.

5.3.3 Scientific institution dimension

Effective emission trading markets require MRV systems to correctly monitor, record, and verify carbon emissions (Dekkers and Oudenes 2007). The emissions trading market in China has concentrated on the power industry due to its data availability and preparedness. As the national emissions market develops to include new industries, a scientifically robust MRV system is needed (Liu, 2019). Here, scientific institutes have numerous crucial duties:

Standardizing Carbon Accounting Methodologies: Tang et al. (2018) called for regional carbon accounting uniformity, especially when integrating regional pilot programs into a national emissions market. Scientific institutions create and standardize carbon accounting criteria to ensure data reporting uniformity and integration.

Innovation in data accuracy technology: Technology for carbon dioxide emission monitoring is crucial to MRV system advancement. In particular, D-MRV based on continuous emission monitoring technology and other information technology improves carbon monitoring and verification accuracy. Scientific institutions must lead research and development to make these methods dependable and effective. Liu (2023) stressed the importance of continuous emission monitoring in MRV systems, and Gunst (2010) stressed the importance of technological innovation, especially in online monitoring, for supporting emissions trading markets.

Capacity Building and Knowledge Sharing: Scientific institutions train and support MRV stakeholders in addition to developing methods and technologies. MRV system integrity and effectiveness depend on well-equipped and knowledgeable emission monitors and reporters (Dekkers and Oudenes 2007).

5.3.4 Third-party dimension

Third-party verification companies are essential to carbon data validity. However, low profit margins may encourage fraud by some agencies. Peak verification times can also result in disappointing results due to verifier shortages and variable ability. Proper regulation and control of these institutions is necessary to assure carbon data integrity.

Set Entry Limits: This includes defining verifier certification standards, enforcing a minimum number of practitioners, and assessing the agency's past performance.

Institutional Constraints: Rules should prevent conflicts of interest, such as preventing verification agencies from managing carbon assets for the same corporation. Poor-performing agencies should also be barred from verification for a while.

Social Supervision: Social supervision increases transparency and responsibility. Commissioners can score verification agencies yearly and publish these rankings to educate stakeholders of their performance and dependability.

5.4 Advantages of this scheme

5.4.1 Completeness

The multi-dimensional carbon emission data quality assurance technique is more thorough than current systems. It first gathers governments, businesses, scientific organizations, and others to collaborate. The government was the major regulator, although enterprises and third parties were regulated. The new approach makes all stakeholders primary actors or accountable parties, fostering ownership. Second, the plan tightens rules and regulations by developing carbon emission trading legislation and setting clear penalties and recommendations for noncompliance with carbon emission data reporting and verification institution misconduct. Finally, MRV guidance science is improved for completeness. Current accounting and verification principles emphasize scientific rigor and practicality. To ensure technical accuracy and clear direction, they use international best practices and the latest scientific research. This method improves data reliability and streamlines deployment across carbon management industries.

5.4.2 Burden sharing

Government regulation has consistently been the primary factor in current carbon emission data quality assurance schemes. Interview results indicate that competent authorities were required to allocate substantial human, material, and financial resources. The emerging multi-dimensional quality assurance scheme has implemented a multi-stakeholder participation model, which includes enterprises, third-party institutions, and scientific institutions. Multi-stakeholder participation facilitates optimal resource allocation, enhancing data management quality and efficiency, while also distributing costs to reduce the government's financial burden.

Scientific institutions are integral to this framework, demonstrated by their advancement of digital technologies that facilitate the implementation of D-MRV. D-MRV integrates data collection and analysis tools, facilitating remote and automated data collection, thereby minimizing the necessity for on-site inspections. D-MRV facilitates the remote collection and verification of data without human intervention, thereby reducing the workload for government entities and advancing innovation in carbon emissions data management.

Enterprises, as essential contributors to the new framework, are urged to maintain elevated standards for carbon data quality. Through education, training, and guidance from government and third-party institutions, enterprises enhance data quality by optimizing internal facilities, strengthening management processes, and standardizing sampling procedures. High-quality data enhance transparency and decrease the necessity for follow-up inspections by improving the initial reliability of the data.

Verification institutions occupy a crucial role within this system. By performing effective preliminary verifications, these institutions guarantee the accuracy and completeness of data. The efficient verification process facilitates early identification and correction of issues, thereby preventing unnecessary iterative inspections, reducing time and costs, and enhancing public trust in carbon data management.

The new carbon data quality assurance scheme optimizes resource allocation and enhances data quality by leveraging the strengths of multiple stakeholders. It also effectively distributes the government's responsibilities and strengthens public trust.

6 Conclusion and research limitation

Since the 1980s, global warming has emerged as a significant problem, eliciting extensive concern across society. The phenomenon of global warming has transcended the realm of climate science and now spans other domains like the environment, economy, agriculture, finance, administration, and law. To combat climate change, reducing greenhouse gas (GHG) emissions to attain carbon neutrality has become a crucial strategy. Emission trading, via market mechanisms, is a crucial policy instrument for attaining the objectives of carbon peaking and carbon neutrality by regulating and diminishing GHG emissions.

Establishing a carbon market is a multifaceted endeavor, and dependable and precise carbon emission data constitutes the essential basis of the carbon market. It serves as essential support for its operation. Carbon allowances, the primary trading instrument in the carbon market, are predicated on carbon emission statistics. Nonetheless, the distortion of carbon emission statistics persists, adversely impacting the evolution of the carbon market. The Chinese government places significant focus on this topic. A reinforced regulatory model is implemented.

While the quality of carbon emission data is assured to some degree under this model, advancements in technology, particularly the advent of digital MRV, will necessitate a transformation of the current sociotechnical system for carbon emission data quality assurance into a more comprehensive and scientifically robust framework.

This shift initially occurs within the realms of science and technology, policy, market, and socio-cultural systems, respectively. Nonetheless, this transition is interconnected, as each regime of the sociotechnical system will engage and exert a co-evolutionary influence. The transformation is enabled by the alignment of activities across three levels (Geels et al. 2017). The initial level is the niche level, characterized by the progression of technology, including information technology and continuous emission monitoring technology, which will be safeguarded and developed inside a specialized or pilot domain. Policy support or user preference prompts a modification in the original regime. The second level, referred to as the regime level, recognizes that the inherent deficiencies of the current sociotechnical system will generate friction and present an opportunity for transformation. The third level examines external factors that will expedite this change. The will of countries to achieve their NDC commitments and the flourishing carbon emission market further facilitate the transition. Ultimately, a novel sociotechnical system will be instituted with the completion of the transformation to ensure the integrity of carbon emission data.

The multi-dimensional carbon emission data assurance system comprises four principal stakeholders: the government, enterprises, scientific institutes, and a third-party agency. The government's function is policy-focused, offering policy assurances and institutional frameworks for superior carbon data. Enterprises are essential in guaranteeing the accuracy of carbon emission statistics by enhancing internal control and disclosing such data. Scientific institutions function as technical support entities, establishing scientific accounting methodologies and monitoring procedures that enhance data authenticity. Third-party organizations offer training and support to firms, facilitating their comprehension and mastery of carbon data accounting and verification techniques for efficient carbon asset management. These consulting firms also function as the primary verification authority. The participants in the four dimensions guarantee the quality of carbon emission data by executing their designated tasks and duties.

This study examines China's national emission market, asserting that carbon neutrality and emission trading are unavoidable global trends. Consequently, the multi-level viewpoint transformation and multi-dimensional carbon emission data quality assurance scheme can be implemented in emission markets globally. Due to the recency of this topic, there is insufficient empirical data to substantiate the approach. Subsequent research ought to concentrate on determining precise numbers for the accuracy rate of carbon data following the implementation of the multi-dimensional carbon emission data quality assurance scheme.

Author's note: This chapter is a revised and expanded English version of the author's article, "Transition to a Multi-Dimensional Carbon Emission Data Quality Assurance Scheme: Analysis Based on MLP Theory", originally published in Policy Science, Vol. 32, No. 3 (2025).

References

Aldy JE (2014) The crucial role of policy surveillance in international climate policy. Clim Chang 126:279–292

Transition to a Multi-Dimensional Carbon Emission Data Quality Assurance Scheme : Analysis Based on MLP Theory", originally published in *Policy Science,* Vol. *32*, No. 3 (2025)

Brian A (2008) Market failure, government failure and externalities in climate change mitigation: the case for a carbon tax. Public administration and development Public Admin. Dev 28:393–401

Autenrieth G (2020) Monitoring, reporting and verification (Digitised-MRV Protocol). Eur Bank Reconstr Dev

Ballassen V, Stephan N, Afriat M et al (2015) Monitoring, reporting and verifying emissions in the climate economy. Nat Clim Chang 5:319–328

Barrett S, Carraro C, Melo J (2015) Towards a workable and effective climate regime. CEPR Press Ferdi

Belenky L, Iyadomi K, Carevic S et al (2022) Digital monitoring, reporting, and verification systems and their application in future carbon markets. World Bank'S Carbon Mark Innov Unit

Bertino E, Merrill S, Nesen A (2019) Christine Utz: Redefining data transparency: A multidimensional approach. Computer 52(1):16–26

Boyd A., Rennkamp B., & Winkler H., et al. (2011). South African approaches to measuring, reporting and verifying: A scoping report. Energy Res Cent, Univ Cape Town. Retrieved from https://www.africaportal.org/publications/south-african-approaches-to-measuring-reporting-and-verifying-a-scoping-report/

Bürgi et al (2019) Automated MRV system: Conceptual design & required changes to existing carbon standards. European Bank for Reconstruction and Development (EBRD)

Christensen CM (1997) The Innovator's Dilemma: When New Technologies Cause Great Firms to Fail. Harvard Business School Press, Boston, MA

Dekkers C, Oudenes ML (2007) The Kyoto Protocol and beyond: Legal and policy challenges of climate change. Asser Press, The Hague, pp 185–200

Freeman C, Louçã F (2001) As Time Goes By: From the Industrial Revolutions to the Information Revolution. Oxford University Press, Oxford

Gao S, Li B (2022) Do energy intensive companies prepare for the national emission trading scheme? Evidence from a nationwide survey in China. Adv Econ, Bus Manag Res 654

Garud R, Karnøe P (2001) Path creation as a process of mindful deviation. In: Garud R, Karnøe P (eds) Path Dependence and Creation. Lawrence Erlbaum Associates Publishers, Mahwah, NJ, pp 1–38

Geels FW (2002) Technological transitions as evolutionary reconfiguration processes: A multi-level perspective and a case-study. Res Policy 31(8–9):1257–1274

Geels FW (2004) From sectoral systems of innovation to socio-technical systems insights about dynamics and change from sociology and institutional theory. Res Policy 33(2004):897–920. https://doi.org/10.1016/j.respol.2004.01.015

Geels FW (2007) Analyzing the breakthrough of rock 'n' roll (1930-1970) multi-regime interaction and reconfiguration in the multi-level perspective. Technol Forecast & Soc Chang 74(2007):1411–1431

Geels FW, Sovacool BK, Schwanen T, Sorrell S (2017) Sociotechnical transitions for deep decarbonization. Science 357(6357):1242–1244. https://doi.org/10.1126/science.aao3760

Gunst A (2010) Carbon pollution (greenhouse gas) measurement and reporting. Asia Pac J Chem Eng 5(4):646–656. https://doi.org/10.1002/apj.498

Hahn R, Reimsbach D, Schiemann F (2015) Organizations, climate change, and transparency: reviewing the literature on carbon disclosure. Organ Environ 28(I):80–102

ICAP (2021). Emissions trading in practice:a handbook on design and implementation. Retrieved from https://icapcarbonaction.com/en/mrv-enforcement

Klaaßen L, Stoll C (2021) Harmon Corp Carbon FootprS Nat Commun 12:6149. https://doi.org/10.1038/s41467-021-26349-x

Kushilevitz E, Lindell Y, Rabin T (2009) Information-Theoretically Secure Protocols and Security Under Composition. STOC

Lee D (2019) Big data quality assurance through data traceability: a case study of the national standard reference data program of Korea. IEEE Access 7:36294–36299. https://doi.org/10.1109/ACCESS.2019.2904286

Leonard-Barton D (1995) Wellsprings of knowledge: building and sustaining the sources of innovation. Harvard Business School Press, Boston, MA

Lin B, Huang C (2022) Analysis of emission reduction effects of carbon trading: market mechanism or government intervention? Sustain Prod Consump 33:28–37

Lin B, Jia Z (2019) Impacts of carbon price level in carbon emission trading market. Appl Energy 239:157–170

Liu J, Hou J, Fan Q, Chen H (2022) China's national ETS: Global and local lessons. Energy Rep 8:428–437. https://doi.org/10.1016/j.egyr.2022.03.097

Liu Z, Deng Z, Huang X (2023) A carbon-monitoring strategy through near-real–time data and space technology. Int J Hydrogen Energ 4(1):100346

Long P.T., Tri D. Q., Hoa V. X., Tam N.T. (2019) Research and propose a legal framework for stakeholders in the national MRV system in Vietnam. 10th NEU-KKU International conference socio-economic and environmental issues in development

Luo L, Tang Q, Fan H, Ayers J (2023) Corporate carbon assurance and the quality of carbon disclosure. Account & Financ 2023(63):657–690

Martillano D, Dita J, Cruz C, Sadhra C (2017) Android based real-time industrial emission monitoring system using IoT technology. J Commun 12(11)

Matisoff DC (2013) Different rays of sunlight: Understanding information disclosure and carbon transparency. Energy Policy 55:579–592

Matsuo, N. (1998). Key elements related to the emissions trading for the Kyoto protocol

Međugorac V, Schuitema G (2023) Why is bottom-up more acceptable than top-down? A study on collective psychological ownership and place-technology fit in the Irish Midlands. Energy Res Soc Sci 96:102924. https://doi.org/10.1016/j.erss.2022.102924

Oberpriller Q, Kohli A, Soini M et al (2022) Principles for best-practice digital verification. INFRAS

Kengo O (2023) Digital MRV can spur growth in voluntary carbon credit market. Lakyara 364

Paranjape Vishwanath (2010). Legal dimensions of cybercrimes and preventive laws with special reference to India. J Indian Law Inst, Vol. 54, No. 3 (July September 2012), pp. 408-412

Park JD, Nishitani K, Kokubu K (2023) Revisiting sustainability disclosure theories: evidence from corporate climate change disclosure in the United States and Japan. J Clean Prod 382(2023):135203

Schaltegger S, Csutora M (2012) Carbon accounting for sustainability and management. status quo and challenges. J Clean Prod 36(2012):1–16

Frederick S (2011) Transparency in three dimensions. Univ Ill Law Rev 2011:1339–1357

South Pole, 2020. Protocol for Digitalized MRV (D-MRV Protocol), South Pole Spain, S. L. Retrieved from http://web.archive.org/web/20220701083436/https://www.ebrd.com/documents/climate-finance/digitised-mrv-protocol.pdf

Tang R, Guo W, Oudenes M, Li P, J., Wang, L., & Wang, H. (2018) Key challenges for the establishment of the monitoring, reporting and verification (MRV) system in China's national carbon emissions trading market. Clim Pol 18(sup1):106–121. https://doi.org/10.1080/14693062.2018.1454882

Tushman, M., Anderson, P. (1986). Technological discontinuities and organization environments. Adm Sci Q 31, 465–493. Retrieved from http://wm.taizhou.gov.cn/art/2023/2/3/art_27_3454497.html

U.K. Government (2013) The Companies Act 2006 (Strategic Report and Directors' Report) Regulations 2013. The Stationery Office, London, England

UNFCCC CDM (2021): SSC_805: Revision to include an option to use remote biogas monitoring in AMS-I.E., AMS-I.I. and AMS-III.R., 2021, Retrieved from https://cdm.unfccc.int/methodologies/SSCmethodologies/clarifications/90305

U.S. Environmental Protection Agency (2012) Inventory of U.S. greenhouse gas emissions and sinks: 1990-2919. Author, Washington, DC

Wang L, Wang L, Li Y, Wang J (2023) A century-long analysis of global warming and earth temperature using a random walk with drift approach. Decis Anal J 7:100237. https://doi.org/10.1016/j.dajour.2023.100237

Welz Bernd (2017). A deficit in digital skills is still among the top barriers to digital transformation. Retrieved from https://blogs.sap.com/2017/07/14/a-deficit-in-digital-skills-is-still-among-the-top-barriers-to-digital-transformation/

World Bank (2022) Digital Monitoring, Reporting, and Verification Systems and Their Application in Future Carbon Markets. World Bank. https://doi.org/10.1596/37622

Zaman I, Pazouki K, Norman R, Younessi S et al (2017) Challenges and opportunities of big data analytics for upcoming regulations and future transformation of the shipping industry. Procedia Eng 194(2017):537–544

Zeng X, Duan M, Yu Z, Li W, Li M, Liang X (2018) Data-related challenges and solutions in building China's national carbon emissions trading scheme. Clim Pol 18(sup1):90–105. https://doi.org/10.1080/14693062.2018.1473239

Zhang C, Randhir TO, Zhang Y (2018) Theory and practice of enterprise carbon asset management from the perspective of low-carbon transformation. Carbon Manag 9(1):87–94. https://doi.org/10.1080/17583004.2018.1426329

Zhang J., Li J., Ye D., Sun C. (2022). The impact of digital economy of resource-based city on carbon emissions trading by blockchain technology. Hindawi Comput Intell Neurosci Vol 2022, Article ID 6366061, 10 pages

Navigating Global Carbon Neutrality and Trade Tensions: The Role of the Carbon Border Adjustment Mechanism (CBAM) and Korea's Strategic Shift

Hyelim You

1 Introduction: The Need for Governance in Achieving Global Carbon Neutrality Goals

Climate change is an unprecedented challenge that transcends borders and demands a collective global response. The Paris Agreement sets a clear goal of limiting the increase in global average temperature to well below 2 °C above pre-industrial levels, preferably to 1.5 °C. Achieving this ambitious goal will require not only individual policy efforts by countries, but also a multi-layered governance framework that harmonizes environmental and economic goals. In this context, "carbon neutrality" has emerged as a key shared global challenge, requiring governments and businesses alike to reset their strategies. The reality, however, is that there are many challenges. Chief among them is the tension between trade liberalization and environmental protection. The proliferation of environmental policies, such as carbon pricing, raises the question of how such schemes can be designed without violating international trade norms. This calls for a new governance framework that can reconcile trade and environmental priorities, ensuring both fairness and sustainability. Countries with export-oriented economies, such as Korea, are particularly sensitive to policy changes in an increasingly carbon-regulated global trade environment and need to carefully consider the impact on their industrial competitiveness and trade structure. Therefore, a strategic response that considers both trade and the environment is becoming more important.

This chapter explores how environmental governance and trade governance can be harmonized, focusing on the Carbon Border Adjustment Mechanism (CBAM), which is increasingly influencing international norms in trade and climate governance. With

H. You (✉)
Faculty of Management, Nagoya University of Commerce & Business, Aichi, Japan
e-mail: hyelim_you@nucba.ac.jp

W. Zhou (ed.), *Decarbonization Policy Engineering*, World Sustainability Series,
https://doi.org/10.1007/978-981-95-5953-4_23

a particular focus on South Korea's response, we examine whether the CBAM can serve as a key mechanism to drive the adoption of a global carbon pricing system. We begin by examining the structural tensions between principles of international trade, such as the World Trade Organization (WTO)'s non-discrimination principle, and increasingly stringent environmental regulations, and analyze the implications of the EU's CBAM as an institutional adjustment mechanism that has emerged in the midst of these tensions. In particular, it explains how the CBAM functions as a regulatory mechanism to prevent carbon leakage and drive the convergence of global carbon pricing policies. Next, using South Korea as a case study, we analyze the policy responses of the South Korean government and strategic changes in industry before and after the introduction of the CBAM. In particular, we examine how the CBAM is being responded to at the corporate level, focusing on POSCO. Finally, we examine the potential impact of the rapidly changing international situation on the global proliferation of CBAMs after the inauguration of the second Trump administration. By comprehensively analyzing domestic and international policy materials and corporate cases, this study aims to propose practical governance strategies that can simultaneously pursue climate goals and a fair trade order.

2 Trade and Environmental Governance in the Era of Carbon Neutrality

2.1 Trade Liberalization Vs. Environmental Protection in the Carbon Neutral Era

As the international community intensifies its response to the climate crisis, environmental regulations are becoming increasingly sophisticated and binding. In particular, with carbon neutrality formalized as a global goal in the wake of the 2015 Paris Agreement, countries are developing medium-and long-term policy roadmaps to achieve carbon neutrality by 2050.[1] However, such increased regulations often significantly increase the likelihood of conflict with existing trade norms, especially free trade regimes.

At its core, the international trading system is based on the principle of removing barriers to trade between countries and ensuring equal treatment for all trading

[1] For example, the European Union (EU) submitted its long-term low-carbon development strategy on March 6, 2020 (European Union, 2020), and South Korea released its 2050 Carbon Neutrality Strategy in December 2020 (Republic of Korea, 2020). Japan submitted its long-term strategy on June 26, 2019 (Japan, 2019), the United States on November 1, 2021 (United States of America, 2021), and China on October 28, 2021 (China, 2021). For more information, visit the UNFCCC's long-term strategy portal (UNFCCC, n.d.). https://unfccc.int/process/the-paris-agreement/long-term-strategies

nations. For example, Article I (most-favored-nation treatment) of the General Agreement on Tariffs and Trade (GATT) under the WTO framework establishes the principle that trade benefits granted by a member to a particular trading partner must be extended to all other members, and Article III (national treatment) ensures that imports can compete on equal terms with domestic products, clarifying that imports should not be subject to taxes or regulations that are less favorable than those applied to domestic products (World Trade Organization (WTO) (n.d.), GATT Articles I & III). On the other hand, regulations aimed at protecting the environment can penalize certain countries or companies based on the carbon emission levels of their products, production methods, sustainability of raw materials, etc. This conflict between the principle of non-discrimination and environmental standards creates a structural tension in international trade governance. For example, if a country imposes additional standards on imported goods on the grounds of protecting its own environment, this can be subject to WTO dispute settlement procedures. This conflict has been repeated in past WTO cases. The 1998 Shrimp-Turtle Case (DS58 1998), a case that redefined the legal boundaries between environmental protection and trade liberalization, was a major turning point. In this case, the United States sought to enforce the use of certain nets to protect endangered marine species, but the WTO ruled that this was procedurally justified and met the principle of non-discrimination (WTO 1998)[2]. Thus, while strengthening international norms for environmental protection is inevitable as we enter the carbon-neutral era, it implies a conflict with the existing trade regime, and new institutional arrangements are urgently needed. Therefore, new normative harmonization and institutional frameworks are needed to reconcile the tensions between environment and trade.

2.2 *Pathways to Complementarity: Environmental Exceptions and Emerging Cooperation*

Although there are inherent tensions between environmental regulations and trade norms, the two areas are not necessarily in conflict. Rather, in recent years, there has been a growing discussion about the policy complementarities between environmental and trade governance. In particular, as net-zero carbon targets have spread across the international community since the Paris Agreement, countries are using trade to promote decarbonization strategies, and environmental regulation is often seen as a policy instrument to create a "level playing field" in international markets (Organisation for Economic Co-operation and Development (OECD) 2025, p. 2). In this context, there are exceptions within the WTO system that allow policy instruments for environmental purposes under certain conditions. One such example is

[2] In this case, the WTO Appellate Body found that while the U.S. measure could be recognized as a measure for the "conservation of exhaustible natural resources" under GATT Article XX(g), it did not satisfy the Chapeau requirement of GATT Article XX because it was arbitrary and discriminatory in its application (WTO, 1998).

GATT Article XX, which permits trade-restrictive measures to "protect human, animal, or plant life or health" and "conserve exhaustible natural resources" in paragraphs (b) and (g), respectively (WTO, n.d., Article XX). While this provision provides an institutional balance to ensure that regulations for environmental purposes do not violate trade liberalization principles, it is also important to note that in order for the exception to apply, the measure must meet the requirements under the chapeau of Article XX: the measure must not cause arbitrary or unjustifiable discrimination in international trade, nor must it operate as a covert protectionist measure. This interpretation means that in order for regulations for environmental purposes to be justified, they must meet international standards of equity and legitimacy in their design and implementation.

In addition, multilateral cooperation models that go beyond individual countries or regions have recently been proposed. For example, new global platforms such as the Climate Club are an attempt to coordinate among countries that have introduced carbon regulations to build decarbonized industrial structures and create a fair trade environment at the same time (Nordhaus 2015; Keohane et al. 2015).[3] It can be understood as an evolutionary form of the interaction between trade governance and environmental governance. In conclusion, in order to address the global challenge of the climate crisis, the twin norms of trade and environment need to be redesigned towards normative consistency rather than conflict, and this trend will play a key role in shaping the international governance system going forward. In particular, in the era of carbon neutrality, trade policy and climate policy are not recognized as independent fields, but as interconnected tasks to achieve the common goal of carbon emission reduction and sustainable industrial structure transformation (Kim 2024).

However, one institutional case that specifically demonstrates the potential for reconciling trade and environmental governance is the CBAM. The CBAM is notable as an institutionalized example of an attempt to simultaneously pursue the goals of environmental protection and fair trade, and in the following section, we provide an in-depth analysis of the concept and design of the CBAM and its implications for global climate governance.

[3] The "Climate Club" concept, proposed by Nordhaus (2015), is a model of international cooperation in which countries implementing carbon regulations apply carbon taxes or border controls to outside groups to encourage cooperative participation and prevent free-riding. See Keohane et al. (2015) for an extended discussion.

3 The Carbon Border Adjustment Mechanism (CBAM) and its Policy Framework

3.1 European CBAM: From Adoption Timeline to Institutional Mechanism

In December 2019, the European Union (EU) launched the European Green Deal, a comprehensive response to climate change and the environmental crisis, as well as a new economic growth strategy. The strategy aims to promote structural transformation in a wide range of sectors, including energy, industry, transportation, and agriculture, with the goal of achieving carbon neutrality by 2050. Notably, the Green Deal first proposed CBAMs as a concrete means to realize the EU's net-zero carbon goal. CBAMs were envisioned as a key policy to prevent carbon leakage that could result from stricter regional carbon regulations and to create a level playing field between domestic and foreign producers (European Union 2023, Article 1).

As shown in Fig. 1, the development and implementation of the EU CBAM has taken place in three main phases: a policy development phase, a legislative and transition period, and full implementation. Following the Green Deal Declaration in 2019, in 2020, the EU Commission presented an initial roadmap and conducted a public consultation from July 22 to October 28 to explore the specific design direction for the CBAM. The main issues discussed during this process included the selection of applicable products (steel, aluminum, cement, etc.), the alignment of the CBAM with free allowances from the EU Emissions Trading System (ETS), and how to minimize trade conflicts with developing countries and free trade agreements (FTAs). Based on these discussions, in July 2021, the EU officially proposed a draft of the CBAM legislation as part of the "Fit for 55" package. The package consists of legislative measures to achieve the goal of reducing greenhouse gas emissions by at least 55% below 1990 levels by 2030, and includes the strengthening of the EU ETS, the raising of renewable energy targets, the improvement of energy efficiency, and the introduction of the CBAM. On April 18, 2023, the European Parliament completed the approval process for the CBAM legislation, and on April 25, 2023, the EU Council gave final approval to the legislation. This is the official endorsement of the political agreement announced last December by the three parties—the European Commission, the European Parliament, and the Council of the European Union. On May 16, 2023, the CBAM Regulation (EU Regulation 2023/956) was officially adopted, completing the detailed institutional design and legislation of the world's first Carbon Border Adjustment Mechanism and paving the way for its implementation. The introduction of the scheme has spurred similar discussions beyond the EU and at the global level (European Parliament, n.d.).

From October 2023 to December 2025, the CBAM will be in a transition phase where importers will only be obliged to report their product-specific carbon emissions on a quarterly basis. From January 2026, the CBAM will enter a full implementation phase where importers will be required to purchase certificates and pay a carbon

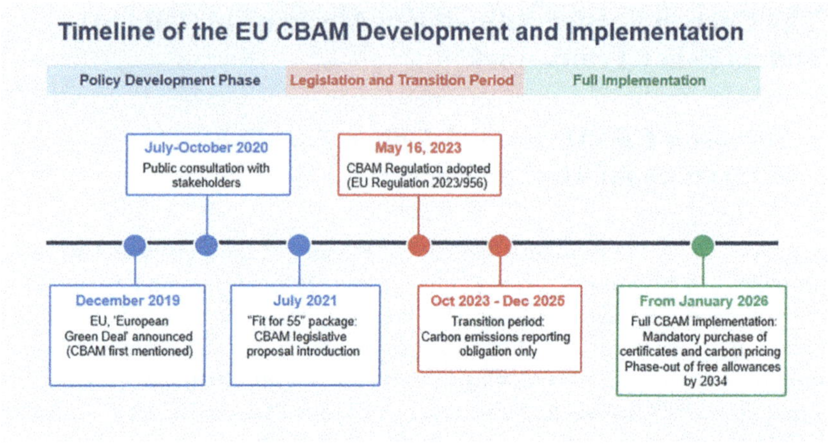

Fig. 1 Timeline of the EU CBAM development and implementation (2019–2034) (Based on European Parliament (n.d.))

price, while the free allocation of EU ETS allowances will be phased out by 2034. Initially, six sectors will be covered: steel, aluminum, cement, fertilizer, power, and hydrogen, all of which are highly carbon-intensive industries and key targets of EU climate policy.

Through its progressive development over time, the CBAM has been designed to ensure that imports from overseas markets with relatively weak carbon emissions regulations bear a carbon cost equivalent to that imposed in the EU's domestic market. By imposing the same level of carbon cost on imports as on products produced within the EU, the scheme aims to close the carbon price gap between domestic and foreign producers and create a 'level playing field' (European Commission 2025). Figure 2 provides a visual representation of this mechanism. An equity issue arises in that in onshore markets covered by the EU ETS, the cost of carbon is already embedded in the price of products, while products imported from offshore markets are not subject to such a cost. CBAMs therefore seek to put imported products on the same footing as onshore products by imposing an additional carbon cost (e.g., through the purchase of certificates). In doing so, it prevents carbon leakage and acts as a regulatory instrument to realize decarbonization through international trade.

The key to CBAM is its link to the EU ETS. In the EU ETS, producers within the EU pay for their carbon credits by purchasing them, and the CBAM applies the same principle to imports to ensure equity between domestic and foreign producers. Specifically, importers are required to purchase and submit "CBAM certificates", the price of which is indexed to the average carbon price under the EU ETS. In addition, if the exporting country has already paid a carbon price, a deduction is provided to avoid a double burden (European Union 2023, Article 9). As such, the CBAM is more than just a trade regulation; it complements the EU ETS and functions as a tool to incentivize carbon reductions through international trade. In this context, the EU claims that CBAM is designed to be in line with WTO norms, emphasizing in

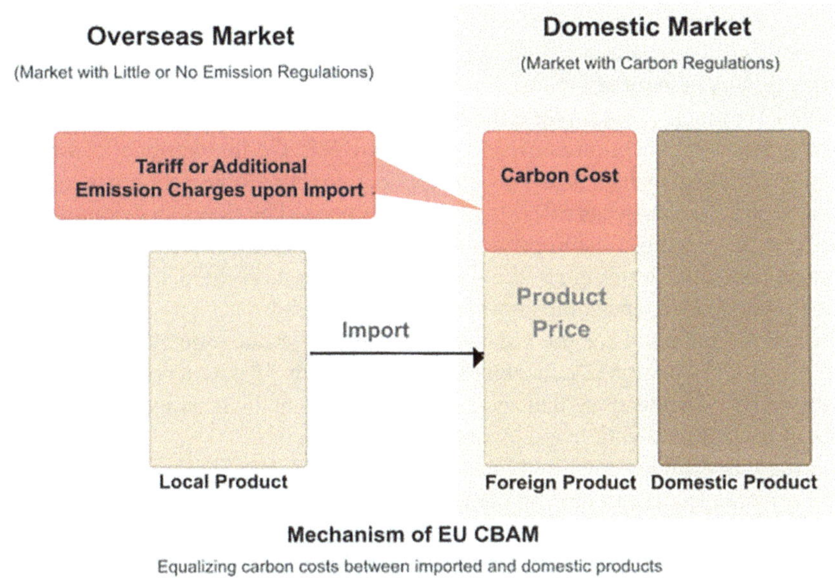

Fig. 2 CBAM Mechanism

particular the balance between the exception for environmental protection (GATT Article XX) and the principle of non-discrimination (European Parliament 2022, pp. 2–3). However, developing countries such as China, India, and South Africa have characterized CBAMs as unilateral trade barriers, and have even raised the possibility of a WTO complaint.[4]

3.2 *CBAM as a Catalyst for Conflict and Coordination: Trade-Environment Tensions and Policy Complementarity*

The CBAM is seen as an innovative attempt to extend the European Green Deal to the international trade system. However, it also creates tensions with the existing WTO system and highlights structural tensions between trade and environmental

[4] China claims that the CBAM is inconsistent with WTO norms and is raising the issue at the WTO. In particular, China criticizes the CBAM for placing a disproportionate burden on developing countries (Reuters, 2023). India considers the CBAM an "unfair and unacceptable measure" and has mentioned a possible WTO complaint. A senior Indian finance ministry official expressed concern that the CBAM could have a negative impact on the country's economy (Reuters, 2024a). South Africa considers the CBAM to be a "unilateral and protectionist measure" and is considering a formal complaint to the WTO. South Africa's Department of Trade, Industry and Competitiveness expressed concern that the CBAM could undermine the industrialization plans of developing countries (Reuters, 2024b).

governance In particular, the CBAM, by imposing costs on imports based on carbon emissions, has the potential to conflict with the core principles of the WTO system, specifically GATT Article I (Most-Favored-Nation treatment, MFN) and Article III (National Treatment, NT).[5] While the EU justifies the CBAM based on the environmental protection exception under GATT Article XX, the international community continues to debate its interpretation and application. Developing countries, as noted earlier, regard the CBAM as a new trade barrier imposed by developed countries and criticize it for placing a disproportionate burden on countries with limited carbon reduction capacity. In doing so, the CBAM is raising more fundamental questions about climate justice, and some international organizations, such as the OECD, have warned of the potential for the CBAM to legitimize protectionism in the name of environmental protection (Dechezleprêtre et al. 2025, p. 10). As a result, the CBAM is at the center of a new governance conflict, highlighting the tension between global norms of trade liberalization and environmental protection.

But the CBAM does not simply exacerbate the clash between trade and environmental governance; it can also be seen as an institutional experiment in embedding a global norm of carbon neutrality into the international trading order. The CBAM is designed to ensure that the EU's domestic emissions reductions translate into real global emissions reductions by containing the carbon leakage that would otherwise occur if strong carbon regulations were introduced. Some studies have shown that the carbon leakage rate, which could be as high as 25.6% under EU abatement alone, can be reduced to 18.4% under the CBAM (Richter and Wanner 2025, p. 24). The CBAM has also been shown to encourage firms in exporting countries to adopt low-carbon production methods, thereby stimulating innovation in decarbonized technologies and the diffusion of green investments (Gérardin and Ferrière 2024; Dechezleprêtre et al. 2025).

The "Brussels Effect" (Bradford 2020), whereby the EU's enforcement of strong regulations induces offshore companies and countries to voluntarily comply, is also materializing through the CBAM. This is in line with the Climate Club initiative centered on the G7, opening up the possibility of multilateral policy convergence. It is worth noting that the CBAM is not just about reducing emissions, but also about creating structural pressures to spread the adoption of carbon pricing at the global level. The CBAM, which was designed to work in conjunction with the EU's ETS, sets a precedent for an integrated approach to international trade and environmental policy, as the system for imposing carbon charges on imports was also designed to be compatible with WTO norms. It can be evaluated as an experimental policy case that attempts to harmonize trade and environmental norms. In conclusion, the CBAM has a complex nature that intensifies the tension between two global norms of trade liberalization and environmental protection, while opening up the possibility of institutional adjustment towards sustainable development through the internalization

[5] While GATT Articles I (MFN) and III (NT) of the WTO Agreement prohibit members from discriminating between domestic and imported goods or between trading partners, concerns have been raised that the CBAM may violate the WTO because they impose differential costs on imports based on carbon emissions.

of a new global norm of carbon neutrality. As such, the CBAM is both a symbol of regulatory conflict and a potentially important driver of the future reshaping of international trade and climate governance regimes.

In particular, since 2019, when discussions on the introduction of the CBAM began in earnest, major trading countries have accelerated their efforts to strengthen their domestic carbon reduction targets and institutional frameworks in response. For example, South Korea announced its 2050 carbon neutrality declaration on October 28, 2020, and enacted the " Framework Act on Carbon Neutrality and Green Growth for Coping with Climate Crisis" (hereinafter referred to as the Carbon Neutrality Framework Act, enforced in March 2022) in September 2021. In October 2022, the second 2050 Carbon Neutrality and Green Growth Committee was established under the Act (Inter-ministerial Joint Committee 2023, p. 3). In addition, as a way to cope with the CBAM, the government has promoted practical measures such as strengthening the carbon emission trading system (K-ETS) and establishing a support system for exporters. In preparation for the CBAM transition in 2023, the Korean government has strengthened consultation channels with the EU and implemented measures to upgrade the carbon emissions monitoring system for domestic companies (Inter-ministerial Joint Committee 2022, pp. 3–4). This series of policy responses demonstrates that the CBAM is more than just a normative debate; it serves as a tangible catalyst for policy and institutional change among trading partners. In addition, major South Korean exporters have responded to the CBAM by investing in decarbonization technologies and strengthening their ESG management. In the next section, we analyze how the Korean government and companies are responding to these global institutional changes in the face of the CBAM, and in particular, how the actual effects of the CBAM have led to structural changes in domestic policies and industrial strategies.

4 Korea's Response to Global Carbon Neutrality and the Rise of CBAM

4.1 Global Climate Neutrality Trends and Korea's Institutional Response

Since the European Union announced the European Green Deal in 2019, carbon neutrality has emerged as a global norm and policy agenda, rather than an optional goal for individual countries. In fact, as of August 2020, 125 countries had either declared or were considering going carbon neutral, according to a report by the International Energy Agency (IEA), with the majority aiming to achieve carbon neutrality by 2050. According to the report, countries accounting for about 10% of global energy-related CO_2 emissions, together with the EU, have already legislated or are in the process of legislating net-zero targets, and major industrialized countries such as Germany, the UK, and France have established legally binding net-zero targets (International Energy Agency (IEA) 2020, pp. 152–153). These data show that carbon neutrality is no longer merely an environmental campaign in a few developed countries, but a full-fledged policy goal for the international community.

The EU has set out a concrete vision of achieving carbon neutrality by 2050, and other major countries, including the United States, China, and Japan, have followed suit. In July 2020, Joe Biden, then a U.S. presidential candidate, pledged to achieve net-zero greenhouse gas emissions and ensure a just transition by 2050 as part of his campaign platform (Biden 2020). Shortly after taking office, President Biden formally rejoined the Paris Climate Agreement and strengthened international climate diplomacy by appointing former Secretary of State John Kerry as Special Presidential Envoy for Climate. President Biden elevated climate issues to the center of the national security and foreign policy agenda, focusing on building multilateral cooperation by supporting vulnerable countries and hosting a global summit to promote climate resilience in the international community (The White House 2021). Furthermore, the Biden administration promoted the linkage between trade and climate policy and pushed for the introduction of a Carbon Border Adjustment (Office of the United States Trade Representative 2021, p. 3). The administration emphasized not only decarbonization but also a just transition toward industrial structural change, set specific climate goals through 2035, and presented a long-term vision for creating high-quality union jobs and securing U.S. leadership in the clean energy economy (The White House 2024). On September 22, 2020, Chinese President Xi Jinping announced at the United Nations General Assembly that China would peak carbon emissions before 2030 and achieve carbon neutrality by 2060 (The State Council Information Office of the People's Republic of China 2020). This marked the first time that China, the world's largest greenhouse gas emitter, declared a long-term reduction target, and it had a significant impact on global climate governance. In Japan, then-Prime Minister Yoshihide Suga declared on October 26, 2020, at the 203rd extraordinary session of the Diet that Japan would achieve carbon neutrality by 2050. To that end, the Japanese government committed to a comprehensive strategy including the expansion of renewable energy and improvements in energy efficiency (Prime Minister's Office of Japan 2020). These developments marked a critical turning point, as major economies began placing climate crisis response at the core of their economic and industrial policy agendas.

In the midst of this global transition, the South Korean government formalized its "2050 Carbon Neutrality" declaration on October 28, 2020, when President Moon Jae-in delivered a policy address to the National Assembly. At that time, the government defined carbon neutrality as a central pillar of its long-term national strategy, moving beyond the traditional domain of environmental policy, and began efforts to institutionalize the goal through concrete action plans. In the same speech, the government presented the "Korean New Deal" as a national growth strategy and announced plans to invest heavily in fostering new industries, including the Digital New Deal and Green New Deal, out of a total of 555.8 trillion won in the 2021 budget. Of this amount, 8 trillion won was allocated for the Green New Deal, reflecting the government's intention to proactively support the goal of carbon neutrality by replacing old buildings and public rental housing with eco-friendly facilities, greening urban spaces and living infrastructure (2.4 trillion won), and expanding the distribution of electric and hydrogen vehicles and building charging infrastructure (4.3 trillion won). To realize its carbon-neutral vision, the government also announced sector-specific strategies, such as reducing coal-fired power generation, expanding renewable energy, and establishing low-carbon industrial clusters within smart industrial zones. President Moon Jae-in reaffirmed Korea's commitment to climate leadership by stating, "We will actively respond to climate change together with the international community and aim for carbon neutrality by 2050" (Office of the President, Republic of Korea 2020).

Although South Korea had previously pursued policies under the banner of "low-carbon green growth," its carbon neutrality strategy since 2020 is grounded in a far more concrete and systematic institutional framework. This shift moved beyond the mere declaration of aspirational goals and laid the foundation for actionable policy implementation. As a key step in legal reform, the Framework Act on Carbon Neutrality and Green Growth for Coping with the Climate Crisis (hereinafter the Carbon Neutrality Act) was enacted in September 2021. This legislation, Korea's first comprehensive climate law, legally codifies the 2050 carbon neutrality goal and establishes a legal foundation for setting national greenhouse gas (GHG) reduction targets, monitoring their implementation, and formulating relevant climate policy (Ministry of Environment 2021, Carbon Neutrality Act, art. 7). In particular, the Act stipulates an interim target of at least a 35% reduction in GHG emissions by 2030 compared to 2018 levels (Carbon Neutrality Act, art. 8), and significantly strengthens institutional accountability by imposing carbon neutrality obligations not only on the central government but also on local governments and public institutions (Carbon Neutrality Act, arts. 4–5).

Meanwhile, institutional changes were also made in terms of organizational structure. On May 29, 2021, the Moon Jae-in administration established the 2050 Carbon Neutrality Committee as a presidential advisory body, launching a joint public–private governance platform that includes government ministries, industry, experts, and civil society. The Committee was created to serve as a central coordinating body for major climate policy initiatives such as Nationally Determined Contribution (NDC) and the formulation of the 2050 carbon neutrality scenario. Compared to its predecessor, the Presidential Committee on Green Growth, it has been granted significantly enhanced authority and functions. In particular, the Committee plays a pivotal role in deliberating and deciding on the national carbon neutrality vision, mid-to long-term reduction targets, and other key plans and policies. It also monitors progress and performance. Through this role, it has facilitated a shift from an expert- and central government–dominated structure to an inclusive governance system that incorporates central and local governments, industry, youth, and labor representatives (2050 Carbon Neutrality and Green Growth Commission 2022). This shift is meaningful not only in that it legitimizes carbon neutrality as a national long-term strategy, but also in that it institutionally embeds a participatory policy-making structure that engages a broad range of social actors.

After the inauguration of the Yoon administration on May 10, 2022, South Korea's carbon neutrality framework remained largely intact. In October of the same year, the government restructured the existing 2050 Carbon Neutrality Committee into the 2050 Carbon Neutrality and Green Growth Committee, aiming to strengthen the foundations of public–private partnership and enhance inter-ministerial coordination. The new administration continued efforts to ensure policy continuity and practical implementation by articulating the national vision of "Carbon Neutrality and Green Growth: Leaping toward a Global Pivotal Country," while advancing specific implementation strategies for the 2030 NDC and increasing investment in carbon reduction infrastructure (2050 Carbon Neutrality and Green Growth Commission 2022).

Since its declaration in 2020, South Korea has comprehensively restructured its legal and institutional systems, maintained cross-administrational policy consistency, and steadily advanced its carbon neutrality implementation mechanisms. Building on these institutional foundations, the country is now preparing to implement more concrete and multilayered policy adjustments in response to the European Union's post-2021 push for the introduction of CBAM.

4.2 CBAM'S Impact on Korea: Policy Adjustments and Corporate Strategic Shifts with a Case Study of POSCO

The EU's CBAM represents an effort to integrate net-zero carbon targets into trade policy by imposing additional costs on products imported from countries with lax or nonexistent carbon regulations. For export-oriented economies such as South Korea, the CBAM presents a potential de facto non-tariff barrier, embodying the tension between trade liberalization and environmental protection discussed in Sect. 2. Since the formal announcement of the CBAM as part of the EU's "Fit for 55" package in 2021, South Korea has initiated a more systematic and strategic response. The CBAM, if implemented without adjustment, would have dealt a significant blow to some of Korea's key export sectors—particularly steel, aluminum, and cement.

The Korean government's response to the CBAM has unfolded in a clear time sequence. First, beginning in the second half of 2021 following the announcement of the CBAM legislation, the Ministry of Trade, Industry and Energy (MOTIE) organized a public–private consultative body to promote the acquisition of carbon emissions data in major sectors such as steel. Simultaneously, it actively engaged in bilateral and multilateral trade diplomacy, emphasizing the importance of WTO compatibility and the principle of non-discrimination against foreign companies. In particular, in December 2022, Ahn Deok-keun, Head of the Trade Negotiations Division, visited Brussels to meet with officials from the European Commission (Directorates-General for Trade, Taxation and Customs Union, and Climate Action) and the European Parliament. During these meetings, the Korean delegation requested that the CBAM not be applied in a discriminatory manner to Korean exporters, and expressed the view that the mechanism should be aligned with international trade norms such as WTO rules and the EU–Korea Free Trade Agreement (Ministry of Trade, Industry and Energy 2023, p. 2). In February 2023, the Ministry of Trade, Industry and Energy (MOTIE) strengthened its CBAM response framework by launching the EU CBAM Task Force and establishing an Inter-ministerial EU CBAM Response TF under its umbrella. The task force comprises major ministries, including the Ministry of Economy and Finance, Ministry of Environment, Ministry of Foreign Affairs, and Ministry of SMEs and Startups, as well as experts from industry and academia—such as trade associations, industrial research institutes, and the Korea Energy Economics Institute. Its main objectives are to monitor EU legislative developments, assess the potential impacts on domestic industries, and develop comprehensive support measures for affected companies. The task force is also formulating broader strategies to address new EU environmental and trade-related regulations, including the Supply Chain Due Diligence Directive and the Battery Regulation. In particular, key policy tasks include supporting the development of carbon reduction technologies and establishing MRV (Monitoring, Reporting, and Verification) infrastructure tailored to specific products, with a strong focus on CBAM-targeted sectors such as steel (Joo 2023). As an extension of its ongoing response efforts, the Ministry of Trade, Industry and Energy, Ministry of Environment, Ministry of SMEs and Startups, Ministry of Foreign Affairs, and the 2050 Carbon Neutrality and Green Growth Committee jointly held the "EU CBAM Corporate Briefing" at the Korea International Trade Association on September 26, 2023. During the event, the government reviewed the status of domestic companies' preparations for the implementation of the CBAM, discussed potential countermeasures, and announced the release of the "EU CBAM Guidelines," a document jointly prepared and published by the five organizing institutions (Ministry of Trade, Industry and Energy, Ministry of Environment, Ministry of SMEs and Startups, Ministry of Foreign Affairs,, and 2050 Carbon Neutrality and Green Growth Committee 2023).

Second, during the CBAM transition period (from October 2023 to December 2025), the Korean government implemented various practical measures to support exporters in complying with the system. These included the establishment of a carbon emissions MRV system for exporters, the publication of industry-specific reporting guidelines, and the organization of regular briefings. For instance, on March 14, 2025, the Ministry of Trade, Industry and Energy, Ministry of Environment, Ministry of SMEs and Startups, and the Korea Customs Service jointly hosted the First Joint Government Briefing on Response to the EU CBAM, where they announced plans to hold a total of six joint briefings. At the event, the government introduced the main elements of the CBAM amendment and provided various support programs, including greenhouse gas (GHG) emissions calculation, reduction consulting, and carbon-neutral facility upgrade assistance (Ministry of Trade, Industry and Energy, Ministry of Environment, Ministry of SMEs and Startups,, and Korea Customs Service 2025). At the same time, to strengthen the response capacity of small and medium-sized enterprises (SMEs), the Ministry of SMEs and Startups launched the "SME CBAM Response Infrastructure Construction" support program in 2025. The initiative aims to alleviate the burden of CBAM compliance for EU-exporting SMEs by providing consulting services for product-specific carbon emissions calculation and reduction, as well as third-party verification by

CBAM-qualified verification bodies (i.e., ISO 14065-accredited organizations). The program targets SMEs that directly or indirectly export CBAM-covered products—such as steel, aluminum, fertilizers, cement, hydrogen, and electricity—to the EU, and offers subsidies covering 70–90% of eligible costs. Support includes consulting on carbon emissions calculation and reduction, assistance with emissions reporting to EU importers, help with preparing verification reports through CBAM-approved verifiers, and the pilot deployment of automated MRV systems for carbon emissions management (Ministry of SMEs and Startups 2025). These policy measures provide tangible support for SMEs—especially those with limited internal capacity—to adapt effectively to the full implementation of the CBAM.

Industry has also accelerated its strategic transformation in response to the CBAM. Even prior to its formal introduction, major conglomerates in Korea had begun managing product-level carbon footprints, investing in low-carbon process transitions, and obtaining green certifications, in line with the global shift toward carbon neutrality. Leading exporters such as Samsung Electronics, Hyundai Motor Company, and LG Energy Solution have already declared their 2050 carbon neutrality targets and joined the RE100 (100% Renewable Energy) initiative. These firms have begun transitioning their production processes to renewable energy, introducing process gas reduction technologies, and responding to growing pressure to reduce supply chain emissions (Samsung Electronics 2022; Hyundai Motor Company 2021; LG Energy Solution (n.d.); Reuters 2022). However, the implementation of the CBAM in Europe is accelerating these strategic changes. In particular, the CBAM requires companies to systematically track and manage carbon emissions across their supply chains. Since the EU mandates disclosure of the carbon footprint of imported products, the ability to manage carbon data across the full value chain—including upstream suppliers—has become a critical factor in global competitiveness. With the European CBAM now in full operation, companies are facing real pressure to integrate carbon performance into core business decision-making processes.

The Korean steel industry is one of the sectors most directly affected by the EU's CBAM, with POSCO serving as a representative example. Under the CBAM, exporters must verify the carbon emissions generated during the production of steel products exported to Europe. This regulatory requirement exerts significant pressure on the cost structure and investment decision-making of affected firms (POSCO 2024a, p. 28). On July 15, 2024, POSCO's Carbon Neutrality Strategy Office estimated that failure to respond effectively to the CBAM could lead to additional annual costs ranging from 200 billion to as much as 1 trillion Korean won (Song 2024). At the same time, European importers are increasingly requesting detailed carbon emission data for POSCO's products. Major clients—including global automakers—have already stated they will not purchase steel unless it demonstrates more than a 30% reduction in carbon intensity. This reflects the growing role of carbon regulation as a determinant of corporate competitiveness. In response to these external pressures, POSCO is pursuing a dual-track strategy that strengthens both carbon reduction technologies and emissions management systems. On the technological front, the company is actively developing its hydrogen-based steelmaking technology,

HyREX,[6] as a replacement for conventional blast furnace methods. This process uses hydrogen instead of coal as a reducing agent, thereby offering the potential for significant reductions in carbon emissions from steel production. POSCO has committed to a large-scale investment plan aimed at transitioning to hydrogen-based steelmaking by 2050—a move it describes as the "second start-up of the steel industry" (Jung 2024).

On the management side, POSCO renewed its Environmental Product Declaration (EPD) certification from the Ministry of Environment in January 2023 for 14 products, including hot rolled, cold rolled, and electrical steel. The company is also actively pursuing low-carbon product certifications for products with improved carbon intensity. In addition, it has developed a product-level carbon tracking system that calculates carbon intensity on a monthly and quarterly basis. This system enables the transparent disclosure of carbon data verified by third-party agencies, facilitating both customer use and compliance with the EU CBAM (POSCO 2024b, p. 23).

The POSCO case is important because it demonstrates that CBAM is not only an inevitable burden for existing companies, but also an opportunity for companies to gain an edge in the global low-carbon market by securing 'proactive ESG management' and 'green technology leadership'. This suggests that corporate competitiveness in the CBAM era directly links to decarbonization innovation capabilities and will likely serve as a key reference case for the Korean steel industry as a whole to accelerate the transition to world-class green processes. POSCO's response demonstrates that CBAM goes beyond a single company and requires structural changes to the entire Korean industry and government policies. As a result, the CBAM has acted as a catalyst to pressure both the Korean government and industry to move from a declaration to a practical implementation system. This shows that the tension between trade and environmental norms has emerged as a real challenge that directly affects the global competitiveness of Korean companies, and suggests that the future trade order may be reorganized around a zero-carbon norm. It also warns that South Korea could face new trade barriers if it does not respond proactively. In sum, CBAMs are emerging as a powerful driver of Korea's net-zero policy and industrial innovation, and a key variable in determining its survival and competitiveness in the future trade environment.

The POSCO case is significant in demonstrating that the CBAM is not merely an unavoidable burden for incumbent firms, but also a strategic opportunity to gain a competitive edge in the global low-carbon market through proactive ESG management and green technology leadership. This case indicates that, in the CBAM era, corporate competitiveness is closely tied to decarbonization innovation capacity, and is likely to serve as a key reference case for the Korean steel industry in accelerating its transition toward world-class green manufacturing processes. POSCO's response shows that the implications of the CBAM extend beyond individual firms,

[6] HyREX (Hydrogen Reduction Ironmaking EXperimental) is POSCO's proprietary hydrogen-based steelmaking process, designed to replace conventional coal-based blast furnaces by using hydrogen as the reducing agent.

highlighting the need for structural transformation across the broader Korean industrial landscape and in national policy frameworks. As a result, the CBAM has acted as a catalyst for both the Korean government and industry to shift from declarative commitments to concrete implementation systems. This shift illustrates that the tension between trade and environmental norms has become a tangible challenge that directly affects the global competitiveness of Korean firms. It also implies that the future trade order may increasingly revolve around zero-carbon standards. Without a proactive response, South Korea risks facing new types of green trade barriers in the near future. In sum, the CBAM is emerging as a powerful driver of Korea's net-zero policy and industrial innovation—and a critical variable that will shape the country's future competitiveness and survival in an evolving global trade environment.

5 Conclusion: CBAM, Strategic Shifts, and the Challenges of a Fragmented Global Order

This chapter has highlighted the CBAM as an example of how trade governance and environmental governance can be made complementary through policy harmonization, rather than conflict. Despite the underlying tensions between WTO non-discrimination principles and environmental regulation, the European Union is pursuing the dual objectives of fair competition and carbon emissions reduction through the CBAM—representing a potential new model of global climate governance.

The CBAM should not be viewed merely as a regional import regulation, but rather as a structural mechanism that can promote international convergence in carbon policies, stimulate investment in decarbonization technologies, and address the persistent issue of free-riding. At the same time, the debate surrounding the CBAM has broadened to include a reinterpretation of international trade norms, renewed attention to climate justice, and concerns over the disproportionate burden placed on developing countries.

In particular, the introduction of the EU's CBAM has served as a decisive catalyst for the Korean government and industry to evolve its carbon neutrality policy from a 'declaration' to a 'practical implementation system'. Specifically, the government has strengthened its institutional response capabilities by checking the carbon emission status of each export industry, establishing a carbon emission calculation, verification, and reporting system, and holding high-level consultations with the EU Commission, while the industry is also realigning its global competitiveness by tracking and managing carbon emissions across the supply chain and investing heavily in low-carbon technology innovation. The case of POSCO, in particular, epitomizes this trend, showing that while CBAM imposes direct and indirect costs on companies, it also provides opportunities to gain competitive advantage by securing "decarbonization innovation capabilities." Furthermore, CBAM has shown that the tension between international trade and environmental norms is not just an abstract discussion, but has elevated carbon neutrality from a peripheral issue to a core issue in South Korea's trade and industrial strategy. It also showed that environmental governance and global trade norms are becoming increasingly intertwined, and that carbon neutrality norms are shaping a new international order that will define competitiveness, market access, and strategic positioning in the global economy.

However, recent shifts in the global political landscape have raised concerns that the proliferation of CBAMs—often described as part of the "Brussels Effect"—may be losing momentum. Most notably, the second Trump administration, which took office in January 2025, formalized the U.S. withdrawal from international climate cooperation on its first day in office by signing Executive Order 14162, titled *Putting America First in International Environmental Agreements*, which declared the U.S. withdrawal from the Paris Agreement (The White House 2025). President Trump justified this move by claiming that the Paris Agreement imposes an unfair burden on the U.S. economy. As a result, the administration has suspended implementation of its NDCs and halted climate finance for developing countries. This reversal of policy is widely interpreted as a sign of weakening U.S. climate leadership and has triggered concerns about negative spillover effects on the global response to climate change (NPR 2025). Furthermore, this policy stance could directly constrain the diffusion of global carbon neutrality norms, including the CBAM. In particular, if the United States intensifies its nationalist trade policies and escalates large-scale tariffs, it could not only undermine the multilateral trading system, but also seriously threaten the integrated evolution of international trade and environmental governance. The structural goal of global carbon pricing convergence envisioned by the CBAM would face major setbacks if major economies—including the United States—withdraw from cooperative climate regimes. The effectiveness of the Brussels Effect depends on the assumption that EU internal norms can spread as de facto global standards. However, if major trading partners reject these norms and pursue divergent paths, norm diffusion is likely to become fragmented and inconsistent. In conclusion, the CBAM remains an important experimental initiative aimed at aligning global trade and environmental norms. Yet, emerging geopolitical shifts may limit or reshape its effectiveness. The future of trade and environmental governance will hinge on the international community's capacity not only to adopt shared norms, but to proactively shape and lead them in a rapidly changing global order.

References

2050 Carbon Neutrality Commission (2022, March 22) Legislation of carbon neutrality vision and GHG reduction commitment: Moving toward a 2050 carbon-neutral society [Press release] (in Korean). https://www.2050cnc.go.kr/base/board/read?boardManagementNo=3&boardNo=631&menuLevel

2050 Carbon Neutrality and Green Growth Commission (2022, October 26) Yoon Administration announces vision and strategy for carbon neutrality and green growth: Official launch of the 2050 Carbon Neutrality and Green Growth Commission [Press release] (in Korean). https://www.opm.go.kr/doc/_attach/file/2022/10/pYUGQXywmxcAHzBTuCol.hwpx.files/Sections1.html

Biden J (2020, July 14) The Biden Plan to Secure Environmental Justice and Equitable Economic Opportunity in a Clean Energy Future. The American Presidency Project. https://www.presidency.ucsb.edu/documents/biden-campaign-press-release-the-biden-plan-secure-environmental-justice-and-equitable

Bradford A (2020) The Brussels Effect: How the European Union Rules the World. Oxford University Press. https://global.oup.com/academic/product/the-brussels-effect-9780190088651?cc=jp&lang=en&utm_source=chatgpt.com

China (2021, October 28) China's mid-century long-term low greenhouse gas emission development strategy. UNFCCC. https://unfccc.int/documents/307765

Dechezleprêtre A, Haramboure A, Kögel C, Lalanne G, Yamano N (2025) Carbon border adjustments: the potential effects of the EU CBAM along the supply chain (OECD science, technology and industry working papers, no, 2025/02). OECD Publishing. https://doi.org/10.1787/e8c3d060-en

European Commission (2025, March 28) Carbon Border Adjustment Mechanism (CBAM). Taxation and Customs Union. https://taxation-customs.ec.europa.eu/carbon-border-adjustment-mechanism_en

European Parliament (n.d.) Carbon border adjustment mechanism as part of the European Green Deal. https://www.europarl.europa.eu/legislative-train/package-fit-for-55/file-carbon-border-adjustment-mechanism. Accessed September 7, 2025

European Parliament (2022) EU carbon border adjustment mechanism: Implications for climate and competitiveness. https://www.europarl.europa.eu/RegData/etudes/BRIE/2022/698889/EPRS_BRI(2022)698889_EN.pdf

European Union (2020, March 6) Long-term low greenhouse gas emission development strategy of the European Union and its member states. UNFCCC. https://unfccc.int/sites/default/files/resource/HR-03-06-2020%20EU%20Submission%20on%20Long%20term%20strategy.pdf

European Union (2023) Regulation (EU) 2023/956 of the European parliament and of the council of 10 May 2023 establishing a carbon border adjustment mechanism. Official Journal of the European Union. https://eur-lex.europa.eu/eli/reg/2023/956/oj/eng

Gérardin M, Ferrière S (2024) Decarbonising steel and other base metals: let's send the right signals. France Stratégie

Hyundai Motor Company (2021) The great shift: Hyundai and the road to carbon neutrality. https://www.hyundai.com/worldwide/en/brand-journal/sustainable-vision/iaa2021-carbon-neutrality

Inter-ministerial Joint Committee (2023, April) National strategy for Carbon neutrality and green growth and the first national basic plan. Seoul: Government of the Republic of Korea (in Korean)

Inter-ministerial Joint Committee (2022, December 26) Status and response plans for the EU carbon border adjustment mechanism (CBAM). Seoul: Government of the Republic of Korea (in Korean)

International Energy Agency (IEA) (2020) World energy outlook 2020. IEA, Paris

Japan (2019, June 26) The long-term strategy under the Paris agreement. UNFCCC. https://unfccc.int/sites/default/files/resource/Japan_LTS2021.pdf

Joo MJ (2023, February 2) Launch of inter-ministerial "EU trade issues control tower"... Responding to EU CBAM. ZDNet Korea. https://zdnet.co.kr/view/?no=20230202151704

Jung S (2024, July 16) CBAM tariffs could cost POSCO up to 1 Trillion won if compliance is not met. Business Korea. https://www.businesskorea.co.kr/news/articleView.html?idxno=221300

Keohane N, Petsonk A, Hanafi A (2015) Toward a club of carbon markets. Clim Chang 144(1):81–95. https://doi.org/10.1007/s10584-015-1506-z

Kim Y (2024, October 18) The era of great transformation: Global supply chain cooperation (KIEP Opinions No. 301). Korea Institute for International Economic Policy. https://www.kiep.go.kr/gallery.es?mid=a20308000000&bid=0008&list_no=11526&act=view

LG Energy Solution (n.d.) Beyond Carbon Neutrality. https://www.lgensol.com/en/carbon-neutrality. Accessed September 7, 2025

Ministry of Environment (2021) Framework Act on Carbon Neutrality and Green Growth for Coping with Climate Crisis (Act No. 18469, Sep. 24, 2021). Korea Legislation Research Institute. https://faolex.fao.org/docs/pdf/kor210354En.pdf

Ministry of SMEs and Startups (2025, February 21) Launch of consulting support for carbon emission measurement, reporting, and verification in preparation for CBAM [Press release]. https://www.2050cnc.go.kr/base/board/read?boardManagementNo=43&boardNo=4917&menuLevel=2&menuNo=92

Ministry of Trade, Industry and Energy (2023) European Union Carbon Border Adjustment Mechanism (CBAM) legislation finalized, implementation scheduled after a transition period starting in 2026 (Press Release). Ministry of Trade, Industry and Energy

Ministry of Trade, Industry and Energy, Ministry of Environment, Ministry of SMEs and Startups, & Korea Customs Service (2025, March 14) Government support programs related to Carbon Border adjustment mechanism (CBAM) at a glance [Press release]. https://www.2050cnc.go.kr/base/board/read?boardManagementNo=43&boardNo=5074&page=1&searchCategory=&searchType=&searchWord=&menuLevel=2&menuNo=92

Ministry of Trade, Industry and Energy, Ministry of Environment, Ministry of SMEs and Startups, Ministry of Foreign Affairs, & 2050 Carbon Neutrality and Green Growth Committee (2023, September 26) Strengthening public-private cooperation in response to EU Carbon Border Adjustment Mechanism [Joint press release]. https://www.mofa.go.kr/www/brd/m_4080/view.do?seq=374152

NPR (2025, January 21) Trump orders U.S. withdrawal from Paris agreement, revokes Biden climate actions. https://www.npr.org/2025/01/21/nx-s1-5266207/trump-paris-agreement-biden-climate-change

Nordhaus WD (2015) Climate clubs: overcoming free-riding in international climate policy. Am Econ Rev 105(4):1339–1370. https://doi.org/10.1257/aer.15000001

Office of the President, Republic of Korea (2020, October 28) President Moon Jae-in's state of the nation address on the 2021 budget proposal (in Korean). https://www.korea.kr/news/policyNewsView.do?newsId=148879196

Office of the United States Trade Representative (2021, March 1) 2021 Trade policy agenda and 2020 annual report of the President of the United States on the trade agreements program. Executive office of the President. https://ustr.gov/sites/default/files/files/reports/2021/2021%20Trade%20Agenda/Online%20PDF%202021%20Trade%20Policy%20Agenda%20and%202020%20Annual%20Report.pdf

Organisation for Economic Co-operation and Development (OECD) (2025, March 13) What to expect from the EU carbon border adjustment mechanism? OECD Publishing. https://www.oecd.org/content/dam/oecd/en/publications/reports/2025/03/what-to-expect-from-the-eu-carbon-border-adjustment-mechanism_a21e9b51/719d2ff9-en.pdf

POSCO (2024a) 2023 POSCO international sustainability report. https://www.poscointl.com/upload/esg/2023_POSCO_INTERNATIONAL_Sustainability_Report_Eng(f).pdf

POSCO (2024b) 2023 POSCO sustainability report: Better world with green steel. https://sustainability.posco.com/assets/file/POSCO_Sustainability_Report_2023_kor.pdf

Prime Minister's Office of Japan (2020, October 26) Policy speech by Prime Minister Suga Yoshihide to the 203rd Session of the Diet (in Japanese). https://www.kantei.go.jp/jp/99_suga/statement/2020/1026shoshinhyomei.html

Republic of Korea (2020, December) 2050 carbon neutral strategy of the Republic of Korea. UNFCCC. https://unfccc.int/sites/default/files/resource/LTS1_RKorea.pdf

Reuters (2022, September 15) FACTBOX-South Korean companies that have joined RE100. https://jp.reuters.com/article/idUSL4N30K0V6/

Reuters (2024a, February 7) India plans to protest EU's carbon tax at WTO meeting. https://www.reuters.com/sustainability/india-plans-protest-eus-carbon-tax-wto-meeting-sources-2024-02-07/

Reuters (2024b, May 23) South Africa considers complaining to WTO against EU carbon border tax. https://www.reuters.com/world/africa/safrica-considers-complaining-wto-against-eu-carbon-border-tax-2024-05-22/

Richter PM, Wanner J (2025) EU climate policy in a globalized world (EconPol policy report No. 53, Vol. 9). EconPol Europe. https://www.econpol.eu/sites/default/files/2025-01/EconPol-PolicyReport_53_EU_Climate_Policy.pdf

Samsung Electronics (2022) Samsung Electronics announces new environmental strategy. https://news.samsung.com/global/samsung-electronics-announces-new-environmental-strategy

Song S (2024, July 16) If the carbon border adjustment mechanism is not followed, POSCO will face 'up to 1 trillion won' in costs. Asia Economy. https://www.asiae.co.kr/article/2024071610145037818

The State Council Information Office of the People's Republic of China (2020, September 23) Full text of Xi's statement at the general debate of the 75th Session of the United Nations general assembly. http://english.scio.gov.cn/topnews/2020-09/23/content_76731466.htm

The White House (2021, January 27) Remarks by President Biden before signing executive actions on tackling climate change, creating jobs, and restoring scientific integrity [Speech transcript]. https://www.whitehouse.gov/briefing-room/speeches-remarks/2021/01/27/remarks-by-president-biden-before-signing-executive-actions-on-tackling-climate-change-creating-jobs-and-restoring-scientific-integrity/

The White House (2024, December 19) FACT SHEET: President Biden sets 2035 climate target aimed at creating good-paying union jobs, reducing costs for all Americans, and securing U.S. leadership in the clean energy economy of the future. https://bidenwhitehouse.archives.gov/briefing-room/statements-releases/2024/12/19/fact-sheet-president-biden-sets-2035-climate-target-aimed-at-creating-good-paying-union-jobs-reducing-costs-for-all-americans-and-securing-u-s-leadership-in-the-clean-energy-economy-of-the-future/

The White House (2025, January 20). Executive order 14162: Putting America first in international environmental agreements. https://www.whitehouse.gov/presidential-actions/2025/01/putting-america-first-in-international-environmental-agreements/

United Nations Framework Convention on Climate Change (UNFCCC) (n.d.) Long-term low greenhouse gas emission development strategies (LT-LEDS). https://unfccc.int/process/the-paris-agreement/long-term-strategies. Accessed September 7, 2025

United States of America (2021, November 1) The long-term strategy of the United States: Pathways to net-zero greenhouse gas emissions by 2050. UNFCCC. https://unfccc.int/sites/default/files/resource/US_accessibleLTS2021.pdf

World Trade Organization (WTO) (n.d.) The general agreement on tariffs and trade (GATT 1947). https://www.wto.org/english/docs_e/legal_e/gatt47_01_e.htm. Accessed September 7, 2025

World Trade Organization (WTO) (1998, October 12) Appellate body report: United States–Import prohibition of certain shrimp and shrimp products (WT/DS58/AB/R). https://www.wto.org/english/tratop_e/envir_e/edis08_e.htm

Policy Implementation and Challenges of Life Cycle Assessment in Japan

Katsuyuki Nakano

1 Life Cycle Thinking

1.1 Life Cycle Assessment (LCA) and Carbon Footprint

To effectively reduce GHG emissions across society, aiming for overall optimization rather than partial optimization is essential. One method that enables this is LCA. LCA is a technique used to assess environmental impacts throughout the entire life cycle of a product—from raw material extraction, manufacturing, and use to disposal. The international standards ISO 14040:2006 (ISO 2006b) and ISO 14044:2006 (ISO 2006c) define the principles, framework, requirements, and guidelines for conducting LCA.

LCA evaluates various environmental impact categories, including climate change, resource consumption, and human toxicity. Among these, assessments that focus solely on the impact of climate change—the amount of GHG emissions and GHG removals—are referred to as Carbon footprint. The standard ISO 14067:2018 (ISO 2018) provides the requirements and guidelines for calculating carbon footprint.

When conducting an LCA, LCA databases are utilized. These databases contain environmental impact data of materials and energy, such as plastics and electricity. AIST-IDEA (AIST Solutions 2025), the process-based LCA database developed by the National Institute of Advanced Industrial Science and Technology (AIST), is widely used in Japan. AIST-IDEA comprises several thousand datasets. Another widely used LCA database is GLIO (Nansai et al. 2012), the environmentally-extended input-output database developed by the National Institute for Environmental Studies (NIES). In AIST-IDEA, the reference flow is generally expressed in physical units, whereas GLIO uses monetary units. While AIST-IDEA contains several thousand datasets, GLIO has approximately 400. AIST-IDEA allows for more

K. Nakano (✉)
College of Policy Science, Ritsumeikan University, Osaka, Japan
e-mail: nakanok@fc.ritsumei.ac.jp

W. Zhou (ed.), *Decarbonization Policy Engineering*, World Sustainability Series,
https://doi.org/10.1007/978-981-95-5953-4_24

detailed analysis at the item level, whereas GLIO covers all domestic industrial activities, including the service sector. Since each database has its own characteristics, they should be selected according to the purpose of the study.

In many cases, LCA studies focus only on GHG emissions. This is because CO_2 emissions can be calculated with relatively high accuracy based on fossil fuel consumption and can be easily aggregated across the life cycle due to their uniform environmental impact, regardless of the emission location (with some exceptions). On the other hand, data on chemical emissions often involve more significant uncertainty in inventory databases, and the environmental impact can vary significantly depending on concentration and emission location. As a result, interpretation of the results requires caution. Climate change mitigation measures should not worsen other environmental issues. Therefore, even in the presence of uncertainties, assessing a broad range of environmental impacts is preferable to identify any trade-offs.

1.2 Environmental Product Declaration (EPD)

An Environmental Product Declaration (EPD) is a method of disclosing product environmental information. A key characteristic of EPDs is that they present life-cycle-based environmental data calculated by LCA and they do not claim environmental superiority. The international standard ISO 14025:2006 (ISO 2006a) defines the principles and procedures for EPDs.

In Japan, the Sustainable Management Promotion Organization (SuMPO) is the sole EPD program operator (SuMPO 2025). Figure 1 illustrates the steps for issuing an EPD under the SuMPO Environmental Labeling Program.

To publish an EPD, an LCA must be conducted in accordance with the relevant Product Category Rule (PCR), and the results must be verified by a third-party EPD

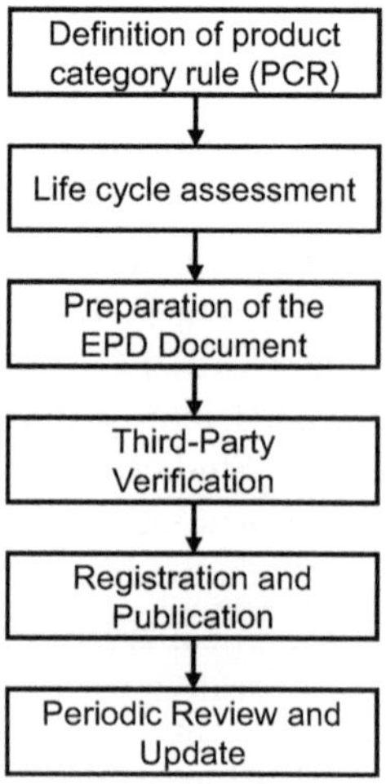

Fig. 1 Procedures for the Publication of an Environmental Product Declaration (EPD)

program operator. PCRs are defined rules for conducting LCAs for specific product categories. They specify what types of data must be collected and what information must be disclosed. Therefore, if there is no existing PCR for a given product, PCR must first be developed.

Under the SuMPO EPD Program, a working group composed of relevant companies and organizations is formed to develop PCR. To ensure industry-wide awareness, the program secretariat invites relevant industry associations and key companies to participate in the PCR working group. The draft PCR created by the working group is then made publicly available for a public comment period. The working group is responsible for responding to comments and revising the draft PCR as necessary. The revised draft is then reviewed by a PCR review panel consisting of LCA experts, who determine whether it meets the necessary criteria. If approved, the PCR is published for general use. As of March 24, 2025, the SuMPO EPD Program has published 65 PCRs and 1963 EPDs.

2 Policies and LCA

2.1 Plan for Global Warming Countermeasures

In October 2020, then-Prime Minister Yoshihide Suga announced Japan's goal to achieve carbon neutrality by 2050. In response, the Cabinet approved the Plan for Global Warming Countermeasures in 2021 (The Government of Japan 2021). This plan is a comprehensive policy framework based on the Act on Promotion of Global Warming Countermeasures. It outlines targets for reducing GHG emissions, as well as the measures to achieve these targets. LCA is positioned as a key element within the plan.

The plan specifies that by 2030, Japan will establish the necessary infrastructure to enable companies and other organizations to quantify GHG emissions throughout the entire life cycle of their products and services—using methods such as carbon footprint—in an objective and standardized way, much like how nutritional information is provided on food packaging. Furthermore, it envisions incorporating such data into product packaging, IC tags, and electronic receipts, thereby facilitating communication between producers, retailers, and consumers. This information could also be used in combination with location data and purchase history to enhance inventory and sales management systems.

In February 2025, the Cabinet approved a new version of the Plan for Global Warming Countermeasures (The Government of Japan 2025). The updated plan includes the development of human resources related to calculating and labeling carbon footprints and other environmental indicators. It also promotes the enhancement of evaluation metrics—such as actual emission reductions and contributions to emission reductions—and encourages the proactive public procurement of low-carbon products and services based on these metrics.

2.2 Ministry of Agriculture, Forestry, and Fisheries (MAFF)

In 2021, Japan's Ministry of Agriculture, Forestry and Fisheries (MAFF) formulated the "Green Food System Strategy" (MAFF 2021). This strategy sets the goal of achieving carbon neutrality in the agriculture, forestry, and fisheries sectors by 2050. One of the key initiatives to reach this goal is enhancing the visibility of efforts to reduce environmental impacts.

In March 2024, MAFF launched a new labeling system called "ChoiSTAR", which stands for "Your Choice boosts SusTainable AgRiculture"—known in Japanese as "Mieru Label" (literally, "visible label") (MAFF 2024). Under this system, producers can calculate the carbon footprint of their agricultural products and report the results to MAFF. If their carbon footprint is lower than the typical benchmark for that product, they are granted permission to use a special star-marked label.

To support producers in these calculations, MAFF has provided an Excel-based calculation tool. The spreadsheet includes the carbon footprint calculation formulas, the necessary background data, and standard carbon footprint values for major products. It is also designed to reflect key environmentally friendly practices, such as the use of organic fertilizers and the application of biochar. As a result, producers can compare their own practices with regional carbon footprint standards in a straightforward manner.

For rice, additional scores are awarded based on biodiversity conservation efforts. These scores are incorporated into the product grading alongside contributions to GHG emissions reduction. For instance, keeping rice fields dry for a longer period during the middle of the growing season is known to reduce methane emissions, thereby contributing to climate change mitigation. However, this practice can negatively affect biodiversity—such as decreasing populations of dragonflies and diving beetles. Therefore, the labeling system assigns points to practices that reduce GHG emissions while minimizing negative biodiversity impacts. This ensures that efforts to mitigate climate change do not come at the expense of other environmental aspects—addressing important trade-offs in a balanced way.

As of December 2025, 25 types of agricultural products are eligible for labeling under this system, including rice, tomatoes, cucumbers, eggplants, spinach, onions, strawberries, and apples. There are also plans to expand the program to include additional products, such as beef, milk, and chicken. Currently, 1756 labeled products are being marketed under the ChoiSTAR system.

In January 2025, Japan's Basic Policy for Procurement of Eco-Friendly Goods, etc., was revised. Under the new policy, the procurement standards for services, such as cafeterias, now include the ChoiSTAR label. This means that food services using ingredients with this label are now prioritized in public procurement processes. As a result, greater adoption of the label is expected in cafeterias located within government facilities.

Looking ahead, MAFF is considering further measures to promote the label in society, including integration with existing private sector point reward systems during FY2025.

2.3 *Ministry of Land, Infrastructure, Transport and Tourism (MLIT)*

Traditionally, the majority of GHG emissions associated with buildings came from the operational phase—such as energy used for heating, cooling, and lighting. However, in recent years, measures such as improved insulation performance and the widespread adoption of photovoltaic panels for on-site power generation have become more common. As a result, an increasing number of buildings now achieve net-zero annual primary energy consumption during operation, thanks to self-generation from renewable energy sources. These buildings are referred to as Zero Energy Houses (ZEH) and Zero Energy Buildings (ZEB). According to Japan's Strategic Energy Plan, approved by the Cabinet in October 2021, newly constructed houses after FY2030 are expected to meet energy efficiency levels equivalent to ZEH standards.

Consequently, the relative contribution of environmental impacts from the production and disposal phases of building materials has increased (Anand and Amor 2017; Röck et al. 2020). As the next target, the concept of Life Cycle Carbon Minus (LCCM) housing has been introduced—referring to houses that achieve a negative carbon balance over their entire life cycle.

Within MLIT, reducing GHG emissions throughout the entire lifecycle of buildings and housing is considered a key policy objective. The Plan for Global Warming Countermeasures calls for the development of systems that promote lifecycle carbon assessment and evaluation, as well as the creation of a market environment in which low-carbon materials and equipment are appropriately recognized and valued.

In response to this, MLIT collaborated with industry and academia to release a new lifecycle carbon assessment tool in October 2024: J-CAT (Japan Carbon Assessment Tool for Building Lifecycle) (Institute for Built Environment and Carbon Neutral for SDGs 2024). This J-CAT has an inventory database based on environmentally-extended input-output analysis. It also allows the use of EPDs of construction materials. Additionally, discussions are underway regarding regulatory frameworks and incentives for applying LCA to buildings. EPD can be used when the Green Purchasing Act is applied to public buildings. Providing incentives for conducting LCAs of buildings is currently being considered.

However, the number of publicly available EPDs remains limited, and few companies currently have hands-on experience with LCA in the construction sector. To address this, in 2025, a subsidy program was launched to support the implementation of LCAs for buildings and building materials. The initiative primarily aims to promote LCA data disclosure and provide incentives for doing so. While the introduction of regulatory limits on GHG emissions is being considered as a future policy tool, no concrete schedule has yet been decided.

EPDs are required to disclose results across multiple environmental impact categories. However, MILT's current use of EPDs is limited solely to GHG emissions. Environmental trade-offs—such as those involving resource depletion or photochemical oxidant formation—may occur when focusing only on climate change mitigation.

Since EPDs include data for these other impact categories, it is desirable that future policies also account for a broader range of environmental impacts.

2.4 Ministry of Economy, Trade, and Industry (METI)

From FY1998 to FY2006, the Ministry of Economy, Trade, and Industry (METI) led a national LCA project, during which it developed LCA methodologies, inventory databases, and impact assessment methods. This national project involved 54 industry associations, including the Japan Iron and Steel Federation and the Japan Chemical Industry Association, and resulted in the establishment of a Japanese inventory database. As part of this initiative, LIME (Life cycle Impact assessment Method based on Endpoint modeling) was developed as an impact assessment method tailored to Japan's regional characteristics (Matsuno et al. 2013; Nakano et al. 2007).

From FY2007 to FY2008, METI launched the Eco-Design Promotion Project for SME, which aimed to promote LCA as a tool for environmentally conscious design in the industry (Nakano and Hirao 2008). Between FY2009 and FY2011, the Supply Chain Resource Efficiency Promotion Project was conducted, promoting environmental impact reduction across manufacturing supply chains through the combined use of LCA and Material Flow Cost Accounting (MFCA) (Nakano and Hirao 2011). In this way, METI has actively supported developing and disseminating LCA as a tool for eco-design in industry.

METI has also supported LCA-based environmental information disclosure. In 2000, METI established the Study Group on Environmental Labels in Green-Oriented Economy and Society, which explored reliable ways to disclose LCA data. As part of this, a trial EPD program was launched. In 2002, the program was transferred to the private sector and began operation under the name Ecoleaf Environmental Label program. By 2006, more than 500 EPDs had been issued under the Ecoleaf program. However, while EPDs disclose LCA information, only a label and certification number appear directly on the product—detailed information, such as GHG emissions, must be accessed online.

To address this limitation, METI launched the Carbon Footprint Pilot Program from FY2009 to FY2011. This pilot was jointly implemented with MLIT, MAFF, and the Ministry of the Environment (MOE). The pilot project focused on GHG emissions, and the project's goal was to display GHG emission values directly on products. The project also supported the development of background databases, the creation of PCRs, and assistance for companies in carbon footprint calculation.

Similar to the Ecoleaf program, the carbon footprint pilot was transferred to the private sector in 2012 and is now operated as the SuMPO Environmental Label (SuMPO 2025). Under the current system, it is no longer possible to issue carbon footprint alone; EPDs must include assessments of multiple environmental impact categories. GHG emissions data can be extracted from EPDs and presented separately as a climate declaration.

In 2023, METI and MOE jointly published the Carbon Footprint Practical Guide (METI and MOE 2023), which outlines methods for companies to calculate, display, disclose, and reduce their carbon footprints. In 2025, the two ministries also released the Carbon Footprint Display Guide (MOE and METI 2025) and the Carbon Footprint Introduction Guide (MOE 2025a).

In FY2023, METI launched a PCR Development Support Program, which assists in creating industry-specific calculation rules aligned with the Carbon Footprint Guidelines. This initiative led to the development of PCRs for categories such as stationery and office supplies, copy and printing paper, office furniture, and software. The program has continued into FY2024.

One notable initiative by METI in the application of LCA is the establishment of an upper limit for GHG emissions from biofuel production. Under the Act on Sophisticated Methods of Energy Supply Structures, biofuels are required to achieve a GHG emissions reduction of at least 55% compared to gasoline on a life cycle basis. As the act defines the life cycle GHG emissions of gasoline as 88.84 g-CO_2 eq/MJ, biofuels must not exceed 39.9 g-CO_2 eq/MJ.

2.5 *Ministry of the Environment (MOE)*

Since 2023, MOE has introduced carbon footprint-based evaluation for selected products under the Green Purchasing Act. This Act mandates government entities to procure products and services that contribute to reducing environmental impact. Under this Act, two levels of environmental performance criteria are defined: Standard 1 (higher environmental performance) and Standard 2 (minimum acceptable performance), along with Consideration Items, which outline desirable environmental practices during procurement. Government entities are encouraged to prioritize products that meet Standard 1 and to set target ratios for such purchases.

Carbon footprint disclosure has been added as a requirement under Standard 1 for certain categories, including imaging equipment, tile carpets, and products made from low-emission steel (with certified emission reductions). Some companies in these sectors disclose EPDs using the SuMPO Environmental Label program. At present, the disclosure of a carbon footprint using EPD alone is sufficient to meet the criteria—GHG emission levels themselves are not evaluated. In the future, policies may evolve to specify maximum GHG emission thresholds or adopt tiered benchmarks, thereby encouraging the procurement of products with lower GHG emissions.

MOE also operates the Green Value Chain Platform (MOE 2025b), which provides information and tools for organizations to calculate, report, and reduce GHG emissions across their supply chains. Internationally, the GHG Protocol (WRI and WBCSD 2013) is the de facto standard for quantifying GHG emissions across an organization's supply chain. The platform offers practical guidance based on the GHG Protocol and provides a database of GHG emission factors necessary for calculations. It also introduces emission reduction strategies for factories, best practices, and details on subsidies and financial incentives. In addition, the platform explains target-setting methodologies such as those used in the Science Based Targets initiative (SBTi) (Science Based Targets Initiative 2025). As of the end of March 2025, 1577 companies in Japan have joined the SBTi, the highest number of participating companies from a single country. MOE's efforts to provide practical information are considered to have contributed to this achievement.

Since 2006, MOE has operated a GHG Emission Inventory, Reporting, and Disclosure System, which requires businesses with significant GHG emissions to calculate and report their GHG emissions to the government. The government aggregates and publishes these reports. This system covers GHGs directly emitted from business operations and those associated with purchased electricity and heat—corresponding to Scope 1 and Scope 2 in the GHG Protocol (WRI and WBCSD 2013). However, there are differences in calculation methods between the MOE system and the GHG Protocol, requiring companies to be especially careful when reporting.

One notable difference is the treatment of J-Credit, which can be used to offset GHG emissions under the Japanese system but is not recognized for this purpose under the GHG Protocol. The J-Credit Scheme allows companies to support domestic emission reduction projects, reflecting the government's policy intent to direct funds toward national climate initiatives. However, due to methodological differences, emission values disclosed under the Japanese system are not directly compatible with those calculated under the GHG Protocol. Going forward, it will be important to design systems that consider domestic policy priorities while maintaining alignment with international initiatives.

2.6 Policy Development of LCA in Japan

LCA-related policies in Japan began with infrastructure development, notably the launch of an LCA database by METI in 1998. It is worth noting that energy analysis research, as a form of energy conservation strategy, had already started in the 1970s in Japan, and supply chain–based assessments similar to LCA were being conducted. However, these earlier efforts are beyond the scope of this discussion.

In 2000, METI began supporting the development of EPDs as a tool for disclosing life cycle-based environmental information. From FY2007, METI expanded its policy to promote LCA as a tool for environmentally conscious design in the manufacturing sector.

From 2009, under the leadership of METI and with MLIT, MAFF, and MOE, a Carbon Footprint Pilot Program was launched. Although the scope was limited to GHG emissions, the program was distinctive in that it sought to communicate numerical values directly to consumers.

The Carbon Footprint Pilot Program continued until FY2011, after which it was transferred to the private sector. Major policy-driven activities related to LCA ceased thereafter. This shift is believed by the author to be partly due to the Great East Japan Earthquake of 2011, which led to the shutdown of many nuclear power plants in Japan.

As nuclear energy—known for its low GHG emissions—was replaced with coal-fired power, GHG emissions from electricity generation increased. Because electricity is widely used across industries, the carbon footprint of many materials and products also rose. Consequently, little policy support was provided for the continued use or development of the carbon footprint scheme after it was handed over to the private sector.

However, in October 2020, the Japanese government announced its goal of achieving net-zero GHG emissions by 2050. In response, the application of the carbon footprint by various ministries was revived. MAFF began using carbon footprints to evaluate agricultural products and introduced environmental labeling for those with lower carbon footprint values. MLIT applied carbon footprint to building evaluation, initially supporting their calculation and disclosure. METI compiled updated guidelines on carbon footprint calculation and disclosure, creating an environment for voluntary actions in the industry while also supporting the development of PCRs by industrial sector. In 2023, MOE started utilizing the Green Purchasing Act to introduce economic incentives for carbon footprint calculation and disclosure. These efforts are seen as laying the groundwork for eventually setting benchmark values for carbon footprints in procurement and policy.

Although the use of carbon footprints in policy is rapidly expanding, it is important to note that the current focus remains limited to GHG emissions. One notable exception is MAFF's system, which explicitly accounts for trade-offs with biodiversity, making it a model worth highlighting. Other initiatives, however, do not yet directly address trade-offs with other environmental aspects. For example, EPDs used in evaluating buildings and building materials include assessments of multiple environmental impact categories beyond climate change, and the results are disclosed accordingly. Therefore, future policy development should consider integrating these broader environmental aspects into climate-based strategies.

3 Utilization of LCA for Evaluations beyond Environmental Aspects

3.1 Climate Risk Assessment

Typhoons and floods can cause disruptions and delays in supply chains, leading to significant economic losses in industrial activities. In severe cases, these disasters may also result in loss of human life, highlighting the importance of considering disaster-related risks for workers. As climate-related disasters are projected to increase in frequency and intensity, companies are expected to assess and manage these risks not only within their own operations but also throughout their supply chains.

While efforts to reduce GHG emissions are referred to as mitigation, activities aimed at reducing the impacts from climate change are known as adaptation. In this context, methods have been proposed to identify vulnerable points in supply chains that are at high risk of climate-related disasters, such as storms, extreme temperatures, floods, and disease outbreaks (Nakano 2021).

The method introduced here involves developing climate-related disaster risk factors, which are then multiplied by the level of economic activity and the number of workers in each supply chain process. This allows for the identification of processes with high climate-related disaster risk (Fig. 2).

As an example, Fig. 3 illustrates the results of an assessment of climate-related disaster risks affecting workers in Japan's automotive supply chain. Approximately 20% of the total risk was located domestically, mainly due to typhoons and extreme heat. However, 80% of the risk was found to be overseas, with the Philippines identified as the highest-risk region. This is likely because labor-intensive auto parts, such as wire harnesses, are produced in the Philippines, and the country also faces a high risk of human-related disaster impacts. Additionally, since Japan imports copper and nickel from the Philippines, risks in the mining sector were also noted.

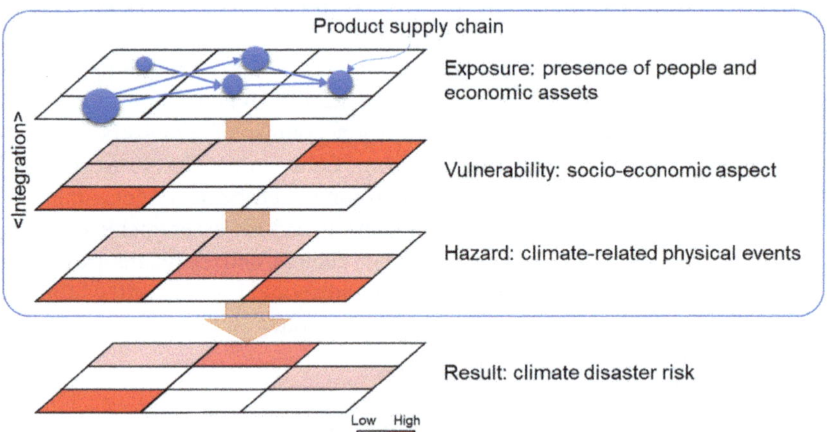

Fig. 2 Concept of the climate-related disaster risk assessment method (Nakano 2021)

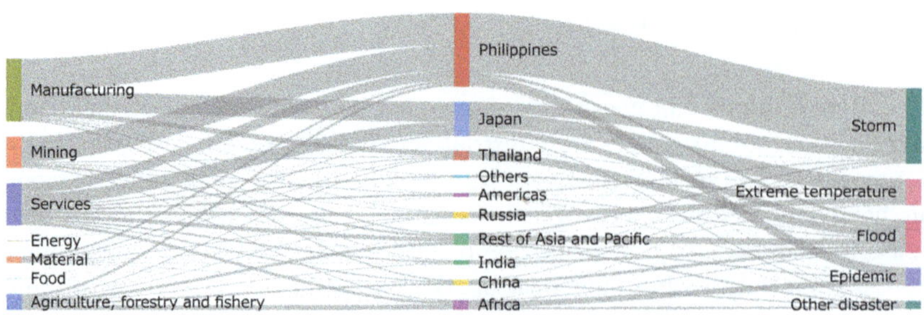

Fig. 3 Work climate-related disaster risk assessment results for Japan (Nakano 2021)

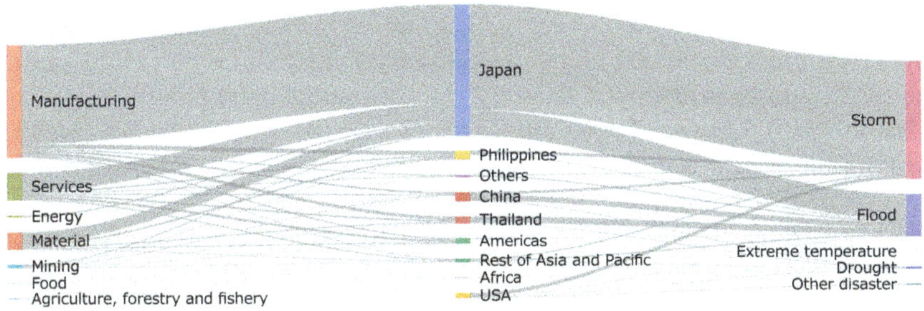

Fig. 4 Economic climate-related disaster risk assessment results for Japan (Nakano 2021)

In terms of economic risk, domestic manufacturing in Japan posed the highest level of risk, due to significant economic losses from recent typhoon-related disasters. Internationally, high-risk regions included the Philippines, Thailand, and China, reflecting broader climate vulnerabilities in Asia. Based on these findings, it is recommended that companies identify specific suppliers in these regions, assess their exposure to disaster risk, and implement appropriate adaptation measures where necessary (Fig. 4).

3.2 Resource Supply Risk

For a country like Japan, which relies heavily on imports for many of its natural resources, ensuring stable resource procurement is a critical issue. LCA enables the quantification of resource use by individual elements, and resource depletion is often evaluated using the inverse of each element's global reserve. However, real-world resource risks are not solely dependent on reserve quantities—they are also influenced by geopolitical factors and, thus, do not always correlate with physical availability.

For instance, a mining risk footprint was calculated for several metals used in electric vehicles and similar technologies (Nansai et al. 2015, 2017). In this assessment, neodymium was identified as having the highest mining risk, compared with platinum and cobalt.

Securing a stable supply of resources is an important policy issue. It is also necessary to consider resource policies using LCA.

3.3 Assessment of Social Aspects

Electric vehicles, which are promoted as a countermeasure to climate change, require significant amounts of rare metals such as copper, nickel, and cobalt. In addition to increased environmental impacts from abiotic resource consumption, there is growing concern about serious human rights violations associated with the mining of these resources, including conflicts with local communities and child labor. In some cases, improper labor practices are still reported in palm plantations (Andika and Ramli 2023). The use of palm oil as a feedstock for biofuels must not exacerbate these social issues. It is essential to ensure that focusing solely on climate change mitigation does not lead to the deterioration of other social problems.

The 2013 Rana Plaza collapse in Bangladesh, which resulted in the deaths of over 1000 workers, serves as a stark reminder of the consequences of neglecting social issues in global supply chains. Despite clear warning signs of the building's structural instability, factory owners forced workers to continue their duties, ultimately leading to the tragedy. Even if not directly related to climate change or environmental concerns, unsafe working conditions and other social issues—such as corruption—must be addressed and avoided throughout the supply chain.

In Japan, the "Guidelines on Respecting Human Rights in Responsible Supply Chains" were established in 2022 by METI in collaboration with related ministries. These guidelines outline methods for identifying and assessing adverse social impacts, preventing or mitigating such impacts, tracking the effectiveness of corporate efforts, and ensuring proper communication and disclosure.

Given Japan's heavy reliance on imported raw materials, identifying social risks at the upstream stages of long and complex global supply chains presents a major challenge. As a result, Social Life Cycle Assessment (S-LCA) is increasingly being used as a tool to support the identification of social risks within supply chains. According to ISO 14075:2024, S-LCA is defined as the "compilation and assessment of the socially relevant inputs and outputs and the potential social impacts of a product system throughout its life cycle"(ISO 2024).

By conducting S-LCA, companies can identify high-risk processes within their supply chains from a social perspective. Prominent databases that support the implementation of S-LCA include the Social Hotspots Database (SHDB) (Benoît Norris et al. 2022) and the Product Social Impact Life Cycle Assessment (PSILCA) (GreenDelta 2020). SHDB is based on GTAP (Aguiar et al. 2016), while PSILCA uses EORA (Lenzen et al. 2012)—both of which are structured using multi-regional input-output (MRIO) tables. However, MRIO-based models typically operate at a low sectoral resolution and do not track element-specific international material flows. Therefore, in the future, more detailed databases—such as those based on process-based LCA approaches—will be necessary to enable finer-grained social risk analyses.

4 Conclusion

LCA has been developed as a methodology aimed at achieving system-wide optimization. In Japan, government ministries are rapidly advancing the use of LCA as a climate change mitigation tool. Many of these efforts focus on promoting information disclosure—for example, MLIT is incentivizing the disclosure of carbon footprints for buildings and construction materials. Similarly, initiatives such as the Green Purchasing Law have begun to prioritize the procurement of products with carbon footprints, albeit in limited cases. In the agricultural sector, labeling schemes have been introduced to identify products with low carbon footprints, thus encouraging environmentally conscious consumption. While regulatory measures such as upper limits on carbon footprints have been implemented in the case of biofuels, no such standards have yet been introduced in other sectors. However, the establishment of such benchmarks may be considered in the future.

Currently, Japan's LCA-related policies primarily focus on the utilization of carbon footprint. However, with increasing EPDs that already quantify and disclose other environmental impact category indicators, it is necessary to expand policy frameworks to include these broader environmental dimensions. Furthermore, incorporating social aspects alongside environmental impacts is essential to prevent unintended adverse effects—such as increased human rights violations in upstream supply chains—and to better manage climate-related disaster risks. Moving forward, a more holistic approach that accounts for environmental, social, and economic dimensions will be critical to advancing truly sustainable decision-making.

Acknowledgments This work was supported by JSPS KAKENHI Grant Number 22 K12491.

References

Aguiar A, Narayanan B, McDougall R (2016) An overview of the GTAP 9 Data Base. Journal of global. Econ Anal 1(1):181–208. Accessed 14 April 2025

AIST Solutions (2025) Service guidance for AIST-IDEA. https://www.aistsolutions.co.jp/english/service/aist_idea/aist_idea_en.html. Accessed 17 December 2025

Anand CK, Amor B (2017) Recent developments, future challenges and new research directions in LCA of buildings: a critical review. Renew Sust Energ Rev 67:408–416

Andika W, Ramli D (2023) Child labor and unfree labor: evidence from the palm oil sector in Sabah (East Malaysia). Journal of human rights 22(3):351–366. Accessed 13 April 2025

Benoît Norris C, Norris GA, Aulisio D (2022) The Social Hotspots Database (Version 5.0) [Data set]. [Online]. https://www.socialhotspot.org/. Accessed 14 April 2025

GreenDelta (2020) PSILCA. [Online]. https://psilca.net/. Accessed 14 April 2025

Institute for Built Environment and Carbon Neutral for SDGs (2024) Whole Life Carbon Assessment Tool for Buildings (J-CAT). [Online]. https://www.ibecs.or.jp/zero-carbon_building/jcat/. Accessed 2 April 2025

ISO (2006a) ISO 14025:2006 Environmental labels and declarations—Type III environmental declarations—Principles and procedures. ISO. [Online]. https://www.iso.org/standard/38131.html. Accessed 30 March 2025

ISO (2006b) ISO 14040. Environmental management—Life cycle assessment—Principles and framework. ISO. https://www.iso.org/standard/37456.html

ISO (2006c) ISO 14044. Environmental management—Life cycle assessment—Requirements and guidelines. ISO. https://www.iso.org/standard/38498.html

ISO (2018) ISO 14067:2018 Greenhouse gases—Carbon footprint of products—Requirements and guidelines for quantification. https://www.iso.org/standard/71206.html

ISO (2024) ISO 14075:2024 Environmental management—Principles and framework for social life cycle assessment. [Online]. https://www.iso.org/standard/61118.html. Accessed 30 March 2025

Lenzen M et al (2012) Mapping the structure of the world economy. Environ Sci Technol 46(15):8374–8381. Accessed 14 April 2025

MAFF (2021) MIDORI Strategy for Sustainable Food Systems. [Online]. https://www.maff.go.jp/e/policies/env/env_policy/midori.html. Accessed 2 April 2025

MAFF (2024) Visualizing Efforts to Reduce the Environmental Impact of Agricultural Products: Contributions to Greenhouse Gas Reduction and Consideration for Biodiversity Conservation. [Online]. https://www.maff.go.jp/j/kanbo/kankyo/seisaku/being_sustainable/mieruka/mieruka.html. Accessed 2 April 2025

Matsuno Y et al (2013) LCA in Japan in the twenty-first century. Int J Life Cycle Assess 18(1):278–284

METI, MOE (2023) Carbon Footprints Practical Guide. [Online]. https://www.meti.go.jp/english/press/2023/0526_002.html. Accessed 2 April 2025

MOE (2025a) Carbon Footprint Introduction Guide. [Online]. https://www.env.go.jp/press/press_04413.html

MOE (2025b) Green Value Chain Platform. [Online]. https://www.env.go.jp/earth/ondanka/supply_chain/gvc/en/. Accessed 30 March 2025

MOE, METI (2025) Carbon Footprint Display Guide. [Online]. https://www.env.go.jp/press/press_04288.html

Nakano K (2021) Risk assessment for adaptation to climate change in the international supply chain. J Clean Prod 319:128785

Nakano K, Hirao M (2008) Introduction of LCA into design process on SME. Journal of Life Cycle Assessment Japan 4(3):237–243

Nakano K, Hirao M (2011) Collaborative activity with business partners for improvement of product environmental performance using LCA. Journal of cleaner production. [Online]. https://www.sciencedirect.com/science/article/pii/S0959652611000837

Nakano K et al (2007) Current activities of the life cycle assessment society of Japan. Int J Life Cycle Assess 12(7):546–546

Nansai K et al (2012) Estimates of embodied global energy and air-emission intensities of Japanese products for building a Japanese input-output life cycle assessment database with a global system boundary. Environ Sci Technol 46(16):9146–9154. Accessed 30 March 2025

Nansai K et al (2015) Global mining risk footprint of critical metals necessary for low-carbon technologies: the case of neodymium, cobalt, and platinum in Japan. Environ Sci Technol 49(4):2022–2031. Accessed 13 April 2025

Nansai K et al (2017) The role of primary processing in the supply risks of critical metals. Econ Syst Res 29(3):335–356. Accessed 13 April 2025

Röck M et al (2020) Embodied GHG emissions of buildings—the hidden challenge for effective climate change mitigation. Appl Energy 258:114107

Science Based Targets Initiative (2025) Ambitious corporate climate action. [Online]. Science Based Targets Initiative. https://sciencebasedtargets.org/. Accessed 3 April 2025

SuMPO (2025) SuMPO EPD. SuMPO EPD. [Online]. https://ecoleaf-label.jp/ Accessed 1 April 2025

The Government of Japan (2021) Cabinet Decision: Plan for Global Warming Countermeasures. [Online]. https://www.env.go.jp/content/000249336.pdf

The Government of Japan (2025) Cabinet Decision: Plan for Global Warming Countermeasures (in Japanese). [Online]. https://www.env.go.jp/earth/ondanka/keikaku/250218.html

WRI, WBCSD (2013) Technical Guidance for Calculating Scope 3 Emissions

Municipal Waste Management for Low Carbon Transition in China

Tao Wang, Zutao Zhang, Guiyan Shen, and Huanzheng Du

1 Introduction

Nearly two gigatonnes (Gt) of municipal solid waste (MSW) are annually generated in the world, in which two-thirds are from developing countries (UNEP 2015). The world's largest MSW producer as from 2004 is China. Its waste generation climbed from 118 megatonnes (Mt) in 2000 to 254 Mt in 2023 (Chaudhry et al. 2005, NBSC 2024). The big quantity of organic content in MSW, along with the growing amounts of waste plastic, increases the risks of climate change (UNEP 2015). Greenhouse gases (GHGs) from MSW mainly consist of carbon dioxide (CO2) from incineration of fossil carbon (primarily in plastic) and methane (CH4) from landfills and biological treatment (Liu et al. 2017). Waste treatment and disposal contribute to some 3% of GHGs emissions in China (Ma et al. 2024). In addition, there is a quantity of carbon resistant to fast degradation, leading to its deposit in landfills and dumping sites. Aging infrastructure and poor management of carbon deposit may provoke chronical or abrupt risks to the environment and public health.

In order to limit global warming within 1.5 °C, net zero emissions and carbon neutrality are to be accomplished by the mid-twenty-first century (IPCC 2018). Many countries from Europe to Asia set up their targets of carbon or climate neutrality

T. Wang (✉) · Z. Zhang · H. Du
Circular Economy Research Institute, Tongji University, Shanghai, China
e-mail: a.t.wang@foxmail.com

Z. Zhang
e-mail: xydst0@gmail.com

H. Du
e-mail: duhuanzheng@tongji.edu.cn

G. Shen
School of Resources and Environmental Engineering, Shanghai Polytechnic University, Shanghai, China
e-mail: 20231516062@sspu.edu.cn

W. Zhou (ed.), *Decarbonization Policy Engineering*, World Sustainability Series,
https://doi.org/10.1007/978-981-95-5953-4_25

(António 2020). China also pledged to achieve peak CO2 emissions before 2030 and to be carbon neutral by 2060 (The State Council of the People's Republic of China 2020). Emissions from the MSW sector should be substantially abated to fulfill the climate targets.

A body of research has been carried out to assess GHGs emissions from municipal waste and to explore mitigation strategies. Whereas sampling and analytical techniques can provide direct and more precise measurement of emissions (Cusworth et al. 2020; Lee et al. 2018), regional-and country-level assessment mainly relies on modeling (IPCC 2006). For instance, Wang and Zhang et al. quantified net GHG emissions from MSW across multiple provinces in China (Wang and Geng 2015, Zhang et al. 2024). Additionally, research has demonstrated that methane emissions from landfills constitute a significant source of GHG emissions (Zhao et al. 2019). However, to identify an effective strategy for emission reduction, it is essential to analyze waste sorting, which facilitates waste recycling, abates landfill volumes, and saves waste management costs (Zhang et al. 2021).

In 2017, under the guidance of China's State Council, 46 major cities, including Beijing, Tianjin, and Shanghai, were selected to initiate a pilot program for MSW sorting with the aim of establishing a standardized sorting model. As from 2019, following the enactment of local legislation, waste sorting was officially launched out in the 46 cities. At present nearly all prefecture-level and above cities in China have taken waste sorting into action. The policy requires mandatory separation of food waste, recyclable materials, and hazardous waste from municipal streams.

The environmental benefits of waste sorting are evident. Taking Shanghai as an example, the implementation of waste sorting had led to >16% reduction in landfill loads, 13.6% decrease of moisture content in waste for incineration, and 16.2% increase of heating value for thermal recovery (Chen et al. 2020). As with the drop of the moisture content in waste, the leachate, which serves as a primary contaminant source for surface water, soil, and groundwater pollution, had been substantially mitigated (Zhang et al. 2023). Sorting also effectively minimized the fly ash and heavy metal from waste incineration (Liu et al. 2023). Furthermore, Zhou et al. (2018) examined the evolution of environmental performance of Hangzhou's household waste management system. It showed that waste-related indicators from water pollution to resource efficiency and recycling had improved. Bian et al. (2022) investigated carbon emissions in Qingdao both before and after the implementation of household waste sorting. It demonstrated that the proper classification of food waste and the enhancement of recycling efficiency for recyclable materials could effectively reduce carbon emissions, and may even enable the waste management sector to achieve carbon neutrality. Ding et al. (2021) compared MSW management between cities in China with their intentional counterparts in Berlin, Singapore, and Tokyo. In contrast to highly developed cities, MSW in Chinese contained higher organic content and generated more GHGs (Ding et al. 2021; Yang et al. 2018).

Put in a nutshell, the MSW property and management practices were found performing a wide diversity, especially in a country as big, heterogenous, and fast changing as China (Dorn et al. 2012). Multiple studies had been conducted to assess the global warming potential of municipal waste in the country as a whole or in specific cities (Kölsch et al. 2007). Nonetheless, a truly comprehensive assessment of carbon metabolism of MSW is still needed. This chapter not only examines MSW and carbon flows across provincial administrative regions in China, but also investigates the influence of socio-economic factors on waste characteristics and management strategies. Through an evaluation of spatial and temporal variations in MSW composition, we can more precisely delineate waste management policies aimed at mitigating climate change challenges.

2 Methods and Data

A material flow analysis was carried out to assess municipal waste flows and carbon content embodied in MSW across 31 provinces in Mainland China in the period 2000–2023. Indirect carbon emissions caused by energy use of waste collection and transportation were not included in the system boundary. Illegal dumping and informal recycling were not counted neither, because dumping was restricted in the new century (Shanghai Chengtou Group Corporation (SCGC) 2021). The informal sector could be active in collecting, processing, and trading of recyclable materials, but the quantity and composition of informal recyclables can hardly be tracked province by province due to data limitations (Linzner and Salhofer 2014, CACE 2021, Shanghai Resource Recycling Association (SRRA) 2021).

2.1 Waste Composition

Nine components in MSW can be identified, i.e., food waste, plastic, textile, paper, metal, wood, stone, glass, and others. Degradable organic carbon (DOC) and fossil carbon content mainly exist in food waste, plastic, textile, paper, and wood (Wang and Geng 2015). Their composition in the gross weight of MSW are presented in Table 1. The moisture content in MSW ranges from 36% to 66% for different components.

Table 1 Composition of MSW in China by gross weight. Mean ± standard deviation of 184 waste samples from 50 cities are computed

Year	Food waste %	Plastic %	Wood %	Paper %	Textile %
1990s	65.4 ± 16.8	9.8 ± 3. 6	1.7 ± 0.8	6.5 ± 2.9	2.4 ± 1.1
2000s	50.1 ± 10.4	13.0 ± 4.5	2.5 ± 2.3	8.7 ± 4.4	3.7 ± 2.9
2010s	50.3 ± 12.6	13. 9 ± 4.5	4.0 ± 3.9	12.8 ± 4.0	3.2 ± 2.9

2.2 Carbon Flows

Carbon coming from paper, textile, food waste, wood, and plastic are computed by Eq. (1). *TC* represents the total carbon content in MSW. W is the gross weight of MSW. α_i represents the percentage of component *i* in MSW. v_i represents the moisture content of composition *i*. $c_{t,\,i}$ represents the carbon content in the dry weight of component *i*.

$$\mathrm{TC} = \mathrm{W} \times \sum_i \alpha_i \times (1 - v_i) \times c_{t,i} \tag{1}$$

The DOC flows are calculated by Eq. (2). *DOC* represents the degradable organic carbon flows in MSW. c_o is the aggregate DOC percentage in gross MSW. $c_{o,\,i}$ represents the DOC percentage in the dry weight of component *i*. W, α_i, and v_i are defined in Eq. (1).

$$\mathrm{DOC} = \mathrm{W} \times \sum_i \alpha_i \times (1 - v_i) \times c_{o,i} \tag{2}$$

The fossil carbon flows are computed by Eq. (3). *FoC* represents the fossil carbon content in MSW. c_f is the fossil carbon content in gross MSW. $c_{f,\,i}$ represents the fossil carbon content in the dry weight of component *i*. W, α_i, and v_i are defined in Eq. (1).

$$\mathrm{FoC} = \mathrm{W} \times \sum_i \alpha_i \times (1 - v_i) \times c_{f,i} \tag{3}$$

2.3 GHGs Emissions

GHGs emissions from municipal waste and food waste treatment processes are estimated as below.

2.3.1 Sanitary Landfills

GHGs emissions from landfills can be computed by two approaches, one relying on mass balance and another employing the FOD model (Kumar et al. 2004). More accurate FOD estimation requires time series landfilling information as long as 50 years, which can hardly be obtained for hundreds of landfills in China. The mass balance method is therefore adopted in this research, and historic waste compositions are considered to mitigate the possible overestimation of emissions.

$E_{L,\,CH4}$ is CH_4 emissions, and $E_{L,\,CO2}$ is CO_2 emissions from sanitary landfills. $\mathrm{W_L}$ W_L is the amount of MSW landfilled. dc_L is decomposed proportion of DOC during landfills. mcf_L is methane correction factors of sanitary landfills. μ_L is the proportion of CH_4 in landfills gas by volume. ox is methane oxidation factor in landfills. 12, 16, 44 are molecular weight of carbon, CH_4, and CO_2, respectively. CH_4 has a global warming potential some 25 times greater than that of CO_2. c_o is the aggregate DOC percentage in gross MSW. $E_{L,\,CH4,\,REC}$ is the recovery amount of CH_4 in landfills. $E_{L,\,CH4,\,FLA}$ is the flaring amount of CH_4 in landfills. $E_{L,\,CH4,\,REL}$ is the released amount of CH_4 in landfills. r_{CH4} is landfills gas recovery ratio. f_{CH4} is landfills gas flaring ratio. The computation is as follows:

$$E_{L,CH4} = W_L \times c_o \times dc_L \times mcf_L \times \mu_L \times (1 - ox) \times \frac{16}{12} \tag{4}$$

$$E_{L,CH4,REC} = E_{L,CH4} \times r_{CH4} \tag{5}$$

$$E_{L,CH4,FLA} = E_{L,CH4} \times f_{CH4} \tag{6}$$

$$E_{L,CH4,REL} = E_{L,CH4} - E_{L,CH4,REC} - E_{L,CH4,FLA} \tag{7}$$

$$E_{L,CO2} = W_L \times c_o \times dc_L \times \left[1 - mcf_L \times \mu_L \times (1 - ox)\right] \times \frac{44}{12} + E_{L,CH4,FLA} \times \frac{44}{16} \tag{8}$$

2.3.2 Simple Landfills and Dumping

GHGs emissions from simple landfills are calculated in Eq. (12 to 13). Emissions from dumps are assumed following the same computation. $E_{D,\,CH4}$ represents CH_4 emissions and $E_{D,\,CO2}$ represents CO_2 emissions from simple landfills and dumping. W_D is the amount MSW entering simple landfills and dumps. dc_D is decomposed proportion of DOC in simple landfills and dumps. mcf_D is methane correction factors of simple landfilling and dumping. μ_D is the proportion of CH_4 in landfill and dumping gas. c_o is the aggregate DOC percentage in gross MSW.

$$E_{D,CH4} = W_D \times c_o \times dc_D \times mcf_D \times \mu_D \times \frac{16}{12} \tag{9}$$

$$E_{D,CO2} = W_D \times c_o \times dc_D \times \left(1 - mcf_D \times \mu_D\right) \times \frac{44}{12} \tag{10}$$

2.3.3 Incineration

CO_2 emissions from MSW incineration are computed in Eq. (9). E_I represents emissions from incineration. W_I is the MSW amount for incineration. δ is oxidation factor in incineration. c_t is the total carbon content in gross MSW.

$$E_I = W_I \times c_t \times \delta \times \frac{44}{12} \tag{11}$$

2.3.4 Composting

The biodegrade carbon mainly turns to CO_2 after composting. The emissions are estimated in Eq. (10 to 11). $E_{C,\,CO2}$ represents CO_2 emissions from composting. $E_{C,\,CH4}$ represents CH_4 emissions from composting. W_C is composting amount of MSW. dc_C is the decomposed proportion of DOC during composting. dc_{CO2} is the decomposed proportion of CO_2 during composting. dc_{CH4} is the decomposed proportion of CH_4 during composting. c_o is the aggregate DOC percentage in gross MSW.

$$E_{C,CO2} = W_C \times c_o \times dc_{CO2} \times \frac{44}{12} \tag{12}$$

$$E_{C,CH4} = W_C \times c_o \times dc_{CH4} \times \frac{16}{12} \tag{13}$$

2.3.5 Anaerobic Digestion

Anaerobic digestion became a major food waste treatment approach since the wide adoption of waste sorting from 2019. The emissions are estimated in Eq. (14) to Eq. (16). $E_{AD,\,CH4,\,REL}$ is CH_4 emissions from anaerobic digestion, $E_{AD,\,CH4,\,REC}$ is CH_4 recovery volume, and $E_{AD,\,CO2}$ is CO_2 emissions. $W_{AD}\mathrm{W_L}$ is the amount of food waste for anaerobic treatment. $\mathrm{c_o}$ is the DOC content in food waste. dca_{CH4} is the proportion of DOC decomposed to CH_4. s_{CH4} is release rate of methane from anaerobic digestion. dca_{CH4} is the proportion of DOC decomposed to CO_2.

$$E_{AD,CH4,REL} = W_{AD} \times c_o \times dca_{CH4} \times s_{CH4} \times \frac{16}{12} \tag{14}$$

$$E_{AD,CH4,REC} = W_{AD} \times c_o \times dca_{CH4} \times (1 - s_{CH4}) \times \frac{16}{12} \tag{15}$$

$$E_{AD,CO2} = W_{AD} \times c_o \times dca_{CO2} \times \frac{44}{12} \tag{16}$$

2.3.6 Other Food Waste Treatment

Other food waste treatment processes count for a very small share in MSW management. The emissions are computed in Eq. (17 to 19). $E_{OFT,\,CH4,\,REL}$ is CH_4 emissions from other food waste treatment, $E_{OFT,\,CH4,\,REC}$ is CH_4 recovery volume, and $E_{OFT,\,CO2}$ is CO_2 emissions. $W_{OFT}\mathrm{W_L}$ is the amount of food waste for other treatment. $\mathrm{c_o}$ is the DOC content in food waste. dco_{CH4} is the proportion of DOC decomposed to CH_4. s_{CH4} is release rate of methane from other food waste treatment. dco_{CH4} is the proportion of DOC decomposed to CO_2.

$$E_{OFT,CH4,REL} = W_{OFT} \times c_o \times dco_{CH4} \times s_{CH4} \times \frac{16}{12} \tag{17}$$

$$E_{OFT,CH4,REC} = W_{OFT} \times c_o \times dco_{CH4} \times (1 - s_{CH4}) \times \frac{16}{12} \tag{18}$$

$$E_{OFT,CO2} = W_{OFT} \times c_o \times dco_{CO2} \times \frac{44}{12} \tag{19}$$

2.3.7 Net and Neutral GHGs Emissions

Net GHGs emissions include CO_2 from burning of fossil carbon and CH_4 from landfills, dumps, and other biological treatment, as computed in Eq. (20). E_n represents net GHGs emissions. $E_{L,\,CH4,\,REL}$ is the released amount of CH_4 in landfills. $E_{D,\,CH4}$ represents CH_4 emissions from dumping. $E_{C,\,CH4}$ represents CH_4 emissions from composting. $E_{AD,\,CH4,\,REL}$ is CH_4 emissions from anaerobic digestion. $E_{OFT,\,CH4,\,REL}$ is CH_4 emissions from other food waste treatment. W_I is the MSW amount for incineration. c_f is the fossil carbon content in gross MSW. δ is oxidation factor in incineration.

$$\begin{aligned} E_n = {} & 25 \times \left(E_{L,CH4,\mathrm{REL}} + E_{D,CH4} + E_{C,CH4} + E_{AD,CH4,REL} + E_{OFT,CH4,REL}\right) \\ & + W_I \times c_f \times \delta \times \frac{44}{12} \end{aligned} \tag{20}$$

CO_2 originating from biological content can be treated as carbon neutral. The quantity of neutral emissions from MSW can be estimated by Eq. (21). E_u represents neutral GHGs emissions. $E_{L,\,CO2}$ is CO_2 emissions from sanitary landfills. W_I is the MSW amount for incineration. c_o is the aggregate DOC percentage in gross MSW. δ is oxidation factor in incineration. $E_{C,\,CO2}$ represents CO_2 emissions from composting. $E_{D,\,CO2}$ represents CO_2 emissions from simple landfills and dumping. $E_{AD,\,CO2}$ represents CO_2 emissions from anaerobic digestion. $E_{OFT,\,CO2}$ is CO_2 emissions from other treatment. $E_{L,\,CH4,\,REC}$ is the recovery amount of CH_4 in landfills. $E_{AD,\,CH4,\,REC}$ is CH_4 recovery from anaerobic digestion, and $E_{OFT,\,CH4,\,REL}$ is the recovery from other treatment. The recycled CH_4 is finally converted to CO_2 after combustion or utilization.

$$E_u = E_{L,CO2} + W_I \times c_o \times \delta \times \frac{44}{12} + E_{C,CO2} + E_{D,CO2} + E_{AD,CO2} + E_{OFT,CO2} + \left(E_{L,CH4,REC} + E_{AD,CH4,REC} + E_{OFT,CH4,REC}\right) \times \frac{44}{16} \tag{21}$$

The total emissions are simply the sum of net plus neutral emissions.

$$E_{tot} = E_n + E_u \tag{22}$$

2.4 Carbon Deposit

The remaining carbon content may be mostly accumulated in landfills and dumps, difficult to be decomposed. The amount of carbon deposit can be estimated by the mass balance of total carbon minus released carbon, as presented in Eq. (23). *DeC* represents deposited carbon. *TC* represents the total carbon content in MSW. Other variables have been depicted in Eq. (4) to (19).

$$DeC = TC - \left(E_{L,CH4,\mathrm{REL}} + E_{L,CH4,\mathrm{REC}} + E_{D,CH4} + E_{C,CH4} + E_{AD,CH4,REL} + E_{AD,CH4,REC} + E_{OFT,CH4,REL} + E_{OFT,CH4,REC}\right) \times \frac{12}{16} - \left(E_{L,CO2} + E_I + E_{C,CO2} + E_{D,CO2} + E_{AD,CO2} + E_{OFT,CO2}\right) \times \frac{12}{44} \tag{23}$$

3 Results

3.1 Change of Carbon Content in MSW

A dataset of MSW composition was developed, containing 184 samples from 50 Chinese cities in 1990–2023. With this dataset, we attempted to identify the relationship between carbon content in MSW and its influential factors from personal income, living expenses, and retail sales of consumer goods. Multiple regressions had been testified, and a sigmoidal growth curve to GDP per capita was proved as more notable and ease of use (Fig. 1). The content of total carbon and DOC increased and finally reached a plateau with the income growth. The corresponding SGompertz model and parameters are displayed in the following equation.

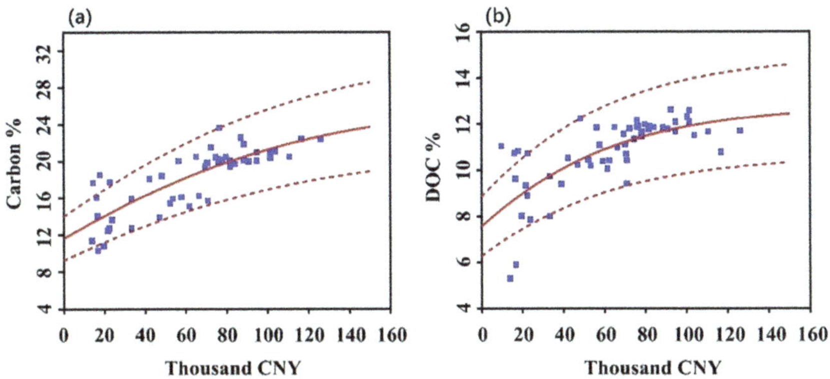

Fig. 1 Regression of carbon content in MSW vs. GDP per capita (in constant 2015 CNY) (a) Total carbon, and (b) DOC. The dashed lines indicate a confidence interval of 90%

$$c = a \times e^{-e^{-k(x-x_c)}} \tag{24}$$

c in Eq. (24) represents carbon content in MSW. x is GDP per capita. a is an asymptote when x tends to infinity. x_c defines the center and k sets the growth rate of the curve.

The regression results are presented in Fig. 1. The most fitted estimate is that $a = 26.796$, $x_c = -1.4218$ and $k = 0.1280$ for total carbon, and $a = 12.797$, $x_c = -3.2905$ and $k = 0.1961$ for DOC. The uncertainty range of the parameters is also analyzed. The biggest impact may be caused by parameter a, which changes from 21.336 to 32.256 for TC and 10.607 to 14.987 for DOC under 90% confidence interval. This regression model was employed to approximate the spatiotemporal variation of carbon flows in provinces of China.

3.2 Carbon Flows in MSW

The carbon content and flows throughout the municipal waste cycle were quantified for the 31 provinces in China as from 2000. The country-level carbon flow diagrams in 2000, 2010, 2018 and 2023 are displayed in Fig. 2. The values representing cumulative carbon flows are in megatonnes carbon (MtC). The figure portrays the carbon embodied in MSW only, indirect carbon emissions associated with MSW collection and transportation were not included.

Figure 2(a), (b), and (c) depict the variations in MSW carbon flow in China prior to extensive implementation of waste sorting. Figure 2(d) demonstrates a new paradigm of carbon flows with a greater amount of food waste been separately processed. The carbon flows embodied in MSW increased from 17.1 MtC in 2000 to 49.0 MtC in 2018. The value further grew up to 50.5 MtC in 2023, with 2.7 MtC contained in food waste separated from gross MSW for specific food waste treatment (mostly in composting and anaerobic digestion). For the country as a whole, the food waste

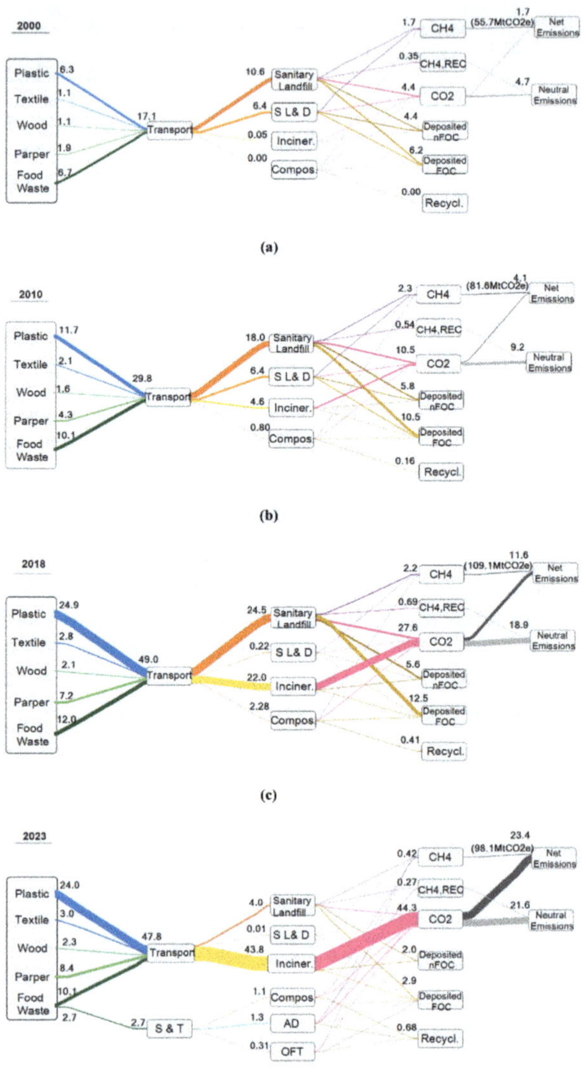

Fig. 2 Carbon Flow of MSW in China in (a) 2000; (b) 2010; (c) 2018; (d) 2023, values are in MtC per annum. S&T, sorting & transport; SL&D, simple landfill & dumping; AD, anaerobic digestion; OFT, other food waste treatments; FoC, fossil carbon; nFoC, non-fossil carbon

separation rate remained low but was increasing in a rapid pace. It enables a more efficient and effective way to recycle nutrients and energy from biological content in MSW.

The waste sorting action also substantially changed the treatment processes of MSW. Until 2018, there was still >50% of MSW been landfilled in China. The landfilling ratio, however, dropped to 7.4% in 2023, and the incineration ratio rocketed to 82.5%. As a result, incineration had become a dominant process influencing carbon metabolism. In addition, the amount of carbon flows deposited in landfills had reduced significantly.

3.3 GHGs Emissions

The carbon emissions from landfilling, incineration, composting, and dumping can be assessed with the equations above. Net GHGs emissions including CH_4 and CO_2 from fossil content increased from 55.7 $MtCO_2e$ in 2000 to 109.1 $MtCO_2e$ in 2018, then slightly sliding to 98.1 $MtCO_2e$ in 2023. The net GHGs emissions per tonne MSW had declined from 0.48 tCO_2e in 2018 to 0.39 tCO_2e in 2023, highlighting the usefulness of waste sorting for climate mitigation. In 2018, non-recycled landfill gases contributed some 68% of net emissions from municipal waste. But with extensive promotion of waste sorting and massive development of incineration facilities, now >80% of the global warming potential of MSW comes from burning of FoC-intensive waste plastics.

Our research found that the conventional approach employing constant composition (IPCC 2006) might overestimate MSW's carbon flows and GHGs emissions. By examining the temporal and spatial changes of waste composition and the influence of economic development, better estimation of carbon metabolism can be obtained. Averagely, per capita net emissions at provincial level could be 14% to 36% lower in our computation (Fig. 3). Gaps bigger than >30% were mainly observed in provinces in western China, including Gansu, Chongqing, Sichuan, Guizhou, and Tibe, where the landfilling rate was greater than the national average. The difference between the two methods was shrinking over time.

3.4 Regional Discrepancy of MSW and Carbon Management

For a country as large and unevenly developed as China, it is no wonder to witness spatial disparities in municipal waste and carbon management. Overall, the regional discrepancy diminished in the past two decades. Under the influence of waste sorting, although the per capita MSW carbon content in most provinces continues to rise (with the exception of Beijing and Shanghai), net GHGs emissions in the majority of provinces exhibit a downward trend. This indicates the positive role of enhanced waste management in reducing emissions. Following the implementation of the waste sorting and incineration policy, per capita net emissions in 25 out of the 31 provinces in China had been in a down trend. Two of the most economically developed municipalities, Beijing and Shanghai rank among the top of the provincial administrations in terms of per capita MSW carbon flow and net emissions. Greater per capita flows were also seen in provinces of Hainan and western China, where the population density is low. This suggest that the carbon metabolism might be affected by economic and demographic forces (Fig. 4).

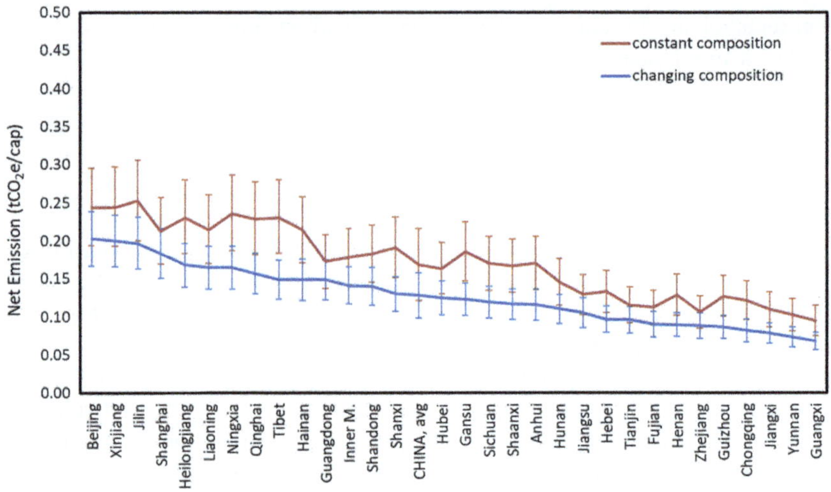

Fig. 3 Estimation of net emissions per capita based on constant and temporal provincial changing waste composition. Annually average per capita emissions in 2000–2023 are displayed for 31 provinces. The upper and lower whiskers define a confidence interval of 90%

4 Discussion and Conclusions

Our study explored the carbon flows of MSW in China from a spatial-temporal evolutionary angle. Despite municipal waste becoming increasingly carbon intensive, GHGs emissions did not increase proportionally in the past two decades. The rate of net emissions per tonne of MSW climbed from 0.47 tCO_2e in 2000 to 0.52 tCO_2e in 2010, but thereafter shrank to 0.48 tCO_2e in 2018 and 0.39 tCO_2e in 2023. Our analysis suggested that net emissions could be 14% to 36% lower than the conventional estimation assuming a static waste composition (Fig. 3). More significant increase in waste and carbon flows were perceived in provinces and cities in southern areas than in northern. The distinct spatial-temporal patterns of municipal waste and carbon metabolism could be interpreted by the following effects.

Increment effect indicates that urbanization and economic development could boost municipal waste generation. MSW kept increasing in China in 2000–2023, with an annualized rate of 3.4%. The increment was more significant in cities in southern China and southeast China, where stronger performance in economic vitality and labor immigration occurred. The trend likely maintains in short to medium term, as potential economic and urban growth is foreseeable in the near future.

Composition effect illuminates a rising carbon intensity (including both DOC and fossil carbon) in MSW. Higher standard of living can lead to use and discard of more carbon intensive materials. Nowadays in many cities, scrap paper and plastic have made some one-third of the municipal waste flows. Another noted composition change has occurred in coal ash from domestic heating. This carbon minimal waste, once constituting more than half of MSW in winter in northern China, has been

Fig. 4 Carbon flows in MSW in provinces in China, including (a) total carbon per capita, kgC/cap and (b) net GHGs emissions per capita, $kgCO_2e$/cap. The three columns from left to right represent the flows in 2000, 2018, and 2023

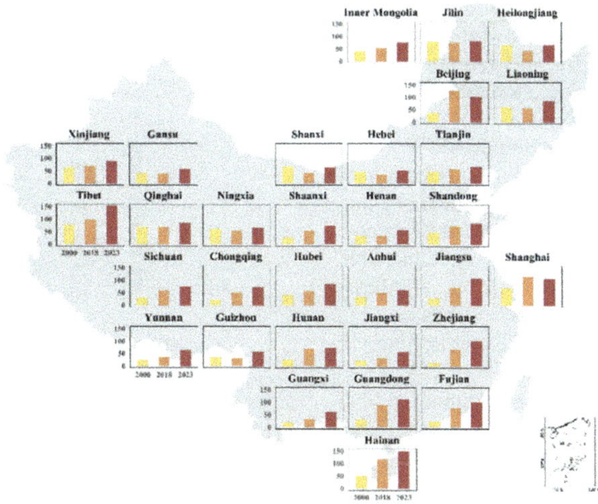

(a)

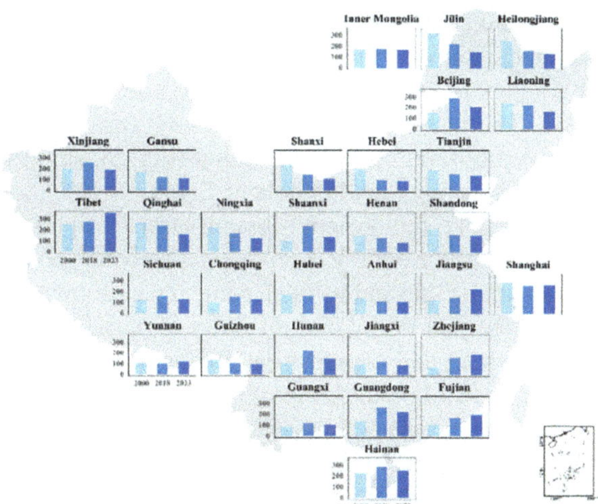

(b)

eliminated after pervasion of central heating and substitution of natural gas boilers for coal-based system. The removal of coal ash from municipal waste management also promoted a decline of MSW in northeast and northern China.

Management effect curbs GHGs emissions with more environmentally friendly treatment of MSW. In the 2000s, >80% of MSW might end up in sanitary or simple landfills (National Statistics Bureau of China (NBSC) 2024). Landfill gas collection and recycling systems were inadequately operated in many landfills (Lan et al. 2011). Methane in landfill gas became a major source of net GHGs emissions, generating

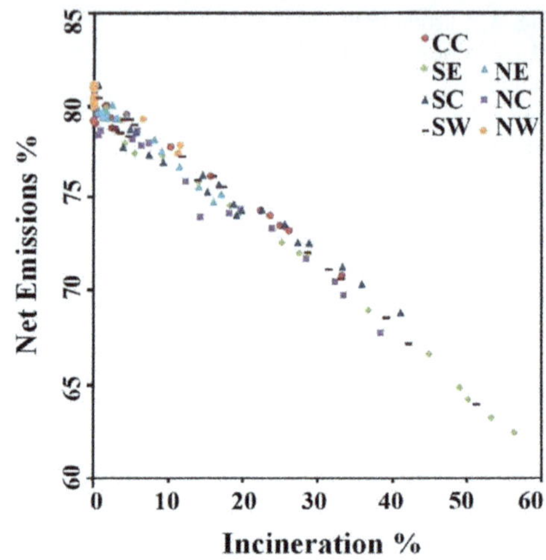

Fig. 5 Relationship between incineration ratio and net GHG emission ratio. Incineration ratio is the percentage of MSW incineration to MSW generation. Net emission ratio is the quotient of E_n to E_{tot}. Provinces are grouped into: CC, Central China; NC, North China; NE, Northeast China; NW, Northwest China; SC, South China; SE, Southeast China; SW, Southwest China

some 90% of the global warming potential. But today incineration has dominated mainstream treatment approaches and food waste biological treatment is rising. The assessment showed that the net emission from incineration was 0.20–0.33 tCO_2e per tonne of MSW treatment, lower than the landfill's emission of 0.50–0.70 tCO_2e/tMSW. The rate of net emission to total (net plus neutral) emission also declined with the proliferation of incineration (Fig. 5), suggesting that incineration was less climate damaging than landfills in China.

In summary, the present research offers insights into municipal waste flows and associated carbon emissions in China. Our future work will keep tracking the changes in municipal waste and its management paradigm and global warming impact. It can be expected that the spatial disparity of waste flows will diminish in the long run. The increment and composition effects may become less influential, because both quantity and carbon intensity of MSW may reach a peak toward a highly developed and material circular society (Emmanuel et al. 2010; Hoornweg et al. 2013).

On the other hand, the management effect will be likely dominant after some fundamental changes occurred in recent years. Compulsory sorting of municipal waste into food waste, recyclables, hazardous waste, and residuals had been implemented or trialed in some 300 cities in China as of 2023. The sorting system will be fully accomplished in almost all of the cities in 2030 (MOHURD 2019). Substantial investments had been pumped into facilities of waste sorting, food waste processing, and incineration (MOHURD 2023). Better waste management paradigm integrating waste sorting, material recycling, nutrient reclamation, energy recovery and incineration will substantially reform the present scheme of municipal waste and carbon metabolism. According to our analysis for the sorted waste in Shanghai, about 21% of the municipal streams were recyclables for material recycling. 32% were food waste from which energy and nutrients can be recovered by means of anaerobic digestion. The remaining 47% were mainly residual waste to be incinerated, and no more primary waste landfilled. Only the fossil carbon in residual waste can generate net emissions after incineration. Thus, a preliminary assessment demonstrated that the net GHGs emissions directly from waste treatment could be mitigated to 0.06–0.10 tCO_2e/t MSW, significantly lower than the national average of 0.47 tCO_2e/t MSW in 2018 and 0.39 tCO_2e/t MSW in 2023.

Elaborate research is to be conducted to investigate future waste flows under the circular economy paradigm and to evaluate its carbon neutrality potential. Reutilization of recyclable materials, reclamation of nutrients from organic waste and recovery of energy from incinerators can create both resource saving and carbon mitigation benefits and should be coherently counted across the country. Furthermore, a circular economy solution is needed to restore and regenerate aging waste landfills and dumps. Some 250 Mt. of carbon deposit in historic landfills must be safely treated or recovered, and the risks be minimized.

Acknowledgments The study was supported by the National Natural Science Foundation of China under Grant 71974144.

This chapter is based on the author's article, "Tracking the carbon flows in municipal waste management in China", published in *Scientific Reports*, Vol. 14, No. 1471 (2024). https://doi.org/10.1038/s41598-024-51698-0. We thank Jing Zhang, Peiyuan Xiao, Sha Lu, Gang Zhao, Jianfu Zhao, and Guangming Li for their permission to reuse their work.

References

António G (2020) Carbon neutrality by 2050: The world's most urgent mission|United Nations Secretary-General. https://www.un.org/sg/en/content/sg/articles/2020-12-11/carbon-neutrality-2050-the-world%E2%80%99s-most-urgent-mission

Bian R, Chen J, Zhang T, Gao C, Niu Y, Sun Y, Zhan M, Zhao F, Zhang G (2022) Influence of the classification of municipal solid wastes on the reduction of greenhouse gas emissions: a case study of Qingdao City. China Journal of Cleaner Production 376:134275. https://doi.org/10.1016/j.jclepro.2022.134275

CACE (2021) Personal communication with experts in CACE in Beijing, China, December 2021. Anonymity requested. China Association of Circular Economy

Chaudhry M, Hoornweg DA, Lam P (2005) Waste management in China: Issues and recommendations. World Bank. https://documents.worldbank.org/en/publication/documents-reports/documentdetail/en/237151468025135801

Chen S, Huang J, Xiao T, Gao J, Bai J, Luo W, Dong B (2020) Carbon emissions under different domestic waste treatment modes induced by garbage classification: case study in pilot communities in Shanghai. China Science of The Total Environment 717:137193. https://doi.org/10.1016/j.scitotenv.2020.137193

Cusworth DH, Duren RM, Thorpe AK, Tseng E, Thompson D, Guha A, Newman S, Foster KT, Miller CE (2020) Using remote sensing to detect, validate, and quantify methane emissions from California solid waste operations. Environ Res Lett 15(5):054012. https://doi.org/10.1088/1748-9326/ab7b99

Ding Y, Zhao J, Liu J-W, Zhou J, Cheng L, Zhao J, Shao Z, Iris Ç, Pan B, Li X, Hu Z-T (2021) A review of China's municipal solid waste (MSW) and comparison with international regions: management and technologies in treatment and resource utilization. J Clean Prod 293:126144. https://doi.org/10.1016/j.jclepro.2021.126144

Dorn T, Flamme S, Nelles M (2012) A review of energy recovery from waste in China. Waste Manag Res 30(4):432–441. https://doi.org/10.1177/0734242X11433530

Emmanuel CG, Anders D, Michael H, Göran F, Ola E, Susan T (2010) Models for waste life cycle assessment: review of technical assumptions. Waste Manag 30(12):2636–2648. https://doi.org/10.1016/j.wasman.2010.06.004

Hoornweg D, Bhada-Tata P, Kennedy C (2013) Environment: waste production must peak this century. Nature 502(7473):615–617. https://doi.org/10.1038/502615a

IPCC (2006) 2006 IPCC Guidelines for National Greenhouse Gas Inventories—IPCC. https://www.ipcc.ch/report/2006-ipcc-guidelines-for-national-greenhouse-gas-inventories/

IPCC (2018) GLOBAL WARMING OF 1.5 °C—an IPCC special report on the impacts of global warming of 1.5 °C above pre-industrial levels and related global greenhouse gas emission pathways, in the context of strengthening the global response to the threat of climate change, sustainable development, and efforts to eradicate poverty. https://www.ipcc.ch/sr15/

Kölsch F, Fricke K, Müller W (2007) Carbon trade project "biological waste treatment" for Gaobeidian (Pr China). CLEAN–soil, air. Water 35(5):504–507. https://doi.org/10.1002/clen.200720024

Kumar S, Gaikwad SA, Shekdar AV, Kshirsagar PS, Singh RN (2004) Estimation method for national methane emission from solid waste landfills. Atmos Environ 38(21):3481–3487. https://doi.org/10.1016/j.atmosenv.2004.02.057

Lan J, Li Y, Zhan L, Chen Y (2011) Landfill gas field pump test in one old landfill, China. International Conference on Electrical and Control Engineering 2011:4552–4556. https://doi.org/10.1109/ICECENG.2011.6058210

Lee H, Yi S-M, Holsen TM, Seo Y-S, Choi E (2018) Estimation of CO2 emissions from waste incinerators: comparison of three methods. Waste Manag 73:247–255. https://doi.org/10.1016/j.wasman.2017.11.055

Linzner R, Salhofer S (2014) Municipal solid waste recycling and the significance of informal sector in urban China. Waste Management & Research: The Journal of the International Solid Wastes and Public Cleansing Association, ISWA 32(9):896–907. https://doi.org/10.1177/0734242X14543555

Liu Y, Sun W, Liu J (2017) Greenhouse gas emissions from different municipal solid waste management scenarios in China: based on carbon and energy flow analysis. Waste Manag 68:653–661. https://doi.org/10.1016/j.wasman.2017.06.020

Liu Z, Fang W, Cai Z, Zhang J, Yue Y, Qian G (2023) Garbage-classification policy changes characteristics of municipal-solid-waste fly ash in China. Sci Total Environ 857:159299. https://doi.org/10.1016/j.scitotenv.2022.159299

Ma S, Deng N, Zhao C, Wang P, Zhou C, Sun C, Guan D, Wang Z, Meng J (2024) Decreasing greenhouse gas emissions from the municipal solid waste sector in Chinese cites. Environ Sci Technol 58(26):11342–11351. https://doi.org/10.1021/acs.est.4c00408

MOHURD (2019) Notice about municipal waste sorting in cities across the country. (in Chinese). Preprint at http://www.gov.cn/xinwen/2019/06/11/content_5399088.htm

MOHURD (2023) China urban construction statistical yearbook, 2022. MOHURD (Ministry of Housing and Urban-Rural Development of China)

National Statistics Bureau of China (NBSC) (2024) China statistical yearbook 2024. China Statistics Press. https://www.stats.gov.cn/sj/ndsj/2024/indexeh.htm

Shanghai Chengtou Group Corporation (SCGC) (2021) Personal communication with experts in SCGC in Shanghai, China, December 2021. Anonymity requested

Shanghai Resource Recycling Association (SRRA) (2021) Personal communication with Chen X in SRRA in Shanghai, China, December 2021

The State Council of the People's Republic of China (2020) Full text of Xi's statement at the General Debate of the 75th Session of the United Nations General Assembly. http://english.scio.gov.cn/topnews/2020-09/23/content_76731466.htm

UNEP (2015, June 3) Global Waste Management Outlook|UNEP—UN Environment Programme. https://www.unep.org/resources/report/global-waste-management-outlook

Wang Z, Geng L (2015) Carbon emissions calculation from municipal solid waste and the influencing factors analysis in China. J Clean Prod 104:177–184. https://doi.org/10.1016/j.jclepro.2015.05.062

Yang N, Damgaard A, Scheutz C, Shao L-M, He P-J (2018) A comparison of chemical MSW compositional data between China and Denmark. J Environ Sci 74:1–10. https://doi.org/10.1016/j.jes.2018.02.010

Zhang J, Du H, Wang T, Xiao P, Lu S, Zhao G, Zhao J, Li G (2024) Tracking the carbon flows in municipal waste management in China. Sci Rep 14(1):1471. https://doi.org/10.1038/s41598-024-51698-0

Zhang J, Yu X, Wang J, Sui Q, Zhao W (2023) Impacts of garbage classification and disposal on the occurrence of pharmaceutical and personal care products in municipal solid waste leachates: a case study in Shanghai. Sci Total Environ 874:162467. https://doi.org/10.1016/j.scitotenv.2023.162467

Zhang J, Zhang Z, Zhang J, Fan G, Wu D (2021) A quantitative study on the benefit of various waste classifications. Advances in Civil Engineering 2021(1):6660927. https://doi.org/10.1155/2021/6660927

Zhao Z, Bian R, Zhao F, Chai X (2019) Implications of municipal solid waste disposal methods in China on greenhouse gas emissions. Environ Prog Sustain Energy 39(3):e13372. https://doi.org/10.1002/ep.13372

Zhou Z, Tang Y, Dong J, Chi Y, Ni M, Li N, Zhang Y (2018) Environmental performance evolution of municipal solid waste management by life cycle assessment in Hangzhou, China. J Environ Manag 227:23–33. https://doi.org/10.1016/j.jenvman.2018.08.083

National Climate Contributions in Context: Greenhouse Gas and Aerosol Forcing from China and Japan, 1850–2020

Xuanming Su and Weisheng Zhou

1 Introduction

To meet the Paris climate reduction targets, it is necessary to conduct detailed attribution studies on climate change in China and Japan, as important economic entities in Northeast Asia. Between 1850 and 2016, China was responsible for 8–14% of all the anthropogenic radiative forcing (Su et al. 2022). Of this, CH4 made up a big chunk (0.1–0.16 Wm-2) compared to CO2's 0.26–0.3 Wm-2, but the negative aerosol forcing cancelled out a lot of the positive forcing. Some earlier studies focused on the evaluation of climate impacts resulting from short-lived climate pollutants (SLCPs), a subset of SLCFs including methane, tropospheric ozone, and black carbon. For example, a simulation study using an online aerosol-climate model found that the SLCPs-induced annual mean surface air temperature accounted for 43% and 55% of the global and Chinese climate impacts from 1850 to 2010, respectively, compared to those caused by CO2 (Xie and Zhang 2022). For future reductions, China needs to achieve a cost-effective reduction potential of 35% for methane, 30% for fluorinated gases, and other GHG emissions in 2030, compared to 2015 levels, to meet the 1.5 ° C climate target (Lin et al. 2022). However, the cooling effects of other aerosols such as sulfur dioxide (SOx) and ammonia (NHx3) are significant and will counterbalance a significant portion of the warming caused by GHG emissions. Reports suggest that reducing aerosol emissions in China between 2013 and 2019 could lead to a 0.1 ° C rise in regional average temperature (Gao et al. 2023).

X. Su (✉)
Research Center for Environmental Modelling and Application/Research Institute for Global Change, Japan Agency for Marine-Earth Science and Technology (JAMSTEC), Yokohama, Japan
e-mail: suxuanming@jamstec.go.jp

W. Zhou
College of Policy Science, Ritsumeikan University, Osaka, Japan
e-mail: zhou@sps.ritsumei.ac.jp

W. Zhou (ed.), *Decarbonization Policy Engineering*, World Sustainability Series,
https://doi.org/10.1007/978-981-95-5953-4_26

Japan accounts for 3.3–4.5% of the present-day global net anthropogenic radiative forcing accumulated since 1850 (Fu et al. 2021). The normalized marginal attribution shows that Japan's historical share of net forcing is modest in absolute terms but driven mainly by LLGHGs, with comparatively little aerosol masking, so further air-pollution controls pose limited short-term warming rebound (Su et al. 2022). Policy priority therefore squarely rests on deep CO_2 cuts (energy/industry) and management of non-CO_2 LLGHGs and F-gases, while scenario results indicate Japan's relative contribution remains low across Paris-aligned pathways—useful framing when contrasting with countries where aerosol declines unmask larger unrealized warming.

Chapter 6 presents the normalized marginal technique that was employed in this investigation. Using this approach, we ascribed the regional climate responses to both LLGHGs and SLCFs for China and Japan. Section 2 presents the emission datasets used in this study. The outcomes of radiative forcing and temperature increases caused by both LLGHGs and SLCFs are then shown for China and Japan in Sect. 3. Sections 4 and 5 presents the discussions and conclusions related to the topic.

2 Emission Datasets Used in This Study

Since the early 1980s, China's emissions have undergone rapid and sustained growth, closely tracking the country's industrial expansion and energy demand. CO_2 from fossil fuel combustion shows the steepest rise, especially post-2000, reflecting coal-dominated electricity generation and heavy industry. CH_4 emissions have also increased, driven by coal mining, agriculture, and waste, with a steady upward trajectory in recent decades. N_2O emissions show a gradual increase, largely from fertilizer use in agriculture. SO_2 emissions grew sharply until the mid-2000s, after which strong air-quality controls led to a pronounced decline, mirrored by decreases in NO_x and black carbon (BC) after peaking. Ammonia (NH_3) emissions remain persistently high, reflecting agricultural sources. Industrial fluorinated gases such as HFC-134a, HFC-23, and perfluorocarbons (CF_4, C_2F_6) have risen steeply in the past two decades, linked to rapid growth in refrigeration, air-conditioning, and electronics manufacturing. Overall, China's profile shows a shift from high aerosol-precursor emissions toward rising LLGHGs, implying an ongoing unmasking of GHG-driven warming, as shown in Fig. 1.

Japan's emissions profile over the last four decades is characterized by stabilization and, for many pollutants, decline. CO_2 emissions peaked in the early 2000s and have trended downward since, reflecting efficiency gains, fuel switching, and slower economic growth. CH_4 emissions have gradually declined since the 1990s due to improvements in waste management and reductions in fugitive emissions. N_2O emissions have remained relatively stable or slightly decreased. SO_2 and NO_x emissions fell sharply from the late 1970s onward, following stringent air-quality regulations and industrial controls, and remain low compared to historical levels. BC and organic carbon (OC) emissions have also declined significantly. In contrast, certain

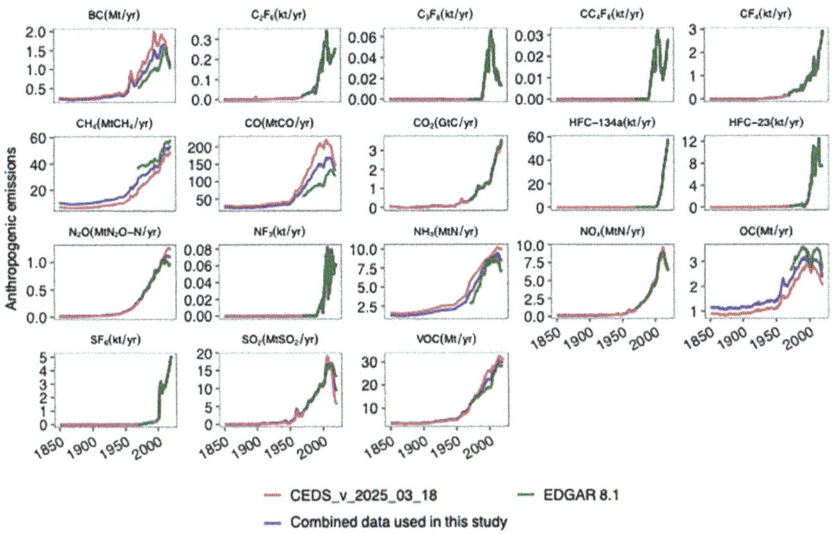

Fig. 1 China's emissions of greenhouse gases and aerosols. Data sources are from the Community Emissions Data System (CEDS_v_2024_07_08) (Hoesly et al. 2018) and the Emission Database for Global Atmospheric Research (EDGAR) v8.0_GHG 1970–2022 and EDGAR v8.1_AP 1970–2022 (Crippa et al. 2021, 2022). For the historical biomass burning, the BB4CMIP (Marle et al. 2017) (not shown) was used when the data was not available

HFCs (notably HFC-125, HFC-143a, HFC-245fa, HFC-32, and HFC-365mfc) have increased over the past two decades, reflecting their use as replacements for phased-out ozone-depleting substances under the Montreal Protocol and growth in electronics manufacturing. Japan's recent emission trends thus reflect a mature industrial economy with strong air-pollution controls, modest but persistent reductions in CO_2 and methane, and growth in some synthetic GHGs, as shown in Fig. 2.

3 Results

Prior to 1980, the emissions of CH_4 from agriculture and CO_2 from land-use change were the primary contributors to China's GHG radiative forcing, as illustrated in Fig. 3a. The significant positive radiative forcing resulting from GHG emissions was largely offset by the negative radiative forcing from aerosols and pollutants, leading to a relatively small total radiative forcing. The rapid increase in GHG emissions in China commenced in 1980, driven by swift economic expansion following the implementation of reforms and open policies (Fu et al. 2021; Su et al. 2022). The associated CO_2 forcing had started to increase until recently. Concurrently, the adverse radiative forcing caused by aerosols and pollutants, especially sulfate, which result from the burning of coal and oil, has also risen markedly. The total radiative forcing attributed

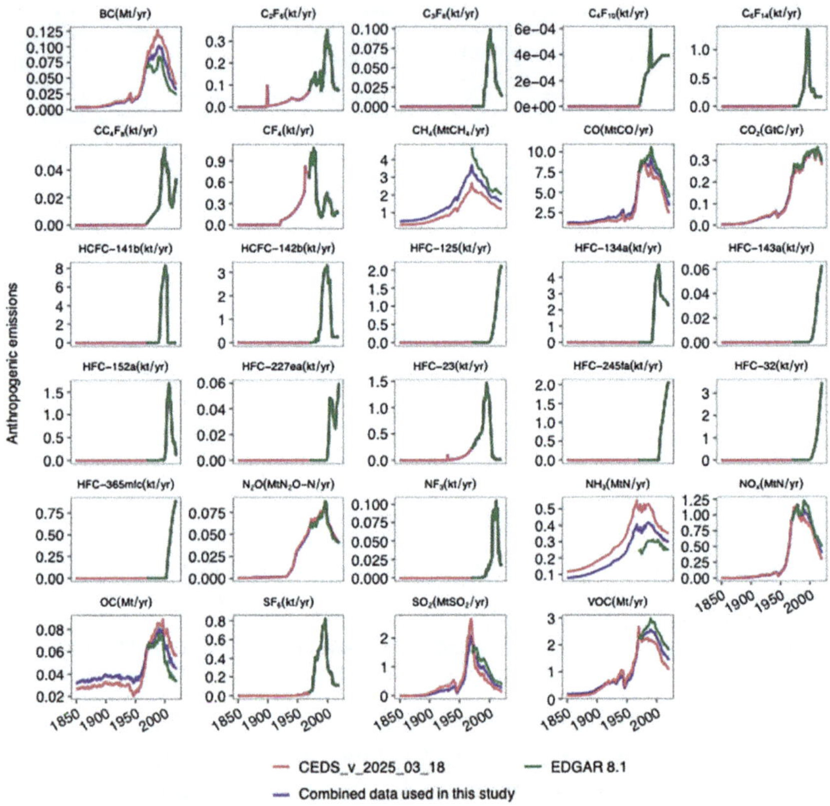

Fig. 2 Japan's emissions of greenhouse gases and aerosols. Data sources are from the Community Emissions Data System (CEDS_v_2024_07_08) (Hoesly et al. 2018) and the Emission Database for Global Atmospheric Research (EDGAR) v8.0_GHG 1970–2022 and EDGAR v8.1_AP 1970–2022 (Crippa et al. 2021, 2022). For the historical biomass burning, the BB4CMIP (Marle et al. 2017) (not shown) was used when the data was not available

to China prior to 2000 was comparatively minor. The sectoral contributions of CO_2 and CH_4 forcing align closely with global patterns, indicating that the energy, residential, industrial, and transportation sectors are the primary contributors to CO_2 forcing. In contrast, the agriculture, energy, and waste treatment sectors are the main drivers of CH_4 forcing (Fig. 3b). In 2020, the radiative forcing associated with LLGHGs was primarily connected to CO_2, while the radiative forcing from SLCFs was also significant, which could cancel a relatively large amount of the positive forcing.

As shown in Fig. 4a, the radiative forcing (ΔF) for Japan peaks near 0.06–0.07 W m^{-2} by 2020. CO_2-FF dominates the positive forcing, with CH_4 and N_2O contributing smaller shares. Negative forcing from aerosols—especially sulfate—partially offsets the GHG-driven positive forcing, although the magnitude of this offset is smaller than in China due to Japan's more diversified energy mix and earlier implementation of air-quality controls. Sectorally, the energy and industrial

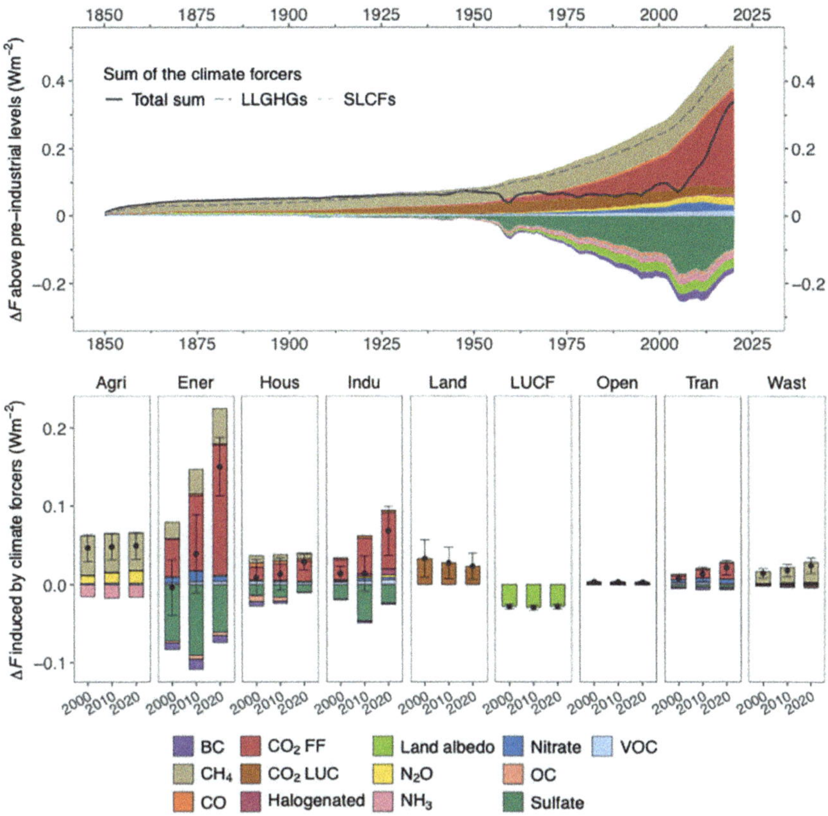

Fig. 3 Radiative forcing (ΔF) induced by climate forcers for China. a Radiative forcing above pre-industrial levels for China. The thick solid line shows the sum of the total; the thin dashed lines show the sums of LLGHGs and SLCFs climate forcers, respectively. **b** Radiative forcing induced by anthropogenic and natural climate forcers in 2000, 2010, and 2020 for China. The points and error bars show the sums of mean and one-sigma obtained from Monte Carlo simulations. Agri, Agriculture; Ener, Energy; Hous, Household; Indu, Industry; Tran, Transport; Wast, Waste; Open, Open burning; Land, Land use related emissions; LUCF, Land albedo; DUST, Dust; NATU, Natural forcing

sectors remain the largest positive forcing contributors, while transport and housing make modest contributions (Fig. 4b). The agricultural sector plays a minor role, and land-use change (CO_2-LUC) is negligible in recent decades.

For Japan, the relatively small absolute contributions do not diminish the importance of continued GHG reductions, particularly in the energy and industrial sectors, which remain the dominant sources of warming and forcing. Because aerosol cooling is modest, further reductions in aerosols will have a smaller short-term warming penalty compared to China, making it easier to align air-quality and climate objectives. In contrast, China must carefully coordinate aerosol and GHG reductions to avoid near-term warming spikes. Globally, the persistence of CO_2-FF as the primary

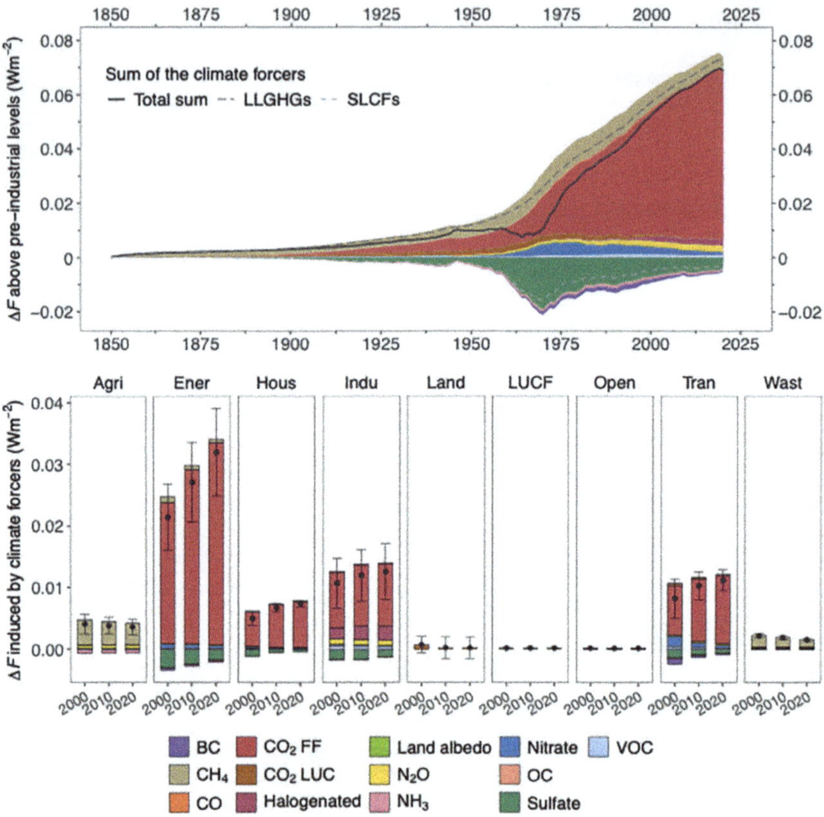

Fig. 4 Japan's radiative forcing (ΔF) induced by climate forcers.**a** Japan radiative forcing above pre-industrial levels. The thick solid line shows the sum of the total; the thin dashed lines show the sums of LLGHGs and SLCFs climate forcers, respectively. **b** Japan's radiative forcing induced by anthropogenic and natural climate forcers in 2000, 2010, and 2020 for Japan. The points and error bars show the sums of mean and one-sigma obtained from Monte Carlo simulations

forcing agent underscores the need for coordinated decarbonization across all major economies, including Japan, whose technological and policy leadership can support broader international mitigation efforts.

Figure 5 decomposes China's contribution to global-mean ΔT since 1850 by forcer and by emitting sector. The temperature increases resulting from China's early industrialization exhibited a gradual growth trend prior to 1980, primarily driven by agricultural CH_4 emissions and CO_2 from land-use changes (Fig. 5a). The overall warming caused by China remained stable until the early 2000s, influenced by the cooling effects of aerosols and pollutants. In Fig. 5b, CO_2-FF is persistently the largest single positive contributor, and CH_4 and N_2O add notable warming. Cooling from sulfate aerosols is strong but appears to weaken from 2000 to 2020, consistent with air-quality controls reducing aerosol burdens; organic carbon (OC) and NH_3-linked aerosol effects add smaller cooling. The sectoral rows indicate that energy and industry dominate the warming ledger (via CO_2-FF), agriculture contributes strongly through CH_4 and N_2O, transport grows in importance in the 2010s–2020s, and waste contributes primarily via CH_4. Land-use change (CO_2-LUC) and land-albedo effects are relatively large but with opposite signs in the recent decades shown. Overall, the net ΔT attributable to China reflects the balance of large, accumulating GHG warming increasingly exposed as aerosol cooling diminishes.

Japan's ΔT attributable to its emissions is small in absolute magnitude compared to China or the global total, reaching on the order of 0.03–0.04 °C by 2020 (Fig. 6a). The long-term trajectory shows a gradual rise from near-zero in 1850, with more rapid increases after the mid-twentieth century coinciding with Japan's post-war industrialization. Positive contributions are dominated by CO_2 from fossil fuel combustion (CO_2-FF), with smaller warming from CH_4 and N_2O. Cooling effects arise mainly

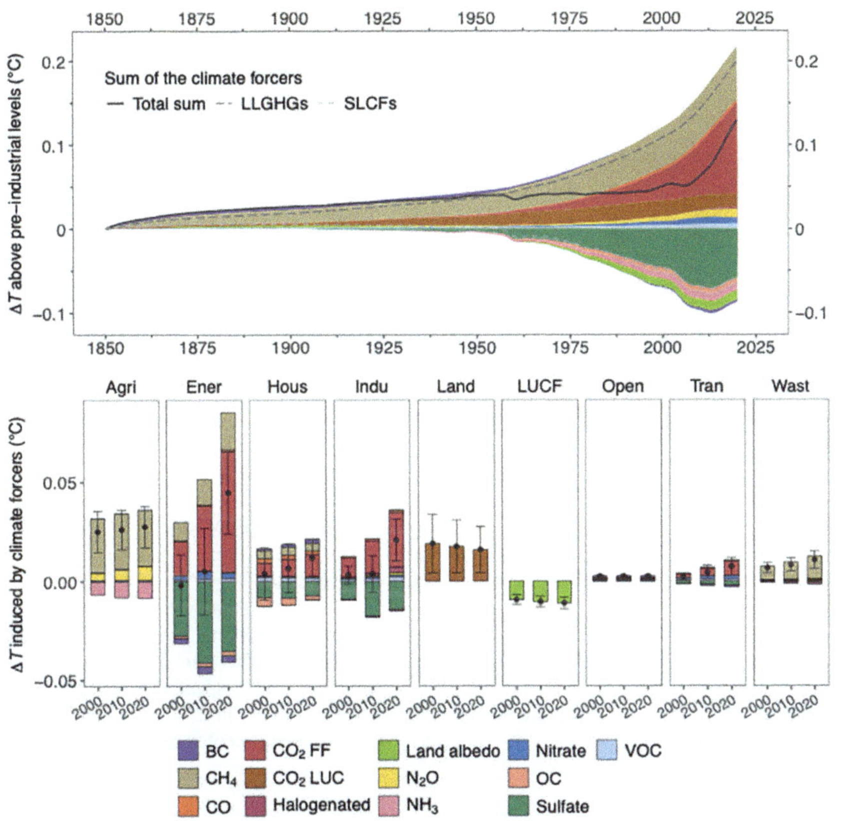

Fig. 5 Temperature increases (ΔT) induced by climate forcers for China. **a** Temperature increases above pre-industrial levels for China. The thick solid line shows the sum of the total; the thin dashed lines show the sums of LLGHGs and SLCFs climate forcers, respectively. **b** Temperature increases induced by anthropogenic and natural climate forcers in 2000, 2010, and 2020 for China. The points and error bars show the sums of mean and one-sigma obtained from Monte Carlo simulations

from sulfate aerosols, with additional minor contributions from organic carbon (OC). Sectoral decomposition indicates that energy and industry are the largest warming sources, with smaller inputs from transport and housing. Agriculture contributes little to the net temperature change, given Japan's relatively low agricultural emissions compared to its energy sector.

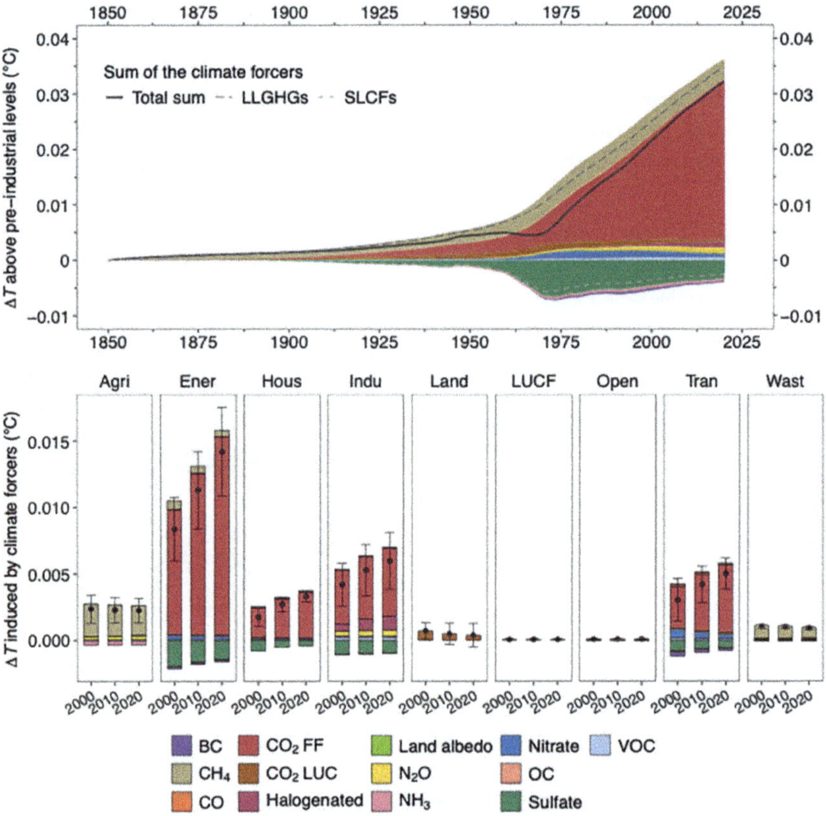

Fig. 6 Japan's temperature increases (ΔT) induced by climate forcers. **a** Japan's temperature increases above pre-industrial levels. The thick solid line shows the sum of the total; the thin dashed lines show the sums of LLGHGs and SLCFs climate forcers, respectively. **b** Japan temperature increases induced by anthropogenic and natural climate forcers in 2000, 2010, and 2020 for Japan. The points and error bars show the sums of mean and one-sigma obtained from Monte Carlo simulations

4 Discussion

The radiative forcing trends highlight a critical dynamic in China's climate influence: while total positive forcing is driven overwhelmingly by LLGHGs, SLCFs cooling from aerosols has historically masked part of this warming potential. As aerosol emissions decline due to air-pollution controls, the net forcing will more closely reflect the underlying GHG forcing, implying a stronger realized warming in the near term. This unmasking effect underscores the importance of sequencing mitigation policies—pairing aerosol reductions with aggressive GHG cuts to avoid abrupt temperature increases. The sectoral patterns suggest targeted opportunities: decarbonization of the energy and industrial sectors to curb CO_2-FF, rapid CH_4 abatement in agriculture, waste, and fossil systems for near-term benefits, and BC controls for both climate and air-quality gains. Uncertainties remain, particularly in aerosol forcing magnitudes and their evolution, as well as in forcer-specific efficacies in translating ΔF to temperature response. Including sensitivity analyses and uncertainty bounds will be essential in interpreting the results. Framed together with the temperature contribution analysis, these forcing results reinforce the conclusion that long-term climate stabilization for China hinges on deep CO_2 reductions, complemented by near-term cuts in CH_4 and co-emitted aerosols.

When comparing China's ΔT trajectory to the global pattern, several key differences emerge. First, China's total ΔT is smaller in magnitude than the global total, reflecting its share of cumulative global emissions and later industrialization onset. However, the shape of the curve is steeper for China after the 2000s, consistent with rapid post-1978 economic growth and industrial expansion. In China, CO_2-FF is also the dominant warming driver, but the relative share of aerosol cooling is larger than the global average, particularly in the late twentieth century, reflecting high aerosol emissions from coal-dominated energy systems. Consequently, the net warming attributable to China in earlier decades was more strongly masked, and the recent weakening of aerosol cooling reveals a sharper uptick in ΔT compared to the global trend.

Relative to China, Japan's ΔT and ΔF magnitudes are much smaller, reflecting its smaller population, lower total emissions, and earlier transition to lower-emission technologies. While both countries' warming is dominated by CO_2-FF, Japan's aerosol cooling is proportionally less important, leading to a smoother long-term warming trajectory without the sharp unmasking effect seen in China. Compared to the global results, Japan's share of total warming and forcing is modest, but its emission profile is more similar to other industrialized economies with mature energy systems and strong air-quality regulations. In all three cases (Japan, China, and the world), CO_2-FF is the leading driver of warming, with CH_4 and N_2O providing secondary contributions and aerosols providing partial offsetting cooling.

5 Conclusions

This study applies the normalized marginal attribution technique to quantify and compare the climate contributions of China and Japan from 1850–2020, focusing on both LLGHGs and SLCFs. Emission inventories were compiled from CEDS, EDGAR, and BB4CMIP, spanning energy, industrial, agricultural, transport, residential, and waste sectors. For China, rapid post-1980 industrialization drove steep increases in CO_2 from fossil fuels, CH_4, and N_2O, with aerosol precursors (SO_2, NO_x, BC) also rising until the mid-2000s before declining due to air-quality controls. This produced historically large aerosol cooling that masked part of the GHG-driven warming; the recent reduction of aerosols has begun to reveal the underlying warming. In 2020, China's climate contribution was dominated by CO_2 and CH_4, with aerosols contributing significant opposing effects. Japan's emissions show stabilization or decline in most pollutants since the late twentieth century. CO_2 emissions peaked in the early 2000s and have since decreased; CH_4 and N_2O have declined or stabilized; SO_2 and NO_x dropped sharply after the 1970s. However, fluorinated gases (HFCs, SF_6, NF_3) have increased in recent decades. Japan's aerosol cooling is modest, meaning further aerosol reductions will have limited short-term warming effects. In 2020, Japan's net radiative forcing peaked at 0.06–0.07 $W\,m^{-2}$, with CO_2-FF as the dominant driver and total attributable warming of 0.03–0.04 °C since 1850. The comparison highlights China's larger absolute contributions and stronger aerosol masking, while Japan's profile resembles other industrialized economies with mature energy systems and stringent air-quality measures. The results underscore the importance of coordinated GHG and aerosol mitigation in China to avoid abrupt warming and continued deep CO_2 reductions in Japan to align climate and air-quality goals.

Acknowledgements This work was supported by the Program for the advanced studies of climate change projection (SENTAN, Grant Number JPMXD0722681344) from the Ministry of Education, Culture, Sports, Science and Technology (MEXT), Japan. The computational resources were provided by the Japan Agency for Marine-Earth Science and Technology (JAMSTEC).

References

Crippa M, Guizzardi D, Banja M, Solazzo E, Muntean M, Schaaf E, Pagani F et al (2022) CO2 emissions of all world countries—JRC/IEA/PBL 2022 report. Publ off Eur Union. https://doi.org/10.2760/730164

Crippa M, Guizzardi D, Solazzo E, Muntean M, Schaaf E, Monforti-Ferrario F, Banja M et al (2021) GHG emissions of all world: 2021 report. Publ off Eur Union. https://doi.org/10.2760/173513

Fu B, Li B, Gasser T, Tao S, Ciais P, Piao S, Balkanski Y et al (2021) The contributions of individual countries and regions to the global radiative forcing. Proc Natl Acad Sci 118(15):e2018211118. https://doi.org/10.1073/pnas.2018211118

Gao J, Yang Y, Wang H (2023) Climate responses in China to domestic and foreign aerosol changes due to clean air actions during 2013–2019. Npj Clim Atmos Sci 6(1):1–11. https://doi.org/10.1038/s41612-023-00488-y

Hoesly RM, Smith SJ, Feng L, Klimont Z, Janssens-Maenhout G, Pitkanen T, Seibert JJ et al (2018) Historical (1750–2014) anthropogenic emissions of reactive gases and aerosols from the Community Emissions Data System (CEDS). Geosci Model Dev 11(1):369–408. https://doi.org/10.5194/gmd-11-369-2018

Lin J, Khanna N, Liu X, Wang W, Gordon J, Dai F (2022) Opportunities to tackle short-lived climate pollutants and other greenhouse gases for China. Sci Total Environ 842:156842. https://doi.org/10.1016/j.scitotenv.2022.156842

van Marle MJE, Kloster S, Magi BI, Marlon JR, Daniau A-L, Field RD, Arneth A et al (2017) Historic global biomass burning emissions for CMIP6 (BB4CMIP) based on merging satellite observations with proxies and fire models (1750–2015). Geosci Model Dev 10(9):3329–3357. https://doi.org/10.5194/gmd-10-3329-2017

Su X, Tachiiri K, Tanaka K, Watanabe M, Kawamiya M (2022) Identifying crucial emission sources under low forcing scenarios by a comprehensive attribution analysis. One Earth 5(12):1–13. https://doi.org/10.1016/j.oneear.2022.10.009

Xie B, Zhang H (2022) A simulation study on the climatic responses to short-lived climate pollutants changes from the pre-industrial era to the present. Front Earth Sci 10:1008164. https://doi.org/10.3389/feart.2022.1008164

Decarbonization of the Residential Sector and Human-Centered Air Conditioning Systems

Tomoyuki Chikamoto

1 The Impact of Activities in Buildings and Houses on Global Warming.

1.1 The Impact of Buildings and Houses on Sectoral CO_2 Emissions.

The movement toward green recovery, which accelerates decarbonization, and carbon neutrality, which means zero greenhouse gas (GHG) emissions, is accelerating, and in April 2021, Japan announced that it would significantly increase its GHG emissions target for fiscal 2030 from the previous target of 26% to a 46% reduction compared to fiscal 2013.

GHG, including CO_2, which causes global warming, are influenced by the very actions of living in buildings and houses. Figure 1 shows the trends in CO_2 emissions by sector.

"Industrial sector", resulting from emissions from factories and other sources, accounted for the largest proportion of Japan's total greenhouse gas emissions in fiscal year 2023, accounting for approximately 34% of the total. However, this has been gradually decreasing, down approximately 33% since 1990. This is due to thorough energy conservation efforts in factories and the globalization of production bases. "Transportation sector" (resulting from emissions from transportation such as cars, trucks, buses, and ships) followed, showing an upward trend in the 1990s, but has been declining since peaking around 2000 due to the spread of hybrid vehicles and the emergence of electric vehicles. Instead, "commercial and other sector", and "residential sector" are showing an increasing ratio. These sectors originate from the energy used in buildings and houses, "commercial and other sector" in offices

T. Chikamoto (✉)
College of Science and Engineering, Ritsumeikan University, Kyoto, Japan
e-mail: tomoyuki@se.ritsumei.ac.jp

W. Zhou (ed.), *Decarbonization Policy Engineering*, World Sustainability Series,
https://doi.org/10.1007/978-981-95-5953-4_27

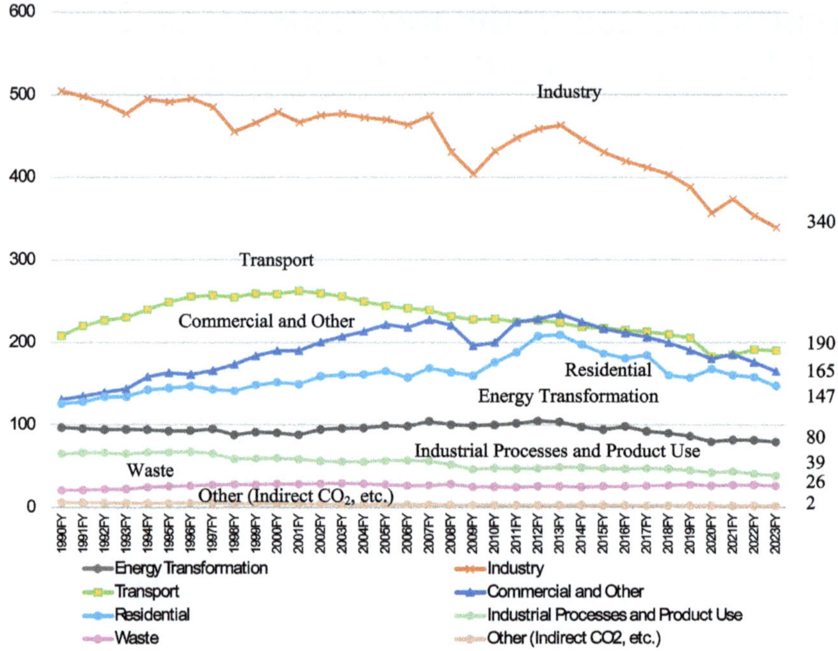

Fig. 1 Trends in CO_2 emissions by sector (after electricity and heat allocation) (FY2023 final, Ministry of the Environment) (Ministry of the Environment 2025)

and commercial facilities and "residential sector" in houses. Together, these two sectors account for approximately 36% of Japan's total, almost the same as "industrial sector". Furthermore, emissions have increased by approximately 38% since 1990, making it the largest target for reducing GHG emissions.

1.2 Characteristics of Japan's CO_2 Emissions.

The results of an evaluation of per capita CO_2 emissions in major OECD countries, including Japan, in terms of energy consumption efficiency and CO_2 intensity is shown in Fig. 2). The results are converted into scores 0 to 10, with 10 being the best and 5 being the average.

In energy consumption efficiency, an indicator of energy demand, Japan scores particularly highly in the residential sector, as well as in the transportation and industry sectors. This is thought to reflect the fact that, compared to other countries, Japan has a warm climate, where heating loads are not as high, and where high-efficiency equipment such as LED lighting and air conditioners are widespread. Furthermore, the transportation sector, where fuel efficiency is improving thanks to

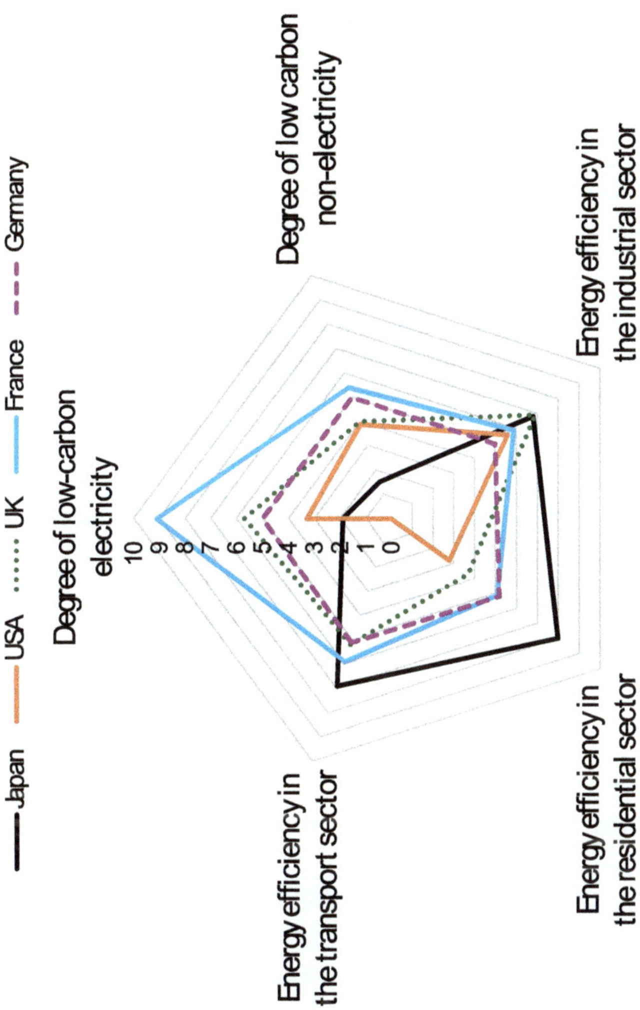

Fig. 2 Comparison of major countries in 2016 CO_2 emissions factor breakdown (Ministry of Economy, Trade and Industry 2019)

the spread of hybrid vehicles, and the industrial sector, where energy-saving measures are progressing, also received high marks.

CO_2 intensity, an indicator of the energy supply side, is notably low. This is due to a slow transition to renewable energy sources (7.6% for traditional hydropower and 15.3% for other power sources such as solar power), a low level of nuclear power plant operation (8.5%), and a reliance on thermal power generation (40.3% for natural gas and oil, and 28.3% for coal-fired power, which emits large amounts of CO_2) (Agency for Natural Resources and Energy 2025). In contrast to Japan, France is highly rated, with roughly 70% of its energy coming from nuclear power and roughly 20% from renewable energy. The UK also receives a high rating, with renewable energy and nuclear power each accounting for more than 20% and thermal power primarily relying on natural gas.

Japan has been promoting the use of renewable energy since 2009 with the Feed-in Tariff (FIT) system, under which electricity generated by renewable sources such as solar power is purchased at a high price for a set period. However, renewable energy, excluding hydroelectric power, still accounts for only around 10% of total electricity generation. The total amount of levies to cover this exceeds 2 trillion yen per year, placing a heavy burden of more than 20,000 yen per person. Shifting to a renewable energy-centered power generation mix and decarbonizing industry will take time, and that we will have no choice but to increase energy consumption efficiency. This means that measures to achieve a 46% reduction by 2030 will require even more emphasis on the residential and commercial sector.

The 6 h Strategic Energy Plan (2021) (Ministry of Economy, Trade and Industry 2021), which outlines the path for energy policy to achieve the goals of carbon neutrality by 2050 and a 46% reduction by 2030 compared to 2013, sets targets for emissions by sector of a 66% reduction in the residential sector and a 51% reduction in the commercial sector compared to 2013 levels by 2030, which is a larger reduction than other sectors (industry: 38% reduction, transportation: 35% reduction), supporting the above statement.

2 Measures for Carbon Neutrality in Buildings and Houses.

2.1 Trends in Architecture toward Carbon Neutrality.

In the field of construction, the Ministry of Land, Infrastructure, Transport and Tourism established the "Committee for Considering Carbon Neutrality in Construction" in 2021 and compiled a report. By 2030, new construction will be subject to the ZEB (net Zero Energy Building) and ZEH (net Zero Energy House) (see below) standards, with 60% of newly constructed detached houses being fitted with PV (photovoltaic power generation equipment). By 2050, the stock average will be guaranteed to achieve energy-saving performance at the ZEB and ZEH standard levels, with renewable energy adoption at a general level.

At the same time, the government will quickly publicize the need for each citizen, including businesses, to take this issue seriously and the specific measures to be taken, and will also publicize and spread information on how to make the most of highly energy-efficient houses, and provide information using methods from behavioral economics (nudge).

2.2 ZEB (net Zero Energy Building), ZEH (net Zero Energy House).

ZEB is defined as "a building whose annual net primary energy consumption is zero or nearly zero, achieved by reducing the primary energy consumption of the building through measures such as improving the energy-saving performance of the building and facilities, using energy on a regional scale, and utilizing on-site renewable energy" (Ministry of Economy, Trade and Industry, "Report of the Study Group on the Realization and Development of ZEB," November 2009).

However, since achieving ZEB all at once is a high hurdle, steps have been taken to promote its widespread adoption. The goals for achieving ZEB are divided into four categories, according to the level of reduction in energy consumption: (1) ZEB Oriented: Lowering the hurdle for widespread adoption, with an energy saving rate of 30% or more; (2) ZEB Ready: Advanced buildings with ZEB in mind, equipped with highly insulated exterior walls and highly efficient energy-saving equipment; (3) Nearly ZEB: Buildings that are as close as possible to ZEB, fulfilling the ZEB Ready requirements and using renewable energy to reduce annual primary energy consumption to nearly zero; and (4) ZEB: Buildings with net zero or negative annual primary energy consumption (Fig. 3).

2.3 ZWB (net Zero Water Building), ZWH (net Zero Water House) for recycling water in addition to energy.

The importance of water resources is being recognized worldwide, and the number of regions experiencing high water stress is rapidly increasing, particularly in Southeast Asia. Figures 4 and 5 show an experimental house initiative that aims to achieve net zero energy consumption, as well as self-sufficiency through the recycling of water (Chikamoto 2016).

The solar panels also serve as eaves, and behind them are installed rainwater storage tanks, solar hot water tanks, membrane modules, and other equipment, aiming to recycle water while using renewable energy.

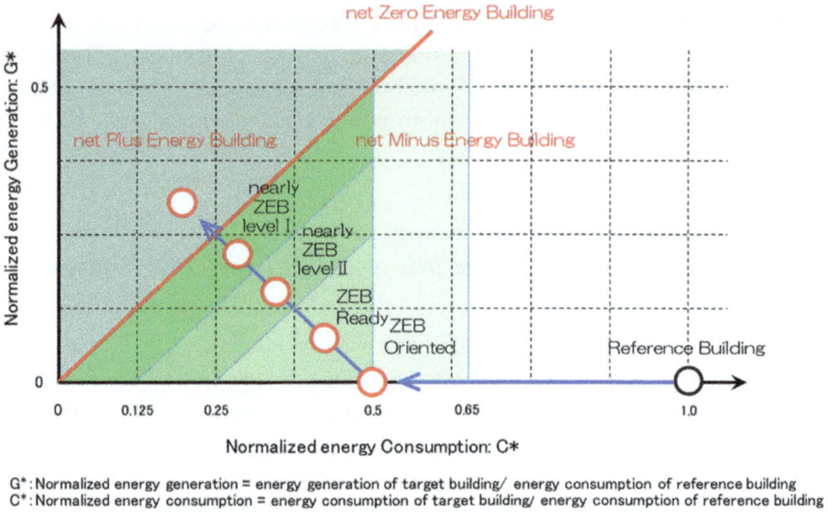

Fig. 3 ZEB Chart (Definitions and Evaluation Method of ZEB, The Society of Heating, Air-Conditioning and Sanitary Engineers of Japan (SHASEJ) 2018)

Fig. 4 An experimental ZWH that aims to recycle water in a ZEH

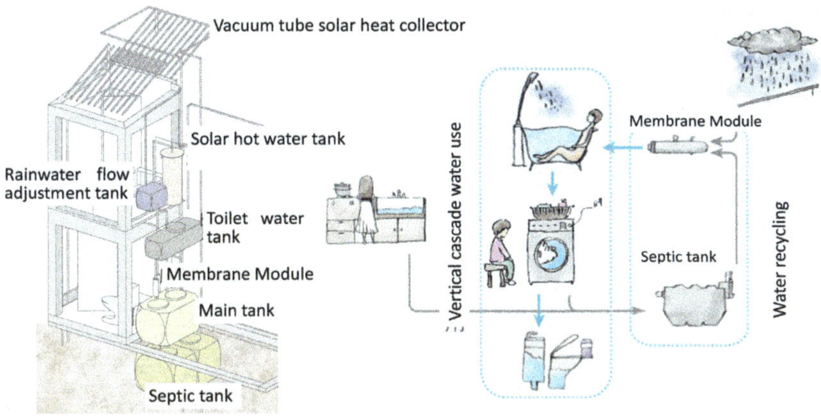

Fig. 5 Outline of water recycling system in ZWH

3 Initiatives toward Sustainability in Buildings and Houses.

3.1 Energy-Saving Methods.

To promote energy conservation, energy-saving and environmentally friendly technologies should be introduced. It is important to carefully consider the energy consumption characteristics of the target building, select effective methods, and reflect them in architectural and facility plans. In a typical office building, Energy consumption related to air-conditioning system accounts for 40 to 50% of the total, making it the largest breakdown. Energy consumption related to lighting system accounts for 20 to 30% of the total. Both are also key to reducing energy consumption. In air conditioning, energy conservation is important not only in heat source equipment, but also in heat transport, which transports hot and cold water and hot and cold air for heating and cooling. In houses, energy conservation is also important in air conditioning (especially heating), hot water supply, and lighting.

3.2 Use of Natural Energy.

The key to reducing energy consumption is to utilize natural energy. Natural ventilation not only cools the room with cool outside air, but also brings fresh air into the room and expels indoor odors, polluted air caused by human breathing and CO_2 from appliances, and humidity, making it essential for good health. Stagnant air due to insufficient ventilation can lead to infectious diseases and chemical sensitivity (sick building syndrome). However, when dealing with nature, we need to find ways to induce wind.

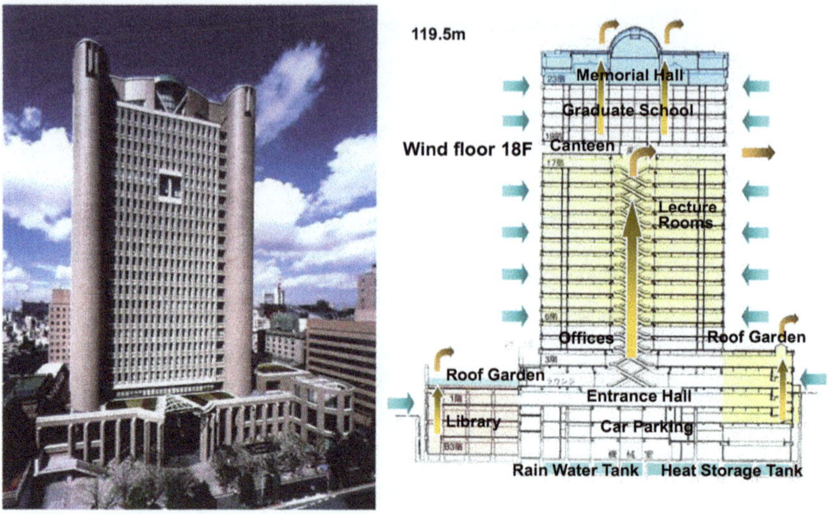

Fig. 6 Natural ventilation in a skyscraper (Meiji University Liberty Tower)

Natural ventilation is enforced in combination with stuck driven forces and cross ventilation in this building (Meiji University Liberty Tower, Tokyo, Japan) (Fig. 6) (Chikamoto et al., 1999). Despite being a skyscraper, the building utilizes natural ventilation. The escalator shaft atrium, which allows for quick movement between classrooms, and the ventilated floor on the 18th floor allows for wind to be drawn from both upwind and downwind.

Natural ventilation is dependent on nature, and does not necessarily create the same comfortable environment. For this reason, when the outside temperature rises slightly, people close the windows and turn on the air conditioner, and conversely, when the outside temperature drops, people often close the windows again because it's cold. This means that the time it can be used is ultimately limited to a very short period. Here, there is the concept of "hybrid ventilation," which uses both natural ventilation and air conditioning. Hybrid ventilation makes it possible to draw out the potential of natural ventilation.

The plan utilizes center core formed by escalator space between first and 17th floor, a major means of mass transit of students, as a course of airflow for natural ventilation, called "wind core". The wind core also causes the stack effect, which promotes natural ventilation. Above the wind core, on 18th level, the "wind floor" is planned to induce the natural ventilation for every floor. It maintains steady condition in the natural ventilation system, regardless of wind direction.

Each room of occupancy has ventilation openings under the glazed windows along perimeter zone to introduce outdoor cool air. The intake air is led to the wind core through a return path duct in which conditioned air also passes through to return. Details of the windows allow natural airflow to come into the rooms via the ventilation openings even when blind screens are shut during lectures using

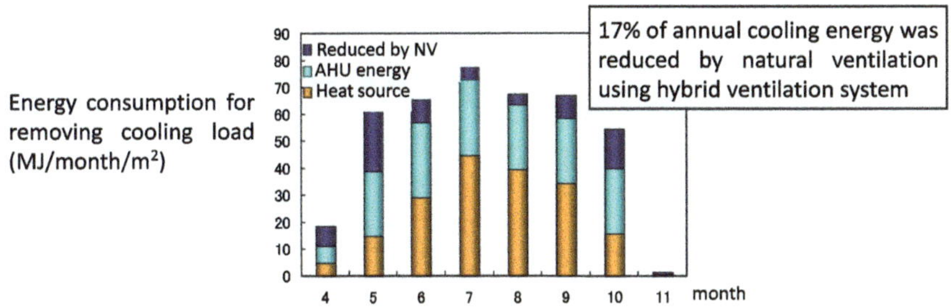

Fig. 7 Energy consumption for removing cooling load (Meiji University Liberty Tower)

audiovisual sets. These openings, with traps to obstruct outer noise, are installed inside window counter units preventing accidents and mischief when the operation is automatically controlled.

The air conditioning system supplements lack of cooling ability by natural ventilation to maintain indoor environment and comfort.

Annual energy for removing cooling load of this building, such as fans in air handling units, heat sources, pumps for carrying cold water from heat source from AHU. Total annual energy for cooling load was 415 MJ/m^2, which is 17% reduction by natural ventilation using hybrid ventilation system (Fig. 7).

3.3 Use of Natural Energy and Individual Control around Human for Comfort and Productivity.

Next building (Institute of Developing Economies, Chiba, Japan) considers the distribution of the indoor thermal environment and IAQ using TAC (task and ambient air-conditioning) system (Fig. 8) (Chang et al., 2004). With this consideration, the opportunity of hybrid ventilation will be extent wider.

Comfort is ensured in occupied areas (task areas) through air conditioning, while in ambient areas outside occupied areas (walking spaces, near ceilings, etc.), excess heat and pollutants are expelled through natural ventilation, saving energy and improving comfort. Windows can be left open to accommodate certain changes in outside temperature. Furthermore, air conditioner outlets for task areas are installed above each desk, and the air volume and direction can be adjusted to suit the occupant's preferences. All airflow is of the type that supplies fresh outside air, so it is supplied directly to occupants without diffusing it, helping to prevent infection and improve freshness for improving not only comfort but also productivity. Air conditioners in ambient areas can be stopped or operated with a relaxed thermal environment, increasing energy efficiency. Taking the outer air, this building uses the cooling/heating tunnel under the ground and solar collector (Fig. 9).

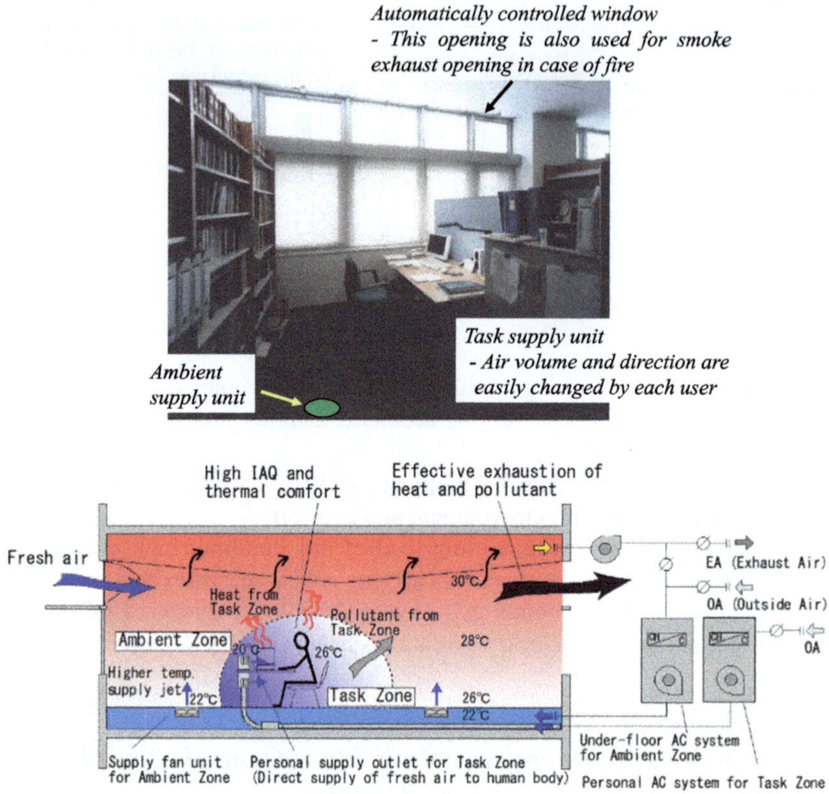

Fig. 8 Hybrid ventilation (Institute of Developing Economies)

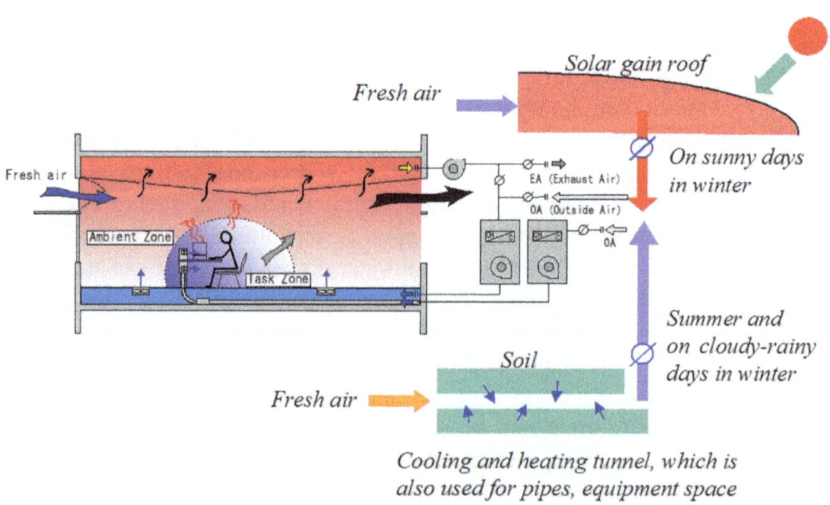

Fig. 9 Method for introducing outer air (Institute of Developing Economies)

3.4 *New TAC System for Individual Control around Human Improving Comfort and Productivity.*

Because of the improving the human comfort and easing the pre-set temperature of the room for the conservation of energy, various TAC systems were proposed. TAC system is a method for providing office workers with control of a local supply of air to control their individual thermal and air environment. TAC system can provide the conditioned air to an individual space, not to all the room, which leads to easy control of the occupied space with low energy.

Controlled variables are air temperature, velocity, direction of the supply air etc., however supply air flow is always jet created by fan. Therefore, vortex ring supplied from air-gun and its pulse flow control technic were paid attention as the new air flow of TAC. Vortex ring is easy to go straight and has more stability and less diffusivity than normal jet. Therefore, wind can be felt even if the TAC device is not set on the office worker's desk etc. and high rotating speed of the vortex ring also can be felt by office worker. The amount of the supplied air can be easily controlled by adjusting the suppling interval of the vortex ring (pulse airflow), which leads to the climate control corresponding to office worker's favor and effective airflow for cooling (Fig. 10). With these control methods with air-gun TAC, personal microclimate with high thermal comfort can be created (Chikamoto and Ueda 2023).

Figure 11 shows an overview of the vortex wheel delivery device newly designed and manufactured to solve these problems. The vortex wheel delivery mechanism and cool air supply method are the same as those of the previous device, but the angle can be adjusted by having an independent driving part. In addition, the type of piston was changed so that it can be installed in the ceiling of many office buildings, making it possible to reduce the height. The ambient airflow was set to 39 [m^3/h].

The vortex ring delivery device is installed on the ceiling of the laboratory, and cold air is supplied to the device from a cold air storage area through a duct.

Table 1 shows experimental cases. For the experimental cases, the room temperature was kept constant at 28 °C, and Case 0, in which the vortex ring was not sent out, and Case 1–3, in which the vortex ring sending air volume was varied, were set.

Fig. 10 Pulse air flow control using vortex ring. Appropriate control for each office worker's comfortable sensation by changing the sending interval of the vortex ring

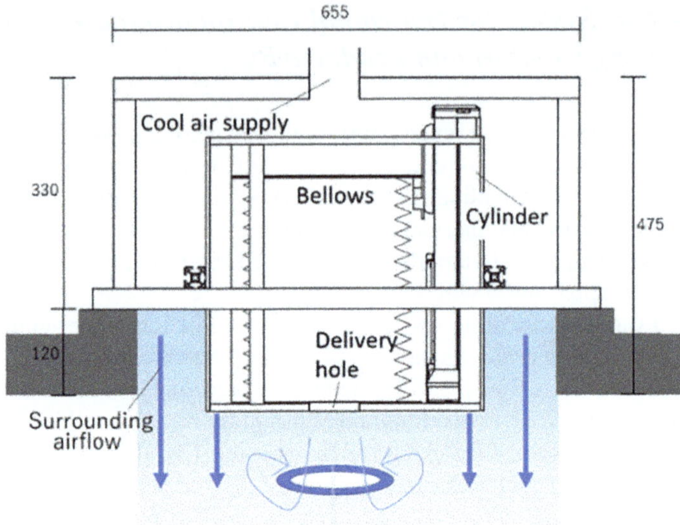

Fig. 11 Outline of vortex ring sending device

The surrounding blowing temperature was set at 18 °C. In addition, in all cases, the delivery angle was set to 30° so that the vortex ring hit the subject's neck, and the delivery interval of the vortex ring was his 1 second.

Thermal sensation result is the average value for all subjects. Figure 12 shows the results of thermal sensation declarations, and Fig. 13 shows the results of comfortable sensation declarations. Compared to Case 0, in which no vortex ring is delivered, Cases 1–3, in which vortex rings are delivered, show a decrease in the number of subjects reporting a hot side. The comfort level was also reported more often in Case 1–3 than in Case 0, indicating the cooling effect of the vortex wheel and its effectiveness as a personal air conditioner. However, Case 2 was superior to Case 3, which had the largest air volume, in terms of both cooling and comfort.

Table 1 Cases analyzed

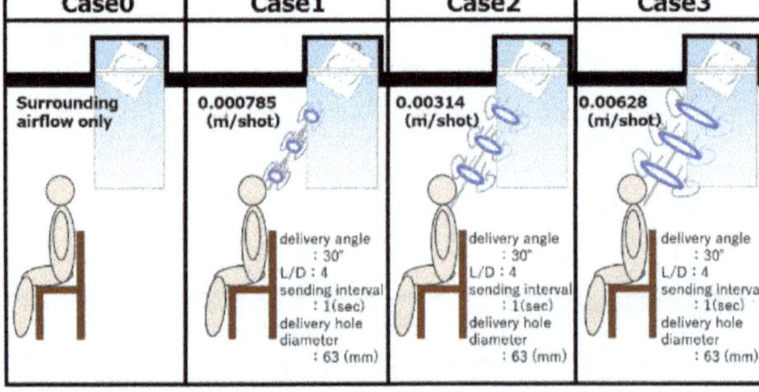

Case0	Case1	Case2	Case3
Surrounding airflow only	0.000785 (m^3/shot) delivery angle : 30° L/D : 4 sending interval : 1(sec) delivery hole diameter : 63 (mm)	0.00314 (m^3/shot) delivery angle : 30° L/D : 4 sending interval : 1(sec) delivery hole diameter : 63 (mm)	0.00628 (m^3/shot) delivery angle : 30° L/D : 4 sending interval : 1(sec) delivery hole diameter : 63 (mm)

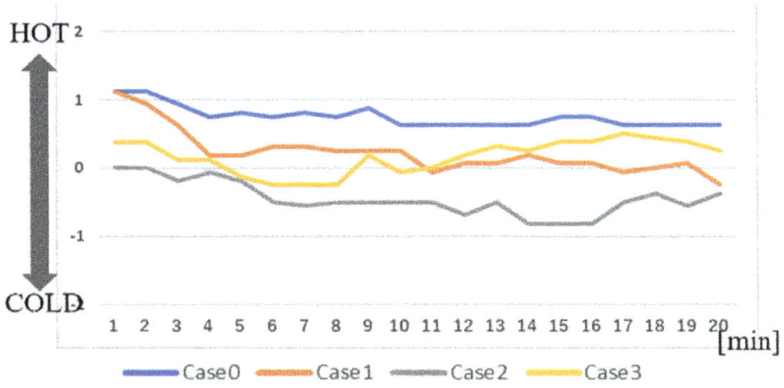

Fig. 12 Comparison of thermal sensation reports

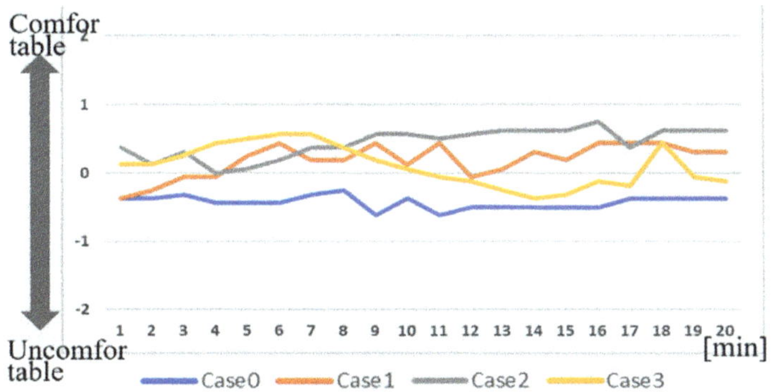

Fig. 13 Comparison of comfort reports

Figure 14 shows the results for airflow strength, and Fig. 15 shows the results for airflow range. With respect to the strength of the airflow, Case 2 is considered to give the greatest airflow sensation and influence the feeling of warmth, coolness, and comfort. In addition, since Case 3, which has the largest airflow, and Case 2 are comparable in terms of the range of airflow, it is considered that the airflow in Case 2 is sufficient for sending air to the neck, where the skin is less exposed. In addition, it is considered that the comfort feeling for airflow differs greatly depending on the individual's likes and dislikes for airflow (Fig. 16).

It was possible to deliver cool air to the subjects efficiently, suggesting that the air conditioner cooled the subjects more efficiently than conventional air conditioning.

In addition, since the range of airflow is similar in Case 3 and Case 2, it is considered that the airflow in Case 2 is sufficient to deliver to the neck, where there is less exposed skin.

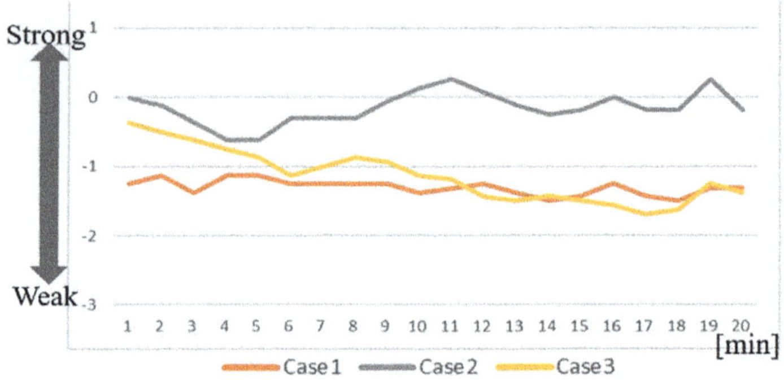

Fig. 14 Airflow strength comparison

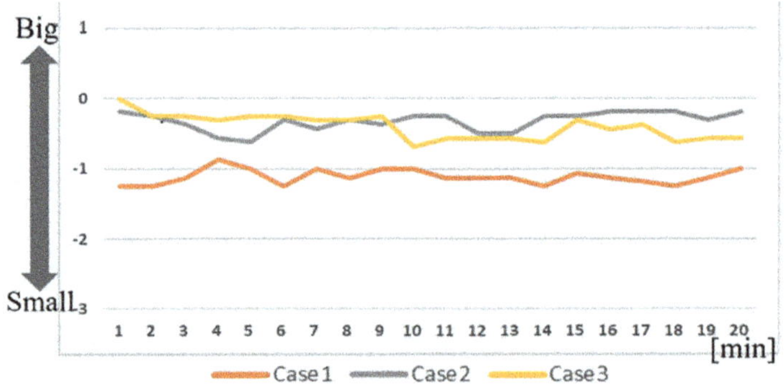

Fig. 15 Comparison of airflow range

4 Towards Sustainable Lifestyle.

4.1 Understanding CO_2 Emissions Associated with Daily Life to Consider Human-Centered Environmental Considerations.

The term "carbon footprint" has been used to indicate the amount of CO_2 emissions throughout a product's life cycle, from raw material procurement and manufacturing to distribution, sales, disposal, and recycling. It is a label used on environmentally friendly products, and was previously us Emissions from personal energy sources and automobiles can be calculated from monthly consumption of electricity, gas, gasoline, etc. This accounts for 20% of the total, with household energy consumption accounting for 15% and private car use for 5%, resulting in an annual average of 1.8 t-CO_2 per person.

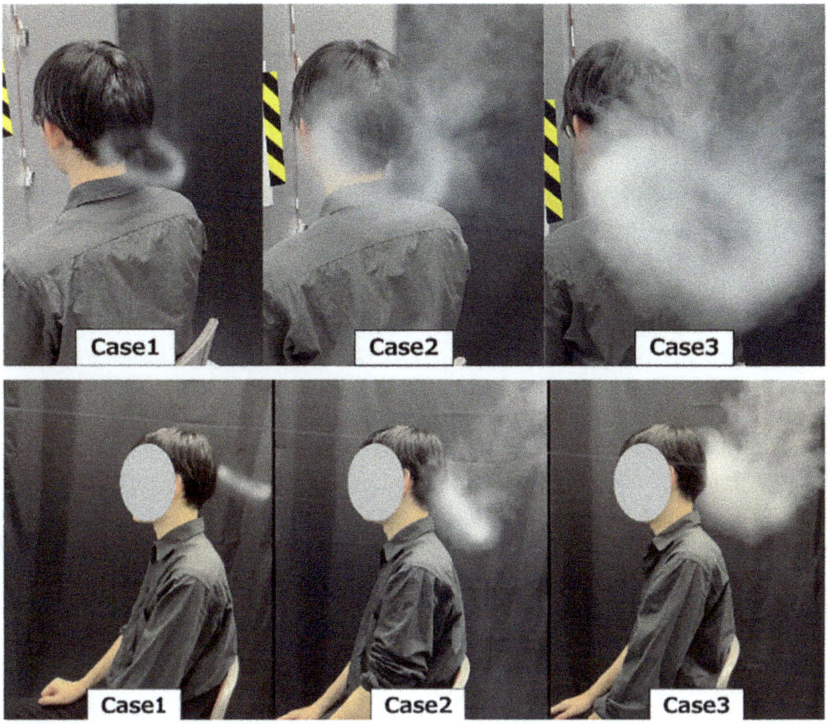

Fig. 16 Flow visualization

Meanwhile, carbon emissions are also involved in the production of food and daily necessities, transportation, and many other industrial and social activities that support our lives. Calculating how much carbon we emit in our daily lives, including emissions beyond the 1.8 t-CO_2, is called "carbon tracking". Apps like "eevie" and "Klima" help support environmentally conscious living by asking simple questions to calculate your carbon footprint, choosing from several actions you want to take, and making them a habit. Knowing the impact of lifestyle improvements on the environment provides incentives to work toward your goals. In addition, you can share your wisdom and struggles in improving your lifestyle through communities with others who are also working on similar initiatives. These efforts also lead to carbon offsetting, such as forestry resource development and support for the construction of solar power plants to offset emissions.

Visualizing CO_2 emissions can raise individual awareness and lead to the creation of a system that changes the awareness and behavior of individuals that tend to be overlooked in everyday life.

4.2 Behavioral Changes through Nudges.

Nudge means to give a gentle nudge with one's elbow, and refers to a policy technique that utilizes knowledge from behavioral science (behavioral insights) to guide people to voluntarily make choices and take actions that are desirable for society and themselves. Specifically, it encourages conformity by comparing people with others, or encourages behavior to avoid losses by emphasizing losses rather than gains.

Tokyo Electric Power Company collaborated with Opower, a U.S. energy information service, to offer a service called "Household Electricity Accounting". Opower uses open data published by the U.S. Energy Information Administration and the Census Bureau to divide customers into detailed segments, rank those with similar characteristics from first to last based on their energy-saving achievements, and advises them on specific energy-saving measures to improve their ranking. The Household Electricity Accounting System also stimulates a sense of competition by comparing households with similar ones, leading to consumers taking energy-saving actions. This comparison is based on behavioral science, which allows consumers to see whether their usage is excessive or whether more energy-efficient households have achieved even greater savings, encouraging them to take voluntary action.

4.3 Human-Centered Design.

Human factors refer to "human factors" or "human characteristics", and refer to the physical, physiological, psychological, and behavioral characteristics of humans. People perceive things differently depending on their race, gender, culture, the climate and region in which they grew up, their experiences and what they learned, their memories, and the situation in which they find themselves.

On the other hand, in architectural spaces, an index is created that averages the comfort and discomfort felt by people, and equipment is installed to adjust the thermal, light, and sound environment to recreate a physical environment that satisfies this index, and efforts are made to control and operate it. Human factors, along with improvements in sensor technology, involve understanding human conditions and shifting to human-centered control.

5 Summary of the current situation and issues raised.

Attention is shifting from considering the performance and functionality required of buildings and houses themselves to the people who live and work there. For example, activity-based working (ABW) responds to the diverse needs of workers by allowing them to freely choose where to work. It offers a relaxing cafeteria, communication spaces, individual booths for focused work, and spaces for group meetings and work.

It also provides places for people to rest and refresh when they feel tired. This aims to enhance creativity and intellectual productivity and lead to new value creation, leading to the expansion of shared offices. While it may seem like "free address" systems, where individuals are free to choose where to work without a desk, ABW is completely different from the concept of effectively utilizing space by eliminating empty seats. ABW focuses on the work style itself, improving individual performance and leading to comfort and health. It seems that the sustainability of buildings and cities also begins with a people-centered approach.

References

Agency for Natural Resources and Energy (2025) General energy statistics FY2023. April 25 2025

Chang H, Kato S, Chikamoto T (2004) Effect of outdoor air conditions on hybrid air conditioning based on task/ambient strategy with natural and mechanical ventilation in office building. Build Environ 39(2):153–164

Chikamoto T (2016) Ritsumeikan university water and living stacking box. Housing and electricity, No.6 (in Japanese)

Chikamoto T, Kato S, Ikaga T (1999) Hybrid air-conditioning system at liberty tower of Meiji university. HybVent Forum '99, Sydney, Australia

Chikamoto T, Ueda K (2023) Subject experiment on personal air-conditioning airflow using a vortex ring. The 11th International Conference on Indoor Air Quality, Ventilation & Energy Conservation in Buildings (IAQVEC2023), E3S. Volume 396

Ministry of Economy, Trade and Industry (2019) Energy white paper FY 2019. June 7 2020

Ministry of Economy, Trade and Industry (2021) The 6h strategic energy plan. October 22 2021

Ministry of the Environment (2025) Japan's national greenhouse gas emissions and removals in FY 2023. April 25 2025

The Society of Heating, Air-Conditioning and Sanitary Engineers of Japan (SHASEJ) (2018) ZEB in Japan

An Empirical Analysis on CO_2 Emissions from International Carbon Flow in East Asia

Hirotaka Haga and Weisheng Zhou

1 Introduction

Climate Change is a global issue. Global efforts are essential in reducing greenhouse gas emissions that cause climate change; however, the Kyoto Protocol adopted at the third session of the Conference of the Parties to the UN Framework Convention on Climate Change (COP3) in 1997 required only developed countries to reduce greenhouse gas emissions. The Paris Agreement was adopted at COP21 held in Paris in December 2015. The Paris Agreement is a fair and effective framework that requires the Parties to submit their nationally determined contributions NDCs to reduce greenhouse gas emissions and undertake efforts toward the achievement of such goals regardless of whether they are developed or developing countries. The Agreement entered into force in November 2016 and has been ratified by more than 190 countries including Japan (as of November 2021). (Ministry of Foreign Affairs of Japan (2022)).

Almost 10 years have passed since the adoption of the Paris Agreement, but the situation surrounding climate change is becoming increasingly serious.

The monthly anomaly of the global average surface temperature in December 2024 (i.e. the average of the near-surface air temperature over land and the SST) was +0.65 °C above the 1991–2020 average (+1.20 °C above the twentieth century average), and was the 2nd warmest since 1891. On a longer time scale, global average surface temperatures have risen at a rate of about 0.79 °C per century (Japan Meteorological Agency 2025).

H. Haga (✉)
Department of Business Economics, Faculty of Regional Design and Development, University of Nagasaki, Nagasaki, Japan
e-mail: hhaga@sun.ac.jp

W. Zhou
College of Policy Science, Ritsumeikan University, Osaka, Japan
e-mail: zhou@sps.ritsumei.ac.jp

W. Zhou (ed.), *Decarbonization Policy Engineering*, World Sustainability Series,
https://doi.org/10.1007/978-981-95-5953-4_28

Due to industrialization of emerging nations and globalization of the economy, the division of labor to provide goods and services spans international borders. Thus, environmental burdens are internationally distributed. There is an "embodied environmental load;" consumption of goods and services induces production of goods in multiple countries, and corresponding environmental loads are embodied in each country (Fujikawa et al. 2016).

CO_2 emissions can be generated by production and consumption. The IPCC Greenhouse Effect Gas Inventory provides a method for estimating production-based emissions, which are the total CO_2 emissions generated inside a country as a result of production in that country. Especially, Consumption is increasingly met by global supply chains often involving large geographical distances and causing emissions in producing countries (Hubacek et al. 2014, 2016; Wiedmann and Lenzen 2018). Therefore, accounting for emissions from production along the entire supply chain to fulfil final demand,—so-called consumption-based emissions (CBEs),—is necessary to understand why emissions occur and to what extent consumption choices and associated supply chains contribute to total emissions, and ultimately how to influence consumption to achieve climate mitigation targets and environmental justice (Vasconcellos 2020).

Production-based emissions (PBEs) and territorial emissions resulting from the production and consumption of goods and services within a region (for both domestic use and export) are often used by authorities to report carbon emissions (Peters 2008). PBEs also include emissions from international activities, which are excluded from territorial emissions (Karstensenet al. 2018; Shan et al. 2018). In contrast, CBEs refer to emissions along the entire supply chains induced by consumption, irrespective of the place of production (Liu et al. 2015). This reflects a shared understanding that a wider system boundary going beyond territorial emissions is important to avoid outsourcing of pollution and to achieve global decarbonization. CBEs allow for the identification of new policy levers through information on a country's trade balance of embodied emissions, households' carbon implications of their lifestyle choices, companies' upstream emissions as input for supply chain management, and cities' footprints outside their administrative boundaries (Davis and Caldeira 2010; Feng et al. 2013).[1]

These are calculated by summing the emissions from production of goods in the country of origin with emissions in the country consuming the goods.

There has been much research on analyzing interdependence of the international division of labor and change of industry and trade structure through input-output analysis, e.g. Shimoda and Watanabe (2005), Shimoda et al. (2009), Ishikawa (2012), Peters GP and Hertwich (2008), Peters et al. (2012), Mózner (2013), Pang et al. (2016), Fujikawa and Ban (2016). Ye et al. (2016), Tamura (2016), Yamdada et al. (2016), Takahashi (2018), Yang (2020),

Shimoda and Watanabe (2005) points out that the following points became clear as a result of our analysis from the viewpoint of value-added attribution. (1) Japan and the U.S. have relatively autonomous economic structures. (2) The rate of domestic

[1] Concerning the details, refer to IPCC (2022)'s previous research below.

production is declining in many East Asian countries. (3) Interdependence within the East Asian region has not necessarily strengthened; in fact, as of the early 2000s, East Asian countries are becoming more dependent on countries outside the region.

In recent research, Ye et al. (2016) focused on the international movement of value added to understand the transformation and current status of the international division of labor structure. In particular, they analyzed the division of labor structure among seven countries (regions): China, Japan, Korea, the US, the EU, and the rest of the world (RoW). In particular, the transformation and current status of the international division of labor in the global value chain were examined using the international input-output table developed by the WIOD. As a result, this chapter confirms the growing dependence on China in the production activities of each country, and this trend is particularly pronounced in the Republic of Korea.

Otherwise, With regard to previous discussions on the interdependence of environmental burdens and international carbon transfer associated with changes in the structure of industrial trade using input-output analysis, Nakamura et al. (2002), Imura et al. (2005), who studied the interdependence of changes in the structure of industrial trade and environmental burdens in Japan, the US, and Asia, and Na (2006), who conducted a study on similar viewpoint like CO_2 attributed emissions, such as Na (2006), and Fujikawa et al. (2016), who used the Asian International Input-Output Table to estimate "embodied energy" and "embodied CO_2" in the Asia-Pacific region.

According to research results of Fujikawa et al. (2016), In the Asia-Pacific region, the 20 year period from 1985 to 2005 has seen an even clearer trend for developed countries like the U.S. and Japan to shoulder the burden of energy consumption and CO_2 emissions on China and other developing countries.

In addition, the World Input-Output Database (WIOD) database of global input-output tables was introduced by Dietzenbacher, E et al., and an increasing number of studies have analyzed the CO_2 emission status using WIOD. For example, Shimoda et al. (2009) and Fujikawa et al. (2016) discussed the interdependence of CO_2 emissions in East Asia, Yoshinaga (2013) attempted to construct and analyze the BRICs international input-output table using the WIOD table, and Haga (2015) analyzed CO_2 emissions associated with industrial activities between regions across national borders. Emissions associated with industrial activities across countries and regions, Haga (2015), and Haga and Zhou (2021), who expanded on these studies to analyze the international transfer of carbon.

As pointed out by Imura et al. (2005), a large part of the flows of environmental impacts associated with imports and exports of goods and services were formed mainly in the United States and China. In particular, the increase in CO_2 flows from China to Japan, the U.S., and other countries was remarkable along with high economic growth. According to Fujikawa et al. (2016), in 1985, China, Japan, and the U.S. were both producers (exporters of goods) and consumers (importers) of goods. By 2005, however, Japan and the U.S. had become more like importers of goods, and they began to depend on China and other Asian countries for energy consumption and CO_2 emissions. On the other hand, China, in contrast to both Japan and the U.S., trans-formed from an importer of goods to an exporter, and began to shoulder the CO_2 emission burden of other countries (Fujikawa et al. (2016) pp. 140–141).

In these studies of the international transfer of carbon, there have been a series of consumption-based and production-based arguments, which were briefly discussed in the background of this study. (Peters GP and Hertwich EG 2008; Peters GP et al. 2012; Mózner Z 2013; Pang J et al. 2016; Fujikawa and Ban 2016; Homma and Akimoto 2018).

Consumption-based CO_2 emissions mean emissions generated inside territory of the country, regardless of the kind of relevant activities. Equivalent to CO_2 emissions in common statistics. On the other hand, consumption-based CO_2 emissions describe CO_2 emissions generated to meet the domestic demand (consumption and investment) of the country, which can be estimated by adding/subtracting CO_2 emissions embodied in imported/exported goods on/from production-based $CO_{2.}$[2]

This paper examines the trend of CO_2 Emissions from International Carbon Flow centering on East Asia like Japan, China, Korea based on input-output analysis using the World Input-Output Database (WIOD). The analytical method in this research will be standard Input-Output analysis like methodology in sect. 2. Furthermore, the author has already published Haga and Zhou (2021) in a similar research field, and this chapter is based on the subsequent development of the research, such as updating the database and conducting analysis based on the integration of sectors in the input-output table as described below.

2 Methodology

2.1 Data

2.1.1 Description of World Input-Output Database

The World Input-Output Database (WIOD) was developed to analyze the effects of globalization on trade patterns, environmental pressures, and socio-economic development across a wide set of countries. The database covers 27 countries of the European Union and 13 other major countries over the period from 1995 to 2009. An important component of a national input-output table (IOT) is that total output by domestic industry is equal to the use of output from domestic industry, such that all flows in the economic system are accounted for[3] (Timmer ed 2012).

In the currently published WIOD 2016 Releases, World Input-Output Tables insist of data, covering 43 countries and ROW (rest of the world), and compositing a model for the rest of the world for the 2000–2014 period. Data for 56 sectors is classified according to the International Standard Industrial Classification revision 4 (ISIC Rev. 4). The tables adhere to the 2008 version of the SNA (University of Groningen 2016).

---

[2] In detail on definition of consumption-based CO_2 emission and production-based, see IPCC (2022), Wiebe and Yamano (2016).

[3] Refer to Timmer (2012) and Dietzenbacher et al. (2013) for details on the history, characteristics, and structure of the WIOD construction.

Still more, while previous studies used the WIOD 2009 Release, this chapter analyzes the WIOT 2014, the most recent global input-output table in the published WIOD 2016 Release.

2.1.2 Environmental Accounts

The core of the environmental database consists of atmospheric emission accounts. Atmospheric emissions resulting in impact categories covered in the WIOD (global warming, acidification, and tropospheric ozone formation) originate from gases emitted in energy-use processes. Inventory data from the UNFCCC and the Convention on Long Range Transboundary Air Pollution (CLRTAP) are compiled using extended energy balances from the International Energy Agency (IEA) (Timmer ed 2012). Environment Accounts: The Joint Research Centre of the European Commission has published useful data on energy use and carbon dioxide emissions by industry and country for 2000–2016, fully consistent with the 2016-release of WIOD (European Commission website (University of Groningen 2016). The author uses CO_2 emissions data from Environmental Accounts WIOD 2016 Release (Corsatea T.D et al. 2019).

2.2 Analysis Method

2.2.1 Creation of WIOT2014 for analysis

As mentioned in Sect. 2.2.1, WIOT 2014 in WIOD2016 Release constitutes 43 countries and areas, 56 sectors in industries. Therefore, the departmental integration was divided into three steps, as follows.

(STEP1) 44 regions divide into 11 country/region-based groups. While basically referring to Homma and Akimoto (2018), Japan, China, Korea are considered independent categories in light of the analytical objective of understanding the status of carbon migration in East Asia. The classification by country/region is shown in Table 1.

(STTP2) Based on the WIOT2014 for processing created in STEP 1, sector consolidation of 56 industry sectors into 16 industry categories. The classification by industry is shown in Table 2.

2.2.2 Model

Production-based and consumption-based emissions are defined as follows. The former case indicates CO_2 emissions from production as the final demand, while the latter case shows CO_2 emissions from the production of each industry like machinery manufacture for the country from which demand stems. Production-based emissions

Table 1 Correspondence table for regions

NO.	EU	GBR	USA	Developed regions	JPN	CHN	KOR	TWN	RUS	IND	ROW
1	Austria	GBR	USA	Australia	Japan	China	Korea	Taiwan	Russia	India	Brasil
2	Belgium			Canada							Indnesia
3	Bulgaria			CHE:Confoederatio Helvetica (Switzerland)							Mexico
4	Cyprus			Norway							Turky
5	Czech Rep.										ROW
6	Denmark										
7	Deutschland										
8	Espania										
9	Estonia										
10	Finland										
11	France										
12	Greece										
13	Republika Hrvatska (Croatia)										
14	Hungary										
15	Ireland										
16	Italy										
17	Latvia										
18	Lithuania										
19	Luxembourg										
20	Malta										
21	Netherlands										

(continued)

Table 1 (continued)

NO.	EU	GBR	USA	Developed regions	JPN	CHN	KOR	TWN	RUS	IND	ROW
22	Poland										
23	Portugal										
24	Romania										
25	Slovakia										
26	Slovenia										
27	Sweden										

(Note) In WIOT, reset countries (around 150 countries) are aggregated as "rest of the world (ROW)". Source: modified correspondence table from Homma and Akimoto (2018)

Table 2 Correspondence table for 56 industrial classifications

Sector classification assumed in this study		WIOT sector classification
Sec01	Agriculture	Crop and animal production, hunting and related service activities
		Forestry and logging
		Fishing and agriculture
Sec02	Mining	Mining and quarrying
Sec03	Food manufacturing	Manufacture of food products, beverages and tobacco products
Sec04	Paper and pulp manufacturing	Manufacture of paper and paper products
		Printing and reproduction of recorded media
Sec05	Textile manufacturing	Manufacturing of textiles, wearing apparel and leather products
Sec06	Coal and petroleum manufacturing	Manufacture of coke and refined petroleum products
Sec07	Chemical manufacturing	Manufacture of chemicals and chemical products
		Manufacture of basic pharmaceutical products and pharmaceutical preparations
		Manufacture of rubber and plastic products
Sec08	Other non-metallic mineral manufacturing	Manufacture of other non-metallic mineral products
Sec09	Basic metal manufacturing	Manufacture of basic metals
Sec10	Machinery manufacturing	Manufacture of fabricated metal products, except machinery and equipment
		Manufacture of computer, electric and optical products
		Manufacture of electrical equipment
		Manufacture of machinery and equipment n.e.c.
Sec11	Transportation equipment manufacturing	Manufacture of motor vehicles, trailers and semi-trailers
		Manufacture of other transport equipment
Sec12	Other manufacturing	Manufacture of wood and products of wood and cork, except furniture: manufacturing articles of straw and plaiting materials
		Manufacture of furniture: other manufacturing
		Repair and installation of machinery and equipment
Sec13	Construction	Construction
Sec14	Electricity and gas	Electricity, gas, steam and air conditioning supply
Sec15	Transport	Land transport and transport via pipe pipelines
		Water transport
		Air transport
		Warehousing and support activities for transportation

(continued)

Table 2 (continued)

Sector classification assumed in this study		WIOT sector classification
Sec16	Service	Water collection, treatment and supply
		Sewerage; waste collection, treatment and disposal activities; materials recovery; remediation activities and other waste management services
		Wholesale and retail trade and repair of motor vehicles and motorcycles
		Wholesale trade, except of motor vehicles and motorcycles
		Retail trade, except of motor vehicles and motorcycles
		Postal and courier activities
		Accommodation and food services activities
		Publishing activities
		Motion picture, video and television programme production, sound recording and music publishing activities
		Telecommunication
		Computer programming, consultancy and related activities, information service activities
		Financial service activities, except insurance and pension funding
		Insurance, reinsurance and pension funding, except compulsory social security
		Activities auxiliary to financial services and insurance activities
		Real estate activities
		Legal and accounting activities: activities of head offices; activities of head offices; management consultancy activities
		Architectural and engineering activities; technical testing and analysis
		Scientific research and development
		Advertising and market research
		Other professional, scientific and technical activities, veterinary activities
		Administrative and support service activities
		Public administration and defence; compulsory social security
		Education
		Human health and social work activities
		Other service activities

(continued)

Table 2 (continued)

Sector classification assumed in this study		WIOT sector classification
		Activities of households as employers; undifferentiated goods- and services- producing activities of households for own use
		Activities of extraterritorial organizations and bodies

Source Homma and Akimoto (2018)

are calculated as the total sum of CO_2 emissions induced in each country/region. Consumption-based emissions estimate the final demand in each area and then calculate the CO_2 emissions induced by this demand. The emissions are calculated as follows:

$$[\mathrm{Y} = \mathrm{C} \cdot \mathrm{X}\ (\text{production - based})] \tag{1}$$

$$[\mathrm{Y} = \mathrm{C} \cdot (\mathrm{I} - \mathrm{A})^{-1} \cdot \mathrm{F}(\text{consumption - based})] \tag{2}$$

where, Y is the estimated matrix of CO_2 emissions; X is the matrix of CO_2 emitted; C is a diagonal matrix containing the coefficient CO_2 emission in each area; I is a unit matrix; A is an input coefficient matrix and F is the final demand vector in each region.

3 Results and Discussions

The results shown in this paper were obtains by utilizing the described above.

As mentioned in 2.2.1, based on the WIOT2014 for processing created in STEP2, 16 industrial sectors are considered as one sector and integrated into one sector to analyze the status of carbon transfer between countries/regions (STEP3).

In the original WIOT2014, the number of countries/regions was 44 and the number of sectors was 56. In the analysis, this number is merged into 11 countries/regions and one sector in a table, and then the status of carbon transfer between countries/regions is determined. In addition, the analysis in this paper will also take into account consumption and production bases in order to grasp interdependence of CO_2 emission and international carbon transfer between countries/regions. The higher the percentage, the more CO_2 is produced as a producer/consumer good in different countries, i.e.

The analysis results show that the proportion of CO_2 induced in each country/region tends to be higher than that of CO_2 induced in the EU.

Table 3 indicates CO_2 emissions transfer from 11 countries/regions: on a production-based CO_2 emissions.

As for production-based CO_2 emissions, some countries, such as India, have a large share of CO_2 emissions induced from one region to the region itself, as well as from production-based emissions. However, for East Asian countries, Japan has 67.15% of its emissions induced from other countries/regions from other than its own induced emissions. Among them, the top countries are ROW and China, in that order. As for China, 87.08% of its emissions come from its own sources, and about 12.92% come from other sources, followed by another countries.

Korea has 71.95% of its own emissions, 13.43% from China, and 7.60% from the ROW.

Table 4 indicates CO_2 emissions transfer from 11 countries/regions: on a consumption-based CO_2 emission. As for consumption-based CO_2 emissions on East Asian countries, Japan has 42.46% of its emissions induced from other than its own induced emissions, while it accounts for 7.23% CO_2 induced from China and 1.95% from Korea, respectively. Among them, the top countries are ROW and China, in that order. As for China, 35.77% of its emissions come from its own sources, and about two-thirds come from other sources, followed by ROW, USA, EU, and other countries.

Korea has 22.63% of its own emissions, 26.81% from China, and 22.47% from the ROW. It used to be said that behind the "import" of consumer goods, Japan "exports" environmental impacts such as CO_2 emissions to China and other countries (Fujikawa et al. 2016). However, the fact that Japan's own inducement on a consumption basis has been decreasing could be due to an increase in the environmental burden on the consumption side, economic conditions, and the relocation of production bases.

Table 5 table calculates the ratio of consumption-based CO_2 emissions divided by production-based CO_2emissions (a value of consumption-based/production

Table 3 CO_2 emissions flow from 11 countries/regions: on a production-based CO_2emissions

	EU	GBR	USA	Developed regions	JPN	CHN	KOR	TWN	RUS	IND	ROW
EU	88.24	0.70	0.89	0.83	0.53	2.77	0.64	0.25	0.13	0.25	4.78
GBR	12.38	62.57	2.61	2.07	1.49	5.75	2.04	0.41	0.15	1.38	9.16
USA	4.34	0.50	63.47	2.52	2.44	11.06	2.43	0.77	0.06	0.71	11.70
Developed regions	11.05	2.17	13.90	23.17	6.24	19.60	3.91	1.45	0.15	2.53	15.84
JPN	2.37	0.20	1.77	0.40	67.15	11.38	5.06	2.60	0.04	0.49	8.53
CHN	1.84	0.17	0.91	0.29	1.28	87.08	1.78	0.69	0.04	0.39	5.53
KOR	2.07	0.13	1.16	0.22	1.80	13.43	71.95	1.12	0.07	0.45	7.60
TWN	2.86	0.27	1.72	0.43	3.61	23.94	2.86	52.40	0.03	0.29	11.61
RUS	12.64	1.15	3.50	1.30	4.68	13.86	2.49	0.76	33.25	1.32	25.04
IND	2.76	0.28	1.40	0.39	1.10	6.20	1.20	0.47	0.05	77.24	8.91
ROW	6.07	0.47	3.69	1.43	4.02	25.31	3.42	1.47	0.15	2.09	51.87

Table 4 CO_2 emissions flow from 11 countries/regions: on a consumption-based CO_2 emissions

	EU	GBR	USA	Developed regions	JPN	CHN	KOR	TWN	RUS	IND	ROW
EU	57.27	3.52	5.41	4.43	2.02	5.65	0.89	0.41	1.35	0.82	18.22
GBR	20.71	33.55	8.63	4.14	2.91	6.29	1.44	0.45	1.13	1.99	18.77
USA	7.13	1.10	51.14	5.57	4.23	8.58	1.39	0.75	0.32	1.25	18.52
Developed regions	14.71	2.40	19.79	12.84	7.23	13.47	2.12	1.08	0.69	3.19	22.47
JPN	6.38	0.70	8.58	1.72	42.46	15.85	3.00	2.46	0.33	1.09	17.43
CHN	9.48	1.33	12.90	3.01	7.23	35.77	1.95	0.88	0.50	1.48	25.48
KOR	7.18	0.64	9.16	1.83	6.48	26.81	22.63	0.96	0.43	1.40	22.47
TWN	8.06	1.31	10.61	2.84	7.15	30.39	1.63	16.86	0.40	1.13	19.63
RUS	12.78	1.55	9.64	2.82	5.48	10.59	1.37	0.73	24.12	2.40	28.53
IND	4.02	0.59	4.51	1.04	1.77	4.29	0.61	0.33	0.21	70.83	11.82
ROW	10.92	1.31	15.22	3.86	7.15	17.52	1.91	1.21	0.71	3.24	36.94

greater than 1) indicates CO_2 emissions abroad through imports, while a value of consumption-based/production less than 1 indicates CO_2 emissions for export emissions abroad through imports, while a value less than 1 indicates high CO_2 emissions for exports (Yasazaki et al. 2012).

Haga and Zhou (2021) also discusses this consumption-based/production-based argument, the results of which are shown in Fig. 1. Although simple comparisons cannot be made because of differences in the regional classification of the input-output table, sector integration, and the timing of the database used, the results of the analysis in that case showed that for China, the production base was greater than

Table 5 Ratio to Consumption-based and Production-based CO_2 emissions

	EU	GBR	USA	Developed regions	JPN	CHN	KOR	TWN	RUS	IND	ROW
EU	0.65	5.06	6.11	5.36	3.79	2.04	1.39	1.64	10.37	3.30	3.81
GBR	1.67	0.54	3.31	2.00	1.96	1.09	0.71	1.08	7.51	1.45	2.05
USA	1.64	2.19	0.81	2.22	1.74	0.78	0.57	0.97	5.41	1.76	1.58
Developed regions	1.33	1.11	1.43	0.56	1.16	0.69	0.54	0.75	4.52	1.26	1.42
JPN	2.69	3.43	4.86	4.30	0.63	1.39	0.59	0.95	8.20	2.22	2.04
CHN	5.14	7.95	14.10	10.56	5.63	0.41	1.10	1.29	12.36	3.79	4.61
KOR	3.47	4.90	7.90	8.38	3.60	2.00	0.31	0.85	6.21	3.09	2.96
TWN	2.82	4.93	6.18	6.58	1.98	1.27	0.57	0.32	12.71	3.91	1.69
RUS	1.01	1.34	2.76	2.16	1.17	0.76	0.55	0.96	0.73	1.82	1.14
IND	1.53	2.20	3.38	2.78	1.69	0.73	0.53	0.74	4.39	0.96	1.39
ROW	1.80	2.78	4.13	2.70	1.78	0.69	0.56	0.83	4.65	1.55	0.71

the consumption base in China and East Asia region, while the consumption base was greater than the production base in the EU and other EU countries, NAFTA, BRIIAT, and ROW. In contrast, the results of this paper's analysis revealed that the consumption base/production base was less than 1 in China and Korea, and that this trend has not changed significantly.

Next, Figure indicates the share of consumption-based and production -based CO_2 emissions from each country and regions in Japan, China, Korea in 2014.

Finally, based on the post-processing input-output table, which merged the WIOT2014 table from 56 sectors to 16 sectors, also described in (STEP2) of 2.2.1, we show the induced CO_2 emissions to each country and regions in Japan, China, Korea for each of the 16 industrial sectors, as shown in Figs. 2, 3, and 4.

The results of the analysis suggested the following:

- **Japan**

Overall, induced CO_2 emissions are seen in the mining and service industries. In addition to Japan's own induced CO_2 emissions, the chemical manufacturing, industrial, and service sectors have high induced CO_2 emissions.

- **China**

China has a large amount of induced CO_2 emissions, with the metal manufacturing industry accounting for 60% of the total. In terms of metal manufacturing and chemicals, China induces its own CO_2 emissions from China to China, but also to the EU, USA, and ROW. As a speculative factor, the impact of automobile manufacturing, including rare metals, is also seen as a possible factor.

- **Korea**

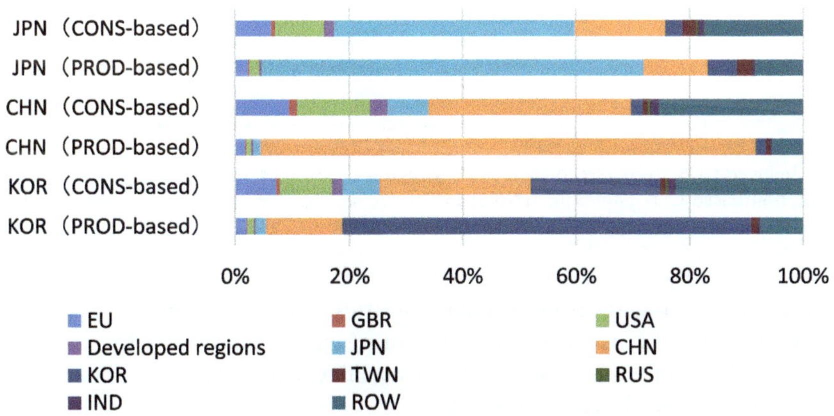

Fig. 1 The share of consumption-based and production -based CO_2 emissions from each country and regions in Japan, China, Korea in 2014

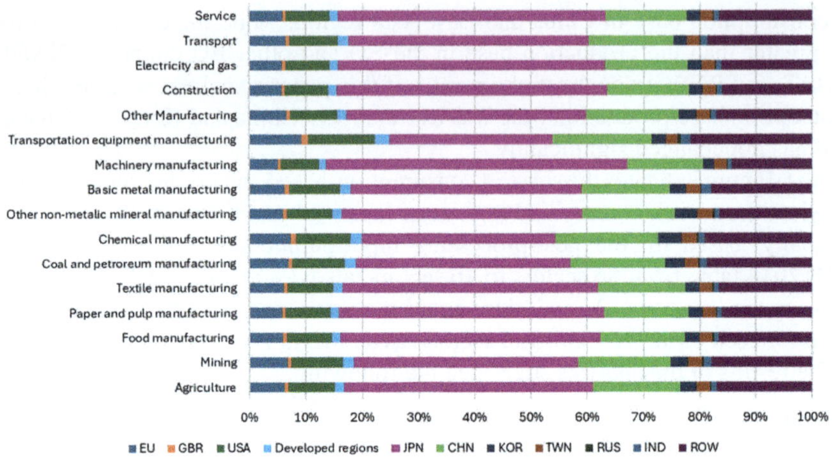

Fig. 2 Induced CO_2 emissions to each country/region for each of 16 industrial sectors in Japan:consumption-based CO_2 emissions

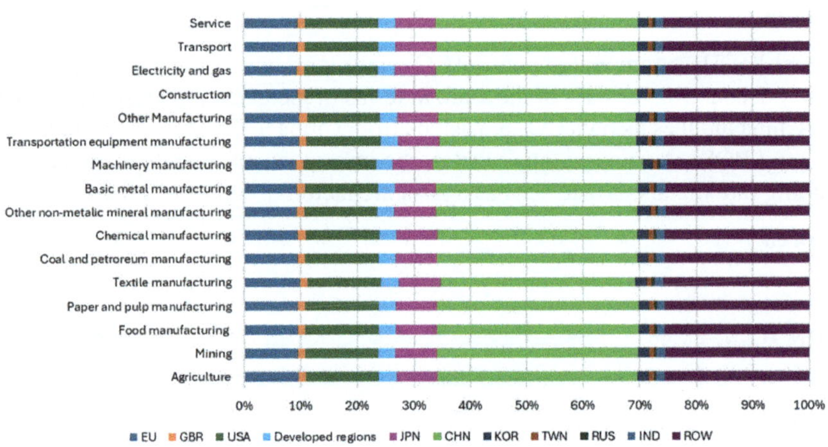

Fig. 3 Induced CO_2 emissions to each country/region for each of 16 industrial sectors in China: consumption-based CO_2 emissions ((Note) In Service section, we omitted the graph for all data value 0)

As for carbon transfer from Korean industries to industries in Korea and other countries/regions, the machinery manufacturing industry accounts for about 50% of the total. In addition, a large number of machinery manufacturing industries are located in China and Korea from Korea.

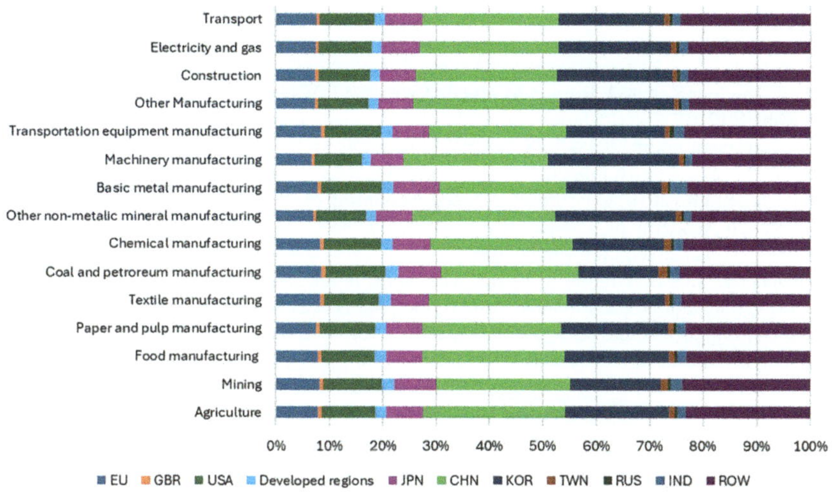

Fig. 4 Induced CO_2 emissions to each country/region for each of 16 industrial sectors in Korea:consumption-based CO_2 emissions

4 Conclusion and Remarks

This paper examines the trend of CO_2 Emissions from International Carbon Flow centering on East Asia like Japan, China, Korea based on input-output analysis using the World Input-Output Database (WIOD).

WIOD2016 was used to identify trends in CO_2 emissions on a production and consumption basis respectively, carbon transfers between countries and regions, and the characteristics of induced CO_2 emissions in Japan, China, Korea in 16 industrial sectors.

In addition, this paper continues to require the creation of highly accurate input-output tables and analysis based on the construction of an input-output model. In addition, we also need to examine the recent trends in the international division of labor, such as the value-added ratio, and the interdependence of international trade, as well as to conduct a detailed factor analysis of the results of this analysis, especially how CO_2 emissions from pre-integration industries, such as machinery manufacturing and chemical services, affect the trends of national and regional carbon transfers, with a view to decarbonizing East Asia. Emissions from premerger industries, especially in the machinery manufacturing and chemical services sectors, and how they affect national and regional carbon transfer trends. We would like to make this an issue for future research as we are required to construct supply chains in the Green Deal era.

References

Corsatea T, Lindner S, Arto I, Roman M, Rueda Cantuche J, Velazquez Afonso A, De Amores HA, Neuwahl F (2019) World input-output database environmental accounts, publications office of the European union, 2019-5-28. https://publications.jrc.ec.europa.eu/repository/handle/JRC116234. Accessed on 16 Feb 2025

Davis SJ, Caldeira K (2010) Consumption-based accounting of CO_2 Emissions, The Proceedings of the National Academy of Sciences (PNAS), 107(12):5687–5692. https://doi.org/10.1073/pnas.0906974107. Accessed on 6 March 2025

Dietzenbacher E, Los B, Timmer M (2013) The world input-output tables in the WIOD database (Japanese translation by Oshita T), J Life Cycle Assess 9(2):91–96. [in Japanese] https://www.jstage.jst.go.jp/article/lca/9/2/9_91/_pdf/-char/ja. Accessed on 13 Feb 2025

Feng K, Davis SJ, Sun L, Hubacek K (2013) Outsourcing CO_2 within China. Proc Natl Acad Sci 110(28):11654–11659. https://doi.org/10.1073/pnas.1219918110. Accessed on 6 March 2025

Fujikawa K, Ban H (2016) An empirical study on the interdependency of energy consumption and CO_2 emissions under the international input-output structure of the Asia-Pacific region. In: Lee S, Pollitt H, Park S-J (eds) (2015) Low-carbon sustainable future in East Asia: improving energy systems, taxation and policy cooperation. Routledge, London, pp 270–287

Fujikawa K, Shimoda M, Watanabe T, Ye Z (2016) Interdependence of CO_2 emissions in East Asia. In: Fujikawa K (eds) Input-output analysis and computable general equilibrium analysis in China's economy. Horitsu-Bunkasha, Kyoto, pp 129–142. [in Japanese]

Haga H (2015). Analysis on CO_2 emission with industrial activities between areas. J Policy Sci (Ritsumeikan University) 9:137–146. http://hdl.handle.net/10367/6565. Accessed on 9 Feb 2025

Haga H, Zhou W (2021) Chapter 16 An empirical analysis of international carbon transfer. In: Zhou W, Qian X, Nakagami K (eds) East Asian low-carbon community realizing a sustainable decarbonized society from technology and social systems. Springer, Singapore, pp 289–305

Homma T, Akimoto K (2018) Decoupling analysis between economic growth and CO_2 emissions: insights from estimation of consumption-based CO_2 emissions, Research Institute of Innovative Technology for the Earth (RITE), September 25, 2018. https://www.rite.or.jp/system/en/global-warming-ouyou/download-data/E-Analysis_Consumption-Based-CO_2.pdf. Accessed on 9 Feb 2025

Hubacek K, Feng K, Minx JC, Pfister S, Zhou N (2014) Teleconnecting consumption to environmental impacts at multiple spatial scales. J Ind Ecol 18(1):7–9. https://doi.org/10.1111/jiec.12082. Accessed on 7 March 2025

Hubacek K, Feng K, Chen B, Kagawa S (2016) Linking local consumption to global impacts. J Ind Ecol 20(3):382–386. https://doi.org/10.1111/jiec.12463. Accessed on 7 March 2025

Imura H, Nakamura E, Morisugi M, (2005) International interdependence of environmental loads associated with trade among Japan, us and Asian Countries. Japanese Journal of JSCE 2005(790):790_11–790_23 [in Japanese] https://www.jstage.jst.go.jp/article/jscej/2005/790/2005_790_790_11/_pdf/-char/en. Accessed on 9 Feb 2025

Intergovernmental Panel on Climate Change (IPCC) (2022), IPCC Sixth assessment report working group III: mitigation of climate change emissions (2 Trends and Drivers), Publication date 8 Jun 2022, https://www.ipcc.ch/report/ar6/wg3/downloads/report/IPCC_AR6_WGIII_Chapter02.pdf. Accessed on 7 March 2025

Ishikawa Y (2012) An analysis of spatial interdependence in Northeast Asia: an Input-Output Approach. Nanzan J Econ Stud 26(3):177–194. [in Japanese]

Japan Meteorological Agency (2025) Monthly anomalies of global average surface temperature in December (1891–2024), publication on the website Last update 7th February 2025. https://www.data.jma.go.jp/tcc/tcc/products/gwp/temp/dec_wld.html. Accessed on 7 March 2025

Karstensen J, Peters GP, Andrew RM (2018) Trends of the EU's territorial and consumption-based emissions from 1990 to 2016. Clim Chang 151(2):131–142. https://doi.org/10.1007/s10584-018-2296-x. Accessed on 6 March 2025

Liu Z, Feng K, Hubacek K, Liang S, Anadon LD, Zhang C, Guan D (2015) Four system boundaries 25 for carbon accounts. Ecol Model 318:118–125. https://doi.org/10.1016/j.ecolmodel.2015.02.001. Accessed on 6 March 2025

Ministry of Environment (2025) Annual report on the environment, the sound material-cycle society and biodiversity in Japan 2024. https://www.env.go.jp/en/index_00003.html. Accessed on 8 March 2025

Ministry of Foreign Affairs of Japan (2022), Diplomatic Bluebook 2022 Chapter 3 Japan Strengthening Its Presence in the International Community, publication on the website 30th September 2022. https://www.mofa.go.jp/policy/other/bluebook/2022/en_html/chapter3/c030202.html. Accessed on 7 March 2025

Mózner ZV (2013) A consumption-based approach to carbon emission accounting—sectoral differences and environmental benefits. Journal Cleaner Prod March 2013, 42:83–95. https://doi.org/10.1016/j.jclepro.2012.10.014. Accessed on 25 December 2025

Na S (2006) Climate change and international cooperation: efficiency, equity, sustainability. Yuhikaku, Tokyo. [in Japanese]

Nakamura E, Morisugi M, Imura H (2002) Does such trading structural change as world-wide globalization affect environmental worse off? A case of study of Japan, U.S.A and Asian countries. Environ Syst Res October 2002, 30:37–43. https://doi.org/10.2208/proer.30.37. Accesed on 25 December 2025

Pang J, Alexandri E, Lin S, Lee S (2016) Measuring both production-based and consumption-based CO_2 emissions of different countries based on the multi-region input-output model. In: Lee S, Pollitt H, Park S (eds) Low-carbon sustainable future in East Asia: improving energy systems, taxation and policy cooperation. Routledge, London, pp 247–269

Peters GP, Hertwich EG (2008) CO_2 embodied in international trade with implications for global climate policy, environmental science & technology, 42(5):1401–1407. https://pubs.acs.org/doi/epdf/10.1021/es072023k?ref=article_openPDF. Accessed on 9 Feb 2025

Peters GP, Davis SJ, Andrew R (2012) A synthesis of carbon in international trade. Biogeosciences 9:3247–3276. http://biogeosciences.net/9/3247/2012/. Accessed on 9 Feb 2025

Shan Y, Guan D, Zheng H, Ou J, Li Y, Meng J, Mi Z, Liu Z, Zhang Q (2018) China CO_2 emission accounts 1997–2015. Sci Data 5(1):170201. https://doi.org/10.1038/sdata.2017.201. Accessed on 6 March 2025

Shimoda M, Watanabe T (2005), A structure of international division of labor in Asia-Pacific region -An empirical study using IDE international IO tables, Input-Output Analysis, Pan Pacific Association of Input-Output Studies (PAPAIOS), 13(2), 2005:42–53 [in Japanese] https://www.jstage.jst.go.jp/article/papaios/13/2/13_42/_pdf/-char/ja. Accessed on 14 Feb 2025

Shimoda M, Watanabe T, Ye Z, Fujikawa K (2009) Interdependence on an environmental load in East Asia: CO_2 attributed emissions, indirect consumption. In: Mori A (ed) Economic development and environmental policy in East Asia, Minerva Shobo pp 40–79 [in Japanese]

Takahashi R (2018), The analysis of global supply chain in East/Southeast Asia: the impacts of China's value-added export in Asia, annual report on Japan academy for Asian market economies, 20, 21 2018:129–137 [in Japanese] https://doi.org/10.20766/jafame.20.21.0_129. Accessed on 13 Feb 2025

Tamura T (2016) The structure of international division of labor in East Asia and China's trade in value added, the 50th founding anniversary commemoration papers, faculty of economics, Ryutsu Keizai University, (1), March 2016:401–435 [in Japanese] https://rku.repo.nii.ac.jp/records/6801. Accessed on 13 Feb 2025

Timmer MP (ed) (2012) The World Input-Output Database (WIOD): contents, souces and methods. Version 09, Working Paper Number:10, April 2012. https://www.rug.nl/ggdc/valuechain/wiod/papers/wiod10.pdf. Accessed on 25 December 2025

University of Groningen (2016) WIOD 2016 Release, Groningen growth and development centre, University of Groningen, https://www.rug.nl/ggdc/valuechain/wiod/wiod-2016-release. Accessed on 25 December 2025

UNFCCC (United Nations Framework on Climate Change) (2015) Adoption of Paris Agreement. FCCC/CP/2015/L.9/Rev.1. https://unfccc.int/resource/docs/2015/cop21/eng/l09r01.pdf. Accessed on 25 December 2025

Vasconcellos Oliveira R (2020) A methodological framework for developing more just footprints: the contribution of footprints to environmental policies and justice. Sci Eng Ethics 26(1):405–429. https://doi.org/10.1007/s11948-019-00100-8.V. Accessed on 7 March 2025

Wiebe KS, Yamano N (2016), Estimating CO_2 emissions embodied in final demand and trade using the OECD ICIO 2015: methodology and results, OECD science, technology and industry working papers, May 2016, 1–39. https://doi.org/10.1787/5jlrcm216xkl-en. Accessed 7 on March 2025

Wiedmann T, M Lenzen (2018) Environmental and social footprints of international trade. Nat Geosci 11(5):314–321. https://doi.org/10.1038/s41561-018-0113-9. Accessed on 7 March 2025

Yamada M, Miyagawa K, Wang Z (2016) Development and challenge of the Japan-Korea-China International Input-Output Table, The Quarterly journal of Rissho Economics Society, Rissho University Economic Society, November, 2016, 89–104. http://hdl.handle.net/11266/5901. Accessed on 13 Feb 2025

Yamazaki M, Su X, Sun F, Zhou W (2012) Estimating international carbon flow using I-O model, 17th Annual Meeting, The Society for Environmental Economics and Policy Studies (SEEPS), Tohoku University, Sendai, Japan [in Japanese] http://www.seeps.org/meeting/2012/submit/abst/2055_MHAG4XRA.pdf

Yang J (2020) Trade structure analysis on China-Japan, US, EU by using WIOD, (Doctoral Dissertation), Graduate School of Humanities and Social Sciences, Saitama University, March 2020 [in Japanese] https://doi.org/10.24561/00019131. Accessed on 14 Feb 2025

Ye Z, Zhang H, Shimoda M, Fujikawa K (2016) International division of labor rate based on value added criteria. In: Fujikawa K (ed). Input-output analysis and computable general equilibrium analysis in China's economy, Horitsu-Bunkasha, Kyoto, pp 117–128 [in Japanese]

Yoshinaga K (2013) Compilation and Analysis of International Input-Output Table for BRICs, The Keizai Ronshu (The economic review of Kansai University). Eco-nomic Society of Kansai University 63(2), September 2013:225–253 [in Japanese] https://kansai-u.repo.nii.ac.jp/records/8539. Accessed on 6 Feb 2025

Zhu Y, Shi Y, Wu J, Wu L, Xiong W (2018) Exploring the characteristics of CO_2 emissions embodied in international trade and the fair share of responsibility. Ecol Econ 146:574–587. https://doi.org/10.1016/j.ecolecon.2017.12.020. Accessed on 9 Feb 2025

Building a Region-Wide Decarbonization Future: Strategic Reflections and Policy Integration

Weisheng Zhou

1 Introduction

In the twenty-first century, as the global climate crisis intensifies, the East Asian region has become one of the most economically dynamic, densely populated, and energy-consuming areas in the world. Consequently, progress in decarbonization in this region significantly influences not only regional sustainability but also the global achievement of carbon neutrality.

The interaction between East Asian countries—primarily China, Japan, and South Korea—and the international community is currently moving toward the formation of a new green order, characterized by both competition and cooperation.

This chapter aims to present a systematic strategic framework and policy design for envisioning and integrating a decarbonized future for the entire East Asian region. National-level efforts alone are insufficient to address complex issues such as transboundary emissions, carbon burdens embedded in supply chains, energy security, and technology transfer.

Therefore, from the perspectives of sustainability science and policy engineering, this chapter proposes the development of a regional coordination mechanism to promote a structural and long-term green transition across East Asia.

2 Essential Conditions for Global Sustainability

The challenges facing humanity in the twenty-first century are wide-ranging, including climate change, resource depletion, energy insecurity, and social inequality. A comprehensive societal vision capable of addressing these interconnected issues

W. Zhou (✉)
College of Policy Science, Ritsumeikan University, Osaka, Japan
e-mail: zhou@sps.ritsumei.ac.jp

W. Zhou (ed.), *Decarbonization Policy Engineering*, World Sustainability Series,
https://doi.org/10.1007/978-981-95-5953-4_29

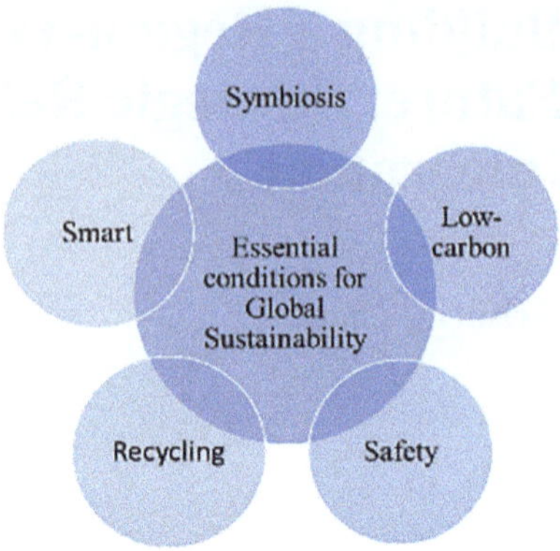

Fig. 1 Essential conditions for human society to achieve sustainable development

is referred to as a sustainable society. However, achieving such a society cannot rely solely on isolated technological innovations or institutional reforms.

What is required is a structural and integrated transition encompassing the social, economic, and environmental dimensions. To realize global sustainability, it is essential to establish five core principles—or necessary conditions—as illustrated in Fig. 1 (Zhou et al. 2021):

1. Low Carbon.
2. Circular.
3. Ecological.
4. Safe and Secure.
5. Smart.

These principles are not independent of one another; rather, they are mutually interconnected and collectively form the foundation for a sustainable society.

2.1 *Low Carbon*

The principle of "Low Carbon" plays a central role in addressing global warming and climate change, and is one of the foundational elements necessary for achieving global sustainability. This principle aims to minimize greenhouse gas (GHG) emissions to the greatest extent possible, requiring a fundamental transformation of both energy supply and consumption structures.

The first and foremost strategy for realizing this goal is the promotion of large-scale energy conservation. It is essential to maximize energy efficiency and reduce overall energy use across all sectors, including industry, buildings, and transportation. In addition, there is a growing shift from fossil fuels such as coal and oil to natural gas, hydrogen, and renewable energy sources such as solar, wind, and geothermal power.

Another key pillar is the expansion of carbon sinks, achieved through activities such as forest restoration and afforestation. Furthermore, attention has increasingly turned to the deployment of innovative environmental technologies. Prominent examples include next-generation technologies that directly capture, store, and utilize carbon, such as CCS (Carbon Capture and Storage), CCUS (Carbon Capture, Utilization and Storage), BECCS (Bioenergy with CCS), and DACCS (Direct Air Carbon Capture and Storage).

On the institutional side, many countries are implementing market-based mechanisms that balance incentives and regulations related to carbon emissions, such as carbon taxes and emissions trading systems (cap-and-trade schemes). In parallel, research and development into future-oriented energy technologies, such as nuclear fusion, is gaining attention as a means of supporting the foundations of a low-carbon society.

Importantly, a low-carbon society does not merely entail switching energy sources to renewables. Rather, it involves a comprehensive transformation of the entire social infrastructure, including urban planning, transportation systems, and industrial structures. Examples include the promotion of compact cities, the enhancement of public transportation systems, and the optimization of electricity management through smart grids.

Moreover, the concept of "low carbon" should be understood as an intermediate step, with the ultimate goal being the realization of carbon neutrality. Approaching zero environmental impact contributes not only to mitigating climate change but also to preserving ecosystems, enhancing societal resilience, and achieving sustainable economic growth.

In this way, the principle of low carbon requires a multi-faceted approach—encompassing technological innovation, institutional design, and structural transformation of society—and serves as an indispensable key to the realization of global sustainability.

2.2 *Circularity*

In the twenty-first century, as resource depletion and environmental crises become increasingly severe, the traditional linear economic model of "take–make–dispose" is proving inadequate for building a sustainable society. Against this backdrop, the principle of circularity has emerged as a core concept for new economic and social development, particularly in the pursuit of the Sustainable Development Goals (SDGs) and the realization of the super-smart society (Society 5.0), as illustrated in Fig. 2 (Keidanren 2025).

The principle of circularity refers to a transition to a circular economy based on the "3Rs": Reduce, Reuse, and Recycle. This model goes beyond simply reusing resources and reducing waste; it emphasizes reducing environmental impact from the design stage of products and services and maximizing the value extracted from resources throughout their life cycle.

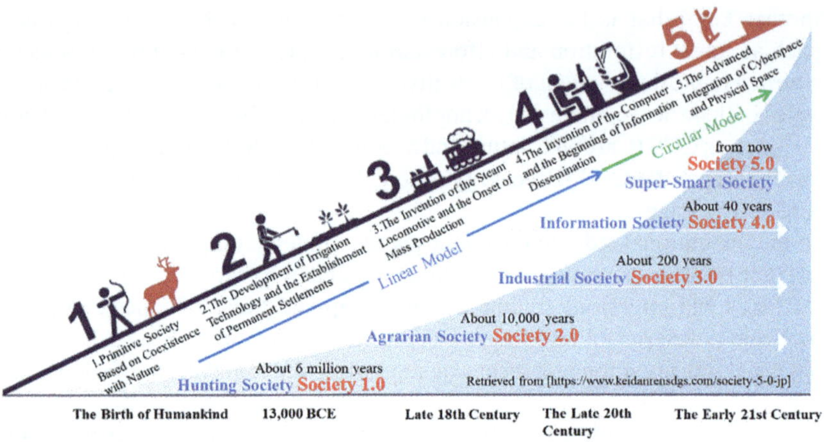

Fig. 2 Transition of human society (Created by the author based on Keidanren 2025)

Based on this approach, practices such as eco-design, which minimizes waste from the initial design phase, and the application of Life Cycle Assessment (LCA), which evaluates and extends the entire lifespan of products, are essential. At the urban level, the development of resource-circulating cities, which promote local recycling of waste, is progressing. Furthermore, under the concept of a Local Circulating and Ecological Sphere (LCS-R), systems that promote local production and consumption of resources and energy—while fostering symbiotic relationships among regions—are gaining importance.

In addition, the circular principle plays a significant role in creating economic value. Through methods such as reuse, remanufacturing, and upcycling, waste is redefined as a new "valuable resource." This contributes not only to effective resource use but also to industrial diversification, job creation, and the enhancement of national and corporate competitiveness. The idea of "creating value from waste" has become a central strategy in the circular economy.

As shown in Fig. 2, the history of human society has evolved step by step—from Society 1.0 (hunting society) to Society 5.0 (super-smart society). In particular, Society 3.0 (industrial society) saw the advent of the steam engine, which led to mass production, mass consumption, and mass disposal, resulting in excessive resource use and environmental degradation. Society 4.0 (information society) brought the development of information technology, yet economic activity remained dependent on the linear model.

In contrast, Society 5.0 clearly positions a shift to a circular model, aiming to achieve a balance between economic growth and environmental conservation by integrating the physical and digital realms. Within this framework, the principle of circularity serves as a foundational philosophy for realizing harmony between human society and nature, as well as between economy and ecology.

In conclusion, the concept of circularity as presented in Section 29.2.2 is no longer merely an environmental measure; rather, it represents a paradigm shift for restructuring the very foundations of our economy, technology, and social systems. In the worldview of Society 5.0, as depicted in the figure, circularity is positioned at the core, and is an indispensable element for realizing a sustainable and competitive future society.

2.3 Ecological Compatibility

"Ecological Compatibility" is positioned as one of the five essential pillars that constitute global sustainability. It aims to reconstruct human activities on the premise of coexistence with nature. This principle goes beyond mere technological responses such as energy conservation or reducing environmental burdens; it offers a perspective that reexamines the very nature of society, based on the fundamental value that "humans are a part of nature."

This principle calls for a shift away from conventional development models centered on human interests and economic efficiency, toward a society designed around the protection and regeneration of ecosystems. A key initiative in this direction is the conservation of biodiversity. In both urban and rural infrastructure development, the adoption of green infrastructure—which helps maintain and enhance ecological diversity—is gaining momentum. This approach leverages the power of nature (e.g., disaster mitigation and water purification) to build sustainable and resilient societal foundations.

Moreover, in terms of land use, the concept of "nature positive" is becoming increasingly important. This perspective does not regard nature as merely an object of development but promotes its active restoration and expansion. There is growing emphasis on conserving and rehabilitating natural areas such as forests, wetlands, and rivers, while harmonizing human activities with the natural environment. In addition, restoring healthy water cycles and soil ecosystems is another critical issue. Rehabilitating water and soil systems that have been degraded by excessive development or chemical use is a fundamental effort directly linked to sustainable agriculture and food security.

In this way, the ecological perspective does not treat environmental issues as isolated "problems to be addressed," but rather as a call to fundamentally reconsider the relationship between humanity and nature. It proposes a comprehensive approach that involves ethical and structural shifts in societal values. This principle complements technology-driven efforts such as low-carbon transitions and circular economies, by providing a philosophical foundation rooted in coexistence with nature.

In other words, "ecological compatibility" is not merely an optional consideration—it is an indispensable principle for envisioning a sustainable future. It urges society to internalize a "symbiotic and cooperative view of nature" that cannot be realized through technological innovation or institutional design alone. Only by incorporating ethics and harmony with nature into society can we achieve a truly balanced and sustainable vision of the future.

2.4 Safe and Secure

"Safe and Secure" is a foundational condition for making sustainability socially legitimate and effective. This principle encompasses not only technical safety aspects such as infrastructure earthquake resistance and nuclear power plant safety but also a broad range of concerns related to citizens' sense of security—for example, the safety of food, water, and air, disaster preparedness, and the reduction of infectious disease risks.

The Fukushima Daiichi nuclear disaster and the COVID-19 pandemic highlighted the reality that "even if something is scientifically safe, it cannot be considered truly sustainable unless society feels secure." These experiences have led to a renewed recognition of the importance of risk governance, transparent information disclosure, and trust-based communication (risk communication).

The principle of "Safe and Secure" functions as a prerequisite for the social acceptance and institutionalization of the other four principles—Low Carbon, Circular, Ecological Compatibility, and Smart.

2.5 Smart

The "Smart" principle refers to the use of advanced technologies such as information and communication technology (ICT) and artificial intelligence (AI) to build efficient, flexible, and optimized systems. It applies to various systems across society, economy, and environment, providing a technological foundation for achieving sustainability.

Typical examples of its application include:

Energy management through smart grids.

Advanced mobility using Mobility as a Service (MaaS).

Optimization of circular logistics through IoT-based waste management.

Design of smart cities that optimize overall urban functions.

Smartification enables the understanding of complex social challenges as systems and realizes "maximum functionality with minimum resources" by controlling supply and demand in real-time, performing predictive analysis, and detecting anomalies.

However, the advancement of smart societies also brings new ethical and institutional challenges such as privacy protection, surveillance concerns, and technological disparities. Therefore, social design based on human-centered design—not merely technology-driven approaches—is indispensable.

In other words, the Smart principle is an essential pillar supporting the realization of sustainability through technology and is closely linked to Society 5.0, as illustrated in Fig. 2, which aims to create a "human-centered super-smart society." By harmonizing humans, nature, and technology, the Smart principle plays a central role in building a fairer and more efficient society.

2.6 *Integration of the Five Principles and Application to Institutional Design*

The five principles described above (necessary conditions) are not independent goals but function in a mutually complementary manner. For example, smart technologies enable greater efficiency in decarbonization and circularity, the ecological principle provides ethical guidance for their implementation, and the safe and secure principle ensures social acceptance.

Based on this integrated understanding, policy design, technology deployment, educational curricula, and regional strategies should be developed with conscious attention to balancing and prioritizing these five principles. A sustainable society is not merely about reducing environmental burdens but is an endeavor to reconstruct the relationship between humans and nature through a new social contract and redefinition of values.

3 China's Emission Reduction Strategy and Comparative Analysis with Japan, the U.S., and South Korea

Achieving decarbonization in the East Asian region requires each country to formulate emission reduction strategies tailored to their respective economic, technological, and institutional contexts while fostering mutual cooperation within the region. Particularly, China's actions, as the world's largest annual emitter of greenhouse gases, significantly influence the success or failure of climate change measures, especially in light of the U.S. Trump administration's withdrawal from the Paris Agreement.

As global climate governance enters the stage of "deep emission reductions" and "carbon neutrality," significant differences among countries emerge in the relationship between per capita greenhouse gas emissions and economic development. By analyzing the historical trends of per capita CO_2 emissions and per capita GDP, it becomes possible to comparatively and systematically understand the strategic turning points and institutional adjustment mechanisms that lead each country to transition toward emission reductions.

China, the United States, Japan, and South Korea are all major global economies and carbon emitters, but their emission trajectories differ markedly in terms of "peak timing," the length of periods with high emission plateaus, and the intensity of decoupling between emissions and economic growth.

Currently, China is at a critical turning point, with per capita carbon emissions approaching their peak and the economy transitioning toward a middle-to upper-income stage. Whether China falls into a "middle carbon trap"—a state in which emissions remain high and fail to decline for an extended period—will determine whether it can successfully shift toward green, high-quality development.

This section comparatively analyzes China, Japan, South Korea, and the U.S., focusing on the evolution paths of per capita carbon emissions, and systematically examines differences in emission reduction policies, energy structures, industrial adjustments, and social acceptance.

3.1 Comparison of CO_2 Emission Pathways among China, Japan, South Korea, and the United States

Figure. 3 depicts the relationship between per capita GDP (horizontal axis) and per capita CO_2 emissions (vertical axis) over time for China, Japan, South Korea, the United States, and other countries. This allows for a visual comparison of how each country has increased or decreased emissions in the course of economic growth—that is, the relationship between carbon emissions and economic growth (the progress of decarbonization).

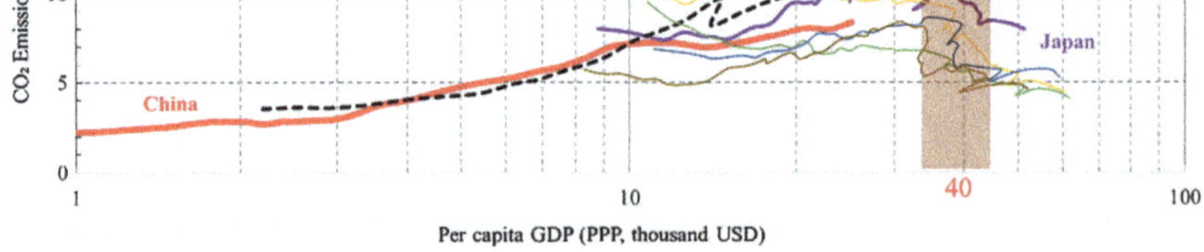

Fig. 3 The Kuznets Curve of carbon dioxide emissions and per capita GDP (PPP, USD) at carbon peak in China, Japan, South Korea, and the United States *Source* Created by the author based on IMF (2025); Ritchie et al. (2023) *Source*: Author's calculations based on IMF (2025); IEA (2023a); and World Bank (n.d.-a). *Note* Nominal GDP converted to US dollars using the Purchasing Power Parity (PPP) for each respective year

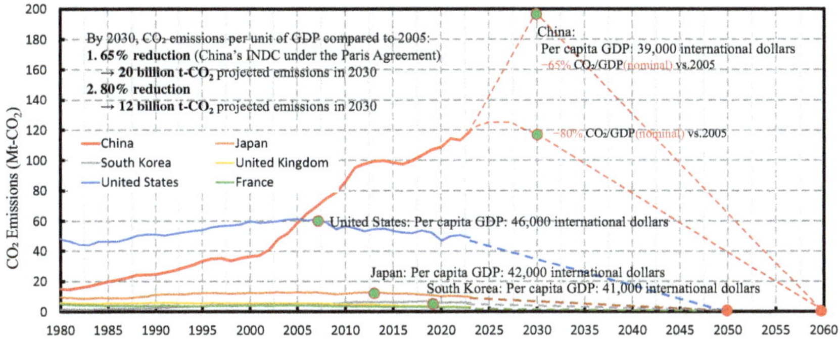

Fig. 4 Actual and Illustrative Trends of Carbon Peak and Carbon Neutrality in China and Selected Developed Countries (International Comparison of Carbon Peaks at a Per-Capita GDP of 40,000 International Dollars) *Source* Created by the author based on IMF (2025); Ritchie et al. (2023); and World Bank (n.d.-a). *Note* Nominal GDP converted to US dollars using the Purchasing Power Parity (PPP) of each year

Figure. 4 shows the carbon dioxide emissions (solid lines) and the progression toward achieving carbon neutrality (dashed lines) in four countries—China, Japan, South Korea, and the United States—as well as the per capita GDP (PPP-adjusted) at the time of the carbon peak.

3.2 *United States: Sharp Decline after High Emissions and a Long Plateau*

The United States, as one of the world's representative high-income countries, currently boasts an economic scale with a per capita GDP exceeding 60,000 dollars. Along with its economic growth, the U.S. has long maintained a very high per capita CO_2 emission level of over 20 tons (Fig. 3). This strongly reflects its high energy-dependent economic structure and a mass-consumption society. However, in recent years, U.S. emissions have declined significantly, marking a transition from a "high-income, high-emission" model to a "high-income, medium-emission" model.

Several factors have contributed in combination to this rapid reduction in emissions. One notable factor is the shift in power generation fuels brought about by the shale revolution. The transition from coal to natural gas, which has a lower CO_2 emission coefficient, has drastically reduced the emission intensity in the power generation sector. Additionally, technological and institutional changes such as the expansion of renewable energy adoption and improvements in energy efficiency have also made significant contributions to emission reductions.

Equally important are changes in industrial structure and the reorganization of global production systems. With parts of the manufacturing sector relocating to emerging economies and the domestic growth of non-energy-intensive service industries such as IT and finance, domestic emissions in the U.S. have statistically declined. This "offshoring of energy-intensive industries" functions as a factor that seemingly reduces emissions, and caution is needed to distinguish this from genuine emission reductions. In fact, international studies have pointed out that consumption-based emissions (carbon footprints) have declined less than production-based emissions (Huang and Jorgenson 2018).

On the policy front, under the Biden administration, the U.S. enacted the Inflation Reduction Act (IRA), which promotes investment in sectors such as renewable energy, electric vehicles, energy storage, and CCUS (carbon capture, utilization, and storage). This legislation aims to simultaneously reduce domestic emissions and strengthen the competitiveness of green industries, forming the core of the U.S. green transition policy.

Meanwhile, debates on climate policy within the country remain diverse. A report released in July 2025 by the U.S. Department of Energy (DOE), titled *A Critical Review of Impacts of Greenhouse Gas Emissions on the U.S. Climate* (Department of Energy 2025), suggested that the warming impact of CO_2 emissions may not be as severe as previously assumed, warning that overly stringent mitigation policies could be economically counterproductive. Such reports encourage critical reevaluation of the scientific consensus and may significantly influence future policy directions.

Overall, the United States is gradually moving away from its high-emission structure, driven by multilayered factors including technological innovation, institutional reform, and industrial structural change. Going forward, it is expected to enter an implementation phase toward building a carbon-neutral society that balances further emission reductions with economic and social equity. However, challenges remain, such as the effects of emission transfers abroad and domestic divisions over climate policy, necessitating a comprehensive and long-term perspective for sustainable decarbonization.

At the time of its carbon emissions peak, the United States had a per capita GDP of 46,000 international dollars (Int$, PPP-adjusted) (Fig. 3, Fig. 4).

3.3 *Japan: Stable Emission Suppression in a Mature Economy*

Japan, following the United States, reached economic maturity relatively early, with its per capita GDP already at a level of around 30,000 to 40,000 US dollars. During its economic development, per capita CO_2 emissions rose to about 10 tons around the 1990s and remained relatively stable at that high level for a long time—effectively maintaining an "emission plateau" (Fig. 3). However, in recent years, emissions have gradually started to decline, falling to below 8 tons at present. This trajectory indicates that Japan has followed a relatively moderate decarbonization path characterized by "economic maturity → emission stabilization → gradual reduction" (Akimoto et al. 2024).

The background of this emission pattern lies in efforts from both technological innovation and institutional responses. Since the period of high economic growth, Japan has actively developed and introduced energy-efficient technologies, and policy measures such as the Energy Conservation Act and industry-specific voluntary action plans have contributed to emission suppression. Notably, the introduction and expansion of nuclear power generation have contributed to CO_2 emission reductions to some extent.

On the other hand, following the Fukushima Daiichi nuclear accident in 2011, the country began reducing its dependence on nuclear power, temporarily increasing reliance on thermal power generation and causing a rebound in emissions. During this period, the dilemma between energy security and decarbonization became evident. However, since then, Japan has returned to an emission reduction trajectory through expanded adoption of renewable energy, further energy-saving efforts, and industrial structural transformation.

Moreover, the shift from manufacturing to service industries and the rapidly advancing population aging have led to lower energy intensity and suppressed overall energy demand, contributing to emission stabilization. These socio-economic changes structurally support Japan's decarbonization efforts.

Overall, Japan's emission reduction pathway is positioned as a sustainable low-carbon trajectory supported by institutional stability and social acceptance rather than dramatic cuts. By accumulating long-term and incremental responses rather than rapid shifts, Japan presents a realistic decarbonization model for mature economies, offering valuable insights for other advanced countries.

At the time of its carbon emissions peak, Japan had a per capita GDP of 42,000 international dollars (Int$, PPP-adjusted) (Fig. 3, Fig. 4).

3.4 South Korea: Rapid Emission Increase among High-Income Countries

South Korea is one of the countries that has experienced rapid economic growth in recent years, with its per capita GDP already reaching between 35,000 and 45,000 dollars. Alongside economic development, CO_2 emissions have also increased significantly, with per capita CO_2 emissions once reaching a very high level of 13 to 14 tons (Fig. 3). This figure is relatively high compared to other OECD countries and reflects the characteristics of South Korea's energy and industrial structures. Although there are currently slight signs of decline in emissions, these trends remain intermittent and unstable, and a clear downward trajectory has not yet been established (IEA, 2023b).

The background to this situation lies in South Korea's advanced industrialization and its supporting energy structure. In particular, the power generation and industrial sectors still rely heavily on fossil fuels such as coal and natural gas, which pose bottlenecks to emission reductions. Additionally, South Korea's decarbonization policies began later than those of other advanced countries, resulting in a delayed structural response.

However, in recent years, the South Korean government has rapidly introduced policy packages aimed at transitioning to a decarbonized society, such as the "Green New Deal" and the "Hydrogen Economy Strategy," accelerating efforts in both energy transition and technological innovation. Especially notable is the positioning of hydrogen as a key next-generation energy source, with active initiatives in infrastructure development and industrial support, indicating promising potential for future emission reductions.

Nevertheless, South Korea's industrial structure remains heavily centered on heavy chemical industries and manufacturing, making it highly emission-sensitive. Challenges also include the limited adoption of renewable energy and difficulties securing large-scale alternative energy sources due to geographic and institutional constraints.

Overall, South Korea is finally intensifying its transition to decarbonization after passing through a high-emission phase associated with economic growth. However, emission reduction outcomes remain unstable, and ensuring policy sustainability and social acceptance will be key moving forward. Attention will focus on how effectively future efforts can link industrial structural transformation with the expanded introduction of renewable energy.

At the time of its carbon emissions peak, South Korea had a per capita GDP of 41,000 international dollars (Int$, PPP-adjusted) (Fig. 3, Fig. 4).

3.5 China: Approaching the "Peak Transition Period" during Economic Growth

In recent years, China has experienced rapid economic growth alongside a steady increase in per capita CO_2 emissions. Currently, the level has reached 7 to 8 tons per capita, approaching levels comparable to those of Japan and South Korea (Fig. 3). Compared to Japan, South Korea, and the United States, China's carbon peak will be higher than that of any of them, and it will also occur later than in those countries. This indicates that China faces a more difficult path toward achieving its carbon neutrality goals (Fig. 3, Fig. 4). However, per capita GDP remains at an upper-middle-income level, and China faces the risk of falling into the so-called "middle carbon trap," where a high-emission structure becomes locked in at the middle-income stage.

This situation indicates that China is now at the critical "peak transition period" for emissions. Once a developing country with relatively low emissions, China's rapid economic expansion since the 2000s has pushed per capita GDP to around \$13,000 (nominal), accompanied by a corresponding increase in emissions. Although current emission levels show signs of stabilization, China has not yet entered a clear decline phase. Future trends will depend heavily on structural economic changes and shifts in energy policy.

Key structural factors behind emissions include the large share of energy-intensive industries such as heavy industry and construction, as well as the continued reliance on coal-fired power as the main source of electricity. Without fundamental energy transition, sustained reductions in CO_2 emissions will be difficult to achieve.

On the other hand, China has already implemented specific policies aimed at decarbonization. Notably, President Xi Jinping's 2020 announcement of the "2030 Carbon Peak and 2060 Carbon Neutrality Goals" (the "dual carbon goals") represents a strong national commitment. Following this, China has introduced a nationwide emissions trading system (ETS) and accelerated investment in low-carbon technologies such as renewable energy, CCUS (carbon capture, utilization, and storage), and hydrogen technologies.

Furthermore, economic upgrading, rising environmental awareness in urban areas, and a shift toward the service sector also hold potential to support emission reductions. Nevertheless, challenges remain for balanced nationwide decarbonization, including regional and industrial disparities and inconsistencies in institutional frameworks.

When assessing China's emissions, it is important to consider both production-based emissions (those associated with activities within the country) and consumption-based emissions (those resulting from domestic consumption). China, as the "world's factory," exports many products, and a significant portion of CO_2 emissions is linked to export-related production activities, while the consumption of these products mainly occurs abroad. Consequently, production-based emissions are very high, but consumption-based emissions partially offset this.

Meanwhile, rising domestic demand and improved living standards have significantly increased consumption-based emissions in recent years. Thus, China's "carbon footprint" (consumption-based emissions) remains high, but its emission trends differ somewhat from production-based emissions due to export-related production. This distinction is important for policy making, requiring efforts to decarbonize export-oriented industries while promoting sustainable domestic consumption.

In September 2025, at the United Nations Climate Summit, China announced a new Nationally Determined Contribution (NDC) target to reduce greenhouse gas emissions by 7–10% from the peak by 2035 (World Resources Institute 2025). This is China's first absolute emission reduction target covering all greenhouse gases. The target is linked to concrete policies including expanding renewable energy deployment, promoting electric vehicle adoption, increasing forest carbon sinks, and expanding the nationwide carbon market. In particular, China aims to increase wind and solar power capacity sixfold from 2020 levels to a total capacity of 3600 GW.

However, the target has been criticized as modest relative to international expectations. The European Union (EU) argues that a reduction of at least 20% is necessary to align with the 1.5 °C target, pointing out that China's goal falls short (Reuters, 2025).

Figure 3 illustrates the historical relationship between per capita GDP and CO_2 emissions, highlighting the trade-offs and complexities of policy choices between economic development and decarbonization. For China, the next 10 to 15 years represent a strategic turning point to simultaneously achieve "carbon peak and high-quality growth." China's future emission reduction strategy must be regarded not merely as environmental policy but as a core national strategy that influences economic competitiveness and international credibility.

Overall, China currently stands at a strategic crossroads of "pursuing both economic growth and emission reduction," where transforming the energy structure, upgrading industries, and sustained policy efforts are essential. The coming 10 to 15 years will determine whether decarbonization is advanced as a "national strategy" crucial for the country's sustainability and international standing, rather than simply an environmental measure.

As shown in Fig. 1.6, it is estimated that China's per capita GDP will be approximately 40,000 international dollars (Int$) in 2030, based on purchasing power parity (PPP), when it reaches its carbon emissions peak.

4 China's Emission Reduction Strategy: Future Focus

4.1 Carbon Peak and 40,000 International Dollars per Capita

China has pledged to achieve its carbon peak (peak carbon emissions) by 2030 and carbon neutrality by 2060. According to the author's forecast, China's per capita GDP (PPP basis) will reach approximately 40,000 international dollars by 2030(Fig. 1.6). This figure is very close to the carbon peak data of the United States (2007), Japan (2013), and South Korea (2019) (see Fig. 3, Fig. 4), and superficially aligns with the "inverted U-shaped" relationship indicated by the Environmental Kuznets Curve (EKC). This similarity raises important questions: Can China simply mimic the emission reduction pathways of developed countries? If not, how will China achieve the same goals under different structural foundations? This section discusses China's structural challenges and the necessity for an autonomous transition strategy, using comparative analysis and charts.

4.2 Structural Characteristics of Developed Countries' Carbon Peak Pathways

4.2.1 Outsourcing of Manufacturing and Emission Transfers

The carbon peak processes of advanced economies, represented by the United States and Japan, are often closely related to "deindustrialization." For example, in Japan, the manufacturing value-added share of GDP has continuously decreased from 27% in 1990 to about 18% in 2013 (World Bank n.d.-a). The U.S. share is even lower, at approximately 11% (World Bank n.d.-b). This decline reflects the relocation of many high energy-consuming industries to China, Southeast Asia, and other regions.

This process reduced the "production-based carbon footprint," but the "consumption-based carbon footprint" did not decrease concurrently. According to OECD data, carbon emissions embedded in U.S. imports account for over 15% of total emissions, indicating that a significant portion of environmental costs is shifted through global value chains (Yamano and Guilhoto 2020).

4.2.2 Energy Transition and the Dominance of the Service Economy

Around the time of carbon peaks, developed countries have almost completed the initial phase of energy transition—for example, switching from coal to natural gas and improving grid integration capacity for renewable energy. At the same time, the service sector has grown to over 70% of the economy, making the consumption sector a focal point for emission reductions.

4.3 China's Structural Differences and the Impossibility of Imitation

4.3.1 Manufacturing's Unshakable Position

China remains in the late industrialization and technological upgrading stage. In 2022, manufacturing value-added accounted for 27.7% of GDP, far exceeding developed countries, and China remains a global manufacturing hub (World Steel Association 2023). High energy-consuming industries such as steel, cement, chemicals, and power maintain world-leading production volumes.

This means China cannot achieve its carbon peak simply through industrial outsourcing; it must advance emission reductions while retaining core manufacturing competitiveness.

4.3.2 Emission Structure Is Primarily Production-Based

Currently, over 70% of China's carbon emissions originate from the industrial and energy production stages, while emissions from the consumption side remain relatively low. This structure indicates that emission reduction should focus on supply-side reforms and restructuring the energy system rather than relying solely on changes in consumer behavior.

4.4 *China's Unique Carbon Peak Pathway: Endogenous Transition and Coordinated Governance*

Under complex structural constraints, China is exploring a unique carbon peak pathway characterized by the following key aspects:

4.4.1 Green Manufacturing Transformation as a Key Pillar

China is promoting low-carbon transformation in traditional industries. For example, it is advancing the adoption of hydrogen metallurgy in steelmaking, piloting CCUS (carbon capture, utilization, and storage) in the cement industry, and actively constructing green industrial parks and smart manufacturing systems. These efforts aim to realize green transformation and high-quality development in the industrial sector.

4.4.2 Deepened Adjustment of the Energy Structure

According to the *China Energy Development Report (2024)*, by 2023 the share of non-fossil energy consumption reached 17.5%, wind and solar power capacity exceeded 130 GW, and nuclear power construction scale became the largest worldwide. Moving forward, China aims to further increase the share of clean energy, raising non-fossil energy's share to over 25% by 2030, and promoting green and low-carbon transformation of energy production and consumption.

4.4.3 Formation of Peak Strategies Based on Regional Differences

The National Development and Reform Commission has proposed a "regional carbon peak" strategy, clearly indicating that economically advanced eastern coastal regions (e.g., Jing-Jin-Ji, Yangtze River Delta) will peak earlier, while central and western regions and resource-based areas will advance their peak sequentially according to development stages. This aims for nationally coordinated peak achievement and balanced development.

4.4.4 Continuous Improvement of Institutional and Market Mechanisms

The national carbon market currently covers the power sector, and expansion to include eight key industries such as steel, cement, and non-ferrous metals is planned. The green finance system is steadily being developed, with the balance of green credits reaching 26 trillion yuan as of 2023. Furthermore, the "dual control" policy on energy consumption is transitioning to an adjustment mechanism centered on carbon intensity and total carbon emissions, establishing a more scientific and effective carbon emissions management system.

Overall, China's carbon peak pathway demonstrates a fusion of endogenous transition and multilayered coordinated governance. It not only promotes the green upgrading of the economic structure but also actively contributes to high-quality development and the achievement of global climate goals.

4.5 China's Path to "Green Modernization"

Table 1 compares per capita GDP (PPP), economic structure (proportions of manufacturing and service industries), and emission reduction pathways of China, the United States, Japan, and South Korea during their carbon peak periods, summarizing the key characteristics and strategic differences among the four.

Although China will reach its carbon peak at a per capita GDP level comparable to that of developed countries, its path is by no means a mere replication. Instead, it represents a distinctive exploration of "green modernization" with unique Chinese characteristics and contemporary significance. Facing the dual pressures of

Table 1 Economic structures and pathways of China, Japan, South Korea, and the United States at the time of carbon peak arrival

Indicator	China (2030, projected)	United States (2007)	Japan (2013)	South Korea (2018)
Per capita GDP (PPP) (World Bank n.d.-c)	\$40,000	\$46,000	\$42,000	\$41,000
Share of manufacturing sector (World Bank n.d.-d)	27%	11%	18%	28%
Share of service sector (World Bank, n.d.-d)	54%	78%	72%	61%
Pathway to carbon peak (International Energy Agency, 2023)	Industrial greening and energy transition	Service-oriented economy and industrial offshoring	Energy efficiency improvement and overseas manufacturing	Energy structure adjustment and digital transformation
Emission reduction mechanism (International Energy Agency, 2023)	National carbon market and dual control (total carbon emission and emission intensity)	Emission trading and federal standards	Energy conservation laws and FIT system	Emission trading and energy efficiency standards

a still-heavy industrial base and rising global responsibilities, China cannot simply follow the carbon peaking experiences of developed countries, which largely rely on outsourcing industries and transferring emissions abroad. China must achieve its green transition domestically. This requires a coordinated approach integrating technology, institutional reforms, structural adjustments, and market mechanisms to forge an independent, coordinated, green, and low-carbon development path. Such a path is not only strategically significant for China itself but also offers a valuable reference model and experience for other developing economies that have yet to reach their carbon peak.

5 Prospects for Decarbonization Policy Engineering in East Asia

As global climate change measures reach a turning point, the East Asia region holds a crucial position in the transition to a decarbonized society due to its economic growth, urbanization, and concentration of energy-intensive industries. The major economic areas of Japan, China, and South Korea each have set targets to achieve carbon neutrality by 2050 or 2060, respectively, and have introduced diverse policy tools such as the adoption of renewable energy, hydrogen societies, and carbon pricing.

Designing and implementing such complex and dynamic decarbonization policies requires evidence-based rational design and a policy engineering perspective that integrates social, technological, and economic factors. This section discusses the prospects for decarbonization policy engineering in East Asia from a multilayered viewpoint.

5.1 Deployment of Decarbonization Technologies and the Role of Policy Support

To achieve carbon neutrality, it is essential to develop and implement a variety of decarbonization technologies in society, including renewable energy (solar, wind, geothermal), fossil fuel clean-up, CCUS (carbon capture, utilization, and storage, including hydrogen production, storage, and use), next-generation batteries, and nuclear power. Appropriately designed policy support is a key factor in accelerating the introduction of these technologies.

From the perspective of policy engineering, the following points are particularly important:

Incentive design according to technology readiness level (TRL): Flexible policy support is required corresponding to the stages of technological development from R&D to market deployment.

Combination of economic instruments: Strategic combinations of subsidies, carbon taxes, emissions trading schemes, and other tools promote the diffusion of decarbonization technologies.

Analysis of barriers and facilitators in social implementation: Identifying factors hindering technology adoption and leveraging accelerators through social acceptance and institutional design.

Optimization of public-private collaboration during market formation: Designing effective public-private partnerships is critical in connecting public resources with private innovation as emerging technologies take root in the market.

Efforts toward decarbonization in East Asian countries reflect diverse approaches based on their institutional and economic characteristics.

China is rapidly advancing large-scale technology deployment and infrastructure development under state leadership, leveraging scale and speed to drive decarbonization.

Japan builds a model that balances private-sector leadership and public support by combining the high technological capabilities of its companies with government institutional support.

South Korea is characterized by a "hybrid model" where private companies lead technology development and market deployment while leveraging public procurement as a lever.

Going forward, if these strengths can be mutually complemented and "policy convergence" (integration and coordination of policies) achieved, the effectiveness of carbon neutrality efforts across East Asia as a whole will be further enhanced.

5.2 Institutional Design for International Cooperation and Regional Integration

Decarbonization is inherently a transboundary challenge, requiring regional institutional design beyond national policy frameworks. In East Asia, the integration of energy infrastructure (power grids, hydrogen supply networks, carbon markets), the sharing of technical standards, and coordination of climate finance are key focal points.

From the five necessary conditions (five principles) for achieving glocal sustainability raised in section 29.2 and from the perspective of policy engineering, the following are important:

5.2.1 Realization of the East Asian Low-Carbon Community

The East Asian Low-Carbon Community is a concept centered around Japan, China, and South Korea that aims to deepen the coordination of technologies, policies, and markets toward decarbonization. Through technological and social innovation, the

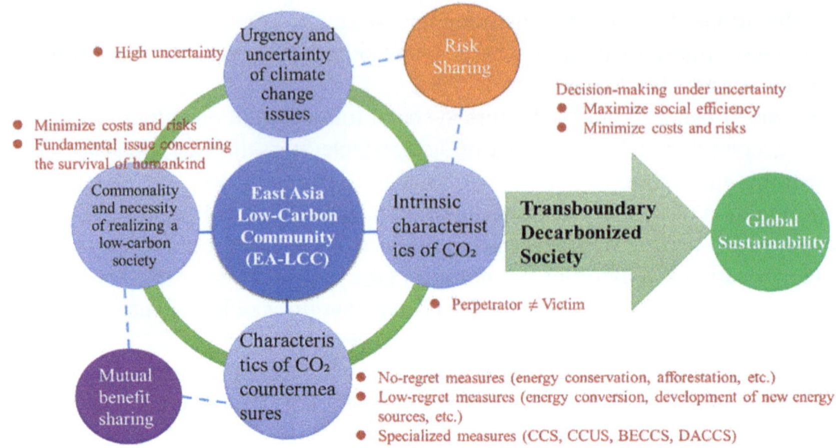

Fig. 5 Background concept of the East Asia Low-Carbon Community (EA-LCC)

goal is to realize a sustainable low-carbon society across the region (Zhou et al. 2021). It is crucial to regard shared energy security and environmental issues as common challenges, promote mutual complementarity in renewable energy, connect power and hydrogen networks regionally, and harmonize policies. Multilateral cooperation is essential to simultaneously pursue economic growth and decarbonization, while building a stable energy order that addresses geopolitical risks.

The author proposed the concept of a broad-area low-carbon community centered on Japan, China, and South Korea under conditions of high uncertainty, based on multi-layered optimization of climate measures along the dimensions of time, space, countermeasures, and outcomes. This is known as the "East Asian Low-Carbon Community" concept. For details, please refer to *East Asian Low-Carbon Community: Realizing a Sustainable Decarbonized Society from Technology and Social Systems* (edited by Zhou et al. 2021. Cham: Springer). This publication provides a substantive analysis method on principles and approaches for best practices in low-carbon policy and introduces significant advances in today's interdisciplinary methodologies. Additionally, it employs policy engineering and scenario design to quantitatively demonstrate the effectiveness of the East Asian Low-Carbon Community, presenting one new pathway toward realizing a cross-border, broad-area decarbonized society (Fig. 5).

5.2.2 Establishing an East Asian Nuclear Power Safety Security Scheme

In utilizing nuclear power generation as part of decarbonization, strengthening nuclear safety and security in the East Asian region is indispensable. In particular, considering natural disaster risks, aging infrastructure, and cross-border impacts in case of accidents, it is necessary to build a regional nuclear safety security scheme that includes information sharing, emergency response coordination, and harmonization of regulatory standards. The East Asian Nuclear Power Safety Security Scheme is a multilateral cooperation framework aimed at ensuring the safety of nuclear power generation in the region and minimizing cross-border risks of radioactive materials.

As illustrated in Fig. 10.12, East Asian countries including Japan, China, and South Korea are geographically close, all operate a significant number of nuclear power plants, and are located in a region prone to frequent natural disasters. Therefore, once an accident occurs, its effects could easily spread beyond national borders. Especially since the Fukushima Daiichi nuclear accident in 2011, the need for international collaboration and information sharing regarding nuclear safety has become widely recognized.

The pillars of this scheme include establishing common safety standards among countries, creating emergency notification systems for nuclear accidents, promoting mutual consultations on environmental impact assessments, as well as professional dialogues and personnel exchanges among regulatory authorities. Ensuring consistency with the International Atomic Energy Agency (IAEA) standards while enhancing transparency and reliability within the region is also required.

However, there remain many challenges to institutional development, such as differences in national sovereignty, energy policies, attitudes toward information disclosure, and immature trust stemming from historical and political backgrounds. Nonetheless, establishing such cooperative schemes is indispensable for the sustainable and safe use of nuclear power. It is expected that institutionalization and ensuring the effectiveness of this framework will contribute to the stability and environmental security of the entire East Asian region. For example, establishing international coordination mechanisms addressing issues of science and trust surrounding the ocean discharge of ALPS-treated contaminated water from the Fukushima Daiichi Nuclear Power Plant is also essential.

5.2.3 Interoperability of the East Asian Carbon Market

Carbon markets (Emission Trading Systems, ETS) are a key economic tool to promote greenhouse gas emission reductions and thus a critical part of decarbonization efforts. Enhancing interoperability among East Asian countries' systems—such as Japan's GX-ETS (Agency for Natural Resources and Energy 2025), China's national ETS (Ministry of Ecology and Environment 2025), and South Korea's K-ETS (Greenhouse Gas Inventory and Research Center of Korea 2020)—will improve carbon price alignment and credit liquidity within the region, enabling companies to reduce emissions more flexibly and efficiently.

By harmonizing trading rules, monitoring, and certification systems and advancing mutual recognition, the integration of regional carbon markets is expected to accelerate the transition to a low-carbon economy.

The East Asian region currently faces triple challenges: economic growth, overcoming pollution, and combating global warming. Japan, China, and South Korea have each declared carbon neutrality by 2050 or 2060—the so-called "5060 goals." Achieving these goals heavily depends on each country's ETS and their international linkage. Together, Japan, China, and South Korea account for over 30% of global CO_2 emissions and are expected to play a central role in achieving global carbon neutrality. Without efficient measures, there is a risk of imposing excessive costs on society.

However, cooperation among the ETS of Japan, China, and South Korea has so far been hampered by three major barriers: "institutional resistance," "governance differences," and "disparities in environmental targets." As stated in Article 6 of the Paris Agreement, the design of ETS systems must sustainably and concurrently address these challenges. For example, if a Cap-and-Trade (C&T) system with mandatory participation and absolute targets is adopted—similar to the EU-ETS—the lack of policy enforcement organizations like government governance in the Japan-China-Korea markets, economic disparities in the region, and the absence of unified reduction targets and total emission caps pose significant barriers to establishing an integrated market. Drawing on the experience of EU-ETS implementation, setting unified reduction targets, resolving allocation issues between countries, and integrating the three countries' distinct climate policies remain critical challenges for the East Asian ETS (EA-ETS).

Therefore, Japan, China, and South Korea should lead innovative technology development, social implementation, and rulemaking for decarbonization under international cooperation. Through forming broad regional markets including emerging economies and achieving scale merits, they can simultaneously enhance domestic industrial competitiveness and climate change responses.

Given Asia's anticipated role as the world's growth engine, international environmental cooperation centered on East Asia, particularly linking the ETS among Japan, China, and South Korea, is indispensable for balancing economic growth and decarbonization. Notably, the establishment of an East Asian Carbon Emissions Trading System (EA-ETS) through international linkage among these countries' ETSs is expected to promote cost equalization by connecting countries with differing reduction costs, thereby lowering overall reduction expenses.

5.2.4 Establishing a Global Resource Recycling System

To achieve carbon neutrality goals, it is essential not only to promote renewable energy and energy conservation but also to maximize resource circulation and minimize environmental impact (zero emissions) by building a "circular carbon-neutral society." A core component of this is the Global Recycling System proposed by the author, consisting of three layers shown in Fig. 6: Local circulation (small scale), Grocal circulation (medium scale), and Global circulation (large scale) (Zhou et al. 2021).

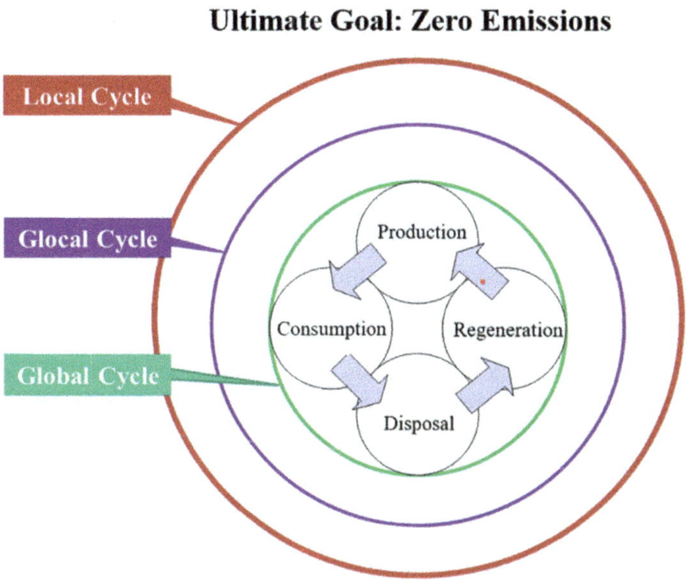

Fig. 6 Conceptual diagram of the Global Resource Recycling System (GRS)

This system comprehensively considers resource characteristics, processing costs, logistics optimization, and environmental impact reduction. By selecting the appropriate circulation scope for recycling, it simultaneously achieves efficient resource utilization and greenhouse gas reduction, as shown in Table 2.

Table 2 Summary of specific application criteria for Local, Grocal, and Global circulation

Type of circulation	Applicable scope	Characteristics of applicable resources	Representative examples
Local circulation (small scale)	Regional level	Lightweight, easily decomposable, can be processed locally	Food waste, construction waste, repairable home appliances and clothing
Grocal circulation (medium scale)	Domestic and neighboring countries	Can be processed domestically but requires specialized facilities	Industrial waste, electronic waste, metal scrap
Global circulation (large scale)	International trade and resource supply networks	Requires advanced processing technologies and meets international demand	Rare metals, plastic waste, remanufacturing, waste paper

Local circulation (small scale): Reusing and recycling resources such as food waste and biomass within a region avoids long-distance transportation and minimizes CO_2 emissions generated locally within the same region. This enables the realization of a community-based low-carbon society.

Grocal circulation (medium scale): Through recycling processes conducted domestically, more advanced resource recovery becomes possible, suppressing CO_2 emissions associated with new resource extraction and processing during manufacturing. For example, reusing scrap metal at domestic steel mills can significantly reduce CO_2 emissions compared to smelting from raw ore.

Global circulation (large scale): Utilizing the most efficient international recycling hubs enables recycling with high energy efficiency and resource recovery rates, contributing to carbon reduction globally. For instance, by using overseas plants capable of highly precise recovery of rare metals, CO_2 emissions related to new mining of scarce resources can be prevented while also contributing to the stable supply of international resources.

Thus, the Global Recycling System goes beyond mere waste disposal and serves as the foundation of a "circular carbon-neutral society" by integrating the circulation of both resources and carbon. It enables balancing economic activities and environmental impacts through reducing raw material use via resource reuse, optimizing transportation to cut emissions, and clearly defining roles at local, domestic, and international levels.

Going forward, strengthening the development and coordination of such circular systems, while promoting technological innovation, policy support, and international cooperation, is crucial for realizing a sustainable and decarbonized society.

6 Transformative Perspective: Viewing Uncertainty as Opportunity

The transformative perspective of seeing uncertainty not as a risk but as an "opportunity" is crucial for innovation in energy policy. Traditionally, uncertainty has been perceived as a barrier or risk to policymaking. However, in today's rapidly changing and unpredictable era, uncertainty can serve as a catalyst to break the inertia and path dependency of existing structures and systems.

Historically, energy systems have relied on centralized fossil fuel-based infrastructures, where numerous stakeholders hold vested interests, limiting scope for transformation. The confusion and unpredictability brought by uncertainty shake these structures and promote new ideas and flexible responses.

From this viewpoint, innovative possibilities emerge in several areas:

First, the decentralization of renewable energy and the reconstruction of regional economies. Introducing distributed energy sources—such as solar, wind, and biomass—tailored to local characteristics goes beyond diversifying electricity supply. It can foster regional employment and energy businesses backed by local capital, leading to economic autonomy and revitalization. New energy governance involving municipalities, regional enterprises, and agriculture and forestry sectors can create a "locally led energy economy" distinct from traditional centralized energy systems.

Second, integration with digital technologies is rapidly advancing new forms of energy supply-demand matching and transactions. The use of IoT (Internet of Things), AI (Artificial Intelligence), and blockchain enables real-time demand forecasting and efficient supply-demand management, promoting the role of "prosumers"—households and companies that both produce and consume energy. Moreover, blockchain-based peer-to-peer (P2P) electricity trading facilitates direct electricity sales between individuals without central utilities, pioneering unprecedented energy distribution methods.

Third, these technological innovations and institutional reforms contribute to addressing energy poverty. Especially in developing regions and low-income households, the spread of decentralized power sources and renewable energy, capable of providing stable and low-cost electricity, reduces disparities in energy access and promotes social inclusion. Microgrids and standalone solar systems that do not rely on central transmission networks enable sustainable electricity supply even in areas where grids are underdeveloped or unstable. This increases the likelihood that energy infrastructure will be built inclusively and fairly, without leaving vulnerable social groups behind.

In summary, actively leveraging uncertainty not as a "crisis" to be avoided but as an "opportunity" to stimulate new ideas and social transformation is vital for future energy policies. This is not merely a matter of technological choices but the starting point for restructuring society, economic systems, and governance frameworks—it is transformation itself.

Climate change policy is long-term, uncertain, and potentially socially divisive. In such a context, policy engineering must emphasize integrating "optimality" and "robustness."

Strategic thinking that uses uncertainty as a chance to pioneer a sustainable and inclusive future is urgently needed today.

References

Agency for Natural Resources and Energy (2025) Energy white paper 2025: Japan's efforts toward carbon neutrality by 2050 (section 1–2). Ministry of Economy, Trade and Industry

Akimoto K, Sano F, Homma T (2024, December 3) Energy supply and demand analysis toward carbon neutrality by 2050 in Japan. The Institute of Applied Energy Systems Analysis Group, RITE

Department of Energy (2025) A critical review of impacts of greenhouse gas emissions on the U.S. climate (Climate Working Group Report).. U.S. Department of Energy

Greenhouse Gas Inventory and Research Center of Korea (2020, May) 2018 Korea emissions trading scheme: summary report. Government of the Republic of Korea

Huang X, Jorgenson AK (2018) The asymmetrical effects of economic development on consumption-based and production-based carbon dioxide emissions, 1990 to 2014. Soc Nat Resour 31(7):814–834

International Energy Agency (2023a) World energy outlook 2023.

International Energy Agency (2023b) Korea 2023 energy policy review.

International Monetary Fund. (2025). GDP per capita, current prices. https://www.imf.org/external/datamapper/PPPPC@WEO/OEMDC/ADVEC/WEOWORLD

Keidanren. (2025). What is society 5.0? Keidanren. https://www.keidanrensdgs.com/society-5-0-jp

Ministry of Ecology and Environment (2025, September) National carbon market development report (2025). People's Republic of China

Reuters (2025, September 25) China's new climate target "falls well short," EU says. Reuters. https://www.reuters.com/sustainability/cop/chinas-new-climate-target-falls-well-short-eu-says-2025-09-25

Ritchie, H., Rosado, P., & Roser, M. (2023). CO_2 and greenhouse gas emissions. Our World in Data. Retrieved from https://ourworldindata.org/co2-and-other-greenhouse-gas-emissions/ Our World in Data

World Bank. (n.d.-a). GDP per capita, PPP (current international $). World Development Indicators. Retrieved December 22, 2025, from https://data.worldbank.org/indicator/NY.GDP.PCAP.PP.CD World Bank Open Data

World Bank. (n.d.-b). Manufacturing, value added (% of GDP) – Japan. World Development Indicators. Retrieved December 22, 2025, from https://data.worldbank.org/indicator/NV.IND.MANF.ZS?locations=JP World Bank Open Data

World Bank. (n.d.-c). Manufacturing, value added (% of GDP) – United States. World Development Indicators. Retrieved December 22, 2025, from https://data.worldbank.org/indicator/NV.IND.MANF.ZS?locations=US World Bank Open Data

World Resources Institute (2025, September 24) China announces new climate target. World Resources Institute

World Steel Association (2023) December 2022 crude steel production and 2022 global crude steel production totals. World Steel Association

Yamano N, Guilhoto J (2020) CO_2 emissions embodied in international trade and domestic final demand: methodology and results using the OECD inter-country input-output database. OECD science, technology and industry working papers, no. 2020/11. OECD Publishing

Zhou W, Qian X, Nakagami K (2021) East Asian low-carbon community: realizing a sustainable decarbonized society from technology and social systems. Springer, pp 61–81

GPSR Compliance
The European Union's (EU) General Product Safety Regulation (GPSR) is a set of rules that requires consumer products to be safe and our obligations to ensure this.

If you have any concerns about our products, you can contact us on

ProductSafety@springernature.com

In case Publisher is established outside the EU, the EU authorized representative is:

Springer Nature Customer Service Center GmbH
Europaplatz 3
69115 Heidelberg, Germany

www.ingramcontent.com/pod-product-compliance
Ingram Content Group UK Ltd.
Pitfield, Milton Keynes, MK11 3LW, UK
UKHW021819280926
475829UK00014B/10

* 9 7 8 9 8 1 9 5 5 9 5 2 7 *